Springer Handbook of Enzymes Volume 29

Dietmar Schomburg and
Ida Schomburg (Eds.)

Springer Handbook of Enzymes

Volume 29
Class 2 · Transferases II
EC 2.1.2.1–2.3.1.59

coedited by Antje Chang

Second Edition

Springer

Professor Dietmar Schomburg
e-mail: d.schomburg@uni-koeln.de

Dr. Ida Schomburg
e-mail: i.schomburg@uni-koeln.de

Dr. Antje Chang
e-mail: a.chang@uni-koeln.de

University to Cologne
Institute for Biochemistry
Zülpicher Strasse 47
50674 Cologne
Germany

ISBN-10 3-540-31918-2 2nd Edition Springer Berlin Heidelberg New York
ISBN-13 978-3-642-42666-7 2nd Edition Springer Berlin Heidelberg New York

The first edition was published as Volume 11 (ISBN 3-540-60295-X) of the "Enzyme Handbook".

Springer is a part of Springer Science+Business Media
springer.com

Softcover re-print of the Hardcover 2nd edition 2006

Cover design: Erich Kirchner, Heidelberg
Typesetting: medionet AG, Berlin

Printed on acid-free paper 2/3141m-5 4 3 2 1 0

Preface

Today, as the full information about the genome is becoming available for a rapidly increasing number of organisms and transcriptome and proteome analyses are beginning to provide us with a much wider image of protein regulation and function, it is obvious that there are limitations to our ability to access functional data for the gene products – the proteins and, in particular, for enzymes. Those data are inherently very difficult to collect, interpret and standardize as they are widely distributed among journals from different fields and are often subject to experimental conditions. Nevertheless a systematic collection is essential for our interpretation of genome information and more so for applications of this knowledge in the fields of medicine, agriculture, etc. Progress on enzyme immobilisation, enzyme production, enzyme inhibition, coenzyme regeneration and enzyme engineering has opened up fascinating new fields for the potential application of enzymes in a wide range of different areas.
The development of the enzyme data information system BRENDA was started in 1987 at the German National Research Centre for Biotechnology in Braunschweig (GBF) and is now continuing at the University at Cologne, Institute of Biochemistry. The present book "Springer Handbook of Enzymes" represents the printed version of this data bank. The information system has been developed into a full metabolic database.
The enzymes in this Handbook are arranged according to the Enzyme Commission list of enzymes. Some 4,000 "different" enzymes are covered. Frequently enzymes with very different properties are included under the same EC-number. Although we intend to give a representative overview on the characteristics and variability of each enzyme, the Handbook is not a compendium. The reader will have to go to the primary literature for more detailed information. Naturally it is not possible to cover all the numerous literature references for each enzyme (for some enzymes up to 40,000) if the data representation is to be concise as is intended.
It should be mentioned here that the data have been extracted from the literature and critically evaluated by qualified scientists. On the other hand, the original authors' nomenclature for enzyme forms and subunits is retained. In order to keep the tables concise, redundant information is avoided as far as possible (e.g. if K_m values are measured in the presence of an obvious cosubstrate, only the name of the cosubstrate is given in parentheses as a commentary without reference to its specific role).
The authors are grateful to the following biologists and chemists for invaluable help in the compilation of data: Cornelia Munaretto and Dr. Antje Chang.

Cologne
Spring 2006 *Dietmar Schomburg, Ida Schomburg*

List of Abbreviations

A	adenine
Ac	acetyl
ADP	adenosine 5'-diphosphate
Ala	alanine
All	allose
Alt	altrose
AMP	adenosine 5'-monophosphate
Ara	arabinose
Arg	arginine
Asn	asparagine
Asp	aspartic acid
ATP	adenosine 5'-triphosphate
Bicine	N,N'-bis(2-hydroxyethyl)glycine
C	cytosine
cal	calorie
CDP	cytidine 5'-diphosphate
CDTA	trans-1,2-diaminocyclohexane-N,N,N,N-tetraacetic acid
CMP	cytidine 5'-monophosphate
CoA	coenzyme A
CTP	cytidine 5'-triphosphate
Cys	cysteine
d	deoxy-
D-	(and L-) prefixes indicating configuration
DFP	diisopropyl fluorophosphate
DNA	deoxyribonucleic acid
DPN	diphosphopyridinium nucleotide (now NAD^+)
DTNB	5,5'-dithiobis(2-nitrobenzoate)
DTT	dithiothreitol (i.e. Cleland's reagent)
EC	number of enzyme in Enzyme Commission's system
E. coli	Escherichia coli
EDTA	ethylene diaminetetraacetate
EGTA	ethylene glycol bis(-aminoethyl ether) tetraacetate
ER	endoplasmic reticulum
Et	ethyl
EXAFS	extended X-ray absorption fine structure
FAD	flavin-adenine dinucleotide
FMN	flavin mononucleotide (riboflavin 5'-monophosphate)
Fru	fructose
Fuc	fucose
G	guanine
Gal	galactose

GDP	guanosine 5'-diphosphate
Glc	glucose
GlcN	glucosamine
GlcNAc	N-acetylglucosamine
Gln	glutamine
Glu	glutamic acid
Gly	glycine
GMP	guanosine 5'-monophosphate
GSH	glutathione
GSSG	oxidized glutathione
GTP	guanosine 5'-triphosphate
Gul	gulose
h	hour
H4	tetrahydro
HEPES	4-(2-hydroxyethyl)-1-piperazineethane sulfonic acid
His	histidine
HPLC	high performance liquid chromatography
Hyl	hydroxylysine
Hyp	hydroxyproline
IAA	iodoacetamide
Ig	immunoglobulin
Ile	isoleucine
Ido	idose
IDP	inosine 5'-diphosphate
IMP	inosine 5'-monophosphate
ITP	inosine 5'-triphosphate
K_m	Michaelis constant
L-	(and D-) prefixes indicating configuration
Leu	leucine
Lys	lysine
Lyx	lyxose
M	mol/l
mM	millimol/l
m-	*meta-*
Man	mannose
MES	2-(N-morpholino)ethane sulfonate
Met	methionine
min	minute
MOPS	3-(N-morpholino)propane sulfonate
Mur	muramic acid
MW	molecular weight
NAD^+	nicotinamide-adenine dinucleotide
NADH	reduced NAD
$NADP^+$	NAD phosphate
NADPH	reduced NADP
NAD(P)H	indicates either NADH or NADPH
NBS	N-bromosuccinimide

NDP	nucleoside 5'-diphosphate
NEM	N-ethylmaleimide
Neu	neuraminic acid
NMN	nicotinamide mononucleotide
NMP	nucleoside 5'-monophosphate
NTP	nucleoside 5'-triphosphate
o-	*ortho-*
Orn	ornithine
p-	*para-*
PBS	phosphate-buffered saline
PCMB	*p*-chloromercuribenzoate
PEP	phosphoenolpyruvate
pH	$-\log 10[H^+]$
Ph	phenyl
Phe	phenylalanine
PHMB	*p*-hydroxymercuribenzoate
PIXE	proton-induced X-ray emission
PMSF	phenylmethane-sulfonylfluoride
p-NPP	*p*-nitrophenyl phosphate
Pro	proline
Q_{10}	factor for the change in reaction rate for a 10°C temperature increase
Rha	rhamnose
Rib	ribose
RNA	ribonucleic acid
mRNA	messenger RNA
rRNA	ribosomal RNA
tRNA	transfer RNA
Sar	N-methylglycine (sarcosine)
SDS-PAGE	sodium dodecyl sulfate polyacrylamide gel electrophoresis
Ser	serine
T	thymine
t_H	time for half-completion of reaction
Tal	talose
TDP	thymidine 5'-diphosphate
TEA	triethanolamine
Thr	threonine
TLCK	N^{α}-*p*-tosyl-L-lysine chloromethyl ketone
T_m	melting temperature
TMP	thymidine 5'-monophosphate
Tos-	tosyl-(*p*-toluenesulfonyl-)
TPN	triphosphopyridinium nucleotide (now $NADP^+$)
Tris	tris(hydroxymethyl)-aminomethane
Trp	tryptophan
TTP	thymidine 5'-triphosphate
Tyr	tyrosine
U	uridine
U/mg	μmol/(mg*min)

UDP	uridine 5'-diphosphate
UMP	uridine 5'-monophosphate
UTP	uridine 5'-triphosphate
Val	valine
Xaa	symbol for an amino acid of unknown constitution in peptide formula
XAS	X-ray absorption spectroscopy
Xyl	xylose

List of Deleted and Transferred Enzymes

Since its foundation in 1956 the Nomenclature Committee of the International Union of Biochemistry and Molecular Biology (NC-IUBMB) has continually revised and updated the list of enzymes. Entries for new enzymes have been added, others have been deleted completely, or transferred to another EC number in the original class or to different EC classes, catalyzing other types of chemical reactions. The old numbers have not been allotted to new enzymes; instead the place has been left vacant or cross-references given to the changes in nomenclature.

Deleted and Transferred Enzymes
For EC class 2.1.2.1–2.3.1.59 these changes are:

Recommended name	Old EC number	Alteration
glutamate formyltransferase	2.1.2.6	deleted, included in EC 2.1.2.5
malonyl-CoA carboxyltransferase	2.1.3.4	deleted
kanamycin 6'-*N*-acetyltransferase	2.3.1.55	transferred to EC 2.3.1.82
6'-deoxychalcone synthase	2.3.1.120	deleted, reaction listed is due to EC 2.3.1.74
diacylglycerol acyltransferase	2.3.1.124	deleted, identical to EC 2.3.1.20

Index of Recommended Enzyme Names

Description of Data Fields

All information except the nomenclature of the enzymes (which is based on the recommendations of the Nomenclature Committee of IUBMB (International Union of Biochemistry and Molecular Biology) and IUPAC (International Union of Pure and Applied Chemistry) is extracted from original literature (or reviews for very well characterized enzymes). The quality and reliability of the data depends on the method of determination, and for older literature on the techniques available at that time. This is especially true for the fields *Molecular Weight* and *Subunits*.

The general structure of the fields is: **Information – Organism – Commentary – Literature**

The information can be found in the form of numerical values (temperature, pH, K_m etc.) or as text (cofactors, inhibitors etc.).

Sometimes data are classified as *Additional Information*. Here you may find data that cannot be recalculated to the units required for a field or also general information being valid for all values. For example, for *Inhibitors, Additional Information* may contain a list of compounds that are not inhibitory.

The detailed structure and contents of each field is described below. If one of these fields is missing for a particular enzyme, this means that for this field, no data are available.

1 Nomenclature

EC number

The number is as given by the IUBMB, classes of enzymes and subclasses defined according to the reaction catalyzed.

Systematic name

This is the name as given by the IUBMB/IUPAC Nomenclature Committee

Recommended name

This is the name as given by the IUBMB/IUPAC Nomenclature Committee

Synonyms

Synonyms which are found in other databases or in the literature, abbreviations, names of commercially available products. If identical names are frequently used for different enzymes, these will be mentioned here, cross references are given. If another EC number has been included in this entry, it is mentioned here.

CAS registry number

The majority of enzymes have a single chemical abstract (CAS) number. Some have no number at all, some have two or more numbers. Sometimes two enzymes share a common number. When this occurs, it is mentioned in the commentary.

2 Source Organism

For listing organisms their systematic name is preferred. If these are not mentioned in the literature, the names from the respective literature are used. For example if an enzyme from yeast is described without being specified further, yeast will be the entry. This field defines the code numbers for the organisms in which the enzyme with the respective EC number is found. These code numbers (form <_>) are displayed together with each entry in all fields of BRENDA where organism-specific information is given.

3 Reaction and Specificity

Catalyzed reaction

The reaction as defined by the IUBMB. The commentary gives information on the mechanism, the stereochemistry, or on thermodynamic data of the reaction.

Reaction type

According to the enzyme class a type can be attributed. These can be oxidation, reduction, elimination, addition, or a name (e.g. Knorr reaction)

Natural substrates and products

These are substrates and products which are metabolized in vivo. A natural substrate is only given if it is mentioned in the literature. The commentary gives information on the pathways for which this enzyme is important. If the enzyme is induced by a specific compound or growth conditions, this will be included in the commentary. In *Additional information* you will find comments on the metabolic role, sometimes only assumptions can be found in the references or the natural substrates are unknown.

In the listings, each natural substrate (indicated by a bold **S**) is followed by its respective product (indicated by a bold **P**). Products are given with organisms and references included only if the respective authors were able to demonstrate the formation of the specific product. If only the disappearance of the substrate was observed, the product is included without organisms of references. In cases with unclear product formation only a ? as a dummy is given.

Substrates and products

All natural or synthetic substrates are listed (not in stoichiometric quantities). The commentary gives information on the reversibility of the reaction,

on isomers accepted as substrates and it compares the efficiency of substrates. If a specific substrate is accepted by only one of several isozymes, this will be stated here.
The field *Additional Information* summarizes compounds that are not accepted as substrates or general comments which are valid for all substrates. In the listings, each substrate (indicated by a bold **S**) is followed by its respective product (indicated by a bold **P**). Products are given with organisms and references included if the respective authors demonstrated the formation of the specific product. If only the disappearance of the substrate was observed, the product will be included without organisms or references. In cases with unclear product formation only a **?** as a dummy is given.

Inhibitors

Compounds found to be inhibitory are listed. The commentary may explain experimental conditions, the concentration yielding a specific degree of inhibition or the inhibition constant. If a substance is activating at a specific concentration but inhibiting at a higher or lower value, the commentary will explain this.

Cofactors, prosthetic groups

This field contains cofactors which participate in the reaction but are not bound to the enzyme, and prosthetic groups being tightly bound. The commentary explains the function or, if known, the stereochemistry, or whether the cofactor can be replaced by a similar compound with higher or lower efficiency.

Activating Compounds

This field lists compounds with a positive effect on the activity. The enzyme may be inactive in the absence of certain compounds or may require activating molecules like sulfhydryl compounds, chelating agents, or lipids. If a substance is activating at a specific concentration but inhibiting at a higher or lower value, the commentary will explain this.

Metals, ions

This field lists all metals or ions that have activating effects. The commentary explains the role each of the cited metal has, being either bound e.g. as Fe-S centers or being required in solution. If an ion plays a dual role, activating at a certain concentration but inhibiting at a higher or lower concentration, this will be given in the commentary.

Turnover number (min^{-1})

The k_{cat} is given in the unit min^{-1}. The commentary lists the names of the substrates, sometimes with information on the reaction conditions or the type of reaction if the enzyme is capable of catalyzing different reactions with a single substrate. For cases where it is impossible to give the turnover number in the defined unit (e.g., substrates without a defined molecular weight, or an undefined amount of protein) this is summarized in *Additional Information*.

Specific activity (U/mg)

The unit is micromol/minute/milligram of protein. The commentary may contain information on specific assay conditions or if another than the natural substrate was used in the assay. Entries in *Additional Information* are included if the units of the activity are missing in the literature or are not calculable to the obligatory unit. Information on literature with a detailed description of the assay method may also be found.

K_m-Value (mM)

The unit is mM. Each value is connected to a substrate name. The commentary gives, if available, information on specific reaction condition, isozymes or presence of activators. The references for values which cannot be expressed in mM (e.g. for macromolecular, not precisely defined substrates) are given in *Additional Information.* In this field we also cite literature with detailed kinetic analyses.

K_i-Value (mM)

The unit of the inhibition constant is mM. Each value is connected to an inhibitor name. The commentary gives, if available, the type of inhibition (e.g. competitive, non-competitive) and the reaction conditions (pH-value and the temperature). Values which cannot be expressed in the requested unit and references for detailed inhibition studies are summerized under *Additional information.*

pH-Optimum

The value is given to one decimal place. The commentary may contain information on specific assay conditions, such as temperature, presence of activators or if this optimum is valid for only one of several isozymes. If the enzyme has a second optimum, this will be mentioned here.

pH-Range

Mostly given as a range e.g. 4.0–7.0 with an added commentary explaining the activity in this range. Sometimes, not a range but a single value indicating the upper or lower limit of enzyme activity is given. In this case, the commentary is obligatory.

Temperature optimum (°C)

Sometimes, if no temperature optimum is found in the literature, the temperature of the assay is given instead. This is always mentioned in the commentary.

Temperature range (°C)

This is the range over which the enzyme is active. The commentary may give the percentage of activity at the outer limits. Also commentaries on specific assay conditions, additives etc.

4 Enzyme Structure

Molecular weight

This field gives the molecular weight of the holoenzyme. For monomeric enzymes it is identical to the value given for subunits. As the accuracy depends on the method of determination this is given in the commentary if provided in the literature. Some enzymes are only active as multienzyme complexes for which the names and/or EC numbers of all participating enzymes are given in the commentary.

Subunits

The tertiary structure of the active species is described. The enzyme can be active as a monomer a dimer, trimer and so on. The stoichiometry of subunit composition is given. Some enzymes can be active in more than one state of complexation with differing effectivities. The analytical method is included.

Posttranslational modifications

The main entries in this field may be proteolytic modification, or side-chain modification, or no modification. The commentary will give details of the modifications e.g.:

- proteolytic modification <1> (<1>, propeptide Name) [1];
- side-chain modification <2> (<2>, N-glycosylated, 12% mannose) [2];
- no modification [3]

5 Isolation / Preparation / Mutation / Application

Source / tissue

For multicellular organisms, the tissue used for isolation of the enzyme or the tissue in which the enzyme is present is given. Cell-lines may also be a source of enzymes.

Localization

The subcellular localization is described. Typical entries are: cytoplasm, nucleus, extracellular, membrane.

Purification

The field consists of an organism and a reference. Only references with a detailed description of the purification procedure are cited.

Renaturation

Commentary on denaturant or renaturation procedure.

Crystallization

The literature is cited which describes the procedure of crystallization, or the X-ray structure.

Cloning

Lists of organisms and references, sometimes a commentary about expression or gene structure.

Engineering

The properties of modified proteins are described.

Application

Actual or possible applications in the fields of pharmacology, medicine, synthesis, analysis, agriculture, nutrition are described.

6 Stability

pH-Stability

This field can either give a range in which the enzyme is stable or a single value. In the latter case the commentary is obligatory and explains the conditions and stability at this value.

Temperature stability

This field can either give a range in which the enzyme is stable or a single value. In the latter case the commentary is obligatory and explains the conditions and stability at this value.

Oxidation stability

Stability in the presence of oxidizing agents, e.g. O_2, H_2O_2, especially important for enzymes which are only active under anaerobic conditions.

Organic solvent stability

The stability in the presence of organic solvents is described.

General stability information

This field summarizes general information on stability, e.g., increased stability of immobilized enzymes, stabilization by SH-reagents, detergents, glycerol or albumins etc.

Storage stability

Storage conditions and reported stability or loss of activity during storage.

References

Authors, Title, Journal, Volume, Pages, Year.

Glycine hydroxymethyltransferase 2.1.2.1

1 Nomenclature

EC number
2.1.2.1

Systematic name
5,10-methylenetetrahydrofolate:glycine hydroxymethyltransferase

Recommended name
glycine hydroxymethyltransferase

Synonyms
EC 4.1.2.6 (formerly)
L-serine hydroxymethyltransferase
L-threonine aldolase
allothreonine aldolase
serine aldolase
serine hydroxymethylase hydroxymethyltransferase, serine
serine hydroxymethyltransferase
serine transhydroxymethylase
threonine aldolase

CAS registry number
9029-83-8

2 Source Organism

<1> *Ovis aries* (lamb [6-8]) [4, 6-8, 25, 27, 31, 36-41, 45]
<2> *Bos taurus* [28, 31]
<3> *Oryctolagus cuniculus* [5, 11-13, 26, 29-31, 34, 38, 45]
<4> *Rattus norvegicus* (Wistar strain [10]) [9, 10, 45]
<5> *Macaca radiata* (Bonnet monkey [14,15]) [14, 15]
<6> *Mus musculus* [16]
<7> *Cricetulus griseus* [31, 45]
<8> *Spinacia oleracea* [2]
<9> *Pisum sativum* (pea [17]) [17, 38, 45]
<10> *Vigna radiata* (mung bean [19]) [19, 45]
<11> *Glycine max* (strain L. Merr. cv. Williams, soybean [18]) [18]
<12> *Euglena gracilis* (strain z) [20, 48]
<13> *Saccharomyces cerevisiae* (strain 3701B [24]) [24, 31, 38]
<14> *Neurospora crassa* [31, 45]

<15> *Hyphomicrobium methylovorum* (strain GM2, a glycine resistant mutant derived from strain KM146 [21]) [21, 22]
<16> *Escherichia coli* [23, 29, 35, 38, 44, 45]
<17> *Salmonella typhimurium* [31]
<18> *Geobacillus stearothermophilus* [1, 42, 43]
<19> *Lactobacillus bulgaricus* (YOP12 [3]) [3]
<20> *Homo sapiens* [32, 38, 45, 46, 47]
<21> *Sulfolobus solfataricus* [33]
<22> *Caenorhabditis briggsae* [49]
<23> *Caenorhabditis elegans* [49]

3 Reaction and Specificity

Catalyzed reaction

5,10-methylenetetrahydrofolate + glycine + H_2O = tetrahydrofolate + L-serine (<1-3, 5-7, 11, 14, 16, 17, 18, 20> mechanism [7, 8, 13-16, 18, 23, 25-31, 35, 41, 43, 45, 47]; <1, 3, 4, 7, 9, 10, 14, 16, 20> enzyme exhibits threonine aldolase activity (EC 4.1.2.5), but the two enzymes are distinct [45]; <18> evidence for a direct transfer mechanism for the enzyme catalysed reaction [42]; <1-3, 7, 14, 17> stereochemistry [28, 30, 31, 35])

Reaction type

hydroxymethyl group transfer

Natural substrates and products

S L-serine + tetrahydrofolate <1-7, 9, 10, 11, 13-18, 20, 21> (<4, 16, 20> major pathway for production of C_1-units of 5,10-methylenetetrahydrofolate [9, 23, 31, 45, 47]; <6, 10, 13, 15, 21> key enzyme of serine pathway for assimilation of C_1-compounds [16, 19, 21, 22, 24, 33]; <3, 4, 16> catalyzes interconversion of serine and glycine [10, 34, 44]; <5, 10> regulatory protein [14, 19]; <6, 11, 20> main glycine source for purine biosynthetic pathway in ureide biogenesis [16, 18, 46]; <1-3, 7, 14, 17> may play important role in central nervous system [31]; <1, 3> enzyme is a component of thymidylate cycle [34, 36]; <1,3,9,13,16,20> enzyme plays a pivotal role in channelling metabolites between amino acid and nucleotide metabolism [38]) [9, 10, 14, 16, 18, 19, 21-24, 26, 31, 33, 34, 36, 38, 41-47]
P glycine + 5,10-methylenetetrahydrofolate + H_2O <1-7, 9, 10, 11, 13-18, 20, 21> [9, 10, 14, 16, 18, 19, 21-24, 26, 31, 33, 34, 36, 38, 41-47]

Substrates and products

S 3-hydroxy-ε-N-trimethyllysine <1-3, 7, 14, 17> (<3> very slow reaction [5]) (Reversibility: r <1-3, 7, 14, 17> [5, 31]) [5, 31]
P glycine + 4-butyrobetaine aldehyde <1-3, 7, 14, 17> [5, 31]
S DL-*p*-hydroxy-*m*-methoxy-erythro-2-phenylserine <1> (Reversibility: ? <1> [8]) [8]
P glycine + *p*-hydroxy-*m*-methoxy-benzaldehyde <1> [8]

S L-3-phenylserine <1, 2, 3, 7, 9, 13-18, 20, 21> (<1, 15> D,L threo-configuration [6,21,22,45]; <1> DL-erythro-configuration [6, 25, 45]) (Reversibility: r <1, 3, 9, 13, 15, 16, 20> [6, 8, 22, 38]; ? <2, 7, 14, 16, 17, 18, 21> [21, 23, 25, 31, 33, 43]) [6, 8, 21, 22, 23, 25, 31, 33, 38, 43, 45]
P glycine + benzaldehyde <1, 2, 7, 9, 13-17, 0, 21> [6, 21, 22, 25, 31, 33, 38]
S L-allothreonine <1-4, 7, 9, 12-18, 20, 21> (<1, 3, 4, 16, 18, 21> tetrahydrofolate independent [5, 6, 9, 11, 22, 23, 33-35, 40, 41, 43]; <16> preferred substrate [35]) (Reversibility: r <1, 3, 9, 13, 16, 20> [5, 6, 13, 30, 35, 38]; ? <2, 4, 7, 12, 14, 15-18, 20, 21> [5, 9, 10, 28, 32, 33, 40, 41, 43, 45, 47, 48]) [5, 6, 9-11, 13, 20, 22, 23, 28, 30-35, 38, 40, 41, 43, 45, 47, 48]
P glycine + acetaldehyde <1-4, 7, 9, 12-17, 20, 21> [5, 6, 9-11, 20, 22, 23, 30, 31, 33, 38, 48]
S L-erythro-3-phenylserine methylester <1> (Reversibility: ? <1> [6]) [6]
P glycine methylester + benzaldehyde <1> [6]
S L-serine + modified folate <21> (<21> not: tetrahydrofolate, synthetic modified folate derivate [33]) (Reversibility: ? <21> [33]) [33]
P glycine + modified methylenefolate <21> [33]
S L-serine + tetrahydrofolate <1-18, 20> (Reversibility: r <1, 3-5, 9, 10-12, 15, 16, 18, 20> [5, 6, 9, 10, 12, 13, 15, 18-23, 26, 30-32, 34-36, 38, 42, 47]; ? <2, 6-9, 13, 14, 16, 17> [2, 4, 11, 14, 16, 17, 24, 27-29, 39-41, 44, 46, 48]) [2, 4-6, 9-24, 26-32, 34-36, 38-48]
P glycine + 5,10-methylenetetrahydrofolate + H_2O <1-3, 6, 7, 9, 10-18, 20> [5, 11, 12, 16, 18, 19, 20-22, 26, 30, 31, 38-43, 45-48]
S L-threo-3,4-dihydroxyphenylserine <15> (Reversibility: ? <15> [22]) [22]
P ?
S L-threo-3-thienylserine <1> (Reversibility: ? <1> [6]) [6]
P glycine + 3-thiophenaldehyde <1> [6]
S L-threonine <1-3, 4, 7, 12, 14-17> (<1-3,7,12,14-17> tetrahydrofolate independent [5,6,11,20-23,30,31]; <4> no substrate [9]) (Reversibility: r <1, 3, 15, 16> [6, 13, 21, 30, 31, 35, 45]; ? <2, 4, 7, 12, 14, 16, 17> [5, 9, 11, 20, 22, 23, 48]) [5, 6, 9, 11, 13, 20-23, 30, 31, 35, 45, 48]
P glycine + acetaldehyde <1-3, 7, 12, 14-17> [6, 11, 13, 20, 22, 23, 30, 31, 45, 48]
S α-methylserine + tetrahydrofolate <1-3, 7, 9, 13, 14, 16, 17, 20> (<3> at high concentrations of enzyme [12]) (Reversibility: ? <1-3, 7, 9, 13, 14, 16, 17, 20> [12, 31, 38, 45]) [12, 31, 38, 45]
P D-alanine + 5,10-methylenetetrahydrofolate <1-3, 7, 14, 17> [12, 31, 38, 45]
S aminomalonate <1, 3, 9, 13, 16, 20> (<1> decarboxylation [6]) (Reversibility: r <1, 3, 9, 13, 16, 20> [31, 38]; ? <1> [6, 45]) [6, 31, 38, 45]
P glycine + CO_2 <1, 3, 9, 13, 16, 20> [6, 31, 38]
S substituted L-erythro-3-phenylserines <1> (Reversibility: ? <1> [8]) [8]
P glycine + substituted benzaldehydes <1> (<1> via pyridoxal 5'-phosphate-stabilized quinoid intermediate [8]) [8]
S Additional information <1-5, 7, 9, 12-18, 20> (<1, 4, 16> enzyme catalyzes the pyridoxal 5'-phosphate dependent reversible cleavage of 3-hydroxy-α-amino acids [6, 10, 35, 39-41]; <1-4, 7, 14, 16, 17> enzyme trans-

aminates D-alanine to pyruvate and pyridoxamine phosphate [6, 10, 23, 31, 35, 38, 40, 45]; <1, 16, 18> enzyme catalyses the racemization of D- and L-alanine [35, 40, 42, 43]; <1> enzym reacts with L-phenylalanine in absence of tetrahydrofolic acid to form an enzyme-bound amino-acid-pyridoxal 5'-phosphate quinonoid intermediate [7]; <1> mutant enzyme H230Y catalyses oxidation of NADH, not wild type enzyme, H230A, H230F, H230N [40]; <5> insoluble enzyme/antibody-complex shows 90% of original activity [14]; <1-4, 7, 12, 15> no: D-serine [9,10,20,22,31,48]; <4, 12> no: D-allothreonine [9, 10, 20, 48]; <1-4, 7, 14, 17> no: L-threonine [9, 10, 31]; <4, 12, 15> no: D-threonine [9, 20, 22, 48]; <1, 3, 4, 7, 9, 10, 14, 16, 20> review and comparison of enzyme activity, threonine aldolase and allothreonine aldolase activity from various sources [45]) [6, 7, 9, 10, 14, 20, 22, 31, 35, 38-43, 45, 47, 48]

P ?

Inhibitors

1,10-phenanthroline <3, 15> [12, 21]
2,2'-dipyridyl <15> [21]
2-aminoisobutyrate <3> [11]
2-mercaptopropionic acid <12> [20]
4-chloro-L-threonine <1, 3, 4, 7, 9, 10, 13, 14, 16, 20> [38, 45]
5,5-dithiobis(2-nitrobenzoic acid) <15, 16> [21, 23]
5-formyltetrahydrofolate monoglutamate <3, 16> [29]
5-methyltetrahydrofolate triglutamate <3, 16> [29]
Ag^+ <15> (<15> strong [21]) [21]
Cibacron blue F3GA <1, 5> (<1> complete inhibition, NAD(H) protects, reversible by tetrahydrofolate [4]; <5> reversible by dialysis [15]) [4, 15]
Cu^{2+} <15> (<15> strong [21]) [21]
D-alanine <1-4, 7, 16> (<4,16> inactivates enzyme by converting the enzyme bound pyridoxal 5'-phosphate to pyridoxamine phosphate in a transamination reaction [10,23,35,36]) [10, 11, 23, 31, 35, 36]
D-β-fluoroalanine <1-3, 7> [31]
D-cycloserine <1-3, 5, 7, 10, 15> (<1> interaction extreme rapidly and irreversible [27]) [4, 14, 19, 21, 22, 27, 31, 38]
DL-2-methylserine <1, 3, 10> [6, 12, 19]
DL-O-methylserine <1, 10> [6, 19]
EDTA <3> (<3> weak [12]) [12]
Hg^{2+} <15> (<15> strong [21]) [21]
KCN <15> [21]
KCl <11, 15> [18, 21]
L-alanine <1, 4, 12> [6, 9, 20, 48]
L-α,β-diaminopropionic acid <12> [48]
L-amino acids <4, 12> (<4> weak, e.g. L-aspartic acid, ornithine, lysine, methionine, phenylalanine, homoserine, threonine, 4-aminobutyric acid [9]) [9, 20, 48]

L-cysteine <3, 4, 12, 21> (<4> mitochondrial enzyme [9]) [5, 9, 20, 33, 48]
L-methionine <12, 13> (<12> not [48]) [31, 48]
L-serine <1, 3, 4, 6, 12> (<3> competitive to glycine [5]; <4> strong [9]) [5, 6, 9, 11, 16, 20, 48]
L-threonine <3> [11]
LiCl <15> [21]
N-ethylmaleimide <3> [12]
NAD^+ <1> (<1> negative effector [4]) [4]
NH_2OH <3, 15> (<3> weak [12]) [12, 21]
$NaCNBH_3$ <21> [33]
NaCl <11, 15> [18, 21]
NaF <15> [21]
O-phosphoserine <10> [19]
PCMB <3, 15> (<15> strong [21]) [12, 21]
Pb^{2+} <15> (<15> strong [21]) [21]
S-adenosyl-L-methionine <13> [31]
allothreonine <3> [11]
aminopterin <5, 10> [14, 19]
antibodies to cytosolic enzyme <3> [5]
antibodies to mitochondrial enzyme <3> (<3> no inhibition of cytosolic enzyme [5]) [5]
β-alanine <4, 12> (<4> weak [9]) [9, 48]
β-aminoalanine <4> [9]
β-chloroalanine <15> (<15> suicide substrate [22]) [22]
β-trifluoroallothreonine <3, 9, 13, 16, 20> [38]
β-trifluorothreonine <1, 3, 9, 13, 16, 20> [38]
bromopyruvate <3> (<3> irreversible inactivation, substrates partially protect [13]; <3> only cytoplasmac not mitochondrial enzyme [30]) [13, 30]
carboxymethoxylamine <8> (<8> strong [2]) [2]
chloroacetaldehyde <3> (<3> irreversible inactivation, substrates partially protect [13]; <3> only cytoplasmic not mitochondrial enzyme [30]) [13, 30]
dichloromethotrexate <5, 10> [14, 19]
dihydrofolate <5> [14]
glycidaldehyde <3> (<3> directed to C_1-binding-site [26]; <3> only cytoplasmic enzyme [30]) [26, 30]
glycine <1, 3-5, 10-12> [6, 9, 11, 14, 18-20, 48]
glycine ethylester <1> [6]
glycine methylester <1> [6]
hydrazine <3> [12]
iodoacetamide <3> (<3> irreversible inactivation, substrates partially protect [13]; <3> only cytoplasmic not mitochondrial enzyme [30]) [13, 30]
methotrexate <5, 10> [14, 19]
methyl methanethiosulfonate <1-3, 7, 16> [23, 31]
penicillamine <15> [21]
phenylhydrazine <15> (<15> partially reversible by pyridoxal 5'-phosphate [21]) [21]
semicarbazide <15> [21]

sodium borohydride <15> [21]
substituted hydroxylamine derivates <1, 3, 9, 13, 16, 20> [38]
sulfonyl fluoride triazine derivates <1, 3, 9, 13, 16, 20> [38]
tetrahydrofolate derivatives <1-3, 5, 6, 7, 11> [14, 16, 18, 31]
thiosemicarbazide <3, 9, 13, 16, 20> [38, 39]
Additional information <1, 3-5, 11-13, 15, 16> (<1,3,5> inhibition kinetics [14,25-27,29]; <1> not inhibitor: sarcosine [6]; <4,12> not: valine, leucine, glutamic acid, 3-hydroxybutyric acid [9,20,48]; <4> not: mercaptopropionic acid, mercaptoethanolamine [9]; <12> not: ethanolamine, ethylenediamine [20]; <3,16> not: iodoacetate [12,13,23]; <3> not: NaN_3, mono- or divalent cations, 2-mercaptoethanol, DTT [12]; <3,13> not: purine nucleoside mono-, di- and triphosphates [12,24]; <3> not: chloroacetamide [13]; not: EDTA, <13,15> [21,24]; <11> not: methionine, S-adenosyl-L-methionine, XMP, IMP, phosphoglycerate [18]; <3> not: N-/O-chloroacetyl and N-/O-bromoacetyl derivatives of glycine and L-serine [30]) [6, 9, 12-14, 18, 20, 21, 23-27, 29, 30, 48]

Cofactors/prosthetic groups

NADH <1, 5, 10> (<10> positive allosteric heterotropic effector [15,19]) [4, 14, 15, 19]
pyridoxal 5'-phosphate <1-5, 7, 9, 10, 12, 13, 15, 16, 18, 20, 21> (<1-5,7> requirement, 4 mol per mol enzyme [6,9-12,14,31]; <12,15> 2 mol per mol enzyme [20,21,48]; <1,3,9,13,16,20> one mol/subunit bound to the ε-amino group of lysine [38,46]; <1> one mol per subunit [39]; <1-3,7> mechanism, active-site structure [31]; <10> not essential [19]; <1,3,9,13,16,20> important role in maintaining the structural integrity of the enzyme by preventing the dissociation of the enzyme into subunits, in addition to its function in catalysis [37-39]) [6-14, 17, 19-21, 23-29, 31, 33, 37-43, 45, 46, 48]
tetrahydrofolate <1-4, 5, 7, 12, 13, 15-16, 20> (<1-4,7,12,13,15-16,20> requirement with L-serine or L-2-methylserine as substrate [5,9,11,20,21-24,30-32,48]; <1-3,7> mechanism [31]; <3> allosteric regulation [5]; <1> not [6]) [5, 6, 9, 10, 11, 14, 20-24, 30-32, 46, 48]
tetrahydrofolate derivatives <16> (<16> requirement, if L-serine is substrate [23]) [23]

Activating compounds

Additional information <5> (<5> in the presence of 1mM dithiothreitol + 1mM pyridoxal 5-phosphate + 20 mM EDTA: 29fold increase in the specific activity [14]) [14]

Metals, ions

Ca^{2+} <3, 13> (<13> activation at 0.5-10 mM [24]; <3> not [12]) [12, 24]
Mg^{2+} <3, 13> (<13> activation at 0.5-10 mM [24]; <3> not [12]) [12, 24]
Mn^{2+} <3, 13> (<13> activation at 0.5-10 mM [24]; <3> not [12]) [12, 24]

Turnover number (min^{-1})

0.6 <1> (L-serine, <1> mutant enzyme E74Q [41]) [41]
4.3 <16> (L-threonine) [35]
5.4 <1> (L-threonine) [6]

5.8 <1> (L-serine, <1> mutant enzyme Y82F [41]) [41]
6.62 <1> (DL-allothreonine, <1> mutant enzyme Y82F [41]) [41]
30 <16> (allothreonine) [23]
33.6 <1> (DL-allothreonine) [6]
42 <18> (DL-allothreonine, <18> only tetrameric form [43]) [43]
42 <1> (L-serine, <1> mutant enzyme [39]) [39]
54 <18> (DL-allothreonine, <18> only dimeric form [43]) [43]
64 <20> (L-allothreonine, <20> recombinant enzyme [32]) [32]
80 <3> (L-allothreonine, <3> recombinant enzyme [34]) [34]
167 <16> (DL-threo-phenylserine) [35]
190.2 <1> (DL-allothreonine, <1> mutant enzyme E74Q [41]) [41]
216 <18> (L-serine, <18> only tetrameric form [43]) [43]
222 <1, 18> (DL-allothreonine) [41, 43]
246 <1> (L-serine) [41]
252 <1> (L-serine) [39]
258 <18> (L-serine) [43]
300 <18> (L-serine, <18> only dimeric form [43]) [43]
420 <1> (DL-threo-phenylserine) [6]
575 <20> (L-serine) [32]
630 <3> (L-serine) [34]
640 <16> (L-serine) [23]
850 <3> (L-serine, <3> recombinant enzyme [34]) [34]
1260 <1> (DL-erythro-3-phenylserine) [6]
1740 <1> (DL-erythro-3-phenylserine methylester) [6]
Additional information <1, 3, 5, 6, 11, 16> [5, 14, 16, 18, 23, 25]

Specific activity (U/mg)

0.012 <1> (<1> mutant enzyme E74Q [41]) [41]
0.02 <1> (<1> crude extract [36]) [36]
0.048 <12> (<12> reaction of L-serine [48]) [48]
0.15 <12> (<12> reaction of L-threonine [48]) [48]
0.16 <1> (<1> mutant enzyme Y82F [41]) [41]
0.52 <1> (<1> mutant enzyme [40]) [40]
0.6 <1> (<1> recombinant enzyme, crude extract [36]; <1> mutant enzyme [39]) [36, 39]
1.388 <12> (<12> reaction of L-allothreonine [48]) [48]
2.3 <1> [6]
2.43 <3> [11]
2.55 <12> [20]
3.2 <1> (<1> purified recombinant enzyme [36]) [36]
3.4 <5> [14]
4.1 <18> (<18> only tetrameric form [43]) [43]
4.2 <1> [41]
4.2 <18> [43]
4.3 <1> [39]
4.8 <1> [40]

6.7 <18> (<18> only dimeric form [43]) [43]
7.19 <4> [10]
7.8 <8> [2]
7.9 <15> (<15> serine synthesis [21]) [21]
9.7 <6> [16]
10 <3, 20> (<3> soluble [12]; <20> recombinant enzyme [32]) [12, 32]
11.9 <11> [18]
13.6 <16> [23]
15.5 <3> (<3> mitochondria [12]) [12]
32.4 <18> (<18> dimeric form at 80°C [43]) [43]
Additional information <1, 3-5, 9-13, 16, 20, 21> (<1,3,4,9,10,16,20> review [45]) [5, 9, 12-14, 17-20, 24, 30, 33, 34, 36, 38, 40, 45, 48]

K_m-Value (mM)

0.02 <3> (tetrahydrofolate, <3> recombinant enzyme [34]) [34]
0.04-0.046 <3, 15, 20> (tetrahydrofolate, <15> glycine [21]; <20> recombinant enzyme [32]) [5, 21, 32]
0.08 <16> (tetrahydrofolate) [23]
0.15-0.9 <1, 3, 4, 6, 8, 12, 13, 16, 20> (L-serine, <3,20> recombinant enzyme [32,34]; <1> mutant enzyme Y82F [41]) [2, 9, 11, 16, 21, 23, 24, 32, 34, 41, 48]
0.25 <11> (tetrahydrofolate) [18]
0.45 <1> (L-allothreonine, <1> mutant enzyme Y82F [41]) [41]
0.6 <20> (L-allothreonine, <20> recombinant enzyme [32]) [32]
0.7 <1, 18> (L-allothreonine, <18> only tetrameric form [43]) [41, 43]
0.7 <18> (L-serine, <18> only tetrameric form [43]) [43]
0.82 <1> (tetrahydrofolate, <1> recombinant enzyme [36]) [36]
0.9 <18> (L-serine, <18> only dimeric form [43]) [43]
0.9 <1> (tetrahydrofolate) [39]
1 <1, 3, 18> (L-serine, <3> mitochondrial enzyme [12]; <1> recombinant enzyme [36]) [12, 36, 39, 41, 43]
1.3 <3> (L-serine, <3> cytosolic enzyme [12]) [12]
1.3-1.5 <1, 3, 16> (allothreonine, <3> recombinant enzyme [34]) [6, 23, 34]
1.4 <3> (L-serine) [5]
1.5 <1, 11> (L-serine, <1> mutant enzyme E74Q [41]) [18, 41]
1.63 <12> (L-allothreonine) [20]
1.8 <10> (L-serine) [19]
2-3.4 <3, 4> (L-allothreonine) [5, 9, 11]
2.1 <1> (tetrahydrofolate, <1> mutant enzyme [39]) [39]
4 <1> (L-serine, <1> mutant enzyme [39]) [39]
4.3 <12> (L-allothreonine) [48]
9.5 <1> (DL-erythro-3-phenylserine) [6]
10.6 <1> (L-allothreonine, <1> mutant enzyme E74Q [41]) [41]
13.9 <12> (L-threonine) [20]
19 <16> (DL-threo-3-phenylserine) [35]
25 <12> (L-serine) [20]
32-40 <1, 3> (L-threonine) [5, 6, 11]

33 <15> (DL-3-phenylserine) [21]
35.7 <12> (L-allothreonine) [48]
43 <16> (L-threonine) [35]
65.3 <12> (L-serine) [48]
70 <1> (DL-erythro-3-phenylserine methylester) [6]
84 <1> (DL-threo-3-phenylserine) [6]
Additional information <1-3, 5-7, 10, 11, 13, 15-16, 19> (<5> allosteric kinetics [14]; <1-3,7> pH-dependence of kinetic parameters [25,31]) [3, 7, 8, 14, 16, 18, 19, 21, 23-25, 31]

K_i-Value (mM)

0.014 <8> (carboxymethoxylamine) [2]
0.3 <12> (L-cysteine) [20]
0.7-0.9 <1, 3> (L-serine) [6, 11]
1.3 <12> (2-aminoalanine) [20]
1.8 <11> (5-methyl-5,6,7,8-tetrahydrofolate) [18]
2.02 <10> (D-cycloserine) [19]
2.3 <10> (glycine) [19]
2.7 <3> (DL-allothreonine) [11]
2.7-6.8 <3> (glycine) [11]
2.8 <12> (L-serine) [20]
2.9 <11> (5,10-methylene-5,6,7,8-tetrahydrofolic acid) [18]
3 <11> (5,10-methenyltetrahydrofolate) [18]
3 <11, 12> (glycine) [18, 20]
3.9 <3> (DL-2-methylserine, <3> soluble fraction [12]) [12]
7.6 <1> (L-alanine) [6]
11 <1> (glycine) [6]
11.2 <3> (DL-2-methylserine, <3> mitochondrial enzyme [12]) [12]
13-23 <3> (aminoisobutyrate) [11]
30 <1> (glycinemethylester) [6]
40-55 <3> (D-alanine) [11]
50 <3> (L-threonine) [11]
60 <1> (glycineethylester) [6]
Additional information <3, 6> (<3> comparison of K_i for allothreonine and threonine [11]) [11, 16, 29]

pH-Optimum

6.5 <19> [3]
6.8 <12> [20, 48]
6.8-8 <4> [9]
7.3 <3> [12]
7.4 <5> [14]
7.5-8 <15> (<15> 3-phenylserine degradation [21]) [21]
8 <15> (<15> serine synthesis [21]) [21]
8-9.5 <13> [24]

8.5 <8, 10, 11> [2, 18, 19]
Additional information <4, 5, 15> (<4> pI, cytosolic enzyme: 4.5, pI, mitochondrial enzyme: 4.8 [9]; <4> pI, cytosolic enzyme: 4.95, pI, mitochondrial enzyme: 5.3 [10]; <5> pI: 4.2 [14]) [9, 10, 14, 21]

pH-Range
6-8.5 <3> (<3> cytosol: about 60% of maximal activity at pH 6, about 70% at 8.5, mitochondria: about 40% of maximal activity at pH 6.0 and 8.5 [12]) [12]
6.5-9.5 <10> (<10> about half-maximal activity at pH 6.5 and 9.5 [19]) [19]
7-9 <15> (<15> serine synthesis: about 80% of maximal activity at pH 7.0 and 9.0 [21]) [21]
7.2-8.6 <15> (<15> 3-phenylserine degradation: about half-maximal activity at pH 7.2 and 8.6 [21]) [21]
7.3-9.5 <11> (<11> about half-maximal activity at pH 7.3 and about 60% of maximal activity at 9.5 [18]) [18]
9.5 <11> (<11> no activity above [18]) [18]
Additional information <1> (<1> pH dependencies of K_m and turnover number [6]) [6, 9]

Temperature optimum (°C)
35 <10> (<10> 2 optima: 35°C and 55°C [19]) [19]
37 <15> (<15> serine synthesis [21]) [21]
37-40 <15> (<15> 3-phenylserine degradation [21]) [21]
40 <19> [3]
55 <10> (<10> 2 optima: 35°C and 55°C [19]) [19]
60 <1, 5> [4, 14]
80 <18> (<18> dimeric form of enzyme [43]) [43]

Temperature range (°C)
25-75 <10> (<10> about half-maximal activity at 25°C and 75°C [19]) [19]
29-42 <15> (<15> serine synthesis, about half-maximal activity at 28.5°C and 90% of maximal activity at 42°C [21]) [21]
30-42 <15> (<15> 2-phenylserine degradation, about half-maximal activity at 29.5°C and 80% of maximal activity at 42°C [21]) [21]
Additional information <1> (<1> thermograms of enzyme and mutant enzyme in the absence and presence of L-serine [41]) [41]

4 Enzyme Structure

Molecular weight
80000 <15> (<15> HPLC gel filtration [21]) [21]
88000 <12> (<12> gel filtration [48]) [48]
90000 <12, 18> (<12, 18> gel filtration [20, 43]; <18> dimeric form [43]) [20, 43]
96000 <16> (<16> sedimentation equilibrium centrifugation, amino acid composition [23]) [23]
98000 <15> (<15> gel filtration [21]) [21]
100000-115000 <16> (<16> HPLC gel filtration [23]) [23]

130000 <1> (<1> apoenzyme, peak 2, gel filtration [37]) [37]
170000 <3> (<3> mitochondria, sedimentation equilibrium centrifugation [12]) [12]
180000 <18> (<18> tetrameric form, gel filtration [43]) [43]
180000-200000 <13> (<13> gel filtration, sucrose density gradient centrifugation [24]) [24]
185000 <3> (<3> soluble, high speed sedimentation equilibrium centrifugation [12]) [12]
190000 <19> (<19> gel filtration [3]) [3]
200000 <1> (<1> holoenzyme, apoenzyme: peak 1, gel filtration [37]) [37]
200000-204000 <4> (<4> mitochondria, gel filtration, PAGE, amino acid composition) [10]
205000 <10> (<10> gel filtration [19]) [19]
208000 <5> (<5> gel filtration [14]) [14]
209000 <8> (<8> gel filtration [2]) [2]
210000 <1> (<1> gel filtration [4]; <1> wild-type and mutant enzyme, gelfiltation [39]) [4, 39]
213000 <1> (<1> gel filtration [37]) [37]
215000 <3, 6> (<3> gel filtration, amino acid composition [5]; <6> HPLC gel filtration [16]) [5, 16]
216000 <4> (<4> cytosol, PAGE [10]) [10]
220000 <1, 4, 9> (<4,9> gel filtration [9,17,41]; <1> mutant enzyme [41]) [9, 17, 41]
227000 <1> (<1> high speed sedimentation equilibrium method by Yphantis [6]) [6]
230000 <4, 11> (<4> cytosol, amino acid composition, gel filtration [10]) [10, 18]
Additional information <1, 3, 4, 9, 10, 16, 20> (<1,3,4,9,10,16,20> review [45]) [45]

Subunits

dimer <1, 9, 12, 13, 15, 16, 18, 20, 21> (<12> 2 * 45000, SDS-PAGE [20]; <12> 2 * 44000, SDS-PAGE [48]; <16> 2 * 46000, SDS-PAGE [23]; <16> 2 * 47000, calculated from DNA-sequence [23]; <15> 2 * 48000, SDS-PAGE [21]; <21> 2 * 53000, SDS-PAGE [33]; <1> removal of the bound pyridoxal 5'-phosphate from the mutant tetrameric enzyme leads to dissociation to a dimer [39, 41]; <18> gelfiltration, crystallographic symmetry [42]) [20, 21, 23, 33, 38, 39, 41, 42, 48]
homotetramer <1-11, 13, 16, 20> (<2> 4 * 45000, SDS-PAGE [28]; <10> 4 * 50000, SDS-PAGE [19]; <5> 4 * 52000, SDS-PAGE [14]; <1,3,8> 4 * 53000, SDS-PAGE [2,5,36]; <1> 4 * 53000, mutant enzyme [41]; <9> lithium(dodecyl sulfate)-PAGE [17]; <4> 4 * 53700, mitochondria, SDS-PAGE [10]; <4> 4 * 54000, SDS-PAGE [9]; <11> 4 * 55000, SDS-PAGE [18]; <6> 4 * 56000, SDS-PAGE [16]; <4> 4 * 56300, cytosol, SDS-PAGE [10]; <1,3,9,13,16,20> actually a dimer of dimers [38]; <1,3,4,9,10,16,20> review [45]) [2, 5, 9, 10, 14, 16-19, 28, 31, 36, 38, 39, 41, 45]
Additional information <18> (<18> 90% dimer and 10% tetramer in the same organism, one subunit: 45000 Da, SDS-PAGE [43]) [43]

Posttranslational modification

side-chain modification <3, 9, 13, 16, 20> (<3,9,13,16,20> in addition to the lysine residue involved in Schiff base formation with the PLP, other residues like arginine, histidine, cysteine and tryptophan essential for catalysis, review [38]) [38]

Additional information <20> (<20> enzyme is a mRNA binding protein [47]) [47]

5 Isolation/Preparation/Mutation/Application

Source/tissue

L-1210 cell <6> (<6> transformed cells [16]) [16]
brain <5> [14]
kidney <5> [14]
leaf <8, 9> [2, 17]
nodule <11> (<11> inoculated with Rhizobium japonicum strain 311b142 [18]) [18]
ovary <7> [31]
pancreas <5> [14]
seed <10> (<10> germinating [19]) [19]
skeletal muscle <5> [14]
Additional information <5> (<5> distribution [14]) [14]

Localization

chloroplast <12> [48]
cytoplasm <1-7, 10, 12-14, 16, 17, 20> (<5> 75% of the enzyme activity [14]) [4, 5, 9, 10, 12-14, 16, 19, 20, 26, 29-32, 34, 36, 38, 46-48]
microsome <5> (<5> 3% of the enzyme activity [14]) [14]
mitochondrion <1-9, 12, 13, 16, 20> (<5> 5% of the enzyme activity [14]; <1-3,7> matrix [31]) [2, 5, 9, 10, 12, 14, 16, 17, 30, 31, 38, 48]
Additional information <3, 4> (<4> antigenically related [10]; <3> immunochemically different [12]; <3,4> comparison of cytosolic and mitochondrial forms of enzyme [5,9]; <3,4> isozymes in cytosol and mitochondria [10,12,30]) [5, 9, 10, 12]

Purification

<1> (recombinant enzyme [36]) [4, 6, 36]
<4> (affinity chromatography on L-adsorbent Sepharose [9]) [9, 10]
<5, 3> (heat-denaturation omitted [14]) [14, 5, 11-13, 34]
<6> [16]
<8> [2]
<9> [17]
<10> [19]
<11> [18]
<12> (affinity chromatography on L-adsorbent [20]) [20, 48]

<13> [24]
<15> [21]
<16> [23]
<18> (gene cloned and expressed in Escherichia coli [1,43]) [1, 43]
<19> [3]
<20> (recombinant enzyme [32]) [32]
<21> (recombinant enzyme [33]) [33]

Crystallization

<3> (mitochondrial enzyme [5,12]) [5, 12, 45]
<18> (internal aldimine form, external aldimine form with bound serine and glycine, ternary complex with glycine and 5-formyl tetrahydrofolate [42]) [42]
<20> [38, 45]

Cloning

<1, 3, 16, 18, 20> (expression in Escherichia coli [1,32,34,36,39-43,47]; <16> Escherichia coli glyA gene [23]) [1, 23, 32, 34-36, 38-43, 47]
<10, 14, 16, 20> [45, 46]
<22, 23> (comparison [49]) [49]

Engineering

E74K <1> (<1> specific activities drastically reduced with serine as substrate, but D-alanine transamination and allothreonine cleavage at rates comparable with wild-type enzyme [41]) [41]
E74Q <1> (<1> specific activities drastically reduced with serine as substrate, but D-alanine transamination and allothreonine cleavage at rates comparable with wild type enzyme [41]) [41]
H230Y <1> (<1> 90% loss of enzyme activity, confers ability to oxidize NADH [40]) [40]
P297R <1> (<1> 85-90% loss of enzyme activity [39]) [39]
Y82F <1> (<1> 95% loss of activity, k_{cat} and K_m decreased, a role in stabilizing the quinonoid intermediate [41]) [41]

Application

medicine <1, 3, 9, 13, 16, 18, 20> (<1, 3, 9, 13, 16, 18, 20> enzyme is a potential target for cancer chemotherapy [38, 41, 42, 45]) [38, 41, 42, 45]

6 Stability

pH-Stability

5.8-8.5 <3> (<3> stable, rabbit cytosol [31]) [31]
6-7.5 <15> (<15> stable [21]) [21]
6-8.5 <16> (<16> stable [23]) [23]
7.2-8.3 <1> (<1> stable, at pH 8.55: loss of activity, tetrameric holoenzyme dissociates into the dimeric form, at pH 9.3: complete dissociation [37]) [37]

Temperature stability

30 <3> (<3> several h, pH 5.8-8.5, cytosol [31]) [31]
40 <15> (<15> below, at least 10 min stable [21]) [21]
45 <15> (<15> above, rapid loss of activity [21]) [21]
50 <5> (<5> 20 min stable [14]) [14]
52 <1> (<1> apoenzyme, absence of pyridoxal 5'-phosphate [37]) [37]
55 <13> (<13> $t_{1/2}$: about 2 min, glycerol stabilizes [24]) [24]
58 <1> (<1> holoenzyme [37]) [37]
60 <1> (<1> subunit interactions retained [4]) [4]
60 <18> (<18> enzyme stable in the absence of any ligand [43]) [43]
65 <5> (<5> 85% loss of activity after 20 min [14]) [14]
70 <18> (<18> 50% loss of activity in the absence of any ligand [43]) [43]
70 <5> (<5> complete loss of activity after 6 min, enzyme-antibody complex: 30% loss of activity after 20 min [14]) [14]
Additional information <1, 5, 13, 18> (<13> glycerol, 30% v/v, enhances thermal stability [24]; <1, 18> L-serine increases the thermal stability [39, 41, 43]; <5> L-serine protects against thermal inactivation [14]; <18> pyridoxal 5'-phosphate increases the thermal stability [43]; <5> enzyme-antibody complex more stable to elevated temperatures than free enzyme, allosteric effectors fail to protect free enzyme [14]; <1> T_m of wild-type and mutant enzyme: 54°C, in the presence of serine T_m of wild-type enzyme: 63°C [39]; <1> T_m: 55°C, increased to 65°C in the presence of L-serine, T_m: 68°C mutant enzyme Y82F in the presence of serine [41]; <18> T_m values of enzyme and dimeric and tetrameric forms in the absence and presence of ligands [43]) [14, 24, 39, 41, 43]

General stability information

<8>, folate stabilizes [2]
<10>, KCl, NaCl or LiCl: denaturation, folate protects [19]
<10>, dialysis inactivates [19]
<10>, tetrahydrofolate stabilizes [19]
<1, 5, 13>, pyridoxal 5'-phosphate stabilizes [6, 14, 24]
<10, 11>, L-serine stabilizes [18, 19]
<5, 10>, 2-mercaptoethanol stabilizes [14, 19]
<5, 10>, EDTA stabilizes [14, 19]
<5, 13>, DTT stabilizes [14, 24]

Storage stability

<1>, 4°C, 5-20 mg protein/ml: indefinitely in the presence of pyridoxal 5'-phosphate [6]
<3>, 5°C, precipitation after several days [12]
<3>, frozen, several months at pH 7.3 [31]
<5>, -20°C, 6-8 months in the presence of pyridoxal 5'-phosphate, 2-mercaptoethanol, DTT and EDTA [14]
<5>, 0°C, at least 2 months in the presence of pyridoxal 5'-phosphate, 2-mercaptoethanol, DTT and EDTA [14]
<8>, 4°C, 8 weeks in the presence of folate [2]
<11>, -20°C, 8 weeks [18]

<13>, -10°C, 50% loss of activity in 10 h [24]
<13>, 4°C, 50% loss of activity in 10 h [24]
<13>, frozen, up to a month in the presence of 30% v/v glycerol [24]
<13>, glycerol, 30% v/v, enhances thermal stability [24]

References

[1] Ide, H.; Hamaguchi, K.; Kobata, S.; Murakami, A.; Kimura, Y.; Makino, K.; Kamada, M.; Miyamoto, S.; Nagaya, T.; et al.: Purification of serine hydroxymethyltransferase from Bacillus stearothermophilus with ion-exchange high-performance liquid chromatography. J. Chromatogr., **596**, 203-209 (1992)
[2] Henricson, D.; Ericson, I.: Serine hydroxymethyltransferase from spinach leaf mitochondria. Purification and characterization. Physiol. Plant., **74**, 602-606 (1988)
[3] Manca de Nadra, M.C.; Raya, R.R.; Holgado, A. Pesce de Ruiz, Oliver, G.: Isolation and properties of threonine aldolase of Lactobacillus bulgaricus YOP12. Milchwissenschaft, **42**, 92-94 (1987)
[4] Manohar, R.; Ramesh, K.S.; Rao, N.A.: Purification, physicochemical and regulatory properties of serine hydroxymethyltransferase from sheep liver. J. Biosci., **4**, 31-50 (1982)
[5] Schirch, L.; Peterson, D.: Purification and properties of mitochondrial serine hydroxymethyltransferase. J. Biol. Chem., **255**, 7801-7806 (1980)
[6] Ulevitch, R.J.; Kallen, R.G.: Purification and characterization of pyridoxal 5-phosphate dependent serine hydroxymethylase from lamb liver and its action upon β-phenylserines. Biochemistry, **16**, 5342-5350 (1977)
[7] Ulevitch, R.J.; Kallen, R.G.: Studies of the reactions of lamb liver serine hydroxymethylase with L-phenylalanine: kinetic isotope effects upon quinonoid intermediate formation. Biochemistry, **16**, 5350-5354 (1977)
[8] Ulevitch, R.J.; Kallen, R.G.: Studies of the reactions of substituted D,L-erythro-β-phenylserines with lamb liver serine hydroxymethylase. Effects of substituents upon the dealdolization step. Biochemistry, **16**, 5355-5363 (1977)
[9] Masuda, T.; Sakamoto, M.; Nishizaki, I.; Hayashi, H.; Yamamoto, M.; Wada, H.: Affinity purification and characterization of serine hydroxymethyltransferases from rat liver. J. Biochem., **101**, 643-652 (1987)
[10] Ogawa, H.; Fujioka, M.: Purification and characterization of cytosolic and mitochondrial serine hydroxymethyltransferases from rat liver. J. Biochem., **90**, 381-390 (1981)
[11] Schirch, L.; Gross, T.: Serine transhydroxymethylase. Identification as the threonine and allothreonine aldolases. J. Biol. Chem., **243**, 5651-5655 (1968)
[12] Fujioka, M.: Purification and properties of serine hydroxymethylase from soluble and mitochondrial fractions of rabbit liver. Biochim. Biophys. Acta, **185**, 338-349 (1969)

[13] Akhtar, M.; El-Obeid, H.A.: Interaction of serine transhydroxymethylase and threonine aldolase activities. Biochim. Biophys. Acta, **791**, 791-799 (1972)
[14] Ramesh, K.S.; Rao, N.A.: Purification and physicochemical, kinetic and immunological properties of allosteric serine hydroxymethyltransferase from monkey liver. Biochem. J., **187**, 623-636 (1980)
[15] Ramesh, K.S.; Rao, N.A.: Inhibition of monkey liver serine hydroxymethyltransferase by Cibacron Blue 3G-A. Biochem. J., **187**, 249-252 (1980)
[16] Strong, W.B.; Tendler, S.J.; Seither, R.L.; Goldman, I.D.; Schirch, V.: Purification and properties of serine hydroxymethyltransferase and C_1-tetrahydrofolate synthase from L1210 cells. J. Biol. Chem., **265**, 12149-12155 (1990)
[17] Bourguignon, J.; Neuburger, M.; Douce, R.: Resolution and characterization of the glycine-cleavage reaction in pea leaf mitochondria. Properties of the forward reaction catalysed by glycine decarboxylase and serine hydroxymethyltransferase. Biochem. J., **255**, 169-178 (1988)
[18] Mitchell, M.K.; Reynolds, P.H.S.; Blevins, D.G.: Serine hydroxymethyltransferase from soybean root nodules. Plant Physiol., **81**, 553-557 (1986)
[19] Rao, D.N.; Rao, N.A.: Purification and regulatory properties of mung bean (Vigna Radiata L.) serine hydroxymethyltransferase. Plant Physiol., **69**, 11-18 (1982)
[20] Sakamoto, M.; Masuda, T.; Yanagimoto, Y.; Nakano, Y.; Kitaoka, S.: Purification and characterization of cytosolic serine hydroxymethyltransferase from Euglena gracilis z. Agric. Biol. Chem., **55**, 2243-2249 (1991)
[21] Miyazaki, S.S.; Toki, S.; Izumi, Y.; Yamada, H.: Purification and characterization of a serine hydroxymethyltransferase from an obligate methylotroph, Hyphomicrobium methylovorum GM2. Eur. J. Biochem., **162**, 533-540 (1987)
[22] Miyazaki, S.S.; Toki, S.; Izumi, Y.; Yamada, H.: Futher characterization of serine hydroxymethyltransferase from a serine-producing methylotroph, Hyphomicrobium methylovorum. Agric. Biol. Chem., **51**, 2587-2589 (1987)
[23] Schirch, V.; Hopkins, S.; Villar, E.; Angelaccio, S.: Serine hydroxymethyltransferase from Escherichia coli: purification and properties. J. Bacteriol., **163**, 1-7 (1985)
[24] Nakamura, K.D.; Trewyn, R.W.; Parks, L.W.: Purification and characterization of serine transhydroxy-methylase from Saccharomyces cerevisiae. Biochim. Biophys. Acta, **327**, 328-335 (1973)
[25] Ching, W.M.; Kallen, R.G.: Mechanism of serine hydroxymethylase catalyzed cleavage of L-erythro-β-phenylserine: pH dependence of elementary kinetic processes from spectroscopic, pre-steady kinetic, and competitive inhibition studies. Biochemistry, **18**, 821-830 (1979)
[26] El-Obeid, H.A.; Akhtar, M.: Glycidaldehyde, an inhibitor directed towards the C_1-units-binding site of serine transhydoxymethylase. Biochem. Soc. Trans., **1**, 1274-1276 (1973)
[27] Manohar, R.; Rao, A.G.A.; Rao, N.A.: Kinetic mechanism of the interaction of D-cycloserine with serine hydroxymethyltransferase. Biochemistry, **23**, 4116-4122 (1984)

[28] Malthouse, J.P.G.; Milne, J.J.; Gariani, L.S.: A comparative study of the kinetics and stereochemistry of the serine hydroxymethyltransferase- and tryptophan synthase-catalysed exchange of the pro-2R and pro-2S protons of glycine. Biochem. J., **274**, 807-812 (1991)
[29] Stover, P.; Schirch, V.: 5-Formyltetrahydrofolate polyglutamates are slow tight binding inhibitors of serine hydroxymethyltransferase. J. Biol. Chem., **266**, 1543-1550 (1991)
[30] Akhtar, M.; El-Obeida, H.A.; Jordan, P.M.: Mechanistic, inhibitory and stereochemical studies on cytoplasmic and mitochondrial serine transhydorxymethylases. Biochem. J., **145**, 159-168 (1975)
[31] Schirch, L.: Serine hydroxymethyltransferase. Adv. Enzymol. Relat. Areas Mol. Biol., **53**, 83-112 (1982)
[32] Kruschwitz, H.; Ren, S.; Di Salvo, M.; Schifch, V.: Expression, purification, and characterization of human cytosolic serine hydroxymethyltransferase. Protein Expr. Purif., **6**, 411-416 (1995)
[33] Fratte, S.D.; White, R.H.; Maras, B.; Bossa, F.; Schirch, V.: Purification and properties of serine hydroxymethyltransferase from Sulfolobus solfataricus. J. Bacteriol., **179**, 7456-7461 (1997)
[34] Di Salvo, M.L.; Delle Fratte, S.; De Biase, D.; Bossa, F.; Schirch, V.: Purification and characterization of recombinant rabbit cytosolic serine hydroxymethyltransferase. Protein Expr. Purif., **13**, 177-183 (1998)
[35] Contestabile, R.; Paiardini, A.; Pascarella, S.; Di Salvo, M.L.; DÁguanno, S.; Bossa, F.: L-Threonine aldolase, serine hydroxymethyltransferase and fungal alanine racemase. A subgroup of strictly related enzymes specialized for different functions. Eur. J. Biochem., **268**, 6508-6525 (2001)
[36] Jagath-Reddy, J.; Ganesan, K.; Savithri, H.S.; Datta, A.; Rao, N.A.: cDNA cloning, overexpression in Escherichia coli, purification and characterization of sheep liver cytosolic serine hydroxymethyltransferase. Eur. J. Biochem., **230**, 533-537 (1995)
[37] Brahatheeswaran, B.; Prakash, V.; Savithri, H.S.; Rao, N.A.: Interaction of sheep liver apo-serine hydroxymethyltransferase with pyridoxal-5'-phosphate: a physicochemical, kinetic, and thermodynamic study. Arch. Biochem. Biophys., **330**, 363-372 (1996)
[38] Appaji Rao, N.; Talwar, R.; Savithri, H.S.: Molecular organization, catalytic mechanism and function of serine hydroxymethyltransferase - a potential target for cancer chemotherapy. Int. J. Biochem. Cell Biol., **32**, 405-416 (2000)
[39] Talwar, R.; Leelavathy, V.; Rao, J.V.K.; Rao, N.A.; Savithri, H.S.: Role of Pro-297 in the catalytic mechanism of sheep liver serine hydroxymethyltransferase. Biochem. J., **350**, 849-853 (2000)
[40] Talwar, R.; Appaji Rao, N.; Savithri, H.S.: A change in reaction specificity of sheep liver serine hydroxymethyltransferase: induction of NADH oxidation upon mutation of His230 to Tyr. Eur. J. Biochem., **267**, 929-934 (2000)
[41] Rao, J.V.K.; Prakash, V.; Appaji Rao, N.; Savithri, H.S.: The role of Glu74 and Tyr82 in the reaction catalyzed by sheep liver cytosolic serine hydroxymethyltransferase. Eur. J. Biochem., **267**, 5967-5976 (2000)

[42] Trivedi, V.; Gupta, A.; Jala, V.R.; Saravanan, P.; Rao, G.S.J.; Rao, N.A.; Savithri, H.S.; Subramanya, H.S.: Crystal structure of binary and ternary complexes of serine hydroxymethyltransferase from Bacillus stearothermophilus: insights into the catalytic mechanism. J. Biol. Chem., **277**, 17161-17169 (2002)
[43] Jala, V.R.; Prakash, V.; Rao, N.A.; Savithri, H.S.: Overexpression and characterization of dimeric and tetrameric forms of recombinant serine hydroxymethyltransferase from Bacillus stearothermophilus. J. Biosci., **27**, 233-242 (2002)
[44] Liu, J.Q.; Ito, S.; Dairi, T.; Itoh, N.; Kataoka, M.; Shimizu, S.; Yamada, H.: Gene cloning, nucleotide sequencing, and purification and characterization of the low-specificity L-threonine aldolase from Pseudomonas sp. strain NCIMB 10558. Appl. Environ. Microbiol., **64**, 549-554 (1998)
[45] Ogawa, H.; Gomi, T.; Fujioka, M.: Serine hydroxymethyltransferase and threonine aldolase: are they identical?. Int. J. Biochem. Cell Biol., **32**, 289-301 (2000)
[46] Chave, K.J.; Snell, K.; Sanders, P.G.: Isolation and characterization of human genomic sequences encoding cytosolic serine hydroxymethyltransferase. Biochem. Soc. Trans., **25**, 53S (1997)
[47] Liu, X.; Reig, B.; Nasrallah, I.M.; Stover, P.J.: Human cytoplasmic serine hydroxymethyltransferase is an mRNA binding protein. Biochemistry, **39**, 11523-11531 (2000)
[48] Sakamoto, M.; Masuda, T.; Yanagimoto, Y.; Nakano, Y.; Kitaoka, S.; Tanigawa, Y.: Purification and characterization of serine hydroxymethyltransferase from mitochondria of Euglena gracilis z. Biosci. Biotechnol. Biochem., **60**, 1941-1944 (1996)
[49] Vatcher, G.P.; Barbazuk, W.B.; O'Neil, N.J.; Marra, M.A.; Ha, T.; Baillie, D.L.: Identification and characterization of a serine hydroxymethyltransferase isoform in Caenorhabditis briggsae. Gene, **230**, 137-144 (1999)

Phosphoribosylglycinamide formyltransferase 2.1.2.2

1 Nomenclature

EC number
2.1.2.2

Systematic name
10-formyltetrahydrofolate:5'-phosphoribosylglycinamide N-formyltransferase

Recommended name
phosphoribosylglycinamide formyltransferase

Synonyms
2-amino-N-ribosylacetamide 5'-phosphate (glycinamide ribotide) transformylase <2> [3]
2-amino-N-ribosylacetamide 5'-phosphate transformylase
5,10-methenyltetrahydrofolate:2-amino-N-ribosylacetamide ribonucleotide transformylase
GAR TFase
GAR formyltransferase
GAR synthetase <1> [1]
GAR transformylase
GARFT <2, 5, 7, 8> [21]
GART <5> [9]
GART synthetase <5, 6, 9, 10, 17> [26]
N^{10}-formyltetrahydrolate:2-amino-N-ribosylacetamide-5'-phosphate transformylase <5> [17]
glycinamide ribonucleotide transformylase
purN transformylase <5> [19]

CAS registry number
9032-02-4

2 Source Organism

<1> *Rattus norvegicus* (rat [1]) [1]
<2> *Gallus gallus* (chicken [2-4, 6, 8, 11-15, 17-19, 21, 24, 25, 28, 32]) [2-4, 6, 8, 11-15, 17-19, 21, 24, 25, 28, 32]
<3> *Columba sp.* (pigeon [3,13]) [3, 13]
<4> *Clostridium cylindrosporum* [3]

<5> *Escherichia coli* (strain B [17]; strains DH5α, BL21(DE3), TX680 [19]; strain TX635 [22]; strains DH5α, BL21(DE3) [23]) [4, 5, 8-10, 17-26, 28-31]
<6> *Saccharomyces cerevisiae* (yeast [5, 6, 8, 12, 13, 19, 24, 26]) [5, 6, 8, 12, 13, 19, 24, 26]
<7> *Homo sapiens* (human [5, 8, 15, 16, 19, 21, 24, 25, 27, 29, 30]) [5, 8, 15, 16, 19, 21, 24, 25, 27, 29, 30]
<8> *Mus musculus* (mouse [5, 7-9, 11, 14-16, 21, 27, 29]) [5, 7-9, 11, 14-16, 21, 27, 29]
<9> *Bacillus subtilis* [5, 8, 19, 24, 26]
<10> *Drosophila melanogaster* (fruit fly [5, 8, 12, 15, 19, 26]) [5, 8, 12, 15, 19, 26]
<11> *Ovis aries* (sheep [6,13]) [6, 13]
<12> *Sus scrofa* (pig [6,13]) [6, 13]
<13> *Pseudomonas aeruginosa* (strain PAO1 [26]) [26]
<14> *Cricetus griseus* (chinese hamster [8]) [8]
<15> *Schizosaccharomyces pombe* (fission yeast [12]) [12]
<16> *Salmonella typhimurium* [17]
<17> *Arabidopsis thaliana* [26]
<18> *Cuniculus sp.* (rabbit [13]) [13]

3 Reaction and Specificity

Catalyzed reaction

10-formyltetrahydrofolate + N^1-(5-phospho-D-ribosyl)glycinamide = tetrahydrofolate + N^2-formyl-N^1-(5-phospho-D-ribosyl)glycinamide

Reaction type

formyl group transfer

Natural substrates and products

S 10-formyltetrahydrofolate + 5-phospho-D-ribosylglycinamide <1-4, 6-17> (<2, 5, 7, 8, 10> essential enzyme in DNA synthesis, third enzyme in de novo purine biosynthesis pathway [5, 8-16, 18, 19, 21, 24, 25, 28, 30, 31]; <2> in uricotelic organisms such as birds, the major role of this active pathway is elimination of waste nitrogen [14]; <2> transformylation reaction of purine biosynthesis [32]; <8> in ureotelic animals such as mammals the role of the far less active pathway is to provide purine nucleotides for further anabolic sequences [14]; <8> carries out the second, third and fifth sequential reactions of de novo purine synthesis pathway [21]; <13> involved in pyoverdine synthesis [26]; <5, 8> third step in de novo purine biosynthetic pathway [29]) (Reversibility: ? <1-4, 6-16> [1-28, 30-32]; r <5> [29]) [1-32]

P tetrahydrofolate + 5'-phosphoribosyl-N-formylglycinamide

S 10-formyltetrahydrofolate + glycinamide ribonucleotide <5, 8> (Reversibility: ? <5, 8> [5, 7]) [5, 7]

P tetrahydrofolate + formylglycinamide ribonucleotide

Substrates and products

S (6R)-N^{10}-formyltetrahydrofolate + 5-phospho-D-ribosylglycinamide <2, 5> (Reversibility: ? <2, 5> [8, 11]) [8, 11]

P (6R)-tetrahydrofolate + 5'-phosphoribosyl-N-formylglycinamide <2, 5> [8, 11]

S 10-formyl-5,8-dideazafolate + carbocyclic β-glycinamide ribonucleotide <5, 7, 8> (Reversibility: ? <5, 7, 8> [7, 8, 14, 21, 25]) [7, 8, 14, 21, 25]

P 5,8-dideazafolate + N-formyl carbocyclic β-glycinamide ribonucleotide <5> [8]

S 10-formyltetrahydrofolate + (α,β)-glycinamide ribonucleotide <2> (Reversibility: ? <2> [28]) [28]

P tetrahydrofolate + formylglycinamide ribonucleotide <2> [28]

S 10-formyltetrahydrofolate + 5-phospho-D-ribosylglycinamide <1-16> (Reversibility: ? <1-4, 6-16> [1-28, 30-32]; r <5> [29]) [1-32]

P tetrahydrofolate + 5'-phosphoribosyl-N-formylglycinamide <1-4, 6-16> [1-32]

S 10-formyltetrahydrofolate + N^5-hydroxyornithine <13> (Reversibility: ? <13> [26]) [26]

P tetrahydrofolate + N^2-formyl-N^5-hydroxyornithine

S 10-formyltetrahydrofolate + O-phosphonate carbocyclic glycinamide ribonucleotide <7> (Reversibility: ? <7> [25]) [25]

P tetrahydrofolate + O-phosphonate-ribosyl-N-formylglycinamide

S 10-formyltetrahydrofolate + β-5-phospho-D-ribosylglycinamide <5> (Reversibility: ? <5> [8]) [8]

P tetrahydrofolate + β-5'-phosphoribosyl-N-formylglycinamide <5> [8]

S 10-formyltetrahydrofolate + carbocyclic β-glycinamide ribonucleotide <2, 7, 8> (Reversibility: ? <2, 7, 8> [15, 16, 25, 28]) [15, 16, 25, 28]

P tetrahydrofolate + formyl glycinamide ribonucleotide

S N^{10}-formyl-5-deazafolate + glycinamide ribonucleotide <5> (Reversibility: ? <5> [8]) [8]

P formylglycinamide ribonucleotide + 5-deazafolate

S N^{10}-formyl-8-deazafolate + glycinamide ribonucleotide <5> (Reversibility: ? <5> [8]) [8]

P formylglycinamide ribonucleotide + 8-deazafolate

S N^{10}-formyldideazafolate + 5-phospho-D-ribosylglycinamide <5> (Reversibility: ? <5> [8]) [8]

P dideazafolate + 5'-phosphoribosyl-N-formylglycinamide

S N^{10}-formylfolate + glycinamide ribonucleotide <5> (Reversibility: ? <5> [8]) [8]

P folate + formylglycinamide ribonucleotide

S N^5,N^{10}-anhydroformyl-5,6,7,8-tetrahydrofolic acid + 2-amino-N-ribosylacetamide 5'-phosphate + H_2O <2> (Reversibility: ir <2> [3]) [3]

P formylglycinamide ribotide + tetrahydrofolic acid + 2-formamido-N-ribosylacetamide 5'-phosphate

S N^5,N^{10}-anhydroformyltetrahydrofolate + glycinamide ribotide + H_2O <4> (Reversibility: ? <4> [3]) [3]

P N^{10}-formyltetrahydrofolate + formylglycinamide ribotide

S N^5,N^{10}-anhydroformyltetrahydrofolic acid + glycinamide ribonucleotide <2> (Reversibility: ? <2> [32]) [32]
P formylglycinamide ribonucleotide + ?
S acetate + ATP <5> (<5> novel non-folate glycinamide ribonucleotide transformylase purT gene product, no significant homology to known EC 2.1.2.2 but significant homology to the purK protein also involved in purine biosynthesis [22,29]) (Reversibility: ? <5> [22, 29]) [22, 29]
P acetyl phosphate + ADP
S formate + ATP + β-glycinamide ribonucleotide <5> (<5> novel non-folate glycinamide ribonucleotide transformylase purT gene product, no significant homology to known EC 2.1.2.2 but significant homology to the purK protein also involved in purine biosynthesis [22]) (Reversibility: ? <5> [22]) [22]
P β-formylglycinamide ribonucleotide + ADP
S inosinic acid + glycinamide ribotide + H_2O <2> (Reversibility: ? <2> [3]) [3]
P 5-amino-4-imidazolecarboxamide ribotide + formylglycinamide ribotide
S Additional information <1, 2, 5, 7-9> (<1, 2, 7, 8> trifunctional enzyme, GAR synthetase, aminoimidazole ribonucleotide synthetase and GAR transformylase [1, 18, 21, 24, 29]; <2> reaction does not occur if glycine, glycinamide or glycinamide riboside is substituted for glycinamide ribotide [3]; <2> GAR transformylase activity copurifies with 5,10-methylene-, 5,10-methenyl, and 10-formyl-H4-folate sythetase (combined) trifunctional enzyme, EC 6.3.4.3, EC 1.5.1.5 and EC 3.5.4.9 [11]; <2> purified along with 10-formyltetrahydrofolate synthetase [13]; <2> uses both diastereomers of 5,10-methenyltetrahydrofolate, (+)-L-diastereomer is used 120 times faster than the (-)-L-diastereomer [13]; <2> glycinamide ribonucleotide nucleoside is no substrate [28]; <5,9> monofunctional enzyme [5, 24]; <2, 7, 8, 11, 12> 3 activities in a multifunctional protein, posessing glycinamide ribonucleotide synthetase, glycinamide ribonucleotidetransformylase and aminoimidazole ribonucleotide synthetase [5, 6, 8, 9, 12]; <5> purT GAR transformylase is the first unknown de novo purine biosynthetic enzyme to be discovered in the last 30 years and represents another step forward in understanding cellular control of purine levels [22]; <5> 10-formyl-dideazaformate, 5-formyltetrahydroformate, formylaminoimidazolecarboxamide ribonucleotide, formylmethionine, formiminoglutamate, GTP, CTP, TTP, ITP and ZTP are no substrates for the purT GAR transformylase [22]; <7> α-glycinamide ribonucleotide and glycinamide ribonucleotide nucleoside are neither substrates for no inhibitors [25]) [1, 3, 5, 6, 8, 9, 12, 21, 22, 24, 25, 28]
P ?

Inhibitors

(6R)-5,10-dideazatetrahydrofolate <7> (<7> lometrexol [27]) [27]
(6R)-dideazatetrahydrofolate diglutamate <7, 8> (<7,8> LY235340 [27]) [27]
(6R)-dideazatetrahydrofolate pentaglutamate <7, 8> (<7,8> LY235542 [27]) [27]

(6R)-dideazatetrahydrofolate tetraglutamate <7, 8> (<7,8> LY266978 [27]) [27]
(6R)-dideazatetrahydrofolate triglutamate <7, 8> (<7,8> LY235337 [27]) [27]
(6S)-N^{10}-formyltetrahydrofolate <5> [8]
(6S)-dideazatetrahydrofolate <7, 8> (<7,8> LY243246 [27]) [27]
(α,β)-hydroxyacetamide ribonucleotide <7> [24]
10-acetyl-5,8-dideazafolate <7, 8, 10> [15, 24]
10-formyl-5,8-dideazafolic acid hexaglutamate <8> [21]
2',3'-dideoxy-2',3'-didehydro-carbocyclic glycinamide ribonucleotide <7> [25]
2',3'-dideoxy-carbocyclic glycinamide ribonucleotide <7> [25]
2',5'-furan-dideazatetrahydrofolate <7> (<7,8> LY222306 [27]) [27]
2',5'-thiophene dideazatetrahydrofolate <7, 8> (<7,8> LY309886, diastereomer A [27]) [27]
2',5'-thiophene dideazatetrahydrofolate <7, 8> (<7,8> LY309887, diastereomer B [21,27]) [21, 27]
2-deaminodideazatetrahydrofolate <8> [21]
5'-phosphonate carbocyclic glycinamide ribonucleotide <7> [25]
5,10-dideaza-5,6,7,8-tetrahydrofolate <5, 7> [5, 9]
5,10-dideazafolic acid <8> [15]
5,10-dideazatetrahydrofolate <5, 7, 8> [5, 9, 21, 25, 27]
5,8-dideazafolate <5> [10, 21]
5,8-dideazafolate hexaglutamate <8> [21]
5,8-dideazapteroyl hexaglutamate <8> [21]
5-deazatetrahydrofolate <8> [21]
Cu^{2+} <2> [3]
L-(+)-10-formyltetrahydrofolate <2> (<2> excellent competitive inhibitor [4]) [4]
N^{10}-(bromoacetyl)-5,8-dideazafolate <5> [8]
N^{10}-formyl-2,4-diaminodideazafolate <5> [8]
α,β-N-(hydroxyacetyl)-D-ribofuranosylamine <8> [15]
α,β-hydroxyacetamideribonucleotide <8> [15]
β-thioglycinamide dideazafolate <5> [8]
dideazafolate <5> [8]
dideazatetrahydrofolate <8> [21]
dideazatetrahydrofolate hexaglutamate <8> [21]
homo-dideazatetrahydrofolate <8> [21]

Cofactors/prosthetic groups

10-formyl-5,8-dideazafolate <5, 8> [7, 14, 19]
5,10-methenyltetrahydrofolate <2> [13]
L-(+-)-5,10-methenyltetrahydrofolate <2> [4, 6]
L-(-)-10-formyltetrahydrofolate <2, 5> [4, 8]
N^{10}-formyltetrahydrofolate <2, 5> (<5> only formyldonor [17]) [13, 17, 19]
N^5,N^{10}-methenyltetrahydrofolate <16> [17]
formate <5> (<5> second GAR transformylase using formate and ATP in place of 10-formyltetrahydrofolate [19]) [19]

Activating compounds

NH_4^+ <2> (<2> activity enhanced by [6]) [6]
glycine <2> (<2> stimulatory effect [3]) [3]
histidine <2> (<2> stimulatory effect [3]) [3]
tryptophan <2> (<2> stimulatory effect [3]) [3]

Metals, ions

Fe^{2+} <13> (expression regulated by amount of iron in the growth medium [26]) [26]
K^+ <2> (<2> activity enhanced by [6]) [6]
Mg^{2+} <5> (<5> purT gene product [22]) [22]

Turnover number (min^{-1})

0.96 <5> (N^{10}-formyltetrahydrofolate) [8]
4.14 <5> (N^{10}-formyl-8-deazafolate) [8]
91.2 <5> (N^{10}-formyl-5-deazafolate) [8]
810 <5> ((6R)-N^{10}-formyltetrahydrofolate, <5> pH 7.5 [8]) [8]
966 <5> (N^{10}-formyl-5,8-dideazafolate, <5> pH 7.5 [8]) [8]
1242 <5> ((6R)-N^{10}-formyltetrahydrofolate, <5> pH 8.5 [8]) [8]
1980 <5> (carbocyclic glycinamide ribonucleotide) [8]
2400 <5> (N^{10}-formyl-5,8-dideazafolate, <5> pH 8.5 [8]) [8]

Specific activity (U/mg)

0.651 <2> [13]
0.95 <2> [12]
1.294 <2> [6]
3.3 <8> [14]
7 <8> [21]
11.6 <5> [8]
14 <5> (<5> purN gene product [22]) [22]
14.8 <2> [11]
53.6 <2> [3]
Additional information <5> (<5> purT gene product, 0.052 mg /min.mg.mL [22]) [22]

K_m-Value (mM)

0.000075 <8> (10-formyl-5,8-dideazafolate, <8> recombinant enzyme [21]) [21]
0.00025 <8> (10-formyl-5,8-dideazafolate) [14]
0.0008 <8> (L-(-)-10-formyltetrahydrofolate) [14]
0.0011 <7> (β-glycinamide ribonucleotide) [24, 25]
0.0013 <8> (10-formyl-5,8-dideazafolate) [21]
0.0015 <7> (10-formyl-5,8-dideazafolate, <7> recombinant enzyme, expressed in E.coli [24]) [24]
0.0019 <2> (10-formyl-5,8-dideazafolate) [14]
0.0036 <7> (carbocyclic β-glycinamide ribonucleotide) [25]
0.00368 <5> (acetate, <5> purT gene product [22]) [22]
0.00576 <2> (L-(+)-5,10-methenyltetrahydrofolate) [4]

0.00594 <2> (L-(-)-10-formyltetrahydrofolate, <2> in absence of trifunctional protein [4]) [4]
0.00676 <2> (L-(-)-10-formyltetrahydrofolate) [4]
0.008 <5> (β-glycinamide ribonucleotide, <5> pH 8.5, tetrahydrofolate as cofactor [8]) [8]
0.01 <5> (10-formyl-5,8-dideazafolate, <5> mutant H108A [19]) [19]
0.01 <8> (glycinamide ribonucleotide) [7]
0.0101 <2, 5> (β-glycinamide ribonucleotide, <5> purT gene product [22]) [22, 25, 28]
0.012 <5> (β-glycinamide ribonucleotide, <5> pH 7.5, tetrahydrofolate as cofactor [22]) [22]
0.0122 <5> (β-glycinamide ribonucleotide, <5> pH 7.5, tetrahydrofolate as cofactor [8]) [8]
0.0123 <5> (10-formyl-5,8-dideazafolate, <5> wild-type [20]) [20, 29]
0.016 <5> (β-glycinamide ribonucleotide, <5> purN/purU hybrid enzyme [23]) [23]
0.0167 <5> (N^{10}-formyldideazafolate, <5> pH 7.5 [8]) [8]
0.017 <5> (10-formyl-5,8-dideazafolate) [19, 23]
0.018 <5> (β-glycinamide ribonucleotide, <5> mutant H108Q [19]) [19]
0.018 <8> (carbocyclic glycinamide ribonucleotide) [7]
0.019 <5> (5,8-dideazafolate) [29]
0.019 <5> (β-glycinamide ribonucleotide) [19, 23]
0.0192 <5> (β-glycinamide ribonucleotide, <5> pH 7.5, dideazafolate as cofactor [8]) [8]
0.0211 <5> (N^{10}-formyl-5,8-dideazafolate, <5> wild-type [31]; <5> strain H108A [31]) [31]
0.0235 <5> (β-glycinamide ribonucleotide, <5> pH 8.5, dideazafolate as cofactor [8]) [8]
0.024 <5> (10-formyl-5,8-dideazafolate, <5> mutant H108Q [19]) [19]
0.025 <5> (10-formyl-5,8-dideazafolate, <5> mutant D144A [19]) [19]
0.0284 <5> (N^{10}-formyl-5,8-dideazafolate, <5> strain D144A [31]) [31]
0.029 <5> (β-glycinamide ribonucleotide, <5> mutant D144A [19]) [19]
0.035 <5> (10-formyl-5,8-dideazafolate, <5> purN/purU hybrid enzyme [23]) [23]
0.0365 <5> (N^{10}-formyldideazafolate, <5> pH 8.5 [8]) [8]
0.0369 <2> (carbocyclic β-glycinamide ribonucleotide) [25, 28]
0.0465 <5> (carbocyclic β-glycinamide ribonucleotide) [8]
0.052 <2> (glycinamide ribotide) [3]
0.054 <5> (β-glycinamide ribonucleotide, <5> mutant H108A [19]) [19]
0.058 <2> (N^5,N^{10}-anhydroformyl tetrahydrofolate) [3]
0.0774 <5> (ATP, <5> purT gene product [22]) [22]
0.0775 <5> ((6R)-N^{10}-formyltetrahydrofolate, <5> pH 8.5 [8]) [8]
0.0848 <5> ((6R)-N^{10}-formyltetrahydrofolate, <5> pH 7.5 [8]) [8]
0.095 <5> (β-glycinamide ribonucleotide, <5> wild-type [31]) [31]
0.099 <5> (N^{10}-formyl-5,8-dideazafolate, <5> strain H121Q [31]) [31]
0.102 <5> (formyl-β-glycinamide ribonucleotide) [29]
0.115 <5> (β-glycinamide ribonucleotide, <5> strain H108A [31]) [31]

0.118 <5> (β-glycinamide ribonucleotide, <5> wild-type [20]) [20, 29]
0.122 <5> (β-glycinamide ribonucleotide, <5> strain D144A [31]) [31]
0.128 <5> (10-formyl-5,8-dideazafolate, <5> mutant 13N + 13C 1:1 [20]) [20]
0.131 <5> (10-formyl-5,8-dideazafolate, <5> mutant B13 [20]) [20]
0.15 <5> (N^{10}-formyl-8-deazafolate) [8]
0.178 <5> (β-glycinamide ribonucleotide, <5> mutant B13 [20]) [20]
0.19 <5> (N^{10}-formyl-5-deazafolate) [8]
0.19 <5> (β-glycinamide ribonucleotide, <5> mutant 13N + 13C 1:1 [20]) [20]
0.21 <5> (N-formylglycinamide ribotide) [17]
0.319 <5> (formate, <5> purT gene product [22]) [22]
0.4 <8> (glycinamide ribonucleotide) [14]
0.414 <5> (β-glycinamide ribonucleotide, <5> strain H121Q [31]) [31]
0.6 <5> (N^{10}-formylfolate) [8]
0.9 <5> (10-formyl-5,8-dideazafolate, <5> mutant N106G [19]) [19]

pH-Optimum

7.8 <2> (<2> Tris buffer, 38°C [3]) [3]
8 <5> (<5> purTgene product [22]) [22]
8.5 <5> [8]
8.5-8.8 <5> [31]

pH-Range

6-9.5 <5> [20]
6-10 <5> [31]

4 Enzyme Structure

Molecular weight

21060 <9> (<9> calculated from peptide sequence deduced from the cDNA [8]) [8]
21800 <7> (<7> recombinant enzyme, predicted from cDNA [24]) [24]
21840 <7> (<7> recombinant enzyme, MALDI-TOF mass spectral analysis [24]) [24]
23000 <5> (<5> purN encoded, DNA sequencing [23]) [5, 10, 23, 31]
23240 <5> (<5> amino acid sequence, deduced from nucleotide sequence [9]) [9]
23240 <5> (<5> calculated from peptide sequence deduced from the cDNA [8]) [8, 19]
24000 <5> (<5> equilibrium sedimentation ultracentrifugation [8]) [8]
25700 <7> (<7> SDS-PAGE [24]) [24]
31000 <13> (<13> DNA sequencing [26]) [26]
41900 <5> (<5> product of the purT gene, SDS-PAGE [22]) [22]
42000 <5> (<5> product of the purT gene, deduced from predicted amino acid sequence [22]) [22]
42400 <5> (<5> product of the purT gene [22]) [22]
70000 <1> (<1> gel filtration [1]) [1]

100000 <2, 8> (<2> protein possessing transformylase activity, sucrose density gradient ultracentrifugation [6]) [6, 9]
110000 <2> (<2> SDS-PAGE [11,12]; <2> GAR transformylase, sucrose density gradient ultracentrifugation [13]) [11-13, 15]
110000-120000 <2> (<2> gel filtration, sucrose density gradient ultracentrifugation, sedimentation velocity [18]) [18]
112000 <7> (<7> HeLa and L1210 cells [15]) [15]
115000 <8> [15]
117000 <2> (<2> ultracentrifugation [11]) [11]
125000 <8> (<8> sucrose density gradient ultracentrifugation [14]) [14]
140000 <2> (<2> sucrose density ultracentrifugation [11]) [11]
145000 <10> [8]
150000 <6> (<6> gel filtration [6]) [6, 13]
160000 <2> (<2> trifunctional protein, sucrose density gradient ultracentrifugation [6]) [6]
190000 <2> (<2> trifunctional protein, sucrose density gradient ultracentrifugation [13]) [13]
201000 <11> (<11> gel filtration [6]) [6, 13]
215000 <18> [13]
218000 <12> (<12> gel filtration [6]) [6, 13]

Subunits

dimer <2, 5, 6, 11> (<2> 2 * 63000, transformylase, homodimer, SDS-PAGE [6]; <2> 2 * 83000, trifunctional protein, SDS-PAGE [6]; <2> 2 * 55000, SDS-PAGE, sucrose density ultracentrifugation, proteolytic cleavage [11]; <2> 2 * 61000, GAR transformylase, SDS-PAGE [13]; <2> 2 * 95000, trifunctional protein, SDS-PAGE [13]; <2> 2 * 55000-57000, SDS-PAGE [18]; <2,6,11> homodimer [6,11]; <2> 2 * 61000, SDS-PAGE [14]; <5> 1 * 12100 + 1 * 11200, heterodimer, SDS-PAGE [20]; <5> pH-dependent dimer-monomer transition near physiological pH might act as a means of regulating enzyme activity [31]) [5, 6, 11, 13, 18, 20, 31]
monomer <1, 2, 5-9, 12> (<1> 1 * 70000, gel filtration [1]; <2> 1 * 117000, ultracentrifugation [11]; <5> 1 * 25000, SDS-PAGE [8]; <5> 1 * 23000 [10, 31]; <5> 1 * 42000, product of the purT gene [22]; <7> 1 * 25700, SDS-PAGE [24]; <8> 1 * 110000, SDS-PAGE [14]; <8> 1 * 102000, SDS-PAGE [14]) [1, 6, 8, 10, 11, 14, 19, 22, 24, 31]

5 Isolation/Preparation/Mutation/Application

Source/tissue

HeLa cell <7> (<7> O cell [8]) [8]
L-1210 cell <8> (<8> murine leukemia [8,11,21]) [8, 11, 21]
L-5178-Y cell <8> (<8> murine lymphoma cell line [7,8,14,15]) [7, 8, 14, 15]
PANC-1 cell <7> (<7> xenograft, pancreatic carcinoma [27]) [27]
colonic cancer cell <7> (xenograft HXGC3, HC1 and VRC5 [27]) [27]
liver <1-3, 11, 12, 18> [1-4, 6, 11-15, 17, 24, 25, 28, 32]

lymphosarcoma cell <8> (<8> 6C3HED [27]) [27]
mammary carcinoma cell <8> (<8> 3CH [27]) [27]
ovary <14> [8]
pancreas <7> [27]

Localization

cytoplasm <5, 7> [30]

Purification

<1> [1]
<2> (purified in combination with formyl-methenyl-methylenetetrahydrofolate synthetase [2,6]; partially [17]; copurification of trifunctional 5,10-methenyl-, 5,10-methylene, and 10-formylH4-folate synthetase, aminoimidazole carboxamide ribotide transformylase and serine transhydroxymethylase [18]) [2, 3, 6, 11-13, 17, 18, 32]
<3> [13]
<5> (transformylase, product of the purT gene [22]) [8-10, 17, 19, 20, 22, 29-31]
<6> [13]
<7> (recombinant enzyme, expressed in E. coli [24,25]) [24, 25]
<8> [7, 14, 15, 21, 27]
<11> [13]
<12> [13]
<18> [13]

Crystallization

<5> (crystals grown by vapor diffusion, space group C_{222}, a = 140.5 A, b = 98.2 A, c = 103.3 A [5]; crystals grown at 20°C by hanging-drop vapor diffusion, space group C_{222}, a = 140.9 A, b = 97.6 A, c = 102.4 [9]; orthorhombic space group C_{222}, a = 141.4 A, b = 98.2 A, c = 103.5 A [10]) [5, 9, 10, 19, 21, 28, 31]

Cloning

<5> (harboring expression vector pJS167, encoding the purN gene [8]; coding region of Escherichia coli GART amplified from chromosomal DNA with PCR method [9]; second GAR transformylase encoded by the Escherichia coli purT gene [19,22,29]; hybrid glycinamide ribonucleotide transformylase, ribonucleotide binding domain from purN, catalytic machinery and formyl tetrahydrofolate binding domain from purU encoded formyl tetrahydrofolate hydrolase [23]) [5, 8-10, 19, 20, 22, 23, 29, 30]
<7> (cloned by functional complementation in Escherichia coli [5]; cloned and expressed in Escherichia coli BL21(DE3)pLysS [24,27]; cloned and expressed in Escherichia coli HMS174(DE3)/pCC12 [25]) [5, 24, 25, 27, 30]
<8> (cDNA, recombinant enzyme expressed in insect cells with a recombinant baculovirus [21]) [21]
<9> (cloning of a 12-gene cluster encoding 9 enzymes for de novo purine nucleotide synthesis [5]) [5]
<10> [12]
<13> (pvdF gene characterization [26]) [26]

Application

medicine <5, 7, 8> (<5, 7, 8> pharmacological importance, target enzyme for chemotherapy, development of novel antifolate drugs, to be used as anti-cancer drugs [5, 9, 14, 15, 24, 28, 30, 31]; <5> target for structure-based drug design [9]; <8> inhibitor 5,10-dideazatetrahydrofolate is antiproliferate and cytotoxic against mammalian tumor cells in culture and have broad spectrum activity against transplantable mouse tumors and human tumor xenografts [21, 25, 27]) [5, 9, 14, 15, 21, 24, 25, 27, 28, 30, 31]

6 Stability

pH-Stability

6-10 <5> [31]

Storage stability

<2>, -10°C, 8months, no appreciable loss of activity [3]

References

[1] Hards, R.G.; Graw, S.L.; Patterson, D.: Purification of mammalian glycinamide ribonucleotide (GAR) synthetase. Adv. Exp. Med. Biol., **195**, 315-319 (1986)

[2] Caperelli, C.A.; Chettur, G.; Lin-Kosley, L.: Purification of glycineamide ribonucleotide transformylase. Dev. Biochem., **4**, 371-376 (1979)

[3] Warren, L.; Buchanan, J.M.: Biosynthesis of the purines. XIX. 2-Amino-N-ribosylacetamide 5'-phosphate(glycinamide ribotide) transformylase. J. Biol. Chem., **229**, 613-626 (1957)

[4] Smith, G.K.; Benkovic, P.A.; Benkovic, S.J.: L(-)-10-formyltetrahydrofolate is the cofactor for glycinamide ribonucleotide transformylase from chicken liver. Biochemistry, **20**, 4034-4036 (1981)

[5] Chen, P.; Schulze-Gahmen, U.; Stura, E.A.; Inglese, J.; Johnson, D.L.; Marolewski, A.; Benkovic, S.J.; Wilson, I.A.: Crystal structure of glycinamide ribonucleotide transformylase from Escherichia coli at 3.0 A resolution. J. Mol. Biol., **227**, 283-292 (1992)

[6] Caperelli, C.A.; Chettur, G.; Lin, L.Y.; Benkovic, S.J.: Purification of glycineamide ribonucleotide transformylase. Biochem. Biophys. Res. Commun., **82**, 403-410 (1978)

[7] Caperelli, C.A.; Price, M.F.: Carbocyclic glycinamide ribonucleotide is a substrate for glycinamide ribonucleotide transformylase. Arch. Biochem. Biophys., **264**, 340-342 (1988)

[8] Inglese, J.; Johnson, D.L.; Shiau, A.; Smith, J.M.; Benkovic, S.J.: Subcloning, characterization, and affinity labeling of Escherichia coli glycinamide ribonucleotide transformylase. Biochemistry, **29**, 1436-1443 (1990)

[9] Almassy, R.J.; Janson, C.A.; Kann, C.C.; Hostomska, Z.: Structures of apo and complexed Escherichia coli glycinamide ribonucleotide transformylase. Proc. Natl. Acad. Sci. USA, **89**, 6114-6118 (1992)

[10] Stura, E.A.; Johnson, D.L; Inglese, J.; Smith, J.M.; Benkovic, S.J.; Wilson, I.A.: Preliminary crystallographic investigations of glycinamide ribonucleotide transformylase. J. Biol. Chem., **264**, 9703-9706 (1989)

[11] Young, M.; Sammons, R.D.; Mueller, W.T.; Benkovic, S.J.: An antibody probe to determine the native species of glycinamide ribonucleotide transformylase in chicken liver. Biochemistry, **23**, 3979-3986 (1984)

[12] Daubner, S.C.; Schrimsher, J.L.; Schendel, F.J.; Young, M.; Henikoff, S.; Patterson, D.; Stubbe, J.; Benkovic, S.J.: A multifunctional protein possessing glycinamide ribonucleotide synthetase, glycinamide ribonucleotide transformylase , and aminoimidazole ribonucleotide synthetase activities in de novo purine biosynthesis. Biochemistry, **29**, 7059-7062 (1985)

[13] Caperelli, C.A.; Benkovic, P.A.; Chettur, G.; Benkovic, S.J.: Purification of a complex catalyzing folate cofactor synthesis and transformylation in de novo purine biosynthesis. J. Biol.Chem., **255**, 1885-1890 (1985)

[14] Caperelli, C.A.: Mammalian glycinamide ribonucleotide transformylase: Purification and some properties. Biochemistry, **24**, 1316-1320 (1985)

[15] Caperelli, C.A.: Mammalian glycinamide ribonucleotide transformylase. Kinetic mechanism and associated de novo purine biosynthesis activities. J. Biol.Chem., **264**, 5053-5057 (1989)

[16] Caperelli, C.A.; Liu, D.: Carbocyclic substrates for de novo purine biosynthesis. J. Biol.Chem., **267**, 9783-9787 (1992)

[17] Dev, I.K.; Harvey, R.J.: N^{10}-Formyltetrahydrofolate is the formyl donor for glycinamide ribonucleotide transformylase in Escherichia coli. J. Biol.-Chem., **253**, 4242-4244 (1978)

[18] Smith, G.K.; Mueller, W.T.; Wasserman, G.F.; Taylor, W.D.; Benkovic, S.J.: Characterization of the enzyme complex involving the folate-requiring enzymes of de novo purine biosynthesis. Biochemistry, **19**, 4313-4321 (1980)

[19] Warren, M.S.; Marolewski, A.E.; Benkovic, S.J.: A rapid screen of active site mutants in glycinamide ribonucleotide transformylase. Biochemistry, **35**, 8855-8862 (1996)

[20] Liu, C.; Shim, J.H.; Benkovic, S.J.: The unexpected catalytic properties of a heterodimer of GAR transformylase. Bioorg. Chem., **28**, 316-323 (2000)

[21] Sanghani, S.P.; Moran, R.G.: Tight binding of folate substrates and inhibitors to recombinant mouse glycinamide ribonucleotide formyltransferase. Biochemistry, **36**, 10506-10516 (1997)

[22] Marolewski, A.; Smith, J.M.; Benkovic, S.J.: Cloning and characterization of a new purine biosynthetic enzyme: a non-folate glycinamide ribonucleotide transformylase from E. coli. Biochemistry, **33**, 2531-2537 (1994)

[23] Nixon, A.E.; Benkovic, S.J.: Improvement in the efficiency of formyl transfer of a GAR transformylase hybrid enzyme. Protein Eng., **13**, 323-327 (2000)

[24] Caperelli, C.A.; Giroux, E.L.: The human glycinamide ribonucleotide transformylase domain: purification, characterization, and kinetic mechanism. Arch. Biochem. Biophys., **341**, 98-103 (1997)

[25] Antle, V.D.; Donat, N.; Hua, M.; Liao, P.L.; Vince, R.; Caperelli, C.A.: Substrate specificity of human glycinamide ribonucleotide transformylase. Arch. Biochem. Biophys., **370**, 231-235 (1999)
[26] McMorran, B.J.; Kumara, H.M.C.S.; Sullivan, K.; Lamont, I.L.: Involvement of a transformylase enzyme in siderophore synthesis in Pseudomonas aeruginosa. Microbiology, **147**, 1517-1524 (2001)
[27] Habeck, L.L.; Leitner, T.A.; Shackelford, K.A.; Gossett, L.S.; Schultz, R.M.; Andis, S.L.; Shih, C.; Grindey, G.B.; Mendelsohn, L.G.: A novel class of monoglutamated antifolates exhibits tight-binding inhibition of human glycinamide ribonucleotide formyltransferase and potent activity against solid tumors. Cancer Res., **54**, 1021-1026 (1994)
[28] Antle, V.D.; Liu, D.; McKellar, B.R.; Caperelli, C.A.; Hua, M.; Vince, R.: Substrate specificity of glycinamide ribonucleotide transformylase from chicken liver. J. Biol. Chem., **271**, 6045-6049 (1996)
[29] Shim, J.H.; Benkovic, S.J.: Evaluation of the kinetic mechanism of Escherichia coli glycinamide ribonucleotide transformylase. Biochemistry, **37**, 8776-8782 (1998)
[30] Gooljarsingh, L.T.; Ramcharan, J.; Gilroy, S.; Benkovic, S.J.: Localization of GAR transformylase in Escherichia coli and mammalian cells. Proc. Natl. Acad. Sci. USA, **98**, 6565-6570 (2001)
[31] Shim, J.H.; Benkovic, S.J.: Evaluation of the kinetic mechanism of Escherichia coli glycinamide ribonucleotide transformylase probed by site-directed mutagenesis and pH-dependent studies. Biochemistry, **38**, 10024-10031 (1999)
[32] Hartman, S.C.; Buchanan, J.M.: Biosynthesis of the purines. XXVI. The Identification of the formyl donors of the transformylation reactions. J. Biol. Chem., **234**, 1812-1816 (1959)

Phosphoribosylaminoimidazolecarboxamide formyltransferase 2.1.2.3

1 Nomenclature

EC number

2.1.2.3

Systematic name

10-formyltetrahydrofolate:5'-phosphoribosyl-5-amino-4-imidazole-carboxamide N-formyltransferase

Recommended name

phosphoribosylaminoimidazolecarboxamide formyltransferase

Synonyms

10-formyltetrahydrofolate:5'-phosphoribosyl-5-amino-4-imidazolecarboxamide formyltransferase
5'-phosphoribosyl-5-amino-4-imidazolecarboxamide formyltransferase
5-amino-1-ribosyl-4-imidazolecarboxamide 5'-phosphate transformylase
5-amino-4-imidazolecarboxamide ribonucleotide transformylase
5-amino-4-imidazolecarboxamide ribotide transformylase
5-aminoimidazole-4-carboxamide ribonucleotide transformylase (AICAR Tfase)-inosine monophosphate cyclohydrolase <6> [13]
5-aminoimidazole-4-carboxamide ribonucleotide transformylase/inosine 5'-monophosphate cyclohydrolase <7> [16]
5-aminoimidazole-4-carboxamide ribotide transformylase-IMP cyclohydrolase <7> (<7> bifunctional enzyme [9,10]) [9, 10]
AICA-ribotide formyltransferase <10> [3]
AICAR TFase <6> [5]
AICAR formyltransferase
AICAR transformylase
AICARFT <7> [10, 17]
AICARFT/IMPCHase <7> [10]
ATIC <6, 7> [13, 16, 17]
IMPCH <6> [13]
aminoimidazolecarboxamide ribonucleotide transformylase

CAS registry number

9032-03-5

2 Source Organism

<1> *Allium cepa* (onion [3]) [3]
<2> *Bacillus subtilis* [9, 10, 12, 15]

<3> *Columba sp.* (pigeon [3]) [3]
<4> *Daucus carota* (carrot [3]) [3]
<5> *Escherichia coli* (wild-type strain 5928, strain PA 15 [2]) [2, 3, 9, 10, 14, 15]
<6> *Gallus sp.* (chicken [1,3-9]; avian [9,10,13,16-18]) [1, 3-11, 13, 15-18]
<7> *Homo sapiens* (human [8-10,12,14-18]) [8-10, 12, 14-18]
<8> *Mus musculus* (mouse, strain MRL/lpr [8]) [5, 7, 8]
<9> *Petroselinum crispum* (parsley [3]) [3]
<10> *Pisum sativum* (pea, L. var. Alaska [3]) [3]
<11> *Saccharomyces cerevisiae* (yeast [16]) [16]
<12> *Salmonella typhimurium* [10, 12, 15]
<13> *Spinacia oleracea* (spinach [3]) [3]
<14> *Trifolium sp.* (trefoil [3]) [3]

3 Reaction and Specificity

Catalyzed reaction

10-formyltetrahydrofolate + 5-amino-1-(5-phospho-D-ribosyl)imidazole-4-carboxamide = tetrahydrofolate + 5-formamido-1-(5-phospho-D-ribosyl)imidazole-4-carboxamide

Reaction type

formyl group transfer

Natural substrates and products

S 10-formyltetrahydrofolate + 5'-phosphoribosyl-5-amino-imidazolecarboxamide <1-14> (<6> precursors for inosinic acid biosynthesis [1]; <6, 7> reduced folate cofactor requiring step of de novo purine biosynthesis [6, 17]; <6,7> penultimate and final steps in the de novo purine biosynthesis pathway [10, 12, 13, 15-18]; <6, 7, 10> de novo synthesis of purine nucleotides [3, 9, 11, 15, 16]) (Reversibility: r <1-14> [1-18]) [1-18]
P tetrahdrofolic acid + 5'-phosphoribosyl-5-formamido-4-imidazolecarboxamide
S N^{10}-formyltetrahydropteroylglutamate + 5'-phosphoribosyl-5-formamido-4-imidazole carboxamide ribonucleotide <6> (Reversibility: ? <6> [4]) [4]
P tetrahydropteroylglutamate + 5'-phosphoribosyl-5-formamido-4-imidazole carboxamide

Substrates and products

S (6R,6S)-10-formyltetrahydrofolic acid + 5-amino-4-imidazole carboxamide ribonucleotide <7> (Reversibility: r <7> [10]) [10]
P tetrahydrofolic acid + 5-formamido-4-imidazole carboxamide ribonucleotide <7> [10]
S 10-formyl-7,8-dihydropteroylpenta-γ-glutamate + 5'-phosphoribosyl-5-amino-4-imidazolecarboxamide <6> (Reversibility: ? <6> [7]) [7]

P 7,8-dihydropteroylpenta-γ-glutarate + 5'-phosphoribosyl-5-formamido-4-imidazolecarboxamide <6> [7]

S 10-formyl-8-deazafolate + 5'-phosphoribosyl-5-amino-4-imidazolecarboxamide <6> (Reversibility: ? <6> [6]) [6]

P 8-deazafolate + 5'-phosphoribosyl-5-formamido-4-imidazolecarboxamide <6> [6]

S 10-formyldihydrofolic acid pentaglutamate + 5'-phosphoribosyl-5-amino-4-imidazolecarboxamide <6> (Reversibility: ? <6> [7]) [7]

P dihydrofolic acid pentaglutamate + 5'-phosphoribosyl-5-formamido-4-imidazolecarboxamide <6> [7]

S 10-formyltetrahydrofolate + 5-amino-imidazole-4-thiocarboxamide ribonucleotide <7> (Reversibility: ? <7> [14]) [14]

P tetrahydrofolate + 6-mercaptopurine ribonucleotide <7> [14]

S N^{10}-formyltetrahydrofolic acid + 5-amino-4-imidazole carboxamide ribonucleotide <1-14> (Reversibility: r <1-14> [1-18]) [1-18]

P tetrahydrofolic acid + 5-formamido-4-imidazole carboxamide ribonucleotide <1-14> [1-18]

S N^{10}-formyltetrahydropteroylglutamate + 5'-phosphoribosyl-5-formamido-4-imidazole carboxamide ribonucleotide <6> (Reversibility: ? <6> [4]) [4]

P tetrahydropteroylglutamate + 5'-phosphoribosyl-5-formamido-4-imidazole carboxamide ribonucleotide <6> [4]

S N^{10}-formyltetrahydropteroylglutamate + 5'-phosphoribosyl-5-formamino-4-imidazole carboxamide ribonucleotide <6> (Reversibility: ? <6> [4]) [4]

P tetrahydropteroylglutamate + 5'-phosphoribosyl-5-formamido-4-imidazole carboxamide ribonucleotide <6> [4]

S N^{10}-formyltetrahydropteroylheptaglutamate + 5'-phosphoribosyl-5-formamino-4-imidazole carboxamide ribonucleotide <6> (Reversibility: ? <6> [4]) [4]

P tetrahydropteroylglutamate + 5'-phosphoribosyl-5-formamido-4-imidazole carboxamide ribonucleotide <6> [4]

S N^{10}-formyltetrahydropteroylhexaglutamate + 5'-phosphoribosyl-5-formamino-4-imidazole carboxamide ribonucleotide <6> (Reversibility: ? <6> [4]) [4]

P tetrahydropteroylglutamate + 5'-phosphoribosyl-5-formamido-4-imidazole carboxamide ribonucleotide <6> [4]

S N^{10}-formyltetrahydropteroylmonoglutamate + 5'-phosphoribosyl-5-formamino-4-imidazole carboxamide ribonucleotide <6> (Reversibility: ? <6> [4]) [4]

P tetrahydropteroylmonoglutamate + 5'-phosphoribosyl-5-formamido-4-imidazole carboxamide ribonucleotide <6> [4]

S N^{10}-formyltetrahydropteroylpentaglutamate + 5'-phosphoribosyl-5-formamino-4-imidazole carboxamide ribonucleotide <6> (Reversibility: ? <6> [4]) [4]

P tetrahydropteroylglutamate + 5'-phosphoribosyl-5-formamido-4-imidazole carboxamide ribonucleotide <6> [4]

S N^{10}-formyltetrahydropteroyltetraglutamate + 5'-phosphoribosyl-5-formamino-4-imidazole carboxamide ribonucleotide <6> (Reversibility: ? <6> [4]) [4]

P tetrahydropteroylglutamate + 5'-phosphoribosyl-5-formamido-4-imidazole carboxamide ribonucleotide <6> [4]

S N^{10}-formyltetrahydropteroyltriglutamate + 5'-phosphoribosyl-5-formamino-4-imidazole carboxamide ribonucleotide <6> (Reversibility: ? <6> [4]) [4]

P tetrahydorpteroylglutamate + 5'-phosphoribosyl-5-formamido-4-imidazole carboxamide ribonucleotide <6> [4]

S N^5,N^{10}-anhydroformyltetrahydrofolic acid + glycinamide ribonucleotide <6> (Reversibility: r <6> [1]) [1]

P N^{10}-formyltetrahydrofolic acid + formylglycinamide ribonucleotide <6> [1]

S tetrahydrofolic acid + 5-formamido-4-imidazole carboxamide ribonucleotide <6> (Reversibility: r <6> [1]) [1]

P N^{10}-formyltetrahydrofolic acid + 5-amino-4-imidazole carboxamide ribonucleotide <6> [1]

S Additional information <1-14> (<1-14> in all organisms studied to date AICARFT activity is accompanied by inosine monophosphate cyclohydrolase located on the same polypeptide encoded by the purH gene [1-18]; <5> no activity with N^5,N^{10}-methenyl analogues [2]) [1-18]

P ?

Inhibitors

(2RS)-N-[4-{1-(acetamideoxy)-3-(2-amino-3,4-dihydro-4-oxo-quinazolin-6-yl)prop-2-yl}benzoyl]-L-glutamic acid <6> [11]
(2RS)-N-[4-{1-(acetoxy)-3-(2-amino-3,4-dihydro-4-oxo-quinazolin-6-yl)prop-2-yl}benzoyl]-L-glutamic acid <6> [11]
(2RS)-N-[4-{3-(2-amino-3,4-dihydro-4-oxo-quinazolin-6-yl)-1,1-(dimethoxy)prop-2-yl}benzoyl]-L-glutamic acid <6> [11]
(2RS)-N-[4-{3-(2-amino-3,4-dihydro-4-oxo-quinazolin-6-yl)-1-(1,1-bromoacetoxy)prop-2-yl}benzoyl]-L-glutamic acid <6> [11]
(2RS)-N-[4-{3-(2-amino-3,4-dihydro-4-oxo-quinazolin-6-yl)-1-(1,1-dimethoxyacetoxy)prop-2-yl}benzoyl]-L-glutamic acid <6> [11]
(2RS)-N-[4-{3-(2-amino-3,4-dihydro-4-oxo-quinazolin-6-yl)-1-(ethoxalyoxy) prop-2-yl}benzoyl]-L-glutamic acid <6> [11]
(2RS)-N-[4-{3-(2-amino-3,4-dihydro-4-oxo-quinazolin-6-yl)-1-(formyloxy)-prop-2-yl}benzoyl]-L-glutamic acid <6> [11]
(2RS)-N-[4-{3-(2-amino-3,4-dihydro-4-oxo-quinazolin-6-yl)-1-(hydroxy)-prop-2-yl}benzoyl]-L-glutamic acid <6> [11]
(2RS)-N-[4-{3-(2-amino-3,4-dihydro-4-oxo-quinazolin-6-yl)-1-(pyruvoyloxy)prop-2-yl}benzoyl]-L-glutamic acid <6> [11]
(6S/R)-5,10-dideaza-5,6,7,8,-tetrahydrofolate <7> [16, 17]
10-formyl-5,6,7,8-tetrahydropteroylpenta-γ-glutamate <6> [7]
10-formyl-5,8,10-trideazafolate <6> [11]
10-formylfolic acid pentaglutamate <6> (<6> competitive inhibitor [7]) [7]

4-N-allyl-5-aminoimidazole-4-carboxamide ribonucleotide <7> [14]
4-N-methyl-5-aminoimidazole-4-carboxamide ribonucleotide <7> [14]
4-acetamidophenol <6> [8]
4-amino-10-methylpteroyl-γ-glutamyl-γ-glutamyl-γ-glutamylglutamic acid <6> (<6> methotrexate polyglutamate, competitive inhibitor [7]) [7]
4-amino-10-methylpteroylglutamic acid <6> (<6,7> methotrexate, non-competitive inhibitor [7,9,10,12,16,18]) [7, 9, 10, 12, 16, 18]
5'-phosphoribosyl-5-formamido-4-imidazolecarboxamide <7> (<7> product inhibitor [14]) [14]
5-amino-1-β-D-ribofuranosylimidazole 5'-phosphate <7> [14]
5-amino-1-β-D-ribofuranosylimidazole-4-carbonitrile 5'-phosphate <7> [14]
5-amino-1-β-D-ribofuranosylimidazole-4-carboxamidoxime 5'-phosphate <7> [14]
5-amino-1-β-D-ribofuranosylimidazole-4-carboxylate 5'-phosphate <7> [14]
5-amino-4-nitro-1-β-D-ribofuranosylimidazole 5'-phosphate <7> [14]
7,8-dihydropteroylpenta-γ-glutamate <6> [7]
AMP <7> [14]
Cu^{2+} <6> [1]
KCl <7> [9]
N^{10}-formyltetrahydropteroylglutamate <5> [2]
XMP <7> [14]
azathioprine <6> (<6> competitive inhibitor [8]) [8]
dihydrofolic acid pentaglutamate <6, 7> [7, 12]
folic acid pentaglutamate <6> (<6> competitive inhibitor [7]) [7]
ibuprofen <6, 8> (<6,8> anti-inflammatory drug [8]) [8]
naproxen <6, 8> (<6,8> anti-inflammatory drug [8]) [8]
paracetamol <6> [8]
pteroylpenta-γ-glutamate <6> [7]
sulphalazine <7> [8]
thioinosinic acid <6> (<6> competitive inhibitor [8]) [8]
tylenol <6> [8]

Cofactors/prosthetic groups

10-formyldihydrofolate <7> [14]
10-formylfolate <7> [14]
N^{10}-formyltetrahydrofolate <6, 7, 10> [3, 5, 14, 17, 18]

Metals, ions

K^+ <6, 10> (<6> required for maximum enzymatic activity [4]; <10> essential component [3]) [3, 4]

Specific activity (U/mg)

0.00067 <7> [9]
0.0371 <6> [4]
0.856 <6> [5]
1.19 <7> [10]
Additional information <10> (<10> 1 unit = removal of 0.0001 mM of AICA-ribotide/0.5 ml/20 min at 37°C [3]) [3]

K_m-Value (mM)

0.00023 <6> (tetrahydrofolic acid) [7]
0.0005 <6> (5-aminoimidazole-4-carboxamide ribonucleotide, <6> mutant K266R [15]) [15]
0.00051 <6> (10-formyldihydrofolic acid pentaglutamate) [7]
0.0009 <7> (5-amino-imidazole-4-thiocarboxamide ribonucleotide, <7> 10-formyldihydrofolate as cofactor [14]) [14]
0.00099 <6> (N^{10}-formyltetrahydropteroyltetraglutamate, <6> 25 mM KCl [4]) [4]
0.00107 <6> (N^{10}-formyltetrahydropteroyltetraglutamate, <6> 150 mM KCl [4]) [4]
0.0015 <7> (5-aminoimidazole-4-carboxamide ribonucleotide, <7> 10-formyldihydrofolate as cofactor [14]) [14]
0.00164 <6> (N^{10}-formyltetrahydropteroylhexaglutamate, <6> 25 mM KCl [4]) [4]
0.00165 <6> (N^{10}-formyltetrahydropteroylmonoglutamate, <6> 25 mM KCl [4]) [4]
0.00183 <6> (N^{10}-formyltetrahydropteroylheptaglutamate, <6> 25 mM KCl [4]) [4]
0.0019 <6> (5-aminoimidazole-4-carboxamide ribonucleotide, <6> wild-type [15]) [15]
0.0019 <7> (5-aminoimidazole-4-carboxamide ribonucleotide, <7> 10-formyltetrahydrofolate as cofactor [14]) [14]
0.00195 <6> (N^{10}-formyltetrahydropteroylhexaglutamate, <6> 150 mM KCl [4]) [4]
0.00206 <6> (N^{10}-formyltetrahydropteroylpentaglutamate, <6> 150 mM KCl [4]) [4]
0.00256 <6> (N^{10}-formyltetrahydropteroylpentaglutamate, <6> 25 mM KCl [4]) [4]
0.00313 <6> (N^{10}-formyltetrahydropteroylheptaglutamate, <6> 150 mM KCl [4]) [4]
0.0043 <6> (10-formyltetrahydrofolate, <6> mutant K266R [15]) [15]
0.00595 <6> (N^{10}-formyltetrahydropteroyltriglutamate, <6> 150 mM KCl [4]) [4]
0.0076 <6> (10-formyltetrahydrofolate, <6> wild-type [15]) [15]
0.01 <7> (10-formyltetrahydrofolate) [17]
0.01 <7> (5-aminoimidazole-4-carboxamide ribonucleotide) [14]
0.0168 <7> (5-aminoimidazole-4-carboxamide ribonucleotide) [10, 18]
0.031 <10> (5'-phosphoribosyl-5-amino-imidazolecarboxamide) [3]
0.0602 <7> ((6R,6S)-10-formyltetrahydrofolic acid) [10]
0.0679 <6> (10-formyltetrahydrofolate) [6]
0.185 <7> (tetrahydrofolate) [17]
0.2 <10> (10-formyltetrahydrofolate) [3]
0.353 <6> (N^{10}-formyltetrahydropteroylmonoglutamate, <6> 150 mM KCl [4]) [4]
0.674 <6> (N^{10}-formyltetrahydropteroyltriglutamate, <6> 25 mM KCl [4]) [4]

K_i-Value (mM)
0.00004 <8> (5-amino-4-nitroimidazole ribonucleotide) [7]
0.00007 <8> (5-formyl-5-aminoimidazole-4-carboxamide ribonucleotide) [7]
0.00008 <6> (folic acid pentaglutamate) [7]
0.00014 <6> (dihydrofolic acid pentaglutamate) [7]
0.0004 <7> (5-formyl-5-aminoimidazole-4-carboxamide ribonucleotide) [14]
0.0007 <7> (5-amino-4-nitro-1-β-D-ribofuranosylimidazole 5'-phosphate) [14]
0.0007 <7> (5-amino-4-nitroimidazole ribonucleotide) [14]
0.0012 <7> (XMP) [14]
0.00137 <6> (10-formylfolic acid pentaglutamate) [7]
0.00315 <6> (methotrexate polyglutamate) [7]
0.006 <7> (5-amino-1-β-D-ribofuranosylimidazole 5'-phosphate) [14]
0.01 <7> (5-amino-1-β-D-ribofuranosylimidazole-4-carboxylate 5'-phosphate) [14]
0.026 <7> (5-amino-1-β-D-ribofuranosylimidazole-4-carbonitrile 5'-phosphate) [14]
0.027 <7> (AMP) [14]
0.039 <6> (thioinosinic acid) [8]
0.075 <6> (methotrexate) [7]
0.09 <8> (azathioprine) [8]
0.11 <8> (thioinosinic acid) [8]
0.12 <6> (azathioprine) [8]

pH-Optimum
7.4 <10> [3]

pH-Range
6-10.5 <6> [15]
6.5-8.5 <10> [3]

Temperature optimum (°C)
37 <10> [3]

Temperature range (°C)
0-50 <10> [3]

4 Enzyme Structure

Molecular weight
64430 <7> (<7> cDNA sequencing [10]) [10]
64450 <7> (<7> electrospray mass spectroscopy [10]) [10]
65000 <7> (<7> recombinant enzyme, SDS-PAGE [12]) [12]
65630 <7> (<7> gel filtration [10]) [10]

Subunits

dimer <6, 7, 11> (<7> 2 * 64400, deduced from nucleotide sequence of cDNA [9]) [9, 16, 18]
monomer <7> (<7> 1 * 64425, calculated from amino acid sequence [10]) [10, 16]

5 Isolation/Preparation/Mutation/Application

Source/tissue

CCRF-CEM cell <7> (<7> leukemia cell, T lymphoblastoid cell line [9,10,12]) [9, 10, 12]
Ehrlich ascites carcinoma cell <8> [5, 7]
HeLa cell <7> (<7> cervical carcinoma cell line [12]) [12]
Hep-G2 cell <7> (<7> hepatocellular carcinoma cell line [12,16]) [12, 16]
MCF-7 cell <7> (<7> breast cancer cell [9,10]) [9, 10]
bulb <1> [3]
embryo <10> [3]
leaf <4, 9, 13, 14> [3]
liver <3, 6, 7> [3-9, 12]
lymphocyte <7> [8]
peripheral blood mononuclear cell <7, 8> [8]
petiole <13, 14> [3]
placenta <7> [12, 16]
root <1, 13, 14> [3]
seed <10> [3]
seedling <10> [3]

Localization

microsome <10> [3]
mitochondrion <10> [3]
nucleus <10> [3]

Purification

<5> [2]
<6> (inosinicase activity copurifies [5]; partially [3,7]; recombinant avian ATIC, purified from E. coli expression system [13]) [1, 3-5, 7, 8, 13, 18]
<7> [9, 10, 12, 17]
<8> [5]
<10> [3]

Crystallization

<6> (ATIC complex crystallized in monoclinic space group P2(1), a = 56.48 A, b = 107.88 A, c = 103.86 A and β = 91.2°; crystallized in a binary complex with methotrexate, single crystal at 96 K to 2.3 A resolution, monoclinic crystals, space group P2(1), with unit-cell dimensions a = 65.17, B = 105.93, c= 103.47 A, β= 108.27 degrees [13]) [4, 13, 17, 18]
<7> [15, 16]

Cloning

<2> (cloned and expressed [10]) [10]
<5> (cloned and expressed [10]) [10]
<6> (purH gene cloning and sequencing of cDNA [10]; ATICcDNA cloned into a pET28a vector and transformed into Escherichia coli BL21.DE3 for overexpression [18]) [10, 18]
<7> (purH gene, cloning and sequencing of cDNA, expressed in Escherichia coli [10]; cloned from a placenta cDNA library, overproduced in Escherichia coli [12,16]; cloned from a hepatoma cDNA library [16]) [10, 12, 16]
<12> (cloned and expressed [10]) [10]

Application

medicine <6-8> (<6> attractive anticancer target [13]; <8> animal model of systemic autoimmune disease [8]; <7> target for development of inhibitors with potential use as chemotherapeutic agents [9,10,12,14-18]; <7> potential target for development of antineoplastic drugs [18]) [8-10, 12-18]

6 Stability

General stability information

<6>, much less stable during storage in maleate buffer than in Tris or phosphate solutions [1]

Storage stability

<6>, 2°C, if the solution of purified enzyme are stored for 3 or 4 months, a requirement for a specific formyl donor can then be shown [1]

References

[1] Hartman, S.C.; Buchanan, J.M.: Biosynthesis of the purines. XXVI. The identification of the formyl donors of the transformylation reactions. J. Biol. Chem., **234**, 1812-1816 (1959)
[2] Patrick, T.W.; Crosbie, G.W.: Specificity of 4-aminoimidazole-5-carboxamide ribotide transformylase of Escherichia coli. Biochem. J., **124**, 31-32 (1971)
[3] Iwai, K.; Fujisawa, Y.; Suzuki, N.: The accumulation of 5'-phosphoribosyl-5-amino-4-imidazolecarboxamide in folate-deficient pea seedlings and the enzymatic reaction in which the compound is involved. Agric. Biol. Chem., **36**, 398-408 (1972)
[4] Baggott, J.E.; Krumdieck, C.L.: Folylpoly-γ-glutamates as cosubstrates of 10-formyltetrahydrofolate:5'-phosphoribosyl-5-amino-4-imidazole-carboxamide formyltransferase. Biochemistry, **18**, 1037-1041 (1979)
[5] Mueller, W.T.; Benkovic, S.J.: On the purification and mechanism of action of 5-aminoimidazole-4-carboxamide-ribonucleotide transformylase from chicken liver. Biochemistry, **20**, 337-344 (1981)

[6] Smith, G.K.; Mueller, T.; Benkovic, P.A.; Benkovic, S.J.: On the cofactor specificity of glycinamide ribonucleotide and 5-aminoimidazole-4-carboxamide ribonucleotide transformylase from chicken liver. Biochemistry, **20**, 1241-1245 (1981)

[7] Baggott, J.E.; Vaughn, W.H.; Hudson, B.B.: Inhibition of 5-aminoimidazole-4-carboxamide ribotide transformylase, adenosine deaminase and 5'-adenylate deaminase by polyglutamates of methotrexate and oxidized folates and by 5-aminoimidazole-4-carboxamide riboside and ribotide. Biochem. J., **236**, 193-200 (1986)

[8] Ha, T.; Morgan, S.L.; Vaughn, W.H; Eto, I.; Baggott, J.E.: Detection of inhibition of 5-aminoimidazole-4-carboxamide ribotide transformylase by thioinosinic acid and azathioprine by a new colorimetric assay. Biochem. J., **272**, 339-342 (1990)

[9] Szabados, E.; Hindmarsh, E.J.; Phillips, L.; Duggleby, R.G.; Christopherson, R.I.: 5-Aminoimidazole-4-carboxamide ribotide transformylase-IMP cyclohydrolase from human CCRF-CEM leukemia cells: purification, pH dependence and inhibitors. Biochemistry, **33**, 14237-14245 (1994)

[10] Rayl, E.A.; Moroson, B.A.; Beardsley, G.P.: The human purH gene product, 5-aminoimidazole-4-carboxamide ribonucleotide formyltransferase/IMP cyclohydrolase. Cloning, sequencing, expression, purification, kinetic analysis, and domain mapping. J. Biol. Chem., **271**, 2225-2233 (1996)

[11] Boger, D.L.; Haynes, N.E.; Warren, M.S.; Gooljarsingh, L.T.; Ramcharan, J.; Kitos, P.A.; Benkovic, S.J.: Functionalized analogues of 5,8,10-trideazafolate as potential inhibitors of GAR Tfase or AICAR Tfase. Bioorg. Med. Chem., **5**, 1831-1838 (1997)

[12] Sugita, T.; Aya, H.; Ueno, M.; Ishizuka, T.; Kawashima, K.: Characterization of molecularly cloned human 5-aminoimidazole-4-carboxamide ribonucleotide transformylase. J. Biochem., **122**, 309-313 (1997)

[13] Reyes, V.M.; Greasley, S.E.; Stura, E.A.; Beardsley, G.P.; Wilson, I.A.: Crystallization and preliminary crystallographic investigations of avian 5-aminoimidazole-4-carboxamide ribonucleotide transformylase-inosine monophosphate cyclohydrolase expressed in Escherichia coli. Acta Crystallogr. Sect. D, **56**, 1051-1054 (2000)

[14] Wall, M.; Shim, J.H.; Benkovic, S.J.: Human AICAR transformylase: Role of the 4-Carboxamide of AICAR in binding and catalysis. Biochemistry, **39**, 11303-11311 (2000)

[15] Shim, J.H.; Wall, M.; Benkovic, S.J.; Diaz, N.; Suarez, D.; Merz, K.M., Jr.: Evaluation of the catalytic mechanism of AICAR transformylase by pH-dependent kinetics, mutagenesis, and quantum chemical calculations. J. Am. Chem. Soc., **123**, 4687-4696 (2001)

[16] Vergis, J.M.; Bulock, K.G.; Fleming, K.G.; Beardsley, G.P.: Human 5-aminoimidazole-4-carboxamide ribonucleotide transformylase/inosine 5'-monophosphate cyclohydrolase. A bifunctional protein requiring dimerization for transformylase activity but not for cyclohydrolase activity. J. Biol. Chem., **276**, 7727-7733 (2001)

[17] Bulock, K.G.; Beardsley, G.P.; Anderson, K.S.: The kinetic mechanism of the human bifunctional enzyme ATIC (5-amino-4-imidazolecarboxamide ribo-

nucleotide transformylase/inosine 5'-monophosphate cyclohydrolase): A surprising lack of substrate channeling. J. Biol. Chem., **277**, 22168-22174 (2002)

[18] Wolan, D.W.; Greasley, S.E.; Beardsley, G.P.; Wilson, I.A.: Structural insights into the avian AICAR transformylase mechanism. Biochemistry, **41**, 15505-15513 (2002)

Glycine formimidoyltransferase 2.1.2.4

1 Nomenclature

EC number
2.1.2.4

Systematic name
5-formimidoyltetrahydrofolate:glycine N-formimidoyltransferase

Recommended name
glycine formimidoyltransferase

Synonyms
FIG formiminotransferase
formiminoglycine formiminotransferase
glycine formiminotransferase

CAS registry number
9029-84-9

2 Source Organism

<1> *Clostridium cylindrosporum* [1, 3]
<2> *Clostridium acidi-urici* [2]

3 Reaction and Specificity

Catalyzed reaction
5-formimidoyltetrahydrofolate + glycine = tetrahydrofolate + N-formimidoylglycine

Reaction type
formimino group transfer

Substrates and products
S 5-formimidoyltetrahydrofolate + glycine <1, 2> (Reversibility: r <1, 2> [1, 2, 3]) [1, 2, 3]
P tetrahydrofolate + N-formimidoglycine <1, 2> [1, 2, 3]

5 Isolation/Preparation/Mutation/Application

Purification

<1> (precipitation at pH 5.0, calcium phosphate, alumina Cγ, 10fold purification [1]) [1]

References

[1] Rabinowitz, J.C.; Pricer, W.E.: Formation, isolation and properties of 5-formiminotetrahydrofolic acid. Fed. Proc., **16**, 236 (1957)

[2] Sagers, R.D.; Beck, J.V.; Gruber, W.; Gunsalus, I.C.: A tetrahydro-folic acid linked formimino transfer enzyme. J. Am. Chem. Soc., **78**, 694-695 (1956)

[3] Rabinowitz, J.C.; Pricer, W.E.: Formimino-tetrahydrofolic acid and methenyltetrahydrofolic acid as intermediates in the formation of N^{10}-formyltetrahydrofolic acid. J. Am. Chem. Soc., **78**, 5702-5704 (1956)

Glutamate formimidoyltransferase 2.1.2.5

1 Nomenclature

EC number
2.1.2.5

Systematic name
5-formimidoyltetrahydrofolate:L-glutamate N-formimidoyltransferase

Recommended name
glutamate formimidoyltransferase

Synonyms
EC 2.1.2.6 (formerly)
formiminoglutamic acid formiminotransferase
formiminoglutamic acid transferase
formiminoglutamic formiminotransferase
formiminotransferase, glutamate
glutamate formiminotransferase
glutamate formyltransferase

CAS registry number
9032-83-1

2 Source Organism

<1> *Sus scrofa* [1, 3-6, 8, 9, 12]
<2> *Oryctolagus cuniculus* [1]
<3> *Bos taurus* (calf [2]) [1, 2]
<4> *Homo sapiens* [1]
<5> *Mus musculus* [1]
<6> *Felis catus* [1]
<7> *Cavia porcellus* [1]
<8> *Rattus norvegicus* (Sprague-Dawley rats [10]) [2, 7, 10]
<9> *Gallus gallus* (chicken, several isoforms of p60, bifunctional formiminotransferase cyclodeaminase [11]) [11]
<10> *Meleagris gallopavo* [11]

3 Reaction and Specificity

Catalyzed reaction

5-formimidoyltetrahydrofolate + L-glutamate = tetrahydrofolate + N-formimidoyl-L-glutamate (<1,3> mechanism [2,5]; <1> catalytic mechanism [9])

Reaction type

formimino group transfer

Natural substrates and products

S N-forminoglutamate + tetrahydrofolate <1-9> (<1-7,9> histidine degradation [1,5,8,9,11]; <8> histidine metabolism pathway, reaction occurs in vivo within the soluble phase of cytoplasm [7]; <8> bifunctional formiminotransferase cyclodeaminase, 58K, is associated with the cytoplasmatic surface of the Golgi apparatus in vivo [7,10]; <8> metabolic enzyme involved in the conversion of histidine to glutamic acid [10]; <1> enzyme serves as additional entry point for the folate pool in liver [8]; <1> physiological substrate is (6S)-tetrahydrofolate [9]; <9> in addition to its enzymatic activity enzyme may have a second physiological function in mediating interaction of Golgi-derived membranes with microtubules [11]) (Reversibility: ? <1-9> [1, 5, 7-11]) [1, 5, 7-11]

P 5-fomiminotetrahydrofolate + glutamate <1-8> (<1> natural biosynthetic product is N^5-formiminotetrahydrofolate [9]) [1, 5, 7-10]

Substrates and products

S 5-formyltetrahydrofolate + L-glutamate <1, 3> (<1> folinic acid binding site, preference for (6R)-enantiomer of folinic acid [9]; <1> N^5-formyl group of Citrovorum factor, specific with respect to substrates, formation of N^5-formiminotetrahydrofolate and anhydro-Citrovorum factor precedes the formation of the N^{10}-formyl folate derivatives [3]; <3> formation of N^5-formyltetrahydrofolate + L-glutamate [2]) (Reversibility: r <1> [3]; ? <1, 3> [2, 9]) [2, 3, 9]

P N-formyl-L-glutamate + tetrahydrofolate <1, 3> [2, 3]

S tetrahydrofolate + N-formimino-L-glutamate <1-10> (<1> relatively specific for formiminoglutamic acid [1]; <1,8> transfer of a single carbon from formiminoglutamate to tetrahydrofolate, product is 5-formiminotetrahydrofolate [7,8]; <3> specificity, equilibrium lies far on formation of formiminotetrahydrofolate and glutamate [2]; <3> α-L-formamidinoglutaric acid, absolute specificity with regard to the isomers of tetrahydrofolate at position 6, reduced pteroylaspartic acid and reduced pteroyltriglutamic acid can replace tetrahydrofolate [2]; <1> tetrahydropteroic acid can replace tetrahydrofolate [5]; <1> mode of substrate binding and channeling, 3 glutamate-binding sites, mechanism for product release, His-82 may act as catalytic base required for the formiminotransferase mechanism, an electrostatic tunnel through the width of the domain facilitates recognition of the substrates [9]) (Reversibility: r <1-7> [1, 2]; ? <1, 8-10> [4-11]) [1, 2, 4-11]

P 5-formiminotetrahydrofolate + L-glutamate <1-10> (<1> labile 5-formiminotetrahydrofolate product [9]) [1, 2, 4-11]

S tetrahydropteroic acid + N-formimino-L-glutamate <1> (<1> tetrahydropteroic acid is a good alternate substrate [5]) (Reversibility: ? <1> [5]) [5]

P 5-formiminotetrahydropteroic acid + L-glutamate <1> [5]

S tetrahydropteroylpentaglutamate + N-formimino-L-glutamate <1> (<1> bifunctional formiminotetrahydrofolate cyclodeaminase: preference for polyglutamylated substrates, polyglutamylation improves binding of both the folate substrates and formiminoglutamate [8]; <1> optimum number of glutamates for channeling of the product to the cyclodeaminase domain is five [9]) (Reversibility: ? <1> [8, 9]) [8, 9]

P 5-formiminotetrahydropteroylpentaglutamate + L-glutamate <1> [8, 9]

S Additional information <1, 3, 8-10> (<1,3> primary reaction of the overall reaction formimino-L-glutamate + tetrahydrofolate = N^{10}-formyltetrahydrofolate + glutamate + NH_3 [1,2]; <3> formation of glutamic acid is reversible while formation of ammonia is not [2]; <1> not: formiminoglycine, very poor substrates: formimino-L-aspartic acid, formyl-L-glutamic acid [1]; <3> not: formamidinoacetic acid, formylisoglutamine, formylglutamine, acetylglutamic acid [2]; <1> twin enzyme EC 2.1.2.5 and formiminotetrahydrofolate cyclodeaminase EC 4.3.1.4: enzyme system forming N^5,N^{10}-methenyltetrahydrofolate from formiminoglutamate and tetrahydrofolate [4]; <1> molecular enzyme complex with formiminotransferase and cyclodeaminase activity, EC 2.1.2.5 and EC 4.3.1.4, both activities are contained in a single polypeptide chain, no substrate: formiminoaspartate [5]; <1> FTCD: bifunctional enzyme with 5-formiminotetrahydrofolate:L-glutamate N-formiminotransferase activity, EC 2.1.2.5, and formiminotetrahydrofolate cyclodeaminase activity, EC 4.3.1.4 [6]; <1> FTCD: bifunctional enzyme formiminotransferase cyclodeaminase efficiently channels polyglutamated folate between catalytic sites, only the octamer is able to directly transfer pentaglutamated intermediate between active sites, subunit structure [8]; <8> 58K: formiminotransferase cyclodeaminase, bifunctional enzyme catalyzing two consecutive steps in the modification of tetrahydrofolate to 5,10-methenyltetrahydrofolate [7, 10]; <1> FTCD: bifunctional enzyme formiminotransferase cyclodeaminase, EC 2.1.2.5 and EC 4.3.1.4, formiminotransferase domain: N-terminal 326 residues, detailed structure [9]; <9,10> p60: bifunctional formiminotransferase cyclodeaminase, can mediate interaction between trans-Golgi network derived vesicles and microtubule in vitro, p60 may bind to the polyglutamate on microtubules, globular hydrophilic protein [11]) [1, 2, 4-11]

P ?

Inhibitors

Ca^{2+} <3> (<3> 0.1 M, strong inhibition in decreasing order: La^{3+}, Ca^{2+}, Mg^{2+}, Li^+, Na^+, K^+ [2]) [2]

$CaCl_2$ <1> (<1> 0.1 M, 72% inhibition [1]) [1]

D-glutamate <3> (<3> 0.015 mM, 47% inhibition [2]) [2]
K^+ <3> (<3> 0.1 M, strong inhibition in decreasing order: La^{3+}, Ca^{2+}, Mg^{2+}, Li^+, Na^+, K^+ [2]) [2]
K_2SO_4 <3> (<3> 0.1 M, 30% inhibition [2]) [2]
KCl <1, 3> (<1> high concentrations [1]; <3> 0.1 M, 30% inhibition [2]) [1, 2]
KNO_3 <3> (<3> 0.1 M, 65% inhibition [2]) [2]
L-aspartate <3> (<3> 0.015 mM, 26% inhibition [2]) [2]
L-glutamate <1, 3> (<3> 0.015 mM, 93% inhibition [2]; <1> product inhibition [5]) [2, 5]
La^{3+} <3> (<3> 0.1 M, strong inhibition in decreasing order: La^{3+}, Ca^{2+}, Mg^{2+}, Li^+, Na^+, K^+ [2]) [2]
Li^+ <3> (<3> 0.1 M, strong inhibition in decreasing order: La^{3+}, Ca^{2+}, Mg^{2+}, Li^+, Na^+, K^+ [2]) [2]
Mg^{2+} <3> (<3> 0.1 M, strong inhibition in decreasing order: La^{3+}, Ca^{2+}, Mg^{2+}, Li^+, Na^+, K^+ [2]) [2]
$MnCl_2$ <1> (<1> 0.1 M: 98% inhibition, 0.01 M: 65% inhibition [1]) [1]
N-formiminoaspartate <1> [5]
NH_4Cl <1> (<1> high concentrations [1]) [1]
Na^+ <3> (<3> 0.1 M, strong inhibition in decreasing order: La^{3+}, Ca^{2+}, Mg^{2+}, Li^+, Na^+, K^+ [2]) [2]
Tris <1> [5]
$ZnCl_2$ <1> (<1> 0.01 M, 99% inhibition [1]) [1]
acetate <3> (<3> inhibition in decreasing order: fumarate, succinate, citrate, acetate [2]) [2]
acetyl-L-glutamate <3> (<3> 0.015 mM, 30% inhibition [2]) [2]
aminopterin <1> (<1> 0.0005 mM, 53% inhibition [1]) [1]
arsenate <3> (<3> weak [2]) [2]
barium acetate <1> (<1> 0.1 M, 90% inhibition [1]) [1]
citrate <1, 3> (<3> inhibition in decreasing order: fumarate, succinate, citrate, acetate [2]) [2, 5]
folic acid <1, 3> (<3> 0.3 mM, 20% inhibition [2]) [2, 5]
formyl-L-glutamate <3> (<3> 0.015 mM, 53% inhibition [2]) [2]
fumarate <3> (<3> inhibition in decreasing order: fumarate, succinate, citrate, acetate [2]) [2]
imidazole <1> [5]
maleate <1> [5]
methotrexate <1> (<1> 0.0005 mM: 33% inhibition, 0.001 mM: 83% inhibition [1]) [1]
p-acetylaminobenzoylglutamate <3> (<3> 0.3 mM, 50% inhibition [2]) [2]
p-aminobenzoylglutamate <3> (<3> 2.5 mM, 50% inhibition [2]) [2]
p-formylaminobenzoylglutamate <3> (<3> 0.3 mM, 50% inhibition [2]) [2]
p-iodobenzoylglutamate <3> (<3> 1.5 mM, 50% inhibition [2]) [2]
phosphate <3> (<3> weak [2]) [2]
proteinase K <8> (<8> degradation of bifunctional formiminotransferase cyclodeaminase [10]) [10]

succinate <3> (<3> inhibition in decreasing order: fumarate, succinate, citrate, acetate [2]) [2]
urea <1, 8> (<8> degradation of bifunctional formiminotransferase cyclodeaminase [10]) [10, 12]
Additional information <3> (<3> not inhibited by formamidinoacetic acid, γ-benzyl-formamidinoglutaric acid, formic acid, formamide, formylglutamine, formylisoglutamine [2]) [2]

Cofactors/prosthetic groups
tetrahydrofolate <1-8> (<1> folate-dependent enzyme [5,6,8]) [1-10]

Metals, ions
KCl <1> (<1> weak stimulation by low concentrations [1]) [1]
NH_4^+ <1> (<3> 0.1 M, 10% stimulation [2]; <1> NH_4Cl: weak stimulation by low concentrations [1]) [1, 2]

Turnover number (min^{-1})
Additional information <1> (<1> 26000 per min per mol or 3250 per min per chain [5]; <1> 3480-4620 per min: bifunctional formiminotransferase cyclodeaminase, 1920 per min: recombinant His-tagged formiminotransferase domain [8]) [5, 8]

Specific activity (U/mg)
7 <8> [7]
10.5 <9> [11]
39-41 <1> (<1> recombinant enzyme, expressed in Escherichia coli [6]) [6]
43-45 <1> [5]
Additional information <1-7> [1-3]

K_m-Value (mM)
0.0017 <1> ((6S)-tetrahydropteroylpentaglutamate, <1> recombinant His-tagged bifunctional formiminotransferase cyclodeaminase [8]) [8]
0.075 <1> ((6S)-tetrahydropteroylpentaglutamate, <1> recombinant His-tagged formiminotransferase domain [8]) [8]
0.1 <1> (tetrahydrofolate) [1, 5]
0.111 <1> ((6R,S)-tetrahydrofolate, <1> recombinant His-tagged formiminotransferase domain [8]) [8]
0.141 <1> ((6R,S)-tetrahydrofolate, <1> wild-type bifunctional formiminotransferase cyclodeaminase [8]) [8]
0.148 <1> ((6R,S)-tetrahydrofolate, <1> recombinant His-tagged bifunctional formiminotransferase cyclodeaminase [8]) [8]
2 <1> (N-formimino-L-glutamate, <1> tetrahydropteroic acid as substrate [5]) [5]
2.5 <1> (tetrahydropteroic acid) [5]
5.8 <1> (N-formimino-L-glutamate, <1> wild-type bifunctional formiminotransferase cyclodeaminase [8]) [8]
6.7 <1> (N-formimino-L-glutamate, <1> recombinant His-tagged bifunctional formiminotransferase cyclodeaminase [8]) [8]

8.3 <1> (N-formimino-L-glutamate, <1> recombinant His-tagged formiminotransferase domain [8]) [8]
11 <1> (N-formimino-L-glutamate) [1]
12 <1> (N-formimino-L-glutamate, <1> tetrahydrofolate as substrate [5]) [5]

K_i-Value (mM)
0.1 <1> (folic acid) [5]
0.9 <1> (L-glutamate) [5]
4 <1> (Tris) [5]
6 <1> (imidazole) [5]
10 <1> (formiminoaspartate) [5]
30 <1> (maleate) [5]
60 <1> (citrate) [5]

pH-Optimum
6.5-7.6 <3> (<3> phosphate buffer [2]) [2]
6.5-9 <3> (<3> unbuffered solution [2]) [2]
7-8 <1> [1]
Additional information <9> (<9> pI: 6.13 [11]) [11]

pH-Range
6-9.8 <3> (<3> about 50% of activity maximum at pH 6 and 9.8 [2]) [2]
6.2-9 <1> (<1> about 50% of activity maximum at pH 6.2 and 9.0 [1]) [1]

Temperature optimum (°C)
25 <1> (<1> assay at [1]) [1]
30 <8> (<8> assay at [7]) [7]
37 <1> (<1> assay at [3,5]) [3, 5]

4 Enzyme Structure

Molecular weight
83000 <1> (<1> recombinant His-tagged N-terminal formiminotransferase domain of bifunctional formiminotransferase cyclodeaminase, gel filtration [8]) [8]
380000 <1> (<1> recombinant His-tagged bifunctional formiminotransferase cyclodeaminase, gel filtration [8]) [8]
438000 <1> (<1> wild-type bifunctional formiminotransferase cyclodeaminase, gel filtration [8]) [8]
480000 <1, 8> (<1,8> approximately, native octameric bifunctional enzyme with formiminotransferase and cyclodeaminase activity, EC 2.1.2.5 and EC 4.3.1.4 [6,7]) [6, 7]

Subunits
? <9> (<9> x * 59100, calculated from the amino acid sequence [11]) [11]
dimer <1, 8> (<1> 2 * 37000, recombinant His-tagged N-terminal formiminotransferase domain of bifunctional formiminotransferase cyclodeaminase, calculated from the amino acid sequence [8]; <1> transferase domain is ex-

pressed in Escherichia coli as monofunctional dimer [12]; <8> 2 * 58000, bifunctional formiminotransferase cyclodeaminase exists as dimeric, tetrameric and octameric complexes, SDS-PAGE, a single 35 kDa protein is cross-linked to each dimer, it may play a role in the association of the protein to the Golgi membrane, functional formiminotransferase activity unit of FTCD is a dimer with binding sites for glutamate and folate [10]) [8, 10, 12]
homodimer <1> (<1> N-terminal formiminotransferase domain of bifunctional formiminotransferase cyclodeaminase, dimer interface is important for the function of the domain [9]) [9]
homooctamer <1, 8> (<8> bifunctional formiminotransferase cyclodeaminase [7]; <1> each identical subunit of octameric bifunctional formiminotransferase cyclodeaminase consists of a transferase and a deaminase domain connected by a short linker sequence [12]) [5, 7, 12]
octamer <1, 8> (<1> 8 * 58926, recombinant bifunctional enzyme with formiminotransferase and cyclodeaminase activity, EC 2.1.2.5 and EC 4.3.1.4, calculated from cDNA sequence [6]; <1> 8 * 59000-60000, bifunctional formiminotransferase cyclodeaminase, calculated from the amino acid sequence, circular tetramer of dimers, each subunit consists of an N-terminal transferase active domain and a C-terminal deaminase active domain separated by a linker sequence of 8-12 residues, both domains self-dimerize: existence of two types of subunit interfaces, transferase and deaminase activities are dependent on the formation of specific subunit interfaces [8]; <1> 8 * 62000, bifunctional formiminotransferase cyclodeaminase, arranged as a circular tetramer of dimers [9]; <8> 8 * 58000, bifunctional formiminotransferase cyclodeaminase exists as dimeric, tetrameric and octameric complexes, SDS-PAGE [10]) [6, 8-10]
tetramer <8> (<8> 4 * 58000, bifunctional formiminotransferase cyclodeaminase exists as dimeric, tetrameric and octameric complexes, SDS-PAGE [10]) [10]

Posttranslational modification

Additional information <9> (<9> p60, bifunctional formiminotransferase cyclodeaminase, is post-translationally modified [11]) [11]

5 Isolation/Preparation/Mutation/Application

Source/tissue

HeLa cell <9> (<9> HeLa spinner cells, contain p60 but no measurable formiminotransferase activity, may be due to the low expression of p60 [11]) [11]
hepatoma cell <9> (<9> DU249 chicken hepatoma cells, contain p60 but no measurable formiminotransferase activity, may be due to the low expression of p60 [11]) [11]
kidney <1-3, 5-8> (<1-3,5-8> low activity [1,2]) [1, 2, 10]
liver <1-10> (<8> richest tissue source of formiminotransferase cyclodeaminase, liver-specific enzyme [7]) [1-8, 10, 11]

Additional information <8> (<8> no activity in the soluble proteins of brain, heart, lung, testis, intestinal mucosa, diaphragm muscle or spleen [2]; <8> undetectable in adrenal gland, blood cells, marrow cells, brain, epididymis, heart, small intestine, kidney, lung, skeletal muscle, spleen and testis [7]) [2, 7]

Localization

Golgi membrane <8, 9> (<8> formiminotransferase cyclodeaminase, 58K, is associated with the cytoplasmic surface of the Golgi apparatus in vivo, cytoplasmically oriented peripheral membrane protein of the Golgi membranes [7]; <8> formiminotransferase cyclodeaminase, 58K, is a peripherically associated Golgi protein, binding is tight but not dependent on presence of intact microtubules, association is likely to be mediated by a protein, dynamic component of the Golgi, a proportion of FTCD molecules cycles between the Golgi and earlier compartments of the secretory pathway [10]; <9> localized on Golgi membranes as well as on some potentially Golgi complex-derived vesicular structures [11]) [7, 10, 11]
Golgi vesicle <9> (<9> localized on Golgi membranes as well as on some potentially Golgi complex-derived vesicular structures [11]) [11]
cytosol <9> [11]
microtubule <8, 9> (<8> formiminotransferase cyclodeaminase binds bovine brain microtubules, but not rat liver microtubules in vitro [7]; <9> microtubule-binding protein [11]) [7, 11]
soluble <8> [2]

Purification

<1> (700fold purification [1]; purification of twin enzyme EC 2.1.2.5 and formiminotetrahydrofolate cyclodeaminase EC 4.3.1.4, combination of affinity chromatography and isoelectric focusing [4]; purification of recombinant bifunctional enzyme with formiminotransferase and cyclodeaminase activity, EC 2.1.2.5 and EC 4.3.1.4, expressed in Escherichia coli [6]; purification of wild type bifunctional formiminotransferase cyclodeaminase, 150fold purification of His-tagged recombinant enzyme, 60fold purification of His-tagged transferase domain, expressed in Escherichia coli BL21/DE3 [8]; purification of recombinant hexahistidine-tagged formiminotransferase domain of FTCD [9]) [1, 3, 4, 6, 8, 9]
<3> (60-80fold purification [2]) [2]
<8> [7]
<9> (purification from liver [11]) [11]
<10> (purification from liver [11]) [11]

Crystallization

<1> (crystal structure of the monofunctional formiminotransferase domain of FTCD, hanging-drop method [9]) [9]

Cloning

<1> (isolation and sequencing of cDNA clones encoding bifunctional enzyme with formiminotransferase and cyclodeaminase activity, EC 2.1.2.5 and EC 4.3.1.4, cDNA encodes an amino acid sequence of 541 residues, expression

in Escherichia coli [6]; expression of full-length formiminotransferase cyclodeaminase and of isolated N-terminal transferase and C-terminal deaminase domains in Escherichia coli BL21/DE3 as His-tagged proteins [8]) [6, 8]
<8> (cDNA encoding formiminotransferase cyclodeaminase is cloned and sequenced [7]; cDNA encoding formiminotransferase cyclodeaminase is cloned, characterized and partially sequenced [10]) [7, 10]
<9> (cloning and sequencing of the cDNA coding for p60, the chicken homologue of bifunctional formiminotransferase cyclodeaminase, a protein of 541 amino acids [11]) [11]

Engineering

Additional information <1> (<1> deletion mutants of formiminotransferase cyclodeaminase with transferase activity [8]) [8]

Application

molecular biology <8> (<8> bifunctional formiminotransferase cyclodeaminase provides a novel marker to study ER-Golgi dynamics [10]) [10]

6 Stability

General stability information

<1>, formiminoglutamate prevents inactivation of enzyme [4]
<3>, no loss of activity after dialysis against phosphate buffer or 0.05 M EDTA, pH 7.4 [2]
<8>, dimeric, tetrameric and octameric complexes are resistant to proteolysis [10]

Storage stability

<1>, -20°C, several weeks, stable [1]
<1>, 2°C, 2 months, 16-45% loss of activity [1]
<1>, 2°C, several weeks, stable [1]
<3>, -15°C, crude extract, 1 year, stable [2]

References

[1] Tabor, H.; Wyngarden, L.: The enzymatic formation of formiminotetrahydrofolic acid, 5,10-methenyltetrahydrofolic acid, and 10-formyltetrahydrofolic acid in the metabolism of formiminoglutamic acid. J. Biol. Chem., **234**, 1830-1846 (1959)
[2] Miller, A.; Waelsch, H.: Formimino transfer from formamidinoglutaric acid to tetrahydrofolic acid. J. Biol. Chem., **228**, 397-417 (1957)
[3] Silverman, M.; Keresztesy, J.C.; Koval, G.J.; Gardiner, R.C.: Citrovorum factor and the synthesis of formylglutamic acid. J. Biol. Chem., **226**, 83-94 (1957)
[4] Slavik, K.; Zizkovsky, V.; Slavikova, V.; Fort, P.: The purification of formiminotransferase and cyclodeaminase by combination of affinity chromatogra-

phy and isoelectric focusing. Biochem. Biophys. Res. Commun., **59**, 1173-1184 (1974)

[5] Beaudet, R.; Mackenzie, R.: Kinetic mechanism of formininotransferase from porcine liver. Biochim. Biophys. Acta, **410**, 252-261 (1975)

[6] Murley, L.L.; Mejia, N.R.; MacKenzie, R.E.: The nucleotide sequence of porcine formiminotransferase cyclodeaminase. Expression and purification from Escherichia coli. J. Biol. Chem., **268**, 22820-22824 (1993)

[7] Bashour, A.M.; Bloom, G.S.: 58K, a microtubule-binding Golgi protein, is a formiminotransferase cyclodeaminase. J. Biol. Chem., **273**, 19612-19617 (1998)

[8] Murley, L.L.; MacKenzie, R.E.: The two monofunctional domains of octameric formiminotransferase-cyclodeaminase exist as dimers. Biochemistry, **34**, 10358-10364 (1995)

[9] Kohls, D.; Sulea, T.; Purisima, E.O.; MacKenzie, R.E.; Vrielink, A.: The crystal structure of the formiminotransferase domain of formiminotransferase-cyclodeaminase: Implications for substrate channeling in a bifunctional enzyme. Structure, **8**, 35-46 (2000)

[10] Gao, Y.S.; Alvarez, C.; Nelson, D.S.; Sztul, E.: Molecular cloning, characterization, and dynamics of rat formiminotransferase cyclodeaminase, a Golgi-associated 58-kDa protein. J. Biol. Chem., **273**, 33825-33834 (1998)

[11] Hennig, D.; Scales, S.J.; Moreau, A.; Murley, L.L.; De Mey, J.; Kreis, T.E.: A formiminotransferase cyclodeaminase isoform is localized to the Golgi complex and can mediate interaction of trans-Golgi network-derived vesicles with microtubules. J. Biol. Chem., **273**, 19602-19611 (1998)

[12] Murley, L.L.; MacKenzie, R.E.: Monofunctional domains of formiminotransferase-cyclodeaminase retain similar conformational stabilities outside the bifunctional octamer. Biochim. Biophys. Acta, **1338**, 223-232 (1997)

Glutamate formyltransferase 2.1.2.6

1 Nomenclature

EC number
2.1.2.6 (deleted, included in EC 2.1.2.5)

Recommended name
glutamate formyltransferase

D-Alanine 2-hydroxymethyltransferase 2.1.2.7

1 Nomenclature

EC number
2.1.2.7

Systematic name
5,10-methylenetetrahydrofolate:D-alanine 2-hydroxymethyltransferase

Recommended name
D-alanine 2-hydroxymethyltransferase

Synonyms
2-methylserine hydroxymethyltransferase
α-methylserine hydroxymethyltransferase
hydroxymethyltransferase, 2-methylserine

CAS registry number
9075-76-7

2 Source Organism

<1> *Pseudomonas sp.* (strain MS, inducible enzyme [2]) [1-3]
<2> *Clostridium tetanomorphum* [4]
<3> *Escherichia coli* (strain K12 [4]) [4]

3 Reaction and Specificity

Catalyzed reaction
5,10-methylenetetrahydrofolate + D-alanine + H_2O = tetrahydrofolate + 2-methylserine

Reaction type
hydroxymethyl group transfer

Substrates and products
S tetrahydrofolate + (+)-2-methylserine <1, 2, 3> (<1> stereospecific, dimedon cannot replace tetrahydrofolate [1-3]) (Reversibility: r <1> [1-3]; ? <2, 3> [4]) [1-4]
P 5,10-methylenetetrahydrofolate + D-alanine + H_2O <1> (<1> 5,10-methylenetetrahydrofolate is non-enzymatically cleaved to tetrahydrofolate and formaldehyde [2]) [1-3]

S tetrahydrofolate + (-)-ethylserine <1> (<1> stereospecific [3]) (Reversibility: r <1> [1-3]) [1-3]
P 5,10-methylenetetrahydrofolate + D-2-aminobutyrate + H_2O <1> [1, 2, 3]
S tetrahydrofolate + 2-hydroxymethylserine <1> (<1> specific removal of one of the two hydroxyl groups [1,3]) (Reversibility: r <1> [1, 3]) [1-3]
P 5,10-methylenetetrahydrofolate + D-serine + H_2O <1> [1, 2]

Inhibitors
Cu^{2+} <1> (<1> at higher concentrations [1,2]) [1, 2]
DL-cycloserine <1> (<1> strong, competitive [1,2]) [1, 2]
Fe^{2+} <1> (<1> at higher concentrations [1,2]) [1, 2]
Fe^{3+} <1> (<1> at higher concentrations [1,2]) [1, 2]
L-serine <1> (<1> weak, competitive [1,2]) [1, 2]
Zn^{2+} <1> (<1> at higher concentrations [1,2]) [1, 2]
p-chloromercuribenzoate <1> [1, 2]
Additional information <1> (<1> no inhibition by Mg^{2+}, Mn^{2+}, Al^{3+}, 2-mercaptoethanol [1,2]; <1> iodoacetate, GSH [2]) [1, 2]

Cofactors/prosthetic groups
pyridoxal 5'-phosphate <1> (<1> requirement [1-3]; <1> K_m: 0.01 mM at pH 7.5 [1,2]) [1-3]

Specific activity (U/mg)
11.1 <1> [1]
17.4 <1> [2]

K_m-Value (mM)
0.25 <1> (L-tetrahydrofolate, <1> pH 7.5, calculated on the basis of one isomer [1,2]) [1, 2]
0.8 <1> (hydroxymethylserine, <1> pH 7.5 [1,2]) [1, 2]
1.5 <1> (2-methyl-DL-serine, <1> pH 7.5 [1,2]) [1, 2]
6.2 <1> (2-ethyl-DL-serine, <1> pH 7.5 [1,2]) [1, 2]

K_i-Value (mM)
0.002 <1> (D-cycloserine) [1]
0.0022 <1> (D-cycloserine) [2]
0.005 <1> (L-cycloserine) [1, 2]
5.6 <1> (L-serine) [1, 2]

pH-Optimum
9 <1> [1, 2]

pH-Range
6-10.5 <1> (<1> about half-maximal activity at pH 6.0 and 10.5 [2]) [2]

Temperature optimum (°C)
45 <1> [1, 2]

Temperature range (°C)
33-48 <1> (<1> about half-maximal activity at 33°C and 90% activity at 48°C [2]) [2]

5 Isolation/Preparation/Mutation/Application

Localization

cytoplasm <1> [1, 2]

Purification

<1> (partial [2]) [1, 2]

Crystallization

<1> [1, 2]

6 Stability

Temperature stability

45 <1> (<1> 96% loss of activity after 10 min, hydroxymethylserine plus tetrahydrofolate and pyridoxal 5'-phosphate prevent [2]) [2]
50 <1> (<1> complete inactivation after 10 min, 55% loss of activity in the presence of tetrahydrofolate, pyridoxal 5'-phosphate plus hydroxymethylserine or D-serine [2]) [2]

General stability information

<1>, prolonged dialysis inactivates [1]
<1>, pyridoxal 5-phosphate restores activity [1]
<1>, pyridoxal 5-phosphate stabilizes during purification [1]
<1>, tetrahydrofolate plus pyridoxal 5-phosphate plus hydroxymethylserine increase temperature-stability [2]

References

[1] Wilson Miles, E.: α-Methylserine transhydroxymethylase (Pseudomonas). Methods Enzymol., **17B**, 341-346 (1971)
[2] Wilson, E.M.; Snell, E.E.: Metabolism of α-methylserine. I. α-methylserine hydroxymethyltransferase. J. Biol. Chem., **237**, 3171-3179 (1962)
[3] Wilson, E.M.; Snell, E.E.: Metabolism of α-methylserine. II. Stereospecificity of α-methylserine hydroxymethyltransferase. J. Biol. Chem., **237**, 3180-3184 (1962)
[4] Sperl, G.T.: Microbial metabolism of 2-methyl amino acids. Curr. Microbiol., **19**, 135-138 (1989)

Deoxycytidylate 5-hydroxymethyltransferase 2.1.2.8

1 Nomenclature

EC number
2.1.2.8

Systematic name
5,10-methylenetetrahydrofolate:deoxycytidylate 5-hydroxymethyltransferase

Recommended name
deoxycytidylate 5-hydroxymethyltransferase

Synonyms
CH
d-cytidine 5'-monophosphate hydroxymethylase
dCMP Hmase
dCMP hydroxymethylase
deoxyCMP hydroxymethylase
deoxycytidylate hydroxymethylase
deoxycytidylate hydroxymethyltransferase
deoxycytidylic hydroxymethylase
hydroxymethyltransferase, deoxycytidylate

CAS registry number
9012-68-4

2 Source Organism

<1> *bacteriophage T6* [4]
<2> *bacteriophage T2* [4]
<3> *bacteriophage T4* (BK 536 and JSW 800, grown on Escherichia coli [4]; Escherichia coli infected with bacteriophage T4 [5,8,9,11]; Escherichia coli strains Cr 63 and B infected with bacteriophage T4 [6]; Escherichia coli B infected with bacteriophage T4D and amber mutants [7]; Escherichia coli JM101 infected with bacteriophage T4 [10]; Escherichia coli BL21 infected with bacteriophage T4 [12]) [3-13]
<4> *bacteriophage T6r+* (Escherichia coli strain B infected with bacteriophage T6r+ [1,2]) [1, 2]

3 Reaction and Specificity

Catalyzed reaction

5,10-methylenetetrahydrofolate + H_2O + deoxycytidylate = tetrahydrofolate + 5-hydroxymethyldeoxycytidylate (<3> mechanism [8,10]; <3> catalytic mechanism [11-13]; <3> ordered bi-bi mechanism [12])

Reaction type

hydroxymethyl group transfer

Natural substrates and products

S 5,10-methylenetetrahydrofolate + H_2O + deoxycytidylate <3> (<3> dCTP depresses the synthesis of 5-hydroxymethyldeoxycytidylate [5]) (Reversibility: ? <3> [5, 7-13]) [5, 7-13]

P tetrahydrofolate + 5-hydroxymethyldeoxycytidylate <3> (<3> 5-hydroxymethyl-dCMP is a virus-specific DNA precursor [7]; <3> enzyme is responsible for production of 5-hydroxymethyl-dCMP found in the DNA of the phage [8,9]; <3> 5-hydroxymethyl-dCMP is essential for DNA synthesis in phage T4 [10,11,13]; <3> involved in biosynthesis of DNA precursor [12,13]; <3> enzyme forms part of phage DNA protection system [13]) [5, 7-13]

S Additional information <3, 4> (<4> biosynthesis of enzyme de novo after infection [1]; <3> enzyme is part of a complex efficiently channeling DNA precursors to the replication apparatus [5,13]; <3> enzyme is part of a complex catalyzing DNA precursor biosynthesis [7]; <3> enzyme is part of the dNTP synthetase complex [9]; <3> enzyme functions in vivo as a component of a multienzyme complex called dNTP synthetase [13]) [1, 5, 7, 9, 13]

P ?

Substrates and products

S 5,10-methylenetetrahydrofolate + H_2O + deoxycytidylate <1-4> (<3> stoichiometry of dCMP binding to the dimeric enzyme is 1:1 and depends upon a reduced thiol group, one thiol-containing active site per mol of dimeric enzyme, sequential order of substrate binding [6]; <3> role of active site Glu-60 is to accelerate bond cleavage between N^5 of tetrahydrofolate and the methylene being transferred [12]; <3> Cys-148 is catalytically essential [10,11]; <3> enzyme structure, dCMP binding site, folate cofactor binding site, way of substrate binding in the active site, key catalytic residue Cys-148 acts as nucleophile to attack the C^6-atom of the cytosine base [13]) (Reversibility: r <3> [6, 10-12]; ? <1-4> [1-5, 7-9, 13]) [1-13]

P tetrahydrofolate + 5-hydroxymethyldeoxycytidylate <1-4> [1-13]

S Additional information <3> (<3> enzyme also catalyzes ^{18}O exchange from solvent water into 5-hydroxymethyl-dCMP in the absence of exogenous tetrahydrofolate and in the presence of tetrahydrofolate analogues that lack N^5, mechanism [11]) [11]

P ?

Inhibitors

10-methyl-5,8-dideazatetrahydrofolate <3> (<3> competitive inhibitor with respect to tetrahydrofolate [11]) [11]
5,5'-dithiobis(2-nitrobenzoic acid) <3> (<3> complete loss of activity, second-order kinetics, inactivation is reversed by dithiothreitol, inactivation involves the modification of one thiol group per mole of dimeric enzyme [6]) [6]
5-deazatetrahydrofolate <3> (<3> competitive inhibitor with respect to tetrahydrofolate [11]; <3> 1 mM, 54% inhibition [12]) [11, 12]
5-fluoro-2'-deoxycytidine 5'-monophosphate <3> (<3> competitive inhibitor with respect to dCMP, not mechanism-based inhibitor [8]) [8]
5-fluorodeoxyuridylate <3> (<3> inactivates by formation of a covalent complex containing enzyme, inhibitor and methylenetetrahydrofolate in a time-dependent manner, 1 mol 5-fluoro-dUMP per enzyme monomer, no substantial inhibition of wild-type enzyme in vivo, inhibition mechanism [10]; <3> mechanism-based inhibitor [12]) [10-12]
5-hydroxymethyl-dCMP <3> (<3> competitor of dCMP, inhibits formation of 5-hydroxymethyl-dCMP [11]) [11]
N-ethylmaleimide <3> (<3> complete loss of activity, second-order kinetics, pH-dependent inactivation, dCMP protects against inactivation, dCMP plus either methotrexate or aminopterin greatly enhances protection, inactivation involves the modification of one thiol group per mole of dimeric enzyme [6]) [6]
aminopterin <3> (<3> competitive inhibitor with respect to tetrahydrofolate [6]) [6]
dihydrofolate <3> (<3> competitive inhibitor with respect to tetrahydrofolate [6]) [6]
methotrexate <3> (<3> competitive inhibitor with respect to tetrahydrofolate [6]) [6]

Cofactors/prosthetic groups

5,10-methylenetetrahydrofolate <1-4> (<3> active cofactor is likely 5,10-methylenetetrahydrofolate [6]; <3> as cofactor [8,13]; <3> weak binding of the folate cofactor to enzyme [13]) [1-13]
tetrahydrofolate <1-4> [1-13]

Activating compounds

dithiothreitol <3> (<3> low specific activity samples can be activated by a 30 min room temperature incubation in 50 mM DTT [6]) [6]

Turnover number (min^{-1})

0.048 <3> (deoxycytidylate, <3> mutant E60Q [12]) [12]
0.42 <3> (deoxycytidylate, <3> mutant E60D [12]) [12]
762 <3> (deoxycytidylate, <3> wild-type enzyme [12]) [12]
Additional information <3, 4> (<4> 320 substrate molecules per min per enzyme molecule [1]; <3> 333 mol substrate per min per 56.8 kg protein [4]) [1, 4]

Specific activity (U/mg)

1 <3> [8]
3.1 <3> (<3> dCMP formation [11]) [11]
4.6 <3> (<3> 5-hydroxymethyl-dCMP formation [11]) [11]
4.9 <3> [7]
5.865 <3> [4]
Additional information <3, 4> [1, 6]

K_m-Value (mM)

0.04 <3> (tetrahydrofolate) [6]
0.36 <3> (deoxycytidylate) [8]

K_i-Value (mM)

0.001 <3> (aminopterin) [6]
0.01 <3> (10-methyl-5,8-dideazatetrahydrofolate) [11]
0.016 <3> (methotrexate) [6]
0.021 <3> (dihydrofolate) [6]
0.025 <3> (5-fluoro-2'-deoxycytidine 5'-monophosphate) [8]
0.1 <3> (5-hydroxymethyl-dCMP) [11]
0.11 <3> (5-deazatetrahydrofolate) [11, 12]

pH-Optimum

7.6 <3> (<3> assay at [6]) [6]

Temperature optimum (°C)

25 <3> (<3> assay at [8]) [8]
30 <3> (<3> assay at [6]) [6]

4 Enzyme Structure

Molecular weight

28400 <3> (<3> calculation from the amino acid sequence [3]) [3]
28450 <1-3> (<1-3> calculation from the amino acid sequence [4]) [4]
51400 <3> (<3> gel filtration [6]) [6]
57000 <3> (<3> gel filtration [7]) [7]
60000 <3> [5]
63000 <3> (<3> sedimentation equilibrium method [7]) [7]
68000 <4> (<4> E. coli B infected with bacteriophage T6r+, gel filtration, sedimentation and diffusion data [1]) [1, 2]

Subunits

? <3> (<3> x * 28600 [10]) [10]
dimer <3> (<3> 2 * 27000, SDS-PAGE [4,7]; <3> 2 * 27000 [5]; <3> 2 * 28000, two equally sized subunits are arranged asymmetrically, SDS-PAGE [6]) [4-7]
homodimer <3> (<3> 2 * 28487 [12]; <3> 2 * 28450, homodimer of 246-residue subunits [13]) [11-13]

5 Isolation/Preparation/Mutation/Application

Localization

Additional information <3> (<3> enzyme is no membrane component or very loosely attached to the membrane [7]) [7]

Purification

<1> [4]
<3> (1026fold purification [6]; Escherichia coli B infected with bacteriophage [6,7]; purification of wild-type enzyme and mutants E60Q and E60D [12]; purification of His6-tagged enzyme [13]) [4, 6-8, 12, 13]
<4> (Escherichia coli infected with bacteriophage [1,2]) [1, 2]

Crystallization

<3> (crystal structure in the presence of substrate dCMP at 1.6 A resolution [13]) [13]

Cloning

<1> (nucleotide sequence of dCMP hydroxymethylase gene encoding a protein of 246 amino acids [4]) [4]
<2> (nucleotide sequence of dCMP hydroxymethylase gene encoding a protein of 246 amino acids [4]) [4]
<3> (nucleotide sequence of g42, the dCMP hydroxymethylase gene, encodes a protein of 246 amino acids [3,4]; cloning and overexpression of g42 in Escherichia coli K38 [4]; structural gene 42 encodes enzyme [5,7,9-12]; overexpression in Escherichia coli BL21 [12]; gene 42 encoding enzyme is sequenced and expressed in Escherichia coli [13]) [3-5, 7, 9-13]

Engineering

C148G <3> (mutant is unable to catalyze a detectable ^{18}O exchange in either the absence or presence of the folate cofactor [11]) [11]
D179A <3> (mutant not inactivated by 5-fluoro-dUMP [10]) [10]
D179N <3> (<3> mutant with altered substrate preference, favoring dUMP rather than dCMP [10,11,13]; mutation of active site, 10fold increased rate of inhibition by 5-fluoro-dUMP [10]) [10, 11, 13]
D179S <3> (mutant not inactivated by 5-fluoro-dUMP [10]) [10]
E60D <3> (<3> mutant with decreased enzyme activity, impaired enzyme function in vitro and in vivo, uncoupling of early steps from later steps in catalysis [12]) [12]
E60Q <3> (<3> mutant with decreased enzyme activity, impaired enzyme function in vitro and in vivo, uncoupling of early steps from later steps in catalysis [12]) [12]
Additional information <3> (<3> transition mutations in gene 42 [4]; <3> gene 42 amber mutants synthesize truncated forms of enzyme [9]) [4, 9]

6 Stability

Temperature stability

20 <3> (<3> room temperature, 24 h, stable [6]) [6]

Storage stability

<3>, frozen, 20 mM KH_2PO_4, pH 6.8, 6 months [6]
<3>, room temperature, 24 h, stable [6]
<4>, 4°C, up to 10 days, only minor loss of activity [1]

References

[1] Mathews, C.K.; Brown, F.; Cohen, S.S.: Virus-induced acquisition of metabolic function. VII. Biosynthesis de novo of deoxycytidylate hydroxymethylase. J. Biol. Chem., **239**, 2957-2963 (1964)
[2] Pizer, L.I.; Cohen, S.S.: Virus-induced acquisition of metabolic function. V. Purification and properties of deoxycytidylate hydroxymethylase and studies on its origin. J. Biol. Chem., **237**, 1251-1259 (1962)
[3] Lamm, N.; Tomaschewski, J.; Ruger, W.: Nucleotide sequence of the deoxycytidylate hydroxymethylase gene of bacteriophage T4 (g42) and the homology of its gene product with thymidylate synthase of E. coli. Nucleic Acids Res., **15**, 3920 (1987)
[4] Lamm, N.; Wang, Y.; Mathews, C.K.; Ruger, W.: Deoxycytidylate hydroxymethylase gene of bacteriophage T4. Nucleotide sequence determination and over-expression of the gene. Eur. J. Biochem., **172**, 553-563 (1988)
[5] Allen, J.R.; Lasser, G.W.; Goldman, D.A.; Booth, J.W.; Mathews, C.K.: T4 phage deoxyribonucleotide-synthesizing enzyme complex. Further studies on enzyme composition and regulation. J. Biol. Chem., **258**, 5746-5753 (1983)
[6] Lee, M.H.; Gautam-Basak, M.; Wooley, C.; Sander, E.G.: Deoxycytidylate hydroxymethylase: purification, properties, and the role of a thiol group in catalysis. Biochemistry, **27**, 1367-1373 (1988)
[7] North, T.W.; Mathews, C.K.: T4 phage-coded deoxycytidylate hydroxymethylase: purification and studies on intermolecular interactions. Biochem. Biophys. Res. Commun., **77**, 898-904 (1977)
[8] Subramaniam, R.; Wang, Y.; Mathews, C.K.; Santi, D.V.: On the inhibition of deoxycytidylate hydroxymethylase by 5-fluoro-2-deoxycytidine 5-monophosphate. Arch. Biochem. Biophys., **275**, 11-15 (1989)
[9] Mathews, C.K.; Wheeler, L.J.; Ungermann, C.; Young, J.P.; Ray, N.B.: Enzyme interactions involving T4 phage-coded thymidylate synthase and deoxycytidylate hydroxymethylase. Adv. Exp. Med. Biol., **338**, 563-570 (1993)
[10] Graves, K.L.; Hardy, L.W.: Kinetic and equilibrium α-secondary tritium isotope effects on reactions catalyzed by dCMP hydroxymethylase from bacteriophage T4. Biochemistry, **33**, 13049-13056 (1994)

[11] Butler, M.M.; Graves, K.L.; Hardy, L.W.: Evidence from ^{18}O exchange studies for an exocyclic methylene intermediate in the reaction catalyzed by T4 deoxycytidylate hydroxymethylase. Biochemistry, **33**, 10521-10526 (1994)
[12] Hardy, L.W.; Graves, K.L.; Nalivaika, E.: Electrostatic guidance of catalysis by a conserved glutamic acid in Escherichia coli dTMP synthase and bacteriophage T4 dCMP hydroxymethylase. Biochemistry, **34**, 8422-8432 (1995)
[13] Song, H.K.; Sohn, S.H.; Suh, S.W.: Crystal structure of deoxycytidylate hydroxymethylase from bacteriophage T4, a component of the deoxyribonucleoside triphosphate-synthesizing complex. EMBO J., **18**, 1104-1113 (1999)

Methionyl-tRNA formyltransferase 2.1.2.9

1 Nomenclature

EC number
2.1.2.9

Systematic name
10-formyltetrahydrofolate:L-methionyl-tRNA N-formyltransferase

Recommended name
methionyl-tRNA formyltransferase

Synonyms
FMT
MTF
N^{10}-formyltetrahydrofolic-methionyl-transfer ribonucleic transformylase
formylmethionyl-transfer ribonucleic synthetase
methionyl ribonucleic formyltransferase
methionyl-tRNA Met formyltransferase
methionyl-tRNA transformylase
methionyl-transfer RNA transformylase
methionyl-transfer ribonucleic transformylase
methyltransferase, methionyl-transfer ribonucleate

CAS registry number
9015-76-3

2 Source Organism

<1> *Escherichia coli* (strain B [1]; strain K12 MO (lot 15-290) [3]; strain K12 EM20031(F32) [2,4]; strain A19 [9]; strain CA274 [16,17]; strain JM109 [16,17,19]; strain DH5αF′ [16]; strains DH5α, Novablue(DE3) pLysS [17]; strain XL1-Blue [19]) [1-6, 9, 12, 13, 16-19, 22]
<2> *Streptococcus faecalis* (strain R [9]) [9]
<3> *Euglena gracilis* (strain Z [7]) [7, 8]
<4> *Triticum aestivum* (wheat [11]) [11]
<5> *Saccharomyces cerevisiae* (strain D243-2B-R1 [10]; strains CMY214, W303, YPALS, DBY2057 [20]) [10, 20]
<6> *Bos taurus* [14, 15]
<7> *Haemophilus influenzae* [21]
<8> *Thermus thermophilus* [21]
<9> *Rickettsia prowazekii* [21]
<10> *Pseudomonas aeruginosa* [21]

3 Reaction and Specificity

Catalyzed reaction

10-formyltetrahydrofolate + L-methionyl-tRNA$_f^{Met}$ + H_2O = tetrahydrofolate + N-formylmethionyl-tRNA$_f^{Met}$

Reaction type

formyl group transfer

Natural substrates and products

S 10-formyltetrahydrofolate + L-methionyl-tRNA$_f^{Met}$ + H_2O <1-5> (<1,2> involved in initiation of protein biosynthesis [2-5,9,16-19,22]; <1> keeps in store initiator tRNA$_f^{Met}$ for initiation of protein synthesis [5]; <3> involved in initiation of protein synthesis in chloroplasts, other mechanism than in cytoplasm of eukaryotes [7]; <4> component of a protein chain initiating system in chloroplasts [11]; <1> enzyme reaction irreversibly commits methionyl-tRNA$_f^{Met}$ to initiation of translation in eubacteria [12]; <1> key step in the targeting of initiator tRNA towards the translation start machinery in prokaryotes [13]; <5> initiation of protein synthesis in mitochondria [20]; <1> formyl group provides a positive determinant for IF2, which can select the initiator fMet-tRNA from other tRNAs and a negative determinant, which blocks the binding of EF-Tu to the initiator tRNA [22]) (Reversibility: ? <1-5> [2-5, 7, 9, 11-13, 16-20, 22]) [2-5, 7, 9, 11-13, 16-20, 22]

P tetrahydrofolate + N-formylmethionyl-tRNA$_f^{Met}$ <1-5> [2-5, 7, 9, 11-13, 16-20, 22]

S Additional information <5, 6> (<5> mitochondrial protein synthesis [10]; <6> translational initiation in mitochondria [14,15]) [10, 14, 15]

P ?

Substrates and products

S 10-formyltetrahydrofolate + L-methionyl-tRNA$_f^{Met}$ + H_2O <1-7> (<1> 1.1 mol 10-L-formyltetrahydrofolate is bound per mol monomeric enzyme [2, 4]; <1> specific for methionyl-tRNA$_f^{Met}$, specificity depends exclusively on the nature of tRNA moiety of aminoacylated tRNA$_f^{Met}$ [3]; <1, 2> specificity is restricted to methionyl-tRNA$_f^{Met}$ [5, 6, 9, 17-19, 22]; <3> high specificity for initiator tRNA^{Met}, strong positive cooperativity for Met-tRNA$_f^{Met}$ and N^{10}-formyltetrahydrofolate [7, 8]; <1, 3-7> formylates methionyl-tRNA$_f^{Met}$ from E. coli [1-7, 11, 14, 20, 21]; <3> formylates methionyl-tRNA$_f^{Met}$ from Euglena gracilis or yeast [7]; <4> formylates methionyl-tRNA$_f^{Met}$ from wheat germ and leaf [11]; <6> formylates bovine mitochondrial and E. coli initiator Met-tRNA with equal efficiency, formylates E. coli elongator Met-tRNA at a lower rate [14, 15]; <5> formylates yeast mitochondrial methionyl-tRNA$_f^{Met}$, specific for initiator tRNAs [20]; <6> broader substrate specificity than that of E. coli enzyme [14, 15]; <2> 10-formyltetrahydropteroyltriglutamate or 10-formyltetrahydrofolate can also act as donor [9]; <1> nucleotide-mutants of tRNA$_f^{Met}$ as substrates [5, 6]; <1> specificity determinants at the end of the acceptor stem of tRNA$^{Met}_{f}$ [5, 6, 18, 19]; <1> affinity of tRNA$^{Met}_{f}$ for enzyme is improved by aminoacylation [2, 4]; <1, 6> recognition mechanism of methionyl-tRNA$^{Met}_{f}$ [12, 15, 16, 19]; <1> tRNA substrate binding [13, 22]; <6> enzyme preferentially recognizes the methionyl moiety of its tRNA substrate, small importance of the tRNA acceptor stem for recognition, E. coli tRNA mutants as substrates [15]; <1> important role of a 16-amino acid insertion loop in enzyme for tRNA recognition [16, 18, 19, 22]; <1> invariant Arg-42 within the 16-amino acid insertion loop, residues 34-49, plays a very important role both in steps of substrate binding and in catalysis, Arg-42 makes functional contacts with the determinants in the acceptor stem of the initiator tRNA [18]; <1> C-terminal region is very important for enzyme activity: basic amino acids contribute mostly toward the nonspecific binding and orientation of the tRNA 3′-end toward the catalytic site [19, 22]; <1> role of the linker region between N- and C-terminal domain [19]; <5> enzyme recognizes nucleotidic features in the acceptor stem of mitochondrial initiator tRNA, recognition mechanism [20]; <7> Asn-107, His-109 and Asp-145 are catalytic residues, Arg-41 is involved in tRNA recognition [21]; <1> determinants for formylation are located in the acceptor stem and D-stem of the initiator tRNA, enzyme binds to the tRNA substrate in an induced fit mechanism, tRNA also undergoes induced structural changes during its binding to enzyme [22]) (Reversibility: ? <1-7> [1-9, 11-22]) [1-9, 11-22]

P tetrahydrofolate + N-formylmethionyl-tRNA$^{Met}_{f}$ <1, 2, 4-7> [1-4, 8, 9, 11-22]

S 10-formyltetrahydrofolate + isoleucyl-tRNA$^{Met}_{f}$ + H_2O <5, 6> (<5, 6> mischarged E. coli tRNA$^{Met}_{f}$ [15,20]; <6> less efficient than with methionyl-tRNA$^{Met}_{f}$ [15]; <5> 250fold reduced catalytic efficiency compared with methionyl-tRNA$^{Met}_{f}$ [20]) (Reversibility: ? <5, 6> [15, 20]) [15, 20]

P tetrahydrofolate + N-formylisoleucyl-tRNA$^{Met}_{f}$

S 10-formyltetrahydrofolate + methionyl-tRNA$^{Val(CAU)}_{1}$ + H_2O <6> (<6> mischarged E. coli tRNA$^{Val(CAU)}_{1}$, less efficient than with methionyl-tRNA$^{Met}_{f}$ [15]) (Reversibility: ? <6> [15]) [15]

P tetrahydrofolate + N-formylmethionyl-tRNA$^{Val(CAU)}_{1}$

S 10-formyltetrahydrofolate + methionyl-tRNA$^{Met}_{m}$ + H_2O <6> (<6> E. coli elongator Met-tRNAmMet, less efficient than with methionyl-tRNA$^{Met}_{f}$ [15]) (Reversibility: ? <6> [15]) [15]

P tetrahydrofolate + N-formylmethionyl-tRNA$^{Met}_{m}$

S 10-formyltetrahydrofolate + phenylalanyl-tRNA$^{Met}_{f}$ + H_2O <1> (<1> mischarged E. coli tRNA$^{Met}_{f}$, reduced velocity compared to the formylation of methionyl-tRNA$^{Met}_{f}$ [3]; <1> 7fold less efficient formylation than of methionyl-tRNA$^{Met}_{f}$ [5]) (Reversibility: ? <1> [3, 5]) [3, 5]

P tetrahydrofolate + N-formylphenylalanyl-tRNA$^{Met}_{f}$ <1> [3]

S 10-formyltetrahydrofolate + valyl-tRNA$^{Met}_{f}$ + H_2O <1, 5, 6> (<1,5,6> mischarged E. coli tRNA$^{Met}_{f}$ [3,5,15,20]; <1> reduced velocity compared to the formylation of methionyl-tRNA$^{Met}_{f}$ [3]; <1> 130fold less efficient formylation than of methionyl-tRNA$^{Met}_{f}$ [5]; <6> less efficiency than with methionyl-tRNA$^{Met}_{f}$ [15]; <5> 25fold reduced catalytic efficiency compared with methionyl-tRNA$_f^{Met}$ [20]) (Reversibility: ? <1, 5, 6> [3, 5, 15, 20]) [3, 5, 15, 20]

P tetrahydrofolate + N-formylvalyl-tRNA$_f^{Met}$ <1> [3]

S 10-formyltetrahydropteroyltriglutamate + L-methionyl-tRNA$_f^{Met}$ + H_2O <2> (<2> 10-formyltetrahydropteroyltriglutamate or 10-formyltetrahydrofolate can act as donor [9]) (Reversibility: ? <2> [9]) [9]

P tetrahydropteroyltriglutamate + N-formylmethionyl-tRNA$_f^{Met}$ <2> [9]

S Additional information <1, 2, 4-6> (<1> broad specificity at the level of tRNA binding, formylation of mischarged E. coli tRNA$_f^{Met}$ [2]; <5> formylates mitochondrial methionyl-tRNA$_f^{Met}$ from yeast, low-level activity with yeast cytoplasmic and E. coli tRNA [10]; <1,2> ATP plus L-methionine and either tRNA$_m^{Met}$ or tRNA$_f^{Met}$ cannot replace methionyl-tRNA$_f^{Met}$ [9]; <1,2,4> no formylation of methionyl-tRNA$_m^{Met}$ [2,4,5,9,11,19,22]; <4> no formylation of cytoplasmic initiating Met-tRNA$_i$ [11]; <6> no formylation of Val-tRNA1Met [15]; <1> no formylation of deacylated heterogenous or deacylated partially purified methionyl-tRNA [1]; <1,5> mutant tRNAs as poor substrates [16,17,20]; <1> no formyl group donor: 5,10-methenyltetrahydrofolate [1,4]; <1> no formyl group donor: 5-formyltetrahydrofolate [4]; <1> enzyme structure [16-19,22]) [1, 2, 4, 5, 9-11, 15-20, 22]

P ?

Inhibitors

10-formylaminopterin <3> (<3> 0.01 mM: 38% inhibition, 0.1 mM: 62% inhibition, competitive [8]) [8]

10-formylpteroylglutamic acid <3> (<3> 0.1 mM: 31% inhibition, 1 mM: 52% inhibition, competitive [8]) [8]

5,10-methylenetetrahydrofolate <3> [8]

5-formyltetrahydrofolate <3> (<3> competitive inhibition [8]) [8]

5-methyltetrahydrofolate <3> (<3> most effective, non-competitive inhibitor [8]) [8]

aminopterin <3> (<3> 1 mM: 26% inhibition, competitive inhibition [8]) [8]

methotrexate <3> (<3> 0.01 mM: 30% inhibition, 0.1 mM, 52% inhibition, competitive [8]) [8]

reduced aminopterin <3> (<3> 1 mM, 34% inhibition, competitive inhibition [8]) [8]

tRNA$_f^{Met}$ <1, 2> (<1> non-aminoacylated tRNA$_f^{Met}$ is a competitive inhibitor [3,9]; <2> tRNA$_f^{Met}$ from Streptococcus faecalis or E. coli K12, strong inhibition [9]; <1>) [3, 9]

tetrahydrofolate <2, 3> (<3> competitive, multisite inhibition [8]; <2> competitive inhibition [9]) [8, 9]

Additional information <1-3> (<3> inhibition studies, not inhibited by trimethoprim [8]; <2,3> not inhibited by folic acid [8,9]; <2> not inhibited by pteroic acid, aminopterin, methionyl-tRNA$_m^{Met}$ and uncharged tRNA$_m^{Met}$ [9]; <1> E. coli A19 enzyme is not inhibited by uncharged tRNAmMet and tRNA$_f^{Met}$ from E. coli K12 or Streptococcus faecalis [9]) [8, 9]

Cofactors/prosthetic groups

10-formyltetrahydrofolate <1-4> (<1> 1.1 mol 10-L-formyltetrahydrofolate is bound per mol monomeric enzyme [2,4]) [1-9, 11]
tetrahydrofolate <1-4> [1-9, 11]
Additional information <1, 2> (<1,2> no cofactors: ribonucleotide mono- or triphosphates [1,9]; <1> no cofactors: ATP, GTP [1]) [1, 9]

Activating compounds

Additional information <1, 2> (<1,2> not stimulated by ATP or GTP [9]) [9]

Metals, ions

Ca^{2+} <1> (<1> can partially replace Mg^{2+} [1]) [1]
K^+ <1, 3, 4> (<1> requirement, maximum activity at 10 mM Mg^{2+} and 150 mM KCl [4]; <1> activity is maximum in presence of 7 mM $MgCl_2$ and 150 mM KCl [2]; <1> can partially replace Mg^{2+} [1]; <3> partial requirement, in presence of 8 mM Mg^{2+} optimum concentration is 20 mM [7]; <4> activates, maximum activity at 5 mM Mg^{2+} and 20 mM KCl [11]) [1, 2, 4, 7, 11]
Mg^{2+} <1-4> (<1,2,4> requirement [1,4,9,11]; <3> partial requirement, optimum concentration is 8 mM, 3fold increased activity [7]; <1,2> optimal concentration: 30 mM [1,9]; <4> maximum activity at 5 mM Mg^{2+} and 20 mM KCl [11]; <1> maximum activity at 10 mM Mg^{2+} and 150 mM KCl [4]; <1> activity is maximum in presence of 7 mM $MgCl_2$ and 150 mM KCl [2]) [1, 2, 4, 7, 9, 11]
NH_4^+ <1, 3, 4> (<1,3> activates, can partially replace Mg^{2+} [1,7]; <4> activates, can replace K^+ [11]) [1, 7, 11]
Na^+ <1, 3, 4> (<1,3> activates, can partially replace Mg^{2+} [1,7]; <4> activates, can replace K^+ [11]) [1, 7, 11]
spermine <3> (<3> activates, can partially replace Mg^{2+} [7]) [7]

Turnover number (min^{-1})

0.03 <5> (E. coli isoleucyl-$tRNA_f^{Met}$) [20]
0.54 <5> (E. coli valyl-$tRNA_f^{Met}$) [20]
1.8 <6> (E. coli methionyl-$tRNA_1^{Val(CAU)}$, <6> above [15]) [15]
2.4 <6> (E. coli methionyl-$tRNA_m^{Met}$, <6> above [15]) [15]
7.8 <5> (E. coli methionyl-$tRNA_f^{Met}$) [20]
12 <1> (methionyl-$tRNA_f^{Met}$, <1> MTFδC20 [19]) [19]
27 <1> (methionyl-$tRNA_f^{Met}$, <1> G41R/R42L double mutant enzyme [18]) [18]
29 <1> (methionyl-$tRNA_f^{Met}$, <1> R42L mutant enzyme [18]) [18]
55.8 <6> (E. coli methionyl-$tRNA_f^{Met}$) [15]
186 <1> (methionyl-$tRNA_f^{Met}$, <1> R42Q mutant enzyme [18]) [18]
496 <1> (methionyl-$tRNA_f^{Met}$, <1> R38L mutant enzyme [18]) [18]
534 <1> (methionyl-$tRNA_f^{Met}$, <1> G178E mutant enzyme [17]) [17]
677.4 <7> (E. coli methionyl-$tRNA_f^{Met}$) [21]
684 <1> (methionyl-$tRNA_f^{Met}$, <1> G41R mutant enzyme [18]) [18]
834 <1> (methionyl-$tRNA_f^{Met}$, <1> wild-type enzyme [17]) [17]
876 <1> (methionyl-$tRNA_f^{Met}$, <1> G41K mutant enzyme [18]) [18]
1200 <1> (methionyl-$tRNA_f^{Met}$, <1> + 10-formyltetrahydrofolate [4]) [2, 4]

1660 <1> (methionyl-tRNA$_f^{Met}$, <1> R42K mutant enzyme [18]) [18]
1890 <1> (methionyl-tRNA$_f^{Met}$, <1> G41L mutant enzyme [18]) [18]
1980 <1> (methionyl-tRNA$_f^{Met}$, <1> G41D mutant enzyme [18]) [18]
2238 <1> (methionyl-tRNA$_f^{Met}$, <1> wild-type enzyme [19]) [19]
2347 <1> (methionyl-tRNA$_f^{Met}$, <1> wild-type enzyme [18]) [18]
2491 <1> (methionyl-tRNA$_f^{Met}$) [16]
2779 <1> (methionyl-tRNA$_f^{Met}$, <1> R38K mutant enzyme [18]) [18]
Additional information <1, 5-7> (<1,5,6> kinetic parameters of the formylation of various mutant methionyl-tRNAs [5,15,20]; <1> values for wild-type and mutant enzyme with wild-type and mutant initiator tRNAs [16,18]; <1> values for several mutant enzymes in the linker region and C-terminal domain [19]; <7> values for several mutant enzymes [21]) [5, 15, 16, 18-21]

Specific activity (U/mg)

0.013 <3> [7]
0.53 <1> [9]
0.64-0.73 <2> [9]
1.14 <1> [1]
1.23 <7> [21]
35.4 <1> [2, 4]
Additional information <1, 3, 4, 7> [7, 11, 18, 19, 21]

K_m-Value (mM)

0.00004 <6> (bovine mitochondrial methionyl-tRNA) [14]
0.000097 <6> (E. coli methionyl-tRNA$_m^{Met}$) [14]
0.0001 <6> (E. coli methionyl-tRNA$_f^{Met}$) [14]
0.0003 <6> (E. coli isoleucyl-tRNA$_f^{Met}$) [15]
0.0003 <5> (E. coli isoleucyl-tRNA$_f^{Met}$) [20]
0.0003 <5> (E. coli methionyl-tRNA$_f^{Met}$) [20]
0.00035 <1> (methionyl-tRNA$_f^{Met}$) [2, 4]
0.0005 <1, 6> (E. coli methionyl-tRNA$_f^{Met}$) [15, 19]
0.0005 <5> (E. coli valyl-tRNA$_f^{Met}$) [20]
0.00051 <1> (methionyl-tRNA$_f^{Met}$, <1> wild-type enzyme [17]) [17]
0.00054 <1> (methionyl-tRNA$_f^{Met}$, <1> wild-type enzyme [18]) [18]
0.00056 <1> (methionyl-tRNA$_f^{Met}$) [16]
0.0006 <1> (methionyl-tRNA$_f^{Met}$) [3]
0.00066 <1> (methionyl-tRNA$_f^{Met}$, <1> G41R mutant enzyme [18]) [18]
0.00072 <1> (methionyl-tRNA$_f^{Met}$, <1> G178E mutant enzyme [17]) [17]
0.00072-0.0009 <2> (methionyl-tRNA$_f^{Met}$) [9]
0.00102 <1> (methionyl-tRNA$_f^{Met}$, <1> G41K mutant enzyme [18]) [18]
0.00112 <7> (E. coli methionyl-tRNA$_f^{Met}$) [21]
0.00121 <1> (methionyl-tRNA$_f^{Met}$, <1> R38K mutant enzyme [18]) [18]
0.0013 <1> (phenylalanyl-tRNA$_f^{Met}$) [3]
0.0014 <6> (E. coli valyl-tRNA$_f^{Met}$) [15]
0.0014 <1> (valyl-tRNA$_f^{Met}$) [3]
0.00162 <1> (methionyl-tRNA$_f^{Met}$, <1> G41L mutant enzyme [18]) [18]
0.0017 <1> (methionyl-tRNA$_f^{Met}$) [6]
0.003 <1> (methionyl-tRNA$_f^{Met}$, <1> G41D mutant enzyme [18]) [18]

0.00324 <1> (methionyl-$tRNA_f^{Met}$, <1> R38L mutant enzyme [18]) [18]
0.0037 <1> (methionyl-$tRNA_f^{Met}$, <1> MTFδC20 [19]) [19]
0.006-0.007 <2> (10-formyltetrahydrofolate) [9]
0.00834 <1> (methionyl-$tRNA_f^{Met}$, <1> G41R/R42L double mutant enzyme [18]) [18]
0.00866 <1> (methionyl-$tRNA_f^{Met}$, <1> R42K mutant enzyme [18]) [18]
0.009 <1> (methionyl-$tRNA_f^{Met}$, <1> R42L mutant enzyme [18]) [18]
0.0096 <1> (methionyl-$tRNA_f^{Met}$, <1> R42Q mutant enzyme [18]) [18]
0.01 <6> (E. coli methionyl-$tRNA_1^{Val(CAU)}$, <6> above [15]) [15]
0.01 <6> (E. coli methionyl-$tRNA_m^{Met}$, <6> above [15]) [15]
0.01 <1> (methionyl-$tRNA^{Met}$) [1]
0.011-0.016 <2> (10-formyltetrahydropteroyltriglutamate) [9]
0.0133 <1> (10-formyltetrahydrofolate) [1]
0.0135 <1> (10-formyltetrahydrofolate) [2, 4]
Additional information <1, 3, 5-7> (<1,5,6> kinetic parameters of the formylation of various mutant methionyl-tRNAs [5,6,15,20]; <3> kinetics [7]; <1> values for wild-type and mutant enzyme with wild-type and mutant initiator tRNAs [16,18]; <1> values for several mutant enzymes in the linker region and C-terminal domain [19]; <7> values for several mutant enzymes [21]) [5-7, 15, 16, 18-21]

K_i-Value (mM)

0.00018-0.00057 <2> ($tRNA_f^{Met}$) [9]
0.0002 <3> (tetrahydrofolate) [8]
0.003 <3> (5-methyltetrahydrofolate) [8]
0.024 <2> (tetrahydrofolate) [9]
0.045 <3> (5-formyltetrahydrofolate) [8]

pH-Optimum

7 <2> (<2> about [9]) [9]
7.5 <1, 4> (<1> with abrupt decrease of reaction rate below pH 7.0 [1]) [1, 11]

pH-Range

6.5-8 <4> (<4> good activity between [11]) [11]
6.8-8.4 <1> (<1> about 60% of maximal activity at pH 6.8 and 8.4 [1]) [1]
7 <1> (<1> below: rapid decrease in reaction rate [1]) [1]

Temperature optimum (°C)

25 <1> (<1> assay at [2,4,5]) [2, 4, 5]
30 <1-3> (<1-3> assay at [7,9]) [7, 9]
33 <5> (<5> assay at [10]) [10]
37 <1, 4, 7> (<1,4,7> assay at [1-3,6,11,18,21]) [1-3, 6, 11, 18, 21]

4 Enzyme Structure

Molecular weight

25000 <1> (<1> high speed equilibrium sedimentation [1]) [1]
29900 <2> (<2> sucrose density gradient centrifugation [9]) [9]

30000 <1> (<1> sedimentation equilibrium method [4]) [4]
33000 <1> (<1> sedimentation equilibrium method, gel filtration, small-angle neutron scattering [2]) [2]
34000 <1> (<1> gel filtration, small-angle neutron scattering [4]) [4]
38000 <3> (<3> gel filtration [7]) [7]
45000 <4> (<4> gel filtration [11]) [11]

Subunits
? <6> (<6> x * 40000, SDS-PAGE [14]) [14]
monomer <1> (<1> 1 * 32000, SDS-PAGE [2,4]; <1> 1 * 32000 [5]) [2, 4, 5]

5 Isolation/Preparation/Mutation/Application

Source/tissue
germ <4> [11]
leaf <4> (<4> higher specific activity in isolated chloroplasts than in the remainder of the leaf extract [11]) [11]
liver <6> [15]

Localization
chloroplast <3, 4> [7, 8, 11]
cytoplasm <1, 2> [1-4, 9]
mitochondrion <5, 6> [10, 14, 15, 20]

Purification
<1> (over 1500fold purification [1]; 10000fold purification [4]; affinity chromatography on Sepharose-bound tRNA [2,4]; purification of wild-type enzyme and G41R, G41K mutant enzymes [16]; 14fold purification of recombinant wild-type enzyme, overexpressed in Escherichia coli Novablue, purification of His-tagged G178E mutant enzyme [17]; purification of wild-type and of R42K, R42L and G41R/R42L mutant recombinant enzymes, expressed in Escherichia coli JM109 [18]; purification of His-tagged wild-type and C-terminal deletion mutant recombinant enzymes, expressed in Escherichia coli [19]) [1, 2, 4, 9, 16-19, 22]
<2> (100-300fold, partial purification [9]) [9]
<3> (1000fold purification from autotrophic cells, 650fold from chloroplasts [7]) [7]
<4> (partial [11]) [11]
<5> (purification of recombinant enzyme, expressed in Escherichia coli [20]) [10, 20]
<6> (purification of His-tagged recombinant enzyme, expressed in Escherichia coli BL21(DE3) [14,15]) [14, 15]
<7> (purification of His6-tagged wild-type and mutant recombinant enzymes, expressed in Escherichia coli BL21(DE3) [21]) [21]

Crystallization
<1> (crystal structure of enzyme complexed with formylmethionyl-$tRNA_f^{Met}$ [12,17,19,22]) [12, 13, 16-19, 22]

Cloning

<1> (fmt gene encoding MTF is cloned and sequenced, expression in Escherichia coli JM109 as His6-tagged enzyme [16]; wild-type and mutant fmt gene encoding MTF is cloned, overexpression of wild-type enzyme in Escherichia coli Novablue [17]; expression of wild-type and of R42K, R42L and G41R/R42L mutant enzymes in Escherichia coli JM109 as His6-tagged proteins [18]; gene encoding MTF is cloned and sequenced, expression of wild-type and mutant enzymes in Escherichia coli [19]) [16-19, 22]
<5> (cloning and expression of the gene encoding mitochondrial enzyme in Escherichia coli Pal13Tr-pMAF and in yeast, several strains lacking gene exhibit in minimal medium a thermosensitive respiratory growth and a slower growth rate [20]) [20]
<6> (gene encoding mitochondrial enzyme is cloned and expressed in Escherichia coli BL21(DE3) as His-tagged enzyme [14,15]) [14, 15]
<7> (cloning of the fmt gene and expression of wild-type and mutant genes in Escherichia coli BL21(DE3) as His6-tagged proteins [21]) [21]
<8-10> (fmt gene encoding enzyme is cloned [21]) [21]

Engineering

A55T <1> (<1> no suppressor mutation [17]) [17]
D145A <7> (<7> mutant enzyme with drastically reduced activity [21]) [21]
D145E <7> (<7> mutant enzyme with 881fold reduced formylation rate, kinetic parameters [21]) [21]
D145N <7> (<7> mutant enzyme with 4031fold reduced formylation rate, kinetic parameters [21]) [21]
G178D <1> (<1> suppressor mutation, compensates for the formylation defect of a mutant elongator lysine-$tRNA_{m}^{Met}$, it compensates for lysine that is attached to the tRNA, G178D, G178K and G178L are progressively weaker suppressors than G178E [17]) [17]
G178E <1> (<1> suppressor mutation, compensates for the formylation defect of a mutant elongator lysine-$tRNA_{m}^{Met}$, G178D, G178K and G178L are progressively weaker suppressors [17]) [17]
G178K <1> (<1> suppressor mutation, compensates for the formylation defect of a mutant elongator lysine-$tRNA_{m}^{Met}$, G178D, G178K and G178L are progressively weaker suppressors than G178E [17]) [17]
G178L <1> (<1> suppressor mutation, compensates for the formylation defect of a mutant elongator lysine-$tRNA_{m}^{Met}$, G178D, G178K and G178L are progressively weaker suppressors than G178E [17]) [17]
G41D <1> (<1> no suppressor mutation, mutant is stably overproduced in E. coli CA274 [16]; <1> mutant enzyme with decreased activity [18]) [16, 18]
G41K <1> (<1> suppressor mutation within a 16-amino acid insertion, compensates for the formylation defect of a U35A36/G72G73 mutant initiator tRNA, lacking a critical determinant in the acceptor stem [16]; <1> mutant enzyme with decreased activity [18]) [16, 18]
G41L <1> (<1> no suppressor mutation, mutant is stably overproduced in E. coli CA274 [16]; <1> mutant enzyme with decreased activity [18]) [16, 18]

G41Q <1> (<1> no suppressor mutation, mutant is stably overproduced in E. coli CA274 [16]) [16]
G41R <1> (<1> suppressor mutation within a 16-amino acid insertion, compensates for the formylation defect of a U35A36/G72G73 mutant initiator tRNA, lacking a critical determinant in the acceptor stem [16]; <1> G41R/R42L double mutant with low activity, G41R: decreased activity [18]) [16, 18]
H109A <7> (<7> mutant enzyme with 4677fold reduced formylation rate, kinetic parameters [21]) [21]
H109K <7> (<7> inactive mutant enzyme [21]) [21]
H109R <7> (<7> mutant enzyme with 7172fold reduced formylation rate, kinetic parameters [21]) [21]
K37A <7> (<7> mutant enzyme with 3.5fold reduced formylation rate, kinetic parameters [21]) [21]
N107A <7> (<7> mutant enzyme with 48fold reduced formylation rate, kinetic parameters [21]) [21]
R38K <1> (<1> mutant enzyme with 2.2fold increased K_m, no effect on k_{cat} [18]) [18]
R38L <1> (<1> mutant enzyme with up to 12fold lower specific activity, 6fold increased K_m and 5fold decreased k_{cat} [18]) [18]
R41A <7> (<7> mutant enzyme with 470fold reduced formylation rate, kinetic parameters [21]) [21]
R41K <7> (<7> mutant enzyme with 140fold reduced formylation rate, kinetic parameters [21]) [21]
R42K <1> (<1> mutant enzyme with lower activity, 16fold increased K_m [18]) [18]
R42L <1> (<1> single mutant and G41R/R42L double mutant with low activity, R42L: 17fold increased K_m and 79fold decreased k_{cat} [18]) [18]
R42Q <1> (<1> mutant enzyme with lower activity, 16-17fold increased K_m [18]) [18]
V31A <7> (<7> mutant enzyme with 1.7fold reduced formylation rate [21]) [21]
Additional information <1, 7> (<1> several C-terminal deletion mutants and substitution mutants in the linker region between the C- and N-terminal domain and the C-terminal domain [19]; <7> mutations of Asp-145 and Arg-41 reduce the affinity for initiator tRNA, mutations of Asn-107 and His-109 affect catalysis but not tRNA binding [21]) [19, 21]

6 Stability

Organic solvent stability

glycerol <1> (<1> 50%, stable to [2]) [2]

General stability information

<1>, EDTA or 2-mercaptoethanol stabilizes [2]
<1>, high ionic strength, e.g. 150 mM KCl stabilizes and prevents adsorption to glassware [2, 4]

Storage stability

<1>, -15°C, 20 mM imidazole-HCl buffer, pH 7.6, 0.1 mM EDTA, 10 mM 2-mercaptoethanol, 50% glycerol, 150 mM KCl, several months, stable [2]
<1>, -20°C, 1 month, 30% loss of activity [1]
<1>, -20°C, 20 mM imidazole, 10 mM 2-mercaptoethanol, 50% glycerol, 150 mM KCl, stable [18]
<1>, -20°C, 20 mM imidazole, pH 7.5, 10 mM 2-mercaptoethanol, 50% glycerol, 150 mM KCl, 2-4 months, stable [18]
<1>, -20°C, 20 mM imidazole-HCl buffer, pH 7.6, 0.1 mM EDTA, 10 mM 2-mercaptoethanol, 50% glycerol, 150 mM KCl, 6 months, stable [4]
<1>, 25°C, 20 mM imidazole-HCl buffer, pH 7.6, 0.1 mM EDTA, 10 mM 2-mercaptoethanol, 150 mM KCl, 2 days, stable [2, 4]
<4>, -20°C, longer than 4 months, stable [11]

References

[1] Dickerman, H.W.; Steers, E.; Redfield, B.G.; Weissbach, H.: Methionyl soluble ribonucleic acid transformylase. I. Purification and partial characterization. J. Biol. Chem., **242**, 1522-1525 (1967)
[2] Blanquet, S.; Dessen, P.; Kahn, D.: Properties and specificity of methionyl-$tRNA_f^{Met}$ formyltransferase from Escherichia coli. Methods Enzymol., **106**, 141-152 (1984)
[3] Giege, R.; Ebel, J.P.; Clarke, B.F.C.: Formylation of mischarged E. coli tRNA Met f. FEBS Lett., **30**, 291-295 (1973)
[4] Kahn, D.; Fromant, M.; Fayat, G.; Dessen, P.; Blanquet, S.: Methionyl-transfer-RNA transformylase from Escherichia coli. Purification and characterisation. Eur. J. Biochem., **105**, 489-497 (1980)
[5] Guillon, J.M.; Meinnel, T.; Mechulam, Y.; Lazennec, C.; Blanquet, S.; Fayat, G.: Nucleotides of tRNA governing the specificity of Escherichia coli methionyl-tRNA(fMet) formyltransferase. J. Mol. Biol., **224**, 359-367 (1992)
[6] Lee, C.P.; Seong, B.L.; RajBhandary, U.L.: Structural and sequence elements important for recognition of Escherichia coli formylmethionine tRNA by methionyl-tRNA transformylase are clustered in the acceptor stem. J. Biol. Chem., **266**, 18012-18017 (1991)
[7] Crosti, P.; Gambini, A.; Lucchini, G.; Bianchetti, R.: Eukaryotic N^{10}-formyl-H4folate:methionyl-$tRNA_f$ transformylase. Some properties of the Euglena gracilis enzyme. Biochim. Biophys. Acta, **477**, 356-370 (1977)
[8] Gambini, A.; Crosti, P.; Bianchetti, R.: Inhibitor studies of methionyl-tRNA transformylase of Euglena gracilis. Biochim. Biophys. Acta, **613**, 73-78 (1980)
[9] Samuel, C.E.; Rabinowitz, J.C.: Initiation of protein synthesis by folate-sufficient and folate-deficient Streptococcus faecalis R: Partial purification and properties of methionyl-transfer ribonucleic acid synthetase and methionyl-transfer ribonucleic acid formyltransferase. J. Bacteriol., **118**, 21-31 (1974)

[10] Halbreich, A.; Rabinowitz, M.: Isolation of Saccharomyces cerevisiae mitochondrial formyltetrahydrofolic acid:methionyl-tRNA transformylase and the hybridization of mitochondrial fMet-tRNA with mitochondrial DNA. Proc. Natl. Acad. Sci. USA, **68**, 294-298 (1971)

[11] Leis, J.P.; Keller, E.B.: N-formylmethionyl-tRNA of wheat chloroplasts. Its synthesis by a wheat transformylase. Biochemistry, **10**, 889-894 (1971)

[12] Schmitt, E.; Panvert, M.; Blanquet, S.; Mechulam, Y.: Crystal structure of methionyl-$tRNA_f^{Met}$ transformylase complexed with the initiator formyl-methionyl-tRNA. EMBO J., **17**, 6819-6826 (1998)

[13] Schmitt, E.; Blanquet, S.; Mechulam, Y.: Structure of crystalline Escherichia coli methionyl-$tRNA^{Met}$ formyltransferase: Comparison with glycinamide ribonucleotide formyltransferase. EMBO J., **15**, 4749-4758 (1996)

[14] Takeuchi, N.; Ueda, T.; Watanabe, K.: Expression and characterization of bovine mitochondrial methionyl-tRNA transformylase. J. Biochem., **124**, 1069-1071 (1998)

[15] Takeuchi, N.; Vial, L.; Panvert, M.; Schmitt, E.; Watanabe, K.; Mechulam, Y.; Blanquet, S.: Recognition of tRNAs by methionyl-tRNA transformylase from mammalian mitochondria. J. Biol. Chem., **276**, 20064-20068 (2001)

[16] Ramesh, V.; Gite, S.; Li, Y.; RajBhandary, U.L.: Suppressor mutations in Escherichia coli methionyl-tRNA formyltransferase: Role of a 16-amino acid insertion module in initiator tRNA recognition. Proc. Natl. Acad. Sci. USA, **94**, 13524-13529 (1997)

[17] Li, Y.; Ramesh, V.; Mangroo, D.; Taneja, C.; RajBhandary, U.L.: Suppressor mutations in Escherichia coli methionyl-tRNA formyltransferase that compensate for the formylation defect of a mutant tRNA aminoacylated with lysine. Biochemistry, **39**, 8039-8046 (2000)

[18] Ramesh, V.; Gite, S.; RajBhandary, U.L.: Functional interaction of an arginine conserved in the sixteen amino acid insertion module of Escherichia coli methionyl-tRNA formyltransferase with determinants for formylation in the initiator tRNA. Biochemistry, **37**, 15925-15932 (1998)

[19] Gite, S.; Li, Y.; Ramesh, V.; RajBhandary, U.L.: Escherichia coli methionyl-tRNA formyltransferase: Role of amino acids conserved in the linker region and in the C-terminal domain on the specific recognition of the initiator tRNA. Biochemistry, **39**, 2218-2226 (2000)

[20] Vial, L.; Gomez, P.; Panvert, M.; Schmitt, E.; Blanquet, S.; Mechulam, Y.: Mitochondrial methionyl-$tRNA_f^{Met}$ formyltransferase from Saccharomyces cerevisiae: Gene disruption and tRNA substrate specificity. Biochemistry, **42**, 932-939 (2003)

[21] Newton, D.T.; Mangroo, D.: Mapping the active site of the Haemophilus influenzae methionyl-tRNA formyltransferase: Residues important for catalysis and tRNA binding. Biochem. J., **339**, 63-69 (1999)

[22] Mayer, C.; RajBhandary, U.L.: Conformational change of Escherichia coli initiator methionyl-$tRNA_f^{Met}$ upon binding to methionyl-tRNA formyl transferase. Nucleic Acids Res., **30**, 2844-2850 (2002)

Aminomethyltransferase 2.1.2.10

1 Nomenclature

EC number
2.1.2.10

Systematic name
S-aminomethyldihydrolipoylprotein:(6S)-tetrahydrofolate aminomethyltransferase (ammonia-forming)

Recommended name
aminomethyltransferase

Synonyms
Amt
T protein
T-protein
glycine synthase
synthase, glycine
tetrahydrofolate aminomethyltransferase
Additional information (a component, with EC 1.4.4.2, of the glycine cleavage system, formerly known as glycine synthase)

CAS registry number
37257-08-2

2 Source Organism

<1> *Gallus gallus* [1]
<2> *Rattus norvegicus* (male Wistar rats [2]) [2]
<3> *Pisum sativum* [3]
<4> *Homo sapiens* [4, 5]
<5> *Mus musculus* [5]
<6> *Escherichia coli* (recombinant enzyme, expressed in Escherichia coli BL21 (DE3)pLysS [6]) [6]

3 Reaction and Specificity

Catalyzed reaction
S-aminomethyldihydrolipoylprotein + (6S)-tetrahydrofolate = dihydrolipoylprotein + (6R)-5,10-methylenetetrahydrofolate + NH_3

Reaction type

C-N bond cleavage
methyl group transfer

Natural substrates and products

S S-aminomethyldihydrolipoylprotein + (6S)-tetrahydrofolate <1-5> (<1, 3, 4> part of the glycine cleavage system [1, 3, 4]; <2> glycine metabolism [2]; <5> important enzyme in glycine metabolism [5]) (Reversibility: ? <1-5> [1-5]) [1-5]

P dihydrolipoylprotein + (6R)-5,10-methylenetetrahydrofolate + NH_3

Substrates and products

S S-aminomethyldihydrolipoylprotein + (6S)-tetrahydrofolate <1-4, 6> (<1, 2> decarboxylated glycine moiety attached to H-protein + tetrahydrofolate as substrates [1, 2]; <6> aminomethyl intermediate bound to the lipoate cofactor of H-protein [6]; <1> strictly dependent on tetrahydrofolate [1]; <3, 6> tetrahydrofolate-dependent enzyme [3,6]; <4> requires tetrahydrofolate [4]; <2> T-protein participates in the formation of the one carbon unit and ammonia or the reverse reaction [2]; <3, 6> T-protein catalyzes the tetrahydrofolate-dependent step of the glycine cleavage reaction [3,6]; <6> folate-binding site: Lys-78, Lys-81 and Lys-352 are involved in binding, Lys-352 may serve as the primary binding site to α-carboxyl group of the first glutamate residue nearest the *p*-aminobenzoic acid ring of 5,10-methylenetetrahydrofolate and 5,10-methylenetetrahydropteroyltetraglutamate, Lys-81 may play a key role to hold the second glutamate residue through binding to its α-carboxyl group, 6.5fold higher affinity for 5,10-methylenetetrahydropteroyltetraglutamate than for 5,10-methylenetetrahydrofolate [6]) (Reversibility: r <1, 2, 6> [1, 2, 6]; ? <3, 4> [3, 4]) [1-4, 6]

P dihydrolipoylprotein + (6R)-5,10-methylenetetrahydrofolate + NH_3 <1-3, 6> [1-3, 6]

S S-aminomethyldihydrolipoylprotein + tetrahydropteroyltetraglutamate <6> (Reversibility: ? <6> [6]) [6]

P dihydrolipoylprotein + 5,10-methylenetetrahydropteroyltetraglutamate + NH_3 <6> (<6> better substrate than 5,10-methylenetetrahydrofolate, 6.5fold higher affinity for 5,10-methylenetetrahydropteroyltetraglutamate than for 5,10-methylenetetrahydrofolate [6]) [6]

S Additional information <1-6> (<1> component of the glycine cleavage system, T-protein is associated with H-protein forming a complex which is composed of one molecule of each of them [1]; <2> enzyme is a component of the reversible glycine cleavage system which is composed of 4 protein components named as P-, H-, L- and T-protein [2]; <3-6> enzyme is a component of the glycine cleavage system which is composed of P-, H-, L- and T-protein, multienzyme complex [3-6]) [1-6]

P ?

Inhibitors

N-ethylmaleimide <2> (<2> 0.1 mM, 70% inhibition [2]) [2]
iodoacetamide <2> (<2> 5 mM, 70% inhibition [2]) [2]

Cofactors/prosthetic groups

tetrahydrofolate <1-4, 6> (<1> strictly dependent on tetrahydrofolate [1]; <3,6> tetrahydrofolate-dependent enzyme [3,6]; <4> requires tetrahydrofolate [4]; <6> folate-binding site [6]) [1-4, 6]

Turnover number (min^{-1})

525 <6> (5,10-methylenetetrahydropteroyltetraglutamate) [6]
864 <6> (5,10-methylenetetrahydrofolate) [6]
Additional information <6> (<6> values for mutant enzymes [6]) [6]

Specific activity (U/mg)

1.08 <3> [3]
Additional information <1> [1]

K_m-Value (mM)

0.0037 <1> (S-aminomethyldihydrolipoylprotein) [1]
0.0104 <6> (5,10-methylenetetrahydropteroyltetraglutamate) [6]
0.0677 <6> (5,10-methylenetetrahydrofolate) [6]
0.17 <1> (tetrahydrofolate) [1]
Additional information <6> (<6> values for mutant enzymes [6]) [6]

pH-Optimum

Additional information <1> (<1> pI: 9.8 [1]) [1]

Temperature optimum (°C)

37 <1, 2, 6> (<1,2,6> assay at [1,2,6]) [1, 2, 6]

4 Enzyme Structure

Molecular weight

33000 <2> (<2> gel filtration [2]) [2]
37000-38000 <1> (<1> sedimentation equilibrium centrifugation, gel filtration [1]) [1]
45000 <3> (<3> gel filtration [3]) [3]

Subunits

monomer <1, 3> (<1> 1 * 41000, SDS-PAGE [1]; <3> 1 * 45000, lithium dodecyl sulfate PAGE [3]) [1, 3]

5 Isolation/Preparation/Mutation/Application

Source/tissue

chorionic villus <4> [4]
leaf <3> [3]
liver <1, 2, 4> [1, 2, 4]

Localization

mitochondrial inner membrane <2> [2]
mitochondrial matrix <3> [3]
mitochondrion <1, 4, 5> [1, 4, 5]

Purification

<1> (641fold purification [1]) [1]
<2> [2]
<3> (8.8fold purification [3]) [3]
<6> (purification of recombinant enzyme, expressed in Escherichia coli BL21 (DE3)pLysS, and of mutants [6]) [6]

Cloning

<4> (Amt gene is located on chromosome 3p21 and contains 9 exons spanning about 6 kb of genomic sequence [5]) [5]
<5> (cloning and sequencing of the Amt gene with 9 closely spaced exons that are contained within about 5 kb of genomic DNA, encoding a protein of 403 amino acids, cis-acting promoter is likely to be very short, Amt gene may be localized on chromosome 9F [5]) [5]

Engineering

D276H <4> (<4> nonketotic hyperglycinemia, rare mutation [4]) [4]
G269D <4> (<4> nonketotic hyperglycinemia, rare mutation [4]) [4]
G47R <4> (<4> nonketotic hyperglycinemia, rare mutation [4]) [4]
H42R <4> (<4> present in many nonketotic hyperglycinemia affected members of an extended Israeli-Arab kindred [4]) [4]
K352E <6> (<6> mutant with 2fold increased K_m-values for folate substrates [6]) [6]
K352Q <6> (<6> mutant with 2fold increased K_m-values for folate substrates [6]) [6]
K352R <6> (<6> no effect on K_m-values [6]) [6]
K75E <6> (<6> mutant with 2.5fold increased K_m-value for 5,10-methylenetetrahydrofolate and 8fold increased K_m-value for 5,10-methylenetetrahydropteroyltetraglutamate [6]) [6]
K78E <6> (<6> mutant with 1.4fold increased K_m-values for folate substrates [6]) [6]
K81E <6> (<6> mutant with 3fold increased K_m-value for 5,10-methylenetetrahydrofolate and 16fold increased K_m-value for 5,10-methylenetetrahydropteroyltetraglutamate [6]) [6]
N145I <4> (<4> nonketotic hyperglycinemia, substitution of conserved N, patient has servere neonatal presentation and died in the newborn period [4]) [4]
Q192X <4> (<4> nonketotic hyperglycinemia, premature stop codon [4]) [4]
R320H <4> (<4> allele frequency of 7% for R320H of T-protein in 50 patients with enzymatic confirmation of their diagnostics of nonketotic hyperglycinemia [4]) [4]

Additional information <4> (<4> allele frequency of 3% for T-protein splice site mutation IVS7-1G-A in 50 patients with enzymatic confirmation of their diagnostics of nonketotic hyperglycinemia, mutation with a one-base deletion 183delC [4]; <4> a subset of nonketotic hyperglycinemia cases is due to mutations in the gene for the T-protein [5]) [4, 5]

Application

medicine <4> (<4> strategy for molecular investigation of patients with nonketotic hyperglycinemia: defective glycine cleavage enzyme system, composed of P-, H-, T- and L-protein, 15% of patients have a T-protein defect [4]; <4> a subset of nonketotic hyperglycinemia cases is due to mutations in the gene for the T-protein [5]) [4, 5]

6 Stability

Temperature stability

50 <2> (<2> 1 min, 20% loss of activity [2]) [2]
70 <2> (<2> 1 min, complete loss of activity [2]) [2]

General stability information

<1>, lability of enzyme under the conditions of its purifications and storage [1]

Storage stability

<1>, -20°C, 20 mM Tris-HCl buffer, pH 8.0, 10 mM 2-mercaptoethanol, 50% v/v glycerol, at least 2 months, stable [1]
<2>, -20°C, 24 h, 60% loss of activity [2]
<2>, 0-4°C, 1 week, stable [2]

References

[1] Okamura-Ikeda, K.; Fujiwara, K.; Motokawa, Y.: Purification and characterization of chicken liver T-protein, a component of the glycine cleavage system. J. Biol. Chem., **257**, 135-139 (1982)

[2] Motokawa, Y.; Kikuchi, G.: Glycine metabolism by rat liver mitochondria. Reconstruction of the reversible glycine cleavage system with partially purified protein components. Arch. Biochem. Biophys., **164**, 624-633 (1974)

[3] Bourguignon, J.; Neuburger, M.; Douce, R.: Resolution and characterization of the glycine-cleavage reaction in pea leaf mitochondria. Properties of the forward reaction catalysed by glycine decarboxylase and serine hydroxymethyltransferase. Biochem. J., **255**, 169-178 (1988)

[4] Toone, J.R.; Applegarth, D.A.; Coulter-Mackie, M.B.; James, E.R.: Recurrent mutations in P- and T-proteins of the glycine cleavage complex and a novel T-protein mutation (N145I): A strategy for the molecular investigation of patients with nonketotic hyperglycinemia (NKH). Mol. Genet. Metab., **72**, 322-325 (2001)

[5] Backofen, B.; Leeb, T.: Genomic organization of the murine aminomethyltransferase gene (Amt). DNA Seq., **13**, 179-183 (2002)
[6] Okamura-Ikeda, K.; Fujiwara, K.; Motokawa, Y.: Identification of the folate binding sites on the Escherichia coli T-protein of the glycine cleavage system. J. Biol. Chem., **274**, 17471-17477 (1999)

3-Methyl-2-oxobutanoate hydroxymethyltransferase 2.1.2.11

1 Nomenclature

EC number
2.1.2.11

Systematic name
5,10-methylenetetrahydrofolate:3-methyl-2-oxobutanoate hydroxymethyltransferase

Recommended name
3-methyl-2-oxobutanoate hydroxymethyltransferase

Synonyms
5,10-methylene tetrahydrofolate:α-ketoisovalerate hydroxymethyltransferase
KHMT
KPHMT
α-ketoisovalerate hydroxymethyltransferase
dehydropantoate hydroxymethyltransferase
hydroxymethyltransferase, ketopantoate
ketopantoate hydroxymethyltransferase
oxopantoate hydroxymethyltransferase

CAS registry number
56093-17-5

2 Source Organism

<1> *Salmonella typhimurium* (strain LT2 [1]) [1]
<2> *Escherichia coli* (strain K12 [1,3,9]; strains W and B [1]; strain ATCC 23783 [5,6]) [1-6, 9]
<3> *Bacillus subtilis* (strains 168, BD170, UR1, UR4 [7]) [7]
<4> *Mycobacterium bovis* (BCG, strain 1173-P2, live vaccine against tuberculosis, used for superficial bladder cancer immunotherapy [8]) [8]
<5> *Mycobacterium tuberculosis* (strain KIT 10468 [8]) [8, 12]
<6> *Aspergillus nidulans* (wild-type strains biA1 and yA2, strain G1765 [10]) [10]
<7> *Salmonella enterica* (serovar Typhimurium [11]) [11]

3 Reaction and Specificity

Catalyzed reaction

5,10-methylenetetrahydrofolate + 3-methyl-2-oxobutanoate = tetrahydrofolate + 2-dehydropantoate (<2> class II aldolase [2-4, 6]; <2, 5> stereochemistry [4-6, 12]; <5> sequential kinetic mechanism, chemical mechanism [12])

Reaction type

hydroxymethyl group transfer

Natural substrates and products

S 5,10-methylenetetrahydrofolate + 3-methyl-2-oxobutanoate <1-7> (<1, 2, 4, 5> first committed step in pantothenate biosynthesis [1-3, 6, 8, 9, 12]; <2> catalytic activity is regulated by the products of the reaction path of which it is one component [2]; <2> biosynthesis of coenzyme A [4, 6]; <2> first step in pantoate biosynthesis [5]; <3, 7> biosynthesis of pantothenate [7, 11]; <6> essential for the biosynthesis of coenzyme A [10]; <5, 7> enzyme may be the rate-limiting reaction in pantothenate biosynthesis [11, 12]) (Reversibility: ? <1-7> [1-12]) [1-12]

P tetrahydrofolate + 2-dehydropantoate <2, 3, 5-7> (<2> enzyme is responsible for catalysis of ketopantoate formation in vivo [1]) [1-7, 10-12]

Substrates and products

S 5,10-methylenetetrahydrofolate + 2-oxobutyrate <2> (<2> also a good substrate [2]) (Reversibility: ? <2> [2, 3]) [2, 3]

P ?

S 5,10-methylenetetrahydrofolate + 2-oxopentanoate <2> (Reversibility: ? <2> [2, 3]) [2, 3]

P ?

S 5,10-methylenetetrahydrofolate + 3-methyl-2-oxobutanoate <1-7> (<2> specificity for α-ketoisovalerate is less rigid than for tetrahydrofolate [2]; <2> forms 2-oxopantoate, syn. 3-hydroxymethyl-3-methyl-2-oxobutanoic acid, from 2-oxoisovalerate with retention of configuration at C-3 [4]; <2> inversion of the configuration at C-3 of 2-ketoisovalerate [5]; <2> condensation of α-ketoisovalerate with the C-1 donor takes place stereospecifically at C-3 and proceeds in a retention mode at C-3, 5,10-methylenetetrahydrofolate-dependent enzyme [6]) (Reversibility: r <1, 2> [1-3, 6, 9]; ? <2-7> [4, 5, 7, 8, 10-12]) [1-12]

P tetrahydrofolate + 2-dehydropantoate <1-7> (<2> formation of ketopantoate, syn. 2-keto-3,3-dimethyl-4-hydroxybutyrate, tetrahydrofolate-dependent enzyme [1]; <2> synthesis of ketopantoate, the following components can replace tetrahydrofolate: tetrahydropteroylmono-, di-, tri-, tetra-, penta-, hexa-, and heptaglutamate, absolute requirement for tetrahydrofolate, only the L-isomer is active [2,3]) [1-12]

S 5,10-methylenetetrahydrofolate + 3-methyl-2-oxopentanoate <2> (Reversibility: ? <2> [2, 3]) [2, 3]

P ?

S formaldehyde + 3-methyl-2-oxobutanoate <5> (<5> enzyme catalyzes methylenetetrahydrofolate-independent hydroxymethyltransferase reaction between free formaldehyde and α-ketoisovalerate, formaldehyde is unlikely to be the natural substrate [12]) (Reversibility: ? <5> [12]) [12]
P 2-dehydropantoate <5> [12]
S formaldehyde + tetrahydrofolate <2> (Reversibility: ? <2> [2]) [2]
P methylentetrahydrofolate <2> [2]
S Additional information <2, 5> (<2> no substrates: pyruvate, isovalerate, D- and L-valine, 3-methyl-2-butanone [2]; <5> enzyme catalyzes deuterium exchange in the methylenetetrahydrofolate-independent enolization of α-ketoisovalerate or other α-keto acids with decreasing efficiency: α-ketoisovalerate, α-ketobutyrate, α-ketovalerate, pyruvate, α-ketomethylthiobutyrate, α-ketoisocaproate, stereochemistry, first step in the reaction leading to ketopantoate is the enolization of α-ketoisovalerate to form the stabilized carbanion [12]) [2, 12]
P ?

Inhibitors

3-methyl-2-butanone <2> (<2> 5 mM, 27% inhibition [2]) [2]
D-valine <2> (<2> 5 mM, 16% inhibition [2]) [2]
L-valine <2> (<2> 5 mM, 23% inhibition [2]) [2]
coenzyme A <2> (<2> above 1 mM [2,3]) [2, 3]
formaldehyde <2> (<2> 0.8 mM, partial inhibition [2,3]; <2> forward reaction [3]) [2, 3]
isovalerate <2> (<2> 5 mM, 39% inhibition [2]) [2]
pantoate <2> (<2> 0.05 mM or above [2,3]) [2, 3]
pantothenate <2> (<2> 0.5 mM or above [2,3]) [2, 3]
pyruvate <2> (<2> 5 mM, 38% inhibition [2]) [2]
tetrahydrofolate <2> (<2> 0.38 mM, partial inhibition [2,3]; <2> forward reaction [3]) [2, 3]
Additional information <2, 5> (<2> not inactivated by borohydride reduction in the presence of excess substrates [2]; <5> not inhibited by methylenetetrahydrofolate, i.e. equimolar formaldehyde and tetrahydrofolate, at concentrations up to 2 mM [12]) [2, 12]

Cofactors/prosthetic groups

5,10-methylenetetrahydrofolate <1-7> (<2> 5,10-methylenetetrahydrofolate-dependent enzyme [6]; <5> 5,10-methylenetetrahydrofolate as cofactor [12]) [1-12]
tetrahydrofolate <1-7> (<2> tetrahydrofolate-dependent reaction [1]; <2> absolute requirement for tetrahydrofolate [2]; <2> requires tetrahydrofolate [3]) [1-12]
Additional information <2> (<2> no requirement of pyridoxal 5′-phosphate as cofactor [6]) [6]

Metals, ions

Ca^{2+} <5> (<5> substrate enolization is divalent metal-dependent with a preference for metal ions in decreasing order: Mg^{2+}, Zn^{2+}, Co^{2+}, Ni^{2+}, Ca^{2+} [12]) [12]
Co^{2+} <2, 5> (<2> activates, Mg^{2+} is most active followed by Mn^{2+}, Ni^{2+}, Co^{2+} and Zn^{2+} [2]; <2> requires Mg^{2+}, Mn^{2+}, Co^{2+} and Zn^{2+} are progressively less active [3]; <5> substrate enolization is divalent metal-dependent with a preference for metal ions in decreasing order: Mg^{2+}, Zn^{2+}, Co^{2+}, Ni^{2+}, Ca^{2+} [12]) [2, 3, 12]
Mg^{2+} <2, 5> (<2> activates and is required for activity, 0.1 mM Mg^{2+} is most active, Mn^{2+}, Ni^{2+}, Co^{2+} and Zn^{2+} are progressively less active, restores activity after dialysis [2]; <2> requires Mg^{2+}, Mn^{2+}, Co^{2+} and Zn^{2+} are progressively less active [3]; <5> substrate enolization is divalent metal-dependent with a preference for metal ions in decreasing order: Mg^{2+}, Zn^{2+}, Co^{2+}, Ni^{2+}, Ca^{2+} [12]) [2, 3, 12]
Mn^{2+} <2> (<2> activates, Mg^{2+} is most active followed by Mn^{2+}, Ni^{2+}, Co^{2+} and Zn^{2+} [2]; <2> requires Mg^{2+}, Mn^{2+}, Co^{2+} and Zn^{2+} are progressively less active [3]) [2, 3]
Ni^{2+} <2, 5> (<2> activates, Mg^{2+} is most active followed by Mn^{2+}, Ni^{2+}, Co^{2+} and Zn^{2+} [2]; <5> substrate enolization is divalent metal-dependent with a preference for metal ions in decreasing order: Mg^{2+}, Zn^{2+}, Co^{2+}, Ni^{2+}, Ca^{2+} [12]) [2, 12]
Zn^{2+} <2, 5> (<2> activates, Mg^{2+} is most active followed by Mn^{2+}, Ni^{2+}, Co^{2+} and Zn^{2+} [2]; <2> requires Mg^{2+}, Mn^{2+}, Co^{2+} and Zn^{2+} are progressively less active [3]; <5> substrate enolization is divalent metal-dependent with a preference for metal ions in decreasing order: Mg^{2+}, Zn^{2+}, Co^{2+}, Ni^{2+}, Ca^{2+} [12]) [2, 3, 12]
Additional information <2, 5> (<2> not activated by Cu^{2+} and Fe^{2+} [2]; <5> metalloenzyme, inactive in absence of divalent metals, enzyme binds metal ions that assist in the polarization of the carbonyl group and stabilize the enolate anion [12]) [2, 12]

Turnover number (min^{-1})

47 <5> (3-methyl-2-oxobutanoate) [12]

Specific activity (U/mg)

0.0025 <3> (<3> strain BD170 [7]) [7]
0.00254 <3> (<3> strain UR1 [7]) [7]
0.00258 <3> (<3> strain UR4 [7]) [7]
0.00262 <3> (<3> strain 168 [7]) [7]
0.007 <2> (<2> strain K12 [9]) [9]
0.01-0.027 <6> (<6> recombinant enzyme expressed in Escherichia coli [10]) [10]
0.019 <4> (<4> recombinant enzyme [8]) [8]
0.162 <2> (<2> strain Hfr3000 YA139/pCEJ02 [9]) [9]
0.348 <2> (<2> strain Hfr3000 YA139/pCEJ01 [9]) [9]
0.666 <2> (<2> strain Hfr3000 YA139/pSAL38 [9]) [9]

7.3 <2> [1, 3]
3890 <2> (<2> purified recombinant enzyme from strain Hfr3000 YA139/ pCEJ01 [9]) [9]

K_m-Value (mM)

0.1 <2> (tetrahydropteroylpentaglutamate) [2]
0.1 <2> (tetrahydropteroyltetraglutamate) [2]
0.15 <2> (ketopantoate) [9]
0.16 <2> (ketopantoate) [2]
0.17 <2> (tetrahydropteroylhexaglutamate) [2]
0.18 <2> (tetrahydrofolate, <2> forward and reverse reaction, methylenetetrahydrofolate: formaldehyde + tetrahydrofolate [2]) [2, 3]
0.18 <2> (tetrahydropteroyltriglutamate) [2]
0.24 <5> (3-methyl-2-oxobutanoate) [12]
0.25 <2> (tetrahydropteroyldiglutamate) [2]
0.29 <2> (tetrahydropteroylheptaglutamate) [2]
0.33 <2> (tetrahydropteroylmonoglutamate) [2]
0.82 <5> (5,10-methylenetetrahydrofolate) [12]
1.1 <2> (3-methyl-2-oxobutanoate) [2, 3]
2.9 <2> (2-oxobutyrate) [2]
5.9 <2> (3-methyl-2-oxopentanoate) [2]
5.9 <2> (formaldehyde, <2> methylenetetrahydrofolate: formaldehyde + tetrahydrofolate [2]) [2, 3]
25 <2> (2-oxopentanoate) [2]
Additional information <2, 5> (<2> low K_m-values for its substrates [1]) [1, 12]

pH-Optimum

7-7.5 <5> (<5> rate of substrate enolization is pH-dependent with optimal activity in the range of [12]) [12]
7-7.6 <2> [2, 3]
8 <2> (<2> at or below [1]) [1]
Additional information <2> (<2> pI: 4.4 [2,3]) [2, 3]

pH-Range

5-9 <2> (<2> about 50% of activity maximum at pH 6 and 9, inactive below pH 5 [2]) [2]
10 <2> (<2> no ketopantoate formation above pH 10 [1]) [1]
Additional information <2> (<2> stable and active over a broad pH-range [3]) [3]

Temperature optimum (°C)

37 <1, 2, 4-6> (<1,2,4-6> assay at [1,3,8,10,12]) [1, 3, 8, 10, 12]
70-80 <2> (<2> reverse reaction, activity decreases rapidly above 80°C [2]) [2]

4 Enzyme Structure

Molecular weight

120000 <5> (<5> gel filtration [12]) [12]
174000 <2> (<2> recombinant enzyme, gel filtration [9]) [9]
255000 <2> (<2> sedimentation equilibrium [2,3]) [2, 3]
285000 <2> (<2> gel filtration [2]) [2]
340000 <6> (<6> recombinant enzyme expressed in Escherichia coli, gel filtration [10]) [10]

Subunits

? <4> (<4> x * 29337, calculated from the amino acid sequence, x * 35000, recombinant enzyme expressed in Escherichia coli BL21, SDS-PAGE [8]) [8]
decamer <2> (<2> 10 * 27000, SDS-PAGE, 10 * 25700, amino acid analysis [2,3]) [2, 3]
hexamer <2> (<2> 6 * 28179, calculated from the amino acid sequence and electrospray mass spectrometry, 6 * 29000, recombinant enzyme expressed in Escherichia coli Hfr3000 YA139/pCEJ01, SDS-PAGE [9]) [9]
octamer <6> (<6> 8 * 37700, calculated from the amino acid sequence, 8 * 40000, recombinant enzyme expressed in Escherichia coli, SDS-PAGE [10]) [10]
tetramer <5> (<5> recombinant enzyme expressed in Escherichia coli BL21 (DE3): 4 * 29000, SDS-PAGE, 4 * 29202, electrospray ionization/mass spectrometry, 4 * 29366, calculated from the amino acid sequence [12]) [12]

Posttranslational modification

no modification <2> [9]

5 Isolation/Preparation/Mutation/Application

Localization

soluble <4, 5> (<4> most of recombinant enzyme is found in a soluble fraction [8]) [8, 12]

Purification

<2> (strain K12, 2450fold purification [1,3]; 19.3fold purification of recombinant enzyme 50fold overexpressed in E. coli panB mutant Hfr3000 YA139/pCEJ01 [9]) [1-3, 9]
<4> (purification of recombinant enzyme overexpressed in Escherichia coli BL21 [8]) [8]
<5> (purification of recombinant enzyme expressed in Escherichia coli BL21 (DE3) [12]) [12]
<6> (affinity purification of recombinant PanB [10]) [10]

Cloning

<2> (panB gene encoding enzyme is cloned, sequenced and 50fold overexpressed in Escherichia coli panB mutant Hfr3000 YA139 containing plasmid

pCEJ01, gene is 792 bp long and encodes a protein of 264 amino acids, gene is likely to be cotranscribed with at least one other gene, panB, panC and panD genes are closely clustered at 3.1 min of the Escherichia coli K12 genetic map [9]) [9]
<3> (panB gene coding for enzyme is located together with panE in the intervall purE-tre of the chromosome [7]) [7]
<4> (panB gene encoding enzyme is cloned, sequenced and overexpressed in Escherichia coli BL21, ORF of 846 bp encodes a protein of 281 amino acids, N-terminal 37 amino acids are essential for enzyme function [8]) [8]
<5> (PCR amplification of panB gene, which is identical to that of Mycobacterium bovis BCG [8]; panB gene encoding enzyme is cloned and overexpressed in Escherichia coli BL21 (DE3) [12]) [8, 12]
<6> (panB gene encoding enzyme is cloned, sequenced and expressed in Escherichia coli as an active octameric enzyme, ORF encodes a protein of 349 amino acids, panB gene is closely linked to nimO gene on Aspergillus nidulans linkage group VII [10]) [10]
<7> (panB gene encoding enzyme is clustered together with other pantothenate biosynthetic genes panC and panD at 3 min of the chromosome map, panBCD operon [11]) [11]

Engineering

Additional information <2, 3, 6, 7> (<3> panB mutant strain UR2 is severely deficient in enzyme, a single mutation is responsible for the lack of transferase [7]; <2> panB mutant Hfr3000 YA139 completely lacks KPHMT [9]; <6> panB100 mutant has a deletion of G-268 [10]; <7> panBp654 mutant with insertion of one GC base pair upstream of the transcription start site results in an optimized panBCD promoter and a 10fold increase in transcription of the pan operon, i.e. an increased expression of panB, which is sufficient to increase pantothenate biosynthesis [11]) [7, 9-11]

Application

pharmacology <5, 6> (<6> enzyme could serve as target for anti-fungal drugs, since it is not present in mammals [10]; <5> enzyme might be an attractive target for inhibitor design [12]) [10, 12]

6 Stability

pH-Stability

4.5 <2> (<2> rapid inactivation below [2]) [2]
5-10 <2> (<2> stable [2]) [2]
5.5 <5> (<5> loss of proper protein folding at pH values lower than 5.5 [12]) [12]
Additional information <2> (<2> stable and active over a broad pH-range [3]) [3]

Temperature stability

55 <2> (<2> treatment at 55°C largely destroys enzyme in crude extract, partially purified enzyme is more heat-stabile [1]) [1]
80 <2> (<2> rapid denaturation above [2]) [2]

General stability information

<2>, ammonium sulfate fractionation inactivates, resistant to urea denaturation [1]
<2>, not inactivated by borohydride reduction in the presence of excess substrates [3]

Storage stability

<2>, 4°C, 100 mM potassium phosphate, pH 6.8, 1 mM EDTA, 0.5 mM dithiothreitol, 6 months, almost no loss of activity [3]

References

[1] Teller, J.H.; Powers, S.G.; Snell, E.E.: Ketopantoate hydroxymethyltransferase. I. Purification and role in pantothenate biosynthesis. J. Biol. Chem., **251**, 3780-3785 (1976)
[2] Powers, S.G.; Snell, E.E.: Ketopantoate hydroxymethyltransferase. II. Physical, catalytic, and regulatory properties. J. Biol. Chem., **251**, 3786-3793 (1976)
[3] Powers, S.G.; Snell, E.E.: Purification and properties of ketopantoate hydroxymethyltransferase. Methods Enzymol., **62**, 204-209 (1979)
[4] Wightman, R.H.: Stereochemistry of 2-oxopantoate formation by oxopantoate hydroxymethyltransferase. J. Chem. Soc. Chem. Commun., **1979**, 818-819 (1979)
[5] Aberhart, D.J.: Stereochemistry of pantoate biosynthesis from 2-ketoisovalerate. J. Am. Chem. Soc., **101**, 1354-1355 (1979)
[6] Aberhart, D.J.; Russell, D.J.: Steric course of ketopantoate hydroxymethyltransferase in E. coli. J. Am. Chem. Soc., **106**, 4902-4906 (1984)
[7] Baigori, M.; Grau, R.; Morbidoni, H.R.; de Mendoza, D.: Isolation and characterization of Bacillus subtilis mutants blocked in the synthesis of pantothenic acid. J. Bacteriol., **173**, 4240-4242 (1991)
[8] Kim, J.K.; Kim, K.D.; Lim, J.S.; Lee, H.G.; Kim, S.J.; Cho, S.H.; Jeong, W.H.; Choe, I.S.; Chung, T.W.; Paik, S.G.; Choe, Y.K.: Cloning and characterization of the Mycobacterium bovis BCG panB gene encoding ketopantoate hydroxymethyltransferase. J. Biochem. Mol. Biol., **34**, 342-346 (2001)
[9] Jones, C.E.; Brook, J.M.; Buck, D.; Abell, C.; Smith, A.G.: Cloning and sequencing of the Escherichia coli panB gene, which encodes ketopantoate hydroxymethyltransferase, and overexpression of the enzyme. J. Bacteriol., **175**, 2125-2130 (1993)
[10] Kurtov, D.; Kinghorn, J.R.; Unkles, S.E.: The Aspergillus nidulans panB gene encodes ketopantoate hydroxymethyltransferase, required for biosynthesis of pantothenate and coenzyme A. Mol. Gen. Genet., **262**, 115-120 (1999)

[11] Rubio, A.; Downs, D.M.: Elevated levels of ketopantoate hydroxymethyltransferase (PanB) lead to a physiologically significant coenzyme A elevation in Salmonella enterica serovar typhimurium. J. Bacteriol., **184**, 2827-2832 (2002)
[12] Sugantino, M.; Zheng, R.; Yu, M.; Blanchard, J.S.: Mycobacterium tuberculosis ketopantoate hydroxymethyltransferase: Tetrahydrofolate-independent hydroxymethyltransferase and enolization reactions with α-keto acids. Biochemistry, **42**, 191-199 (2003)

Methylmalonyl-CoA carboxytransferase 2.1.3.1

1 Nomenclature

EC number

2.1.3.1

Systematic name

(S)-methylmalonyl-CoA:pyruvate carboxytransferase

Recommended name

methylmalonyl-CoA carboxytransferase

Synonyms

carboxyltransferase, methylmalonyl coenzyme A
methylmalonyl CoA carboxyltransferase
methylmalonyl coenzyme A carboxyltransferase
methylmalonyl-CoA transcarboxylase
oxalacetic transcarboxylase
transcarboxylase

CAS registry number

9029-86-1

2 Source Organism

<1> *Propionibacterium sp.* [1, 6]
<2> *Propionibacterium shermanii* (i.e. Propionibacterium freudenreichii [8]; strain 52W [2]) [2-5, 7, 8, 10-17, 19]
<3> *Canis familiaris* [2]
<4> *Veillonella alcalescens* (activity is located on the α-subunit of membrane-bound methylmalonyl-CoA decarboxylase EC 4.1.1.41 [9]) [9]
<5> *Streptomyces fradiae* (strain T1558 [18]) [18]

3 Reaction and Specificity

Catalyzed reaction

(S)-methylmalonyl-CoA + pyruvate = propanoyl-CoA + oxaloacetate (mechanism <1,2> [1,5,6,8,11,12]; <2> partial reaction 1 is catalysed specifically by the 12S subunit, partial reaction 2 is catalysed specifically by the 5S subunit [12-16])

Reaction type

carboxyl group transfer

Natural substrates and products

S propionyl-CoA + oxaloacetate <1> (Reversibility: r <1> [1]) [1]
P (S)-methylmalonyl-CoA + pyruvate <1> [1]
S Additional information <5> (<5> enzyme involved in the biosynthesis of the tylosin [18]) [18]
P ?

Substrates and products

S 3-fluoropropionyl-CoA + oxaloacetate <1> (<1> no F^-release if pyruvate or malate are substituted for oxaloacetate [6]) [6]
P acrylyl-CoA + pyruvate + F^- + CO_2 <1> [6]
S acetoacetyl-CoA + oxaloacetate <1, 2> (<1> carboxylation at about 2.5% the rate of propionyl-CoA [1]) [1, 2]
P 3-oxoglutaryl-CoA + pyruvate <1, 2> [1, 2]
S acetyl-CoA + oxaloacetate <1, 2, 4> (<1> carboxylation at about 50% the rate of propionyl-CoA [1]) (Reversibility: r <1> [1, 3]) [1, 2, 3, 9]
P malonyl-CoA + pyruvate <1, 2, 4> [1, 2, 9]
S butyryl-CoA + oxaloacetate <1, 2> (<1> carboxylation at about 10% the rate of propionyl-CoA [1]) [1, 2, 11]
P ethylmalonyl-CoA + pyruvate <1, 2> [1, 2]
S propionyl-CoA + oxaloacetate <1-5> (<2> two partial reactions [8,11-17,19]; <2> spontaneous decarboxylation follows [10]) (Reversibility: r <1-4> [1-4, 7-17, 19]) [1-4, 7-19]
P (S)-methylmalonyl-CoA + pyruvate <1-5> (<1> enzyme specific for the S isomer [1]) [1-4, 7-19]
S Additional information <1-3, 5> (<2> structural requirements of the C_1-donor: carbonyl group in β-position to carboxyl group being donated and adjacent either to another carboxyl group or to coenzyme A [2]; <1> 2-oxobutyrate, 2-oxovalerate, 2-oxoglutarate, 3-oxoglutarate cannot replace pyruvate and is no C_1-donor [1]; <5> enzyme catalyses the formation of the precursor of tylosin, protylonoide [18]) [1, 2, 18]
P ?

Inhibitors

(S)-methylmalonyl-CoA <1> [1]
2-oxobutyrate <1> [1]
3-methyloxaloacetate <1> (<1> strong [1]) [1]
SH-reagents <1> (<1> weak [1]) [1]
avidin <1, 2> (<1> strong [1]; <2> biotin restores [2]) [1, 2]
coenzyme A <1> [1]
guanidinium chloride <2> (<2> Co^{2+} protects [3,5]) [3, 5]
oxalate <1> [1]
oxaloacetate <1> [1]
propionyl pantetheine <1> [1]
propionyl-CoA <1> [1]

pyruvate <1> [1]
Additional information <1> (<1> no inhibition by metal chelating agents, such as EDTA [1]; <2> loss of assembly-promoting factor activity after digestion with trypsin, V8 protease, pronase [17]) [1, 17]

Cofactors/prosthetic groups
biotin <1, 2> (<1,2> requirement [1-6,8,10,12-16]; <1> biotinyl-enzyme, 6 mol per 18S and 4 mol per 16S enzyme species [1]; <2> covalently linked to the 1.3 subunit to the ε-amino group of Lys-89, which lies in the conserved sequence Ala-Met-Lys-Met [7,10,14]; <2> biotin containing subunit of enzyme acts as a carboxyl carrier between the CoA ester sites on the central 12S subunit of enzyme and keto acid sites on outer 5S subunit of enzyme and links the 12S and 5S subunits together to form a 26S multisubunit enzyme complex, biotin content: 28% [13]) [1-6, 7, 8, 10, 12-16]
Additional information <1, 2> (<1,2> no ATP requirement [1,2]) [1, 2]

Metals, ions
Co^{2+} <1-3> (<1-3> requirement [1-8]; <2> necessary for 5S activity [16]; <2> 3 types of Co^{2+}: tightly i.e. catalytic, not so tightly, and weakly bound, above 2 mM, stabilizes quarternary structure, no exchange between catalytic and stabilizing Co^{2+} [5]; <2> no formation of Co-thiol bonds [5]) [1-8, 16]
Zn^{2+} <1, 2> (<1> requirement, can replace Co^{2+} [1]; <2> slightly stabilizing [5]; <2> necessary for 5S activity [16]) [1, 5, 16]
Additional information <1> (<1> no additional metal requirement [1]) [1]

Turnover number (min^{-1})
Additional information <2> (<2> 2530 nmol oxaloacetate per nmol biotin per min [7]) [7]

Specific activity (U/mg)
0.075 <5> (<5> carbon source: glucose [18]) [18]
0.077 <2> [3]
0.15 <5> (<5> carbon source: starch [18]) [18]
0.19 <5> (<5> carbon source: rapeseed oil [18]) [18]
2.3 <2> (<2> recombinant 5S subunit, overall reaction [16]) [16, 17]
17 <2> [8]
33.6 <2> [5]
40 <1> [1]
43 <2> (<2> recombinant 5S subunit + assembly-promotin factor, overall reaction [17]) [17]
48 <2> (<2> 5S subunit, overall reaction [16]) [16, 17]
323 <2> (<2> 5S subunit, partial reaction 2 [16]) [16]
330 <2> (<2> recombinant 5S subunit, partial reaction 2 [16]) [16]
Additional information <2, 5> (<2> comparison of wild-type and mutant enzymes [7]; <2> comparison of wild-type and truncated forms [15]; <2> the recombinant 5S subunit is active in partial reaction 2 but not in overall enzyme reaction and not able to form a enzyme complex [16]; <5> specific activity depends on the carbon source and increases during culture time [18]) [7, 15, 16, 17, 18]

K_m-Value (mM)

0.0044 <2> (pyruvate, <2> 5S-subunit [7]) [7]
0.0077 <1, 2> (S-methylmalonyl-CoA) [1, 7]
0.034 <1> (propionyl-CoA) [1]
0.035 <1> (malonyl-CoA) [1]
0.063 <1> (oxaloacetate) [1]
0.25 <1> (butyryl-CoA) [1]
0.5 <1> (acetyl-CoA) [1]
0.77 <1> (pyruvate) [1]
Additional information <1, 2> (<1> kinetics and mechanism [1]; <2> kinetic parameters of wild-type and mutant enzyme [7]; <2> kinetic deuterium and ^{13}C-isotope effects [8]) [1, 7, 8]

K_i-Value (mM)

0.000018 <1> (oxalate) [1]
0.00004 <1> ((S)-methylmalonyl-CoA) [1]
0.00049 <1> (propionyl-CoA) [1]
0.0007 <1> (oxaloacetate) [1]
0.003 <1> (3-methyloxaloacetate) [1]
0.003-0.007 <1> (propionyl pantetheine) [1]
0.0063 <1> (coenzyme A) [1]
0.0089 <1> (pyruvate) [1]
0.025 <1> (2-oxobutyrate) [1]

pH-Optimum

5.5-7.8 <1> (<1> broad [1]) [1]

pH-Range

7-7.5 <5> (<5> assay at [18]) [18]

Temperature optimum (°C)

25 <2> (<2> assay at [5,7]) [5, 7]
28 <1> (<1> assay at [6]) [6]
30 <2> (<2> assay at [2,8]) [2, 8]

4 Enzyme Structure

Molecular weight

Additional information <1, 2> (<1> two active forms of enzyme: 18S and 16S, the 16S species arises from 18S species by loss of a 6S biotin, Co, Zn subunit [1]; <2> two active forms of enzyme: 18S and 26S [4]) [1, 4]

Subunits

Additional information <1, 2> (<1,2> structure [1,11]; <2> the 26S holoenzyme consists of a hexameric central 12S subunit, 360000 Da and six peripheral dimeric 5S subunits, 6 * 120000 Da each of which is linked to the central subunit by 2 biotinyl 1.3S subunits, 12 * MW 12000 [3,4,7,13-17]; <2> amino acid sequence of biotinyl subunit [4,7,13]; <2> MW of 12S hexamer:

338000Da [11]; <2> 5S subunit is a dimer of identical monomers, contains Co^{2+} and Zn^{2+} and assembles with two 1.3 subunits to form a stable complex termed the 6S subunit [12,16]; <2> 5S subunit with a native MW 120000 Da, gel filtration and a subunit MW 60000 Da, SDS-PAGE, [16]; <2> 1.3 S is homogen, SDS-PAGE [13]; <2> existence of a novel, previously unidentified subunit, a assemply-promoting factor, APF, which is nessessary for the assemply of enzyme, MW: 60000 Da, SDS-PAGE [17]) [1, 3, 4, 7, 11-17]

Posttranslational modification

side-chain modification <2> (<2> a series of truncated forms of the 1.3S subunit is prepared and the effect on the first and the second partial reaction and the overall enzyme reaction measured, result: the sequence 59-78 of 1.3S subunit is necessary for the carboxylation of biotin of the 1.3 S subunit, the inability of the truncated peptides to form enzyme complexes explains the activity loss in the overall reaction [12]; <2> 97 amino acids deleted of the carboxyl-terminal region of the monomer of the 12S subunit, 46% loss of activity because of a reduces stability of the 12S subunit and a decrease in binding of 6S subunit to the 12 subunit [15]; deletion of residues 10 to 48 of 1.3S subunit, not necessary for catalysis of the partial reaction [19]) [12, 15, 19]

5 Isolation/Preparation/Mutation/Application

Source/tissue

skeletal muscle <3> (<3> not heart muscle [2]) [2]
Additional information <3> (<3> not ox liver [2]) [2]

Localization

cytoplasm <1-3> [1, 2]
membrane <4> [9]

Purification

<1> [1]
<2> (partially [2]; separation of transcarboxylase complexes from uncombined 12S, 5S and 1.3S subunits by gel filtration [7]; recombinant enzyme and mutant 1.3S subunit [7,10,12]; recombinant enzyme and mutant 1.3S subunit, affinity chromatography on avidin(monomeric)-agarose, copurification of apo and biotinylated 1.3S forms [13]; recombinant 5S subunit [16]; a deletion mutant and recombinant enzyme [19]) [2-4, 7, 10, 12, 13, 16, 19]
<2, 4> (affinity chromatography on avidin-Sepharose [3,9]) [3, 9]

Crystallization

<2> (Raman spectroscopy [11]) [11, 17]

Cloning

<2> (1.3S subunit cloned and expressed in Escherichia coli [7,10,12-14,17]; 12S subunit cloned and expressed in Escherichia coli [15]; 5S subunit cloned and expressed in Escherichia coli [16,17]; a prokaryotic expression system for

the production of a ^{13}C- and ^{15}N-labeled 1.3S subunit mutant that is suitable for structure determination and functional studies by NMR [19]) [7, 10, 12-17, 19]

Engineering

A87G <2> (<2> K_m not significantly changed, significantly reduced k_{cat} [7]) [7]
M88A <2> (<2> K_m not significantly changed, significantly reduced k_{cat} [7]) [7]
M88C <2> (<2> K_m not significantly changed, significantly reduced k_{cat} [7]) [7]
M88L <2> (<2> K_m not significantly changed, significantly reduced k_{cat} [7]; <2> partial reaction 1: 99% loss activity, partial reaction 2: 65% loss of activity, possibly alterations in the microenvironment of the biocytin [14]) [7, 14]
M88T <2> (<2> K_m not significantly changed, significantly reduced k_{cat} [7]) [7]
M90L <2> (<2> K_m and k_{cat} not significantly changed [7]; <2> partial reaction 1: 50% loss of activity, partial reaction 2: 115% activity compared to the 1.3 wild-type enzyme [14]) [7, 14]
Additional information <2> (<2> double mutant A87M and M88A [7]) [7]

Application

medicine <5> (<5> industrial antibiotic production [18]) [18]

6 Stability

pH-Stability

5.5 <2> (<2> 26S-form stable [4,5]) [4, 5]
6.5 <2> (<2> stable, at pH 8 enzyme dissociates [3]) [3]
6.8 <2> (<2> 18S form stable [4]) [4]
8 <2> (<2> dilute solution, 50 mM buffer, 25°C: enzyme dissociates into subunits and rapidly loses activity, Co^{2+} protects, Zn^{2+} protects slightly [5]) [5]
9 <2> (<2> enzyme dissociates into 12SH, 5SE, 6SE and 6SH subunits, in the presence of Co^{2+}: the 6SE subunits binds to 12SH subunit [5]) [5]
Additional information <2> (<2> effect of pH on the overall conformation of the 1.3 subunit, below 3.5 and above 9.0: the N-terminus of the 1.3S protein undergoes a transition into a random coil or unordered conformation [10]) [10]

Temperature stability

25 <2> (<2> dissociation and rapid loss of activity in dilute solution in alkaline 50 mM buffer ($t_{1/2}$: 12 min), Co^{2+} protects, Zn^{2+} protects slightly ($t_{1/2}$: 22 min) [5]) [5]
50 <2> (<2> denaturation within 1 min, Co^{2+} protects, Zn^{2+} protects slightly [5]) [5]
Additional information <2> (<2> assembly-promoting factor heat stable [17]) [17]

Oxidation stability

<2>, oxygen prevents reassembly of dissociated subunits [5]

General stability information

<1>, glycerol, 20% retards dissociation of 12S and 6S-biotinyl subunit [1]
<1>, polyvalent anions stabilize [1]
<2>, Co^{2+} protects [5]
<1, 2>, alkaline pH, low ionic strength, monovalent ions, low protein concentration and elevated temperatures favor dissociation to inactive 5S, 6S and 1.3S subunits [1, 3]

Storage stability

<2>, -12°C, partially purified at least 1 month [2]

References

[1] Wood, H.G.: Transcarboxylase. The Enzymes, 3rd Ed. (Boyer, P.D., ed.), **6**, 83-115 (1972)
[2] Swick, R.W.; Wood, H.G.: The role of transcarboxylase in propionic acid ermentation. Proc. Natl. Acad. Sci. USA, **46**, 28-41 (1960)
[3] Berger, M.; Wood, H.G.: Purification of the subunits of transcarboxylase by affinity chromatography on avidin-sepharose. J. Biol. Chem., **250**, 927-933 (1975)
[4] Maloy, W.L.; Bowien, B.U.; Zwolinski, G.K.; Kumar, K.G.; Wood, H.G.; Ericsson, L.H.; Walsh, K.A.: Amino acid sequence of the biotinyl subunit from transcarboxylase. J. Biol. Chem., **254**, 11615-11622 (1979)
[5] Harmon, F.R.; Goss, N.H.; Wood, H.G.: Stabilization of the quaternary structure of transcarboxylase by cobalt (II) ions. Biochemistry, **21**, 2847-2852 (1982)
[6] Stubbe, J.; Fish, S.; Abeles, R.H.: Are carboxylations involving biotin concerted or nonconcerted?. J. Biol. Chem., **255**, 236-242 (1980)
[7] Shenoy, B.C.; Xie, Y.; Park, V.L.; Kumar, G.K.; Beegen, H.; Wood, H.G.; Samols, D.: The importance of methionine residues for the catalysis of the biotin enzyme, transcarboxylase. Analysis by site-directed mutagenesis. J. Biol. Chem., **267**, 18407-18412 (1992)
[8] O'Keefe, S.J.; Knowles, J.R.: Biotin-dependent carboxylation catalyzed by transcarboxylase is a stepwise process [published erratum appears in Biochemistry 1987 Sep 8;26(18):5952]. Biochemistry, **25**, 6077-6084 (1986)
[9] Hoffmann, A.; Hilpert, W.; Dimroth, P.: The carboxyltransferase activity of the sodium-ion-translocating methylmalonyl-CoA decarboxylase of Veillonella alcalescens. Eur. J. Biochem., **179**, 645-650 (1989)
[10] Rivera-Hainaj, R.E.; Pusztai-Carey, M.; Reddy, D.V.; Choowongkomon, K.; Soennichsen, F.D.; Carey, P.R.: Characterization of the carboxylate delivery module of transcarboxylase: following spontaneous decarboxylation of the 1.3S-CO_2 subunit by NMR and FTIR spectroscopies. Biochemistry, **41**, 2191-2197 (2002)

[11] Zheng, X.; Rivera-Hainaj, R.E.; Zheng, Y.; Pusztai-Carey, M.; Hall, P.R.; Yee, V.C.; Carey, P.R.: Substrate binding induces a cooperative conformational change in the 12S subunit of transcarboxylase: raman crystallographic evidence. Biochemistry, **41**, 10741-10746 (2002)
[12] Shenoy, B.C.; Kumar, G.K.; Samols, D.: Dissection of the biotinyl subunit of transcarboxylase into regions essential for activity and assembly. J. Biol. Chem., **268**, 2232-2238 (1993)
[13] Shenoy, B.C.; Magner, W.J.; Kumar, G.K.; Phillips, N.F.B.; Haase, F.C.; Samols, D.: The nonbiotinylated form of the 1.3 S subunit of transcarboxylase binds to avidin (monomeric)-agarose: Purification and separation from the biotinylated 1.3 S subunit. Protein Expr. Purif., **4**, 85-94 (1993)
[14] Shenoy, B.C.; Samols, D.; Kumar, G.K.: The conserved methionines of the 1.3 S biotinyl subunit of transcarboxylase: effect of mutations on conformation and activity. Arch. Biochem. Biophys., **304**, 359-366 (1993)
[15] Woo, S.B.; Shenoy, B.C.; Wood, H.G.; Magner, W.J.; Kumar, G.K.; Beegen, H.; Samols, D.: Effect of deletion from the carboxyl terminus of the 12 S subunit on activity of transcarboxylase. J. Biol. Chem., **268**, 16413-16419 (1993)
[16] Xie, Y.; Shenoy, B.C.; Magner, W.J.; Hejlik, D.P.; Samols, D.: Purification and characterization of the recombinant 5 S subunit of transcarboxylase from Escherichia coli. Protein Expr. Purif., **4**, 456-464 (1993)
[17] Shenoy, B.C.; Xie, Y.; Sha, D.; Samols, D.: Identification and characterization of a factor which is essential for assembly of transcarboxylase. Biochemistry, **32**, 10750-10756 (1993)
[18] Choi, D.B.; Park, Y.; Okabe, M.: Effects of rapeseed oil on activity of methylmalonyl-CoA carboxyltransferase in culture of Streptomyces fradiae. Biosci. Biotechnol. Biochem., **62**, 902-906 (1998)
[19] Jank, M.M.; Bokorny, S.; Rohm, K.H.; Berger, S.: Expression and biotinylation of a mutant of the transcarboxylase carrier protein from Propioni shermanii. Protein Expr. Purif., **17**, 123-127 (1999)

Aspartate carbamoyltransferase 2.1.3.2

1 Nomenclature

EC number
2.1.3.2

Systematic name
carbamoyl-phosphate:L-aspartate carbamoyltransferase

Recommended name
aspartate carbamoyltransferase

Synonyms
(S)-2-methyl-3-oxopropanoyl-CoA:pyruvate carboxyltransferase
ATC domain of CAD (in mammalia aspartate transcarbamylase activity is carried by a large multifunctional protein that catalyzes the first 3 steps of the de novo pyrimidine biosynthetic pathway, the same molecule also carries glutamine-dependent carbamoyl phosphate synthetase, EC 6.3.5.5, dihydroorotase, EC 3.5.2.3 and regulatory binding sites binding 5-phosphoribosyldiphosphate and UTP)
ATCase
L-aspartate transcarbamoylase
L-aspartate transcarbamylase
aspartate carbamyltransferase
aspartate transcarbamoylase
aspartate transcarbamylase
aspartic acid transcarbamoylase
aspartic carbamyltransferase
aspartic transcarbamylase
carbamoylaspartotranskinase
carbamoyltransferase, aspartate
carbamylaspartotranskinase

CAS registry number
9012-49-1

2 Source Organism

<1> *Oryctolagus cuniculus* [3]
<2> *Mus musculus* [3, 11]
<3> *Mytilus edulis* (mussel [20]) [20]
<4> *Escherichia coli* [1, 3-12, 19, 36, 37, 38, 40, 43, 47, 48, 49, 50, 51]

<5> *Rattus norvegicus* [2, 3, 33]
<6> *pigeon* [2]
<7> *Neurospora crassa* [3]
<8> *Saccharomyces cerevisiae* [3, 46]
<9> *lettuce* [3]
<10> *Phaseolus aureus* (mung bean [3]) [3, 14, 21, 22, 28, 29]
<11> *Triticum vulgare* (wheat [3]) [3, 13, 16, 17, 39, 44]
<12> *Salmonella typhimurium* [3]
<13> *Pseudomonas vulgaris* [3]
<14> *Pseudomonas aeruginosa* [3, 52]
<15> *Pseudomonas fluorescens* [3, 25, 31]
<16> *Azotobacter vinelandii* [3]
<17> *Serratia marcescens* [3, 43]
<18> *Aerobacter aerogenes* [3]
<19> *Bacillus subtilis* [3, 15, 23, 38]
<20> *Citrobacter freundii* [3]
<21> *Streptococcus faecalis* [3, 24, 26, 27]
<22> *Halobacterium cutirubrum* [3]
<23> *Lathyrus sativus* [14]
<24> *Eleusine corocana* [14]
<25> *Trigonella foenum-graecum* [14]
<26> *Mesocricetus auratus* [11, 18, 33, 34]
<27> *Drosophila melanogaster* [30]
<28> *Mycobacterium smegmatis* [32]
<29> *Pyrococcus abyssi* [35, 53]
<30> *Helicobacter pylori* [41]
<31> *Vibrio sp.* (strain 2693 isolated from the deep atlantic [42]) [42]
<32> *Vibrio natriegens* [42]
<33> *Sulfolobus acidocaldarius* [45]
<34> *Methanococcus jannaschii* [48]

3 Reaction and Specificity

Catalyzed reaction

carbamoyl phosphate + L-aspartate = phosphate + N-carbamoyl-L-aspartate (<1,2,4,5,7-22> reaction mechanism [3,4,22-24,27,29]; <4> enzyme exhibits homotropic cooperativity for aspartate, is heterotropically activated by ATP and is heterotropically inhibited by CTP and UTP [4,38]; <4> reaction proceeds via a nucleophilic attack by the free amino group of L-aspartate on the carbon of carbamoylphosphate [9]; <10> kinetic data suggest an ordered bi bi mechanism [29])

Reaction type

carbamoyl group transfer

Natural substrates and products

S carbamoylphosphate + L-aspartate <1-34> (<4> catalyzes the formation of carbamoyl-L-aspartate, the first compound unique to the biosynthetic pathway for pyrimidine nucleotides [1,3]; <27> second enzyme of pyrimidine synthesis [30]) (Reversibility: ? <1-34> [1-53]) [1-53]

P phosphate + N-carbamoyl-L-aspartate <1-34> [1-53]

Substrates and products

S carbamoyl phosphate + L-aspartate <1-34> (<10> specific for L-aspartate [28]) (Reversibility: ? <1-34> [1-53]) [1-53]

P phosphate + N-carbamoyl-L-aspartate <1-34> [1-53]

Inhibitors

2-thioUMP <4> [3]
5'-UMP <10> (<10> 2'-UMP and 3'-UMP have no effect [28]) [28]
ATP <10, 14, 15> (<10> 1 mM, 40% inhibition [28]; <14> excess $MgCl_2$ abolishes inhibition [52]) [25, 28, 31, 52]
CDP <15, 30> (<30> 10 mM, 61% inhibition [41]) [25, 31, 41]
CTP <4, 12, 14, 15, 22, 29, 30, 31> (<4> synergistic inhibition by CTP and UTP [6,8]; <4> 800-1000 mM urea lower or eliminate CTP inhibition [3]; <4> competitive vs. carbamoyl phosphate [7]; <30> 10 mM, 79% inhibition [41]) [3-8, 25, 31, 35, 41, 42, 52]
Ca^{2+} <28> [32]
Cl^- <4> [19]
Cu^{2+} <3> (<3> strong inhibition [20]) [20]
GTP <4, 10, 14, 30> (<10> 1 mM, 20% inhibition [28]) [3, 25, 28, 31, 41, 52]
$HgCl_2$ <21> (<21> 0.001 mM, 50% inhibition [24]) [24]
$HgSO_4$ <28> [32]
ITP <4> (<4> weak inhibition [3]) [3]
L-malate <4> [19]
Mn^{2+} <28> [32]
N-(phosphonoacetyl)-L-aspartate <2, 4, 10, 19, 29, 30> (<10> competitive vs. carbamoyl phosphate, noncompetitive vs. aspartate [22]; <29> 0.011 mM, 50% inhibition, low concentrations activate [35]; <30> 0.0001 mM, 50% inhibition [41]; <29> 0.002 mM, 50% inhibition of catalytic subunit [53]) [3, 9, 11, 22, 23, 35, 41, 53]
N-diphosphoryl-L-aspartate <4> [9]
N-methyl phosphonoacetamide <4> [3]
Ni^{2+} <28> [32]
UDP <10> (<10> 1 mM, 45% inhibition [28]) [28]
UMP <10, 11, 23-25, 28> (<11> 0.1 mM, 25% inhibition, inhibition increases to 80% and 90% in the presence of 0.2 mM and 0.6 mM deoxycholate respectively [13]; <10> 0.8 mM, 85% inhibition [14]; <10> 1 mM, 75% inhibition [28]; <11> fatty acids with chains of C_8 or longer, dodecylsulfate and decylsulfonate potentiate inhibition [39]) [13, 14, 21, 28, 29, 32, 39]
UTP <4, 8, 10, 14, 15, 29> (<4> weak inhibition [3]; <4> synergistic inhibition by CTP and UTP, no inhibition unless CTP is present [6,8]; <10> 1mM, 40% inhibition [28]) [3, 5, 6, 8, 25, 28, 29, 31, 35, 52]

Zn^{2+} <28> [32]
acetate <4> [19]
acetyl phosphate <10, 30> (<30> 3 mM, 50% inhibition [41]) [22, 41]
aspartate <26> (<26> at high concentrations [18]) [18]
carbamoyl aspartate <10> (<10> noncompetitive vs. carbamoyl phosphate and aspartate [22]) [22, 29]
cytidine <4> [3]
diphosphate <4, 15, 21> [3, 27, 31]
diphosphate analogues <4> [9]
guanidine-HCl <10> (<10> 800 mM, almost complete inhibition [29]) [29]
inosine <4> [3]
iodoacetate <5, 28> [2, 32]
maleate <10, 21, 28, 30> (<30> 15 mM, 50% inhibition [41]) [27, 28, 32, 41]
mersalyl <21> (<21> 0.01 mM, 50% inhibition [24]) [24]
mesotartrate <4> [19]
nucleotides <4, 14, 15, 20> [3]
p-chloromercuribenzoate <5, 10> (<10> 6 mM, almost complete inhibition [29]) [2, 29]
p-hydroxymercuribenzoate <3, 28> (<3> strong inhibition [20]) [20, 32]
p-mercuribenzoate <21> (<21> 0.01 mM, 50% inhibition [24]) [24]
phenylglyoxal <31> [42]
phosphate <10, 15, 28> (<10> competitive vs. carbamoyl phosphate, noncompetitive vs. aspartate [22]) [22, 29, 31, 32]
phosphonoacetamide <4> [3]
phosphonoacetic acid arsenate <30, 5> (<30> 1 mM, 50% inhibition [41]) [41, 2]
succinate <4, 10, 28, 30, 31> (<10> activator at low concentrations of both succinate and aspartate, inhibitor at high succinate concentrations and at high aspartate concentrations [29]; <30> 10 mM, 50% inhibition [41]) [19, 22, 28, 29, 32, 41, 42]
urea <10> (<10> 2.5 M, almost complete inhibition [29]) [29]
Additional information <3, 29> (<3> not inhibited by CTP [20]; <29> not inhibited by phosphonoacetate, diphosphate or phosphate [35]; <29> 50% inhibition at 80 MPa [35]) [20, 35]

Cofactors/prosthetic groups

ATP <4> (<4> activates [3,5]; <3> no effect [20]) [3, 5]
dATP <4> (<4> stimulates [3]) [3]

Activating compounds

4-aminobutyrate <21> (<21> 400% activation [24]) [24, 26]
ATP <4, 29, 33> (<33> holoenzyme, catalytic subunits alone are inhibited [45]) [4, 35, 45]
CTP <17, 33> (<33> holoenzyme, catalytic subunits alone are inhibited [45]) [43, 45]
D-aspartate <21> (<21> 5700% activation [24]) [24]
GTP <33> (<33> holoenzyme, catalytic subunits alone are inhibited [45]) [45]

L-2-aminobutyrate <21> (<21> 1400% activation [24]) [24, 26]
L-alanine <21> (<21> 800% activation [24]) [24]
L-malate <4> (<4> activation in presence of excess carbamoyl phosphate and limiting L-aspartate [19]) [19]
N-(phosphonoacetyl)-L-aspartate <14, 29> (<29> at concentrations below 0.001 mM [35]; <14> at concentrations below 0.000002 mM, inhibition above [52]) [35, 52]
NH_4Cl <21> (<21> 30 mM, 2710% activation [26]) [26]
NaCl <21> (<21> 30 mM, 1730% activation [26]) [26]
Tris-HCl <21> (<21> 30 mM, 2620% activation [26]) [26]
UTP <33> (<33> holoenzyme, catalytic subunits alone are inhibited [45]) [45]
acetate <19> (<19> 25 mM, 2.3fold activation [38]) [38]
adenosine <4> (<4> very weak activation [3]) [3]
aminoethanol <21> (<21> activates [26]) [26]
β-alanine <21> (<21> 600% activation [24]) [24]
butyrate <21> (<21> 5700% activation [24]) [24, 26]
carbamoyl-β-alanine <21> (<21> 5000% activation [24]) [24, 26]
dimethyl sulfoxide <3> (<3> strong stimulation [20]) [20]
dimethylformamide <3> (<3> strong stimulation [20]) [20]
formate <21> (<21> activates [24]) [24, 26]
hydroxylamine-HCl <21> (<21> 30 mM, 2120% activation [26]) [26]
imidazole-HCl <21> (<21> 30 mM, 2710% activation [26]) [26]
isobutyrate <21> (<21> 5800% activation [24]) [24, 26]
mesotartrate <4> (<4> activation in presence of excess carbamoyl phosphate and limiting L-aspartate [19]) [19]
propionate <21> (<21> 5600% activation [24]) [24, 26]
succinate <4, 29> (<4> activation in presence of excess carbamoyl phosphate and limiting L-aspartate [19]; <29> low concentrations, 35% activation in the presence of 2 mM aspartate [35]) [19, 35]
triethanolamine-HCl <21> (<21> 30 mM, 1950% activation [26]) [26]
Additional information <3, 19, 21> (<21> every type of anion tested, including ATP and CTP can stimulate the reaction to a certain extent [24]; <21> at high aspartate levels addition of an activator has no effect [26]; <3> not activated by ATP [20]; <19> activity is not regulated by nucleotide triphosphates [38]) [20, 24, 26, 38]

Metals, ions

Zn^{2+} <4> (<4> regulatory subunit contains 6 Zn^{2+} per 300000 Da protein, Zn^{2+} may be replaced by Hg^{2+} or Cd^{2+} [3]) [3, 47]
Additional information <28> (<28> no requirement for any metal ion [32]) [32]

Turnover number (min^{-1})

2153 <26> (carbamoyl phosphate, <26> aspartate transcarbamoylase activity domain of the multienzyme complex [18]) [18]
2830 <11> (carbamoyl phosphate) [44]
3446 <26> (carbamoyl phosphate, <26> multienzyme complex [18]) [18]

4780 <26> (aspartate, <26> aspartate transcarbamoylase activity domain of the multienzyme complex [18]) [18]
9378 <26> (L-aspartate, <26> multienzyme complex [18]) [18]
25000 <4> (L-aspartate) [1]

Specific activity (U/mg)
0.2 <10> [14]
0.35 <10> [29]
0.44 <25> [14]
0.68 <27> [30]
0.7 <23> [14]
0.8 <24> [14]
2.7 <10> [28]
3.16 <15> [31]
10.21 <11> [44]
14 <11> [17]
29.95 <33> (<33> at 55°C [45]) [45]
60.7 <11> [16]
116.7 <4> [1]
148.3 <19> (<19> recombinant enzyme [38]) [38]
238.3 <4> (<4> recombinant enzyme, overexpressed in E. coli [10]) [10]
333 <19> (<19> recombinant enzyme, in the presence of 50 mM Tris-acetate [38]) [38]
350 <21> [24, 26]
450 <14> (<14> recombinant enzyme [52]) [52]
560 <19> [23]
7900 <28> [32]
Additional information <4> (<4> specific activity of the catalytic subunit alone is about 50% higher than that of the holoenzyme [5]) [5]

K_m-Value (mM)
0.004 <26> (carbamoyl phosphate, <26> in crude extracts of fibroblasts [11]) [11]
0.0077 <21> (carbamoyl phosphate) [27]
0.009 <26> (L-aspartate, <26> multienzyme complex [18]) [18]
0.014 <15> (carbamoyl phosphate) [31]
0.02 <26> (carbamoyl phosphate, <26> multienzyme complex and aspartate transcarbamoylase activity domain of the multienzyme complex alone [18]) [18]
0.0207 <26> (carbamoyl phosphate) [18]
0.021 <26> (L-aspartate, <26> aspartate transcarbamoylase activity domain of the multienzyme complex alone [18]) [18]
0.02134 <26> (L-aspartate) [18]
0.023 <11> (carbamoyl phosphate, <11> in the absence of UMP [44]) [44]
0.038 <29> (carbamoyl phosphate, <29> recombinant catalytic subunit, at 37°C [53]) [53]
0.091 <10> (carbamoyl phosphate) [29]
0.11 <19> (carbamoyl phosphate) [23]

0.11 <19> (carbamoyl phosphate) [38]
0.16 <11> (L-aspartate) [44]
0.2 <21> (carbamoyl phosphate, <21> at high salt concentrations [24]) [24]
0.2 <33> (carbamoyl phosphate, <33> holoenzyme [45]) [45]
0.3 <31> (carbamoyl phosphate) [42]
0.49 <14> (carbamoyl phosphate) [52]
0.6 <33> (carbamoyl phosphate, <33> catalytic subunits [45]) [45]
0.61 <30> (carbamoyl phosphate, <30> in cell extracts [41]) [41]
1.7 <10> (L-aspartate) [29]
2.3 <4> (L-aspartate, <4> C47A/A241C mutant enzyme [51]) [51]
2.6 <14> (L-aspartate) [52]
2.75 <15> (L-aspartate) [31]
7 <19> (L-aspartate) [23]
7 <19> (aspartate) [38]
7 <3> (carbamoyl phosphate) [20]
7.9 <4> (L-aspartate, <4> complex of catalytic and regulatory subunits, C3R6 [7]) [7]
10 <21> (L-aspartate) [27]
10.9 <34> (L-aspartate, <34> at 55°C [48]) [48]
11.6 <30> (L-aspartate, <30> in cell extracts [41]) [41]
12.3 <29> (L-aspartate, <29> recombinant catalytic subunit, at 30°C [53]) [53]
14.1 <29> (L-aspartate, <29> recombinant catalytic subunit, at 45°C [53]) [53]
15.5 <29> (L-aspartate, <29> recombinant catalytic subunit, at 37°C [53]) [53]
18 <3> (L-aspartate) [20]
18.2 <34> (L-aspartate, <34> at 37°C [48]) [48]
19.7 <29> (L-aspartate, <29> recombinant catalytic subunit, at 55°C [53]) [53]
34.6 <34> (L-aspartate, <34> at 25°C [48]) [48]
40.4 <34> (L-aspartate, <34> at 15°C [48]) [48]
44.7 <4> (L-aspartate, <4> complex of catalytic subunits, C3 [7]) [7]
88 <34> (carbamoyl phosphate, <34> at 25°C [48]) [48]
162 <34> (carbamoyl phosphate, <34> at 37°C [48]) [48]
Additional information <4, 19, 29, 33> (<29> sigmoidal saturation curve for carbamoylphosphate and aspartate [35]; <19> enzyme exhibits Michaelis-Menten kinetics for both of its substrates [38]; <33> sigmoidal saturation curve for aspartate [45]; <4> sigmoidal saturation curve for aspartate [47]) [35, 38, 45, 47]

K_i-Value (mM)

0.000001 <26> (N-(phosphonacetyl)-L-aspartate) [11]
0.000006 <19> (N-(phosphonoacetyl)-L-aspartate) [23]
0.00001 <26> (4,5-dicarboxy-2-ketopentylphosphonate) [11]
0.000027 <4> (N-(phosphonacetyl)-L-aspartate) [3, 9]
0.00024 <4> (N-pyrophosphoryl-L-aspartate) [9]

0.0004 <15> (diphosphate, <15> when carbamoyl phosphate is limiting [25]) [25, 31]
0.00049 <11> (UMP) [44]
0.002 <4> (O-phosphonoacetyl-oxosuccinate) [9]
0.0035 <4> (carbonyldiphosphonate) [9]
0.0055 <4> (S-phosphonoacetyl-mercaptosuccinate) [9]
0.007 <15> (CTP, <15> when carbamoyl phosphate is limiting [25]) [25, 31]
0.0085 <15> (ATP, <15> when carbamoyl phosphate is limiting [25]) [25, 31]
0.0085 <15> (UTP, <15> when carbamoyl phosphate is limiting [25]) [25, 31]
0.017 <4> (S-diphosphoryl-mercaptosuccinate) [9]
0.02 <4> (hypophosphate) [9]
0.022 <4> (phosphonoformate) [9]
0.032 <4> (dichlormethylenediphosphonate) [9]
0.037 <4> (CTP) [3]
0.037 <4> (methylenediphosphonate) [9]
0.05 <4> (N-phosphoryl-L-aspartate) [9]
0.051 <4> (diphosphate) [9]
0.071 <4> (imidodiphosphonate) [9]
0.09 <4> (diphosphate) [3]
0.1 <15> (CDP, <15> when carbamoyl phosphate is limiting [25]) [25, 31]
0.1 <15> (phosphate, <15> when carbamoyl phosphate is limiting [25]) [25, 31]
0.125 <4> (S-phosphoryl-mercaptosuccinate) [9]
0.15 <4> (phosphonoacetate) [9]
0.17 <4> (maleate, <4> inhibition of catalytic subunits, C3 [19]) [19]
0.175 <4> (peroxydiphosphonate) [9]
0.66 <4> (phosphonoacetamide) [3]
0.7 <4> (mesotartrate, <4> inhibition of catalytic subunits, C3 [19]) [19]
0.9 <21> (diphosphate, <21> competitive vs. carbamoyl phosphate [27]) [27]
1 <21> (maleate, <21> competitive vs. L-aspartate [27]) [27]
1.1 <4> (chloride, <4> inhibition of catalytic subunits, C3 [19]) [19]
1.5 <15> (ribose 5-phosphate, <15> when carbamoyl phosphate is limiting [25]) [25, 31]
1.54 <4> (N-methyl phosphonoacetamide) [3]
1.6 <10> (succinate) [14]
2 <4> (ethylamidocarbonylphosphonate) [9]
2.2 <4> (2,2-dimethylsuccinate, <4> inhibition of catalytic subunits, C3 [19]) [19]
3.5 <4> (succinate, <4> inhibition of catalytic subunits, C3 [19]) [19]
4.59 <26> (L-aspartate, <26> aspartate transcarbamoylase activity domain of the multienzyme complex [18]) [18]
5 <21> (phosphate, <21> competitive vs. carbamoyl phosphate [27]) [27]
5.3 <4> (fumarate, <4> inhibition of catalytic subunits, C3 [19]) [19]
7 <21> (carbamoyl aspartate, <21> competitive vs. carbamoyl phosphate [27]) [27]
7.5 <4> (phosphonopropinate) [9]

7.9 <4> (L-aspartate, <4> substrate inhibition, complex of catalytic and regulatory subunits, C3R6 [7]) [7]
13 <21> (carbamoylaspartate, <21> noncompetitive vs. carbamoyl phosphate [27]) [27]
13.56 <26> (L-aspartate, <26> multienzyme complex [18]) [18]
14 <21> (phosphate, <21> competitive vs. L-aspartate [27]) [27]
15 <21> (carbamoyl aspartate, <21> competitive vs. L-aspartate [27]) [27]
80 <4> (aspartate, <4> substrate inhibition, complex of catalytic subunits, C3 [7]) [7]

pH-Optimum
7.8 <28> [32]
8 <10, 30> (<10> maxima at pH 8.0 and pH 10.2 [28]; <30> sharp decline in activity below and above [41]) [28, 41]
8.5 <4, 14, 15, 19> (<4> maxima at pH 8.5 and pH 10.2 [3]) [3, 15, 23, 31, 52]
9 <29> [35]
9-10 <31> [42]
10 <4> [7]
10.2 <4, 10> (<4> maxima at pH 8.5 and pH 10.2 [3]; <10> maxima at pH 8.0 and pH 10.2 [28]) [3, 28]
10.5 <10> [29]

pH-Range
7-8.5 <14> (<14> steep increase in activity until pH 8.5, decrease above [52]) [52]
7.5-9.5 <15> [31]

Temperature range (°C)
2-50 <31> (<31> 20% of maximal activity at 2°C, 30% at 6°C [42]) [42]
30-75 <10> (<10> approx. 70% of maximal activity at 30°C, approx. 30% of maximal activity at 75°C [28]) [28]

4 Enzyme Structure

Molecular weight
32000 <34> (<34> regulatory subunit, gel filtration [48]) [48]
83000 <10> (<10> gel filtration [29]) [29]
93000 <10, 20> (<20> 250000 Da and 93000 Da isoforms [3]; <10> glycerol gradient [29]) [3]
100000 <13> (<13> 100000 Da and 300000 Da isoforms [3]) [3]
102000 <19> (<19> sedimentation velocity and sedimentation equilibrium analysis [15]) [15]
120000-140000 <21> [3]
125000 <21> (<21> sucrose density gradient centrifugation [24]) [24]
128000 <10, 21> (<10> gel filtration, sedimentation data [28]; <21> sedimentation equilibrium [24,26]) [14, 24, 26, 28]

129000 <34> (<34> catalytic subunit, gelfiltration [48]) [48]
220000 <4> (<4> analytical ultracentrifugation [1]) [1]
246000 <28> (<28> gel filtration [32]) [32]
250000 <20> (<20> 250000 Da and 93000 Da isoforms [3]) [3]
300000 <12-14, 16, 17, 32> (<12-14,16,17> 100000 Da and 300000 Da isoforms [3]) [3, 42]
310000 <4> (<4> sedimentation equilibrium [3]) [3]
320000 <31> (<31> gel filtration [42]) [42]
340000 <33> (<33> gel filtration [45]) [45]
360000 <15> (<15> sedimentation equilibrium [25]) [3, 25, 31]

Subunits

? <10, 34> (<10> x * 17000 + x * 34000, SDS-PAGE [14]; <34> x * 37000 + x * 17000, most probably a dodecamer, SDS-PAGE [48]) [14, 48]
dimer <15, 28> (<15> 2 * 180000, SDS-PAGE [25,31]; <28> 1 * 138000 + 1 * 108000, SDS-PAGE [32]) [25, 31, 32]
dodecamer <4, 14, 33> (<4> 6 * 33000 + 6 * 17000, 2C3/3R2 holoenzyme [38]; <33> 6 * 36500 + 6 * 18000, SDS-PAGE [45]; <14> 6 * 45000 + 6 * 36000, SDS-PAGE [52]) [38, 45, 52]
hexamer <10> (<10> 6 * 20600 [21]) [21]
tetramer <21> (<21> 4 * 32500, SDS-PAGE [24]) [24, 26]
trimer <19> (<19> 3 * 33500, SDS-PAGE, sedimemtation equilibrium analysis of protein dissolved in 6 M guanidine hydrochloride [15]) [15, 23]
Additional information <4, 17, 26> (<26> hamster aspartate carbamoyltransferase domain is an oligomer consisting of 2 or 3 identical copies of the 40000 Da proteolytic fragment of CAD, 2 or 3 * 40000, SDS-PAGE [18]; <4> 2 distinct kinds of separable subunits, a catalytically active C subunit of 33000 Da and a regulatory R subunit of 17000 Da, native structure: (C3)2(R2)3 i.e. a dimer of C subunits and a trimer of R subunits [3]; <4> model for homotrophic cooperativity [5]; <4> characterization of a complex between catalytical and regulatory subunit [7]; <4> crosslinking of subunits with tartryl diazide [12]; <17> dodecameric structure, 2C3:3R2 i.e. 2 trimers of the catalytic subunit and 3 dimers of the regulatory subunit [43]; <4> holoenzyme is a dodecamer composed of 2 catalytic trimers and 3 regulatory dimers, the catalytic chain is composed of 2 structural domains, the aspartate domain and the carbamoylphosphate domain which are involved in the binding of aspartate and carbamoylphosphate respectively, the regulatory chain is also composed of 2 domains, the allosteric and the zinc domains, which are involved in the binding of allosteric effectors and zinc [47]) [3, 5, 7, 12, 18, 33, 34, 43, 47]

5 Isolation/Preparation/Mutation/Application

Source/tissue

cell culture <2, 26, 27> (<27> 165-23 cell line [30]; <26> simian virus 40 transformed syrian hamster cell line [11,18]; <2> fibroblast cell line MA-134 [11]) [11, 18, 30, 34]
erythrocyte <1> [3]
germ <10, 11> [3, 13, 17]
hepatoma ascites cell <5> [33]
liver <5, 6, 26> [2, 33]
mantle <3> [20]
second instar larva <27> [30]
seedling <9, 10> [3, 14, 21, 22, 28]
spleen <1, 2, 5> [3, 11]

Localization

endoplasmic reticulum <8> [46]
mitochondrial membrane <8> [46]
nuclear envelope <8> [46]
plasmalemma <8> [46]

Purification

<4> (rapid procedure [10]; heat, ammonium sulfate, DEAE-cellulose [1]; recombinant His6-tagged enzyme [49]) [1, 3, 10, 47, 49]
<5> (multienzyme complex [33]) [33]
<10> (ammonium sulfate, DEAE-cellulose, DEAE-Sephadex, Sephadex G-200, partially purified [29]; ammonium sulfate, aspartate-Sepharose [14]; $MnSO_4$, ammonium sulfate, DEAE-cellulose, Sephadex G-100 [28]) [14, 21, 28, 29]
<11> ($CaCl_2$, heat, ethanol, 10-carboxydecylamino-Sepharose, DEAE-cellulose [16]; purification steps at room temperatur [17]; ethanol, 10-carboxydecyl-Sepharose, DEAE-cellulose [44]) [16, 17, 44]
<14> (recombinant enzyme, Ni^{2+}-affinity column [52]) [52]
<15> (Streptomycin sulfate, ammonium sulfate, Sephadex G-200, hydroxylapatite, preparative gel electrophoresis [25]) [25, 31]
<19> (protamine sulfate, heat treatment, ammonium sulfate, Sephadex G-150, DEAE-Sephadex, DEAE-cellulose [23]; Q-Sepharose, Matrex gel Red A, Matrex Phenyl Cellufine [38]) [15, 23, 38]
<21> (pH 4.8, hydroxylapatite, ammonium sulfate, DEAE-Sephadex [24]) [24, 26, 27]
<26> (multienzyme complex [33]; aspartate transcarbamoylase activity domain of the multienzyme complex [18]) [18, 33, 34]
<27> [30]
<28> (streptomycin sulfate, ammonium sulfate, DE-52 column, Sephadex G-200, aspartate linked CNBr [32]) [32]
<29> (ion-exchange, gel filtration [35]) [35]
<31> (native and recombinant enzyme [42]) [42]
<34> (heat, Q-Sepharose, Phenyl-Sepharose [48]) [48]

Renaturation

<15, 21> (dialysis after SDS-denaturation, 1% of original activity is restored [25,31]; <21> 4% of original activity is restored [26]) [25, 26, 31]

Crystallization

<4> (approx. 30% ammonium sulfate, crystals appear after about 2 weeks at 4°C [1]; X-ray structure at 5.5 A resolution [3]; X-ray structure of the holoenzyme in the presence of the substrate analog N-phosphonoacetyl-L-aspartate, carbamoylphosphate and succinate [5]; cocrystallization of catalytic subunit with N-(phosphonoacetyl)-L-aspartate by vapor diffusion, crystals grow from 0.005 ml of 100 mM Tris-HCl, pH 6.8, 20 mM calcium acetate, 5.8% polyethylene glycol 8000 and 0.005 ml of catalytic subunit in 10 mM Tris-HCl, pH 7.5, 1 mM 2-mercaptoethanol and 2 mM N-(phosphonoacetyl)-L-aspartate [49]) [1, 3, 5, 23, 49]

<14> (hanging drop method, equal volumes of 1 mg/ml enzyme in 100 mM Tris-HCl, pH 8.5 is mixed with 100 mM citrate, pH 5, and 10% 6K polyethylene glycol [52]) [52]

<19> (50 mM Tis-HCl, pH 8.1, 70% ammonium sulfate [23]) [23]

<29> (hanging-drop vapour-diffusion, 0.002 ml of enzyme solution, 7 mg/ml, in 20 mM Tris-HCl, pH 8.2, 2 mM 2-mercaptoethanol, 300 mM NaCl, is mixed with 0.002 ml reservoir solution consisting of 1.4 M citrate, pH 6.8, X-ray structure to 1.8 A resolution [53]) [53]

Cloning

<4> (overexpression in Escherichia coli [6]) [6, 10]

<14> (expression in Escherichia coli [52]) [52]

<19> (overexpression in Escherichia coli [38]) [38]

<29> (expression of catalytic subunit in Escherichia coli [53]) [53]

<31> (expression in Escherichia coli [42]) [42]

<33> (expression in Escherichia coli [45]) [45]

<34> (expression in Escherichia coli [45]) [48]

Engineering

A241C <4> (<4> reduced affinity for aspartate, hyperbolic aspartate saturation curve [51]) [51]

C47A <4> (<4> Hill coefficient 1.3 as compared to 2.4 for wild-type [51]) [51]

C47A/A241C <4> (<4> non-reducing conditions, reduced affinity for aspartate, hyperbolic aspartate saturation curve [51]) [51]

D162A <4> (<4> 7700fold reduction in specific activity, 2fold decrease in affinity for aspartate, loss of homotropic cooperativity and decreased activation by ATP [50]) [50]

D19A <4> (<4> loss of the synergistic inhibition of UTP in the presence of CTP [8]) [8]

E50A <4> (<4> mutant enzyme shows a low activity, low affinity state, only 2fold activation with N-(phosphonacetyl)-L-aspartate, kinetic mechanism is changed [37]) [37, 39]

L151Q <4> (<4> strongly reduced stimulation by ATP, synergistic inhibition by UTP is decreased [36]) [36]
L151V <4> (<4> stimulation by ATP is reduced by 50% [36]) [36]
L32A <4> (<4> stimulation by ATP is reduced by 25% [36]) [36]
L76A <4> (<4> synergistic inhibition by UTP is decreased [36]) [36]
P268A <4> (<4> 40fold reduction in activity, concentration of N-(phosphonacetyl)-L-aspartate for maximal activation is increased 233fold as compared to the wild-type, less activation by ATP, stronger inhibition by CTP [47]) [47]
Q73E <4> (<4> stimulation by ATP is reduced by 80% [36]) [36]
V106A <4> (<4> synergistic inhibition by UTP is decreased [36]) [36]
V106W <4> (<4> strongly reduced stimulation by ATP [36]) [36]
V106W/Y77F <4> (<4> strongly reduced stimulation by ATP [36]) [36]
Y165F <4> (<4> mutant enzyme shows greatly reduced affinity for aspartate and activity [40]) [40]
Y240F <4> (<4> mutant enzyme shows higher affinity for aspartate and increased activity [40]) [40]
Y77F <4> (<4> synergistic inhibition by UTP is decreased [36]) [36]
Additional information <17> (<17> chimeric enzyme consisting of E. coli catalytic subunit and Serratia marcescens regulatory subunit [43]) [43]

6 Stability

pH-Stability

6 <19> (<19> unstable below [23]) [23]
6.6 <21> (<21> stable [24]) [24, 26]

Temperature stability

0-25 <19> (<19> quite stable [23]) [23]
45 <28, 30> (<28> loss of activity after 10 min, carbamoyl phosphate protects [32]) [32, 41]
50 <10> (<10> 5 min, no loss of activity [28]) [28]
55 <10> (<10> 5 min, 20% loss of activity [28]) [28]
60 <10, 15, 31> (<10> 5 min, 50% loss of activity [28]; <15> loss of approx. 60% activity after 20°C [31]; <31> more than 90% inactivation after 15 min, half-life: 4 min [42]) [28, 31, 42]
75 <4, 34> (<34> catalytic subunit looses 25% activity after 60 min [48]; <4> catalytic subunit looses 50% activity after 1 min [48]) [48]
85 <33> (<33> no loss of activity after 15 min [45]) [45]
90 <29, 33> (<29> no loss of activity after 6 h [35]; <33> native and recombinant enzyme, loss of less than 10% activity after 40 min [45]; <29> catalytic subunit, half-life: 80 min, half-life of holenzyme: 240 min [53]) [35, 45, 53]
98 <29> (<29> half-life of the holoenzyme is 2.5 times higher than that of the catalytic subunit [53]) [53]
Additional information <19> (<19> carboxylic acids e.g. acetic acid, citric acid and high ionic strength, 200 mM KCl, protect against thermal denaturation at 65-70°C [23]) [23]

General stability information

<4>, chemical stabilization of conformational states of enzyme [12]
<4>, thermal denaturation [4]
<10>, phosphate required for maximum stability [29]
<10>, presence of 20 mM potassium phosphate as well as 10% v/v glycerol is required for maintenance of activity of the purified enzyme [29]
<10>, purified enzyme is inactivated if stored frozen [28]
<11>, 50% glycerol stabilizes [16]
<11>, cold-labile when highly purified, must be stored at room temperature, complete loss of activity after 24 h at pH 7.0 and 0°C [17]
<19>, carbamoyl phosphate protects against proteolysis [23]
<19>, carboxylic acids, e.g. acetic acid, citric acid and high ionic strength, 200 mM KCl, protect against thermal denaturation at 65-70°C [23]
<19>, unstable in very dilute solutions [23]
<26>, dimethylsulfoxid and glycerol are required to stabilize the enzyme [18]
<28>, carbamoyl phosphate protects against thermal denaturation [32]
<28>, freezing and thawing in 50 mM Tris-HCl buffer, pH 7.4, 1 mM 2-mercaptoethanol has no effect on stability [32]

Storage stability

<4>, 4°C, mercaptoethanol-imidazole buffer, several months [1]
<10>, 0-4°C, 1-2 days [28]
<11>, 4°C, 50 mM Tris-HCl, pH 7.5, 25% glycerol, 3 mM sodium azide, indefinitely, no loss of activity [44]
<19>, -10°C -20°C, many months, no loss of activity [23]
<28>, -20°C, 2 mM carbamoyl phosphate, 1 mM 2-mercaptoethanol, at least a week, no loss of activity [32]
<21, 21>, -20°C, 0.1-1 mg protein/ml, at least 2 years, no loss of activity 4°C, 0.1-1 mg/ml, at least 2 months, no loss of activity [24, 26, 24, 26]

References

[1] Sheperdson, M.; Pardee, A.B.: Production and crystallization of aspartate transcarbamylase. J. Biol. Chem., **235**, 3233-3237 (1960)
[2] Lowenstein, J.M.; Cohen, P.P.: Studies on the biosynthesis of carbamylaspartic acid. J. Biol. Chem., **220**, 57-70 (1956)
[3] Jacobson, G.R.; Stark, G.R.: Aspartate transcarbamylases. The Enzymes, 3rd Ed. (Boyer, P.D., ed.), **9**, 225-308 (1973)
[4] Allewell, N.M.: Escherichia coli aspartate transcarbamoylase: structure, energetics, and catalytic and regulatory mechanisms. Annu. Rev. Biophys. Biophys. Chem., **18**, 71-92 (1989)
[5] Kantrowitz, E.R.; Lipscomb, W.N.: Escherichia coli aspartate transcarbamoylase: the molecular basis for a concerted allosteric transition. Trends Biochem. Sci., **15**, 53-59 (1990)

[6] England, P.; Herve, G.: Synergistic inhibition of Escherichia coli aspartate transcarbamylase by CTP and UTP: binding studies using continuous-flow dialysis. Biochemistry, **31**, 9725-9732 (1992)

[7] Mort, J.S.; Chan, W.W.C.: Subunit interactions in aspartate transcarbamylase. Characterization of a complex between the catalytic and the regulatory subunits. J. Biol. Chem., **250**, 653-660 (1975)

[8] Zhang, Y.; Kantrowitz, E.R.: The synergistic inhibition of Escherichia coli aspartate carbamoyltransferase by UTP in the presence of CTP is due to the binding of UTP to the low affinity CTP sites. J. Biol. Chem., **266**, 22154-22158 (1991)

[9] Laing, N.; Chan, W.W.C.; Hutchinson, D.W.; Íberg, B.: Phosphorus-containing inhibitors of aspartate transcarbamoylase from Escherichia coli. FEBS Lett., **260**, 206-208 (1990)

[10] Nowlan, S.C.; Kantrowitz, E.R.: Superproduction and rapid purification of Escherichia coli aspartate transcarbamylase and its catalytic subunit under extreme derepression of the pyrimidine pathway. J. Biol. Chem., **260**, 14712-14716 (1985)

[11] Swyryd, E.A.; Seaver, S.S.; Stark, G.R.: N-(phosphonacetyl)-L-aspartate, a potent transition state analog inhibitor of aspartate transcarbamylase, blocks proliferation of mammalian cells in culture. J. Biol. Chem., **249**, 6945-6950 (1974)

[12] Enns, C.A.; Chan, W.W.C.: Chemical stabilization of conformational states of aspartate transcarbamoylase. Methods Enzymol., **135**, 569-577 (1987)

[13] Yon, R.J.: Wheat-germ aspartate transcarbamoylase: potentiation of end-product inhibition in vitro by deoxycholate. Biochem. Soc. Trans., **1**, 676-678 (1973)

[14] Rao, G.S.J.; Savithri, H.S.; Seethalakshmi, S.; Rao, N.A.: Plant aspartate transcarbamylase: an affinity chromatographic method for the purification of the enzyme from germinated seedings. Anal. Biochem., **95**, 401-405 (1979)

[15] Brabson, J.S.; Switzer, R.L.: Purification and properties of Bacillus subtilis aspartate transcarbamylase. J. Biol. Chem., **250**, 8664-8669 (1975)

[16] Yon, R.J.: Versatility of mixed-function adsorbents in biospecific protein desorption: accidental affinity and an improved purification of aspartate transcarbamoylase from wheat germ. Anal. Biochem., **113**, 219-228 (1981)

[17] Grayson, J.E.; Yon, R.J.; Butterworth, P.J.: Wheat-germ aspartate transcarbamoylase. Purification and cold-lability. Biochem. J., **183**, 239-245 (1979)

[18] Grayson, D.R.; Evans, D.R.: The isolation and characterization of the aspartate transcarbamylase domain of the multifunctional protein, CAD. J. Biol. Chem., **258**, 4123-4129 (1983)

[19] Jacobson, G.R.; Stark, G.R.: Aspartate transcarbamylase of Escherichia coli. Mechanisms of inhibition and activation by dicarboxylic acids and other anions. J. Biol. Chem., **250**, 6852-6860 (1975)

[20] Mathieu, M.: Partial characterization of aspartate transcarbamylase from the mantle of the mussel Mytilus edulis. Comp. Biochem. Physiol. B Comp. Biochem., **82**, 667-674 (1985)

[21] Prasad, P.V.; Rao, N.A.: Purification and regulation of aspartate transcarbamylase from germinated mung bean (Vigna radiata) seedlings. J. Biosci., **6**, 233-248 (1984)
[22] Savithri, H.S.; Vaidyanathan, C.S.; Rao, N.A.: Plant aspartate transcarbamylase: kinetic properies of the enzyme from mung bean (Phaseolus aureus) seedlings. Proc. Indian Acad. Sci. Sect. B, **87B**, 81-94 (1978)
[23] Brabson, J.S.; Maurizi, M.R.; Switzer, R.L.: Aspartate transcarbamylase from Bacillus subtilis. Methods Enzymol., **113**, 627-635 (1985)
[24] Chang, T.Y.; Prescott, L.M.; Jones, M.E.: Aspartate carbamyltransferase (Streptococcus faecalis). Methods Enzymol., **51**, 41-50 (1978)
[25] Adair, L.B.; Jones, M.E.: Aspartate carbamyltransferase (Pseudomonas fluorescens). Methods Enzymol., **51**, 51-58 (1978)
[26] Chang, T.Y.; Jones, M.E.: Aspartate transcarbamylase from Streptococcus faecalis. Purification, properties, and nature of an allosteric activator site. Biochemistry, **13**, 629-638 (1974)
[27] Chang, T.Y.; Jones, M.E.: Aspartate transcarbamylase from Streptococcus faecalis. Steady-state kinetic analysis. Biochemistry, **13**, 638-645 (1974)
[28] Achar, B.S.; Savithri, H.S.; Vaidyanathan, C.S.; Rao, N. A.: Studies on plant aspartate transcarbamylase. Purification and properties of the enzyme from mung-bean (Phaseolus aureus) seedlings. Eur. J. Biochem., **47**, 15-22 (1974)
[29] Ong, B.L.; Jackson, J.F.: Aspartate transcarbmoylase from Phaseolus aureus. Biochem. J., **129**, 571-581 (1972)
[30] Jarry, B.P.: Purification of aspartate transcarbamylase from Drosophila melanogaster. Eur. J. Biochem., **87**, 533-540 (1978)
[31] Adair, L.B.; Jones, M.E.: Purification and characteristics of aspartate transcarbamylase from Pseudomonas fluorescens. J. Biol. Chem., **247**, 2308-2315 (1972)
[32] Masood, R.; Venkitasubramanian, T.A.: Purification and properties of aspartate transcarbamylase from Mycobacterium smegmatis. Biochim. Biophys. Acta, **953**, 106-113 (1988)
[33] Mori, M.; Tatibana, M.: A multienzyme complex of carbamoyl-phosphate synthase (glutamine):aspartate carbamoyltransferase:dihydroorotase (rat ascites hepatoma cells and rat liver). Methods Enzymol., **51**, 111-121 (1978)
[34] Coleman, P.F.; Suttle, D.P.; Stark, G.R.: Purification of a multifunctional protein bearing carbamyl-phosphate synthase, aspartate transcarbamylase, and dihydroorotase enzyme activities from mutant hamster cells. Methods Enzymol., **51**, 121-134 (1978)
[35] Purcarea, C.; Erauso, G.; Prieur, D.; Herve, G.: The catalytic and regulatory properties of aspartate transcarbamoylase from Pyrococcus abyssi, a new deep-sea hyperthermophilic archaeobacterium. Microbiology, **140**, 1967-1975 (1994)
[36] Xi, X.G.; De Staercke, C.; Van Vliet, F.; Triniolles, F.; Jacobs, A.; Stas, P.P.; Ladjimi, M.M.; Simon, V.; Cunin, R.; Herve, G.: The activation of Escherichia coli aspartate transcarbamylase by ATP. Specific involvement of helix H2' at the hydrophobic interface between the two domains of the regulatory chains. J. Mol. Biol., **242**, 139-149 (1994)

[37] Lee, B.H.; Ley, B.W.; Kantrowitz, E.R.; O'Leary, M.H.; Wedler, F.C.: Domain closure in the catalytic chains of Escherichia coli aspartate transcarbamoylase influences the kinetic mechanism. J. Biol. Chem., **270**, 15620-15627 (1995)
[38] Baker, D.P.; Aucoin, J.M.; Williams, M.K.; DeMello, L.A.; Kantrowitz, E.R.: Overexpression and purification of the trimeric aspartate transcarbamoylase from Bacillus subtilis. Protein Expr. Purif., **6**, 679-684 (1995)
[39] Khan, A.; Chowdhry, B.Z.; Yon, R.J.: Effects of lipids on nucleotide inhibition of wheat-germ aspartate transcarbamoylase: evidence of an additional level of control?. Biochem. J., **313**, 669-673 (1996)
[40] Yuan, X.; LiCata, V.J.; Allewell, N.M.: Effects of assembly and mutations outside the active site on the functional pH dependence of Escherichia coli aspartate transcarbamylase. J. Biol. Chem., **271**, 1285-1294 (1996)
[41] Burns, B.P.; Mendz, G.L.; Hazell, S.L.: In situ properties of Helicobacter pylori aspartate carbamoyltransferase. Arch. Biochem. Biophys., **347**, 119-125 (1997)
[42] Xu, Y.; Zhang, Y.; Liang, Z.; Van de Casteele, M.; Legrain, C.; Glansdorff, N.: Aspartate carbamoyltransferase from a psychrophilic deep-sea bacterium, Vibrio strain 2693: properties of the enzyme, genetic organization and synthesis in Escherichia coli. Microbiology, **144**, 1435-1441 (1998)
[43] Liu, L.; Wales, M.E.; Wild, J.R.: Temperature effects on the allosteric responses of native and chimeric aspartate transcarbamoylases. J. Mol. Biol., **282**, 891-901 (1998)
[44] Khan, A.I.; Chowdhry, B.Z.; Yon, R.J.: Wheat germ aspartate transcarbamoylase: revised purification, stability and re-evaluation of regulatory kinetics in terms of the Monod-Wyman-Changeux model. Eur. J. Biochem., **259**, 71-78 (1999)
[45] Durbecq, V.; Thia-Toong, T.L.; Charlier, D.; Villeret, V.; Roovers, M.; Wattiez, R.; Legrain, C.; Glansdorff, N.: Aspartate carbamoyltransferase from the thermoacidophilic archaeon Sulfolobus acidocaldarius cloning, sequence analysis, enzyme purification and characterization. Eur. J. Biochem., **264**, 233-241 (1999)
[46] Vorisek, J.; Noaillac-Depeyre, J.; Denis-Duphil, M.: Life-cycle-dependent changes of aspartate carbamoyltransferase localization in membranes of Saccharomyces cerevisiae - centrifugal elutriation and ultracytochemical study. Folia Microbiol., **44**, 289-294 (1999)
[47] Jin, L.; Stec, B.; Lipscomb, W.N.; Kantrowitz, E.R.: Insights into the mechanisms of catalysis and heterotropic regulation of Escherichia coli aspartate transcarbamoylase based upon a structure of the enzyme complexed with the bisubstrate analogue N-phosphonacetyl-L-aspartate at 2.1 Å Proteins Struct. Funct. Genet., **37**, 729-742 (1999)
[48] Hack, E.S.; Vorobyova, T.; Sakash, J.B.; West, J.M.; Macol, C.P.; Herve, G.; Williams, M.K.; Kantrowitz, E.R.: Characterization of the aspartate transcarbamoylase from Methanococcus jannaschii. J. Biol. Chem., **275**, 15820-15827 (2000)
[49] Endrizzi, J.A.; Beernink, P.T.; Alber, T.; Schachman, H.K.: Binding of bisubstrate analog promotes large structural changes in the unregulated catalytic

trimer of aspartate transcarbamoylase: implications for allosteric regulation. Proc. Natl. Acad. Sci. USA, **97**, 5077-5082 (2000)
[50] Fetler, L.; Tauc, P.; Baker, D.P.; Macol, C.P.; Kantrowitz, E.R.; Vachette, P.: Replacement of Asp-162 by Ala prevents the cooperative transition by the substrates while enhancing the effect of the allosteric activator ATP on E. coli aspartate transcarbamoylase. Protein Sci., **11**, 1074-1081 (2002)
[51] West, J.M.; Tsuruta, H.; Kantrowitz, E.R.: Stabilization of the R allosteric structure of Escherichia coli aspartate transcarbamoylase by disulfide bond formation. J. Biol. Chem., **277**, 47300-47304 (2002)
[52] Vickrey, J.F.; Herve, G.; Evans, D.R.: Pseudomonas aeruginosa aspartate transcarbamoylase. Characterization of its catalytic and regulatory properties. J. Biol. Chem., **277**, 24490-24498 (2002)
[53] Van Boxstael, S.; Cunin, R.; Khan, S.; Maes, D.: Aspartate transcarbamylase from the hyperthermophilic archaeon Pyrococcus abyssi: thermostability and 1.8 A resolution crystal structure of the catalytic subunit complexed with the bisubstrate analogue N-phosphonacetyl-L-aspartate. J. Mol. Biol., **326**, 203-216 (2003)

Ornithine carbamoyltransferase 2.1.3.3

1 Nomenclature

EC number
2.1.3.3

Systematic name
carbamoyl-phosphate:L-ornithine carbamoyltransferase

Recommended name
ornithine carbamoyltransferase

Synonyms
L-ornithine carbamoyltransferase
L-ornithine carbamyltransferase
L-ornithine transcarbamylase
OTC
carbamoyltransferase, ornithine
carbamylphosphate-ornithine transcarbamylase
citrulline phosphorylase
ornithine carbamyltransferase
ornithine transcarbamoylase

CAS registry number
9001-69-8

2 Source Organism

<1> *Streptococcus sp.* (D10 [2]) [2]
<2> *Lathyrus sativus* (putrescine synthase has inherent activities of agmatine iminohydrolase, putrescine carbamoyltransferase, ornithine carbamoyltransferase and carbamate kinase [1]) [1]
<3> *Streptococcus faecalis* [3-5, 21, 37]
<4> *Bos taurus* [3-5, 7, 22, 37]
<5> *Mycobacterium smegmatis* [6]
<6> *Squalus acanthias* [35]
<7> *Suncus murinus* (house musk shrew [7]) [7]
<8> *Rattus norvegicus* [7, 15, 25, 27, 30, 36]
<9> *Oryctolagus cuniculus* [7, 48]
<10> *ray* [7]
<11> *shark* [7]
<12> *Gallus gallus* [7, 32]

<13> *Arundinaria sp.* (sugar cane, cytoplasmic and mitochondrial enzyme form [8]) [8]
<14> *Nostoc muscorum* (blue-green alga [9]) [9]
<15> *Geotrichum candidum* [10]
<16> *Escherichia coli* (strain W [11]) [11, 33, 53]
<17> *Bacillus subtilis* (strain 168 [34]) [34]
<18> *Saccharomyces cerevisiae* (plasmid-containing enzyme-overproducing strain [38]) [12, 38]
<19> *Homo sapiens* (normal liver enzyme and enzyme from a liver of a patient with Reye's syndrome [39]; ornithine transcarbamoyltransferase deficiency is the most common cause of inherited urea cycle disorders [46]) [39, 40, 46, 55]
<20> *Phaseolus vulgaris* (bean [13]) [13, 48]
<21> *Chlorella autotrophica* [14]
<22> *Chlorella saccharophila* [14]
<23> *Dunaliella tertiolecta* [14]
<24> *Pisum sativum* [16, 19, 23, 24, 48]
<25> *Pseudomonas syringae* (pv. phaseolicola [17]) [17]
<26> *Vitis vinifera* [18]
<27> *Streptomyces fradiae* [20]
<28> *Salmonella typhimurium* [24]
<29> *Alnus glutinosa* (isoenzyme 1 [26]) [26]
<30> *Neisseria gonorrhoeae* [28]
<31> *Daucus carota* (carrot [29]) [29]
<32> *Neurospora crassa* [31]
<33> *Sus scrofa* [41]
<34> *Pseudomonas aeruginosa* (catabolic ornithine carbamoyltransferase [42,50,51,52]) [42, 50, 51, 52]
<35> *Halobacterium halobium* (also known as Halobacterium salinarium, catabolic ornithine carbamoyltransferase [43]) [43]
<36> *Nocardia lactamdurans* (3'-hydroxymethylcephem O-carbamoyltransferase [44]) [44]
<37> *Streptomyces clavuligerus* (3'-hydroxymethylcephem O-carbamoyltransferase [44]) [44]
<38> *Pyrococcus furiosus* [45, 49, 56]
<39> *Thermus thermophilus* [47]
<40> *Dolichos lablab* [48]
<41> *Glycine max* [48]
<42> *Perilla frutescens* [48]
<43> *Solanum melongena* [48]
<44> *Lycopersicon esculentum* [48]
<45> *Brassica juncea* [48]
<46> *Canavalia lineata* (2 isoforms [57]) [48, 54, 57]
<47> *Canavalia maritima* [48]
<48> *Canavalia gladiata* [48]
<49> *Canavalia brasiliensis* [48]
<50> *Dioclea megacarpa* [48]

<51> *Wisteria floribunda* [48]
<52> *Mus musculus* [48]
<53> *Caretta caretta* (loggerhead turtle [58]) [58]
<54> *Spinacia oleracea* [59]

3 Reaction and Specificity

Catalyzed reaction

carbamoyl phosphate + L-ornithine = phosphate + L-citrulline (<3,5,6,16,28> mechanism [6,11,21,24,35]; <3> ping pong mechanism [21]; <31> ordered sequential mechanism, carbamoylphosphate binds first, followed by L-ornithine, L-citrulline leaves first, followed by phosphate [29]; <8> ordered mechanism, carbamoylphosphate binds first [27]; <54> ordered bi-bi-sequential reaction mechanism, carbamoylphosphate binds first followed by L-ornithine, L-citrulline leaves first followed by phosphate [59])

Reaction type

carbamoyl group transfer

Natural substrates and products

S carbamoyl phosphate + L-ornithine <6, 13, 14, 16, 18, 21-23, 25, 32> (<13> cytoplasmic form has predominantly catabolic function, mitochondrial form has anabolic function [8]; <16> enzyme belongs to the arginine biosynthesis pathway [11]; <21-23,25> anabolic function [14,17]; <32> enzyme is involved in the biosynthesis of arginine in many organisms and participates in the urea cycle of mammals [31]; <6> catalyzes the last step of mitochondrial citrulline synthesis [35]; <14> enzyme seems to be involved in a bicarbonate-fixing pathway [9]; <18> first step in arginine biosynthesis [38]) (Reversibility: r <16> [11]; ir <25> [17]; ? <6, 13, 14, 18, 21, 23, 32> [8, 9, 14, 31, 35, 38]) [8, 9, 11, 14, 17, 31, 35, 38]

P phosphate + L-citrulline <6, 13, 16, 18, 21-23, 25, 32> [8, 11, 14, 17, 31, 35, 38]

S carbamoyl phosphate + canaline <46> (<46> 66fold higher activity with canaline than with ornithine, enzyme may play a role in canavanine synthesis [54]) (Reversibility: ? <46> [54]) [54]

P O-ureidohomoserine + phosphate <46> [54]

Substrates and products

S 3'-hydroxymethylcephem + carbamoyl phosphate <36, 37> (Reversibility: ? <36, 37> [44]) [44]

P cephamycin C + phosphate <36, 37> [44]

S arsenate + L-citrulline <34> (<34> pseudo-reverse reaction, arsenate is first coupled to citrulline, followed by elimination of carbamoylarsenate [51]) (Reversibility: r <34> [51]) [51]

P carbamoylarsenate + L-ornithine <34> [51]

S carbamoyl phosphate + L-ornithine <1-54> (<1> enzyme catalyzes the synthesis of δ-acetyl ornithine at 0.3-0.5% the rate of citrulline synthesis at pH 8.5 when acetyl phosphate replaces carbamoyl phosphate [2]; <2> enzyme also catalyzes the reactions of EC 2.1.3.6, EC 2.7.2.2 and EC 3.5.3.12, thus acting as putrescine synthase converting agmatine and ornithine into putrescine and citrulline [1]; <3,4> highly specific for L-ornithine at pH 8.0 [4]; <34,35> catabolic ornithine carbamoyltransferase, catalyzes the reverse reaction, i.e. the cleavage of citrulline [42,43]; <20> very low activity with canaline [48]; <46> isoform 1 shows 14times higher ornithine-dependent transcarbamoylase activity than canaline-dependent transcarbamoylase activity [57]) (Reversibility: r <2, 16, 20> [1, 11, 13]; ir <25> [17]; ? <3-15, 17-19, 20, 21-24, 26-54> [2-10, 12, 13-16, 18-48, 58, 59]) [1-48, 58, 59]

P phosphate + L-citrulline <1-54> [1-48, 58, 59]

S carbamoyl phosphate + canaline <46> (<46> 66fold higher activity with canaline than with ornithine [54]; <46> isoform 2 shows 13times higher canaline-dependent transcarbamoylase activity than ornithine-dependent transcarbamoylase activity [57]) (Reversibility: ? <46> [54]) [54, 57]

P O-ureidohomoserine + phosphate <46> [54]

S carbamoyl phosphate + lysine <3, 4, 8> (<3,4> at alkaline pH [4]; <8> 4.3% of the rate with ornithine [36]) (Reversibility: ? <3, 4, 8> [4, 36]) [4, 36]

P phosphate + homocitrulline <3, 4, 8> [4, 36]

Inhibitors

1,4-diaminobutane <31, 34> (<31> 100 mM, 50% inhibition [29]; <34> 10 mM; 45% inhibition [42]) [29, 42]

2,4,6-trinitrobenzenesulfonate <3, 4> (<3> 10 mM, almost complete inactivation [5]) [5]

2-mercaptoethanol <19> [46]

4-hydroxy-1-aminovaleric acid <3> (<3> competitive vs. ornithine, uncompetitive vs. carbamoylphosphate [21]) [21]

5,5'-dithiobis(2-nitrobenzoate) <3, 4> (<3,4> complete inactivation at alkaline pH [37]) [5, 37]

ADP <34, 35> (<35> 5 mM, 90% inhibition [43]) [43, 51]

AMP <34, 35> (<35> 5 mM, 90% inhibition [43]) [43, 51]

ATP <34, 35> (<35> 5 mM, 90% inhibition [43]) [43, 51]

CMP <34> [51]

CTP <34, 35> (<35> 5 mM, 90% inhibition [43]) [43, 51]

Cd^{2+} <21> (<21> 1 mM, 99% inhibition [14]) [14]

Cu^{2+} <21> (<21> 1 mM, 75% inhibition [14]) [14]

DL-2-amino-5-hydroxypentanoic acid <3, 4> [4, 21]

GMP <34> [51]

GTP <34, 35> (<35> 5 mM, 90% inhibition [43]) [43, 51]

L-2,4-diaminobutyric acid <3, 4, 16, 19> [4, 11, 39]

L-2-amino-4-pentenoic acid <19> [39]

L-2-aminobutyrate <3, 4, 8, 19, 27> (<3,4,19,27> L-isomer [4,20,39]; <27> 5-20 mM [20]) [4, 20, 36, 39]

L-alanine <3, 4, 16> [4, 11]

L-asparagine <16> (<16> weak [11]) [11]
L-cysteine <20> (<20> 10 mM, 90% inhibition, noncompetitive vs. ornithine [48]) [48]
L-isoleucine <8, 16> [11, 36]
L-leucine <3, 4, 8, 16, 19> [4, 11, 36, 39]
L-lysine <8, 16> (<8,16> weak [11,36]) [11, 36]
L-norleucine <3, 4> [4]
L-norvaline <3, 4, 5, 6, 8, 16, 19, 31> (<5> competitive vs. ornithine [6]; <3> competitive vs. ornithine, uncompetitive vs. carbamoylphosphate [21]; <31> 2 mM, 50% inhibition [29]) [4, 6, 11, 21, 29, 35, 36, 39]
L-serine <16> (<16> weak [11]) [11]
L-valine <8, 16> [11, 36]
$MgCl_2$ <6> (<6> 10 mM, 14% inhibition [35]) [35]
N-δ-(phosphosulfamyl)ornithylalanylhomoarginine <16, 25> (<16> trivial name phaseolotoxin, 0.1 mM, 93% inhibition [33]; <25> inhibition of isoform 1 only [17]) [17, 33]
NaCl <6> (<6> 50 mM, 25% inhibition [35]) [35]
S-carbamoyl cysteine <20> (<20> 10 mM, 90% inhibition [48]) [48]
Tris <24> (<24> 50 mM, more than 80% inhibition [48]) [48]
UTP <34, 35> (<35> 5 mM, 90% inhibition [43]) [43, 51]
Zn^{2+} <19, 21, 27> (<21> 1 mM, 44% inhibition [14]; <27> 5-20 mM [20]) [14, 20, 46]
acetylimidazole <3> (<3> 50 mM, 95% inhibition at 25°C in 100 mM imidazole-HCl, pH 7.5 [5]) [5]
agmatine <31> (<31> 13 mM, 50% inhibition [29]) [29]
arginine <5, 28, 31> (<31> L-isomer, 8 mM, 50% inhibition [29]; <5> several arginine analogs have no effect [6]; <28> 5 mM, 42% inhibition [24]) [6, 24, 29]
arsenate <16> (<16> competitive vs. carbamoylphosphate, noncompetitive vs. ornithine [11]) [11]
carbamoyl phosphate <3> [4]
chlorosis-inducing toxin of Pseudomonas phaseolicola <20> (<20> no more than 85% inhibition [13]) [13]
citrulline <6> (<6> competitive to carbamoyl phosphate [35]) [35]
cystamine <3, 4> (<3,4> complete inactivation, activity is completely recovered by incubation with 20 mM dithiothreitol [5]) [5]
δ-N-phosphonoacetyl-L-ornithine <4, 8, 17, 33, 38, 39> (<4,8> transition state analogue [22,25]; <8> synthesis and properties [27]; <8> 0.001 mM, 86% inhibition, 0.0025 mM, 95% inhibition [15]; <4> competitive vs. carbamoylphosphate [22]) [15, 22, 25, 27, 34, 41, 45, 47]
dithiodiglycol <4> (<4> 100 mM, inactivation at pH 9,0 [5]) [5]
dithiothreitol <19> [46]
glutamine <8> [36]
glycine <16> [11]
histidine <8> [36]
iodoacetate <3, 4> [5]
methionine <8, 16> (<16> L-isomer [11]) [11, 36]

ornithine <3, 5, 6, 14, 16, 19, 26, 27, 29, 54> (<16> noncompetitive substrate inhibition vs. carbamoylphosphate [11]; <26> above 10 mM [18]; <27> above 8 mM [20]; <3,5,14,16> at high concentrations [5,6,9,11]; <19> recombinant wild-type enzyme [46]; <54> competitive vs. carbamoylphosphate in the presence of phosphate [59]) [5, 6, 9, 11, 18, 20, 26, 35, 46, 59]
ornithine-δ-P(O)(NH_2)-NH-SO_3H <25> (<25> inhibition of isoform 1 only [17]) [17]
p-chloromercuribenzoate <21> [14]
p-hydroxymercuribenzoate <3, 4> (<3,4> inhibition is completely reversed by 2-mercaptoethanol [5]) [5]
phenylglyoxal <33> (<33> complete inactivation after 1 h incubation [41]) [41]
phosphate <3, 5, 6, 8, 14, 16, 31, 33, 39> (<16> competitive vs. carbamoylphosphate, noncompetitive vs. ornithine [11]; <3> competitive vs. carbamoylphosphate, uncompetitive vs. ornithine [21]; <31> competitive vs. carbamoylphosphate [29]; <6> competitive vs. carbamoylphosphate [35]; <35> slight inhibition [43]) [6, 9, 11, 15, 21, 29, 35, 36, 41, 43, 47]
spermidine <34> (<34> 10 mM, 93% inhibition, 3.7 mM, 50% inhibition [42]; <34> E105G mutant enzyme, competitive vs. carbamoylphosphate [50]; <34> 10 mM, more than 98% inhibition of arsenolytic cleavage activity [51]) [42, 50, 51]
Additional information <8, 21-23, 25, 31> (<3,4> effects of chemical modifications of specific residues on ligand binding and enzymatic activity [5]; <21-23> not inhibited by 10 mM 2-oxoglutarate, proline, glutamate, aspartate, acetylglutamate, pyruvate, oxaloacetate, glycine, acetylornithine, ATP, GTP, CTP, L-asparagine, arginine and glutamine [14]; <31> not inhibited by lysine and α-methyl-L-ornithine [29]; <8> not inhibited by arginine [36]; <25> isoform 2 is not inhibited by phaseolotoxin and ornithine-δ-P(O)(NH_2)-NH-SO_3H [17]) [14, 17, 29, 36]

Activating compounds

ADP <34> (<34> 10 mM, 350% arsenolytic cleavage activity [51]) [51]
AMP <34> (<34> strong allosteric activator, 1.5 mM, half-maximal activation [42]; <34> 10 mM, 550% arsenolytic cleavage activity, activation by phosphate and AMP correspond to different mechanisms [51]) [42, 50, 51]
CMP <34> (<34> strong allosteric activator [42]; <34> 10 mM, 340% arsenolytic cleavage activity [51]) [42, 51]
CTP <34> (<34> 10 mM, 340% arsenolytic cleavage activity [51]) [51]
GMP <34> (<34> allosteric activator [42]; <34> 10 mM, 340% arsenolytic cleavage activity [51]) [42, 51]
UMP <34> (<34> allosteric activator [42]; <34> 10 mM, 600% arsenolytic cleavage activity [51]) [42, 51]
arginine <35> (<35> 50 mM, 2fold activatin [43]) [43]
arsenate <34> (<34> potent activator in the concentration range of 0-10 mM [50]) [50]
phosphate <34> (<34> presumably via interaction with the carbamoyl phosphate-binding site [42]; <34> potent activator in the concentration range of

0-10 mM [50]; <34> 10 mM, 225% arsenolytic cleavage activity [51]) [42, 50, 51]
Additional information <19> (<19> activity increases with ionic strength up to 100 mM [46]) [46]

Turnover number (min^{-1})

1080 <29> (ornithine) [26]
5400 <3> (L-citrulline) [4]
11220 <4> (ornithine) [4]
22200 <38> (ornithine, <38> at 30°C [56]) [56]
30000 <38> (ornithine, <38> at 55°C [56]) [56]
132000 <38> (ornithine, <38> Y227C/E277G double mutant, at 30°C [56]) [56]
174000 <38> (ornithine, <38> A240D/E277G double mutant, at 30°C [56]) [56]
204000 <3> (ornithine) [4]
210000 <38> (ornithine, <38> Y227C/E277G double mutant, at 55°C [56]) [56]
258000 <38> (ornithine, <38> A240D/E277G double mutant, at 55°C [56]) [56]

Specific activity (U/mg)

0.0928 <46> (<46> isoenzyme 2 with ornithine [57]) [57]
0.47 <2> (<2> multifunctional enzyme [1]) [1]
0.574 <46> (<46> isoenzyme 1 with canaline [57]) [57]
1.24 <46> (<46> isoenzyme 2 with canaline [57]) [57]
3.33 <39> (<39> recombinant enzyme [47]) [47]
4.02 <25> [17]
5.56 <29> (<29> isoenzyme 1 [26]) [26]
8.31 <46> (<46> isoenzyme 1 with ornithine [57]) [57]
26.4 <33> [41]
27.1 <6> [35]
40.2 <19> (<19> recombinant enzyme [46]) [46]
50 <27> [20]
77 <12> [32]
83.3 <19> (<19> enzyme from a liver of a Reye's syndrome patient [39]) [39]
86.5 <19> (<19> enzyme from normal liver [39]) [39]
139 <24> [16]
157.8 <24> [19]
198.6 <18> [12]
224 <53> [58]
233 <19> [40]
250-260 <8> [25]
251 <32> [31]
265 <17> [34]
436 <54> [59]
500 <27> [20]
526 <20> [48]

561 <18> [38]
780 <4> [3]
885 <8> [36]
1000 <1> [2]
1901 <28> [24]
2833 <16> [11]
2914 <16> [33]
3200 <3> [3]
Additional information <35> (<35> highest activity at 4 M KCl [43]) [43]

K_i-Value (mM)

0.0000047 <4> (δ-N-phosphonoacetyl-L-ornithine, <4> at pH 8.5 [22]) [22]
0.000007 <4> (δ-N-phosphonoacetyl-L-ornithine, <4> at pH 7.2 [22]) [22]
0.00001 <39> (δ-N-(phosphonacetyl)-L-ornithine) [47]
0.00013 <33> (δ-N-phosphonoacetyl-L-ornithine) [41]
0.0002 <16> (N-δ-(phosphosulfamyl)ornithylalanylhomoarginine, <16> mixed or competitive vs. carbamoyl phosphate [33]) [33]
0.00025 <8> (δ-N-phosphonoacetyl-L-ornithine) [15]
0.0009 <16> (N-δ-(phosphosulfamyl)ornithylalanylhomoarginine, <16> non-competitive vs. ornithine [33]) [33]
0.01 <38> (δ-N-phosphonoacetyl-L-ornithine) [45]
0.01 <17> (phosphonacetyl-L-ornithine, <17> between pH 7.5-9.0 [34]) [34]
0.012 <3> (carbamoyl phosphate, <3> at 25°C and pH 8.4 [4]) [4]
0.06 <6> (ornithine) [35]
0.071 <8> (L-norvaline) [36]
0.09 <21> (norvaline) [14]
0.25 <8> (phosphate) [15, 36]
0.27 <8> (δ-N-(phosphonoacetyl)-L-ornithine, <8> at 30°C and pH 7.4, competitive vs. carbamoyl phosphate, noncompetitive vs. ornithine [27]) [27]
0.4 <16> (L-norvaline) [11]
0.58 <8> (L-α-aminobutyrate) [36]
1 <21> (phosphate) [14]
1.1-1.3 <24> (phosphate, <24> chloroplast enzyme [19]) [19]
1.2 <8> (L-leucine) [36]
1.7 <29> (ornithine, <29> at 2 mM carbamoyl phosphate [26]) [26]
3.6 <29> (ornithine, <29> at 10 mM carbamoyl phosphate [26]) [26]
5.5 <16> (L-leucine) [11]
7 <16> (2,4-diaminobutyrate) [11]
8 <3> (phosphate, <3> at 25°C and pH 7.1 [4]) [4]
10.5 <8> (L-isoleucine) [36]
12.5 <8> (L-valine) [36]
13 <16> (L-alanine) [11]
13.5 <16> (L-valine) [11]
14 <16> (L-isoleucine) [11]
15 <16> (glycine) [11]
40 <8> (L-histidine) [36]

40 <16> (L-methionine) [11]
63.5 <16> (L-serine) [11]
110 <16> (L-asparagine) [11]
Additional information <3> (<3> K_i for norvaline varies with pH, carbamoyl phosphate or phosphate concentration [4]; <8> approximate K_i of 1.0-2.0 mM for phosphate in permeabilized mitochondria [15]) [4, 15]

pH-Optimum

6.5 <38> (<38> at 55°C [45]) [45]
6.7 <12> (<12> in potassium phosphate buffer 5 mM ornithine, pH optimum shifts to more alkaline values with decreasing ornithine concentrations [32]) [32]
7 <20, 24> (<20> ornithine synthesis [13]; <24> enzyme 2 [23]) [13, 23]
7.1 <3> (<3> ornithine synthesis [4]) [4]
7.2 <34> (<34> at 10 mM carbamoyl phosphate, sharp decrease of activity above [42]) [42]
7.5 <12, 31> (<31> half-maximal activity at pH 9.0 [29]; <12> in triethanolamine-HCl [32]) [29, 32]
7.6 <8> (<8> triethanolamine/acetate buffer [36]) [36]
7.7 <19> (<19> triethanolamine buffer, 85% of maximal activity at pH 8.5, 25% of maximal activity at pH 6.8 [40]) [40]
7.8 <29> [26]
7.9 <8> (<8> HEPES/NaOH buffer [36]) [36]
8 <2, 19, 24, 28, 33, 39, 46> (<33> sharp decrease in activity between pH 8.4 and 8.8 [41]; <19> gradual loss of activity above pH 8.5 or below pH 6.5 [46]; <39> at an ornithine concentration around the K_m value, at higher concetration the pH optimum shifts to pH 7.0 [47]; <46> with canaline [54]) [1, 24, 41, 46, 47, 48, 54]
8.3 <17, 24> (<17> bicine-NaOH buffer [34]) [19, 34]
8.4 <3, 25> (<25> phaseolotoxin insensitive isoform 2 [17]) [4, 17]
8.4-8.8 <26> [18]
8.5 <3, 4, 20, 22, 23, 46> (<3,20> citrulline synthesis [4,13]; <46> with ornithine [54]) [4, 13, 14, 54]
8.5-9.5 <25> (<25> phaseolotoxin sensitive isoform 1 [17]) [17]
8.8 <35> (<35> 30% of maximal activity at pH 7.0 [43]) [43, 43]
9 <21, 24> (<24> enzyme 1 [23]) [14, 23]
9.5 <14, 32> (<14> citrulline synthesis [9]) [9, 31]
9.5-10 <27> [20]

pH-Range

6-7.2 <34> (<34> at 10 mM carbamoyl phosphate, sharp decrease in activity above [42]) [42]
6-8.3 <17> (<17> approx. 50% of maximal activity at pH 6.0 [34]) [34]
6-9 <29> [26]
6.7-8.5 <8> (<8> approx. 50% of maximal activity at pH 6.7 and 8.5, triethanolamine/acetate buffer [36]) [36]
6.8-8.5 <19> (<19> approx. 85% of maximal activity at pH 6.8, approx. 25% of maximal activity at pH 8.5 [40]) [40]

7-9 <28> (<28> 50% of maximal activity at pH 9.0, 70% at pH 7.0 [24]) [24]
7-10 <27> (<27> sharp decrease in activity below pH 9.0 [20]) [20]
7.2-8.3 <8> (<8> 71% of maximal activity at pH 7.2, 94% of maximal activity at pH 8.3, HEPES/NaOH buffer [36]) [36]
7.5-9 <31> (<31> approx. 50% of maximal activity at pH 9.0 [29]) [29]
7.5-10 <17> (<17> nearly constant activity in this range, half-maximal activity at pH 7.0, negliable activity at pH 6.0 [34]) [34]

Temperature optimum (°C)

35-45 <29> [26]
38-40 <26> [18]
40 <20> [13]
40-45 <27> [20]
65 <39> [47]

4 Enzyme Structure

Molecular weight

56000 <2> (<2> multifunctional enzyme with agmatine iminohydrolase, putrescine carbamoyltransferase, ornithine carbamoyltransferase and carbamate kinase activity, gel filtration [1]) [1]
75000 <14> [9]
77600 <24> (<24> gel filtration [19]) [19]
79000 <13> (<13> cytoplasmic form [8]) [8]
87000 <27> (<27> gel filtration [20]) [20]
90000 <17> (<17> dimeric aggregation state, sucrose density gradient centrifugation [34]) [34]
95000 <16> (<16> gel filtration [33]) [33]
98000 <8> (<8> sedimentation equilibrium [25]) [25]
105000 <16> [11]
107000 <33> (<33> gel filtration, sedimentation equilibrium [41]) [41]
108000 <4, 8> (<4> sedimentation equilibrium measurement [3]; <8> high performance hydrophobic interaction chromatography [30]) [3, 30]
108200 <24> (<24> gel filtration [16]) [16]
109000 <20, 46> (<20> gel filtration [48]) [48, 54]
110000 <12, 18, 19, 25, 30, 32, 39> (<25> gel filtration [17]; <30> gel filtration [28]; <32> SDS-PAGE of dimethyl suberimidate cross-linked enzyme [31]; <12> gel filtration [32]; <18> equilibrium sedimentation [38]; <19> gel filtration [39]; <39> gel filtration [47]) [17, 28, 31, 32, 38, 39, 47]
112000 <8, 53> (<8> sedimentation equilibrium analysis [36]; <53> gel filtration [58]) [36, 58]
113000 <6> (<6> gel filtration [35]) [35]
114000 <19> (<19> gel filtration, glycerol density gradient centrifugation [40]) [40]
115000 <32> (<32> gel filtration [31]) [31]

116000 <5, 28> (<28> sucrose density gradient ultracentrifugation [24]) [6, 24]
118000 <54> (<54> gel filtration [59]) [59]
125000 <18> (<18> gel filtration [12]) [12]
158000 <31> (<31> gel filtration [29]) [29]
167000 <17> (<17> tetrameric aggregation state, sucrose density gradient centrifugation [34]) [34]
200000 <29, 35> (<29> gradient PAGE [26]; <35> gel filtration [43]) [26, 43]
223000 <3> (<3> sedimentation equilibrium measurement [3]) [3]
224000 <13> (<13> mitochondrial enzyme [8]) [8]
236000 <17> (<17> hexameric aggregation state, sucrose density gradient centrifugation [34]) [34]
260000 <17> (<17> gel filtration [34]) [34]
400000 <38> (<38> gel filtration [45]) [45]
420000 <34> (<34> gel filtration [42]) [42]

Subunits

? <7, 17, 27, 36> (<7> x * 35000, SDS-PAGE [7]; <17> x * 44000, SDS-PAGE, enzyme exists as a mixture of dimeric, tetrameric and hexameric forms, enzyme is converted to a dimer in presence of 2-mercaptoethanol [34]; <27> x * 40000, enzyme may be a dimer, SDS-PAGE [20]; <36> x * 57148, deduced from nucleotide sequence [44]) [7, 20, 34, 44]
dodecamer <34> (<34> 12 * 38000, SDS-PAGE [42]) [42]
hexamer <3, 35> (<3> 6 * 39600, SDS-PAGE [3]; <35> 6 * 32000, deduced from nucleotide sequence [43]) [3, 43]
monomer <2> (<2> 1 * 55000, multifunctional enzyme, SDS-PAGE [1]) [1]
trimer <4, 6, 8, 12, 16, 18, 19, 20, 24, 28, 30, 32, 33, 39, 46, 53, 54> (<4> 3 * 37800, SDS-PAGE [3]; <8> 3 * 39600, SDS-PAGE, sedimentation equilibrium analysis in 6 M guanidine-HCl [36]; <8> 3 * 39900, amino acid sequence [36]; <6> 3 * 38000, SDS-PAGE [35]; <19> SDS-PAGE under reducing conditions [40]; <12> 3 * 36000, SDS-PAGE [32]; <30> 3 * 36500, SDS-PAGE [28]; <32> 3 * 37000, SDS-PAGE [31]; <16> 3 * 35000, SDS-PAGE [11]; <18> 3 * 37500, SDS-PAGE [12]; <18> 3 * 37000, SDS-PAGE [38]; <24> 3 * 36500, SDS-PAGE [16]; <28> 3 * 39000, SDS-PAGE [24]; <8> 3 * 35300, SDS-PAGE [25]; <19> 3 * 36500, SDS-PAGE [39]; <16> 3 * 37500, SDS-PAGE [33]; <33> 3 * 37000, SDS-PAGE [41]; <39> 3 * 36000, SDS-PAGE [47]; <20> 3 * 36000, SDS-PAGE [48]; <46> 3 * 36000, SDS-PAGE [54]; <53> 3 * 38000, SDS-PAGE [58]; <54> 3 * 40000, SDS-PAGE [59]) [3, 11, 12, 16, 24, 25, 28, 31, 32, 33, 35, 36, 39, 40, 41, 47, 48, 54, 58, 59]
Additional information <18> (<18> structural properties [38]) [38]

5 Isolation/Preparation/Mutation/Application

Source/tissue

cell suspension culture <31> [29]
fruit <26> [18]

kidney <12> [7, 32]
leaf <20, 24, 26, 40-51, 46> [16, 18, 19, 48, 57]
liver <4, 6-11, 19, 33, 53> (<4,7-11> ureotelic vertebrates [7]) [3, 4, 7, 15, 22, 25, 30, 35-37, 39, 40, 41, 58]
root nodule <29> [26]
seed <26> (<26> germinating [18]) [18]
seedling <2, 24, 26> [1, 18, 23]
seedling <24> (<24> shoot [19]) [19]
spore <15> (<15> of conidia [10]) [10]

Localization

cytoplasm <2, 13> (<13> 2 enzyme forms in sugar cane: cytoplasmic and mitochondrial [8]) [1, 8]
mitochondrion <6, 8, 13, 21-23> (<13> 2 enzyme forms in sugar cane: cytoplasmic and mitochondrial [8]; <21-23> soluble matrix [14]) [8, 14, 30, 35]

Purification

<1> [2]
<2> (putrescine synthase has inherent activities of agmatine iminohydrolase, putrescine carbamoyltransferase, ornithine carbamoyltransferase and carbamate kinase, ammonium sulfate, putrescine-Sepharose affinity column [1]) [1]
<3> (ammonium sulfate, streptomycin, DEAE-cellulose, heating, DEAE-Sephadex [3]) [3]
<4> (ammonium sulfate [3]) [3]
<5> [6]
<6> (DEAE-Sephadex, hydoxyapatite C, δ-N-(phosphonacetyl)-L-ornithine affinity chromatography [35]) [35]
<7> (immunoadsorbent column [7]) [7]
<8> (affinity chromatography on δ-N-(phosphonacetyl)-L-ornithine, Sephacryl S-200 [25]; high-performance hydrophobic interaction chromatography [30]) [25, 30, 36]
<12> (60-65°C for 5 min, ammonium sulfate, DEAE-cellulose, Bio-Gel P-300 [32]) [32]
<14> [9]
<15> [10]
<16> (heat, ammonium sulfate, DEAE-Sephadex, Sephadex G-200, aminohexyl-Sepharose [11]; N-δ-(phosphonoacetyl)ornithine affinity chromatography, Sephadex G-100 [33]; recombinant enzyme [53]) [11, 33, 53]
<17> (heat, ammonium sulfate, DEAE-cellulose, hydroxylapatite, DEAE-Sephacel, Bio-Gel A 1.5 [34]) [34]
<18> (plasmid-containing enzyme-overproducing strain [38]; ammonium sulfate, DEAE-Sephadex, Sephadex G-200 [12]) [12, 38]
<19> (enzyme from normal liver and liver of a Reye's syndrome patient, DEAE-cellulose, hydroxylapatite, Sephadex G-200 [39]; recombinant wild-type and R277W mutant enzyme, ammonium sulfate, δ-N-(phosphonacetyl)-L-ornithine affinity chromatography [46]) [39, 40, 46]

<20> (partially purified [13]; δ-N-(phosphonoacetyl)-L-ornithine-Sepharose 6B affinity chromatography [48]) [13, 48]
<24> (single step procedure using δ-N-(phosphonoacetyl)-L-ornithine-Sepharose 6B affinity chromatography [16]; ammonium sulfate, heat treatment, celite, DEAE-Sephadex A-50, Sephacryl S-200, partial purification [19]; ammonium sulfate, DEAE-cellulose, isoforms 1 and 2 [23]) [16, 19, 23]
<25> (isoforms 1 and 2 [17]) [17]
<27> (ammonium sulfate, DEAE-cellulose, aminohexyl-agarose [20]) [20]
<28> (protamine sulfate, ammonium sulfate, diethylaminoethyl-cellulose, Sephadex G-150, hydroxylapatite [24]) [24]
<29> (DEAE-Sephacel, hydroxylapatite, preparative PAGE, isoenzyme 1 [26]) [26]
<30> (DEAE-cellulose, isoelectric focusing [28]) [28]
<31> [29]
<32> (ammonium sulfate, 70°C for 25 min, gel filtration, ion-exchange chromatography [31]) [31]
<33> (ammonium sulfate, Sephacryl S-200, δ-N-phosphonoacetyl-L-ornithine affinity chromatography [41]) [41]
<35> (gel filtration, ammonium sulfate-mediated hydrophobic chromatography [43]) [43]
<38> (ammonium sulfate, DEAE-Sepharose, ornithine[AcPO$(OH)_2$]-Sepharose [45]; recombinant enzyme, Mono Q, arginine-Sepharose column [56]) [45, 56]
<39> (recombinant enzyme, thermodenaturation at 85°C, δ-N-(phosphonacetyl)-L-ornithine affinity chromatography [47]) [47]
<53> (heat treatment, δ-N-(phosphonacetyl)-L-ornithine affinity chromatography [58]) [58]
<54> (ammonium sulfate, heat, activated charcoal, δ-N-(phosphonacetyl)-L-ornithine affinity chromatography [59]) [59]

Renaturation

<3, 4> (effect of pH and temperature on the rate of renaturation of guanidine-HCl dissociated enzyme, approx. 90% recovery at 25°C and pH 7.0 [3]) [3]

Crystallization

<1> [2]
<16> (ornithine carbamoyltransferase-N^{δ}-(N'-sulfodiaminophosphinyl)-L-ornithine complex, hanging-drop vapour diffusion, equal volumes of protein are combined with a solution containing 17.8% polyethylene glycol 8000 and 2.2% polyethylene glycol 1000, pyramidal crystals after 1-2 weeks [53]) [53]
<19> (ornithine transcarbamoylase-carbamoylphosphate complex, X-ray structure at 2.4-2.6 A resolution [55]) [55]
<34> (T allosteric form, hanging-drop vapour diffusion against a reservoir solution containing 100 mM glycylglycine, pH 9, 12% polyethylene glycol 6000 and 1 mM dithiothreitol, X-ray analysis, 4.5 A resolution, R allosteric form, hanging-drop vapour diffusion against a reservoir solution containing

1.9 M ammonium sulfate, 50 mM HEPES, pH 7.2, 1 mM DTT, 1 mM EDTA, 3% polyethylene glycol 400 and 10 mM spermidine [52]) [52]
<38> (hanging-drop vapor diffusion, 1/1 mixture of 12 mg/ml enzyme solution and a reservoir solution containing 1 M NaCl, 100 mM acetate buffer, pH 4.0, crystals are obtained after 7-10 days, X-ray structure at 2.7 A resolution [49]) [49]

Cloning
<16> (expression in Escherichia coli [53]) [53]
<19> (coexpression with GroES and GroEL in Escherichia coli [46]) [46]
<35> [43]
<38> (expression in Saccharomyces cerevisiae [45]; expression of wild-type, Y227C, A240D, E277G, Y227C/E277G and A240D/E277G mutant enzyme in Saccharomyces cerevisiae [56]) [45, 56]
<39> (expression in Escherichia coli [47]) [47]
<46> (expression of isoform 1 and 2 in Escherichia coli [57]) [57]
<36, 37> [44]

Engineering
A240D <38> (<38> 10fold increase of K_m at 55°C [56]) [56]
E105G <34> (<34> no cooperativity towards carbamoyl phosphate, follows Michaelis-Menten kinetics [50]) [50]
E277G <38> (<38> 14fold increase of K_m at 55°C [56]) [56]
H121R <46> (<46> isoform 2, 80% increase in ornithine dependent activity, 64% decrease in canaline-dependent activity [57]) [57]
L118M <46> (<46> isoform 2, 515% increase in ornithine dependent activity, 54% decrease in canaline-dependent activity [57]) [57]
L118M <46> (<46> isoform 2, 617% increase in ornithine dependent activity, 55% decrease in canaline-dependent activity [57]) [57]
R277W <19> (<19> shows no substrate inhibition by ornithine, 70fold lower affinity for L-ornithine [46]) [46]
Y227D <38> (<38> slight increase of K_m at 55°C [56]) [56]

6 Stability

pH-Stability
5.8-8.2 <19> (<19> 120 min, 37°C, stable [40]) [40]
6.8-9 <8> (<8> 60 min, stable [36]) [36]
8.3 <17> (<17> highest stability, bicine-NaOH buffer [34]) [34]
9 <3> (<3> stable at low ionic strength, gradual loss of activity in 1 M KCl [3]) [3]

Temperature stability
-50-60 <19> (<19> stable [40]) [40]
45 <26> (<26> rapid inactivation above [18]) [18]
51 <19> (<19> R277W mutant enzyme, 50% loss of activity [46]) [46]
56 <19> (<19> recombinant enzyme, 50% loss of activity [46]) [46]

56 <54> (<54> half-live: 99 min [59]) [59]
58 <54> (<54> half-life: 60 min [59]) [59]
60 <53> (<53> half-life: 116 min [58]) [58]
64 <53> (<53> half-life: 39 min [58]) [58]
64 <54> (<54> half-life: 11 min [59]) [59]
66 <53> (<53> half-life: 10 min [58]) [58]
67 <53> (<53> half-life: 7 min [58]) [58]
68 <16> (<16> half-life: 5 min, 8 min with 500 mM KCl, 10 min with 10 mM citrulline, 20 min with 100 mM citrulline, 43 min with 20 mM phosphate, 48 min with 100 mM ornithine, 108 min with 100 mM phosphate, 120 min with 10 mM phosphate and 10 mM ornithine, 122 min with 10 mM phosphate and 100 mM citrulline, more than 200 min with 100 mM phosphate and 100 mM citrulline or with 10 mM phosphate and 100 mM ornithine [11]) [11]
68 <54> (<54> half-life: 10.5 min, half-life increases to 45 min in the presence of 50 mM ornithine [59]) [59]
70 <53> (<53> half-life: 5 min [58]) [58]
85 <39> (<39> half-life: 18 min, 10 mM ornithine and phosphate increase half-life 5fold and 7fold respectively [47]) [47]
100 <38> (<38> half-life, 30-40 min, presence of ornithine and/or phosphate protects [45]) [45]

General stability information

<2>, prolonged dialysis and freeze-thawing causes loss of activity [1]
<2>, purified enzyme is highly unstable even in the presence of glycerol, dithiothreitol and Mg^{2+} [1]
<8>, loss of activity by freezing and thawing [25]
<19>, cetyltrimethylammonium stabilizes enzyme in crude extract against heat inactivation in presence of the detergent [40]
<19>, loss of activity after the second thawing [40]
<30>, 1 mM carbamoyl phosphate or 20% glycerol stabilize the enzyme in cell-free extracts, a minimal enzyme concentration of 0.2 mg/ml is obligatory for stability [28]
<32>, glycerol, 20% v/v, stabilizes during freeze-thawing [31]
<35>, unstable below 1 M KCl [43]

Storage stability

<2>, 4°C, 48 h, complete loss of activity [1]
<3>, enzyme suspension in 3.0 M ammonium sulfate, pH 5.0, 3 years, no loss of activity [3]
<6>, -20°C, 0.05 potassium phosphate buffer, pH 7.6, 1 month, no loss of activity [35]
<8>, -20°C or 4°C, no loss of activity [25]
<8>, -70°C, 2% loss of activity per month [36]
<16>, -80°C, prolonged storage, no loss of activity [33]
<17>, -20°C, protein concentration 1-2 mg/ml, several months, no loss of activity after repeated freezing and thawing [34]
<19>, -20°C, 20 mM Tris acetate, 2 mM EDTA, pH 7.4, 50% glycerol, no loss of activity [46]

<19>, -20°C, 50% glycerol, 10 mM potassium phosphate, pH 7.0, 1 mM 2-mercaptoethanol, 4 months, no loss of activity [40]
<19>, -50°C, indefinitely [40]
<27>, 0-4°C, 2 months, no loss of activity [20]
<33>, -20°C, several weeks, no loss of activity [41]
<35>, 25°C, high-salt solution, 1 year, no loss of activity [43]

References

[1] Srivenugopal, K.S.; Adiga, P.R.: Enzymic conversion of agmatine to putrescine in Lathyrus sativus seedlings. Purification and properties of a multifunctional enzyme (putrescine synthase). J. Biol. Chem., **256**, 9532-9541 (1981)
[2] Bishop, S.H.; Grisolia, S.: Crystalline ornithine transcarbamylase. Biochim. Biophys. Acta, **139**, 344-348 (1967)
[3] Marshall, M.; Cohen, P.P.: Ornithine transcarbamylase from Streptococcus faecalis and bovine liver. I. Isolation and subunit structure. J. Biol. Chem., **247**, 1641-1653 (1972)
[4] Marshall, M.; Cohen, P.P.: Ornithine transcarbamylase from Streptococcus faecalis and bovine liver. II. Multiple binding sites for carbamyl-P and L-norvaline, correlation with steady state kinetics. J. Biol. Chem., **247**, 1654-1668 (1972)
[5] Marshall, M.; Cohen, P.P.: Ornithine transcarbamylase from Streptococcus faecalis and bovine liver. 3. Effects of chemical modifications of specific residues on ligand binding and enzymatic activity. J. Biol. Chem., **247**, 1669-1682 (1972)
[6] Ahmad, S.; Bhatnagar, R.K.; Venkitasubramanian, T.A.: Ornithine transcarbamylase from Mycobacterium smegmatis ATCC 14468: purification, properties, and reaction mechanism. Biochem. Cell Biol., **64**, 1349-1355 (1986)
[7] Tsuji, S.; Kanazawa, S.; Fukushina, T.; Oda, S.: Some characteristics of hepatic ornithine transcarbamylase from the house musk shrew, Suncus murinus. Nippon Chikusan Gakkai Ho, **56**, 962-970 (1985)
[8] Glenn, E.; Maretzki, A.: Properties and subcellular distribution of two partially purified ornithine transcarbamoylases in cell suspensions of sugarcane. Plant Physiol., **60**, 122-126 (1977)
[9] Boggess, S.F.; Naylor, A.W.: Partial purification and properties of ornithine transcarbamoylase from Nostoc muscorum. Plant Physiol., **56**, 640-644 (1975)
[10] Barash, I.; Mor, H.: Control and properties of onithine transcarbamylase in germinated spores of Geotrichum candidum. Plant Cell Physiol., **13**, 119-130 (1972)
[11] Legrain, C.; Stalon, V.: Ornithine carbamoyltransferase from Escherichia coli W. Purification, structure and steady-state kinetic analysis. Eur. J. Biochem., **63**, 289-301 (1976)
[12] Penninckx, M.; Simon, J.P.; Wiame, J.M.: Interaction between arginase and L-ornithine carbamoyltransferase in Saccharomyces cerevisiae. Purification

of S. cerevisiae enzymes and evidence that these enzymes as well as rat-liver arginase are trimers. Eur. J. Biochem., **49**, 429-442 (1974)
[13] Tam, L.Q.; Patil, S.S.: Mode of action of the toxin from Pseudomonas phaseolicola. II. Mechanism of inhibition of bean ornithine carbamoyltransferase. Plant Physiol., **49**, 808-812 (1972)
[14] Laliberte, G.; Hellebust, J.A.: Partial characterization of ornithine carbamoyltransferase in three microalgae. Plant Physiol., **93**, 62-66 (1990)
[15] Cohen, N.S.; Cheung, C.W.; Sijuwade, E.; Raijman, L.: Kinetic properties of carbamoyl-phosphate synthase (ammonia) and ornithine carbamoyltransferase in permeabilized mitochondria. Biochem. J., **282**, 173-180 (1992)
[16] Siocum, R.D.; Richardson, D.P.: Purification and characterization of ornithine transcarbamoylase from pea (Pisum sativum L.). Plant Physiol., **96**, 262-268 (1991)
[17] Jahn, O.; Sauerstein, J.; Reuter, G.: Characterization of two ornithine carbamoyltransferases from Pseudomonas syringae pv. phaseolicola, the producer of phaseolotoxin. Arch. Microbiol., **147**, 174-178 (1987)
[18] Roubelakis, K.A.; Kliewer, W.M.: Enzymes of Krebs-Henseleit cycle in Vitis vinifera L.. Plant Physiol., **62**, 337-339 (1978)
[19] De Ruiter, H.; Kollöffel, C.: Properties of ornithine carbamoyltransferase from Pisum sativum L.. Plant Physiol., **77**, 695-699 (1985)
[20] Szilagyi, I.; Vargha, G.; Szabo, G.: Ornithine carbamoyltransferase of Streptomyces fradiae: purification and properties. FEMS Microbiol. Lett., **48**, 115-119 (1987)
[21] Kurtin, W.E.; Bishop, S.H.; Himoe, A.: Ornithine transcarbamylase: steady-state kinetic properties. Biochem. Biophys. Res. Commun., **45**, 551-556 (1971)
[22] Mori, M.; Aoyagi, K.; Tatibana, M.; Ishikawa, T.; Ishii, H.: N δ-(phosphonacetyl)-L-ornithine, a potent transition state analogue inhibitor of ornithine carbamoyltransferase. Biochem. Biophys. Res. Commun., **76**, 900-904 (1977)
[23] Eid, S.; Waly, Y.; Abdelal, A.T.: Separation and properties of two ornithine carbamoyltransferases from Pisum sativum seedlings. Phytochemistry, **13**, 99-102 (1974)
[24] Abdelal, A.T.H.; Kennedy, E.H.; Nainan, O.: Ornithine transcarbamylase from Salmonella typhimurium: purification, subunit composition, kinetic analysis, and immunological cross-reactivity. J. Bacteriol., **129**, 1387-1396 (1977)
[25] Hoogenraad, N.J.; Sutherland, T.M.; Howlett, G.J.: Purification of ornithine transcarbamylase from rat liver by affinity chromatography with immobilized transition-state analog. Anal. Biochem., **101**, 97-102 (1980)
[26] Martin, F.; Hirel, B.; Gadal, P.: Purification and properties of ornithine carbamoyltransferase 1 from Alnus glutinosa root nodules. Z. Pflanzenphysiol., **111**, 413-422 (1983)
[27] Hoogenraad, N.J.: Synthesis and properties of δ-N-(phosphonacetyl)-L-ornithine. A transition-state analog inhibitor of ornithine transcarbamylase. Arch. Biochem. Biophys., **188**, 137-144 (1978)

[28] Powers, C.N.; Pierson, D.L.: Stabilization and purification of ornithine transcarbamylase from Neisseria gonorrhoeae. J. Bacteriol., **141**, 544-549 (1980)

[29] Baker, S.R.; Yon, R.J.: Characterization of ornithine carbamoyltransferase from cultured carrot cells of low embryogenic potential. Phytochemistry, **22**, 2171-2174 (1983)

[30] Miller, N.T.; Feibush, B.; Corina, K.; Powers-Lee, S.; Karger, B.L.: High-performance hydrophobic interaction chromatography: purification of rat liver carbamoylphosphate synthetase I and ornithine transcarbamoylase. Anal. Biochem., **148**, 510-517 (1985)

[31] Bates, M.; Weiss, R.L.; Clarke, S.: Ornithine transcarbamylase from Neurospora crassa: purification and properties. Arch. Biochem. Biophys., **239**, 172-183 (1985)

[32] Tsuji, S.: Chicken ornithine transcarbamylase: purification and some properties. J. Biochem., **94**, 1307-1315 (1983)

[33] Templeton, M.D.; Sullivan, P.A.; Sheperd, M.G.: The inhibition of ornithine transcarbamoylase from Escherichia coli W by phaseolotoxin. Biochem. J., **224**, 379-388 (1984)

[34] Neway, J.O.; Switzer, R.L.: Purification, characterization, and physiological function of Bacillus subtilis ornithine transcarbamylase. J. Bacteriol., **155**, 512-521 (1983)

[35] Xiong, X.; Anderson, P.M.: Purification and properties of ornithine carbamoyl transferase from liver of Squalus acanthias. Arch. Biochem. Biophys., **270**, 198-207 (1989)

[36] Lusty, C.J.; Jilka, R.L.; Nietsch, E.H.: Ornithine transcarbamylase of rat liver. Kinetic, physical, and chemical properties. J. Biol. Chem., **254**, 10030-10036 (1979)

[37] Marshall, M.; Cohen, P.P.: The essential sulfhydryl group of ornithine transcarbamylases. Reaction with anionic, aromatic disulfides and properties of its cyano derivative. J. Biol. Chem., **255**, 7291-7295 (1980)

[38] Eisenstein, E.; Osborn, J.C.; Chaiken, I.M.; Hensley, P.: Purification and characterization of ornithine transcarbamoylase from Saccharomyces cerevisiae. J. Biol. Chem., **259**, 5139-5145 (1984)

[39] Pierson, D.L.; Cox, S.L.; Gilbert, B.E.: Human ornithine transcarbamylase. Purification and characterization of the enzyme from normal liver and the liver of a Reyes syndrome patient. J. Biol. Chem., **252**, 6464-6469 (1977)

[40] Kalousek, F.; Francois, B.; Rosenberg, L.E.: Isolation and characterization of ornithine transcarbamylase from normal human liver. J. Biol. Chem., **253**, 3939-3944 (1978)

[41] Koger, J.B.; Howell, R.G.; Kelly, M.; Jones, E.E.: Purification and properties of porcine liver ornithine transcarbamylase. Arch. Biochem. Biophys., **309**, 293-299 (1994)

[42] Tricot, C.; Schmid, S.; Baur, H.; Villeret, V.; Dideberg, O.; Haas, D.; Stalon, V.: Catabolic ornithine carbamoyltransferase of Pseudomonas aeruginosa. Changes of allosteric properties resulting from modifications at the C-terminus. Eur. J. Biochem., **221**, 555-561 (1994)

[43] Ruepp, A.; Mueller, H.N.; Lottspeich, F.; Soppa, J.: Catabolic ornithine transcarbamylase of Halobacterium halobium (salinarium): purification, characterization, sequence determination, and evolution. J. Bacteriol., **177**, 1129-1136 (1995)
[44] Coque, J.J.R.; Perez-Llarena, F.J.; Enguita, F.J.; Fuente, J.L.; Martin, J.F.; Liras, P.: Characterization of the cmcH genes of Nocardia lactamdurans and Streptomyces clavuligerus encoding a functional 3'-hydroxymethylcephem O-carbamoyltransferase for cephamycin biosynthesis. Gene, **162**, 21-27 (1995)
[45] Legrain, C.; Villeret, V.; Roovers, M.; Gigot, D.; Dideberg, O.; Pierard, A.; Glansdorff, N.: Biochemical characterization of ornithine carbamoyltransferase from Pyrococcus furiosus. Eur. J. Biochem., **247**, 1046-1055 (1997)
[46] Morizono, H.; Tuchman, M.; Rajagopal, B.S.; McCann, M.T.; Listrom, C.D.; Yuan, X.; Venugopal, D.; Barany, G.; Allewell, N.M.: Expression, purification and kinetic characterization of wild-type human ornithine transcarbamylase and a recurrent mutant that produces 'late onset' hyperammonemia. Biochem. J., **322**, 625-631 (1997)
[47] Sanchez, R.; Baetens, M.; Van De Casteele, M.; Roovers, M.; Legrain, C.; Glansdorff, N.: Ornithine carbamoyltransferase from the extreme thermophile Thermus thermophilus. Analysis of the gene and characterization of the protein. Eur. J. Biochem., **248**, 466-474 (1997)
[48] Lee, Y.; Jun, B.O.; Kim, S.G.; Kwon, Y.M.: Purification of ornithine carbamyltransferase from kidney bean (Phaseolus vulgaris) leaves and comparison of the properties of the enzyme from canavanine-containing and -deficient plants. Planta, **205**, 375-379 (1998)
[49] Villeret, V.; Clantin, B.; Tricot, C.; Legrain, C.; Roovers, M.; Stalon, V.; Glansdorff, N.; Van Beeumen, J.: The crystal structure of Pyrococcus furiosus ornithine carbamoyltransferase reveals a key role for oligomerization in enzyme stability at extremely high temperatures. Proc. Natl. Acad. Sci. USA, **95**, 2801-2806 (1998)
[50] Tricot, C.; Villeret, V.; Sainz, G.; Dideberg, O.; Stalon, V.: Allosteric regulation in Pseudomonas aeruginosa catabolic ornithine carbamoyltransferase revisited: association of concerted homotropic cooperative interactions and local heterotropic effects. J. Mol. Biol., **283**, 695-704 (1998)
[51] Sainz, G.; Tricot, C.; Foray, M.F.; Marion, D.; Dideberg, O.; Stalon, V.: Kinetic studies of allosteric catabolic ornithine carbamoyltransferase from Pseudomonas aeruginosa. Eur. J. Biochem., **251**, 528-533 (1998)
[52] Sainz, G.; Vicat, J.; Kahn, R.; Tricot, C.; Stalon, V.; Dideberg, O.: Purification, crystallization and preliminary X-ray analysis of catabolic ornithine carbamoyltransferase from Pseudomonas aeruginosa. Acta Crystallogr. Sect. D, **55**, 1591-1593 (1999)
[53] Langley, D.B.; Templeton, M.D.; Fields, B.A.; Mitchell, R.E.; Collyer, C.A.: Mechanism of inactivation of ornithine transcarbamoylase by N^{δ}-(N'-sulfodiaminophosphinyl)-L-ornithine, a true transition state analogue? Crystal structure and implications for catalytic mechanism. J. Biol. Chem., **275**, 20012-20019 (2000)

[54] Lee, Y.; Kwon, Y.M.: Identification of an isoform of ornithine carbamoyltransferase that can effectively utilize canaline as a substrate from the leaves of Canavalia lineata. Plant Sci., **151**, 145-151 (2000)
[55] Shi, D.; Morizono, H.; Yu, X.; Tong, L.; Allewell, N.M.; Tuchman, M.: Human ornithine transcarbamylase: crystallographic insights into substrate recognition and conformational changes. Biochem. J., **354**, 501-509 (2001)
[56] Roovers, M.; Sanchez, R.; Legrain, C.; Glansdorff, N.: Experimental evolution of enzyme temperature activity profile: selection in vivo and characterization of low-temperature-adapted mutants of Pyrococcus furiosus ornithine carbamoyltransferase. J. Bacteriol., **183**, 1101-1105 (2001)
[57] Lee, Y.; Choi, Y.A.; Hwang, I.D.; Kim, S.G.; Kwon, Y.M.: cDNA cloning of two isoforms of ornithine carbamoyltransferase from Canavalia lineata leaves and the effect of site-directed mutagenesis of the carbamoyl phosphate binding site. Plant Mol. Biol., **46**, 651-660 (2001)
[58] Bellocco, E.; Di Salvo, C.; Lagana, G.; Leuzzi, U.; Tellone, E.; Kotyk, A.; Galtieri, A.: Purification and properties of ornithine carbamoyltransferase from loggerhead turtle liver. Physiol. Res., **51**, 151-158 (2002)
[59] Bellocco, E.; Di Salvo, C.; Lagana, G.; Galtieri, A.; Ficarra, S.; Kotyk, A.; Leuzzi, U.: Ornithine carbamoyltransferase from Spinacia oleracea: purification and characterization. Biol. Plant., **45**, 533-538 (2002)

Malonyl-CoA carboxyltransferase 2.1.3.4

1 Nomenclature

EC number
2.1.3.4 (deleted)

Recommended name
malonyl-CoA carboxyltransferase

Oxamate carbamoyltransferase 2.1.3.5

1 Nomenclature

EC number
2.1.3.5

Systematic name
carbamoyl-phosphate:oxamate carbamoyltransferase

Recommended name
oxamate carbamoyltransferase

Synonyms
oxamic transcarbamylase

CAS registry number
62213-52-9

2 Source Organism

<1> *Streptococcus allantoicus* [1]
<2> *Streptococcus faecalis* (strain ATCC 11700, enzyme is induced in cells growing on oxalureate [2]) [2]

3 Reaction and Specificity

Catalyzed reaction
carbamoyl phosphate + oxamate = phosphate + oxalureate

Reaction type
carbamoyl group transfer

Natural substrates and products
S carbamoyl phosphate + oxamate <2> (<2> involved in the oxalureate catabolic pathway [2]) (Reversibility: ? <2> [2]) [2]
P phosphate + oxalureate <2> [2]

Substrates and products
S carbamoyl phosphate + oxamate <1> (Reversibility: r <1> [1]) [1]
P phosphate + oxalureate <1> [1]

Metals, ions

Ca^{2+} <1> (<1> supports activity [1]) [1]
Mg^{2+} <1> (<1> required for activity [1]) [1]
Mn^{2+} <1> (<1> required for activity [1]) [1]

Specific activity (U/mg)

0.11 <1> (<1> crude extract, in the presence of Ca^{2+} [1]) [1]
0.59 <1> (<1> crude extract, in the presence of 2.5 mM Mn^{2+} [1]) [1]
0.6 <1> (<1> crude extract, in the presence of 2.5 mM Mg^{2+} [1]) [1]

pH-Optimum

8.3-8.7 <1> [1]

pH-Range

7.5-9.5 <1> [1]

References

[1] Bojanowski, R.; Gaudy, E.; Valentine, R.C.; Wolfe, R.S.: Oxamic transcarbamylase of Streptococcus allantoicus. J. Bacteriol., **87**, 75-80 (1964)
[2] Wauwen, C.V.; Simon, J.P.; Slos, P.; Stalon, V.: Control of enzyme synthesis in the oxalureate catabolic pathway of Streptococcus faecalis ATCC 11700: evidence for the existence of a third carbamate kinase. Arch. Microbiol., **145**, 386-390 (1986)

Putrescine carbamoyltransferase 2.1.3.6

1 Nomenclature

EC number
2.1.3.6

Systematic name
carbamoyl-phosphate:putrescine carbamoyltransferase

Recommended name
putrescine carbamoyltransferase

Synonyms
PTCase
putrescine synthase
putrescine transcarbamylase

CAS registry number
9076-55-5

2 Source Organism

<1> *Streptococcus faecalis* (strain 10C1 [1]) [1, 2, 5]
<2> *Pediococcus acidilactici* [2]
<3> *Lathyrus sativus* [3, 4]
<4> *Cucumis sativus* [6]

3 Reaction and Specificity

Catalyzed reaction
carbamoyl phosphate + putrescine = phosphate + N-carbamoylputrescine (<3,4> multifunctional enzyme that catalyzes the reactions of EC 2.1.3.3 ornithine carbamoyltransferase, EC 2.7.2.2 carbamate kinase and EC 3.5.3.12 agmatine deiminase, thus acting as putrescine synthase, converting agmatine and ornithine into putrescine and citrulline, respectively [4,6])

Reaction type
transfer of carbamoyl phosphate

Natural substrates and products
S carbamoyl phosphate + putrescine <1> (Reversibility: r <1> [1]) [1]
P phosphate + N-carbamoylputrescine <1> (<1> phosphorolysis of N-carbamoylputrescine is possibly the physological role of the enzyme [1]) [1]

Substrates and products

S arsenate + N-carbamoylputrescine <1, 3> (<1> carbamoyl arsenate as intermediate is not formed [1]) (Reversibility: ir <1> [1]; <3> ? [4]) [1, 4]
P ?
S carbamoyl phosphate + 1,3-diaminopropane <1> (Reversibility: ? <1> [2]) [2]
P phosphate + N^1-carbamoyl-3-aminohexane <1> [1, 2]
S carbamoyl phosphate + 1,6-diaminohexane <1> (Reversibility: ? <1> [2]) [2]
P phosphate + N^1-carbamoyl-6-aminohexane <1> [1, 2]
S carbamoyl phosphate + cadaverine <1> (Reversibility: ? <1> [2]) [2]
P phosphate + N-carbamoylcadaverine <1> [1, 2]
S carbamoyl phosphate + ornithine <1> (Reversibility: ? <1> [2]) [2]
P phosphate + citrulline <1> [1, 2]
S carbamoyl phosphate + putrescine <1, 3> (Reversibility: r <1> [1]; ? <3> [3]) [1, 2, 3]
P phosphate + N-carbamoylputrescine <1, 3> [1, 2, 3]
S carbamoyl phosphate + spermidine <1> (Reversibility: ? <1> [2]) [2]
P phosphate + N-carbamoylspermidine <1> [1, 2]
S carbamoyl phosphate + spermine <1> (Reversibility: ? <1> [2]) [2]
P phosphate + N-carbamoylspermine <1> [1, 2]

Inhibitors

4-aminobutyrate <1> [2]
arginine <1> [2]
arsenate <1> (<1> competitive vs. carbamoylphosphate [2]) [2]
diphosphate <1> (<1> competitive vs. carbamoylphosphate [2]) [2]
glutamate <1> [2]
guanidobutyrate <1> [2]
norvaline <1> [2]
oxalurate <1> [2]
phosphate <1> (<1> competitive vs. carbamoylphosphate [2]) [2, 5]
spermidine <1> [5]

Specific activity (U/mg)

81 <1> [1]
460 <1> [2]
508.3 <1> [5]

K_m-Value (mM)

0.02 <1> (putrescine, <1> at pH 7.0 [2]) [2]
0.029 <1> (putrescine) [5]
0.17 <1> (carbamoyl phosphate, <1> at pH 7.0 [2]) [2]
0.35 <1> (spermine) [5]
2.9 <1> (1,3-diaminopropane) [5]
4 <1> (arsenate, <1> arsenolytic cleavage [2]) [2]
6.6 <1> (1,6-diaminohexane) [5]
7.7 <1> (cadaverine) [5]

10 <1> (N-carbamoylputrescine, <1> arsenolytic cleavage [2]) [2]
13 <1> (ornithine) [5]
19.7 <1> (spermidine) [5]

K_i-Value (mM)
0.06 <1> (diphosphate) [2]
2.5 <1> (spermidine, <1> vs. putrescine [5]) [5]
4 <1> (phosphate) [2]
4.5 <1> (spermidine, <1> vs. carbamoylphosphate [5]) [5]
6 <1> (phosphate, <1> vs. carbamoylphosphate [5]) [5]
8.5 <1> (phosphate, <1> vs. putrescine [5]) [5]
12 <1> (arsenate) [2]
13.5 <1> (phosphate, <1> vs. ornithine [5]) [5]

pH-Optimum
6.7 <1> (<1> in potassium 3,3-dimethylglutarate [1]) [1]
7 <1> (<1> in Tris-HCl [1]) [1]
7-9 <1> (<1> at 10 mM putrescine and carbamoyl phosphate [2]) [2, 5]
7.8 <1> (<1> at at 0.5 mM putrescine and carbamyol phosphate [2]) [2]
9 <1> (<1> at 0.5 mM putrescine [2]) [2, 5]

pH-Range
5-9 <1> [1]

4 Enzyme Structure

Molecular weight
56000 <3> (<3> gel filtration [4]) [4]
140000 <1> (<1> gel filtration [5]) [5]

Subunits
? <4> (<4> x * 150000, SDS-PAGE, nonreducing [6]; <4> x * 15000 + x * 18000 + x * 44000, SDS-PAGE, in the presence of 2-mercaptoethanol [6]) [6]
monomer <3> (<3> 1 * 55000, SDS-PAGE [4]) [4]
trimer <1> (<1> 3 * 40000, treatment with glutaraldehyde, SDS-PAGE [2,5]) [2, 5]

5 Isolation/Preparation/Mutation/Application

Source/tissue
seedling <3, 4> [3, 4, 6]

Purification
<1> (ammonium sulfate, heat treatment, Sephadex G-100 [1]; 65°C, ammonium sulfate, DEAE-Sephadex, Sephadex G-200 [2]) [1, 2, 5]
<3> ($MnCl_2$, ammonium sulfate, putrescine-CH-Sepharose affinity column [4]) [4]

6 Stability

Temperature stability

55 <1> (<1> carbamoylphosphate, diphosphate, ornithine, 2,4-diaminobutyrate and norvaline protect against denaturation [5]) [5]

General stability information

<3>, highly unstable even in the presence of glycerol, dithiothreitol and Mg^{2+}, activity is stabilized in dilute solutions, less than 0.05 mg protein/ml for about 3-4 h at 37°C by 0.25 mg/ml bovine serum albumine [4]

Storage stability

<1>, -20°C, 2.5 mg/ml, 50 mM potassium phosphate, pH 7.5, 3 years, no loss of activity [2]

<3>, 4°C, 48 h, complete loss of activity [4]

References

[1] Roon, R.J.; Barker, H.A.: Fermentation of agmatine in Streptococcus faecalis: occurence of putrescine transcarbamoylase. J. Bacteriol., **109**, 44-50 (1972)

[2] Stalon, V.: Putrescine carbamoyltransferase (Streptococcus faecalis). Methods Enzymol., **94**, 339343 (1983)

[3] Venugopal, K.S.S.; Adiga, P.R.: A simple procedure for purification of N-carbamoylputrescine: application to assays of putrescine transcarbamoylase and agmatine iminohydrolase activities. Anal. Biochem., **104**, 440-444 (1980)

[4] Srivenugopal, K.S.; Adiga, P.R.: Enzymic conversion of agmatine to putrescine in Lathyrus sativus seedlings. Purification and properties of a multifunctional enzyme (putrescine synthase). J. Biol. Chem., **256**, 9532-9541 (1981)

[5] Wargnies, B.; Lauwers, N.; Stalon, V.: Structure and properties of the putrescine carbamoyltransferase of Streptococcus faecalis. Eur. J. Biochem., **101**, 143-12 (1979)

[6] Prasad, G.L.; Adiga, P.R.: Purification and characterization of putrescine synthase from cucumber seedlings. A multifunctional enzyme involved in putrescine biosynthesis. J. Biosci., **10**, 373-391 (1986)

3-Hydroxymethylcephem carbamoyltransferase — 2.1.3.7

1 Nomenclature

EC number
2.1.3.7

Systematic name
carbamoyl-phosphate:3-hydroxymethylceph-3-em-4-carboxylate carbamoyltransferase

Recommended name
3-hydroxymethylcephem carbamoyltransferase

Synonyms
carbamoyltransferase, 3-hyroxymethylcephem

CAS registry number
74315-96-1

2 Source Organism

<-2> no activity in *Streptomyces lipmanii* [3]
<-1> no activity in *Streptomyces griseus* [3]
<1> *Streptomyces clavuligerus* [1]
<2> *Nocardia lactamdurans* [2, 3]
<3> *Streptomyces cattleya* [3]

3 Reaction and Specificity

Catalyzed reaction
carbamoyl phosphate + a 3-hydroxymethyl ceph-3-em-4-carboxylate = phosphate + a 3-carbamoyloxymethylcephem

Reaction type
carbamoyl group transfer

Natural substrates and products
S carbamoyl phosphate + a 3-hydroxymethyl ceph-3-em-4-carboxylate <1, 2> (<1>, catalyzes the synthesis of a wide range of carbamoyloxymethylcephems, e.g. cefuroxime [1]; <2>, the cmcH gene - encoding 3-hydroxymethylcephem carbamoyltransferase - upstream region does not contain a functional promoter, suggesting that cmcH is transcribed as a part of a

polycistronic mRNA [2]; <2>, enzyme is involved in biosynthesis of cephamycin [3]) (Reversibility: ? <1, 2> [1, 2, 3]) [1, 2, 3]

P phosphate + a 3-carbamoyloxymethylcephem

Substrates and products

S carbamoyl phosphate + 7-aminodeacetylcephalosporanic acid <1> (Reversibility: ? <1> [1]) [1]

P 7-β-acylamino-3-carbamoyloxymethylcephem + phosphate

S carbamoyl phosphate + 7β-acylamino-3-hydroxymethylcephem <1> (Reversibility: ? <1> [1]) [1]

P 7β-acylamino-3-carbamoyloxymethylcephem + phosphate

S carbamoyl phosphate + [3-hydroxymethyl]deacetylcephalosporin C <1> (<1>, ATP may be a substrate or an effector [1]) (Reversibility: ? <1> [1]) [1]

P 7β-(5-D-aminoadipamido)-3-carbamoyloxymethylceph-3-em-4-carboxylic acid + phosphate <1> [1]

S carbamoyl phosphate + decarbamoylcefuroxime <1> (Reversibility: ? <1> [1]) [1]

P carbamoylcefuroxime + phosphate <1> [1]

Inhibitors

$(NH_4)_2SO_4$ <1> [1]
ATP <1> (<1>, above 1 mM [1]) [1]
Ba^{2+} <1> (<1>, weak [1]) [1]
Cu^{2+} <1> [1]
Fe^{2+} <1> (<1>, weak [1]) [1]
NaCl <1> [1]
Pb^{2+} <1> (<1>, weak [1]) [1]
Sn^{2+} <1> [1]
Zn^{2+} <1> [1]
adenosine-α,β-methylene triphosphate <1> [1]
adenosine-β,γ-methylene triphosphate <1> [1]
diphosphate <1> [1]

Cofactors/prosthetic groups

ATP <1> (<1>, activation, up to 1 mM, may be substrate or effector, maximal activity in the presence of Mg^{2+} and Mn^{2+} [1]) [1]
dATP <1> (<1>, activation, can replace ATP to some extent [1]) [1]

Metals, ions

Mg^{2+} <1> (<1>, activation, maximal activity in the presence of Mn^{2+} and ATP [1]) [1]
Mn^{2+} <1> (<1>, activation, maximal activity in the presence of Mg^{2+} and ATP [1]) [1]

pH-Optimum

6.8 <1> [1]

pH-Range

5.9-7.4 <1> (<1>, about 75% of maximal activity at pH 5.9 and 7.4 [1]) [1]

4 Enzyme Structure

Subunits

? <2> (<2>, x * 57149, calculation from nucleotide sequence [3]) [3]

5 Isolation/Preparation/Mutation/Application

Localization

cytoplasm <1> [1]

Purification

<1> [1]

Cloning

<2> (cmcH subcloned in pIJ702 and expressed in Streptomyces lividans [3]) [3]

6 Stability

General stability information

<1>, phosphate anions stabilize during purification [1]

Storage stability

<1>, -10°C, 50 mM phosphate buffer, pH 7.0, stable for several weeks [1]
<1>, 4°C, in the presence of ATP and carbamoyl phosphate, stable for at least 3 days [1]

References

[1] Brewer, S.J.; Taylor, P.M.; Turner, M.K.: An adenosine triphosphate-dependent carbamoylphosphate-3-hydroxymethylcephem O-carbamoyltransferase from Streptomyces clavuligerus. Biochem. J., **185**, 555-564 (1980)

[2] Chary, V.K.; de la Fuente, J.L.; Liras, P.; Martin, J.F.: Amy as a reporter gene for promoter activity in Nocardia lactamdurans: comparison of promoters of the cephamycin cluster. Appl. Environ. Microbiol., **63**, 2977-2982 (1997)

[3] Coque, J.J.R.; Perez-Llarena, F.J.; Enguita, F.J.; Fuente, J.L.; Martin, J.F.; Liras, P.: Characterization of the cmcH genes of Nocardia lactamdurans and Streptomyces clavuligerus encoding a functional 3'-hydroxymethylcephem O-carbamoyltransferase for cephamycin biosynthesis. Gene, **162**, 21-27 (1995)

Lysine carbamoyltransferase 2.1.3.8

1 Nomenclature

EC number
2.1.3.8

Systematic name
carbamoyl-phosphate:L-lysine carbamoyltransferase

Recommended name
lysine carbamoyltransferase

Synonyms
lysine transcarbamylase <1> (<1> not identical with EC 2.1.3.3. ornithine carbamoyltransferase [1]) [1]

CAS registry number
86352-19-4

2 Source Organism

<1> *Bos taurus* [1]
<2> *Mus musculus* [2]
<3> *Rattus norvegicus* [2]

3 Reaction and Specificity

Catalyzed reaction
carbamoyl phosphate + L-lysine = phosphate + L-homocitrulline

Reaction type
transfer of carbamoyl phosphate

Substrates and products
S carbamoyl phosphate + L-lysine <1> (Reversibility: ? <1> [1]) [1]
P phosphate + L-homocitrulline <1> [1]

K_m-Value (mM)
6.3 <1> (L-lysine, <1> in mitochondrial extracts [1]) [1]

5 Isolation/Preparation/Mutation/Application

Source/tissue

liver <1, 2, 3> [1, 2]

Localization

mitochondrion <1, 2, 3> (<2,3> localized outside the inner mitochondrial membrane [2]) [1, 2]

References

[1] Hommes, F.A.; Eller, A.G.; Scott, D.F.; Carte, A.L.: Separation of ornithine and lysine activities of the ornithine-transcarbamylase-catalyzed reaction. Enzyme, **29**, 271-277 (1983)

[2] Carter, A.L.; Eller, A.G.; Rufo, S.; Metoki, K.; Hommes, F.A.: Further evidence for a separate enzymic entity for the synthesis of homocitrulline, distinct from the regular ornithine transcarbamylase. Enzyme, **32**, 26-36 (1984)

Glycine amidinotransferase 2.1.4.1

1 Nomenclature

EC number
2.1.4.1

Systematic name
L-arginine:glycine amidinotransferase

Recommended name
glycine amidinotransferase

Synonyms
EC 2.6.2.1 (formerly)
GAT <16> [11]
arginine-glycine amidinotransferase
arginine-glycine transamidinase
glycine transamidinase

CAS registry number
9027-35-4

2 Source Organism

<-2> no activity in *Streptomyces sp.* [17]
<-1> no activity in *Leuconostoc mesenteroides* [1]
<1> *Bos taurus* (cow [5]) [5]
<2> *Canavalia ensiformis* (jack bean [5]) [5]
<3> *Cuniculus sp.* (rabbit [13]) [13]
<4> *Felis domestica* (cat [13]) [13]
<5> *Gallus gallus* (cDNA clone, sequence of the chicken AT gene, GenBank Accession code AF237950 [19]) [19]
<6> *Gallus gallus* (chicken [5,8,19]; White Leghorn [19]) [5, 8, 19]
<7> *Glycine max* (soybean [21]) [21]
<8> *Homo sapiens* (mitochondrial enzyme [14,16]) [14, 16]
<9> *Homo sapiens* (cytosolic enzyme [14,16]) [14, 16]
<10> *Homo sapiens* (human [5,6,8,9,12-20]) [5, 6, 8, 9, 12-20]
<11> *Lacerta sp.* (lizard [5]) [5]
<12> *Lampetra planeri* (lamprey [17]) [17]
<13> *Mus musculus* (mouse [13]) [13]
<14> *Rana sp.* (frog [5,19]) [5, 19]

<15> *Rattus norvegicus* (cDNA sequence, EMBL accession number U07971 [14,16]) [14, 16]
<16> *Rattus norvegicus* (rat [2,6-13,15-17,19]; strain Sprague-Dawley [8,13]; strain Wistar [11]) [2, 5-13, 15-17, 19]
<17> *Sus scrofa* (pig, porcine, hog [1-9,12,14-17,19]) [1-9, 12, 14-17, 19]

3 Reaction and Specificity

Catalyzed reaction

L-arginine + glycine = L-ornithine + guanidinoacetate (canavanine can act instead of arginine)

Reaction type

amidine group transfer
transamidination

Natural substrates and products

S L-arginine + glycine <1-17> (<9,15> first reaction in the de novo biosynthesis of creatine [5,7,8,11,12,14-16,18,20]; <7> nitrogen metabolism, putrescine biosynthesis [21]) (Reversibility: r <1-17> [1-21]) [1-21]
P L-ornithine + guanidinoacetate
S L-ornithine + guanidinoacetate <6, 7, 9, 15, 17> (Reversibility: r <6, 7, 9, 15, 17> [1-21]) [1-21]
P L-arginine + glycine

Substrates and products

S L-alanine + glycine <10> (Reversibility: ? <10> [18]) [18]
P ?
S L-aminobutyric acid + glycine <10> (Reversibility: ? <10> [18]) [18]
P ?
S L-hydroxyguanidine + glycine <10> (Reversibility: ? <10> [18]) [18]
P hydroxylamine + guanidinoacetate
S L-norvaline + glycine <10> (Reversibility: ? <10> [5, 18]) [5, 18]
P ?
S arginine + 3-aminopropionic acid <10, 16> (Reversibility: ? <10, 16> [5, 13, 18]) [5, 13, 18]
P 3-guanidinopropionic acid + ornithine
S arginine + 4-aminobutyric acid <3, 10, 16> (Reversibility: ? <3, 10, 16> [5, 13, 18]) [5, 13, 18]
P 4-guanidinobutyric acid + ornithine
S arginine + 5-aminovaleric acid <16> (Reversibility: ? <16> [13]) [13]
P 5-guanidinovaleric acid + ornithine
S arginine + δ-aminovaleric acid <10> (Reversibility: ? <10> [18]) [18]
P δ-guanidinovaleric acid + ornithine
S arginine + ethanolamine <3, 10, 16> (Reversibility: ? <3, 10, 16> [13]) [13]
P 2-guanidinoethanol + ornithine

S arginine + γ-aminobutyric acid <10> (Reversibility: ? <10> [18]) [18]
P γ-guanidinobutyric acid + ornithine
S arginine + hydroxylamine <16, 17> (Reversibility: ? <16, 17> [2]) [2]
P ornithine + hydroxyguanidine
S arginine + lysine <16> (Reversibility: ? <16> [13]) [13]
P homoarginine + ornithine
S arginine + ornithine <16, 17> (Reversibility: ? <16, 17> [2]) [2]
P ornithine + arginine
S arginine + taurine <13, 16> (Reversibility: ? <13, 16> [13]) [13]
P 2-guanidinoethanesulfonic acid + ornithine
S canavanine + canaline <10, 16, 17> (Reversibility: ? <10, 16, 17> [2, 6]) [2, 6]
P canaline + canvanine
S canavanine + glycine <10, 16, 17> (Reversibility: r <10, 16, 17> [1, 2, 5, 6, 13, 18]) [1, 2, 5, 6, 13, 18]
P guanidinoacetate + canaline
S canavanine + hydroxylamine <16, 17> (Reversibility: ? <16, 17> [2]) [2]
P canaline + hydroxyguanidine
S canavanine + ornithine <1-17> (Reversibility: r <1-17> [1-21]) [1-21]
P arginine + canaline
S glycine + guanidionoacetate <16, 17> (Reversibility: r <16, 17> [2]) [2]
P guanidionoacetate + glycine
S guanidinoacetate + canaline <16, 17> (Reversibility: ? <16, 17> [2]) [2]
P glycine + canavine
S guanidinoacetate + hydroxylamine <10, 16, 17> (Reversibility: ? <10, 16, 17> [2, 3, 6]) [2, 3, 6]
P hydroxyguanidine + glycine
S guanidinoacetate + ornithine <16, 17> (Reversibility: ? <16, 17> [2, 3, 5]) [2, 3, 5]
P glycine + arginine
S homoarginine + glycine <10, 16, 17> (Reversibility: r <10, 16, 17> [3, 6, 13]) [3, 6, 13]
P homoornithine + guanidinoacetate
S Additional information <17> (<17> lysine will not substitute for ornithine [1]; <17> possible amidino group acceptors γ-aminobutyric acid, β-alanine, hydroxylamine [3]; <16> L-glutamic acid, L-aspartic acid and histidine are no substrates [13]; <10> homoarginine, α-amino-γ-guanidinobutyric acid and α-amino-β-guanidinopropionic acid are not substrates [18]) [1, 3, 13, 18]
P ?

Inhibitors

2-thio-6-aminouracil <16, 17> (<16,17> plus ferricyanide, strong inhibition [2]) [2]
2-thio-6-oxypurine <16, 17> (<16,17> plus ferricyanide, strong inhibition [2]) [2]
2-thiouracil <16, 17> (<16,17> plus ferricyanide, strong inhibition [2]) [2]

2-thiouracil-5-carboxylic acid <16, 17> (<16,17> plus ferricyanide, strong inhibition [2]) [2]
5,5'-dithiobis-2-nitrobenzoate <17> [5]
CO_2 <10, 16, 17> (<10,16,17> in presence of amidino group donor [5]) [5]
Cu^{2+} <16, 17> [2]
Hg^{2+} <9> [14]
L-norvaline <1, 6, 11, 14, 16, 17> (<1,6,11,14,16,17> good competitive inhibitor [5]) [5]
Ni^{2+} <9> [14]
Zn^{2+} <9> [14]
creatine <9> (<9> represses enzyme expression [14,16]) [14, 16]
dithiobiuret <16, 17> (<16,17> plus ferricyanide, strong inhibition [2]) [2]
glycine <16> (<16> inhibits in presence of low amounts of arginine [7]) [7]
guanidine hydrochloride <16> [11]
ornithine <10, 16, 17> (<16,17> above 1 mg per ml guanidinoacetate-ornithine transamidination is inhibited [2]; <10,16> competitive inhibition [8,18]) [2, 8, 14, 16, 18]
p-chloromercuribenzoate <16, 17> [2, 6]
thiolhistidine <16, 17> (<16,17> plus ferricyanide, strong inhibition [2]) [2]

Activating compounds

cysteine <16, 17> (<16,17> relative activity 103% [2]) [2]
thiolhistidine <16, 17> (<16,17> relative activity 102% [2]) [2]
thiourea <16, 17> (<16,17> relative activity 104% [2]) [2]
thyroxine <9> (<9> induces enzyme expression [16]) [16]

Specific activity (U/mg)

0.34 <16> [7]
1.25 <17> [3, 9]
23.33 <9> [9, 14]
Additional information <17> (<17> specific activity 3400 cpm per mg of protein [4]) [4]

K_m-Value (mM)

0.011 <16> (ornithine) [11]
0.89 <7> (glycine) [21]
1.8 <17> (arginine) [7]
2.27 <17> (arginine) [8]
2.3 <17> (arginine) [9]
2.3 <17> (glycine) [7]
2.4 <16> (arginine, <16> β-transamidinase [7,9]) [7, 9]
2.5 <10> (arginine) [9, 15]
2.5 <10> (glycine) [9]
2.8 <16> (arginine, <16> α-transamidinase [7,9]) [7, 9]
3 <16> (glycine, <16> α-transamidinase [7,9]) [7, 9]
3.1 <16> (glycine, <16> β-transamidinase [7,9]) [7, 9]
3.8 <7> (arginine) [21]

5 <16> (glycine, <16> α-and β-transamidinase [13]) [13]
9.21 <16> (arginine) [8]
14 <16> (arginine) [13]
14.3 <16> (arginine) [13]
18 <17> (glycine) [9]
23 <16> (lysine, <16> α-transamidinase [13]) [13]
23.5 <16> (5-aminovaleric acid, <16> β-transamidinase [13]) [13]
23.5 <16> (lysine, <16> β-transamidinase [13]) [13]
23.9 <16> (5-aminovaleric acid, <16> α-transamidinase [13]) [13]
24.9 <16> (4-aminobutyric acid, <16> β-transamidinase [13]) [13]
27.2 <16> (4-aminobutyric acid, <16> α-transamidinase [13]) [13]
57.4 <16> (3-aminopropionic acid, <16> α-transamidinase [13]) [13]
57.5 <16> (3-aminopropionic acid, <16> β-transamidinase [13]) [13]
163 <16> (ethanolamine) [13]
171 <16> (ethanolamine, <16> β-transamidinase [13]) [13]
174 <16> (ethanolamine, <16> α-transamidinase [13]) [13]
392 <16> (taurine, <16> α-transamidinase [13]) [13]
450 <16> (ethanolamine, <16> β-transamidinase [13]) [13]

K_i-Value (mM)

0.253 <8> (ornithine) [16]
0.253 <15> (ornithine) [8, 16]

pH-Optimum

7-7.5 <17> [6]
7.2 <17> [1]
7.4 <16, 17> (<16> synthesis of guanidinoacetic acid [13]) [3, 13]
7.5 <17> [9]
8.3 <16> (<16> synthesis of 4-guanidinobutyric acid [13]) [13]
8.4 <16> (<16> synthesis of homoarginine [13]) [13]
8.6 <16> (<16> synthesis of 5-guanidinovaleric acid [13]) [13]
9.2 <16> (<16> synthesis of 3-guanidinopropionic acid, 2-guanidinoethanol and 2-guanidinoethanesulfonic acid [13]) [13]
9.5 <7> [21]

pH-Range

6-9 <17> [3]
6.2-8.2 <17> [1]

Temperature optimum (°C)

37 <7> [21]

4 Enzyme Structure

Molecular weight

46000 <16> (<16> preprotein, size predicted from cDNA [12]) [12]
52000-56000 <10> (<10> wild-type enzyme AT, mutant ATΔ11, mutant ATΔM302, gel filtration [18]) [18]

57000 <16> (<16> in vitro synthesized preprotein, SDS-PAGE [12]) [12]
82600 <16> (<16> α-transamidinase, sedimentation equilibrium [7,9]) [7, 9]
83000 <16> (<16> gel filtration [7]) [7]
83300 <16> (<16> β-transamidinase, sedimentation equilibrium [7,9]) [7, 9]
84000 <16> [12]
89000 <10> (<10> gel filtration, equilibrium sedimentation [9]) [9]
89200 <10> (<10> mutant ATΔ11, calculated molecular mass [18]) [18]
90000 <10> (<10> mutant ATΔM302, sedimentation analysis [18]) [18]
90100 <10> (<10> mutant ATΔ11, sedimentation analysis [18]) [18]
91600 <10> (<10> mutant ATΔM302, calculated molecular mass [18]) [18]
91800 <10> (<10> wild-type enzyme AT, calculated molecular mass [18]) [18]
100000 <17> (<17> sucrose density gradient sedimentation [3]; <17> gel filtration [5]) [3, 5, 7, 9]
240000 <7> [21]

Subunits

dimer <10, 16, 17> (<16> 2 * 42000, α-transamidinase, SDS-PAGE, 2 * 44000, β-transamidinase, SDS-PAGE [7,9,12]; <10> 2 * 44000, SDS-PAGE [9]) [7, 9, 12, 16]
monomer <17> [16]
tetramer <7> [21]

5 Isolation/Preparation/Mutation/Application

Source/tissue

blood <10, 17> [6]
brain <3, 4, 10, 16, 17> [6, 8, 9, 12, 13, 20]
decidua <16> [5]
deciduoma cell <16> [5]
embryo <6> [5]
erythrocyte <17> [6]
fibroblast <10> [8]
heart <9, 17> [6, 15, 16]
kidney <6, 9, 15, 17> [1-15, 17, 18]
liver <1, 6, 9, 11, 15, 17> [5, 8, 9, 14, 15, 19, 20]
lung <17> [6]
lymphocyte <10> [8]
muscle <9, 15, 17> [5, 6, 8, 9, 12, 14-16, 18]
neuron <9> [5, 14, 16]
pancreas <6, 9, 15, 17> [5-9, 12-15, 17, 19, 20]
photoreceptor <9> [15, 16, 18]
proximal tubule <10, 16> [9, 12]
seed <2> [5]
seedling <7> [21]
semen <9> [15, 16]

spleen <10, 16, 17> [6, 9]
testis <17> [6]
thymus <17> [6]

Localization

cytoplasm <6, 9> [14, 15, 19]
cytosol <9> [14, 16]
mitochondrial inner membrane <9, 15> [8, 12, 16]
mitochondrial intermembrane space <6, 9> [14-16, 19]
mitochondrion <6, 9, 15, 17> [8, 12, 14-16, 19]

Purification

<7> [21]
<9> [9, 14, 15, 18]
<16> (2 fractions of enzyme, designated α- and β transamidase individually purified [7,13]) [7, 9, 12, 13]
<17> [3, 5, 15, 16]

Crystallization

<9> (recombinant human enzyme, bipyramidal, tetragonal crystals, belonging to space group P4(3)2(1)2, lattice constants a = b = 83.6 A, c = 200.4 A and $\alpha = \beta = \gamma = 90°$ [16,18]) [15, 16, 18]

Cloning

<6> (cDNA clone isolated [19]) [19]
<9> (cDNA, nucleotide sequence determined, cloned and expressed in Escherichia coli BL21(DE3)-pLysS [14-17]) [14-17]
<12> (cDNA, nucleotide sequence determined [17]) [17]
<16> (cDNA, nucleotide sequence determined [17]) [12, 15, 17]
<17> (cDNA, nucleotide sequence determined [17]) [15, 17]

Application

medicine <10> (<10> presence of the enzyme in serum or urine may prove useful in development of kidney disease diagnosis, hyperornithinemia, an autosomal recessive disease caused by decreased enzyme activity, enzyme activity is also dimished in thyrotoxicosis and myotonic muscular dystrophy [6,8,9]; <10> enzyme is a target of the estrogen receptor, is involved in carcinogenesis, mental disorder, osteoporosis and cardiovascular disease [19]; <10> diagnostic tool for biochemical diagnosis of creatine metabolism disorders [20]) [6, 8, 9, 19, 20]

6 Stability

Temperature stability

100 <17> (<17> inactivation by incubation for 5 min [1]) [1]

Storage stability

<16>, -18°C, homogenized tissues stored during 1 month without any loss of enzyme activity [8]

References

[1] Walker, J.B.: Biosynthesis of arginine from canavanine and ornithine in kidney. J. Biol. Chem., **218**, 549-556 (1956)

[2] Walker, J.B.: Studies on the mechanism of action of kidney transamidinase. J. Biol. Chem., **224**, 57-66 (1957)

[3] Conconi, F.; Grazi, E.: Transamidinase in hog kidney. I. Purification and properties. J. Biol. Chem., **240**, 2461-2463 (1965)

[4] Grazi, E.; Conconi, F.; Vigi, V.: Transamidinase in hog kidney. II. Isolation of a stable enzyme-amidine complex. J. Biol. Chem., **240**, 2465-2467 (1965)

[5] Walker, J.B.: Amidinotranferases. The Enzymes, 3rd Ed. (Boyer, P.D., ed.), **9**, 497-509 (1973)

[6] Dubach, U.C.: Transamidinase. Methods Enzym. Anal., 3rd Ed. (Bergmeyer, H.U., ed.), **1**, 740-743 (1974)

[7] McGuire, D.M.; Tormanen, C.D.; Segal, I.S.; Van Pilsum, J.F.: The effect of growth hormone and thyroxine on the amount of L-arginine:glycine amidinotransferase in kidneys of hypophysectomized rats. J. Biol. Chem., **255**, 1152-1159 (1980)

[8] Sipilae, I.: Inhibition of arginine-glycine amidinotransferase by ornithine. A possible mechanism for the muscular and chorioretinal atrophies in gyrate atrophy of the choroid and retina with hyperornithinemia. Biochim. Biophys. Acta, **613**, 79-84 (1980)

[9] Gross, M.D.; Eggen, M.A.; Simon, A.M.; Van Pilsum, J.F.: The purification and charcterization of human kidney L-arginine:glycine amidinotransferase. Arch. Biochem. Biophys., **251**, 747-755 (1986)

[10] Gross, M.D.; Simon, A.M.; Jenny, R.J.; Gray, E.D.; McGuire, D.M.; Van Pilsum, J.F.: Multiple forms of rat kidney L-arginine:glycine amidinotransferase. J. Nutr., **118**, 1403-1409 (1988)

[11] Raman Rao, S.V.R.; Indira, K.: Effects of guanidine hydrochloride on catalytic efficiency of glycine amidinotransferase of rat. Biochem. Mol. Biol. Int., **29**, 63-67 (1993)

[12] Guthmiller, P.; Van Pilsum, J.F.; Boen, J.R.; McGuire, D.M.: Cloning and sequencing of rat kidney L-arginine:glycine amidinotransferase. Studies on the mechanism of regulation by growth hormone and creatine. J. Biol. Chem., **269**, 17556-17560 (1994)

[13] Watanabe, Y.; Van Pilsum, J.F.; Yokoi, I.; Mori, A.: Synthesis of neuroactive guanidino compounds by rat kidney L-arginine:glycine amidinotransferase. Life Sci., **55**, 351-358 (1994)

[14] Humm, A.; Fritsche, E.; Mann, K.; Goehl, M.; Huber, R.: Recombinant expression and isolation of human L-arginine:glycine amidinotransferase and identification of its active-site cysteine residue. Biochem. J., **322**, 771-776 (1997)

[15] Humm, A.; Fritsche, E.; Steinbacher, S.: Structure and reaction mechanism of L-arginine:glycine amidinotransferase. Biol. Chem., **378**, 193-197 (1997)

[16] Humm, A.; Fritsche, E.; Steinbacher, S.; Huber, R.: Crystal structure and mechanism of human L-arginine:glycine amidinotransferase: a mitochon-

drial enzyme involved in creatine biosynthesis. EMBO J., **16**, 3373-3385 (1997)
[17] Bedekar, A.; Zink, R.M.; Sherman, D.H.; Line, T.V.; Van Pilsum, J.F.: The comparative amino acid sequences, substrate specificities and gene or cDNA nucleotide sequences of some prokaryote and eukaryote amidinotransferases: implications for evolution. Comp. Biochem. Physiol. B, **119B**, 677-690 (1998)
[18] Fritsche, E.; Humm, A.; Huber, R.: The ligand-induced structural changes of human L-arginine:glycine amidinotransferase. A mutational and crystallographic study. J. Biol. Chem., **274**, 3026-3032 (1999)
[19] Zhu, Y.; Evans, M.I.: Estrogen modulates the expression of L-arginine:glycine amidinotransferase in chick liver. Mol. Cell. Biochem., **221**, 139-145 (2001)
[20] Carducci, C.; Birarelli, M.; Leuzzi, V.; Carducci, C.; Battini, R.; Cioni, G.; Antonozzi, I.: Guanidinoacetate and creatine plus creatinine assessment in physiologic fluids: an effective diagnostic tool for the biochemical diagnosis of arginine:glycine amidinotransferase and guanidinoacetate methyltransferase deficiencies. Clin. Chem., **48**, 1772-1778 (2002)
[21] Lee, G.T.; Kim, W.J.; Cho, Y.D.: Polyamine synthesis in plants. Purification and properties of amidinotransferase from soybean (Glycine max) axes. Phytochemistry, **61**, 781-789 (2002)

scyllo-Inosamine-4-phosphate amidinotransferase 2.1.4.2

1 Nomenclature

EC number
2.1.4.2

Systematic name
L-arginine:1-amino-1-deoxy-scyllo-inositol-4-phosphate amidinotransferase

Recommended name
scyllo-inosamine-4-phosphate amidinotransferase

Synonyms
L-arginine:inosamine phosphate amidinotransferase
L-arginine:inosamine-P-amidinotransferase
aminocyclitolamidinotransferase <14> [6]
arginine:Xamidinotransferase <14> [1]
inosamine-P-amidinotransferase
inosamine-phosphate amidinotransferase

CAS registry number
52227-63-1

2 Source Organism

<-1> no activity in *Streptomyces griseus ATCC 10971* [1]
<1> *Streptomyces bikiniensis* (ATCC 11062 [1-4]) [1, 2, 4, 5]
<2> *Streptomyces bluensis* (genebank acc. No. X78972 [8]) [8]
<3> *Streptomyces bluensis* (var. bluensis NRRL 2876 [1,2]) [1, 2]
<4> *Streptomyces consensus* [8]
<5> *Streptomyces galbus* (ATCC 14077 [2,4]; strB1 [9]) [2, 4, 8, 9]
<6> *Streptomyces gallous* (genbank acc. No. S01837 [8]) [8]
<7> *Streptomyces glaucescens strB1* (strB1, genebank acc. No. X78974 [8,9]) [8, 9]
<8> *Streptomyces glaucescens strB2* (strB2, genebank acc. No. X78975 [8,9]) [8, 9]
<9> *Streptomyces griseocarneus* (ATCC 12628 [1,2,4]) [1, 2, 4]
<10> *Streptomyces griseus strB1* (EMBL accession number S55493 [9]) [9]
<11> *Streptomyces griseus strB1* (EMBL accession number P29780 [9]) [9]
<12> *Streptomyces griseus strB2* [8]
<13> *Streptomyces griseus* (strB1, genbank seq.acc. number Y00459 [8,9]) [8, 9]
<14> *Streptomyces griseus* (strain W4 [1]; ATCC 12475 [1,2,4]; ATCC10137 [4,8]; ATCC 11429 [4]; ATCC 11984 [2,4]; ATCC 27001 [4]; strB2 [8]; strain SPH [8]) [1, 2, 5-10]

<15> *Streptomyces humidus* (ATCC 12760 [2,4]) [2, 4]
<16> *Streptomyces hygroscopicus* (forma glebosus, ATCC 14607 [2-4]) [2-4]
<17> *Streptomyces kanamyceticus* (ATCC 12853 [1]) [1]
<18> *Streptomyces ornatus* (ATCC 23265 [4]) [4]

3 Reaction and Specificity

Catalyzed reaction

L-arginine + 1-amino-1-deoxy-scyllo-inositol 4-phosphate = L-ornithine + 1-guanidino-1-deoxy-scyllo-inositol 4-phosphate (1D-1-guanidino-3-amino-1,3-dideoxy-scyllo-inositol 6-phosphate, streptamine phosphate and 2-deoxystreptamine phosphate can also act as acceptors; canavanine can act as donor)

Reaction type

amidine group transfer

Natural substrates and products

S L-arginine + 1-amino-1-deoxy-scyllo-inositol 4-phosphate <1-18> (<1,2,13,16> biosynthesis of streptomycin [1-10]) (Reversibility: ? <1-18> [1-10]) [1-10]
P L-ornithine + 1-guanidino-1-deoxy-scyllo-inositol 4-phosphate

Substrates and products

S 2-deoxystreptidine 6-phosphate + scyllo-inosaminephosphate <13> (Reversibility: ? <13> [9]) [9]
P ?
S L-arginine + 1-amino-1-deoxy-scyllo-inositol-4-phosphate <1-18> (Reversibility: ? <1-18> [1-10]) [1-10]
P L-ornithine + 1-guanidino-1-deoxy-scyllo-inositol 4-phosphate
S L-canavanine + L-ornithine <1, 13, 16> (Reversibility: r <1, 13, 16> [2, 4, 9]) [2, 4, 9]
P L-canaline + L-arginine
S N-amidino-scyllo-inosamine 4-phosphate + scyllo-inosaminephosphate <13> (<13> weak amidino donor [9]) (Reversibility: ? <13> [9]) [9]
P ?
S arginine + 1,4-diaminobutylphosphonic acid <1> (Reversibility: ? <1> [2, 4, 9]) [2, 4, 9]
P ornithine + ?
S arginine + 1-D-1,3-diamino-1,2,3-trideoxy-scyllo-inositol 6-phosphate <16> (Reversibility: ? <16> [3]) [3]
P ornithine + ?
S arginine + 1-D-1,3-diamino-1,3-dideoxy-scyllo-inositol 6-phosphate <16> (Reversibility: ? <16> [3]) [3]
P ornithine + ?
S arginine + 1-D-1-O-carbamoyl-3-amino-3-deoxy-scyllo-inositol 6-phosphate <16> (Reversibility: ? <16> [3]) [3]

P ornithine + ?

S arginine + 1-D-1-guanidino-3-amino-1,3-dideoxy-scyllo-inositol 6-phosphate <16> (Reversibility: ? <16> [3]) [3]

P ornithine + ?

S arginine + 2-amino-2-deoxy-neo-inositol 5-phosphate <13> (Reversibility: ? <13> [9]) [9]

P ?

S arginine + 2-amino-2-deoxy-neoinositol 5-phosphate <1, 16> (Reversibility: ? <1, 16> [2, 4]) [2, 4]

P ornithine + ?

S arginine + 2-deoxystreptamine 6-phosphate <1, 3, 16> (Reversibility: ? <1, 3, 16> [1, 4]) [1, 4]

P N-amidino-2-deoxystreptamine 6-phosphate + ornithine

S arginine + 2-deoxystreptidine <1, 16> (Reversibility: ? <1, 16> [4]) [4]

P ornithine + ?

S arginine + 3-amino-3-deoxy-scyllo-inosamine 4-phosphate <13> (Reversibility: ? <13> [9]) [9]

P ?

S arginine + N-amidinostreptamine 6-phosphate <13> (Reversibility: ? <13> [9]) [9]

P ?

S arginine + NH_2OH <16> (Reversibility: ? <16> [3]) [3]

P ornithine + ?

S arginine + canaline <13> (Reversibility: ? <13> [9]) [9]

P ?

S arginine + hydroxylamine <1, 13> (Reversibility: ? <1, 13> [2, 5, 9]) [2, 5, 9]

P ornithine + hydroxyguanidine

S arginine + inosamine phosphate <14> (Reversibility: ? <14> [6]) [6]

P ornithine + ?

S arginine + ornithine <13> (Reversibility: ? <13> [9]) [9]

P ?

S arginine + scyllo-inosamine 4-phosphate <13> (Reversibility: ? <13> [9]) [9]

P ?

S arginine + scyllo-inosamine 4-phosphate <1, 3, 13, 16> (<3> reversibility of transamidinations with the monoamidinated inosamine phosphate has not yet been demonstrated [1]) (Reversibility: ? <1, 3, 13, 16> [1, 4, 9]) [1, 4, 9]

P ornithine + N-amidino-scyllo-inosamine-4-phosphate

S arginine + streptamine 6-phosphate <3> (Reversibility: ? <3> [1]) [1]

P ornithine + N-amidinostreptamine 6-phosphate

S streptidine 6-phosphate + scyllo-inosaminephosphate <13> (Reversibility: ? <13> [9]) [9]

P ?

S streptidine phosphate + glycylglycine <1, 3, 9, 13, 17> (Reversibility: ? <1, 3, 9, 13, 17> [1, 2, 4, 9]) [1, 2, 4, 9]

P ?

S Additional information <1, 14> (<1> inosamine derivatives must be phosphorylated to serve as substrates, phosphates must be esterified with the hydroxyl group para to thc amino group to bc transamidinated [4]; <14> enzyme cannot utilize glycine as guanidino acceptor [6,8]) [4, 6]

P ?

Inhibitors

cystamine <1, 16> [4]
cystine <1, 16> [4]
formamidine disulfidedihydrochloride <1, 16> [4]
ornithine <1, 16> (<1,16> product inhibition, most potent competitive inhibitor [4]) [4]
p-chloromercuribenzoate <1, 16> [4]

Specific activity (U/mg)

0.17 <1> [4]

4 Enzyme Structure

Molecular weight

42000 <14> (<14> SDS-PAGE [5]) [5]
68900 <14> (<14> sedimentation analysis [10]) [10]
80300 <14> (<14> calculated molecular mass [10]) [10]

5 Isolation/Preparation/Mutation/Application

Source/tissue

mycelium <1, 2, 5, 9, 13, 15-18> [1-10]

Purification

<1> (partially [2]) [1, 2, 4]
<3> [1]
<13> (recombinant enzyme, expressed in Streptomyces lividans [5]; recombinant enzyme [9]) [1, 5, 9]
<16> [4]

Crystallization

<13> (tetragonal-bipyramidal shaped crystals, belonging to the tetragonal space group P4(3), a = b = 121.3 A, c = 63.74 A [9]) [9]

Cloning

<13> (cloned in Streptomyces lividans using multicopy vector pIJ702 [5]; strain strB1, cloned as N-terminal hexa-histidine fusion protein [9]) [5, 9]

References

[1] Walker, M.S.; Walker, J.B.: Enzymic studies on the biosynthesis of streptomycin. J. Biol. Chem., **241**, 1262-1270 (1966)
[2] Walker, J.B.: Amidinotranferases. The Enzymes, 3rd Ed. (Boyer, P.D., ed.), **9**, 497-509 (1973)
[3] Walker, J.B.: Biosynthesis of the monoguanidinated inositol moiety of bluensomycin, a possible evolutionary precursor of streptomycin. J. Biol. Chem., **249**, 2397-2404 (1974)
[4] Walker, J.B.: L-Arginine:inosamine-P amindinotransferase (s). Methods Enzymol., **43**, 451-458 (1975)
[5] Tohyama, H.; Okami, Y.; Umezawa, H.: Nucleotide sequence of the streptomycinphosphotransferase and amidinotransferase from Streptomyces griseus. Nucleic acids Res., **15**, 1819-1833 (1987)
[6] Guthmiller, P.; Van Pilsum, J.F.; Boen, J.R.; McGuire, D.M.: Cloning and sequencing of rat kidney L-arginine:glycine amidinotransferase. Studies on the mechanism of regulation by growth hormone and creatine. J. Biol. Chem., **269**, 17556-17560 (1994)
[7] Humm, A.; Fritsche, E.; Mann, K.; Goehl, M.; Huber, R.: Recombinant expression and isolation of human L-arginine:glycine amidinotransferase and identification of its active-site cysteine residue. Biochem. J., **322**, 771-776 (1997)
[8] Bedekar, A.; Zink, R.M.; Sherman, D.H.; Line, T.V.; Van Pilsum, J.F.: The comparative amino acid sequences, substrate specificities and gene or cDNA nucleotide sequences of some prokaryote and eukaryote amidinotransferases: implications for evolution. Comp. Biochem. Physiol. B, **119B**, 677-690 (1998)
[9] Fritsche, E.; Bergner, A.; Humm, A.; Piepersberg, W.; Huber, R.: Crystal Structure of L-arginine:inosamine-phosphate amidinotransferase StrB1 from streptomyces griseus: an enzyme involved in streptomycin biosynthesis. Biochemistry, **37**, 17664-17672 (1998)
[10] Fritsche, E.; Humm, A.; Huber, R.: The ligand-induced structural changes of human L-arginine:glycine amidinotransferase. A mutational and crystallographic study. J. Biol. Chem., **274**, 3026-3032 (1999)

Transketolase 2.2.1.1

1 Nomenclature

EC number
2.2.1.1

Systematic name
sedoheptulose-7-phosphate:D-glyceraldehyde-3-phosphate glycolaldehydetransferase

Recommended name
transketolase

Synonyms
glycolaldehydetransferase

CAS registry number
9014-48-6

2 Source Organism

<1> *Oryctolagus cuniculus* [1, 21, 36]
<2> *Rattus norvegicus* [3, 12, 17, 18, 27, 36]
<3> *Mus musculus* [22, 36]
<4> *Sus scrofa* [12, 13, 36]
<5> *Homo sapiens* [14-16, 19, 23, 36]
<6> *Triticum aestivum* (wheat, cv. Sappo) [11]
<7> *Spinacia oleracea* (spinach) [1, 3, 10, 11, 31, 32, 36]
<8> *Candida utilis* [5, 36]
<9> *Saccharomyces cerevisiae* (baker's yeast) [1, 4, 6, 9, 10, 12, 20, 24-26, 29, 30, 35, 36]
<10> *Saccharomyces carlsbergensis* (brewer's yeast, strain Sapporo [7]) [1, 7, 36]
<11> *Torula sp.* [1]
<12> *yeast* [1, 8, 28]
<13> *Lactobacillus pentosus* [1]
<14> *Alcaligenes faecalis* [2]
<15> *Escherichia coli* (K12 [33]; JM 107/pQR 700 [34]) [33-36]

3 Reaction and Specificity

Catalyzed reaction

sedoheptulose 7-phosphate + D-glyceraldehyde 3-phosphate = D-ribose 5-phosphate + D-xylulose 5-phosphate

Reaction type

keto group transfer

Natural substrates and products

S D-xylulose 5-phosphate + D-ribose 5-phosphate <1, 2, 4-15> (<1, 7, 9-13> ubiquitous enzyme, involved in carbohydrate and nucleic acid metabolism, bacterial pentose biosynthesis, microbial shikimic acid biosynthesis, formaldehyde metabolism [1]; <7, 9> pathway in non-oxidative sequence of pentose cycle [10]; <15> ping pong bi bi mechanism [34]) (Reversibility: r <1, 2, 4-15> [1-6, 9, 11, 12, 14, 15, 17-19, 21]) [1-6, 9-12, 14, 15, 17-19, 21, 33, 34]

P sedoheptulose 7-phosphate + D-glyceraldehyde 3-phosphate

Substrates and products

S 2-deoxy-D-erythrose 4-phosphate + ? <12> (Reversibility: ? <12> [28]) [28]

P ?

S 2-deoxy-D-ribose 5-phosphate + ? <12> (Reversibility: ? <12> [28]) [28]

P ?

S D,L-glyceraldehyde + ? <15> (Reversibility: ? <15> [33]) [33]

P ?

S D,L-glyceraldehyde 3-phosphate + ? <9, 15> (Reversibility: ? <9, 15> [33, 36]) [33, 36]

P ?

S D-arabinose 5-phosphate + ? <12> (Reversibility: ? <12> [28]) [28]

P ?

S D-erythrose + ? <15> (Reversibility: ? <15> [33]) [33]

P ?

S D-erythrose 4-phosphate + ? <5, 12, 15> (Reversibility: ? <5, 12, 15> [28, 33, 36]) [28, 33, 36]

P ?

S D-ribose + ? <7> (Reversibility: ? <7> [36]) [36]

P ?

S D-threose 4-phosphate + ? <12> (Reversibility: ? <12> [28]) [28]

P ?

S D-xylose + ? <7> (Reversibility: ? <7> [36]) [36]

P ?

S D-xylulose 5-phosphate + D-erythrose 4-phosphate <2, 5-7> (<5> best acceptor with D-xylulose 5-phosphate as donor [14]) (Reversibility: r <2, 5-7> [3, 11, 14]) [3, 11, 14]

P fructose 6-phosphate + D-glyceraldehyde 3-phosphate

S D-xylulose 5-phosphate + D-ribose 5-phosphate <1, 2, 4-15> (<5> best donor with D-ribose 5-phosphate as acceptor [14]; <5> catalyzes reversible transfer of a keto group between several donor and acceptor substrates [14]; <1, 2, 4-7, 9-14> wide specificity for both reactants: donor substrates may be: fructose 6-phosphate [1-4, 6, 11, 12, 14, 18, 21]; <1, 2, 4, 6, 7, 9-13> sedoheptulose 7-phosphate [1, 11, 12]; <1, 7, 9-13> octulose 8-phosphate [1]; <1, 2, 4, 7, 9-13> L-erythrulose [1, 3, 4, 12]; <1, 7, 9-13> D-xylulose [1]; <1, 2, 4, 7, 9-13> and acceptor substrates may be: D-glyceraldehyde 3-phosphate [1, 3, 4, 10, 12]; <1, 2, 4-7, 9-13> D-erythrose 4-phosphate [1, 3, 11, 12, 14]; <1, 2, 4, 5, 7, 9-13> D-ribose 5-phosphate [1, 3, 4, 12, 14, 18, 21]; <1, 7, 9-13> D-deoxyribose 5-phosphate [1, 4]; <1, 7, 10-13> allose 6-phosphate, glucose 6-phosphate, formaldehyde [1]; <1, 7, 9-13> D-arabinose 5-phosphate [1]; <1, 2, 4, 7, 9-13> glycolaldehyde [1, 4, 12]; <1, 7, 9-13> DL-glyceraldehyde [1, 4]) (Reversibility: r <1, 2, 4-15> [1-6, 9, 11, 12, 14, 15, 17-19, 21, 25, 28-36]) [1-6, 9, 11, 12, 14, 15, 17-19, 21, 25, 28-36]

P sedoheptulose 7-phosphate + D-glyceraldehyde 3-phosphate

S L-arabinose + ? <7> (Reversibility: ? <7> [36]) [36]

P ?

S L-erythrulose + ribose 5-phosphate <9> (Reversibility: ? <9> [4]) [4]

P ?

S L-lyxose + ? <7> (Reversibility: ? <7> [36]) [36]

P ?

S formaldehyde + ? <15> (Reversibility: ? <15> [33]) [33]

P ?

S fructose 6-phosphate + ? <5, 7, 9, 15> (Reversibility: ? <5, 7, 9, 15> [33, 36]) [33, 36]

P ?

S fructose 6-phosphate + $Fe(CN)_6^{3-}$ <12> (Reversibility: ? <12> [8]) [8]

P glycolic acid + erythrose 4-phosphate + $Fe(CN)_6^{4-}$ + H^+

S fructose 6-phosphate + ribose 5-phosphate <1, 2, 5, 9> (<5> 40% as effective as D-xylulose 5-phosphate [14]) (Reversibility: ? <1, 2, 5, 9> [4, 14, 18, 21]) [4, 14, 18, 21]

P ?

S glycolaldehyde + ? <15> (Reversibility: ? <15> [33]) [33]

P ?

S hydroxyethylthiamine diphosphate <9> (Reversibility: ? <9> [24]) [24]

P ethanal + thiamine diphosphate

S hydroxypyruvate + ? <15> (Reversibility: ? <15> [33]) [33]

P ?

S hydroxypyruvate + D-glyceraldehyde 3-phosphate <1, 2, 4, 7, 9-13> (<7,9> enantioselective [10]) (Reversibility: ir <2, 4, 9> [12]) [1, 4, 10, 12]

P CO_2 + ribulose 5-phosphate

S hydroxypyruvate + ribose 5-phosphate <9, 15> (Reversibility: ? <9, 15> [4, 36]) [4, 36]

P sedoheptulose 7-phosphate + ?

S ribulose 5-phosphate + ribose 5-phosphate <9> (<9> ribulose is cleaved and ribose acts as acceptor [4]) (Reversibility: ? <9> [4]) [4]
P a heptulose phosphate + glyceraldehyde 3-phosphate
S sedoheptulose 7-phosphate + ? <15> (Reversibility: ? <15> [33]) [33]
P ?
S Additional information <2, 5, 9> (<9> no acceptors are D-ribose, formaldehyde, acetaldehyde, glucose 6-phosphate [4]; <2,5> no acceptors with D-xylulose 5-phosphate as donor substrate: glucose 6-phosphate [14,18]; <2> D-arabinose 5-phosphate [18]) [4, 14, 18]
P ?

Inhibitors

D-arabinose 5-phosphate <2, 15> (<15> competitive inhibition [33]) [18, 33]
EDTA <9, 10, 15> (<10> reversible by cofactors [7]) [7, 9, 33]
Hg^{2+} <5> [16]
L-erythrulose <15> (<15> competitive inhibition [34]) [34]
N-acetylimidazole <9> (<9> inhibition kinetics [20]) [20]
PCMB <5> (<5> reversible by cysteine [16]) [16]
arsenate <8> [5]
oxythiamine diphosphate <9, 12> (<9> strong [1]) [1, 12]
phenylglyoxal <10> (<10> prevents reconstitution of apotransketolase [7]) [7]
phosphate <8, 9> (<5> not [16]) [5, 12]
rabbit antibodies <5, 9> (<9> polyclonal [10]; <7> not [10]; <5> anti-transketolase IgG, 20% inhibition [16]; <2> not [17]) [10, 16]
sulfate <8, 9> [5, 12]

Cofactors/prosthetic groups

thiamine diphosphate <1-15> (<7,9> i.e. functional form of vitamin B1 [10]; <3,5> requirement, thiamine diphosphate protein, tightly bound [22,23]; <1-3,5> additional thiamine does not enhance activity [16,18,21-23]; <1,5,7,9-13> not thiamine, thiamine mono-or triphosphate, [1,15,16]; <2,4,9> 2 mol thiamine diphosphate per mol enzyme [12,13,17]; <1> 0.9 mol thiamine diphosphate per mol subunit [21]; <9> 2 catalytic sites per enzyme molecule [10]; <2,4,9> with different affinities for the cofactor [12]; <7> 1 catalytic site per enzyme molecule [10]) [1-12, 14-19, 21-23, 33]

Activating compounds

cysteine <5> (<5> activation, 10 mM [16]) [16]
ethanol <5> (<5> activation, 1-10% v/v [16]) [16]

Metals, ions

Ca^{2+} <9, 15> (<9> can replace Mg^{2+} to some extent [9]) [9, 15, 33]
Mg^{2+} <1, 2, 4-15> (<1,2,4-14> requirement [1-12,14]; <2,4,9> indispensable for thiamine diphosphate/enzyme binding [12]; chloroplastic, <6,7> not cytosolic enzyme [11]; <2,4,5,9> no exogenous Mg^{2+}-addition necessary [12,15,16]) [1-12, 14-16, 19, 33]
Additional information <2, 4, 9> (<4> no addition of exogenous metal ions required [12]) [12, 13]

Turnover number (min^{-1})

0.89 <7> (D-ribose 5-phosphate) [32]
1.12 <7> (xylulose 5-phosphate) [32]
3 <12> (D-arabinose 5-phosphate, <12> D477A [28]) [28]
4.2 <12> (D-threose 4-phosphate, <12> D477A [28]) [28]
18 <12> (2-deoxy-D-ribose 5-phosphate, <12> wild-type [28]) [28]
18 <12> (D-threose 4-phosphate, <12> wild-type [28]) [28]
24 <12> (2-deoxy-D-ribose 5-phosphate, <12> D477A [28]) [28]
24 <12> (D-arabinose 5-phosphate, <12> wild-type [28]) [28]
48 <12> (D-ribose 5-phosphate, <12> D477A [28]) [28]
78 <12> (2-deoxy-D-erythrose 4-phosphate, <12> D477A [28]) [28]
84 <12> (D-erythrose 4-phosphate, <12> D477A [28]) [28]
294 <12> (2-deoxy-D-erythrose 4-phosphate, <12> wild-type [28]) [28]
2760 <12> (D-ribose 5-phosphate, <12> wild-type [28]) [28]
3400 <12> (D-ribose 5-phosphate) [1]
3400 <12> (xylulose 5-phosphate) [1]
4140 <12> (D-erythrose 4-phosphate, <12> wild-type [28]) [28]

Specific activity (U/mg)

0.88 <4> [12, 13]
1.24 <14> [2]
1.5 <2> [12]
1.6 <6> [11]
3.2-3.9 <5> [16]
4 <7> [31]
8.1 <5> [19]
10.27 <5> [14]
12 <7> [32]
18 <9> [10]
20 <9> [12]
24 <12> [1]
25 <8> [5]
40 <7> [10]
50.4 <15> [33]

K_m-Value (mM)

0.0003 <9> (thiamine diphosphate, <9> H481A [26]) [26]
0.0004 <5> (thiamine diphosphate, <5> in the presence of Mg^{2+} [15]) [15]
0.0005 <9> (thiamine diphosphate, <9> H469A [25]; <9> H30N, H418S [26]) [25, 26]
0.0006 <9> (thiamine diphosphate, <9> wild-type [25,26]) [25, 26]
0.0007 <9> (thiamine diphosphate, <9> H481Q [26]) [26]
0.0008 <9> (thiamine diphosphate, <9> H30A [26]) [26]
0.0012 <9> (thiamine diphosphate, <9> D477A [25]; <9> H263A [26]) [25, 26]
0.0018 <9> (thiamine diphosphate, <9> H69A [26]) [26]
0.0019 <9> (thiamine diphosphate, <9> R359A [25]) [25]
0.0024 <9> (thiamine diphosphate, <9> R528A [25]) [25]
0.0054 <8> (thiamine diphosphate) [5]

0.007 <3> (thiamine diphosphate) [22]
0.023 <9> (D-xylulose 5-phosphate, <9> H263A [26]) [26]
0.025 <2> (D-xylulose 5-phosphate) [18]
0.0362 <2> (D-ribose 5-phosphate, <2> wild-type [27]) [27]
0.038 <1> (D-xylulose 5-phosphate) [21]
0.0399 <2> (D-xylulose 5-phosphate, <2> wild-type [27]) [27]
0.067 <7> (D-xylulose 5-phosphate) [32]
0.07 <9> (D-xylulose 5-phosphate, <9> wild-type [26]) [26]
0.073 <9> (D-xylulose 5-phosphate, <9> wild-type [25]) [25]
0.09 <15> (D-erythrose 4-phosphate) [33, 36]
0.093 <9> (D-ribose 5-phosphate, <9> H263A [26]) [26]
0.095 <1> (D-ribose 5-phosphate) [21]
0.12 <3> (D-xylulose 5-phosphate, <3> (+ D-ribose 5-phosphate) [22]) [22]
0.146 <9> (D-ribose 5-phosphate, <9> wild-type [25,26]) [25, 26]
0.15 <9> (D-ribose 5-phosphate, <9> H481A [26]) [26]
0.15 <5> (D-xylulose 5-phosphate, <5> (+ D-ribose 5-phosphate) [16]) [16]
0.16 <15> (D-xylulose 5-phosphate) [33, 36]
0.163 <9> (D-xylulose 5-phosphate, <9> R359A [25]) [25]
0.18 <5> (D-xylulose 5-phosphate, <5> (+ D-ribose 5-phosphate) [15]) [15]
0.193 <9> (D-ribose 5-phosphate, <9> H69A [26]) [26]
0.21 <9, 12> (D-xylulose 5-phosphate, <12> (+ D-ribose 5-phosphate) [1,12]) [1, 12, 36]
0.29 <9> (D-ribose 5-phosphate, <9> H30A [26]) [26]
0.3 <5, 5> (D-ribose 5-phosphate, (+ D-xylulose 5-phosphate) [15]) [15]
0.318 <9> (D-xylulose 5-phosphate, <9> R528A [25]) [25]
0.33 <3, 7> (D-ribose 5-phosphate, <3> (+ D-xylulose 5-phosphate) [22]) [22, 32]
0.36 <5> (D-erythrose 4-phosphate) [36]
0.4 <7, 9, 12> (D-ribose 5-phosphate, <12> (+ D-xylulose 5-phosphate) [1,12]) [1, 12, 16, 36]
0.4 <9> (D-xylulose 5-phosphate, <9> H69A [26]) [26]
0.44 <2> (D-xylulose 5-phosphate, <2> (+ D-ribose 5-phosphate) [17]) [17]
0.4579 <2> (D-xylulose 5-phosphate, <2> R433A [27]) [27]
0.49 <5> (D-xylulose 5-phosphate) [36]
0.53 <5> (D-ribose 5-phosphate) [36]
0.595 <9> (D-xylulose 5-phosphate, <9> D477A [25]) [25]
0.63 <5> (D-ribose 5-phosphate, <5> (+ D-xylulose 5-phosphate) [14]) [14]
0.72 <2> (D-ribose 5-phosphate, <2> (+ D-xylulose 5-phosphate) [17]) [17]
0.829 <9> (D-xylulose 5-phosphate, <9> H469A [25]) [25]
0.879 <2> (D-ribose 5-phosphate, <2> R433A [27]) [27]
1.01 <9> (D-xylulose 5-phosphate, <9> H30A [26]) [26]
1.1 <15> (fructose 6-phosphate) [33, 36]
1.24 <9> (D-xylulose 5-phosphate, <9> H481A [26]) [26]
1.4 <15> (D-ribose 5-phosphate) [33, 36]
1.75 <9> (D-ribose 5-phosphate, <9> D477A [25]) [25]
1.8 <9, 12> (fructose 6-phosphate, <12> (+ ribose 5-phosphate) [1,12]) [1, 12, 36]

2.1 <15> (D,L-glyceraldehyde 3-phosphate) [33, 36]
3.2 <7> (fructose 6-phosphate) [36]
3.35 <9> (D-xylulose 5-phosphate, <9> H481S [26]) [26]
4 <15> (sedoheptulose 7 phosphate) [33]
4.08 <9> (D-xylulose 5-phosphate, <9> H481Q [26]) [26]
4.9 <12> (L-erythrulose, <12> (+ glyceraldehyde 3-phosphate) [1,12]) [1, 12]
4.9 <9> (glyceraldehyde 3-phosphate) [36]
5.65 <9> (D-ribose 5-phosphate, <9> R359A [25]) [25]
5.97 <9> (D-ribose 5-phosphate, <9> H469A [25]) [25]
7 <9> (D-ribose 5-phosphate, <9> R528A [25]) [25]
7 <5> (fructose 6-phosphate) [36]
10 <15> (D,L-glyceraldehyde) [33]
11 <1> (fructose 6-phosphate) [21]
14 <15> (glycolaldehyde) [33]
18 <15> (hydroxypyruvate) [33, 36]
31 <15> (formaldehyde) [33]
33 <9> (hydroxypyruvate) [36]
45 <7> (D-xylose) [36]
55 <7> (L-arabinose) [36]
120 <7> (L-lyxose) [36]
150 <15> (D-erythrose) [33]
230 <7> (D-ribose) [36]
Additional information <9> (<9> kinetic data concerning the lag phase of transketolase reaction [9]) [9]

pH-Optimum

6.2-7.8 <9> (<9> R359A, R528A [25]) [25]
6.6-7.4 <9> (<9> H481Q [26]) [26]
6.6-7.8 <9> (<9> H469A [25]) [25]
6.8-7.8 <9> (<9> D477A [25]) [25]
6.8-8 <9> (<9> H30A [26]) [26]
7-7.5 <15> (<15> in phosphate buffer [36]) [36]
7-7.8 <9> (<9> H481A [26]) [26]
7.5 <9> [10]
7.5-7.6 <7, 9, 10> [36]
7.5-7.8 <9> (<9> H69A [26]) [26]
7.5-8.5 <6, 7> [11]
7.5-8.6 <5> [23]
7.6 <2, 4, 9, 12> (<12> sedoheptulose 7-phosphate formation [1]) [1, 12]
7.6-7.8 <5> [15]
7.6-8 <5, 9> (<9> H263A [26]) [16, 26]
7.6-8.2 <9> (<9> wild-type [25,26]) [25, 26]
7.7 <5, 8> [36]
7.8 <8> [5]
7.8-8 <1> [21]
7.8-8.2 <4> [12, 13]
8 <2, 7> [10, 17, 18]

8-8.5 <5, 15> (<15> in glycylglycine buffer [36]) [14, 33, 36]
8.5-9 <6, 7> [11]
Additional information <2, 4, 5, 8, 9> (<8> pI: 4.8 [5]; <5> pI: 6.5-8.5 (6-8 distinct isozymes, isoelectric focusing) [14]; <4> pI: 7.6-7.8 [12,13]; <5> pI: 8.5 [23]; <5> pI: 8.6 [16]) [5, 12-14, 16, 23]

pH-Range

5.8-9.5 <8> (<8> about half-maximal activity at pH 5.8 and 9.5 [5]) [5]
6.5-10.2 <2> (<2> about half-maximal activity at pH 6.5 and 10.2 [17]) [17]
6.5-10.5 <2> (<2> about half-maximal activity at pH 6.5 and about 60% of maximal activity at pH 10.5 [18]) [18]
6.8-9 <5> (<5> about half-maximal activity at pH 6.8 and 9.0 [16]) [16]
7.1-9.5 <5> (<5> about half-maximal activity at pH 7.1 and 9.5 [14]) [14]

Temperature optimum (°C)

20 <4> (<4> assay at [13]) [13]
25 <2, 7, 9> (<2,7,9> assay at [9,17,18,31]) [9, 17, 18, 29, 31]
30 <4, 9, 15> (<4,9,15> assay at [4,13,33]) [4, 13, 33]
37 <5, 8, 14> (<5,8,14> assay at [2,5,14-16]) [2, 5, 14-16]
40 <1> [21]
45 <2> (<2> assay at, low enzyme concentration [17]) [17]

4 Enzyme Structure

Molecular weight

100000 <10> (<10> gel filtration [7]) [7]
104000 <5> (<5> gel filtration [16]) [16]
130000 <2> (<2> sedimentation equilibrium method [12]) [12]
135000 <5> (<5> gel filtration [14]) [14]
136000 <5> (<5> [23]) [23]
138000 <4> (<4> meniscus depletion method [12,13]) [12, 13]
139000 <2> (<2> gel filtration [17,18]) [17, 18]
140000 <12> (<12> low speed sedimentation without reaching equilibrium [1]) [1]
145000 <15> (<15> gel filtration [33]) [33]
148000 <2> (<2> PAGE [27]) [27]
150000 <6> (<6> gel filtration [11]) [11]
158000-159000 <9> (<9> sedimentation equilibrium data [12]) [12]
160000 <7> (<7> gel filtration [32]) [32]
162000 <1> (<1> gel filtration [21]) [21]
163000 <8> (<8> gel filtration [5]) [5]

Subunits

? <5> (<5> ? * 70000, SDS-PAGE) [19]
dimer <2, 5-10, 15> (<10> 2 * 50000, SDS-PAGE after overnight exposition to 2-mercaptoethanol [7]; <5> α_2, 2 * 66000, SDS-PAGE [14,16,17]; <2> α_2, 2 * 69000, SDS-PAGE [17,18]; <2> α_2, 2 * 74000, SDS-PAGE [27]; <9> α_2, 2 *

74200, SDS-PAGE [10,25]; <6> apoenzyme, gel filtration after 2-mercaptoethanol incubation [11]; <8> α_2, 2 * 78000, SDS-PAGE [5]; <7> α_2, 2 * 74000, SDS-PAGE [32]; <15> α_2, 2 * 73000, SDS-PAGE [33]) [5, 7, 10, 11, 14, 16-18, 27, 32, 33]
monomer <7> (<7> 1 * 100000, SDS-PAGE) [10]
tetramer <4, 6> (<6> 4 * 37600, holoenzyme, SDS-PAGE [11]; <4> $\alpha_2\beta_2$, 2 * 27000-29000 + 2 * 52000-56000, SDS-PAGE [12,13]; <4> $\alpha_2\beta_2$, 2 * 30000 + 2 * 54000, pretreated with 1% SDS, gel filtration in the presence of 0.1% SDS [13]) [11-13]
Additional information <6, 9> (<6> catalytically active apoenzyme dimer and holoenzyme tetramer [11]; <9> at sufficiently low concentration the apo-, not the holoenzyme dissociates reversibly into 2 subunits of equal molecular weight, individual subunits are equally catalytically active [12]; <6> tetramer dissociates into dimers in the absence of thiamine diphosphate or in dilute solution [11]) [11, 12]

5 Isolation/Preparation/Mutation/Application

Source/tissue
brain <3> [22, 36]
erythrocyte <5> [12, 15, 16, 23, 36]
leaf <6, 7> [3, 10-12]
leukocyte <5> [14, 36]
liver <1, 2, 4> [1, 3, 12, 13, 17, 18, 21, 36]
muscle <1> [1]

Localization
chloroplast <6, 7, 11> (<6,7> no stimulation by light [11]) [1, 11, 31, 32]
cytosol <1, 2, 4, 6, 7, 9-13> [1, 11, 12, 17]

Purification
<1> [21]
<2> (immunoaffinity chromatography [17,18]; partial, 5 multiple forms by isoelectric focussing [17]) [12, 17, 18]
<4> [12, 13]
<5> (partial [15]) [14-16, 19, 23]
<6> [11]
<7> (immunoaffinity chromatography [10]) [3, 10, 11, 31, 32]
<8> (no isozymes, inseparable from triosephosphate isomerase EC 5.3.1.1 [5]) [5]
<9> (immunoaffinity chromatography [10]) [4, 10, 12]
<10> [7]
<14> [2]
<15> [33]

Renaturation

<6, 7> (thiamine diphosphate reconstitutes dissociated holoenzyme [11]) [11]

Crystallization

<9> (cocrystallization of apotransketolase with 5 mM thiamine diphosphate, 5 mM $CaCl_2$, 50 mM fructose-6-phosphate, 13-16% (w/W) polyethylenglycol 6000 in 50 mM glycyl-glycine buffer, pH 7.6, 0.0075 ml of a 20 mg/ml solution mixed with the same amount of mother liquid, space group: P212121 [25]) [1, 4, 12, 25, 29]
<15> [35]

Cloning

<2> (expression in Escherichia coli [27]) [27]
<7> (expression in Escherichia coli [31]) [31]
<9> (expression in H402 yeast cells [30]) [30]
<15> (overexpression in Escherichia coli [34]) [34]

Engineering

D477A <12> (<12> lower activity than wild-type enzyme [28]) [28]
D477A <9> (<9> lower activity than wild-type enzyme [25]) [25]
H263A <9> (<9> lower activity than wild-type enzyme [26]) [26, 30]
H30A <9> (<9> lower activity than wild-type enzyme [26]) [26]
H30N <9> (<9> lower activity than wild-type enzyme [26]) [26]
H469A <9> (<9> lower activity than wild-type enzyme [25]) [25]
H481A <9> (<9> lower activity than wild-type enzyme [26]) [26]
H481Q <9> (<9> lower activity than wild-type enzyme [26]) [26]
H481S <9> (<9> lower activity than wild-type enzyme [26]) [26]
H69A <9> (<9> lower activity than wild-type enzyme [26]) [26]
R102A <2> (<2> similar catalytyc activity like wild-type enzyme [27]) [27]
R350A <2> (<2> lower activity than wild-type enzyme [27]) [27]
R359A <9> (<9> lower activity than wild-type enzyme [25]) [25]
R433A <2> (<2> lower activity than wild-type enzyme, less stable at 55°C than wild-type [27]) [27]
R506A <2> (<2> lower activity than wild-type enzyme [27]) [27]
R528A <9> (<9> lower activity than wild-type enzyme [25]) [25]

6 Stability

pH-Stability

5 <4, 8> (<8> denaturation below [5]; <4> thiamine diphosphate dissociates from holoenzyme [12,13]) [5, 12, 13]
6-10 <8> (<8> stable in the presence of cofactors [5]) [5]
6.5 <2> (<2> below, unstable [17]) [17]
6.5-8.7 <4> (<4> 20°C, 20 min stable [12,13]) [12, 13]
Additional information <8> (<8> thiamine diphosphate enhances pH-stability) [5]

Temperature stability

20 <4> (<4> at pH 6.5-8.7, 20 min stable [12,13]) [12, 13]

40 <4> (<4> pH 8.0, activity hardly changes for 60 min [12,13]) [12, 13]

50 <4> (<4> $t_{1/2}$: 10 min, [12,13]; <4> 1 h stable in the presence of 4 mM thiamine diphosphate [13]) [12, 13]

55 <5> (<4> $t_{1/2}$: 5 min, albumin stabilizes) [16]

Additional information <4, 5> (<4> thiamine diphosphate enhances thermal stability [12]; <5> albumin enhances thermal stability [16]) [12, 16]

General stability information

<2>, glycerol, 10% v/v, stabilizes [17]

<5>, dilution, stable to [14]

<5>, freezing and thawing, looses activity [36]

<5>, freezing and thawing, stable to [16]

<5>, purification and storage lead to modification [16]

<8>, Mg^{2+} or thiamine diphosphate stabilizes [5]

Storage stability

<2>, -20°C, more than a week [17]

<2>, -20°C, up to 2 months with 10% v/v glycerol [17]

<2>, 0-4°C, partially purified, several months as precipitate in 80% ammonium sulfate [18]

<2>, 0-4°C, purified, several weeks as precipitate in 80% ammonium sulfate [18]

<2>, 0-4°C, several h [17]

<2>, frozen, at least a week [12]

<4>, -20°C, several days [12, 13]

<5>, -20°C, 2 months [14]

<5>, -20°C, stable [19]

<5>, -90°C, several weeks [16]

<5>, 4°C, at least 16 days [15]

<5>, 4°C, several weeks [14]

<5>, room temperature, after 13 days at 4°C, at least 4 h stable [15]

<9>, 0°C, crystalline, several weeks in 0.008 M glycylglycine buffer, pH 7.6 [4]

<9>, 4°C, several weeks in 8 M glycylglycine buffer, pH 7.4 [12]

<9>, room temperature, crystalline, in alkaline ammonium sulfate suspension [4]

<12>, prolonged storage of 10 months and longer leads to loss of solubility without loss of activity [1]

<14>, frozen, several months [2]

<1, 7, 9-13>, 2°C, crystalline, several months, after 10 months and longer the enzyme loses its solubility without losing activity [1]

<2, 7>, -16°C, several months in the presence of cofactors [3]

References

[1] Racker, E.: Transketolase. The Enzymes, 2nd Ed. (Boyer, P.D., Lardy, H., Myrbäck, K., eds.), **5**, 397-406 (1961)

[2] Domagk, G.F.; Horecker, B.L.: Fructose in erythrose metabolism in Alcaligensis faecalis. Arch. Biochem. Biophys., **109**, 342-349 (1965)

[3] Horecker, B.L.; Smyrniotis, P.Z.; Hurwitz, J.: The role of xylulose 5-phosphate in the transketolase reaction. J. Biol. Chem., **223**, 1009-1019 (1956)

[4] De La Haba, G.; Leder, I.G.; Racker, E.: Crystalline transketolase from bakers yeast: isolation and properties. J. Biol. Chem., **214**, 409-426 (1955)

[5] Klein, H.; Brand, K.: Purification and properties of transketolase from Candida utilis. Hoppe-Seyler's Z. Physiol. Chem., **358**, 1325-1337 (1977)

[6] Kochetov, G.A.; Philippov, P.P.; Razjivin, A.P.; Tikhomirova, N.K.: Kinetics of reconstruction of holo-transketolase. FEBS Lett., **53**, 211-212 (1975)

[7] Saitou, S.; Ozawa, T.; Tomita, I.: The purification and some properties of brewers yeast apotransketolase. FEBS Lett., **40**, 114-118 (1974)

[8] Kochetov, G.A.: Determination of transketolase activity via ferricyanide reduction. Methods Enzymol., **89**, 43-44 (1982)

[9] Egan, R.M.; Sable, H.Z.: Transketolase kinetics. The slow reconstitution of the holoenzyme is due to rate-limiting dimerization of the subunits. J. Biol. Chem., **256**, 4877-4883 (1981)

[10] Demuynck, C.; Fisson, F.; Bennani-Baiti, I.; Samaki, H.; Mani, J.C.: Immunoaffinity purification of transketolases from yeast and spnich leaves. Agric. Biol. Chem., **54**, 3073-3078 (1990)

[11] Murphy, D.J.; Walker, D.A.: The properties from transketolase from photosynthetic tissue. Planta, **155**, 316-320 (1982)

[12] Kochetov, G.A.: Transketolase from yeast, rat liver, and pig liver. Methods Enzymol., **90**, 209-223 (1982)

[13] Philippov, P.P.; Shestakova, I.K.; Tikhomirova, N.K.; Kochetov, G.A.: Characterization and properties of pig liver transketolase. Biochim. Biophys. Acta, **613**, 359-369 (1980)

[14] Mocali, A.; Paoletti, F.: Transketolase from human leukocytes. Isolation, properties and induction of polyclonal antibodies. Eur. J. Biochem., **180**, 213-219 (1989)

[15] Warnock, L.G.; Prudhomme, C.R.: The isolation and preliminary characterization of apotransketolase from human erythrocytes. Biochem. Biophys. Res. Commun., **106**, 719-723 (1982)

[16] Takeuchi, T.; Nishino, K.; Itokawa, Y.: Purification and characterization of, and preparation of an antibody to, transketolase from human red blood cells. Biochim. Biophys. Acta, **872**, 24-32 (1986)

[17] Paoletti, F.; Aldinucci, D.: Immunoaffinity purification of rat liver transketolase: evidence for multiple forms of the enzyme. Arch. Biochem. Biophys., **245**, 212-219 (1986)

[18] Paoletti, F.: Purification and properties of transketolase from fresh rat liver. Arch. Biochem. Biophys., **222**, 489-496 (1983)

[19] Schellenberg, G.D.; Wilson, N.M.; Copeland, B.R.; Furlong, C.E.: Transketolase from human red blood cells. Methods Enzymol., **90**, 223-228 (1982)
[20] Kuimov, A.N.; Kovina, M.V.; Kochetov, G.A.: Inhibition of transketolase by N-acetylimidazole. Biochem. Int., **17**, 517-521 (1989)
[21] Masri, S.W.; Ali, M.; Gubler, C.J.: Isolation of transketolase from rabbit liver and comparison of some of its kinetic properties with transketolase from other sources. Comp. Biochem. Physiol. B Comp. Biochem., **90**, 167-172 (1988)
[22] Blass, J.P.; Piacentini, S.; Boldizsar, E.; Baker, A.: Kinetic studies of mouse brain transketolase. J. Neurochem., **39**, 729-733 (1982)
[23] Heinrich, P.C.; Wiss, O.: Transketolase from human erythrocytes. Purification and properties. Helv. Chim. Acta, **54**, 2658-2668 (1971)
[24] Meshalkina, L.E.; Neef, H.; Tjaglo, M.V.; Schellenberger, A.; Kochetov, G.A.: The presence of a hydroxyl group at the C-1 atom of the transketolase substrate molecule is necessary for the enzyme to perform the transferase reaction. FEBS Lett., **375**, 220-222 (1995)
[25] Nilsson, U.; Meshalkina, L.; Lindqvist, Y.; Schneider, G.: Examination of substrate binding in thiamin diphosphate-dependent transketolase by protein crystallography and site-directed mutagenesis. J. Biol. Chem., **272**, 1864-1869 (1997)
[26] Wikner, C.; Nilsson, U.; Meshalkina, L.; Udekwu, C.; Lindqvist, Y.; Schneider, G.: Identification of catalytically important residues in yeast transketolase. Biochemistry, **36**, 15643-15649 (1997)
[27] Soh, Y.; Song, B.J.; Jeng, J.; Kallarakal, A.T.: Critical role of Arg433 in rat transketolase activity as probed by site-directed mutagenesis. Biochem. J., **333**, 367-372 (1998)
[28] Nilsson, U.; Hecquet, L.; Gefflaut, T.; Guerard, C.; Schneider, G.: Asp477 is a determinant of the enantioselectivity in yeast transketolase. FEBS Lett., **424**, 49-52 (1998)
[29] Kovina, M.V.; Tikhonova, O.V.; Solov'eva, O.N.; Bykova, I.A.; Ivanov, A.S.; Kochetov, G.A.: Influence of transketolase substrates on its conformation. Biochem. Biophys. Res. Commun., **275**, 968-972 (2000)
[30] Fiedler, E.; Golbik, R.; Schneider, G.; Tittmann, K.; Neef, H.; Konig, S.; Hubner, G.: Examination of donor substrate conversion in yeast transketolase. J. Biol. Chem., **276**, 16051-16058 (2001)
[31] Flechner, A.; Dressen, U.; Westhoff, P.; Henze, K.; Schnarrenberger, C.; Martin, W.: Molecular characterization of transketolase (EC 2.2.1.1) active in the Calvin cycle of spinach chloroplasts. Plant Mol. Biol., **32**, 475-484 (1996)
[32] Teige, M.; Melzer, M.; Suss, K.H.: Purification, properties and in situ localization of the amphibolic enzymes D-ribulose 5-phosphate 3-epimerase and transketolase from spinach chloroplasts. Eur. J. Biochem., **252**, 237-244 (1998)
[33] Sprenger, G.A.; Schoerken, U.; Sprenger, G.; Sahm, H.: Transketolase A of Escherichia coli K12. Purification and properties of the enzyme from recombinant strains. Eur. J. Biochem., **230**, 525-532 (1995)

[34] Gyamerah, M.; Willetts, A.J.: Kinetics of overexpressed transketolase from Escherichia coli JM 107/pQR 700. Enzyme Microb. Technol., **20**, 127-134 (1997)
[35] Schneider, G.; Lindqvist, Y.: Crystallography and mutagenesis of transketolase: mechanistic implications for enzymic thiamin catalysis. Biochim. Biophys. Acta, **1385**, 387-398 (1998)
[36] Schenk, G.; Duggleby, R.G.; Nixon, P.F.: Properties and functions of the thiamin diphosphate dependent enzyme transketolase. Int. J. Biochem. Cell Biol., **30**, 1297-1318 (1998)

Transaldolase 2.2.1.2

1 Nomenclature

EC number
2.2.1.2

Systematic name
sedoheptulose-7-phosphate:D-glyceraldehyde-3-phosphate glyceronetransferase

Recommended name
transaldolase

Synonyms
dihydroxyacetone synthase
dihydroxyacetonetransferase
formaldehyde transketolase

CAS registry number
9014-46-4

2 Source Organism

<1> *Saccharomyces cerevisiae* [1]
<2> *Saccharomyces carlsbergensis* (brewer's yeast) [1, 3]
<3> *Escherichia coli* (K12 [15,18]) [1, 3, 15-18]
<4> *Spinacia oleracea* [1-3]
<5> *Candida utilis* (3 isoenzymes: I, II, III [1]; I and II [6,11,12]) [1, 5-7, 10-12]
<6> *Bos taurus* [1, 9]
<7> *Oryctolagus cuniculus* [1]
<8> *Rattus norvegicus* [1, 3]
<9> *Homo sapiens* [1, 14, 19]
<10> *Dictyostelium discoideum* [4]
<11> *Arthrobacter sp.* (P1, 2 isoenzymes: C1-inducible enzyme, C2-constitutive enzyme) [8]
<12> *Musca domestica* [1]
<13> *Tetranychus telarius* [1]
<14> *Euglena sp.* [1]
<15> *Chlorella sp.* [1]
<16> *Chromatium sp.* [1]
<17> *Chlorobium thiosulfatophilum* [1]
<18> *Carcinus maenas* [13]

3 Reaction and Specificity

Catalyzed reaction

sedoheptulose 7-phosphate + D-glyceraldehyde 3-phosphate = D-erythrose 4-phosphate + D-fructose 6-phosphate (mechanism, <1-9,12-17> [1-3]; ping-pong kinetics with sedoheptulose 7-phosphate adding first, <10> [4])

Reaction type

aldehyde group transfer

Natural substrates and products

S sedoheptulose 7-phosphate + D-glyceraldehyde 3-phosphate <1-17> (Reversibility: r <1-9, 12-17> [1, 5]) [1-12]
P D-erythrose 4-phosphate + D-fructose 6-phosphate
S Additional information <1-9, 12-17> (<1-9,12-17> enzyme participates in the interconversion of metabolically significant triose, tetrose, pentose, hexose and heptose phosphates, enzyme is involved in the metabolism of carbohydrates via the nonoxidative part of the pentose phosphate pathway [1]) [1]
P ?

Substrates and products

S 2,5-D-threo-diketohexose phosphate + D-glyceraldehyde 3-phosphate <1-9, 12-17> (Reversibility: ? <1-9, 12-17> [1, 5, 15]) [1, 5, 15]
P ?
S D-erythrose 4-phosphate + sedoheptulose 7-phosphate <1-9, 12-17> (Reversibility: ? <1-9, 12-17> [1, 5, 15]) [1, 5, 15]
P ?
S D-erythrose+ sedoheptulose 7-phosphate <1-9, 12-17> (Reversibility: ? <1-9, 12-17> [1, 5, 15]) [1, 5, 15]
P ?
S D-fructose 6-phosphate + D-glyceraldehyde 3-phosphate <1-9, 12-17> (Reversibility: ? <1-9, 12-17> [1, 5, 15]) [1, 5, 15]
P ?
S D-glyceraldehyde + sedoheptulose 7-phosphate <1-9, 12-17> (Reversibility: ? <1-9, 12-17> [1, 5, 15]) [1, 5, 15]
P ?
S L-glyceraldehyde 3-phosphate + sedoheptulose 7-phosphate <1-9, 12-17> (Reversibility: ? <1-9, 12-17> [1, 5, 15]) [1, 5, 15]
P ?
S L-sorbose 6-phosphate + D-glyceraldehyde 3-phosphate <1-9, 12-17> (Reversibility: ? <1-9, 12-17> [1, 5, 15]) [1, 5, 15]
P ?
S hydroxypyruvic aldehyde + sedoheptulose 7-phosphate <1-9, 12-17> (Reversibility: ? <1-9, 12-17> [1, 5, 15]) [1, 5, 15]
P ?
S octulose 8-phosphate phosphate + D-glyceraldehyde 3-phosphate <1-9, 12-17> (Reversibility: ? <1-9, 12-17> [1, 2]) [1, 2]

P ?

S ribose 5-phosphate + sedoheptulose 7-phosphate <1-9, 12-17> (Reversibility: ? <1-9, 12-17> [1, 2]) [1, 2]

P ?

S sedoheptulose 7-phosphate + D-glyceraldehyde 3-phosphate <1-18> (<5> maximal forward rate is about one third of the maximal reverse rate [5]; <1-9,12-17> not: glycoaldehyde, glycoaldehyde phosphate [1,2]) (Reversibility: r <1-9, 12-18> [1, 5, 13, 15]) [1-15]

P D-erythrose 4-phosphate + D-fructose 6-phosphate

S sedoheptulose phosphate + D-glyceraldehyde 3-phosphate <1-9, 12-17> (Reversibility: ? <1-9, 12-17> [1, 5, 15]) [1, 5, 15]

P ?

Inhibitors

D-arabinose-5-phosphate <3> [15]
D-erythrose 4-phosphate <10> [4]
L-glyceraldehyde <3> [15]
Tris-HCl <3> [15]
diphosphate <1-9, 12-17> [1]
phosphate <1-10, 12-17> [1, 2, 4, 7, 15]
sedoheptulose 7-phosphate <10> [4]
sodium borohydride <6> [9]
sulfate <1-9, 12-17> [1]
tetranitromethane <5> (<5> isoenzyme III, only in presence of the substrates D-fructose 6-phosphate and sedoheptulose 7-phosphate [10]) [10]

Cofactors/prosthetic groups

Additional information <1-9, 12-17> (<1-9,12-17> no cofactor requirement [1,2,9]) [1, 2, 9]

Metals, ions

Additional information <4> (<4> no metal requirement [2]) [2]

Specific activity (U/mg)

0.9 <18> [13]
3.84 <6> [9]
15 <9> [14]
60 <3> [15]

K_m-Value (mM)

0.007 <6> (D-erythrose 4-phosphate) [9]
0.018 <1> (D-erythrose 4-phosphate) [1]
0.02 <5> (D-erythrose 4-phosphate) [6]
0.038 <3> (glyceraldehyde 3-phosphate) [15]
0.072 <10> (glyceraldehyde 3-phosphate) [4]
0.078 <3> (D-erythrose 4-phosphate, <3> T156A [18]) [18]
0.09 <3> (D-erythrose 4-phosphate) [15]
0.1 <10> (D-erythrose 4-phosphate) [4]
0.1 <3> (D-erythrose 4-phosphate, <3> wild-type, N35A [18]) [18]

0.11 <3> (D-erythrose 4-phosphate, <3> D17N [18]) [18]
0.111 <3> (D-erythrose 4-phosphate, <3> S176A [18]) [18]
0.117 <3> (D-erythrose 4-phosphate, <3> E96Q [18]) [18]
0.122 <3> (D-erythrose 4-phosphate, <3> E96A [18]) [18]
0.18 <1> (sedoheptulose 7-phosphate) [1-3]
0.2 <11> (D-fructose 6-phosphate) [8]
0.22 <1> (glyceraldehyde 3-phosphate) [1-3]
0.265 <3> (D-erythrose 4-phosphate, <3> R300A [17]) [17]
0.285 <3> (sedoheptulose 7-phosphate) [15]
0.295 <3> (D-erythrose 4-phosphate, <3> wild-type [17]) [17]
0.32 <1> (D-fructose 6-phosphate) [1]
0.35 <3> (D-erythrose 4-phosphate, <3> R300E [17]) [17]
0.46 <10> (sedoheptulose 7-phosphate) [4]
0.555 <3> (D-erythrose 4-phosphate, <3> D17A [18]) [18]
0.6 <3> (D-fructose 6-phosphate, <3> E96A [18]) [18]
0.7 <3> (D-fructose 6-phosphate, <3> D17A [18]) [18]
0.8 <5> (D-fructose 6-phosphate) [6]
0.94 <3> (D-fructose 6-phosphate, <3> wild-type [17]) [17]
0.96 <3> (D-fructose 6-phosphate, <3> D17N [18]) [18]
1.02 <3> (D-fructose 6-phosphate, <3> T156A [18]) [18]
1.1 <3> (D-fructose 6-phosphate, <3> N35A [18]) [18]
1.14 <3> (D-fructose 6-phosphate, <3> E96Q [18]) [18]
1.2 <3> (D-fructose 6-phosphate, <3> R300E [17]) [17]
1.2 <3> (D-fructose 6-phosphate, <3> wild-type [18]) [15, 18]
1.35 <3> (D-fructose 6-phosphate, <3> R300A [17]) [17]
1.6 <10> (D-fructose 6-phosphate) [4]
4.9 <3> (D-fructose 6-phosphate, <3> S176A [18]) [18]

pH-Optimum

7.3-8.1 <2, 4> [2, 3]
7.8 <5> [6]
8 <5> (<5> isoenzyme I, II, III [1]) [1, 7]
8.2 <6> [9]
8.5-9.5 <3> [15]

pH-Range

5-11 <3> [15]
6.3-9.5 <5> (<5> about 50% of activity maximum at pH 6.3 and 9.5, isoenzyme I, II, III [1,6]) [1, 6]

Temperature optimum (°C)

15-40 <3> [15]

4 Enzyme Structure

Molecular weight

36200 <18> (<18> computer analysis of a 2-D PAGE [13]) [13]
47500 <11> (<11> gel filtration, C1-inducible enzyme [8]) [8]
52000 <11> (<11> gel filtration, constitutive enzyme [8]) [8]
63600-65200 <5> (<5> ultracentrifugation and gel filtration, isoenzyme III [1]) [1]
65000 <6> (<6> gel filtration [9]) [9]
68100 <5> (<5> ultracentrifugation and gel filtration, isoenzyme II [1]) [1]
70000 <3> (<3> gel filtration [15]) [15]
76100-77400 <5> (<5> ultracentrifugation and gel filtration, isoenzyme I, sedimentation equilibrium ultracentrifugation [1]) [1]

Subunits

dimer <3, 5, 9> (<5> α_2, 2 * 38000, SDS-PAGE isoenzyme I [12]; <5> β_2, 2 * 35500, SDS-PAGE, isoenzyme III [12]; <5> $\alpha\beta$, 1 * 38000 + 1 * 35500, SDS-PAGE, isoenzyme II [1]; <3> α_2, 2 * 35000, SDS-PAGE [15]; <9> the crystal forms a dimer [19]) [1, 12, 15, 19]
monomer <11> (<11> 1 * 50000, SDS-PAGE [8]) [8]

5 Isolation/Preparation/Mutation/Application

Source/tissue

Y-organ <18> [13]
artery <9> (<9> wall [1]) [1]
brain <6-8> [1]
commercial preparation <6> [9]
erythrocyte <9> [1]
kidney <6-8> [1]
leaf <4> [3]
liver <6-8> [1, 3]
mammary gland <7, 8> [1, 9]

Localization

mitochondrion <18> [13]

Purification

<1> [1]
<2> (brewer's yeast [1,3]) [1, 3]
<3> [15, 16]
<5> (isoenzymes: I [1,6,11,12]; II [1]; III [1,6,11,12]) [1, 6, 11, 12]
<6> [9]
<10> [4]
<11> [8]

Crystallization

<3> (crystallization with a combination of micro- and macro-seeding techniques, space group: P212121 [16]) [16]
<9> (hanging drop method, mother liquid consits of: 27% w/v polyethylenglycol 4000, 0.08% NaN_3, 0.3 M NH_4COO, pH 4.4 [19]) [19]
<1-3,5> [1, 2, 6, 7, 12, 16, 18]

Cloning

<3> (overexpression in Escherichia coli [15]) [15]
<9> (expression as a fusion protein with glutathione S-transferase [14]) [14]
<9> (expression in Escherichia coli [19]) [19]

Engineering

D17A <3> (<3> lower specific activity than wild-type [18]) [18]
E96A <3> (<3> lower specific activity than wild-type [18]) [18]
N35A <3> (<3> lower specific activity than wild-type [18]) [18]
R300A <3> (<3> little lower activity than the wild type, but same stability against urea and thermal inactivation [17]) [17]
R300E <3> (<3> little lower activity than the wild type, but same stability against urea and thermal inactivation [17]) [17]
S176A <3> (<3> lower specific activity than wild-type [18]) [18]
T156A <3> (<3> lower specific activity than wild-type [18]) [18]

6 Stability

pH-Stability

4-9 <6> (<6> stable [9]) [9]
7 <5> (<5> room temperature, stable for a few hours [6]) [6]

Temperature stability

22 <5> (<5> room temperature, at neutral pH stable for a few hours [6]) [6]
50 <3> (<3> above 50°C rapid loss in activity [15]) [15]
50 <4, 11> (<4> heat-labile [2]; <11> 2 min, C1-inducible enzyme loses 90% of activity, constitutive enzyme remains fully active [8]) [2, 8]
55 <3> (<3> total inactivation of enzyme [15]) [15]

General stability information

<4>, highly purified preparations can be dialyzed with little loss of activity [2]
<4>, stable to lyophilization [2]
<6>, complete inactivation by treatment with sodium borohydride at pH 6.0 and 2°C in absence of substrate D-fructose 6-phosphate [9]

Storage stability

<2>, -16°C, 10-20% loss of activity after 1 week [3]
<4>, -20°C, stable for several months [2]
<5>, isoenzymes I and III are stable in the cold in the crystalline form for over 1 year [6]

<6>, 2°C, for at least 8 weeks [9]
<11>, -20°C, stable for at least 2 months [8]

References

[1] Tsolas, O.; Horecker, B.L.: Transaldolase. The Enzymes, 3rd Ed. (Boyer, P.D., ed.), **7**, 259-280 (1972)
[2] Racker, E.: Transaldolase. The Enzymes, 2nd.Ed. (Boyer, P.D., Lardy, H., Myrbäck, K., eds.), **5**, 407-412 (1961)
[3] Horecker, B.L.; Smyrniotis, P.Z.: Purification and properties of yeast transaldolase. J. Biol. Chem., **212**, 811-825 (1955)
[4] Albe, K.R.: Partial purification and kinetic characterization of transaldolase from Dictyostelium discoideum. Exp. Mycol., **15**, 255-262 (1991)
[5] Wood, T.: The forward and reverse reactions of transaldolase. FEBS Lett., **25**, 153-155 (1972)
[6] Tsolas, O.; Joris, L.: Transaldolase. Methods Enzymol., **42 C**, 290-297 (1975)
[7] Horecker, B.L.; Tsolas, O.: Purification and crystallization of transaldolase from Candida utilis. Methods Enzymol., **182**, 788-793 (1990)
[8] Levering, P.R.; Dijkhuizen, L.: Transaldolase isoenzymes frm Arthrobacter P1. Methods Enzymol., **188**, 405-411 (1990)
[9] Kuhn, E.; Brand, K.: Purification and properties of transaldolase from bovine mammary gland. Biochemistry, **11**, 1767-1772 (1972)
[10] Kobashi, K.; Brand, K.: Substrate-dependent inactivation of transaldolase activity with tetranitromethane. Arch. Biochem. Biophys., **148**, 169-179 (1972)
[11] Schutt, H.; Brand, K.: Active center and subunit structure of transaldolase from Candida utilis. Arch. Biochem. Biophys., **169**, 287-297 (1975)
[12] Sun, S.C.; Joris, L.; Tsolas, O.: Purification of crystallization of transaldolase isozyme I and evidence for different genetic origin of isozymes I and III in Candida utilis. Arch. Biochem. Biophys., **178**, 69-78 (1977)
[13] Lachaise, F.; Somme, G.; Carpentier, G.; Granjeon, E.; Webster, S.; Baghdassarian, D.: A transaldolase. An enzyme implicated in crab steroidogenesis. Endocrine, **5**, 23-32 (1996)
[14] Banki, K.; Perl, A.: Inhibition of the catalytic activity of human transaldolase by antibodies and site-directed mutagenesis. FEBS Lett., **378**, 161-165 (1996)
[15] Sprenger, G.A.; Schoerken, U.; Sprenger, G.; Sahm, H.: Transaldolase B of Escherichia coli K-12: cloning of its gene, talB, and characterization of the enzyme from recombinant strains. J. Bacteriol., **177**, 5930-5936 (1995)
[16] Jia, J.; Schorken, U.; Lindqvist, Y.; Sprenger, G.A.; Schneider, G.: Crystal structure of the reduced Schiff-base intermediate complex of transaldolase B from Escherichia coli: mechanistic implications for class I aldolases. Protein Sci., **6**, 119-124 (1997)
[17] Schorken, U.; Jia, J.; Sahm, H.; Sprenger, G.A.; Schneider, G.: Disruption of Escherichia coli transaldolase into catalytically active monomers: evidence against half-of-the-sites mechanism. FEBS Lett., **441**, 247-250 (1998)

[18] Schorken, U.; Thorell, S.; Schurmann, M.; Jia, J.; Sprenger, G.A.; Schneider, G.: Identification of catalytically important residues in the active site of Escherichia coli transaldolase. Eur. J. Biochem., **268**, 2408-2415 (2001)
[19] Thorell, S.; Gergely, P., Jr.; Banki, K.; Perl, A.; Schneider, G.: The three-dimensional structure of human transaldolase. FEBS Lett., **475**, 205-208 (2000)

Formaldehyde transketolase 2.2.1.3

1 Nomenclature

EC number
2.2.1.3

Systematic name
D-xylulose-5-phosphate:formaldehyde glycolaldehydetransferase

Recommended name
formaldehyde transketolase

Synonyms
dihydroxyacetone synthase
synthase, dihydroxyacetone
transketolase, formaldehyde
Additional information (not identical with EC 2.2.1.1)

CAS registry number
124566-23-0
76774-46-4

2 Source Organism

<1> *Candida boidinii* (strain KD1 [3,5]; Kloeckera sp. No. 2201 [2]; methanol grown [1-7]) [1-10]
<2> *Hansenula polymorpha* (strain DL1, methanol-grown [3,4]) [3, 4, 8, 11]
<3> *Acinetobacter sp.* (JC1 DSM 3803 [12]) [12]

3 Reaction and Specificity

Catalyzed reaction
D-xylulose 5-phosphate + formaldehyde = glyceraldehyde 3-phosphate + glycerone (mechanism (yeast, <1> [1]), <1,2> [1-3, 5])

Reaction type
keto group transfer

Natural substrates and products
S D-xylulose 5-phosphate + formaldehyde <1, 2> (<1> pathway for formaldehyde fixation [2]; <1,2> first step in formaldehyde assimilation via xy-

lulose monophosphate (dihydroxyacetone) cycle [3]; <1> inducible [7]) (Reversibility: r <1> [2]; ir <1> [5]) [1-11]

P glyceraldehyde 3-phosphate + dihydroxyacetone <1, 2> (<1,2> i.e. 1,3-dihydroxy-2-propanone [1-7]) [1-7]

Substrates and products

S D-erythrose 4-phosphate + D-xylulose 5-phosphate <3> (Reversibility: ? <3> [12]) [12]

P D-fructose 6-phosphate + glyceraldehyde 3-phosphate

S D-xylulose 5-phosphate + D-erythrose <1> (Reversibility: ? <1> [6]) [6]

P glyceraldehyde 3-phosphate + D-fructose

S D-xylulose 5-phosphate + D-glucose <1> (Reversibility: ? <1> [6]) [6]

P glyceraldehyde 3-phosphate + D-glycero-D-ido-oct-2-ulose

S D-xylulose 5-phosphate + D-glyceraldehyde <1, 3> (Reversibility: ? <1, 3> [3, 12]) [3, 12]

P glyceraldehyde 3-phosphate + D-xylulose

S D-xylulose 5-phosphate + D-ribose <1> (Reversibility: ? <1> [6]) [6]

P glyceraldehyde 3-phosphate + sedoheptulose

S D-xylulose 5-phosphate + butyraldehyde <1> (Reversibility: ? <1> [6]) [6]

P glyceraldehyde 3-phosphate + 1,3-dihydroxyhexane-2-one

S D-xylulose 5-phosphate + formaldehyde <1, 2> (<1> best substrate [6]; <1> D-xylulose 5-phosphate cannot be replaced by D-ribulose 5-phosphate, D-glucose 6-phosphate, or D-ribose 5-phosphate [6]; <1> formaldehyde is the only acceptor [5]; <1> no acceptors are glyceraldehyde, glyceraldehyde 3-phosphate, erythrose 4-phosphate, ribose 5-phosphate, glucose 6-phosphate, D-ribose, glutaraldehyde, dihydroxyacetone [5,6]) (Reversibility: r <1> [2]; ir <1> [5]) [1-11]

P glyceraldehyde 3-phosphate + dihydroxyacetone <1, 2> (<1,2> i.e. 1,3-dihydroxy-2-propanone [1-7]) [1-7]

S D-xylulose 5-phosphate + glycolaldehyde <1> (Reversibility: ? <1> [3, 6]) [3, 6]

P glyceraldehyde 3-phosphate + 1,3,5-trihydroxypentane-2-one

S D-xylulose 5-phosphate + heptaldehyde <1> (Reversibility: ? <1> [6]) [6]

P glyceraldehyde 3-phosphate + 1,3-dihydroxynonane-2-one

S D-xylulose 5-phosphate + propionaldehyde <1> (Reversibility: ? <1> [6]) [6]

P glyceraldehyde 3-phosphate + 1,3-dihydroxypentane-2-one

S D-xylulose 5-phosphate + ribose 5-phosphate <1> (Reversibility: ? <1> [3]) [3]

P glyceraldehyde 3-phosphate + sedoheptulose 7-phosphate

S acetaldehyde + D-xylulose 5-phosphate <3> (Reversibility: ? <3> [12]) [12]

P glyceraldehyde 3-phosphate + 1,3-dihydroxybutane-2-one

S fructose 6-phosphate + formaldehyde <1-3> (<1> fructose 6-phosphate can replace D-xylulose 5-phosphate with 10% efficiency as glycolaldehyde donor [6]) (Reversibility: ? <1-3> [3-6, 12]) [3-6, 12]

P dihydroxyacetone + erythrose 4-phosphate
S glucose 6-phosphate + formaldehyde <3> (Reversibility: ? <3> [12]) [12]
P ?
S glycolaldehyde + D-xylulose 5-phosphate <3> (Reversibility: ? <3> [12]) [12]
P glyceraldehyde 3-phosphate + 1,3-dihydroxypentane-2-one
S hydroxypyruvate + formaldehyde <1, 3> (<1> hydroxypyruvate can replace D-xylulose 5-phosphate with 35% [1]; <1> hydroxypyruvate can replace D-xylulose 5-phosphate with 41.4%, hydroxypyruvate can replace D-xylulose 5-phosphate with 56% [6]; <1> efficiency as glycolaldehyde donor [2]) (Reversibility: ? <1, 3> [1-3, 6, 12]) [1-3, 6, 12]
P CO_2 + dihydroxyacetone
S ribose 5-phosphate + formaldehyde <3> (Reversibility: ? <3> [12]) [12]
P ?
S ribulose 5-phosphate + formaldehyde <3> (Reversibility: ? <3> [12]) [12]
P glyceraldehyde 3-phosphate + 1,3-dihydroxyacetone
S sedoheptulose 7-phosphate + formaldehyde <1> (<1> sedoheptulose 7-phosphate can replace D-xylulose 5-phosphate with 10% efficiency as glycolaldehyde donor [6]) [3, 5, 6]
P dihydroxyacetone + ribose
S Additional information <1> (<1> with xylulose 5-phosphate as glycolyl donor: aliphatic aldehydes (C1-C7), glycolaldehyde and glyceraldehyde are better acceptors than ribose 5-phosphate [2]) [2]
P ?

Inhibitors

1,10-phenanthroline <1> [2]
As_2O_3 <1> [2]
Co^{2+} <1, 2> (<1,2> 0.1 mM [3]) [2, 3, 6]
$CuSO_4$ <1> [2]
EDTA <1> [2, 5]
$FeCl_3$ <1> [2]
GSH <1> [3, 5, 6]
$HgCl_2$ <1> [2]
Mn^{2+} <1, 2> (<1,2> 0.1 mM [3]) [3]
NEM <1> (<1> irreversible, GSH partially protects [5]) [3, 5]
Ni^{2+} <1> [6]
PCMB <1> [2]
Zn^{2+} <1, 2> (<1,2> 0.1 mM [3]) [2, 3]
formaldehyde <1> (<1> above 1 mM, xylulose 5-phosphate protects [5]) [3, 5]
iodoacetate <1> [2]
ribose <1> [6]
sulfate <1> [6]

Cofactors/prosthetic groups

thiamine diphosphate <1, 3> (<1,3> requirement, thiamine diphosphate protein, 0.5 mM [5,12]; <1> tightly bound [3,5]) [2, 3, 5, 12]

Metals, ions

Mg^{2+} <1, 3> (<1,3> requirement [2,6,12]; <1> 5 mM, essential for effective binding of thiamine diphosphate to enzyme molecule [5]; <1> activation [3]) [2, 3, 5, 6, 12]
Mn^{2+} <1> (<1> activation, replaces Mg^{2+} to some extent [6]) [6]
Additional information <1, 2> (<1> no metal cations [3]; <1> no activation by Ca^{2+} [3,5]) [3, 5]

Specific activity (U/mg)

0.285 <1> [4]
0.766 <1> [6]
1.12 <3> [12]
2.22 <1> [8]
3.2 <1> [7]
3.88 <1> [2]
4 <1> [5]
4.1 <1> [3]

K_m-Value (mM)

0.0333 <3> (D-xylulose 5-phosphate) [12]
0.42 <1> (glyceraldehyde 3-phosphate) [2]
0.43 <1> (formaldehyde) [2]
0.52 <1> (dihydroxyacetone) [2]
1 <1> (D-xylulose 5-phosphate) [2]
1.86 <3> (formaldehyde) [12]

pH-Optimum

6.8-7.1 <1> [3, 5]
7 <1> [2]
7.4-7.6 <1, 2> (<1,2> markedly less activity in Tris than in phosphate or glycylglycine buffer [4]) [4]
Additional information <1> (<1> pI: 7.1, [3]) [3, 5]

pH-Range

6.5-7.5 <1> (<1> about 80% of maximal activity at pH 6.5 and 7.5 [2]) [2]
7 <3> [12]

Temperature optimum (°C)

30 <3> [12]
35 <1> [2]

Temperature range (°C)

35-45 <1> (<1> maximal activity at 35°C and about 80% of maximal activity at 45°C [2]) [2]

4 Enzyme Structure

Molecular weight

105000-110000 <1> (<1> gel filtration [6]) [6]
135000 <1> (<1> gel filtration [7]) [7]
140000 <3> (<3> gel filtration [12]) [12]
145000 <1> (<1> gel filtration [3,5]; <1> PAGE [3]) [3, 5]
190000 <1> (<1> gel filtration [2]) [2]

Subunits

dimer <1, 3> (<1> α_2, 2 * 62500, non-equilibrium pH-gradient electrophoresis followed by SDS-PAGE [6]; <1> α_2, 2 * 65500, SDS-PAGE [6]; <1> α_2, 2 * 76000, SDS-PAGE [3,5]; <1> α_2, 2 * 79000, SDS-PAGE [7]; <3> α_2, 2 * 73000, SDS-PAGE [12]) [3, 5-7, 12]
tetramer <1> (<1> α_4, 4 * 55000, SDS-PAGE [2]) [2]

Posttranslational modification

Additional information <1> (<1> no evidence for glycosylation [7]) [7]

5 Isolation/Preparation/Mutation/Application

Localization

peroxisomal matrix <1> [10]
peroxisome <1, 2> [3, 7]

Purification

<1> (partial [1,4]; spheroblast and peroxisome lysis [7]) [1-5, 7]
<2> (partial [4]) [3, 4]

Cloning

<1> (expression in Escherichia coli JM109 [9]) [9]

6 Stability

Temperature stability

20 <1> (<1> $t_{1/2}$: about 100 min [1]) [1]
30 <1> (<1> $t_{1/2}$: about 40 min [1]) [1]
35 <1> (<1> $t_{1/2}$: about 10 min [1]; <1> 5 min without stabilizing agents [6]; <1> 15 min in stabilizing buffer [6]) [1, 6]
40 <1> (<1> 10 min stable [2]; <1> $t_{1/2}$: about 5 min [1]) [1, 2]
50 <1> (<1> inactivation after 10 min [2]) [2]
70 <3> (<3> rapidly inactivated [12]) [12]

General stability information

<1>, Ca^{2+} does not stabilize [5]
<1>, Mg^{2+} or NH_4^+ stabilizes for at least 80 h [5]

<1>, Mg^{2+}, 5 mM, thiamine diphosphate 0.5 mM, DTT, 1 mM, EDTA 1 mM stabilize [2]
<1>, PMSF, 0.024% w/v, stabilizes [2]
<1>, thiamine diphosphate, 0.005% w/v, stabilizes slightly [1]
<1>, unstable, formaldehyde, EDTA, ammonium sulfate, methanol, ethanol, DTT, xylulose 5-phosphate and PMSF do not stabilize [1]
<1, 2>, Mg^{2+} or EDTA stablizes [3]
<1, 2>, dodecanethiol stabilizes most effectively [3]
<1, 2>, glycerol, 20% w/v, stabilizes during purification [4]

Storage stability
<1>, 0°C, $t_{1/2}$: 45 h [1]
<1>, 4°C, partially purified, inactivation within 12 h [4]
<1>, 5°C, at least 1 month in buffer with stabilizing agents [2]
<1>, 5°C, inactivation within 10 days with one of the stabilizing agents of the buffer omitted [2]
<1, 2>, 4°C, at least 2-3 weeks under 1 ml dodecanethiol [3]
<1, 2>, 4°C, inactivation within a few h in the absence of dodecanethiol [3]

References

[1] Waites, M.J.; Quayle, J.R.: The interrelation between transketolase and dihydroxyacetone synthase activities in the methylotrophic yeast Candida boidinii. J. Gen. Microbiol., **124**, 309-316 (1981)
[2] Kato, N.; Higuchi, T.; Sakazawa, C.; Nishizawa, T.; Tani, Y.; Yamada, H.: Purification and properties of a transketolase responsible for formaldehyde fixation in a methanol-utilizing yeast, candida boidinii (Kloeckera sp.) No. 2201. Biochim. Biophys. Acta, **715**, 143-150 (1982)
[3] Bystrykh, L.V.; De Koning, W.; Harder, W.: Dihydroxyacetone synthase from Candida boidinii KD1. Methods Enzymol., **188**, 435-445 (1990)
[4] O'Connor, M.L.; Quayle, J.R.: Pentose phosphate-dependent fixation of formaldehyde by methanol-grown Hansenula polymorpha and Candida boidinii. J. Gen. Microbiol., **120**, 219-225 (1980)
[5] Bystrykh, L.V.; Sokolov, A.P.; Trotsenko, Y.A.: Purification and properties of dihydroxyacetone synthase from methylotrophic yeast Candida boidinii. FEBS Lett., **132**, 324-328 (1981)
[6] Waites, M.J.; Quayle, J.R.: Dihydroxyacetone synthase: a special transketolase for formaldehyde fixation from the methylotrophic yeast Candida boidinii CBS 5777. J. Gen. Microbiol., **129**, 935-944 (1983)
[7] Goodman, J.M.: Dihydroxyacetone synthase is an abundant constituent of the methanol-induced peroxisome of Candida boidinii. J. Biol. Chem., **260**, 7108-7113 (1985)
[8] Yanase, H.; Okuda, M.; Kita, K.; Sato, Y.; Shibata, K.; Sakai, Y.; Kato, N.: Enzymic preparation of [1,3-^{13}C]dihydroxyacetone phosphate from [13C]methanol and hydroxypyruvate using the methanol-assimilating sys-

tem of methylotrophic yeasts. Appl. Microbiol. Biotechnol., **43**, 228-234 (1995)

[9] Sakai, Y.; Nakagawa, T.; Shimase, M.; Kato, N.: Regulation and physiological role of the DAS1 gene, encoding dihydroxyacetone synthase, in the methylotrophic yeast Candida boidinii. J. Bacteriol., **180**, 5885-5890 (1998)

[10] Stewart, M.Q.; Esposito, R.D.; Gowani, J.; Goodman, J.M.: Alcohol oxidase and dihydroxyacetone synthase, the abundant peroxisomal proteins of methylotrophic yeasts, assemble in different cellular compartments. J. Cell Sci., **114**, 2863-2868 (2001)

[11] Salomons, F.A.; Kiel, J.A.K.W.; Faber, K.N.; Veenhuis, M.; van der Klei, I.J.: Overproduction of Pex5p stimulates import of alcohol oxidase and dihydroxyacetone synthase in a Hansenula polymorpha pex14 null mutant. J. Biol. Chem., **275**, 12603-12611 (2000)

[12] Ro, Y.T.; Eom, C.Y.; Song, T.; Cho, J.W.; Kim, Y.M.: Dihydroxyacetone synthase from a methanol-utilizing carboxydobacterium, Acinetobacter sp. strain JC1 DSM 3803. J. Bacteriol., **179**, 6041-6047 (1997)

Acetoin-ribose-5-phosphate transaldolase 2.2.1.4

1 Nomenclature

EC number
2.2.1.4

Systematic name
3-hydroxybutan-2-one:D-ribose-5-phosphate aldehydetransferase

Recommended name
acetoin-ribose-5-phosphate transaldolase

Synonyms
1-deoxy-D-altro-heptulose-7-phosphate synthase
1-deoxy-D-altro-heptulose-7-phosphate synthetase
synthase, 1-deoxy-D-altro-heptulose 7-phosphate

CAS registry number
87843-76-3

2 Source Organism

<1> *Bacillus pumilus* (transketolase mutant strain BG2532 [1]) [1]

3 Reaction and Specificity

Catalyzed reaction
3-hydroxybutan-2-one + D-ribose 5-phosphate = acetaldehyde + 1-deoxy-D-altro-heptulose 7-phosphate

Reaction type
aldehyde group transfer

Natural substrates and products
- **S** 3-hydroxybutan-2-one + D-ribose 5-phosphate <1> (<1> i.e. acetoin [1]) (Reversibility: ? <1> [1]) [1]
- **P** acetaldehyde + 1-deoxy-D-altro-heptulose 7-phosphate

Substrates and products
- **S** 3-hydroxybutan-2-one + D-ribose 5-phosphate <1> (<1> i.e. acetoin [1]) (Reversibility: ? <1> [1]) [1]
- **P** acetaldehyde + 1-deoxy-D-altro-heptulose 7-phosphate <1> [1]
- **S** methylacetoin + D-ribose 5-phosphate <1> (Reversibility: ? <1> [1]) [1]

P acetone + 1-deoxy-D-altro-heptulose 7-phosphate <1> [1]

S Additional information <1> (<1> no substrates are acetaldehyde, pyruvate, acetone, hydroxyacetone, acetylacetone i.e. 2,4-pentanedione, propanetriol, diacetyl i.e. 2,3 butanedione, 2,3 butanediol, 1,2,3 butanetriol, 2 butanone [1]) [1]

P ?

Cofactors/prosthetic groups

thiamine diphosphate <1> (<1> requirement, thiamine diphosphate protein [1]) [1]

Metals, ions

Mg^{2+} <1> (<1> requirement [1]) [1]

Specific activity (U/mg)

0.0145 <1> [1]

pH-Optimum

8.5-9 <1> [1]

Temperature optimum (°C)

37 <1> (<1> assay at [1]) [1]

4 Enzyme Structure

Molecular weight

120000 <1> (<1> gel filtration [1]) [1]

5 Isolation/Preparation/Mutation/Application

Localization

cytoplasm <1> [1]

Purification

<1> (partial [1]) [1]

6 Stability

pH-Stability

6-8 <1> (<1> stable for 1 h at 37°C in the presence of cofactors and 2-mercaptoethanol [1]) [1]

Temperature stability

37 <1> (<1> stable for 1 h at pH 6.0-8.0 in the presence of cofactors and 2-mercaptoethanol [1]) [1]

80 <1> (<1> inactivation after 3 min [1]) [1]

General stability information

<1>, thiamine diphosphate, or Mg^{2+}, or 2-mercaptoethanol stabilizes [1]

References

[1] Yokota, A.; Sasajima, K.: Enzymatic formation of a new monosaccharide, 1-deoxy-D-altro-heptulose phosphate, from DL-acetoin and D-ribose 5-phosphate by a transketolase mutant of Bacillus pumilus. Agric. Biol. Chem., **47**, 1545-1553 (1983)

2-Hydroxy-3-oxoadipate synthase 2.2.1.5

1 Nomenclature

EC number
2.2.1.5

Systematic name
2-oxoglutarate:glyoxylate succinaldehydetransferase (decarboxylating)

Recommended name
2-hydroxy-3-oxoadipate synthase

Synonyms
2-hydroxy-3-oxoadipate glyoxylate-lyase (carboxylating)
2-oxoglutarate: glyoxylate carboligase
EC 4.1.3.15 (formerly)
α-ketoglutarate:glyoxylate carboligase
α-ketoglutaric-glyoxylic carboligase
carboligase
glyoxylate-2-oxoglutarate carboligase
oxoglutarate: glyoxylate carboligase

CAS registry number
9054-72-2

2 Source Organism

<1> *Rattus norvegicus* (rat) [1, 5, 8, 10, 13]
<2> *Oryctolagus cuniculus* (rabbit) [1]
<3> *Tetrahymena pyriformis* [2, 3]
<4> *Mycobacterium phlei* [4]
<5> *Bos taurus* (bovine) [5, 6]
<6> *Sus scrofa* (pig) [7]
<7> *Escherichia coli* [9]
<8> *Nicotiana rustica* (tobacco) [12]
<9> *Saccharomyces cerevisiae* [11]

3 Reaction and Specificity

Catalyzed reaction

2-oxoglutarate + glyoxylate = 2-hydroxy-3-oxoadipate + CO_2 (<4>, mechanism [4]; <6> various alternative mechanisms proposed [7])

Reaction type

decarboxylation (synergistic)

Natural substrates and products

S 2-oxoglutarate + glyoxylate <1-3, 8> (<1,2>, linked to citric acid cycle [1]; <3> hypothesized 2-oxoglutarate regeneration cycle [2]; <8>, possible 1-carbon source for nicotine synthesis [12]; <1>, important as diversion of glyoxylate from oxalate formation [13]) [1, 2, 12, 13]

P ?

Substrates and products

S 2-oxoglutarate <3-5> (<3>, identical with 2-oxoglutarate decarboxylase [2]; <5>, decarboxylation [5]) [2, 4, 5]

P ? + CO_2 <4> [4]

S 2-oxoglutarate + acetaldehyde <4, 5, 9> [4, 5, 11]

P 5-hydroxy-4-oxohexanoate + CO_2 <4> [4]

S 2-oxoglutarate + glyoxylate <1-9> [1-13]

P 2-hydroxy-3-oxohexandioate + CO_2 <3-5, 7> (<3, 5> i. e. 2-hydroxy-3-oxoadipic acid [3, 6]; <3,4> further spontaneous decarboxylation in the presence of acid [3, 4]; <5>, also 2,3-dihydroxy-4-ketopimelic acid [5]; further decarboxylation leads to 5-hydroxy-4-ketovaleric acid [6]) [3-6, 9]

Inhibitors

EDTA <3, 4, 6> (<4>, 93% inhibition at 1 mM [4]; <6>, complete inhibition at 10 mM [7]) [3, 4, 7]
N-ethylmaleimide <6> (<6>, 74% inhibition at 0.01 mM [7]) [7]
Zn^{2+} <4> (<4>, 97% inhibition at 1 mM [4]) [4]
iodoacetic acid <4> (<4>, 83% inhibition at 3.3 mM [4]) [4]
p-chloromercuriphenylsulfonic acid <4> (<4>, complete inhibition at 1 mM [4]) [4]

Cofactors/prosthetic groups

thiamine diphosphate <1-9> (<1>, deficiency may lead to hyperoxaluria and glyoxylate excretion [10]) [1-13]

Activating compounds

avidin <6> [7]
mercaptoethanol <6> (<6>, stimulates [7]) [7]
perchloric acid <7> [9]
sucrose <6> (<6>, stimulates [7]) [7]

Metals, ions

Cl^- <6> (<6>, stimulates [7]) [7, 12]
Mg^{2+} <3, 4, 7, 8> (<4>, 31% activation at 1mM [4]; <1>, Mg^{2+}-deficient food leads to 27-35% loss of activity in vivo [8]) [3, 4, 8, 9, 12]
Mn^{2+} <3, 4, 8> (<3>, substitutes for Mg^{2+} [3]; <4>, 175% activation at 1 mM [4]) [3, 4, 12]

Specific activity (U/mg)

0.128 <3> (<3>, maximum, cell-cycle-dependent [3]) [3]
0.8 <5> [5]
2.21 <4> [4]
4.1 <1> [13]
85 <6> [7]
Additional information <1, 3> (<1>, thiamine deficiency decreases specific activity [10]) [3, 10]

K_m-Value (mM)

1.8 <1> (2-oxoglutarate, <1>) [13]
2 <4> (2-oxoglutarate, <4>) [4]
3.2 <4> (glyoxylate, <4>) [4]
3.6 <6> (glyoxylate, <6>) [7]
6.8 <1> (glyoxylate, <1>) [13]
16 <6> (2-oxoglutarate, <6>) [7]

pH-Optimum

6.3 <4> [4]
6.5 <1, 2, 4> (<1,2,4>, enzyme assay at [1, 4, 8]) [1, 4, 8]
7 <1, 3, 6, 8, 9> (<1,3,9>, enzyme assay at [3, 11, 13]) [3, 7, 11-13]
7.1 <5> (<5>, enzyme assay at [5, 6]) [5, 6]

pH-Range

6-7.5 <6> [7]

Temperature optimum (°C)

30 <3, 9> (<3,9>, enzyme assay at [3, 11]) [3, 11]
37 <1, 2, 4, 5> (<1,2,4,5>, enzyme assay at [1, 4-6, 8, 13]) [1, 4-6, 8, 13]

5 Isolation/Preparation/Mutation/Application

Source/tissue

brain <1, 5> [1, 5]
heart <1, 5> (<1>, distribution in rat tissues [5]) [1, 5, 6]
kidney <1, 5> [5, 8, 10]
liver <1, 2, 5, 6> [1, 5, 7, 8, 10]
muscle <5> [5]
root <8> [12]

Localization

mitochondrion <1-3, 6, 8> (<1,2>, exclusively [1]; <1,2,9> part of α-ketoglutarate dehydrogenase complex [1, 11, 13]) [1-3, 7, 10-13]

Purification

<1> [13]
<4> [4]
<5> [5]
<6> (partial [7]) [7]

Application

medicine <1> (<1>, deficiency leads to hyperoxaluria, relative decrease in liver biomass [8]) [8]
medicine <1> (<1>, involved in renal stone formation, decreased activity due to thiamine deficiency may increase oxalate excretion by 35% [10]) [10]

6 Stability

Temperature stability

47 <8> (<8>, 50% loss of activity after 3 min [12]) [12]

Oxidation stability

<6>, access of air destabilises [7]

General stability information

<6>, Mg^{2+} stabilises [7]
<6>, mercaptoethanol stabilises [7]
<6>, sucrose stabilises [7]

Storage stability

<1>, -20°C, potassium phosphate buffer, pH 7.1, 10% glycerol, 1 mM dithiothreitol, at least several days [13]
<6>, -20°C, potassium phosphate buffer, pH 7.0, 5 days [7]

References

[1] O'Fallon, J.V.; Brosemer, R.W.: Cellular localization of α-ketoglutarate: glyoxylate carboligase in rat tissues. Biochim. Biophys. Acta, **499**, 321-328 (1977)
[2] Zaror-Behrens, G.; Mavrides, C.: Further studies on 2-oxoglutarate:glyoxylate carboligase activity in Tetrahymena pyriformis. Int. J. Biochem., **7**, 111-116 (1976)
[3] Zaror-Behrens, G.; Mavrides, C.: Demonstration of 2-oxoglutarate:glyoxylate carboligase activity in Tetrahymena pyriformis. Can. J. Biochem., **51**, 1505-1508 (1973)

[4] Yamasaki, H.; Moriyama, T.: Purification, general properties and two other catalytic activities of α-ketoglutarate:glyoxylate carboligase of Mycobacterium phlei. Biochim. Biophys. Acta, **242**, 637-647 (1971)
[5] Schlossberg, M.A.; Bloom, R.J.; Richert, D.A.; Westerfeld, W.W.: Carboligase activity of α-ketoglutarate dehydrogenase. Biochemistry, **9**, 1148-1153 (1970)
[6] Schlossberg, M.A.; Richert, D.A.; Bloom, R.J.; Westerfeld, W.W.: Isolation and identification of 5-hydroxy-4-ketovaleric acid as a product of α-ketoglutarate:glyoxylate carboligase. Biochemistry, **7**, 333-337 (1968)
[7] Steward, P.R.; Quale, J.R.: The synergistic decarboxylation of glyoxylate and 2-oxoglutarate by an enzyme system from pig-liver mitochondria. Biochem. J., **102**, 885-897 (1967)
[8] Rattan, V.; Thind, S.K.; Jethi, R.J.; Sidhu, H.; Nath, R.: Oxalate metabolism in magnesium-deficient rats. Magnesium Res., **6**, 127-133 (1993)
[9] Yamasaki, H.; Mariyama, T.: α-Ketoglutarate:glyoxylate carboligase activity in Escherichia coli. Biochem. Biophys. Res. Commun., **39**, 790-795 (1970)
[10] Sidhu, H.; Gupta, R.; Thind, S.K.; Nath, R.: Oxalate metabolism in thiamine-deficient rats. Ann. Nutr. Metab., **31**, 354-361 (1987)
[11] Hirabayashi, T.; Harada, T.: Inhibition of the α-ketoglutarate dehydrogenase complex from baker's yeast by acetaldehyde and glyoxylate. Agric. Biol. Chem., **36**, 1249-1251 (1972)
[12] Prather, C.W.; Sisler, E.C.: Glycine and glyoxylate decarboxylation in Nicotiana rustica roots. Phytochemistry, **11**, 1637-1647 (1972)
[13] Bats, R.; Rofe, A.M.; Conyers, R.A.J.: Investigations into the effect of glyoxylate decarboxylation and transamination on oxalate formation in rat. Nephron, **57**, 460-469 (1991)

Acetolactate synthase 2.2.1.6

1 Nomenclature

EC number
2.2.1.6

Systematic name
pyruvate:pyruvate acetaldehydetransferase (decarboxylating)

Recommended name
acetolactate synthase

Synonyms
AHAS
ALS
EC 4.1.3.18 (formerly)
GST-mALS
GST-wALS
acetohydroxy acid synthase
acetohydroxy acid synthetase
acetohydroxyacid synthase
acetolactate pyruvate-lyase (carboxylating)
acetolactate synthetase
acetolactic synthetase
α-ALS
α-acetohydroxy acid synthetase
α-acetohydroxyacid synthase
α-acetolactate synthase
α-acetolactate synthetase
synthase, acetolactate

CAS registry number
9027-45-6

2 Source Organism

<1> *Zea mays* (black mexican sweet, two enzyme forms: AHAS I and AHAS II [1]) [1, 2, 7, 8, 10, 36]
<2> *Arabidopsis thaliana* (herbicide sensitive cell line [3]; mutant enzymes A122V, W574S, W574L and S653N [33]) [3, 32, 33, 39]
<3> *Nicotiana tabacum* [3, 41, 50]
<4> *Spirulina platensis* (<1>, strain C1 [4]) [4]

<5> *Methanococcus aeolicus* [5, 51]
<6> *Methanococcus deltae* [5]
<7> *Methanococcus maripaludis* [5]
<8> *Methanococcus thermolithotrophicus* [5]
<9> *Methanococcus voltae* [5]
<10> *Escherichia coli* (K-12 [11,12,14,17,31]; isoenzyme I [11,12,15]; isoenzyme III [12]; isoenzyme II [34]; 3 isoenzymes: I, II, and III [6,14]; isoenzyme I and III are expressed in wild type cells, isoenzyme II is cryptic [14]; wild type and two isogenic strains PS1035, containing only acetohydroxy acid synthase III, and strain PS1036, containing only acetohydroxy acid synthase I [17]; 2 enzyme form AHS I and AHS II [30]; isoenzyme AHS III [31]) [6, 11, 12, 14, 15, 17, 30, 31, 34, 47]
<11> *Salmonella typhimurium* (3 isoenzymes: I, II, and III [6,14]; isoenzyme II, recombinantly produced in Escherichia coli HB101/pDU9 [9]; isoenzyme I and II are expressed in wild type cells, isoenzyme III is cryptic [14]; 2 enzyme forms [29]; 2 enzyme form AHS I and AHS II [30]) [6, 9, 14, 29]
<12> *Euglena gracilis* [13]
<13> *Rhodopseudomonas sphaeroides* [16]
<14> *Aerobacter aerogenes* (strain 1-12, 3 types of enzyme: sensitive pH 8.0 enzyme, insensitive pH 6.0 enzyme, insensitive pH 8.0 enzyme [18]; pH 6 acetolactate forming enzyme [27,28]) [18, 27, 28]
<15> *Bacillus subtilis* [19]
<16> *Neurospora crassa* [20, 26]
<17> *Saccharomyces cerevisiae* [21, 45]
<18> *Pseudomonas aeruginosa* [22, 23]
<19> *Thermus aquaticus* [24]
<20> *Bacillus sp.* [24]
<21> *Schizosaccharomyces pombe* [25]
<22> *Hordeum vulgare* [35, 40, 46]
<23> *Spinacia oleracea* [35]
<24> *Brassica napus* (2 enzyme forms [37]) [37]
<25> *Triticum aestivum* [38]
<26> *Lactococcus lactis* (subsp. lactis bv, diacetylactis) [52]
<27> *Serratia marcescens* (ATCC 25419 [42]) [42, 43, 53]
<28> *Claviceps purpurea* [44]
<29> *Leuconostoc mesenteroides* (subsp. cremonis [48]) [48]
<30> *Leuconostoc lactis* (strain NCW1 [49]) [49]

3 Reaction and Specificity

Catalyzed reaction

2 pyruvate = 2-acetolactate + CO_2 (<1>, Breslow's mechanism [36])

Reaction type

C-C bond formation
decarboxylation

Natural substrates and products

S pyruvate <1-4, 10-13, 15, 17, 22, 26, 27, 29> (<1-3, 10, 13, 11, 22, 25>, first enzyme unique to biosynthesis of the branched chain amino acids Val, Leu, and Ile [1, 3, 10, 14, 16, 32-34, 38, 39, 46, 50]; <12>, key enzyme in synthesis of branched-chain amino acids [13]; <4>, the expression is negatively controlled by Val. Leu and Ile slightly stimulate the enzyme production [4]; <10,11>, isoenzyme AHAS I enables a bacterium to cope with poor carbon sources, which lead to low endogenous pyruvate concentrations [6]; <10,11>, isoenzyme I is regulated by Leu and Val [14]; <10,11>, isoenzyme II is regulated by Leu, Ile and Val [14]; <10,11>, isoenzyme II is regulated by Leu [14]; <10>, production of isoenzyme AHS I is under multivalent control by Val and Leu, production of isoenzyme AHS II is under multivalent control by Ile, Val and Leu [30]; <27,29>, catabolic enzyme is involved in 2,3-butanediol pathway [42,48]; <27>, the enzyme plays a role in not only preventing intracellular acidification but also supplying α-acetolactate as an intermediate of branched chain amino acids biosynthesis [42]; <26>, constitutive high expression, the enzyme is active only under conditions of pyruvate excess [52]) [4, 6, 10, 13, 14, 16, 19, 21, 30, 32-34, 38, 39, 41, 42, 46, 48, 50, 52]

P ?

Substrates and products

S 2-oxobutanoate + lactylthiamine diphosphate <1> [36]

P acetohydroxybutanoate + ? <1> [36]

S pyruvate <1-30> (Reversibility: ? <1-30> [1-53]) [1-53]

P 2-acetolactate + CO_2 <1-30> [1-53]

S pyruvate + 2-oxobutanoate <10-12, 22, 23> (<10,11>, isoenzyme I shows no product preference, isoenzymes II and III form acetohydroxybutanoate at 180fold and 60fold faster rates, respectively than acetolactate [6]; <12>, much higher affinity for 2-oxobutanoate than for pyruvate [13]; <22,23>, preference for 2-ketobutanoate at the second substrate site [35]; <27>, no activity [53]) (Reversibility: ?, <10-12, 22, 23> [6, 13, 35, 47]) [6, 13, 35, 47]

P acetohydroxybutanoate <10-12, 22, 23> [6, 13, 35, 47]

Inhibitors

2-oxobutanoate <10, 12, 16> (<12>, inhibits formation of acetolactate from pyruvate [13]; <10> isoenzyme I has lower sensitivity to inhibition than isoenzyme III [17]) [13, 17, 26]

3-phosphoglycerate <29> (<29>, noncompetitive [48]) [48]

ATP <29> (<29>, noncompetitive [48]) [48]

Cadre <22> (<22>, imidazolinone) [46]

EDTA <26> [52]

Ile <1, 7, 9, 10-12, 18-22, 24, 25, 27> (<22>, slight [46]; <1>, enzyme form AHS I is inhibited. Enzyme form AHS II is not inhibited [1]; <12>, 5 mM, 32% inhibition [13]; <10>, mild inhibition of isoenzyme I and III [14]; <11>, inhibition of isoenzyme I, no inhibition of isoenzyme II [14]; <21>, less potent, noncompetitive [25]; <27>, 1 mM, 50% inhibition [53]; <10>,

isoenzyme AHS I is sensitive to feed-back inhibition, isoenzyme AHS II is insensitive [30]; <2>, insensitive to [32]; <5,27>, no inhibition [5,42,44]) [1, 5, 13, 14, 22, 24, 25, 30, 37, 38, 46, 53]
L-2-aminobutanoate <12> (<12>, 5 mM, 21% inhibition) [13]
L-Thr <12> (<12>, 5 mM, 6.6% inhibition) [13]
L-Leu <1, 2, 7, 10-12, 18, 22, 24, 25, 27> (<1>, enzyme form AHS I is inhibited. Enzyme form AHS II is not inhibited [1]; <1>, cooperative effect with Val [8]; <10>, mixed noncompetitive inhibition of isoenzyme, pH-independent inhibition of isoenzyme III [12]; <12>, 5 mM, 23% inhibition [13]; <11>, inhibition of isoenzyme I, no inhibition of isoenzyme II [14]; <10>, no inhibition of isoenzyme I and III [14]; <10>, isoenzyme AHS I is sensitive to feed-back inhibition, isoenzyme AHS II is insensitive [30]; <27>, 1 mM, 50% inhibition [53]; <2>, inhibition of the enzyme from Arabidopsis thaliana, no inhibition of the enzyme expressed in E. coli [39]; <2>, insensitive to [32]; <5,9,19,20,27>, no inhibition [5,24,42,44]) [1, 5, 8, 12-14, 22, 24, 30, 37, 38, 39, 46, 53]
N-phthalyl anilide <1> (<1>, and related compounds [8]) [8]
NEM <28> [44]
PCMB <16, 28> [26, 44]
SO_4^{2-} <15> (<15>, competitive when assayed in phosphate buffer, mixed type, when assayed in acetate buffer [19]) [19]
L-Val <1, 2, 5, 7, 9, 10-13, 16, 18-22, 24, 25, 27> (<1>, enzyme form AHS I is inhibited. Enzyme form AHS II is not inhibited [1]; <5,7,9,12,19-21>, feedback inhibition [5,13,24,25]; <1>, cooperative effective with Leu [8]; <12,16,19,21>, noncompetitive [13,24-26]; <13,20>, competitive [16,24]; <12>, 5 mM, 89% inhibition [13]; <10>, isoenzyme I and III inhibited [15]; <10,11>, isoenzymes I and II are inhibited, isoenzyme II is not inhibited [14]; <10>, isoenzyme I is more resistant to inhibition than isoenzyme III [17]; <10>, isoenzyme AHS I is sensitive to feed-back inhibition, isoenzyme AHS II is insensitive [30]; <16>, pH-dependent inhibition [26]; <11>, two enzyme forms: one is very sensitive to inhibition by Val, the second is not subject to feedback inhibition [29]; <27>, 0.1 mM, 50% inhibition [53]; <2>, inhibition of the enzyme from Arabidopsis thaliana, no inhibition of the enzyme expressed in E. coli [39]; <2,27>, no inhibition [32,42,44]) [1, 5, 8, 13-17, 20, 22, 24-26, 29, 30, 37-39, 46, 53]
chlorimuron ethyl <10> [34]
chlorsulphuron <2, 10, 24, 25> (<2>, inhibition of the enzyme from Arabidopsis thaliana and of the enzyme expressed in E. coli [39]) [32, 34, 37-39]
imazapyr <1, 2, 10, 25> (<1>, enzyme form AHS I and AHS II [1]; <1>, slow, tight-binding inhibitor [7]) [1, 7, 32, 34, 38]
imazaquin <25, 27> (<27>, 1 mM, 50% inhibition [43]) [38, 43]
imazethapyr <25> [38]
imidazolinones <1> (<1>, potent [10]) [10]
metsulfuron-methyl <3, 25> [38, 50]
phosphate <15> (<15>, inhibits activity of enzyme assayed in acetate buffer) [19]
phosphoenolpyruvate <29> (<29>, noncompetitive [48]) [48]

primisulfuron-methyl <27> (<27>, 1 mM, 50% inhibition [43]) [43]
pyrimidylsalicylate <27> (<27>, 0.025 mM, 60% inhibition) [43]
sulfometuron-methyl <1, 22, 23, 27> (<1>, enzyme form AHS I and AHS II [1]; <27>, 2.5 mM, 50% inhibition [43]) [1, 35, 43]
thiamine thiazolone diphosphate <1> [36]
trisulfuron <27> (<27>, 1 mM, 50% inhibition [43]) [43]

Cofactors/prosthetic groups

ATP <12> (<12>, activates) [13]
FAD <1-3, 5, 10, 11, 17, 19, 27> (<17, 19, 27>, required [24, 45, 53]; <2>, required as cofactor [32]; <1>, 0.005 mM, 35-40% enhancement of activity [2]; <11>, contains bound FAD, 0.88 mol per mol of protomer of 69000 Da [9]; <10>, absolutly required for isoenzyme I and II, no requirement for isoenzyme III [14]; <11>, partial requirement for isoenzyme III [14]; <10, 13>, required [15, 16]; <10>, K_m: 0.0003 mM [15]; <10>, required by isoenzyme I, no requirement by isoenzyme III [17]; <5>, K_m: 0.0004 mM [51]; <10>, K_m: 0.00016 mM for isoenzyme I [17]; <17>, K_m: 0.0003 mM [45]; <18>, no stimulation [22]; <20, 28>, no requirement [24, 44]) [2, 5, 9, 14-17, 24, 32, 41, 45, 51, 53]
FMN <5> (<5>, stimulates, K_m: 0.0013 mM) [51]
riboflavin <5> (<5>, plus 10 mM phosphate, stimulates) [51]
thiamine diphosphate <1-3, 5-11, 13, 16-20, 26, 28> (<1, 2, 5-11, 16, 17, 19, 20, 26, 28>, required as cofactor [2, 3, 5, 9, 15, 17, 24, 26, 32, 36, 44, 45, 51, 52]; <5, 7, 9>, required after dialysis [5]; <18>, stimulates [22]; <1>, saturated at 0.5 mM thiamine diphosphate [2]; <7>, K_m: 0.0033 mM [5]; <9>, K_m: 0.0031 mM [5]; <11>, K_m: 0.0015 mM [9]; <10>, K_m: 0.0012 mM [15]; <10>, K_m: 0.0087 mM for isoenzyme I, 0.026 mM for isoenzyme III [17]; <28>, K_m: 0.0012 mM [44]; <17>, K_m: 0.11 mM [45]; <26>, K_m: 0.0032 mM [52]; <5>, K_m: 0.0016 [51]) [2, 3, 5, 9, 15-17, 22, 24, 26, 32, 36, 41, 43, 44, 45, 51, 52]

Activating compounds

2-mercaptoethanol <12> (<12>, stimulates) [13]
NH_4^+ <5> (<5>, 200 mM, 1.6fold stimulation) [51]
acetate <15> (<15>, stimulates) [19]
acetate <29> (<29>, increases activity) [48]
dithiothreitol <12> (<12>, stimulates) [13]
ethanol <29> (<29>, increases activity) [48]

Metals, ions

Ca^{2+} <14> (<14>, stimulates) [28]
Mg^{2+} <1, 2, 5, 10, 11, 13, 14, 16-20, 27, 28> (<2, 10, 11, 13, 17-19, 20, 27>, required [9, 15-17, 22, 24, 28, 32, 45, 53]; <18>, stimulates [22]; <1>, saturated at 1 mM $MgCl_2$ [2]; <5>, strong requirement [5]; <11>, K_m: 0.022 mM [9]; <10>, K_m: 0.06 mM [15]; <17>, K_m: 0.28 mM [45]; <13>, K_m: 1.01 mM [16]; <5>, K_m: 0.3 mM $MgCl_2$ [51]; <10>, K_m: 0.3 mM for isoenzyme I, 0.4 mM for isoenzyme III [17]; <5, 16, 28>, Mn^{2+} or Mg^{2+} required [26, 44, 51]) [2, 5, 9, 15-17, 22, 24, 26, 28, 32, 44, 45, 51, 53]

Mn^{2+} <5, 14, 16, 21, 28> (<21>, can substitute for Mg^{2+} [25]; <5, 16>, Mn^{2+} or Mg^{2+} required [26, 51]; <14>, stimulates [28]; <5>, K_m: 0.016 mM $MnCl_2$ [51]) [25, 26, 28, 32, 44, 51]
Ni^{2+} <14> (<14>, stimulates) [28]
Zn^{2+} <14> (<14>, stimulates) [28]
Additional information <29> (<29>, the enzyme does not require a divalent cation for activity) [48]

Turnover number (min^{-1})
2844 <10> (pyruvate, <10>, fusion protein containing an N-terminal oligohistidine sequence on the large subunit) [34]
4002 <10> (pyruvate, <10>, native enzyme) [34]

Specific activity (U/mg)
0.0616 <25> [38]
0.135 <28> [44]
0.4075 <16> [26]
0.66 <1> [7]
0.844 <29> [48]
1.59 <22> [40]
2.67 <22> [46]
4.4 <2> (<2>, mutant enzyme W574L) [33]
6.25 <27> [42]
7.88 <2> [32]
8.28 <15> [19]
8.51 <2> (<2>, mutant S653N) [33]
11 <2> (<2>, mutant W574) [33]
12 <2> (<2>, mutant enzyme A122V) [33]
25.3 <11> [9]
25.6 <18> [22]
49 <10> [11]
53 <10> (<10>, fusion protein with N-terminal oligohistidine at the large subunit) [34]
Additional information <1, 10, 17, 26, 30> (<1>, spectrophotometric assay involving an indirect detection of the product acetolactate [2]; <10>, assay methods [47]) [2, 45, 47, 49, 52]

K_m-Value (mM)
0.5 <9> (pyruvate) [5]
1 <10> (2-oxobutanoate, <10>, at 1.5 mM pyruvate) [47]
1 <10> (pyruvate, <10>, pH 7.6, isoenzyme AHAS I) [6]
1.3 <10> (pyruvate, <10>, isoenzyme I) [17]
2.1 <12> (pyruvate) [13]
2.3 <5> (pyruvate) [5]
2.6 <10> (pyruvate, <10>, native protein) [34]
2.9 <10> (pyruvate, <10>, fusion protein containing an N-terminal oligohistidine sequence on the large subunit) [34]
3 <22> (pyruvate) [46]

3.2 <16> (pyruvate) [26]
3.93 <17> (pyruvate) [45]
4.2 <7> (pyruvate) [5]
4.5 <13> (pyruvate) [16]
5 <10> (2-oxobutanoate, <10>, at 2 mM pyruvate) [47]
5 <1> (pyruvate, <1>, AHAS I, [1]) [1, 2]
5.5 <22> (pyruvate) [40]
6.8 <5> (pyruvate) [51]
7 <10, 24> (pyruvate, <10>, pH 7.6, isoenzyme AHAS III [6]; <24>, enzyme form II [37]) [6, 37]
7.6 <10> (pyruvate, <10>, isoenzyme III) [17]
8 <1, 24> (pyruvate, <1>, AHAS II, [1]; <24>, enzyme form I [37]) [1, 37]
8.01 <2> (pyruvate, <2>, K_m-value of the first active site) [32]
10 <29> (pyruvate) [48]
10.6 <11> (pyruvate, <11>, pH 7.6, isoenzyme AHAS II, [6]) [6]
12.1 <3> (pyruvate, <3>, wild-type recombinant enzyme) [50]
13 <15> (pyruvate) [19]
17 <16, 28> (pyruvate) [26, 44]
22.6 <27> (pyruvate) [42]
50 <26> (pyruvate) [52]
100 <2> (pyruvate, <2>, K_m-value of the second active site) [32]
Additional information <2> (<2>, non-hyperbolic substrate-saturation curve, involving interaction between the active sites of the dimer) [32]

K_i-Value (mM)

0.1 <27> (Val) [53]
1 <27> (Ile) [53]
1 <27> (Leu) [53]
1 <27> (imazaquin) [43]
1 <27> (primisulfuron-methyl) [43]
1 <27> (trisulfuron) [43]
2.5 <27> (sulfometuron-methyl) [43]

pH-Optimum

5.3 <29> [48]
5.5 <27, 30> [42, 49]
5.8 <15> (<15>, acetate buffer) [19]
6 <20, 26, 28> [24, 44, 52]
6-7 <1> (<1>, AHAS I [1]) [1, 2]
6-8.5 <22> (<22>, enzyme from green barley extract) [40]
6.3-6.6 <21> [25]
6.5 <17, 22> (<22>, enzyme from etiolated preparations [40]) [40, 45]
6.5-8.5 <2> [32]
7 <1, 10, 11, 15, 22> (<1>, AHAS II [1]; <10,11>, isoenzyme I [14]; <15>, phosphate buffer [19]) [1, 14, 19, 46]
7-7.4 <24> (<24>, enzyme form I) [37]
7.2-7.4 <13> [16]
7.2-7.6 <24> (<24>, enzyme form II) [37]

7.5 <2, 10, 12, 16, 27> (<10>, isoenzyme I [15]; <2>, enzyme from mutant A122V, W574S, W574L and S653N [33]) [13, 15, 26, 33, 53]
7.6 <5> [51]
8 <10, 11, 19> (<10,11>, isoenzyme II [14]) [14, 24]
8-8.8 <18> [22]
8.5 <10> (<10>, isoenzyme III) [15]
9 <10, 11> (<10,11>, isoenzyme III [14]) [14]

pH-Range
5.5-8 <1> (<1>, pH 5.5: about 80% of maximal activity, pH 8.0: about 55% of maximal activity, AHAS I, [1]) [1, 2]
6-6.9 <21> (<21>, pH 6.0: about 50% of maximal activity, pH 6.9: about 30% of maximal activity) [25]
6-7.5 <17> (<17>, pH 6.0: about 25% of maximal activity, pH 7.5: about 65% of maximal activity) [45]
6-8 <15> (<15>, pH 6.0: about 55% of maximal activity, pH 8.0: about 50% of maximal activity) [19]
6.5-7.5 <1> (<1>, about 60% of maximal activity at pH 6.5 and pH 7.5, AHAS II) [1]

Temperature optimum (°C)
40 <28> [44]
46-50 <1> [2]
50 <12> [13]
70 <19> [24]
75 <20> [24]

Temperature range (°C)
Additional information <19, 20> [24]

4 Enzyme Structure

Molecular weight
55000 <1> (<1>, AHAS II, gel filtration) [1]
109000 <2> (<2>, gel filtration in absence of FAD) [32]
113000 <2> (<2>, gel filtration in presence of FAD) [32]
119000 <10> (<10>, gel filtration) [34]
124000 <24> (<24>, gel filtration) [37]
128000 <25> (<25>, gel filtration) [38]
130000 <10> (<10>, sucrose density gradient sedimentation, glycerol density gradient sedimentation) [15]
132000 <24> (<24>, glycerol density gradient sedimentation) [37]
140000 <11> (<11>, gel filtration) [9]
150000 <17, 27, 29> [42, 45, 48]
172000 <26> (<26>, gel filtration) [52]
178000 <27> (<27>, non-denaturing PAGE) [53]

190000-210000 <22> (<22>, another enzyme species with MW of 430000-460000 Da is detected, gel filtration) [40]
193000 <1> (<1>, AHAS I, gel filtration) [1]
200000 <14> (<14>, calculation from sedimentation and diffusion data) [28]
204000 <27> (<27>, gel filtration) [53]
220000 <14> (<14>, equilibrium sedimentation) [27]
225000-250000 <22> (<22>, nondenaturing PAGE) [46]
250000 <15> (<15>, gel filtration) [19]
277000 <30> (<30>, gel filtration) [49]
430000-460000 <22> (<22>, another enzyme species with MW of 190000-210000 Da is detected, gel filtration) [40]
440000 <1> (<1>, gel filtration) [7]
456000 <16> (<16>, predominant form, an enzyme form of 110000 Da also detected, gel filtration) [26]
600000 <18> (<18>, equilibrium sedimentation) [22]

Subunits

? <10, 22, 25, 26, 30> (<10>, x * 9500 + x * 60000, isoenzyme I, SDS-PAGE [11]; <10>, x * 11000 + x * 60000 [47]; <25>, x * 15000 + x * 57000 + x * 58000, SDS-PAGE [38]; <10>, x * 16200 + x * 67200, SDS-PAGE [34]; <10>, x * 17500 + x * 61800, calculation from nucleotide sequence [31]; <30>, 3 major bands detected by SDS-PAGE: 26000 Da, 35000 Da and 46000 Da [49]; <22>, x * 58000, SDS-PAGE with urea [40]; <26>, x * 62000, SDS-PAGE [52]) [11, 31, 34, 38, 40, 47, 49, 52]
dimer <2, 10, 17, 22, 24, 27> (<10>, 2 * 60000, SDS-PAGE [15]; <2>, 2 * 61000, SDS-PAGE [32]; <2>, 2 * 63864, ion spray MS analysis [32]; <22,27>, 2 * 64000, SDS-PAGE [42,46]; <24>, 2 * 65000 and/or 66000, SDS-PAGE [37]; <17>, 2 * 75000, SDS-PAGE [45]) [15, 32, 37, 42, 45, 46]
oligomer <18> (<18>, 8 * 15000 + 8 * 60000, SDS-PAGE) [23]
tetramer <11, 14> (<11>, 2 * 9800 + 2 * 59000, SDS-PAGE [9]; <14>, 4 * 58000, SDS-PAGE [27]) [9, 27]
trimer <29> (<29>, 3 * 55000, SDS-PAGE) [48]

Posttranslational modification

Side-chain modification <14> (<14>, 6 atoms of phosphorus tightly bound to one molecule of enzyme, 3 molecules of cocarboxylase tightly bound per molecule of enzyme) [28]

5 Isolation/Preparation/Mutation/Application

Source/tissue

cell suspension culture <1> [1, 2, 7]
cotyledon <24> [37]
leaf <23, 25> [35, 38]
root <25> [38]
seedling <25> [38]
shoot <22> (<22>, etiolated [46]) [35, 40, 46]

Localization

chloroplast <25> [38]
mitochondrion <12, 16, 17> [13, 20, 21, 26]
plastid stroma <24> [37]

Purification

<1> [7]
<2> (mutant enzymes A122V, W574S, W574L and S653N expressed in Escherichia coli [33]) [32, 33]
<3> [50]
<5> [51]
<10> (isoenzyme I [15,47]; isoenzyme II [34]) [11, 15, 34, 47]
<11> (isoenzyme II, recombinantly produced) [9]
<15> [19]
<16> [26]
<17> [45]
<18> [22]
<21> (partial [25]) [25]
<22> [40, 46]
<24> (partial [37]) [37]
<25> [38]
<27> [42, 53]
<28> (partial [44]) [44]
<29> [48]
<30> [49]

Crystallization

<14> [28]

Cloning

<2> (cloned into the bacterial expression vector pT7-7 and expressed in Escherichia coli BL21(DE3) [32]; cloned into the bacterial expression plasmid pKK233-2, expression in AHAS-deficient Escherichia coli strain MF2000 [39]) [3, 32, 39]
<3> [3]
<3> (cloned into the bacterial expression vector pGEX-2T and used to transform Escherichia coli strain pGEX-2T [41,50]; wild-type recombinant enzyme fused with glutathione S-transferase [41,50]; GST-wALS, and sulfonylurea-resistant enzyme fused with glutathione S-transferase, GST-mALS [41]) [41, 50]
<10> (expression is considerably higher from a vector that introduces a 50 residue N-terminal fusion containing an oligohistidine sequence on the large subunit [34]) [34]
<11> [9]
<17> (expression in Escherichia coli) [45]

Engineering

A122V <2> (<2>, reduced affinity for all Mg^{2+}, thiamine diphosphate and FAD) [33]

S653N <2> (<2>, binds FAD more strongly than the wild-type enzyme [33]) [33]
W574L <2> (<2>, insensitive to sulfonurea herbicides) [33]
W574S <2> (<2>, reduction in sensitivity to sulfonurea herbicides compared to the wild-type enzyme) [33]

6 Stability

Temperature stability

0 <16> (<16>, freezing and subfreezing temperatures are severely damaging) [26]
37 <16> (<16>, half-life: 7.5 min, without substrate and cofactors [20]) [20]
45 <10> (<10>, 15 min, 69% loss of activity, isoenzyme I, slight stimulation of activity of isoenzyme III) [17]
50 <27> (<27>, 30 min, stable) [53]
60 <27> (<27>, 30 min, stable) [42]
65 <28> (<28>, half-life: 88 seconds) [44]
70 <28> (<28>, half-life: 65 seconds) [44]

Oxidation stability

<5>, 5% O_2, half-life: 35 min [51]
<11>, oxygen-dependent inactivation of the enzyme complexed with thiamine diphosphate, FAD and magnesium acetolactate, no inactivation of the enzyme-FAD complex [9]
<5, 7>, 10% O_2, 37°C, 30 min, 90% loss of activity [5]

General stability information

<4>, FAD protects the enzyme during dialysis [4]
<10>, dialysis causes 38% loss of activity of isoenzyme I, stimulates activity of isoenzyme III [17]
<16>, stabilized by pyruvate, 2-mercaptoethanol or dithiothreitol [26]

Storage stability

<1>, -20°C or -80°C, not stable [7]
<2>, -70°C, 21 days, 15% loss of activity of mutant enzyme A122V, less than 1% loss of activity of mutant enzymes W574S and S653N, 3% loss of activity of mutant enzyme W574L [33]
<2>, -70°C, 4% loss of activity after 3 weeks with three freeze-thaw cycles, 8% loss of activity after 5 weeks with five freeze-thaw cycles, 9-10% loss of activity after 6 months with no intervening freeze-thaw cycle [32]
<2>, 4°C, 10% loss of activity after 3 weeks, 23% loss of activity after 5 weeks [32]
<2>, 4°C, 21 days, 30% loss of activity of mutant enzyme A122V, 6% loss of activity of mutant enzyme W574S, 20% loss of activity of mutant enzyme W574L, 18% loss of activity of mutant enzyme S653N [33]
<10>, 4°C, stable for several days [15]
<11>, -80°C, concentrated enzyme solution is indefinitely stable [9]

<12>, 4°C, 24 h, 36% loss of activity [13]
<16>, -20°C, stable for over 1 month [20]
<18>, 0-4°C, 50 mM potassium phosphate, pH 7.0, 10 mM mercaptoethanol, 1 mM $MgCl_2$, 0.01 mM thiamine diphosphate, 1 month, 20% loss of activity [22]
<25>, -20°C, 50 mM phosphate, pH 8, 50% v/v glycerol, 3 weeks, less than 25% loss of activity [38]
<26>, -20°C, 20% glycerol, stable for 2 months [52]
<28>, -25°C, no significant loss of activity after 6 months [44]

References

[1] Singh, B.K.; Stidham, M.A.; Shaner, D.L.: Separation and characterization of two forms of acetohydroxy acid synthase from black mexican sweet corn cells. J. Chromatogr., **444**, 251-261 (1988)
[2] Singh, B.K.; Stidham, M.A.; Shaner, D.L.: Assay of acetohydroxyacid synthase. Anal. Biochem., **171**, 173-179 (1988)
[3] Mazur, B.J.; Chui, C.F.; Smith, J.K.: Isolation and characterization of plant genes coding for acetolactate synthase, the target enzyme for two classes of herbicides. Plant Physiol., **85**, 1110-1117 (1987)
[4] Riccardi, G.; DeRossi, E.; Nielsen, E.; DeFelice, M.: Detection and characterization of acetohydroxy acid synthase in Spirulina platensis. FEMS Microbiol. Lett., **49**, 13-17 (1988)
[5] Xing, R.; Whitman, W.B.: Sulfometuron methyl-sensitive and -resistant acetolactate synthases of the archaebacteria Methanococcus spp.. J. Bacteriol., **169**, 4486-4492 (1987)
[6] Barak, Z.; Chipman, D.M.; Gollop, N.: Physiological implications of the specificity of acetohydroxy acid synthase isozymes of enteric bacteria. J. Bacteriol., **169**, 3750-3756 (1987)
[7] Muhitch, M.J.; Shaner, D.L.; Stidham, M.A.: Imidazolinones and acetohydroxyacid synthase from higher plants. Plant Physiol., **83**, 451-456 (1987)
[8] Huppatz, J.L.; Casida, J.E.: Acetohydroxyacid synthase inhibitors: N-phthalyl-L-valine and related compounds. Z. Naturforsch. C, **40**, 652-656 (1985)
[9] Schloss, J.V.; VanDyk, D.E.; Vasta, J.F.; Kutny, R.M.: Purification and properties of Salmonella typhimurium acetolactate synthase isozyme II from Escherichia coli HB101/pDU9. Biochemistry, **24**, 4952-4959 (1985)
[10] Shaner, D.L.; Anderson, P.C.; Stidham, M.A.: Imidazolinones. Potent inhibitors of acetohydroxyacid synthase. Plant Physiol., **76**, 545-546 (1984)
[11] Eoyang, L.; Silverman, P.M.: Purification and subunit composition of acetohydroxyacid synthase I from Escherichia coli K-12. J. Bacteriol., **157**, 184-189 (1984)
[12] Gollop, N.; Chipman, D.M.; Barak, Z.: Inhibition of acetohydroxy acid synthase by leucine. Biochim. Biophys. Acta, **748**, 34-39 (1983)
[13] Oda, Y.; Nakano, Y.; Kitaoka, S.: Properties and regulation of valine-sensitive acetolactate synthase from mitochondria of Euglena gracilis. J. Gen. Microbiol., **128**, 1211-1216 (1982)

[14] DeFelice, M.; Lago, C.T.; Squires, C.H.; Calvo, J.M.: Acetohydroxy acid synthase isoenzymes of Escherichia coli K12 and Salmonella typhimurium. Ann. Microbiol., **133A**, 251-256 (1982)

[15] Grimminger, H.; Umbarger, H.E.: Acetohydroxy acid synthase I of Escherichia coli: purification and properties. J. Bacteriol., **137**, 846-853 (1979)

[16] Semeraro, R.J.; Wixom, R.L.: Studies in valine biosynthesis. X. The acetolactate synthase from Rhodopseudomonas spheroides. Microbios, **20**, 7-14 (1977)

[17] DeFelice, M.; Squires, C.; Levinthal, M.: A comparative study of the acetohydroxy acid synthase isoenzymes of Escherichia coli K-12. Biochim. Biophys. Acta, **541**, 9-17 (1978)

[18] Asada, Y.; Okuzawa, Y.; Yamaguchi, K.: The existence of three types of acetohydroxy acid synthetase in an isoleucine-requiring mutant of Aerobacter aerogenes. Biochim. Biophys. Acta, **429**, 1029-1035 (1976)

[19] Holtzclaw, W.D.; Chapman, L.F: Degradative acetolactate synthase of Bacillus subtilis: purification and properties. J. Bacteriol., **121**, 917-922 (1975)

[20] Kuwana, H.; Date, M.: Solubilization of valine-sensitive acetohydroxy acid synthetase from Neurospora mitochondria. J. Biochem., **77**, 257-259 (1975)

[21] Ryan, E.D.; Kohlhaw, G.B.: Subcellular localization of isoleucine-valine biosynthetic enzymes in yeast. J. Bacteriol., **120**, 631-637 (1974)

[22] Arfin, S.M.; Koziell, D.A.: Acetolactate synthase of Pseudomonas aeruginosa I. Purification and allosteric properties. Biochim. Biophys. Acta, **321**, 348-355 (1973)

[23] Arfin, S.M.; Koziell, D.A.: Acetolactate synthase of Pseudomonas aeruginosa. II. Evidence for the presence of two nonidentical subunits. Biochim. Biophys. Acta, **321**, 356-360 (1973)

[24] Chin, N.W.; Trela, J.M.: Comparison of acetohydroxy-acid synthetases from two extreme thermophilic bacteria. J. Bacteriol., **114**, 674-678 (1973)

[25] McDonald, R.A.; Satyanarayana, T.; Kaplan, J.G.: Biosynthesis of branched-chain amino acids in Schizosaccharomyces pombe: properties of acetohydroxy acid synthetase. J. Bacteriol., **114**, 332-340 (1973)

[26] Glatzer, L.; Eakin, E.; Wagner, R.P.: Acetohydroxy acid synthetase with a pH optimum of 7.5 from Neurospora crassa mitochondria: characterization and partial purification. J. Bacteriol., **112**, 453-464 (1972)

[27] Huseby, N.E.; Christensen, T.B.; Olsen, B.R.; Störmer, F.C.: The pH6 acetolactate-forming enzyme from Aerobacter aerogenes. Subunit structure. Eur. J. Biochem., **20**, 209-214 (1971)

[28] Störmer, F.C.; Solberg, Y.; Hovig, T.: The pH 6 acetolactate-forming enzyme from Aerobacter aerogenes. Molecular properties. Eur. J. Biochem., **10**, 251-260 (1969)

[29] O'Neill, J.P.; Freundlich, M.: Two forms of biosynthetic acetohydroxy acid synthetase in Salmonella typhimurium. Biochem. Biophys. Res. Commun., **48**, 437-443 (1972)

[30] Blatt, J.M.; Pledger, W.J.; Umbarger, H.E.: Isoleucine and valine metabolism in Escherichia coli. XX. Multiple forms of acetohydroxy acid synthetase. Biochem. Biophys. Res. Commun., **48**, 444-450 (1972)

[31] Squires, C.H.; DeFelice, M.; Devereux, J.; Calvo, J.M.: Molecular structure of ilvIH and its evolutionary relationship to ilvG in Escherichia coli K12. Nucleic Acids Res., **11**, 5299-5313 (1983)
[32] Chang, A.K.; Duggleby, R.G.: Expression, purification and characterization of Arabidopsis thaliana acetohydroxyacid synthase. Biochem. J., **327**, 161-169 (1997)
[33] Chang, A.K.; Duggleby, R.G.: Herbicide-resistant forms of Arabidopsis thaliana acetohydroxyacid synthase: characterization of the catalytic properties and sensitivity to inhibitors of four defined mutants. Biochem. J., **333**, 765-777 (1998)
[34] Hill, C.H.; Pang, S.S.; Duggleby, R.G.: Purification of Escherichia coli acetohydroxyacid synthase isoenzyme II and reconstitution of active enzyme from its individual pure subunits. Biochem. J., **327**, 891-898 (1997)
[35] Delfourne, E.; Bastide, J.; Badon, R.; Rachon, A.; Genix, P.: Specificity of plant acetohydroxyacid synthase: formation of products and inhibition by herbicides. Plant Physiol. Biochem., **32**, 473-477 (1994)
[36] Roux, C.; Delfourne, E.; Bastide, J.: Activation and inhibition of maize acetohydroxyacid synthase by thiamine analogues. Plant Physiol. Biochem., **34**, 293-299 (1996)
[37] Bekkaoui, F.; Schorr, P.; Crosby, W.L.: Acetolactate synthase from Brassica napus: immunological characterization and quarternary structure of the native enzyme. Physiol. Plant., **88**, 475-484 (1993)
[38] Southan, M.D.; Copeland, L.: Physical and kinetic properties of acetohydroxyacid synthase from wheat leaves. Physiol. Plant., **98**, 824-832 (1996)
[39] Singh, B.; Szamosi, I.; Hand, J.M.; Misra, R.: Arabidopsis acetohydroxyacid synthase expressed in Escherichia coli is insensitive to the feedback inhibitors. Plant Physiol., **99**, 812-815 (1992)
[40] Durner, J.; Böger, P.: Acetolactate synthase from barley (Hordeum vulgare L.): purification and partial characterization. Z. Naturforsch. C, **43**, 850-856 (1988)
[41] Kil, M.W.; Chang, S.I.: Expression in Escherichia coli, purification, and characterization of the tobacco sulfonylurea herbicide-resistant recombinant acetolactate synthase and its interaction with the triazolopyrimidine herbicides. J. Biochem. Mol. Biol., **31**, 287-295 (1998)
[42] Joo, H.S.; Kim, S.S.: Purification and characterization of the catabolic α-acetolactate synthase from Serratia marcescens. J. Biochem. Mol. Biol., **31**, 37-43 (1998)
[43] Yang, J.; Kim, S.: Effect of pyrimidylsalicylate on the valine sensitive acetolactate synthase purified from Serratia marcescens. J. Biochem. Mol. Biol., **30**, 13-17 (1997)
[44] Maier, W.; Luthra, R.; Gröger, D.: Acetohydroxyacid synthase from Claviceps purpurea: partial purification and characterization. J. Basic Microbiol., **29**, 483-490 (1989)
[45] Poulsen, C.; Stougaard, P.: Purification and properties of Saccharomyces cerevisiae acetolactate synthase from recombinant Escherichia coli. Eur. J. Biochem., **185**, 433-439 (1989)

[46] Chong, C.K.; Chang, S.I.; Choi, J.D.: Purification and characterization of acetolactate synthase from barley. J. Biochem. Mol. Biol., **30**, 274-279 (1997)
[47] Eoyang, L.; Silverman, P.M.: Purification and assays of acetolactate synthase I from Escherichia coli K-12. Methods Enzymol., **166**, 435-445 (1988)
[48] Phalip, V.; Schmitt, P.; Divies, C.: Purification and characterization of the catabolic α-acetolactate synthase from Leuconostoc mesenteroides subsp. cremonis. Curr. Microbiol., **31**, 316-321 (1995)
[49] Carroll, N.M.; Sheehan, D.; Cogan, T.M.: Purification of α-acetolactate synthase from Leuconostoc lactis NCW1. Biochem. Soc. Trans., **23**, 366S (1995)
[50] Chang, S.I.; Kang, M.K.; Choi, J.D.; Namgoong, S.K.: Soluble overexpression in Escherichia coli, and purification and characterization of wild-type recombinant tobacco acetolactate synthase. Biochem. Biophys. Res. Commun., **234**, 549-553 (1997)
[51] Xing, R.; Whitman, W.B.: Purification and characterization of the oxygen-sensitive acetohydroxy acid synthase from the archaebacterium Methanococcus aeolicus. J. Bacteriol., **176**, 1207-1213 (1994)
[52] Snoep, J.L.; Teixeira de Mattos, M.J.; Starrenburg, M.J.C.; Hugenholtz, J.: Isolation, characterization, and physiological role of the pyruvate dehydrogenase complex and α-acetolactate synthase of Lactococcus lactis subsp. lactis bv. diacetylactis. J. Bacteriol., **174**, 4838.-4841 (1992)
[53] Yang, J.H.; Kim, S.S.: Purification and characterization of the valine sensitive acetolactate synthase from Serratia marcescens ATCC 25419. Biochim. Biophys. Acta, **1157**, 178-184 (1993)

1-Deoxy-D-xylulose-5-phosphate synthase 2.2.1.7

1 Nomenclature

EC number
2.2.1.7

Systematic name
pyruvate:D-glyceraldehyde-3-phosphate acetaldehydetransferase (decarboxylating)

Recommended name
1-deoxy-D-xylulose-5-phosphate synthase

Synonyms
1-D-deoxy-D-threo-2-pentulose 5-phosphate synthetase
1-D-deoxyxylulose 5-phosphate synthase
1-deoxy-D-threo-pentulose synthase
1-deoxy-D-xylulose 5-phosphate synthetase
1-deoxy-D-xylulose-5-phosphate pyruvate lyase
1-desoxy-D-xylulose-5-phosphate synthase
D-1-deoxyxylulose 5-phosphate synthase
DXP-synthase
DXPS <3, 4> [3]
DXPase <6> [5]
DXS <8> [8]
EC 4.1.3.37 (formerly)
deoxyxylulose 5-phosphate synthase
deoxyxylulose-5-phosphate synthetase
glyceraldehydes 3-phosphate-pyruvate ligase
synthase, 1-deoxy-D-xylulose 5-phosphate

CAS registry number
202218-79-9

2 Source Organism

<1> *Streptomyces sp.* (strain CL190 [1]) [1]
<2> *Escherichia coli* [1, 2, 7]
<3> *Bacillus subtilis* (strain PY79 [3]) [3]
<4> *Synechocystis sp.* (PCC 6803 [3]) [3]
<5> *Pseudomonas aeruginosa* [4]
<6> *Rhodobacter capsulatus* [5]

<7> *Arabidopsis thaliana* (transgenic lines with increased or decreased enzyme levels [6]) [6]
<8> *Lycopersicon esculentum* (tomato [8]) [8]
<9> *Mentha x piperita* (peppermint [9]) [9]

3 Reaction and Specificity

Catalyzed reaction

pyruvate + D-glyceraldehyde 3-phosphate = 1-deoxy-D-xylulose 5-phosphate + CO_2 (requires thiamine diphosphate; the enzyme forms part of an alternative nonmevalonate pathway for terpenoid biosynthesis)

Reaction type

condensation (<2> acyloin-type condensation reaction [2])

Natural substrates and products

S pyruvate + D-glyceraldehyde 3-phosphate <7> (<7> plastidic isoprenoid synthesis in plants, catalyses one of the rate-limiting steps [6]) [6]
P 1-deoxy-D-xylulose 5-phosphate + CO_2

Substrates and products

S pyruvate + D-glyceraldehyde <1, 2, 6> (Reversibility: ? <1, 2, 6> [1, 5]) [1, 5]
P 1-deoxy-D-xylulose + CO_2
S pyruvate + D-glyceraldehyde 3-phosphate <1-7> (Reversibility: ? <1-7> [1-6]) [1-6]
P 1-deoxy-D-xylulose 5-phosphate + CO_2
S pyruvate + DL-glyceraldehyde 3-phosphate <2, 5> (Reversibility: ? <2, 5> [2, 4]) [2, 4]
P 1-deoxy-D-xylulose 5-phosphate + CO_2

Inhibitors

EDTA <1> (<1> recombinant protein [1]) [1]
fluoropyruvate <5> (<5> recombinant protein, 50% inhibition at 0.4 mM, fluoropyruvate is supposed to bind covalently to the active site [4]) [4]

Cofactors/prosthetic groups

thiamine diphosphate <1> (<1> failure of DXP producting in the absence of thiamine diphosphate indicates its absolute requirement for the enzymatic reaction [1]) [1]

Metals, ions

Mg^{2+} <1, 2, 6> (<1> recombinant protein, optimum concentration 1 mM [1]; <6> recombinant isoenzyme A [5]) [1, 5]
Mn^{2+} <1, 2, 6> (<1> recombinant protein, optimum concentration 1 mM [1]; <6> recombinant isoenzyme A [5]) [1, 5]

Specific activity (U/mg)

1.6 <6> (<6> histidine-tagged DXP-synthase A, expressed in Escherichia coli [5]) [5]

K_m-Value (mM)

0.065 <1> (pyruvate, <1> recombinant protein [1]) [1]
0.096 <2> (pyruvate) [1]
0.12 <1> (D-glyceraldehyde 3-phosphate, <1> recombinant protein [1]) [1]
0.12 <6> (D-glyceraldehyde 3-phosphate, <6> recombinant isoenzyme B [5]) [5]
0.15 <6> (D-glyceraldehyde 3-phosphate, <6> recombinant isoenzyme A [5]) [5]
0.24 <2> (D-glyceraldehyde 3-phosphate) [1]
0.61 <6> (pyruvate, <6> recombinant isoenzyme A, cosubstrate D-glyceraldehyde 3-phosphate [5]) [5]
3 <6> (pyruvate, <6> recombinant isoenzyme B, cosubstrate D-glyceraldehyde 3-phosphate [5]) [5]
4.3 <6> (pyruvate, <6> recombinant isoenzyme A, cosubstrate D-glyceraldehyde [5]) [5]
10 <6> (D-glyceraldehyde, <6> recombinant isoenzyme B [5]) [5]
14 <6> (D-glyceraldehyde, <6> recombinant isoenzyme A [5]) [5]
14 <6> (pyruvate, <6> recombinant isoenzyme B, cosubstrate D-glyceraldehyde [5]) [5]
35 <1> (D-glyceraldehyde, <1> recombinant protein [1]) [1]
38 <2> (D-glyceraldehyde) [1]

pH-Optimum

7-7.5 <6> (<6> recombinant isoenzyme A [5]) [5]
7.5-8 <2> [1]
9 <1> (<1> recombinant protein [1]) [1]

Temperature optimum (°C)

42-44 <1, 2> (<1> recombinant protein [1]) [1]

4 Enzyme Structure

Molecular weight

65000-69000 <2> (<2> SDS-PAGE [1,2]) [1, 2]
68000 <1> (<1> predicted size from gene sequence [1]) [1]
68000 <6> (<6> histidine-tagged recombinant isoenzyme A, SDS-PAGE [5]) [5]
69000 <4> (<4> predicted size from gene sequence [3]) [3]
70000 <1> (<1> recombinant protein, SDS-PAGE [1]) [1]
70000 <3> (<3> predicted size from gene sequence [3]) [3]
71000 <9> (<9> predicted size from gene sequence, mature protein [9]) [9]
77600 <8> (<8> predicted size from gene sequence [8]) [8]

Subunits

dimer <1> (<1> 2 * 65000-70000, gel filtration, native PAGE [1]) [1]
dimer <2> (<2> 2 * 60000, gel filtration [1]) [1]
dimer <6> (<6> 2 * 70500, histidine-tagged recombinant isoenzyme A, homodimer, gel filtration [5]) [5]

5 Isolation/Preparation/Mutation/Application

Source/tissue

fruit <8> (<8> level of mRNA increases greatly during fruit ripening, highest level detectable in orange fruit, predominantly in the outer cell-layers of the pericarp [8]) [8]
inflorescence <8> (<8> RNA blot hybridization [8]) [8]
leaf <8, 9> (<8> young, developing and fully expanded leaves, RNA blot hybridisation [8]; <9> RNA blot analysis of oil gland secretory cells from leaves of different developmental stages [9]) [8, 9]
stem <8> (<8> RNA blot hybridization [8]) [8]
Additional information <8> (<8> mRNA not detectable in root [8]) [8]

Localization

oleosome <9> [9]
plastid <8, 9> (<8,9> N-terminal domain shows general features of plastidial targeting sequences [8,9]) [8, 9]

Purification

<1> (recombinant protein, expressed in Escherichia coli [1]) [1]
<2> (overexpression [1]) [1]
<5> (histidine-tagged recombinant protein, expressed in Escherichia coli [4]) [4]
<6> (histidine-tagged recombinant proteins DXP-synthase A and DXP-synthase B, expressed in Escherichia coli [5]) [5]

Cloning

<1> (expressed in Escherichia coli [1]) [1]
<2> (overexpression [2]) [2]
<3> (expressed in Escherichia coli [3]) [3]
<4> (expressed in Escherichia coli [3]) [3]
<5> (expressed in Escherichia coli [4]) [4]
<6> (two genes dxsA and dxsB, dxsA is located in the photosynthesis cluster, both isoenzymes expressed in Escherichia coli [5]) [5]
<8> (isolation of cDNA, expression analysis indicates developmental and organ-specific regulation of mRNA accumulation [8]; complementation assay in Escherichia coli [8]) [8]
<9> (expressed in Escherichia coli [9]) [9]

Engineering

H49Q <2> (<2> mutant enzyme shows no detectable DXP-synthase activity, this demonstrates the key role of H49 for enzyme activity [7]) [7]

6 Stability

Storage stability

<6>, -20°C, 50 mM Tris-HCl pH 7.5, 5 mM β-mercaptoethanol, 0.15 mM phenylmethysulfonyl fluoride, 30% glycerol, 4-6 weeks [5]

References

[1] Kuzuyama, T.; Takagi, M.; Takahashi, S.; Seto, H.: Cloning and characterization of 1-deoxy-D-xylulose 5-phosphate synthase from Streptomyces sp. strain CL190, which uses both the mevalonate and nonmevalonate pathways for isopentenyl diphosphate biosynthesis. J. Bacteriol., **182**, 891-897 (2000)

[2] Lois, L.M.; Campos, N.; Putra, S.R.; Danielsen, K.; Rohmer, M.; Boronat, A.: Cloning and characterization of a gene from Escherichia coli encoding a transketolase-like enzyme that catalyzes the synthesis of D-1-deoxyxylulose 5-phosphate, a common precursor for isoprenoid, thiamin, and pyridoxol biosynthesis. Proc. Natl. Acad. Sci. USA, **95**, 2105-2110 (1998)

[3] Harker, M.; Bramley, P.M.: Expression of prokaryotic 1-deoxy-D-xylulose-5-phosphatases in Escherichia coli increases carotenoid and ubiquinone biosynthesis. FEBS Lett., **448**, 115-119 (1999)

[4] Altincicek, B.; Hintz, M.; Sanderbrand, S.; Wiesner, J.; Beck, E.; Jomaa, H.: Tools for discovery of inhibitors of the 1-deoxy-D-xylulose 5-phosphate (DXP) synthase and DXP reductoisomerase: an approach with enzymes from the pathogenic bacterium Pseudomonas aeruginosa. FEMS Microbiol. Lett., **190**, 329-333 (2000)

[5] Hahn, F.M.; Eubanks, L.M.; Testa, C.A.; Blagg, B.S.J.; Baker, J.A.; Poulter, C.D.: 1-Deoxy-D-xylulose 5-phosphate synthase, the gene product of open reading frame (ORF) 2816 and ORF 2895 in Rhodobacter capsulatus. J. Bacteriol., **183**, 1-11 (2001)

[6] Estevez, J.M.; Cantero, A.; Reindl, A.; Reichler, S.; Leon, P.: 1-Deoxy-D-xylulose-5-phosphate synthase, a limiting enzyme for plastidic isoprenoid biosynthesis in plants. J. Biol. Chem., **276**, 22901-22909 (2001)

[7] Querol, J.; Rodriguez-Concepcion, M.; Boronat, A.; Imperial, S.: Essential role of residue H49 for activity of Escherichia coli 1-deoxy-D-xylulose 5-phosphate synthase, the enzyme catalyzing the first step of the 2-C-methyl-D-erythritol 4-phosphate pathway for isoprenoid synthesis. Biochem. Biophys. Res. Commun., **289**, 155-160 (2001)

[8] Lois, L.M.; Rodriguez-Concepcion, M.; Gallego, F.; Campos, N.; Boronat, A.: Carotenoid biosynthesis during tomato fruit development: Regulatory role of 1-deoxy-D-xylulose 5-phosphate synthase. Plant J., **22**, 503-513 (2000)

[9] Lange, B.M.; Wildung, M.R.; McCaskill, D.; Croteau, R.: A family of transketolases that directs isoprenoid biosynthesis via a mevalonate-independent pathway. Proc. Natl. Acad. Sci. USA, **95**, 2100-2104 (1998)

Fluorothreonine transaldolase 2.2.1.8

1 Nomenclature

EC number
2.2.1.8

Systematic name
fluoroacetaldehyde:L-threonine aldehydetransferase

Recommended name
fluorothreonine transaldolase

CAS registry number
403503-13-9

2 Source Organism

<1> *Streptomyces cattleya* (NRRL 8057 [1,2]) [1, 2]

3 Reaction and Specificity

Catalyzed reaction
L-threonine + fluoroacetaldehyde = acetaldehyde + 4-fluoro-L-threonine

Reaction type
aldehyde group transfer

Natural substrates and products
S L-threonine + fluoroacetaldehyde <1> (<1>, enzyme is important in biosynthesis of 4-fluorothreonine [1,2]) (Reversibility: ? <1> [1]) [1, 2]
P acetaldehyde + 4-fluoro-L-threonine <1> [1]

Substrates and products
S L-threonine + chloroacetaldehyde <1> (Reversibility: ? <1> [1,2]) [1]
P acetaldehyde + 4-chloro-L-threonine <1> [1, 2]
S L-threonine + fluoroacetaldehyde <1> (<1>, C3 and C4 of threonine are replaced with the aldehyde carbons, generating a new molecule of threonine [1]; <1> the enzyme-PLP complex accepts acetaldehyde as a substrate and generates a new molecule of threonine in a neutral crossover reaction [2]) (Reversibility: ? <1> [1,2]) [1, 2]
P acetaldehyde + 4-fluoro-L-threonine <1> [1, 2]

Cofactors/prosthetic groups

pyridoxal 5'-phosphate <1> (<1>, pyridoxal 5'-phosphate protein [1,2]) [1, 2]

4 Enzyme Structure

Molecular weight

120000 <1> (<1> gel filtration [2]) [2]

Subunits

dimer <1> (<1> 2 * 60000, SDS-PAGE [2]) [2]

5 Isolation/Preparation/Mutation/Application

Purification

<1> [2]

References

[1] Murphy, C.D.; Schaffrath, C.; O'Hagan, D.: Fluorinated natural products: the biosynthesis of fluoroacetate and 4-fluorothreonine in Streptomyces cattleya. Chemosphere, **52**, 455-461 (2003)
[2] Murphy, C.D.; O'Hagan, D.; Schaffrath, C.: Identification of a PLP-dependent threonine transaldolase: a novel enzyme involved in 4-fluorothreonine biosynthesis in Streptomyces cattleya. Angew. Chem., **40**, 4479-4481 (2001)

Amino-acid N-acetyltransferase 2.3.1.1

1 Nomenclature

EC number
2.3.1.1

Systematic name
acetyl-CoA:L-glutamate N-acetyltransferase

Recommended name
amino-acid N-acetyltransferase

Synonyms
AGAS
N-acetyl-L-glutamate synthetase
N-acetylglutamate synthase
N-acetylglutamate synthetase
acetylglutamate synthase
acetylglutamate synthetase
acetylglutamic synthetase
acetyltransferase, amino acid
amino acid acetyltransferase

CAS registry number
9029-88-3

2 Source Organism

<1> *Escherichia coli* (enzyme is repressed by arginine [6]) [1, 3, 6, 17]
<2> *Salmonella typhimurium* [1]
<3> *Rattus norvegicus* [1, 2, 5, 8, 10-12]
<4> *Homo sapiens* [1, 10, 13]
<5> *Pseudomonas aeruginosa* [1, 9]
<6> *Neurospora crassa* [7, 16]
<7> *Chlorella vulgaris* [14]
<8> *Pisum sativum* [15]
<9> *Bos taurus* [4]
<10> *Mus musculus* [18]
<11> *Saccharomyces cerevisiae* [19, 21]
<12> *Mus musculus* [20]

3 Reaction and Specificity

Catalyzed reaction

acetyl-CoA + L-glutamate = CoA + N-acetyl-L-glutamate (<4> rapid-equilibrium random bi bi mechanism [13])

Reaction type

acyl group transfer

Natural substrates and products

S acetyl-CoA + L-glutamate <1-3, 5, 6, 11, 12> (<1,2,5> enzyme catalyzes the first step in the biosynthesis of arginine [1,6,9]; <6> indispensible enzyme of arginine biosynthesis [7]; <3> enzyme activates carbamoyl-phosphate synthase ammonia in small intestine mucosa, to allow citrulline synthesis in the tissue [8]; <11> enzyme catalyzes the first step in arginine biosynthesis [19]; <12> first enzyme in urea cycle [20]) (Reversibility: ? <1-3, 5, 6, 11, 12> [1, 6-9, 19, 20]) [1, 6-9, 19, 20, 21]

P CoA + N-acetyl-L-glutamate <1-3, 5, 6, 11, 12> [1, 6-9, 19, 20, 21]

Substrates and products

S acetyl-CoA + DL-2-aminopimelate <1, 3> (<3> 5.2% of activity with L-glutamate [1]) (Reversibility: ? <3> [1]) [1, 11]

P CoA + 2-acetylaminoheptanedioate <1, 3> [1, 11]

S acetyl-CoA + L-2-aminoadipate <1, 3> (<1> 1.3% of activity with L-glutamate [1]; <3> DL-2-aminoadipate, 5.2% of activity with L-glutamate [11]) (Reversibility: ? <1, 3> [1, 11]) [1, 11]

P CoA + 2-acetylaminohexanedioate <1, 3> [1, 11]

S acetyl-CoA + L-aspartate <3> (<3> 3.0% of activity with L-glutamate [8]) (Reversibility: ? <3> [8]) [8]

P CoA + N-acetyl-L-aspartate <3> [1, 2, 8, 11]

S acetyl-CoA + L-glutamate <1-12> (Reversibility: ? <1-12> [1-21]) [1-21]

P CoA + N-acetyl-L-glutamate <1-12> [1-21]

S acetyl-CoA + L-glutamate-γ-hydroxamate <3> (<3> 15.5% of activity with glutamate [11]) (Reversibility: ? <3> [11]) [11]

P CoA + N-acetyl-L-glutamate-γ-hydroxamate <3> [11]

S acetyl-CoA + L-glutamine <1, 3> (<3> 6.8% of activity with L-glutamate [1]; <1> 5% of activity with acetyl-CoA [1]) (Reversibility: ? <1, 3> [1, 11]) [1, 2, 11]

P ?

S acetyl-CoA + glycine <3> (<3> 2.9% of activity with L-glutamate [1]) (Reversibility: ? <3> [1]) [1, 2, 8, 11]

P CoA + acetylaminoacetate <3> [1, 2, 8, 11]

S benzoyl-CoA + L-glutamate <3> (<3> very low activity [2]) (Reversibility: ? <3> [2]) [2]

P CoA + N-benzoyl-L-glutamate <3> [2]

S butyryl-CoA + L-glutamate <3> (<3> very low activity [2]) (Reversibility: ? <3> [2]) [2]

P CoA + N-butyryl-L-glutamate <3> [2]

S propionyl-CoA + L-glutamate <1, 3, 6> (<3> 4.3% of activity with acetyl-CoA [1]; <1> 4% of activity with acetyl-CoA [1]; <6> 18% of activity with acetyl-CoA [7]) (Reversibility: ? <1, 3, 6> [1, 7]) [1, 2, 7, 8, 11]

P CoA + N-propionyl-L-glutamate <1, 3, 6> [1, 2, 7, 8, 11]

Inhibitors

1,3-diaminopropane <5> (<5> 10 mM, 88% inhibition in the presence of 0.5 mM N-acetyl-L-glutamate [9]) [9]
5,5'-dithiobis(2-nitrobenzoate) <3> (<3> strong inhibition [2]) [2]
$AgNO_3$ <3> (<3> 0.1 mM, 70-90% inhibition [2]) [2]
$BaCl_2$ <3> (<3> 1 mM, 30-50% inhbition [2]) [2]
$CaCl_2$ <3> (<3> 1 mM, 30-50% inhbition [2]) [2]
Cd^{2+} <1> (<1> 0.1 mM, 94% inhibition [3]) [3]
$CoCl_2$ <3> (<3> 0.1 mM, 30-50% inhbition [2]) [2]
Cu^{2+} <1, 3> (<3> $CuSO_4$, 0.1 mM, 70-90% inhibition [2]; <1> 0.1 mM, 68% inhibition [3]) [2, 3]
EDTA <3> (<3> weak inhibition [2]) [2]
$FeCl_3$ <3> (<3> 0.1 mM, 30-50% inhbition [2]) [2]
$FeSO_4$ <3> (<3> 0.1 mM, 30-50% inhbition [2]) [2]
Hg^{2+} <1, 3> (<3> $Hg(CHCOO)_2$, 0.1 mM, 70-90% inhibition [2]; <1> 0.1 mM; 50% inhibition [3]) [2, 3]
K_2SO_4 <3> [2]
KCl <1, 3> (<3> 100 mM, 30% inhibition, 200 mM, 50% inhibition [2]; <1> 100 mM, 36% inhibition after 1 h, 300 mM, 74% inhibition after 1 h, 500 mM, 87% inhibition after 1 h [3]) [2, 3]
L-α-acetoxylglutamate <3> (<3> 2 mM, 17% inhibition [2]) [2, 2]
L-citrulline <1> (<1> 10 mM, 75% inhibition [3]) [3]
L-indospicine <1> (<1> 0.2 mM; 50% inhibition [6]) [6]
$MgCl_2$ <3> (<3> 1 mM, 30-50% inhbition [2]) [2]
$MnCl_2$ <3> (<3> 0.1 mM, 30-50% inhbition [2]) [2]
N-acetyl-D-glutamate <3> (<3> 2 mM, 30% inhibition [2]) [2, 2]
N-acetyl-DL-α-aminoadipate <3> (<3> 2 mM, 78% inhibition [2]) [2]
N-acetyl-L-aspartate <3> (<3> 2 mM, 25% inhibition [2]) [2, 2]
N-acetyl-L-glutamine <3, 3> (<3> 2 mM, 46% inhibition [2]) [2, 2]
N-acetylglutamate <1, 3> (<3> 2 mM, 88% inhibition [1]; <1> 25 mM, 50% inhibition [3]) [1, 2, 3, 9, 13]
N-benzoyl-L-glutamate <3> (<3> 2 mM, 29% inhibition [2]) [2, 2]
N-butyryl-L-glutamate <3> (<3> 2 mM, 19% inhibition [2]) [2, 2]
N-carbamoyl-L-glutamate <3> (<3> 2 mM, 31% inhibition [2]) [2, 2]
N-ethylmaleimide <1> [3]
N-propionyl-L-glutamate <3, 3> (<3> 2 mM, 63% inhibition [2]) [2, 2]
$Ni(NO_3)_2$ <3> (<3> 0.1 mM, 30-50% inhbition [2]) [2]
O-(L-norvalyl-5)-isourea <1> (<1> 0.02 mM, 50% inhibition [6]) [6]
$Pb(NO_3)_2$ <3> (<3> 0.1 mM, 70-90% inhibition [2]) [2]
Zn^{2+} <1, 3> (<3> $ZnCl_2$, 0.1 mM, 70-90% inhibition [2]; <1> 0.1 mM, 43% inhibition [3]) [2, 3]

arginine <1, 5, 6, 7> (<1,6> L-arginine [3,6,7]; <1> 0.02 mM, 50% inhibition [3]; <1> 0.02 mM, 50% inhibition [6]; <6> feed back inhibition, 0.16 mM, 50% inhibition [7]; <5> 0.2 mM, 73% inhibition, 97% inhibition in the presence of 0.5 mM N-acetyl-L-glutamate [9]) [3, 6, 7, 9, 14]
cadaverine <5> (<5> 10 mM, 80% inhibition in the presence of 0.5 mM N-acetyl-L-glutamate [9]) [9]
coenzyme A <1, 3, 6> (<1> 2.5 mM, 50% inhibition [1,3]; <3> 1 mM, 20% inhibition [1]; <6> 1 mM, 30% inhibition [7]) [1, 2, 3, 7, 13]
high ionic strength <1, 3> [1, 3]
oxaloacetate <6> (<6> 1 mM, 68% inhibition [7]) [7]
p-chloromercuribenzoate <1> [3]
p-hydroxymercuribenzoate <1, 3> (<3> strong inhibition [2]) [2, 3]
polyamines <5> [9]
potassium phosphate <3> [2]
propionyl-CoA <3> [1]
putrescine <5> (<5> 10 mM, 74% inhibition in the presence of 0.5 mM N-acetyl-L-glutamate [9]) [9]
spermidine <5> (<5> 1 mM, 78% inhibition in the presence of 0.5 mM N-acetyl-L-glutamate [9]) [9]
spermine <5> (<5> 1 mM, 88% inhibition in the presence of 0.5 mM N-acetyl-L-glutamate [9]) [9]
succinate <3> (<3> 2 mM, 21% inhibition [2]) [2, 2]

Activating compounds

L-argininic acid <3> (<3> activates to a lesser extent than L-arginine [11]) [11]
Triton X-100 <3> (<3> 0.1%, 4fold activation [1]) [1]
arginine <3, 4, 10, 12> (<3> 0.02 mM, 4.2fold activation [1]; <3> 1 mM, 2-6fold activation in mitochondria, 1.6fold activation of partially purified enzyme [8]; <3> rapid rate of synthesis of acetyl glutamate requires a concentration of arginine of about 0.025 mM and either relatively high glutamate concentrations or relatively high pH [5]; <3> 0.1 mM, 2fold activation at pH 7.2, 8fold activation at pH 8.2 [5]; <3> 5.1fold activiation of matrix enzyme [11]; <10> activation is high if the enzyme is purified from fed mice and low if purified from fasted mice, thiol/disulfide interchange may be responsible for arginine sensitivity [18]; <12> 1 mM, 2-6fold activation of recombinant enzyme [20]) [1, 2, 5, 8, 11, 18, 20]
cationic polypeptides <3> (<3> 4fold activation [1]) [1]
culpein <3> [11]
polyarginine <3> [11]
polylysine <3> [11]
salmin <3> [11]
Additional information <11> (<11> enzyme must interact stoichoimetrically in vivo with acetylglutamate kinase to be active [19,21]) [19, 21]

Specific activity (U/mg)

0.0000135 <3> (<3> activity in liver of fetus, activity increases strongly after eating external diets until week 10 after birth [12]) [12]

0.0000136 <3> (<3> activity in intestine of fetus [12]) [12]
0.000014 <3> [1]
0.0000203 <3> (<3> activity in intestine at day 3 after birth [12]) [12]
0.0000455 <3> (<3> activity in intestine of adult [12]) [12]
0.0000643 <3> (<3> activity in intestine at day 1 after birth [12]) [12]
0.00013 <1, 2> [1]
0.000268 <3> (<3> activity in liver of adult [1]) [12]
0.00286 <4> [13]
0.012 <3> [2]
0.346 <12> (<12> mature recombinant enzyme in the absence of arginine [20]) [20]
0.803 <12> (<12> mature recombinant enzyme in the presence of 1 mM arginine [20]) [20]
31.6 <3> [8]
92.4 <3> (<3> in the presence of 1 mM L-arginine [11]) [11]
133 <1> [3]
Additional information <3, 4, 8> (<3,4,8> assay procedure [10,12,15]; <9> rapid and specific assay [4]; <3> 0.00000017 mmol N-acetylglutamate/min/g mucosa [8]) [4, 8, 10, 12, 15]

K_m-Value (mM)
0.38 <3> (acetyl-CoA) [5]
0.7 <3> (acetyl-CoA) [2]
0.76 <3> (acetyl-CoA) [1, 11]
0.8 <3> (acetyl-CoA) [8]
1.6 <6> (glutamate) [7]
2.1 <3> (glutamate) [5]
2.5 <3> (L-glutamate) [8]
3 <3> (L-glutamate) [2]
3.2 <7> (acetyl-CoA) [14]
3.7 <3> (L-glutamate) [1, 11]
4.4 <4> (acetyl-CoA) [13]
4.6 <2> (acetyl-CoA) [1]
4.7 <4> (L-glutamate, <4> partially purified enzyme [1]) [1]
6.3 <6> (acetyl-CoA) [7]
8.1 <4> (acetyl-CoA, <4> partially purified enzyme [1]) [1]
8.1 <4> (glutamate) [13]

K_i-Value (mM)
0.07 <3> (acetyl-glutamate) [2]
0.49 <4> (N-acetyl-L-glutamate) [13]
0.71 <3> (propionyl-CoA) [1]

pH-Optimum
8 <3> (<3> liver enzyme, in the absence of arginine [1,2]; <3> in the absence of arginine [2]) [1, 2]
8-8.5 <7> [14]

8.5 <3, 4> (<4> liver enzyme, in the presence of arginine [1,13]; <3> in the presence of arginine [2]) [1, 2, 8, 13]
8.8-9 <6> [7]
10 <1> [1, 3]

pH-Range
8-9.5 <6> (<6> 38% of maximal activity at pH 8.0., 25% of maximal activity at pH 9.5 [7]) [7]

4 Enzyme Structure

Molecular weight
30000-300000 <3> (<3> gel filtration, mixture of multiple forms) [2]
160000 <3> (<3> gel filtration, sucrose density gradient centrifugation [1,11]) [1, 11]
190000 <4> (<4> gel filtration [1,13]) [1, 13]

Subunits
? <6> (<6> x * 73000, deduced from nucleotide sequence of the mature enzyme [16]) [16]
hexamer <1> (<1> 6 * 51700, SDS-PAGE, cross-linking studies with dimethylsuberimidate, in the maximally aggregated state the enzyme exists as a hexamer [3]) [3]
trimer <3> (<3> 3 * 57000, SDS-PAGE [1,11]) [1, 11]

5 Isolation/Preparation/Mutation/Application

Source/tissue
colon <3> [10]
cotyledon <8> [15]
duodenum <4> [10]
liver <3, 4> [1, 2, 5, 10-13]
lung <3> [10]
placenta <3> [10]
small intestine <3, 4> (<3> mucosa [1,8]; <3> little activity in muscle layer and serosa [8]) [1, 8, 10, 12]
spleen <3> [10]
stomach <4> (<4> mucosa [10]) [10]
submaxillary gland <3> [10]
testis <3> [10]
thymus <3> [10]
uterus <3> [10]

Purification
<1> (DEAE-cellulose, hydroxylapatite [3]) [3]
<3> (ammonium sulfate, DEAE Biogel, Sephacryl S-200, aminoacetyl biogel A, affi-gel blue, sucrose gradient, isoelectric focusing [1,11]; ammonium sul-

fate, hydroxyapatite, DEAE-cellulose, Sephadex G-100 [2]; ammonium sulfate, hydroxyapatite [8]) [1, 2, 8, 11]
<4> (ammonium sulfate, hydroxyapatite, DEAE-cellulose, Sephacryl-300 [13]) [13]
<5> (ammonium sulfate, hydroxyapatite, 10fold purification [9]) [9]
<7> [14]
<12> (recombinant enzyme, Ni^{2+}-affinity column [20]) [20]

Cloning

<1> (cloning of N-acetylglutamate synthase gene argA and a mutant gene fbr-argA responsible for arginine feedback resistance [17]) [17]
<6> [16]
<11> (coexpression of N-acetylglutamate synthase gene ARG2 together with ARG5 and ARG6 coding for acetylglutamate kinase and acetylglutamyl-phosphate reductase in Saccharomyces cerevisiae [19]) [19, 21]
<12> (expression in Escherichia coli [20]) [20]

Engineering

G287S <1> (<1> arginine feedback resistant mutant enzyme [17]) [17]
H15Y <1> (<1> arginine feedback resistant mutant enzyme [17]) [17]
Q432R <1> (<1> arginine feedback resistant mutant enzyme [17]) [17]
R58H <1> (<1> arginine feedback resistant mutant enzyme [17]) [17]
S54N <1> (<1> arginine feedback resistant mutant enzyme [17]) [17]
Y19C <1> (<1> arginine feedback resistant mutant enzyme [17]) [17]

6 Stability

pH-Stability

7 <1> (<1> highest stability [1,3]) [1, 3]

Temperature stability

38 <3> (<3> strong decrease in activity after 5 min in the absence of arginine, presence of 1 mM arginine stabilizes [10]) [10]
60 <3> (<3> 5 min, complete loss of activity [8]) [8]

General stability information

<1>, L-arginine or N-acetyl-L-glutamate stabilizes the high molecular weight form of 300000 Da [3]
<1>, glycerol stabilizes [3]
<3>, Triton X-100 stabilizes [11]
<3>, Triton X-100, 0.1%, stabilizes liver enzyme [1]
<1-5>, use of silicone-treated glassware or plastic tubes, e.g. polyethylene, polycarbonate or polypropylene stabilizes the enzyme [1]

Storage stability

<3>, -80°C, 100 mM phosphate buffer, pH 7.5, 6 months, 50% loss of activity [2]

<3>, 4°C, 100 mM potassium phosphate buffer, pH 7.5, 24 h, 10% loss of activity [2]
<4>, -20°C, stable for 2 months [13]

References

[1] Powers-Lee, S.G.: N-Acetylglutamate synthase. Methods Enzymol., **113**, 27-35 (1985)
[2] Shigesada, K.; Tatibana, M.: N-Acetylglutamate synthetase from rat-liver mitochondria. Partial purification and catalytic properties. Eur. J. Biochem., **84**, 285-291 (1978)
[3] Marvil, D.K.; Leisinger, T.: N-Acetylglutamate synthase of Escherichia coli: purification, characterization, and molecular properties. J. Biol. Chem., **252**, 3295-3303 (1977)
[4] Driessen, H.P.C.; De Jong, W.W.; Tesser, G.I.; Bloemendal, H.: A rapid and specific assay for N-α-acetyltransferase activity. Anal. Biochem., **125**, 352-357 (1982)
[5] Kamemoto, E.S.; Atkinson, D.E.: Modulation of the activity of rat liver acetylglutamate synthase by pH and arginine concentration. Arch. Biochem. Biophys., **243**, 100-107 (1985)
[6] Leisinger, T.; Haas, D.: N-Acetylglutamate synthase of Escherichia coli: regulation of synthesis and activity by arginine. J. Biol. Chem., **250**, 1690-1693 (1975)
[7] Hinde, R.W.; Jacobson, J.A.; Weiss, R.L.; Davis, R.H.: N-Acetyl-L-glutamate synthase of Neurospora crassa. Characteristics, localization, regulation, and genetic control. J. Biol. Chem., **261**, 5848-5852 (1986)
[8] Uchiyama, C.; Mori, M.; Tatibana, M.: Subcellular localization and properties of N-acetylglutamate synthase in rat small intestinal mucosa. J. Biochem., **89**, 1777-1786 (1981)
[9] Haas, D.; Leisinger, T.: Multiple control of N-acetylglutamate synthetase from Pseudomonas aeruginosa: synergistic inhibition by acetylglutamate and polyamines. Biochem. Biophys. Res. Commun., **60**, 42-47 (1974)
[10] Wakabayashi, Y.; Iwashima, A.; Yamada, E.; Yamada, R.: Enzymological evidence for the indispensability of small intestine in the synthesis of arginine from glutamate. II. N-acetylglutamate synthase. Arch. Biochem. Biophys., **291**, 9-14 (1991)
[11] Sonoda, T.; Tatibana, M.: Purification of N-acetyl-L-glutamate synthetase from rat liver mitochondria and substrate and activator specificity of the enzyme. J. Biol. Chem., **258**, 9839-9844 (1983)
[12] Yamada, E.; Wakabayashi, Y.: Development of pyrroline-5-carboxylate synthase and N-acetylglutamate synthase and their changes in lactation and aging. Arch. Biochem. Biophys., **291**, 15-23 (1991)
[13] Bachmann, C.; Krahenbuhl, S.; Colombo, J.P.: Purification and properties of acetyl-CoA:L-glutamate N-acetyltransferase from human liver. Biochem. J., **205**, 123-127 (1982)

[14] Morris, C.J.; Thompson, J.F.: Acetyl coenzyme A-glutamate acetyltransferase and N^2-acetylornithine-glutamate acetyltransferase of Chlorella. Plant Physiol., **55**, 960-967 (1975)

[15] Jain, J.C.; Shargool, P.D.: A modified assay system for enzymes involved in N-acetyl group transfer reactions: its use to study enzymes involved in ornithine biosynthesis in plants. Anal. Biochem., **138**, 25-29 (1984)

[16] Yu, Y.G.; Turner, G.E.; Weiss, R.L.: Acetylglutamate synthase from Neurospora crassa: structure and regulation of expression. Mol. Microbiol., **22**, 545-554 (1996)

[17] Rajagopal, B.S.; Deponte, J.; Tuchman, M.; Malamy, M.H.: Use of inducible feedback-resistant N-acetylglutamate synthetase (argA) genes for enhanced arginine biosynthesis by genetically engineered Escherichia coli K-12 strains. Appl. Environ. Microbiol., **64**, 1805-1811 (1998)

[18] Kawamoto, S.; Sonoda, T.; Ohtake, A.; Suzuki, Y.; Okuda, K.; Tatibana, M.: Alteration in arginine activation of N-acetylglutamate synthetase in vitro by disulfide or thiol compounds. J. Mol. Catal. B, **10**, 191-197 (2000)

[19] Abadjieva, A.; Pauwels, K.; Hilven, P.; Crabeel, M.: A new yeast metabolon involving at least the two first enzymes of arginine biosynthesis. Acetylglutamate synthase activity requires complex formation with acetylglutamate kinase. J. Biol. Chem., **276**, 42869-42880 (2001)

[20] Caldovic, L.; Morizono, H.; Yu, X.; Thompson, M.; Shi, D.; Gallegos, R.; Allewell, N.M.; Malamy, M.H.; Tuchman, M.: Identification, cloning and expression of the mouse N-acetylglutamate synthase gene. Biochem. J., **364**, 825-831 (2002)

[21] Pauwels, K.; Abadjieva, A.; Hilven, P.; Stankiewicz, A.; Crabeel, M.: The N-acetylglutamate synthase/N-acetylglutamate kinase metabolon of Saccharomyces cerevisiae allows co-ordinated feedback regulation of the first two steps in arginine biosynthesis. Eur. J. Biochem., **270**, 1014-1024 (2003)

Imidazole N-acetyltransferase 2.3.1.2

1 Nomenclature

EC number
2.3.1.2

Systematic name
acetyl-CoA:imidazole N-acetyltransferase

Recommended name
imidazole N-acetyltransferase

Synonyms
acetyltransferase, imidazole
imidazole acetylase
imidazole acetyltransferase

CAS registry number
9029-89-4

2 Source Organism

<1> *Clostridium kluyveri* (not detected in other sources [1]) [1]

3 Reaction and Specificity

Catalyzed reaction
acetyl-CoA + imidazole = CoA + N-acetylimidazole

Reaction type
acyl group transfer

Natural substrates and products
S acetyl-CoA + imidazole <1> (Reversibility: ? <1> [1]) [1]
P CoA + N-acetylimidazole

Substrates and products
S acetyl-CoA + imidazole <1> (Reversibility: ? <1> [1]) [1]
P CoA + N-acetylimidazole <1> [1]
S propanoyl-CoA + imidazole <1> (Reversibility: ? <1> [1]) [1]
P CoA + N-propanoylimidazole <1> [1]

S Additional information <1> (<1> no naturally occurring compound tested can replace imidazole [1]) [1]

P ?

Specific activity (U/mg)

5.02 <1> [1]

5 Isolation/Preparation/Mutation/Application

Purification

<1> [1]

References

[1] Kinsky, S.C.: Assay, purification, and properties of imidazole acetylase. J. Biol. Chem., **235**, 94-98 (1960)

Glucosamine N-acetyltransferase 2.3.1.3

1 Nomenclature

EC number
2.3.1.3

Systematic name
acetyl-CoA:D-glucosamine N-acetyltransferase

Recommended name
glucosamine N-acetyltransferase

Synonyms
acetyltransferase, glucosamine
glucosamine acetylase

CAS registry number
9032-94-4

2 Source Organism

<1> *pigeon* [1]
<2> *Avena sativa* [2]

3 Reaction and Specificity

Catalyzed reaction
acetyl-CoA + D-glucosamine = CoA + N-acetyl-D-glucosamine

Reaction type
acyl group transfer

Natural substrates and products
S acetyl-CoA + D-glucosamine <2> (<2> the first enzyme in the metabolism of D-glucosamine [2]) (Reversibility: ? <2> [2]) [2]
P CoA + N-acetyl-D-glucosamine <2> [2]

Substrates and products
S acetyl-CoA + D-glucosamine <1, 2> (Reversibility: ? <1, 2> [1, 2]) [1, 2]
P CoA + N-acetyl-D-glucosamine <1, 2> [1, 2]
S acetyl-CoA + galactosamine <1, 2> (<2> not [2]) (Reversibility: ? <1> [1]) [1, 2]
P CoA + N-acetyl-D-galactosamine

Inhibitors

ADP <2> [2]
AMP <2> [2]
ATP <2> [2]
Co^{2+} <2> [2]
Cu^{2+} <2> [2]
EDTA <2> (<2> 1 mM EDTA inhibits enzyme by 20% [2]) [2]
Fe^{2+} <2> [2]
Fe^{3+} <2> [2]
GDP <2> [2]
GMP <2> [2]
GTP <2> [2]
Zn^{2+} <2> [2]
Additional information <2> (<2> not: Mg^{2+}, Mn^{2+}, Ca^{2+}, mono-, di and triphosphates of uridine, cytosine and tymidine [2]) [2]

K_m-Value (mM)

0.24 <2> (D-glucosamine, <2> membrane-bound enzyme [2]) [2]
0.26 <2> (D-glucosamine, <2> CHAPS-solubilized enzyme [2]) [2]

5 Isolation/Preparation/Mutation/Application

Source/tissue

liver <1> [1]

Localization

membrane <2> [2]

References

[1] Chou, T.C.; Soodak, M.: The acetylation of D-glucosamine by pigeon liver extracts. J. Biol. Chem., **196**, 105-109 (1957)
[2] Piro, G.; Buffo, M.; Dalessandro, G.: Membrane-bound glucosamine acetyltransferase in coleoptile segments of Avena sativa. Physiol. Plant., **90**, 181-186 (1994)

Glucosamine 6-phosphate N-acetyltransferase 2.3.1.4

1 Nomenclature

EC number
2.3.1.4

Systematic name
acetyl-CoA:D-glucosamine-6-phosphate N-acetyltransferase

Recommended name
glucosamine 6-phosphate N-acetyltransferase

Synonyms
D-glucosamine-6-P N-acetyltransferase
N-acetylglucosamine-6-phosphate synthase
acetyltransferase, glucosamine phosphate
aminodeoxyglucosephosphate acetyltransferase
glucosamine 6-phosphate N-acetyltransferase
glucosamine 6-phosphate acetylase
glucosamine phosphate acetyltransferase
glucosamine-6-phosphate acetylase
phosphoglucosamine N-acetylase
phosphoglucosamine acetylase
phosphoglucosamine transacetylase

CAS registry number
9031-91-8

2 Source Organism

<1> *Phaseolus aureus* (mung bean [6]) [6]
<2> *Streptococcus sp.* (hemolytic, group A [2]) [2]
<3> *Ovis aries* [1]
<4> *Neurospora crassa* (strain 5297a [3]) [1-3]
<5> *Penicillium sp.* [2]
<6> *Oryctolagus cuniculus* [2]
<7> *Canis familiaris* [2]
<8> *Homo sapiens* (3 isoenzymes [8]) [2, 8]
<9> *Rattus norvegicus* [5]
<10> *Blastocladiella emersonii* [4]
<11> *Sus scrofa* [7]
<12> *Mus musculus* [9]

3 Reaction and Specificity

Catalyzed reaction

acetyl-CoA + D-glucosamine 6-phosphate = CoA + N-acetyl-D-glucosamine 6-phosphate (<9> ping pong bi bi mechanism [5])

Reaction type

acyl group transfer

Natural substrates and products

S acetyl-CoA + D-glucosamine 6-phosphate <10> (<10> catalyzes a step in the biosynthesis of uridine 5'-diphosphate-N-acetyl D-glucosamine [4]) (Reversibility: ? <10> [4]) [4]
P CoA + N-acetyl-D-glucosamine 6-phosphate <10> [4]

Substrates and products

S acetyl-CoA + D-glucosamine 6-phosphate <1-12> (<1> highly specific for D-glucosamine-6-phosphate [6]) (Reversibility: ? <1-12> [1-9]) [1-9]
P CoA + N-acetyl-D-glucosamine 6-phosphate <1-12> [1-9]
S propionyl-CoA + D-glucosamine 6-phosphate <3> (<3> 15% of activity with acetyl-CoA [1]) (Reversibility: ? <3> [1]) [1]
P CoA + N-propionyl-D-glucosamine 6-phosphate <3> [1]

Inhibitors

CoA <9> [5]
D-glucosamine-6-phosphate <11> (<11> feedback inhibition [7]) [7]
EDTA <1, 4> (<1> 20 mM, 50% inhibition [6]) [1, 6]
Mn^{2+} <1> [6]
Ni^{2+} <3> [1]
Zn^{2+} <1> [6]
acetyl-CoA <3, 9> (<3> high concentrations [1]; <9> complex inhibition pattern [5]) [1, 5]
p-chloromercuribenzoate <3, 4> (<3> 0.001 mM, 50% inhibition, 0.01 mM, complete inhibition, reversed by cysteine [1]; <4> 1 mM, complete inhibition [2]) [1, 2, 3]
p-hydroxymercuribenzoate <1, 8> (<1> 0.05 mM, 91% inhibition, 0.1 mM, complete inhibition 1 mM L-cysteine restores activity [6]; <8> 2.5 mM, 81% inhibition of isoenzyme 1, 92% inhibition of isoenzyme 2 and 92.5% inhibition of isoenzyme 3, inhibition by 1.25 mM is reversible upon addition of 2.5 mM dithiothreitol [8]) [6, 8]
Additional information <4, 10> (<4> not inhibited by EDTA [2]; <10> not inhibited by EDTA [4]; <4> not inhibited by iodoacetate [2]) [2, 4]

Cofactors/prosthetic groups

Additional information <4> (<4> no cofactor required [1-3]) [1-3]

Activating compounds

EDTA <3> (<3> activates [1]) [1]
free sulfhydryl groups <3> (<3> activate [1]) [1]

Metals, ions

Mg^{2+} <1, 3> (<3> slight but consistent enhancing effect [1]; <1> required for maximal activity [6]) [1, 6]
Mn^{2+} <3> (<3> slight but consistent enhancing effect [1]) [1]
Additional information <11> (<11> not stimulated by K^+, Na^+, NH_4^+ and acetate [7]) [7]

Specific activity (U/mg)

0.00447 <8> (<8> isoenzymes 1 and 2 [8]) [8]
0.0054 <3> [1]
0.0379 <8> (<8> isoenzyme 3 [8]) [8]
2.41 <10> [4]
11.7 <4> [3]
12.21 <4> [2]
Additional information <4> (<4> assay method [3]) [3]

K_m-Value (mM)

0.012 <9> (acetyl-CoA) [5]
0.038 <10> (Na_2S-acetyl coenzyme A) [4]
0.04 <12> (D-glucosamine 6-phosphate, <12> recombinant enzyme [9]) [9]
0.06 <9> (D-glucosamine 6-phosphate) [5]
0.12 <12> (acetyl-CoA, <12> recombinant enzyme [9]) [9]
0.157 <8> (acetyl-CoA, <8> isoenzyme 1 [8]) [8]
0.187 <8> (acetyl-CoA, <8> isoenzyme 2 [8]) [8]
0.24 <10> (D-glucosamine 6-phosphate) [4]
0.25 <3> (acetyl-CoA) [1]
0.28 <8> (acetyl-CoA, <8> isoenzyme 3 [8]) [8]
0.45 <3> (D-glucosamine 6-phosphate) [1]
0.5 <4> (acetyl-CoA) [3]
0.7 <1, 4> (D-glucosamine 6-phosphate, <1> acetyl-CoA [6]) [3, 6]
0.78 <4> (D-glucosamine 6-phosphate) [2]
0.78 <4> (acetyl-CoA) [2]
1.51 <1> (D-glucosamine 6-phosphate) [6]
3 <8> (D-glucosamine 6-phosphate, <8> isoenzymes 2 and 3 [8]) [8]
7.1 <8> (D-glucosamine 6-phosphate, <8> isoenzyme 1 [8]) [8]

K_i-Value (mM)

0.05-0.15 <9> (acetyl-CoA) [5]

pH-Optimum

5.2 <1> [6]
6-7.1 <4> [1-3]
7.4 <3> [1]
7.5 <11> [7]
8.1 <10> [4]
9.7 <8> (<8> in glycine-NaOH [8]) [8]

pH-Range

6-8 <4> (<4> approx. 90% of maximal activity at pH 6.0, approx. 95% of maximal activity at pH 8.0 [1]) [1]

Temperature optimum (°C)

37 <11> [7]

4 Enzyme Structure

Molecular weight

24000 <11> [7]

Subunits

? <12> (<12> x * 21000, SDS-PAGE, immunoblotting, deduced from nucleotide sequence [9]) [9]

5 Isolation/Preparation/Mutation/Application

Source/tissue

brain <3> [1]
kidney <6, 7> [2]
liver <5-9, 11> [2, 5, 7]
muscle <6> [2]
seed <1> [6]
zoospore <10> [4]

Localization

cytosol <9> [5]
membrane <12> (<12> attached to cytoplasmic membrane leaflet of Golgi and other intracellular membranes [9]) [9]

Purification

<1> (protamine sulfate, pH 4.9, ammonium sulfate, acetone, CM-Sphadex, partially purified [6]) [6]
<3> (ammonium sulfate, calcium phosphate gel, heat treatment at 65°C [1]) [1]
<4> (calcium phosphate gel, heat treatment, 60°C, mixed bed resin, alumina Cγ gel, ammonium sulfate [2]; protamine sulfate, 60°C, pH 4.8, mixed-bed resin, ammonium sulfate [3]) [2, 3]
<8> (isoenzymes 1-3, CM-Sephadex [8]) [8]
<9> (ammonium sulfate, Ultragel AcA54, CoA-Sepharose [5]) [5]
<10> (acetone, DEAE-cellulose [4]) [4]
<11> (hydroxylapatite, ammonium sulfate, cellulose phosphate chromatography, electrofocusing, gel electrophoresis [7]) [7]
<12> (recombinant His-tagged enzyme [9]) [9]

Cloning

<12> (expression in Sf9 insect cells and Escherichia coli [9]) [9]

6 Stability

pH-Stability

6 <4> (<4> 60°C, 10 min, stable [3]) [3]

Temperature stability

60 <4> (<4> at pH 6, 10 min, stable [3]) [3]

General stability information

<10>, loss of activity with successive freezings and thawings [4]
<11>, ampholine or glycine protect against inactivation [7]

Storage stability

<1>, over liquid N_2, more than a month, little loss of activity [6]
<3>, frozen state, 2 weeks, 50% loss of activity [1]
<10>, -20°C [4]
<11>, unstable during storage [7]
<12>, -70°C, 6 months, several freeze/thaw cycles, no loss of activity [9]

References

[1] Pattabiraman, T.N.; Bachhawat, B.K.: Purification of glucosamine-6-phosphate N-acetylase from sheep brain. Biochim. Biophys. Acta, **59**, 681-689 (1962)

[2] Davison, E.A.; Blumenthal, H.J.; Roseman, S.: Glucosamine metabolism. II. Studies on glucosamine 6-phosphate N-acetylase. J. Biol. Chem., **226**, 125-133 (1957)

[3] Davison, E.A.: Glucoseamine 6-phosphate N-acetylase. Methods Enzymol., **9**, 704-707 (1966)

[4] Giddings, T.H.; Cantino, E.C.: Partial purification and properties of D-glucosamine 6-phosphate N-acetyltransferase from zoospores of Blastocladiella emersonii. J. Bacteriol., **120**, 976-979 (1974)

[5] Corfield, A.P.; Mortimer, D.J.; Winterburn, P.J.: Purification and properties of acetyl-CoA:glucosamine-6-phosphate N-acetytransferase from rat liver. Biochem. Soc. Trans., **12**, 565 (1984)

[6] Vessal, M.; Hassid, W.Z.: Partial purification and properties of D-glucosamine 6-phosphate N-acetyltransferase from Phaseolus aureus. Plant Physiol., **51**, 1055-1060 (1973)

[7] Porowski, T.S.; Porowska, H.; Galasinski, W.: Isolation, purification, and characterization of glucosamine-6-phosphate-N-acetylase from pig liver. Biochem. Med. Metab. Biol., **44**, 1-12 (1990)

[8] Vessal, M.; Jaberi-Pour, M.: Partial purification and kinetic properties of three different D-glucosamine 6-P:N-acetyltransferase forms from human placenta. Comp. Biochem. Physiol. B, **121B**, 379-384 (1998)
[9] Boehmelt, G.; Fialka, I.; Brothers, G.; McGinley, M.D.; Patterson, S.D.; Mo, R.; Hui, C.C.; Chung, S.; Huber, L.A.; Mak, T.W.; Iscove, N.N.: Cloning and characterization of the murine glucosamine-6-phosphate acetyltransferase EMeg32. Differential expression and intracellular membrane association. J. Biol. Chem., **275**, 12821-12832 (2000)

Arylamine N-acetyltransferase 2.3.1.5

1 Nomenclature

EC number
2.3.1.5

Systematic name
acetyl-CoA:arylamine N-acetyltransferase

Recommended name
arylamine N-acetyltransferase

Synonyms
2-naphthylamine N-acetyltransferase
4-aminobiphenyl N-acetyltransferase
N-acetyltransferase
NAT
NAT1
NAT2
acetyl CoA-arylamine N-acetyltransferase
acetyltransferase, 2-naphthylamine N-
acetyltransferase, 4-aminobiphenyl
acetyltransferase, arylamine
acetyltransferase, *p*-aminosalicylate N-
acetyltransferase, procainamide N-
acetyltransferase, serotonin N-
arylamine acetylase
arylamine acetyltransferase
β-naphthylamine N-acetyltransferase
indoleamine N-acetyltransferase
p-aminosalicylate N-acetyltransferase
serotonin N-acetyltransferase
serotonin acetyltransferase
Additional information (cf. EC 2.3.1.87)

CAS registry number
9027-33-2

2 Source Organism

<-1> no activity in *Canis familiaris* [10, 21, 29]
<1> *Rattus norvegicus* (Wistar [16]; Sprague-Dawley [19]; hepatocyte primary culture [30]) [6, 7, 10, 16, 17, 19, 30]

<2> *Mus musculus* (fast acetylator strain C57BL6, slow acetylator strain A/J [4]) [4, 10, 29]
<3> *Oryctolagus cuniculus* (New Zealand white [17,24]; slow and fast acetylator phenotype [23]; rapid acetylator type III/J [24]) [6, 8, 10, 17, 21, 23, 24, 29]
<4> *Mesocricetus auratus* (recombinant protein, fusion with dihydrofolate reductase [31]; recombinant protein, NAT1 [32]; recombinant NAT1, NAT2 [35]) [9, 29, 31, 32, 35]
<5> *monkey* [17]
<6> *Homo sapiens* (genetic polymorphism [2]; two isozymes NAT1, NAT2 [28]; recombinant NAT2, several genetic variants [34]; NAT1 [36]; NAT1 and NAT2 [48]) [2, 10, 17, 20, 27-29, 34, 36, 46, 48, 50, 51]
<7> *Bos taurus* [1]
<8> *Gallus gallus* [3, 5, 8, 22, 25, 26]
<9> *Columba sp.* (pigeon) [8, 13, 14, 18, 24]
<10> *Meleagris gallopavo* [14]
<11> *Anas sp.* (duck) [14]
<12> *Oncorhynchus mykiss* (trout) [14]
<13> *Hansenula ciferrii* [12]
<14> *Cordyceps militaris* (insect pathogen fungus [15]) [15]
<15> *Aerobacter aerogenes* (strain N5-36 [11]) [11]
<16> *Bacillus cereus* (distribution among soil microorganisms, overview [15]) [15]
<17> *Salmonella typhimurium* [29, 47]
<18> *Mycobacterium tuberculosis* [33]
<19> *Mycobacterium bovis* (strain Bacille Calmette-Guerin [33]) [33]
<20> *Ovis ammon* [37]
<21> *Enterobacter aerogenes* [38]
<22> *Lactobacillus acidophilus* [39]
<23> *Enterobius vermicularis* [40]
<24> *Candida albicans* [41]
<25> *Helicobacter pylori* [42]
<26> *Anisakis simplex* [43]
<27> *Pseudomonas aeruginosa* [44]
<28> *Escherichia coli* [45]
<29> *Musca domestica* (housefly [49]) [49]

3 Reaction and Specificity

Catalyzed reaction

acetyl-CoA + an arylamine = CoA + an N-acetylarylamine (<3,9> mechanism and characterization of active site [14,24]; <1-6,17> ping-pong bi-bi mechanism [17,19,29]; <4> mechanism and inactivation [35])

Reaction type

acyl group transfer

Natural substrates and products

S acetyl-CoA + an arylamine <1, 8, 9, 14-16> (<1, 9, 14-16> constitutive enzyme [11, 13, 15, 19]; <1> pineal gland: inducible [19]; <8> involved in metabolism of drugs and endogenous substances containing an amine or hydrazine group [25]) (Reversibility: ? <1, 8, 9, 14-16> [11, 13, 15, 19, 25]) [11, 13, 15, 19, 25]

P CoA + a N-acetylarylamine

Substrates and products

S N-hydroxy-2-acetylaminofluorene + 4-aminoazobenzene <4> (<4> much better acetyl donor for NAT1 than for NAT2 [35]) (Reversibility: ? <4> [35]) [35]

P CoA + ?

S acetyl-CoA + 2-(4-aminobenzamido)pyridine <1-3, 6> (Reversibility: ? <1-3, 6> [10]) [10]

P CoA + 2-(4-acetylamidobenzamido)pyridine

S acetyl-CoA + 2-aminobenzoic acid <15> (<15> i.e. anthranilate, ir [11]) (Reversibility: ir <15> [11]) [11]

P CoA + N-acetylanthranilate <15> [11]

S acetyl-CoA + 2-aminofluorene <3, 4, 6, 8, 21-28> (<6> N-acetyltransferase 2 [27]; <4> NAT1, poor substrate [32]) (Reversibility: ? <3, 4, 6, 8, 21-28> [2, 3, 21-23, 27, 32, 38-46]) [2, 3, 21-23, 27, 32, 38-46]

P CoA + 2-acetylaminofluorene

S acetyl-CoA + 3,5-dimethyl-4-nitroaniline <16> (Reversibility: ? <16> [15]) [15]

P CoA + ?

S acetyl-CoA + 3-aminobenzoic acid <15> (Reversibility: ? <15> [11]) [11]

P CoA + N-acetyl-3-aminobenzoic acid <15> [11]

S acetyl-CoA + 4-amino-4'-nitroazobenzene <16> (Reversibility: ? <16> [15]) [15]

P CoA + ?

S acetyl-CoA + 4-aminobenzoic acid <1-4, 6, 15, 24-26, 28> (<6> N-acetyltransferase 1 [2]; <15> poor [11]; <8> no substrate [22]) (Reversibility: ? <1-4, 6, 15, 24-26, 28> [2, 11, 17, 21, 23, 28-30, 32, 41-43, 45]) [2, 11, 17, 21, 23, 28-30, 32, 41-43, 45]

P CoA + 4-acetlyaminobenzoic acid

S acetyl-CoA + 4-aminosalicylic acid <2-4, 6> (<6> N-acetyltransferase 1 [2]; <15> not [11]) (Reversibility: ? <2-4> [2, 17, 21, 27-29]) [2, 17, 21, 27-29]

P CoA + 4-acetylamino-2-hydroxybenzoate

S acetyl-CoA + 4-ethylaniline <3, 9> (Reversibility: ? <3, 9> [24]) [24]

P CoA + 4-ethylacetanilide

S acetyl-CoA + 4-methylaniline <3, 8, 9> (Reversibility: ? <3, 8, 9> [8]) [8]

P CoA + N-acetyl-4-methylaniline

S acetyl-CoA + 4-nitroaniline <8, 9, 14, 16> (<3,13> not [12, 17]) (Reversibility: ? <8, 9, 14, 16> [8, 13, 15, 18]) [8, 13, 15, 18]

P CoA + N-acetyl-4-nitroaniline

S acetyl-CoA + 4-nitrobenzalhydrazone <16> (Reversibility: ? <16> [15]) [15]
P CoA + ?
S acetyl-CoA + 5-aminosalicylic acid <6> (Reversibility: ? <6> [28]) [28]
P CoA + 5-acetylamino-2-hydroxybenzoate
S acetyl-CoA + 5-methoxytryptamine <8> (<8> poor substrate [3,25]) (Reversibility: ? <8> [3, 22, 25, 26]) [22, 25, 26]
P CoA + N-acetyl-5-methoxytryptamine
S acetyl-CoA + D-glucosamine <13> (<3> not [17]) (Reversibility: ? <13> [12]) [12]
P CoA + N-acetyl-D-glucosamine
S acetyl-CoA + aniline <3, 8, 14-16> (Reversibility: ? <3, 8, 14-16> [3, 11, 15, 17, 21, 25]) [3, 11, 15, 17, 21, 25]
P CoA + acetanilide <14-16> [11, 15]
S acetyl-CoA + benzidine <3> (Reversibility: ? <3> [21]) [21]
P CoA + ?
S acetyl-CoA + β-phenylethylamine <1, 8> (<8> poor substrate [3]; <8> not chicken pineal gland and brain [22]) (Reversibility: ? <1, 8> [3, 6, 16, 22, 25]) [3, 6, 16, 22, 25]
P CoA + N-(2-phenylethyl)-acetaminde
S acetyl-CoA + decylamine <13> (Reversibility: ? <13> [12]) [12]
P CoA + N-acetyldecylamine
S acetyl-CoA + diaminodiphenylsulfone <3> (Reversibility: ? <3> [17, 21]) [17, 21]
P CoA + ?
S acetyl-CoA + diaminodiphenylsulfone <6> (<6> i.e. dapsone, predominantly acetylated by NAT2 [51]) (Reversibility: ? <6> [51]) [51]
P monoacetyldiaminodiphenylsulfone + CoA
S acetyl-CoA + hexylamine <13> (Reversibility: ? <13> [12]) [12]
P CoA + N-acetylhexylamine
S acetyl-CoA + histamine <1, 3, 13> (Reversibility: ? <1, 3, 13> [6, 12, 17, 21]) [6, 12, 17, 21]
P CoA + N-acetylhistamine
S acetyl-CoA + hydralazine <3> (Reversibility: ? <3> [17, 21]) [17, 21]
P CoA + ?
S acetyl-CoA + hydroxytyramine <13> (Reversibility: ? <13> [12]) [12]
P CoA + ?
S acetyl-CoA + isoniazide <3, 6, 18, 19> (<3> i.e. isonicotinyl hydrazide [17]) (Reversibility: ? <3, 6, 18, 19> [17, 20, 21, 33]) [17, 20, 21, 33]
P CoA + acetylniazide <3> [17]
S acetyl-CoA + long-chain primary amines <13> (<13> C6-C_{16} [12]) (Reversibility: ? <13> [12]) [12]
P CoA + ?
S acetyl-CoA + methyl bis-2-chloroaniline <3> (Reversibility: ? <3> [21]) [21]
P CoA + ?
S acetyl-CoA + naphthylamine <3> (Reversibility: ? <3> [21]) [21]
P CoA + N-acetylnaphthylamine

S acetyl-CoA + noradrenaline <13> (Reversibility: ? <13> [12]) [12]
P CoA + ?
S acetyl-CoA + normetanephrine <1> (Reversibility: ? <1> [6]) [6]
P CoA + N-acetylnormetanephrine
S acetyl-CoA + octylamine <13> (Reversibility: ? <13> [12]) [12]
P CoA + N-acetyloctylamine
S acetyl-CoA + *p*-octopamine <1> (Reversibility: ? <1> [6, 16]) [6, 16]
P CoA + N-acetyloctopamine
S acetyl-CoA + *p*-phenetidine <3, 8> (<8> i.e. 4-ethoxybenzenamine, best substrate [25]) (Reversibility: ? <3, 8> [3, 17, 21, 22, 25, 26]) [3, 17, 21, 22, 25, 26]
P CoA + N-(4-ethoxyphenyl)-acetamide
S acetyl-CoA + *p*-tyramine <1, 3, 29> (<1> ie. 4-(2-aminoethyl)phenol [6]) (Reversibility: ? <1, 3, 29> [6, 16, 21, 49]) [6, 16, 21, 49]
P CoA + N-acetyl-*p*-tyramine <1> [16]
S acetyl-CoA + phenelzine <3> (Reversibility: ? <3> [21]) [21]
P CoA + ?
S acetyl-CoA + procainamide <3, 6> (<6> N-acetyltransferase 2 [2, 27]) (Reversibility: ? <3, 6> [2, 21, 27, 28]) [2, 21, 27, 28]
P CoA + N-acetylprocainamide
S acetyl-CoA + serotonin <1, 3, 9, 13> (Reversibility: ? <1, 3, 9, 13> [6, 12, 17, 18, 21]) [6, 12, 17, 18, 21]
P CoA + N-acetylserotonin <1, 9> [6, 18]
S acetyl-CoA + sulfadiazine <3, 6> (Reversibility: ? <3, 6> [17, 21, 28]) [17, 21, 28]
P CoA + ?
S acetyl-CoA + sulfamerazine <3, 6> (Reversibility: ? <3, 6> [17, 28]) [17, 28]
P CoA + ?
S acetyl-CoA + sulfamethazine <1, 3, 6> (<6> N-acetyltransferase 2 [2,27,30]) (Reversibility: ? <1, 3, 6> [2, 17, 21, 23, 27, 28, 30]) [2, 17, 21, 23, 27, 28, 30]
P CoA + N-4-acetylsulfamethazine <3> [17]
S acetyl-CoA + sulfanilamide <3> (Reversibility: ? <3> [17, 21]) [17, 21]
P CoA + ?
S acetyl-CoA + sulfapyridine <3> (Reversibility: ? <3> [21]) [21]
P CoA + ?
S acetyl-CoA + tetradecylamine <13> (Reversibility: ? <13> [12]) [12]
P CoA + N-acetyltetradecylamine
S acetyl-CoA + thiazolsulfone <3> (Reversibility: ? <3> [21]) [21]
P CoA + ?
S acetyl-CoA + tryptamine <1, 7, 8, 13> (<8> not chicken pineal gland and brain [22]) (Reversibility: ? <1, 7, 8, 13> [1, 6, 12, 16, 19, 22]) [1, 6, 12, 16, 19, 22]
P CoA + N-acetyltryptamine
S *p*-nitroacetanilide + aniline <9> (Reversibility: ? <9> [14]) [14]
P *p*-nitroaniline + acetanilide <9> [14]

S Additional information <-1, 1-6, 9-17> (<2-4,6,9> the reaction proceeds via a covalent acetyl-enzyme intermediate [14,29]; <15> propionyl-CoA can replace acetyl-CoA [11]; <9> 4-nitroacetanilide, 4-nitrophenylacetate can replace acetyl-CoA [14]; <3> acetylthiocholine can replace acetyl-CoA [17]; <3> N-diacetyl cysteamine can replace acetyl-CoA with less than 10% efficiency [17,21]; <15> not butyryl-CoA [11]; <3> the enzyme exhibits arylhydroxamic acid N,O-acetyltransferase activity [21]; <15> no substrate: glucosamine 6-phosphate [11]; <13> no substrate: polyamines, amino acids, sphingosine bases, ethanolamine, C_1-C_4 primary amines, some secondary and tertiary amines [12]; <9,16> no substrate: *o*-nitroaniline [14,15]; <9> no substrate: 2-amino-6-nitropyridine [14]; <16> no substrate: 2-methyl-4-nitroaniline, 2,4-dinitroaniline, 4-nitrophenylhydrazine, carbopyridine [15]; <3> no substrate: phenylalanine, cyclohexylamine [17]) [11, 12, 14, 15, 17, 21, 29]

P ?

Inhibitors

3',5'-ADP 4-bromoacetaminobenzoic acid <9, 3> (<9> kinetics [14]) [14, 24]
AMP <9> (<9> kinetics [14]) [14]
$AgNO_3$ <13> [12]
Ca^{2+} <1, 21-23, 27> (<1> at increased concentrations [7]) [7, 38-40, 44]
Co^{2+} <27> [44]
Cu^{2+} <1, 3, 4, 22, 26-28> (<4> reversible by EDTA or histidine [9]) [7, 9, 17, 21, 39, 43-45]
DTNB <9, 13> (<9> partly reversible by dithioerythritol [14]) [12-14]
EDTA <13> (<13> partly reversible by $MgCl_2$ [12]) [12]
Fe^{2+} <21-24, 27> [38-41, 44]
K^+ <1> (<1> at increased concentrations [7]) [7]
Mg^{2+} <1, 22, 27> (<1> at increased concentrations [7]) [7, 39, 44]
Mn^{2+} <3, 27> [17, 21, 44]
N-ethylmaleimide <1, 3, 5, 6, 15, 20> [11, 17, 21, 37]
N-hydroxy-2-acetylaminofluorene <4, 6> (<4,6> mechanism-based inactivator, kinetics [35,36]) [35, 36]
NH_4^+ <24> [41]
Na^+ <1> (<1> at increased concentrations [7]) [7]
Ni^{2+} <3> [17, 21]
SCN^- <24> [41]
Triton X-100 <13> [12]
Zn^{2+} <3, 4, 21-23, 26, 28> (<4> reversible by dimercaptosuccinate [9]) [9, 17, 21, 38-40, 43, 45]
amethopterin <9> (<9> kinetics [13]) [13]
bromo-β-naphthalide butyryl-CoA <3, 13, 15> [24, 11, 12]
bromoacetanilide <3, 9> (<3,9> product analogue, kinetics [24]) [24]
chloro-β-naphthalide <3> [24]
cholate <13> [12]
coenzyme A <3, 8, 9, 13> [8, 12, 14, 24]

deoxycholate <13> [12]
diethyldicarbonate <22, 27> [39, 44]
dithiothreitol <9> (<13> above 0.5 mM, not [12]) [18]
glucosamine <15> [11]
glucosamine 6-phosphate <15> [11]
iodoacetamide <9, 21, 22, 24-28> [14, 38-45]
iodoacetic acid <3, 9, 15> [11, 14, 24]
methotrexate <1> (<1> inhibits reaction with 4-aminobenzoic acid, but with sulfamethazine, possibly due to two different enzymes [30]) [30]
p-chloromercuribenzoate <3, 9, 15> [11, 14, 17, 21]
p-hydroxymercuribenzoate <13> [12]
p-mercuribenzoate <13, 29> [12, 49]
palmitoyl-CoA <13> [12]
phenylglyoxal <22, 27> [39, 44]
propionyl-CoA <15> [11]
salicylic acid <15> [11]
taurocholate <13> [12]
Additional information <1, 3, 5, 6, 13> (<13> no inhibition by acetylated products and 2-mercaptoethanol [12]; <3> no inhibition by $CdCl_2$, arsenite, diisopropylfluorophosphate [17, 21]; <3> no inhibition by paraoxon [21]) [12, 17, 21]

Activating compounds

2-mercaptoethanol <4> (<4> 1 mM, activation [9]) [9]
albumin <13> (<13> activation, with long-chain primary amines as substrate [12]) [12]
cysteine <4> (<4> activation [9]) [9]
dimercaptosuccinate <4> (<4> 1 mM, activation [9]) [9]
penicillamine <4> (<4> 1 mM, activation [9]) [9]
Additional information <4> (<4> no activation by EDTA or histidine [9]) [9]

Metals, ions

Ca^{2+} <1> (<1> activation [7]) [7]
K^+ <1> (<1> activation [7]) [7]
Mg^{2+} <1> (<1> activation [7]) [7]
Na^+ <1> (<1> activation [7]) [7]

Turnover number (min^{-1})

1500 <3> (4-ethylaniline) [24]

Specific activity (U/mg)

0.145 <3> (<3> 2-aminofluorene [23]) [23]
0.16 <3> [17]
0.177 <3> (<3> sulfamethazine [23]) [23]
0.198 <3> [21]
0.41 <23> (<23> 2-aminofluorene [40]) [40]
0.46 <28> (<28> 4-aminobenzoic acid [45]) [45]
0.55 <6> [28]
0.67 <28> (<28> 2-aminofluorene [45]) [45]

0.75 <25> (<25> 4-aminobenzoic acid [42]) [42]
0.91 <25> (<25> 2-aminofluorene [42]) [42]
2.54 <26> (<26> 4-aminobenzoic acid [43]) [43]
2.89 <26> (<26> 2-aminofluorene [43]) [43]
3.3 <9> [18]
14.02 <9> [13]
Additional information <1-4, 6, 8, 9> [14, 16, 25, 29]

K_m-Value (mM)

0.0084 <29> (acetyl-CoA) [49]
0.0088 <29> (tyramine) [49]
0.013 <8> (acetyl-CoA) [8]
0.02 <3> (acetyl-CoA) [17]
0.028 <7> (acetyl-CoA) [1]
0.031 <7> (tryptamine) [1]
0.04 <16> (4-amino-4'-nitroazobenzene) [15]
0.05 <1, 1> (acetyl-CoA) [16, 19]
0.07 <13> (tetradecylamine) [12]
0.08 <1> (*m*-tyramine) [16]
0.08 <1, 13> (octylamine) [12, 16]
0.11-0.125 <1> (acetyl-CoA) [19]
0.11-0.125 <1> (serotonin, <1> brain [16]) [16]
0.14 <13> (hexylamine) [12]
0.14-0.145 <1, 13> (*p*-tyramine) [12, 16]
0.17 <24> (2-aminofluorene) [41]
0.18 <3> (isoniazid) [17]
0.18 <1> (phenylethylamine) [16]
0.24-0.25 <1, 9> (tryptamine, <1> liver [19]) [18, 19]
0.25 <9> (serotonin) [18]
0.28-0.37 <1, 13, 16> (4-nitroaniline) [12, 15, 16]
0.3 <16> (4-nitrobenzalhydrazone) [15]
0.3 <13> (dodecylamine) [12]
0.3 <1> (phenylethylamine) [16]
0.38-0.58 <6> (isoniazide) [20]
0.42 <13> (serotonin) [12]
0.5 <16> (3,5-dimethyl-4-nitroaniline) [15]
0.504 <27> (2-aminofluorene) [44]
0.53 <13> (tryptamine, <13> pineal gland [12]) [12]
0.71-0.72 <13> (hydroxytyramine) [12]
0.72 <21> (2-aminofluorene) [38]
0.72 <13> (histamine) [12]
0.77 <16> (acetyl-CoA) [15]
0.77-0.8 <13, 16> (decylamine) [12, 15]
0.81 <23> (2-aminofluorene) [40]
0.842 <22> (2-aminofluorene) [39]
1.06 <26> (2-aminofluorene) [43]
1.1 <25> (2-aminofluorene) [42]

1.3 <1> (*p*-octopamine, <1> liver [16]) [16]
2.25 <26> (4-aminobenzoic acid) [43]
2.34 <25> (4-aminobenzoic acid) [42]
3.5 <6> (sulfamethazine, <6> isozyme NAT1 [27]) [27]
12.5 <15> (3-aminobenzoic acid) [11]
13 <15> (2-aminobenzoic acid) [11]
45 <6> (procainamide, <6> isozyme NAT1 [27]) [27]
Additional information <6, 9, 13> (<13> reaction kinetics [12]; <9> kinetic data of various acetyl donors with 4-nitroaniline or aniline as acceptor amine and of various acceptor amines with acetyl-CoA as donor [14]; <9> assay based on HPLC [18]; <6> kinetic data of isozymes NAT1 and 2 [28]) [12, 14, 18, 28]

K_i-Value (mM)
0.67 <3> (bromoacetanilide) [24]

pH-Optimum
5-9 <3> (<3> isoniazide, sulfanilamide [21]) [17, 21]
5.6 <3> (<3> sulfamethazine, 2 optima [21]) [21]
5.7 <1, 3, 5> (<3> sulfadiazine [21]) [17, 21]
5.7-7.4 <3> (<3> sulfamethazine [17]) [17]
5.9 <9> (<9> 4-nitroaniline [14]) [14]
6 <25> [42]
6.4 <9> (<9> 1,2-phenylenediamine [14]) [14]
7 <22, 28> [39, 45]
7 <9> (<9> 1,3-phenylenediamine [14]) [14]
7.2 <3, 29> (<3> sulfamethazine, 2 optima [21]) [21, 49]
7.3 <13> (<13> tricine buffer [12]) [12]
7.5 <13, 20, 21, 23> (<13> potassium phosphate or piperazine-N,N-bis(2-ethanolsulfonic acid) buffer [12]) [12, 37, 38, 40]
7.5-8.5 <1> [6]
7.7-8.6 <8> [8]
7.8 <1> (<1> tyramine in phosphate buffer [16]) [16]
7.9 <9> (<9> 1,4-phenylenediamine [14]) [14]
8 <24, 26> [41, 43]
8.3 <13> (<13> N-tris(hydroxymethyl)methyl-2-aminoethane buffer [12]) [12]
8.5 <27> [44]
9.1 <9> (<9> benzylamine [14]) [14]
9.5 <9, 13> (<13> cyclohexylamineopropanesulfonic acid buffer [12]; <9> 0.1 M glycine/KOH buffer [18]) [12, 18]
9.8 <13> (<13> trisethanolamine buffer [12]) [12]
Additional information <9-12> (<6,9> pI: 4.8 [13,28]; <9-12> pH-optimum depends on the nature of the acetyl acceptor amine [14]; <13> pH-optimum depends on buffer [12]) [13, 14, 28]

pH-Range

6.2-7.3 <16> (<16> about half-maximal activity at pH 6.2 and 7.3 [15]) [15]
9.3-9.5 <9> (<9> about half-maximal activity at pH 9.3, maximal at 9.5 [18]) [18]

Temperature optimum (°C)

32 <16> [15]
35 <9> [18]
37 <21, 25-27> [38, 42-44]
37 <22, 27> [39, 44]
40 <24> [41]

Temperature range (°C)

16-36 <16> (<16> about half-maximal activity at 16°C and 36°C [15]) [15]
25-40 <9> (<9> nearly maximal activity in this range [18]) [18]

4 Enzyme Structure

Molecular weight

25000 <1> (<1> gel filtration [16]) [16]
26000 <1> (<1> liver, gel filtration [19]) [19]
26500 <6> (<6> gel filtration [20]) [20]
27600 <29> (<29> gel filtration [49]) [49]
30000 <2> (<2> strain C57BL6, gel filtration [4]) [4]
32900 <9> (<9> high-speed sedimentation equilibrium analysis [13]) [13]
33000 <2, 3, 9> (<2> strain A/J, gel filtration [4]; <3> gel filtration [23]; <9> gel filtration [13]) [4, 13, 23]
34000 <8> (<8> gel filtration [8]) [8]
39000 <1> (<1> pineal gland, gel filtration [19]) [19]
Additional information <9> (<9> amino acid analysis [13]) [13]

Subunits

? <1, 8, 21, 22, 23> (<1> x * 11000, SDS-PAGE [6]; <8> x * 34000, SDS-PAGE [25]; <8> x * 34000, SDS-PAGE, recombinant protein [26]; <21> x * 44900, SDS-PAGE [38]; <22> x * 44900, SDS-PAGE [39]; <23> x * 44900, SDS-PAGE [40]) [6, 25, 26, 38-40]
dimer <1> (<1> 2 * 12000, liver, gel filtration in the presence of cysteamine [19]) [19]
monomer <2, 3, 6, 9, 29> (<2> 1 * 30000, strain C57BL6 [4]; <6> 1 * 31000, SDS-PAGE [28]; <9> 1 * 31900, SDS-PAGE [13]; <2> 1 * 33000, strain A/J [4]; <3> 1 * 33000, SDS-PAGE [21,23]; <29> 1 * 26000, SDS-PAGE [49]) [4, 13, 21, 23, 28, 49]
tetramer <1> (<1> 4 * 10000, pineal gland, gel filtration in the presence of cysteamine [19]) [19]
Additional information <9> (<9> enzyme forms dimer under conditions of catalysis [13]) [13]

5 Isolation/Preparation/Mutation/Application

Source/tissue

bladder <4> [29]
brain <1, 8> (<1> extrapineal distribution [16]) [16, 22]
intestine <3, 4, 6> [21, 29]
kidney <1, 3, 8, 20> (<8> distribution [3]) [3, 6, 16, 22, 37]
liver <1-6, 8-12, 20> [2, 4, 6, 8, 10, 13, 14, 16-20, 23-30, 37, 51]
lung <4, 20> [29, 37]
pineal gland <1, 3, 7, 8> [1, 3, 5-7, 19, 22]
umbilical cord <6> [50]
Additional information <1-4, 6, 17, 20> (<2-4,6> distribution in mammalian tissues [21, 29]; <1> in rat tissues [16]; <20> activity in different tissues [37]) [16, 21, 29, 37]

Localization

cytosol <1-4, 6, 13, 17> [2, 6, 12, 21, 23, 28, 29]

Purification

<1> (partial [6]) [6, 16]
<2> (partial, 3 isozymes from 2 strains [4]) [4]
<3> [23]
<4> (recombinant protein, fusion with dihydrofolate reductase [31]) [31]
<6> [20, 28]
<8> (cDNA-isolation [3]; monoclonal antibodies [25]) [3, 8, 25]
<9> (affinity chromatography [13]) [13]
<13> (partial [12]) [12]
<15> [11]
<21> [38]
<22> [39]
<23> [40]
<29> [49]

Crystallization

<17> [47]

Cloning

<4> (isozyme NAT1 [32]) [32]
<6> (2 genes, NAT1 and 2, encoding enzyme proteins, transiently expressed in cultured monkey kidney COS-1 cells [2]; NAT1 and NAT2, expressed in Escherichia coli XA90 [27]) [2, 27]
<8> (liver cDNA clones introduced in Chinese hamster ovary cells [22,26]) [22, 26]
<18> [33]

Engineering

C223G <6> (<6> NAT2, enzymatically active, markedly reduced in vitro stability [27]) [27]

C44G <6> (<6> NAT2, enzymatically active, markedly reduced in vitro stability [27]) [27]
C68G <6> (<6> NAT2, no enzymatic activity [27]) [27]
Additional information <6, 17> (<17> truncation mutants with either C-terminal undecapeptide or C-terminal 85 amino acids missing, in contrast to complete protein, mutants hydrolyse acetyl-CoA even in the absence of arylamine substrate [47]; <6> study of NAT1/NAT2 chimerea [48]) [47, 48]

Application

medicine <6> (<6> kinetics of several substrates for different genetic variants of enzyme [34]; <6> activity in benign prostatic hyperplasia tissue [46]; <6> activity in umbilical cord tissue from different samples [50]; <6> inactivation of antiopportunistic infection drugs [51]) [34, 46, 50, 51]

6 Stability

pH-Stability

6-8.5 <9> (<9> in several buffer systems, at 4°C, several days stable [13]) [13]
6.5 <3> (<3> irreversible inactivation below [17]) [17]

Temperature stability

34 <16> (<16> inactivation above [15]) [15]
37 <22> (<22> half life of 32 min [39]) [39]
37 <6> (<6> $t_{1/2}$: 35 h NAT2 wild-type, 3.5 h Cys44Gly-mutant, 0.5 h Cys223Gly-mutant [27]) [27]
55 <3> (<3> inactivation after 2 min [17]) [17]

General stability information

<1>, cysteamine stabilizes [19]
<6>, 2-mercaptoethanol stabilizes [20]
<6>, cysteine stabilizes [20]
<6>, glycerol and albumin stabilize [28]
<6>, thioglycolate stabilizes [20]
<8>, EGTA stabilizes [5]
<9>, EDTA stabilizes [13]
<9>, repeated freeze-thawing inactivates [13]
<13>, Tris-buffer cannot replace phosphate buffer during purification [12]
<13>, glycerol, 10% v/v, stabilizes [12]
<13>, sucrose or mannitol, 0.1 M, cannot fully replace glycerol [12]
<3, 9, 16>, dithiothreitol stabilizes [13, 15, 21]

Storage stability

<3>, 4°C, 20% loss of activity within 4 days in the presence of 1 mM EDTA [17]
<3>, frozen, over a long-time range as ammonium sulfate precipitate [21]
<6>, -60°C, 60-80% loss of activity after 3 weeks [20]
<6>, -60°C, at least 3 weeks in the presence of 2-mercaptoethanol, thioglycolate or cysteine [20]

<9>, -70°C, frozen at neutral pH in liquid nitrogen, no loss of activity over a long-time range [13]
<9>, 4°C, several days in various buffer systems at pH 6.0-8.5 [13]
<13>, -20°C, at least 2 years in the presence of 10% glycerol v/v [12]
<21>, -70°C, stable for several months [38]

References

[1] Fajardo, N.; Abreu, P.; Alonso, R.: Determination of kinetic properties of serotonin-N-acetyltransferase in bovine pineal gland using HPLC with fluorimetric detection. J. Pineal Res., **13**, 80-84 (1992)
[2] Grant, D.M.; Blum, M.; Beer, M.; Meyer, U.A.: Monomorphic and polymorphic human arylamine N-acetyltransferases: a comparison of liver isozymes and expressed products of two cloned genes. Mol. Pharmacol., **39**, 184-191 (1991)
[3] Ohtomi, M.; Tomoyuki, I.; Deguchi, T.: Arylamine N-acetyltransferase from the chicken kidney: characterization of the enzyme and localization of the mRNA by in situ hybridization. Biomed. Res., **11**, 263-267 (1990)
[4] Watson, A.P.; Wang, P.D.; Sim, E.: Arylamine N-acetyltransferase from fast (C57BL6) and slow (A/J) N-acetylating strains of mice. Biochem. Pharmacol., **39**, 647-654 (1990)
[5] Rodriguez-Cabello, J.C.; Agapito, M.T.; Garcia-Herrero, I.; Recio, J.M.: Effects of EGTA and calmodulin, neutral thiol proteinases and protein kinase C inhibitors on loss of chicken pineal serotonin N-acetyltransferase activity. J. Comp. Physiol. B, **159**, 583-588 (1989)
[6] Namboodiri, M.A.A.; Brownstein, M.J.; Weller, J.L.; Voisin, P.; Klein, D.C.: Multiple forms of arylalkylamine N-acetyltransferases in the rat pineal gland: purification of one molecular form. J. Pineal Res., **4**, 235-246 (1987)
[7] Morton, D.J.: Effect of various cations on the activity of pineal gland N-acetyltransferase (EC 2.3.1.5) and hydroxyindole-O-methyl transferase (EC 2.1.1.4) in vitro. J. Pineal Res., **4**, 161-167 (1987)
[8] Rougraff, P.M.; Paxton, R.: Purification and partial characterization of checken liver acetyl coenzymeA:arylamine N-acetyltransferase. Comp. Biochem. Physiol. B Comp. Biochem., **86**, 601-606 (1987)
[9] Jendryczko, A.; Drozdz, M.: Effects of some complexing agents on acetyl-CoA:arylamine N-acetyltransferase. Rev. Roum. Biochim., **22**, 201-203 (1985)
[10] Gollamudi, R.; Rackley, R.J.; Autian, J.: A new substrate for the measurement of N-acetyltransferase activity. Enzyme, **30**, 155-161 (1983)
[11] Paul, R.C.; Ratledge, C.: Further studies on anthranilate N-acetyltransferase and the metabolism of N-acetylanthranilic acid in Aerobacter aerogenes. Biochim. Biophys. Acta, **320**, 9-15 (1973)
[12] Barenholz, Y.; Edelman, I.; Gatt, S.: Amine:acetyl CoenzymeA acetyltransferase from the soluble fraction of Hansenula ciferri: isolation and properties. Biochim. Biophys. Acta, **358**, 262-274 (1974)

[13] Andres, H.H.; Kolb, H.J.; Weiss, L.: Purification and physical-chemical properties of acetyl-CoA:arylamine N-acetyltransferase from pigeon liver. Biochim. Biophys. Acta, **746**, 182-192 (1983)

[14] Andres, H.H.; Kolb, H.J.; Schreiber, R.J.; Weiss, L.: Characterization of the active site, substrate specificity and kinetic properties of acetyl-CoA:arylamine N-acetyltransferase from pigeon liver. Biochim. Biophys. Acta, **746**, 193-201 (1983)

[15] Hasmann, M.J.; Seidl, P.H.; Engelhardt, G.; Schleifer, K.H.: Acetyl-coenzyme A:arylamine N-acetyltransferases in microorganisms: screening and isolation of an enzyme from Bacillus cereus. Arch. Microbiol., **146**, 275-279 (1986)

[16] Yu, P.H.; Boulton, A.A.: N-Acylation of tyramines: purification and characterization of an arylamine N-acetyltransferase from rat brain and liver. Can. J. Biochem., **57**, 1204-1209 (1979)

[17] Weber, W.W.: N-Acetyltransferase (mammalian liver). Methods Enzymol., **17B**, 805-811 (1971)

[18] Mannens, G.; Siegers, G.; Claeys, A.: Assay for acetyl-CoA:arylamine N-acetyltransferase by high-performance liquid chromatography applied to serotonin N-acetylation. Biochim. Biophys. Acta, **1037**, 1-6 (1990)

[19] Morrissey, J.J.; Edwards, S.B.; Lovenberg, W.: Comparison of rat pineal gland and rat liver serotonin-N-acetyltransferase. Biochem. Biophys. Res. Commun., **77**, 118-123 (1977)

[20] Schulte, E.H.; Schloot, W.; Goedde, H.W.: Purification of human liver serotonin/isoniazid N-acetyltransferase by preparative polyacrylamide gel electrophoresis and determination of molecular weight. Z. Naturforsch. C, **29c**, 661-666 (1974)

[21] Weber, W.W.; King, C.M.: N-Acetyltransferase and arylhydroxamic acid acyltransferase. Methods Enzymol., **77**, 272-280 (1981)

[22] Ohtomi, M.; Sasaki, M.; Deguchi, T.: Two arylamine N-acetyltransferases from chicken pineal gland as identified by cDNA cloning. Eur. J. Biochem., **185**, 253-261 (1989)

[23] Glowinski, I.B.; Weber, W.W.; Fysh, J.M.; Vaught, J.B.; King, C.M.: Evidence that arylhydroxamic acid N,O-acyltransferase and the genetically polymorphic N-acetyltransferase are properties of the same enzyme in rabbit liver. J. Biol. Chem., **255**, 7883-7890 (1980)

[24] Andres, H.H.; Klem, A.J.; Schopfer, L.M.; Harrison, J.K.; Weber, W.W.: On the active site of liver acetyl-CoA. Arylamine N-acetyltransferase from rapid acetylator rabbits (III/J). J. Biol. Chem., **263**, 7521-7527 (1988)

[25] Deguchi, T.; Sakamoto, Y.; Sasaki, Y.; Uyemura, K.: Arylamine N-acetyltransferase from chicken liver. I. Monoclonal antibodies, immunoaffinity purification, and amino acid sequences. J. Biol. Chem., **263**, 7528-7533 (1988)

[26] Ohsako, S.; Ohtomi, M.; Sakamoto, Y.; Uyemura, K.; Deguchi, T.: Arylamine N-acetyltransferase from chicken liver II. Cloning of cDNA and expression in Chinese hamster ovary cells. J. Biol. Chem., **263**, 7534-7538 (1988)

[27] Dupret, J.M.; Grant, D.M.: Site-directed mutagenesis of recombinant human arylamine N-acetyltransferase expressed in Escherichia coli. Evidence for

direct involvement of Cys68 in the catalytic mechanism of polymorphic human NAT2. J. Biol. Chem., **267**, 7381-7385 (1992)
[28] Grant, D.M.; Lottspeich, F.; Meyer, U.A.: Evidence for two closely related isozymes of arylamine N-acetyltransferase in human liver. FEBS Lett., **244**, 203-207 (1989)
[29] Hein, D.W.: Acetylator genotype and arylamine-induced carcinogenesis. Biochim. Biophys. Acta, **948**, 37-66 (1988)
[30] Zaher, H.; Lindblad, W.J.; Jiang, W.; Doll, M.A.; Hein, D.W.; Svensson, C.K.: Acetyl CoA:arylamine N-acetyltransferase activity in rat hepatocytes cultured on different extracellular matrixes. Toxicol. in Vitro, **11**, 271-283 (1997)
[31] Sticha, K.R.; Sieg, C.A.; Bergstrom, C.P.; Hanna, P.E.; Wagner, C.R.: Overexpression and large-scale purification of recombinant hamster polymorphic arylamine N-acetyltransferase as a dihydrofolate reductase fusion protein. Protein Expr. Purif., **10**, 141-153 (1997)
[32] Bergstrom, C.P.; Wagner, C.R.; Ann, D.K.; Hanna, P.E.: Hamster monomorphic arylamine N-acetyltransferase: expression in Escherichia coli and purification. Protein Expr. Purif., **6**, 45-55 (1995)
[33] Upton, A.M.; Mushtaq, A.; Victor, T.C.; Sampson, S.L.; Sandy, J.; Smith, D.M.; Van Helden, P.V.; Sim, E.: Arylamine N-acetyltransferase of Mycobacterium tuberculosis is a polymorphic enzyme and a site of isoniazid metabolism. Mol. Microbiol., **42**, 309-317 (2001)
[34] Hickman, D.; Palamanda, J.R.; Unadkat, J.D.; Sim, E.: Enzyme kinetic properties of human recombinant arylamine N-acetyltransferase 2 allotypic variants expressed in Escherichia coli. Biochem. Pharmacol., **50**, 697-703 (1995)
[35] Sticha, K.R.K.; Bergstrom, C.P.; Wagner, C.R.; Hanna, P.E.: Characterization of hamster recombinant monomorphic and polymorphic arylamine N-acetyltransferases. Bioactivation and mechanism-based inactivation studies with N-hydroxy-2-acetylaminofluorene. Biochem. Pharmacol., **56**, 47-59 (1998)
[36] Butcher, N.J.; Ilett, K.F.; Minchin, R.F.: Inactivation of human arylamine N-acetyltransferase 1 by the hydroxylamine of *p*-aminobenzoic acid. Biochem. Pharmacol., **60**, 1829-1836 (2000)
[37] Guray, T.; Guvenc, T.: Sheep tissue acetyl coenzyme A-dependent arylamine N-acetyltransferases. Comp. Biochem. Physiol. C, **118C**, 305-310 (1997)
[38] Tsou, M.F.; Chung, J.G.; Wu, L.T.; Cheng, K.S.; Hung, C.F.: Characterization of arylamine N-acetyltransferase in Enterobacter aerogenes. Microbios, **94**, 133-143 (1998)
[39] Chen, G.W.; Hung, C.F.; Chang, S.H.; Lin, J.G.; Chung, J.G.: Purification and characterization of an arylamine N-acetyltransferase from Lactobacillus acidophilus. Microbios, **98**, 159-174 (1999)
[40] Chung, J.G.: Purification and characterization of an arylamine N-acetyltransferase in the nematode Enterobius vermicularis. Microbios, **98**, 15-25 (1999)

[41] Fang, S.H.; Chung, J.G.; Chang, W.C.; Chang, S.S.: Evidence for arylamine N-acetyltransferase activity in the fungi Candida albicans. Toxicol. Lett., **92**, 109-116 (1997)
[42] Chung, J.G.; Wang, H.H.; Tsou, M.F.; Hsieh, S.E.; Lo, H.H.; Yen, Y.S.; Chang, S.S.; Wu, L.T.; Lee, J.H.; Hung, C.F.: Evidence for arylamine N-acetyltransferase activity in the bacterium Helicobacter pylori. Toxicol. Lett., **91**, 63-71 (1997)
[43] Chung, J.G.; Kuo, H.M.; Lin, T.H.; Ho, C.C.; Lee, J.H.; Lai, J.M.; Levy, G.N.; Weber, W.W.: Evidence for arylamine N-acetyltransferase in the nematode Anisakis simplex. Cancer Lett., **106**, 1-8 (1996)
[44] Hsieh, S.E.; Lo, H.H.; Chung, J.G.: The characteristics of arylamine N-acetyltransferase in Pseudomonas aeruginosa. Curr. Microbiol., **36**, 353-360 (1998)
[45] Chang, F.C.; Chung, J.G.: Evidence for arylamine N-acetyltransferase activity in the Escherichia coli. Curr. Microbiol., **36**, 125-130 (1998)
[46] Yeh, C.C.; Hung, C.F.; Wang, W.L.; Chung, J.G.: Kinetics of acetyl coenzyme A:arylamine N-acetyltransferase from rapid and slow acetylator human benign prostatic hyperplasia tissues. Urol. Res., **29**, 311-316 (2001)
[47] Mushtaq, A.; Payton, M.; Sim, E.: The COOH terminus of arylamine N-acetyltransferase from Salmonella typhimurium controls enzymic activity. J. Biol. Chem., **277**, 12175-12181 (2002)
[48] Goodfellow, G.H.; Dupret, J.M.; Grant, D.M.: Identification of amino acids imparting acceptor substrate selectivity to human arylamine acetyltransferases NAT1 and NAT2. Biochem. J., **348**, 159-166 (2000)
[49] Whitaker, D.P.; Goosey, M.W.: Purification and properties of the enzyme arylamine N-acetyltransferase from the housefly Musca domestica. Biochem. J., **295**, 149-154 (1993)
[50] Chung, J.G.; Chen, G.W.; Yeh, H.N.; Hung, C.F.; Huang, D.S.: Kinetics of acetyl coenzyme A:arylamine N-acetyltransferase from the human umbilical cord. Res. Commun. Pharmacol. Toxicol., **2**, 193-204 (1997)
[51] Palamanda, J.R.; Hickman, D.; Ward, A.; Sim, E.; Romkes-Sparks, M.; Unadkat, J.D.: Dapsone acetylation by human liver arylamine N-acetyltransferases and interaction with antiopportunistic infection drugs. Drug Metab. Dispos., **23**, 473-477 (1995)

Choline O-acetyltransferase 2.3.1.6

1 Nomenclature

EC number
2.3.1.6

Systematic name
acetyl-CoA:choline O-acetyltransferase

Recommended name
choline O-acetyltransferase

Synonyms
acetyl CoA:choline-O-acetyltransferase
acetyl-CoA:choline-O-acetyltransferase
acetyltransferase, choline
chAcT
choline acetylase
choline acetyltransferase

CAS registry number
9012-78-6

2 Source Organism

<1> *Homo sapiens* [1-3, 14, 18, 24, 26, 27, 37, 39, 44, 54]
<2> *Bos taurus* (2 enzyme forms [28]) [1-4, 7, 19, 25, 28, 31, 33, 38, 50]
<3> *Rattus norvegicus* [1-3, 9, 12, 19, 21, 25, 29, 34, 40, 41, 42, 46, 47, 48, 52, 53, 54, 55]
<4> *Gallus gallus* [1, 17, 43, 45]
<5> *Torpedo californica* [2]
<6> *Drosophila melanogaster* [2, 8, 16, 42]
<7> *Drosophila sp.* [1, 11]
<8> *Musca domestica* (housefly) [3]
<9> *Columba livia* [3]
<10> *Cavia porcellus* [3]
<11> *Sus scrofa* [5, 42]
<12> *Torpedo marmorata* [6]
<13> *Torpedo sp.* [1, 30]
<14> *Rhizoglyphus echinopus* (bulb-mite) [10]
<15> *Schistocerca gregaria* (locust) [13]
<16> *Caenorhabditis elegans* (nematode) [15, 42]

<17> *catfish* [20]
<18> *Loligo pealei* (squid) [22, 23]
<19> *Loligo opalescens* (squid) [22, 35]
<20> *squid* [38]
<21> *Mus musculus* (3 enzyme forms) [30, 49]
<22> *monkey* (3 enzyme forms) [30]
<23> *Oryctolagus cuniculus* [3, 30]
<24> *Helix aspersa* (snail) [32]
<25> *Periplaneta americana* (cockroach) [32]
<26> *Limulus polyphemus* (horseshoe crab) [32]
<27> *Carassius auratus* (goldfish) [36]
<28> *Electrophorus electricus* [51]

3 Reaction and Specificity

Catalyzed reaction

acetyl-CoA + choline = CoA + O-acetylcholine

Reaction type

acyl group transfer

Natural substrates and products

S acetyl-CoA + choline <1-4, 7, 13> (Reversibility: r <1-3, 8-10> [1, 3, 24, 26]) [1, 3, 24, 26]
P acetylcholine + CoA

Substrates and products

S acetyl-CoA + carnitine <28> (<28> a mutant enzyme that incorporates four amino acid substitutions from wild type, shows a substantial increase in catalytic efficiency toward L-carnitine [51]) (Reversibility: r <28> [51]) [51]
P acetylcarnitine + CoA
S acetyl-CoA + choline <1-28> (<1> kinetic data indicate that choline acetyltransferase operates by forming an acetylated enzyme as an intermediate in the reversible reaction between acetyl-CoA and choline [37]) (Reversibility: r <1-3, 8-10, 23, 27, 28> [3, 24, 26, 33, 36, 37, 51, 52]) [1-55]
P acetylcholine + CoA

Inhibitors

3-bromoacetonyltrimethylammonium bromide <1> [2]
3-bromoacetonyltrimethylammonium ion <24-26> [32]
3-chloro-4-stilbazole <1-3, 8-10, 23> [3]
3-trimethylammoniomethylcatechol <2> [4]
4-(1-naphthylvinyl)pyridine <1-3, 8-10, 23> [3]
5,5'-dithiobis(2-nitrobenzoate) <1, 2, 7, 14, 18, 19, 24-26> [10, 11, 22, 24, 32, 33]
AMPA <4> [43]
$CdCl_2$ <2> [31]

CoA <24-26> (<24-26> competitive to acetyl-CoA, non-competitive to choline [32]) [32]
Cu^+ <4> [17]
$CuSO_4$ <2> [31]
H_2O_2 <3> [48]
$HgCl_2$ <2> [31]
N,N-dimethylaminoethyl acrylate <3> [29]
N,N-dimethylaminoethyl bromoacetate <3> [29]
N,N-dimethylaminoethyl chloroacetate <3> [29]
N-ethylmaleimide <1, 2, 4, 18, 19> (<2> the enzyme is completely protected against N-ethylmaleimide inactivation by acetyl coenzyme A and is substantially protected by acetyl choline [33]) [17, 22, 24, 33]
N-hexamethylene-4-(1-naphthylvinyl)pyridinium-6-trimethylammonium ion <2, 8> [3]
N-methyl-4-(1-naphthylvinyl)pyridine <1-3, 5, 6, 8> [2, 3]
[2-[3-(2-ammonioethoxy)-benzoyl]ethyl]trimethylammonium bromide <12> [6]
acetylaminocholine <1> [26]
acetylcholine <1, 24-26> (<24-26> competitive to choline, non-competitive to acetyl-CoA [32]) [26, 32]
acyloylcholine <1-3, 5, 6, 24-26> [2, 32]
amyloid β-peptide <3> (<3> the aggregated form of A-β 25-35 decreased significantly enzyme activity only in the aged striatum [48]) [48]
arachidonic acid <3> [48]
bromoacetyl-S-CoA <2> [2]
bromoacetylcholine <1-3> [3]
catecholine <2> [4]
chloroacetylcholine <24-26> [32]
chlorocholine <1> [26]
choline <1> [26]
choline mustard aziridinium salt <1-3, 5, 6> [2, 12]
dexamethasone <3> [47]
diethylaminoethanol <1> [26]
diethyldicarbonate <7> [11]
domoic acid <4> [43]
ethanol <3> [47]
ethylenimine <2> [33]
flubenzimine <14> (<14> i.e. N-[3-phenyl-4,5-bis[(trifluoromethyl)immino]-2-thiazolidinylidene]benzenamine [10]) [10]
glutamate <4> [43]
iodoacetamide <2> [33]
iodoacetate <2, 24-26> [32, 33]
juglone <23> [3]
kainate <4> [43]
nerve growth factor <21> (<21> nerve growth factor, under specific development conditions, leads to a paradoxical down-regulation of the enzyme [49]) [49]

p-chloromercuribenzoate <1, 2> [24, 33]
phenylacetate <1> [39]
phenylacetyl-CoA <1> [39]
phenylbutyrate <1> [39]
phenylmethanesulfonylfluoride <18, 19> [22]
quisqualate <4> [43]
reactive blue 2 <18, 19> [22]
styrylpyridinium salts <1-3, 5, 6, 8> [2, 3, 19]
thioctic acid <2> (<2> it is proposed that dihydrolipoic acid serves an essential role in the regulation of the activity of the the enzyme and that the ratio of reduced to oxidized lipoic acid in the cell may play an important role in the regulation of the activity of the enzyme [50]) [50]
veratridine <4> [43]
zinc acetate <2> [31]
Additional information <1-3, 8-10, 23> [3]

Activating compounds

Br^- <1> [27, 37]
Cl^- <1> (<1> enzyme activity increases in the presence of ionic strength, Cl^- is the most potent activator in comparison with Br^- and I^- [27]) [27, 37]
I^- <1> [27]
KCl <1, 2, 4, 24-26> [17, 31, 32, 37]
LiCl <1> [37]
NGFI-A <1> (<1> NGF activates transcriptional factors which influence the promotor region of the enzyme [44]) [44]
NGFI-C <1> (<1> NGF activates transcriptional factors which influence the promotor region of the enzyme [44]) [44]
NH_4Cl <1> [37]
Na_2HPO_4 <1> [37]
Na_2SO_4 <1> [37]
NaBr <1> [37]
NaCl <1, 2, 4, 18, 19, 24-26> [17, 22, 31, 32, 37]
NaF <1> [37]
creatinine-HCl <2> [31]
dihydrolipoic acid <2> (<2> it is proposed that dihydrolipoic acid serves an essential role in the regulation of the activity of the the enzyme and that the ratio of reduced to oxidized lipoic acid in the cell may play an important role in the regulation of the activity of the enzyme [50]) [50]
guanidine-HCl <2> [31]
ocadaic-acid <3> (<3> increases the activity of water-soluble and nonionically membrane-bound enzyme [46]) [46]
sarin (O-isopropyl-methylphosphonofluoridate) <3> (<3> choline acetyltransferase activity is increased in cortex, brainstem and midbrain 6h after LD50 treatment, and the elevated enzyme activity persisted up to 20h after treatment [40]) [40]
sodium acetate <1> [37]

Metals, ions

K^+ <2> [31]

Specific activity (U/mg)

0.07742 <28> [51]
0.19 <3> [41]
73 <3, 12> [6, 21]
142 <2> [7]
500 <6> [8]
Additional information <1-7, 11, 13, 15, 16, 18, 19> (<1-7,13> overview: assay methods [1, 2]) [1, 2, 5, 13-15, 17, 22, 28, 34, 37]

K_m-Value (mM)

0.00025 <3> (CoA) [55]
0.0004 <3> (CoA, <3> enzyme mutant R458A [55]) [55]
0.0004 <3> (acetylcholine) [55]
0.001 <3> (CoA, <3> enzyme mutant R453A [55]) [55]
0.0011 <3> (acetylcholine, <3> enzyme mutant R463A [55]) [55]
0.0015 <3> (acetylcholine, <3> enzyme mutant R453A [55]) [55]
0.0016 <3> (CoA, <3> enzyme mutant R463A [55]) [55]
0.002 <3> (acetylcholine, <3> enzyme mutant R458A [55]) [55]
0.00201 <3> (acetyl-CoA, <3> cosubstrate carnitine [52]) [52]
0.00284 <3> (acetyl-CoA, <3> enzyme mutant ChAT-R/TET, cosubstrate carnitine [52]) [52]
0.0082-0.032 <18> (acetyl-CoA) [22]
0.0092 <3> (acetylcholine, <3> enzyme mutant R452A [55]) [55]
0.012 <3> (acetyl-CoA, <3> enzyme mutant ChAT-R, cosubstrate carnitine [52]) [52]
0.014 <6> (CoA, <6> enzyme mutant H268N [42]) [42]
0.0148 <3> (acetyl-CoA, <3> enzyme mutant ChAT-TET, cosubstrate carnitine [52]) [52]
0.015 <3> (acetyl-CoA, <3> cosubstrate choline [52]) [52]
0.0154 <3> (CoA, <3> enzyme mutant R452A [55]) [55]
0.0155 <3> (acetyl-CoA, <3> enzyme mutant ChAT-R, cosubstrate choline [52]) [52]
0.016 <6> (CoA) [42]
0.016 <25> (acetyl-CoA) [32]
0.017 <6> (CoA, <6> enzyme mutant H393N [42]) [42]
0.0225 <3> (acetyl-CoA, <3> enzyme mutant ChAT-TET, cosubstrate choline [52]) [52]
0.025 <7> (CoA) [11]
0.029 <7> (acetyl-CoA) [11]
0.04 <6> (CoA, <6> enzyme mutant H268L [42]) [42]
0.051 <24> (acetyl-CoA) [32]
0.0516 <3> (acetyl-CoA, <3> enzyme mutant ChAT-R/TET, cosubstrate acetyl-CoA [52]) [52]
0.055 <1> (acetyl-CoA) [39]
0.068 <26> (acetyl-CoA) [32]

0.12 <14> (acetyl-CoA) [10]
0.192 <3> (choline, <3> cosubstrate acetyl-CoA [52]) [52]
0.221 <3> (acetyl-CoA) [41]
0.221 <3> (acetyl-CoA) [9]
0.299 <3> (choline) [9, 41]
0.33 <7> (choline) [11]
0.37 <24> (choline) [32]
0.55 <25> (choline) [32]
0.66 <22> (choline) [30]
0.67 <6> (acetylcholine, <6> enzyme mutant H393N [42]) [42]
0.88 <21> (choline) [30]
0.92 <23> (choline) [30]
1.14 <14> (choline) [10]
1.7 <6> (acetylcholine, <6> enzyme mutant H268L [42]) [42]
1.88 <3> (choline, <3> enzyme mutant ChAT-R, cosubstrate acetyl-CoA [52]) [52]
2 <7> (acetylcholine) [11]
2 <20> (choline) [38]
2.1 <6> (acetylcholine) [42]
2.7 <6> (acetylcholine, <6> enzyme mutant H268N [42]) [42]
3.5 <13> (choline) [30]
4.4 <1> (choline) [39]
4.4-19 <18> (choline) [22]
4.8 <3> (carnitine, <3> enzyme mutant ChAT-R/TET, cosubstrate acetyl-CoA [52]) [52]
5 <2> (acetyl-CoA) [38]
5.79 <3> (choline, <3> enzyme mutant ChAT-TET, cosubstrate acetyl-CoA [52]) [52]
27 <26> (choline) [32]
100 <3> (carnitine, <3> cosubstrate acetyl-CoA [52]) [52]
100 <3> (carnitine, <3> enzyme mutant ChAT-R, cosubstrate acetyl-CoA [52]) [52]
100 <3> (carnitine, <3> enzyme mutant ChAT-TET, cosubstrate acetyl-CoA [52]) [52]
128 <3> (choline, <3> enzyme mutant ChAT-TET, cosubstrate acetyl-CoA [52]) [52]
Additional information <1-4, 7-10, 13, 16, 17, 23, 27> (<1-4,7-10,13,23> overview [1,3]; <16> dependence on salt concentration [15]; <27> temperature dependence [36]) [1, 3, 4, 15, 20, 26, 33, 36]

K_i-Value (mM)

0.00031 <1> (phenylacetyl-CoA) [39]
0.008 <2, 3> (coenzyme A) [3]
0.8 <2> (thioctic acid) [50]
15 <2> (3-trimethylammoniomethyl catechol) [4]
25 <2> (catecholine) [4]

pH-Optimum

6.3 <1> (<1> without NaCl in assay medium [37]) [37]
7-8 <1> (<1> presence of NaCl [37]) [37]
7.2 <17> [20]
7.4 <4> [17]
8.3 <1> [14]
Additional information <18, 19> (<18,19> broad optimum in alkaline range [22]) [22]

Temperature range (°C)

5-45 <27> [36]

4 Enzyme Structure

Molecular weight

30000-300000 <18> (<18> self-association to large aggregates, gel filtration, sedimentation equilibrium) [23]
66800 <15> (<15> gel filtration) [13]
67000-69000 <1, 3, 4, 6> (<6> gel filtration, sucrose density gradient centrifugation [8]; <4> native PAGE [17]; <1> gel filtration [24]; <3> gel filtration [25]) [8, 17, 24, 25]
71000 <16> (<16> gel filtration) [15]
80000 <28> (<28> gel filtration) [51]
106000 <2> (<2> largest aggregate, formation after ammonium sulfate fractionation, gel filtration) [31]
120000-125000 <19> (<19> peaks A, B1, gel filtration, gel electrophoresis) [35]
128000 <17> (<17> gel filtration) [20]
140000 <17> (<17> gradient polyacrylamide electrophoresis) [20]
200000 <19> (<19> peak B2, gel filtration, gel electrophoresis) [35]
Additional information <1-7, 13> (<1-7,13> overview [1,2]) [1, 2]

Subunits

? <1, 2, 6, 12, 17, 18> (<1> x * 62000, form 1, x * 67000, form 2, SDS-PAGE [14]; <18> x * 37000 + x * 56000, SDS-PAGE, ultracentrifugation in presence of guanidine-HCl [23]; <12> x * 68300, SDS-PAGE [6]; <2> x * 72000, isozyme CAT-A, x * 76000, isozyme CAT-B, SDS-PAGE [7]; <6> x * 54000 + x * 67000, SDS-PAGE [8]; <2> x * 69000 + x * 34000, SDS-PAGE [28]; <17> dimer or tetramer: 2 * 68000 or 4 * 34000, SDS-PAGE [20]) [6-8, 14, 20, 23, 28]
dimer <4, 6, 28> (<6> 1 * 54000 + 1 * 13000, major form isolated, may be generated from monomeric form by limited proteolysis, SDS-PAGE [16]; <4> chicken enzyme consists of 2 subunits of identical molecular weight [17]; <28> 2 * 42000, SDS-PAGE [51]) [16, 17, 51]
monomer <1, 3, 6, 15> (<15> 1 * 65000, SDS-PAGE [13]; <6> 1 * 67000, SDS-PAGE [16]; <1> 1 * 66000, SDS-PAGE [24]; <3> 1 * 66000, SDS-PAGE [25]; 1 * 68500, SDS-PAGE [41] ; 1 * 68000, enzyme is initially synthesized as 75 kDa precursor-protein, SDS-PAGE [42]) [13, 16, 24, 25, 41, 42]

Posttranslational modification

phosphoprotein <1, 3> (<3> rat choline acetyltransferase is phosphorylated in Sf9 cells [9,41]; <1> serine 440 of the recombinant human 69-kDa protein is the phoshorylation site [54]) [9, 41, 54]

5 Isolation/Preparation/Mutation/Application

Source/tissue

brain <1-6, 8, 11, 17, 21-23, 27> (<4> regional distribution [17]; <3> cortex [40,48]; <1,2> caudate nuleus [1,7,14,19,31]; <3> cerebrum [19]; <3> brainstem, midbrain [40]; <3> triatum [48]; <3> cerebral cortex [47]) [1-3, 5, 7, 12, 14, 17, 19-21, 28-31, 33, 34, 36, 40, 47, 48]
electric organ <12, 13, 28> [1, 6, 30, 51]
eye <4> (<4> uvea [45]) [45]
ganglion <1, 4, 18-20> (<1> autonomic [3]; <18-20> head [22,35,38]; <4> ciliary [45]) [3, 22, 35, 38, 45]
head <15, 19> [13, 35]
hippocampus <3> [48]
motor spinal root <1-3, 8-10, 23> [3]
nerve trunk <1-3, 8-10, 23> (<1-3,8-10,23> preganglionic [3]) [3]
neuron <21> (<21> transgenic [49]) [49]
placenta <1> [2, 18, 24, 26, 27, 37, 39]
retina <1-3, 8-10, 23> [3, 42, 45]
whole body <14> [10]

Localization

membrane <1, 3> (<1> bound [14]; <3> associated [12]) [12, 14, 54]

Purification

<1> (placental acidic and basic form [18]) [14, 18, 24, 37]
<2> (2 isozymes [25]) [2, 7, 25, 28, 33]
<3> [41]
<3, 15> [9, 21, 25, 34, 13]
<4> [17]
<4> (overview procedures) [1]
<5> (overview procedures) [2]
<6> [8, 16]
<7> (overview procedures) [1]
<11> (large scale [5]) [5]
<12> [6]
<13> (overview procedures) [1]
<16> (partial [15]) [15, 15]
<17> [20]
<18> (2 isozymes) [22]
<19> (2 isozymes) [22, 35]
<20> [38]
<24> [32]

<25> [32]
<26> [32]
<28> [51]

Cloning

<1> (expression of wild-type- and S440A mutant-enzyme in HEK293 cells [54]) [54]
<1> (two fragments of the hChAT gene are used for functional analysis carrying 944 bp (P1) and 4000 bp (P2) of the 5' flanking region in front of the chloramphenicol acetyltransferase (CAT) reporter gene. They are transfected in NG108-15, SN6 and COS-1 cells [44]) [44]
<3> (expression in FR3T3 fibroblasts, SN6 cells and transgenic mice, a 3800-bp 5'-flanking segment from the rat ChAT gene promotor directed cell type-specific expression of a reporter gene in cholinergic cells. A 2342-bp segment from the 5' flanking region of the ChAT gene behaves as an enhancer in cholinergic cells but as a repressor in noncholinergic cells in an orientation-independent manner [53]) [53]
<3> (expression in Spodoptera frugiperda Sf9 cells) [9]
<3> (expression in Spodoptera frugiperda Sf9 cells using a baculovirus expression system [41]) [41]
<7> (expression in Escherichia coli [11]) [11]

Engineering

H268L <6> (<6> an active site histidine of the enzyme is believed to act as general acid/base catalyst, a comparison of the deduced amino acid sequence of the enzyme from Drosophila, pig, rat and Caenorhabditis elegans reveales three conserved histidines: His268, His393 and His426 [42]) [42]
H268N <6> (<6> an active site histidine of the enzyme is believed to act as general acid/base catalyst, a comparison of the deduced amino acid sequence of the enzyme from Drosophila, pig, rat and Caenorhabditis elegans reveales three conserved histidines: His268, His393 and His426 [42]) [42]
H393L <6> (<6> an active site histidine of the enzyme is believed to act as general acid/base catalyst, a comparison of the deduced amino acid sequence of the enzyme from Drosophila, pig, rat and Caenorhabditis elegans reveales three conserved histidines: His268, His393 and His426 [42]) [42]
H393N <6> (<6> an active site histidine of the enzyme is believed to act as general acid/base catalyst, a comparison of the deduced amino acid sequence of the enzyme from Drosophila, pig, rat and Caenorhabditis elegans reveales three conserved histidines: His268, His393 and His426 [42]) [42]
H426L <6> (<6> an active site histidine of the enzyme is believed to act as general acid/base catalyst, a comparison of the deduced amino acid sequence of the enzyme from Drosophila, pig, rat and Caenorhabditis elegans reveales three conserved histidines: His268, His393 and His426 [42]) [42]
H426N <6> (<6> an active site histidine of the enzyme is believed to act as general acid/base catalyst, a comparison of the deduced amino acid sequence of the enzyme from Drosophila, pig, rat and Caenorhabditis elegans reveales three conserved histidines: His268, His393 and His426 [42]) [42]

N514R <3> (<3> is described as ChAT-R, the introduction of an Arg at position 514 in rat enzyme is predicted to provide the ionic charge required to interact with, and neutralize, the carboxyl group of carnitine [52]) [52]

R452A <3> (<3> kinetic as well chemical modification studies have implicated the presence of an active site arginine in enzyme, whose function is to stabilize coenzyme binding, conserved arginines are converted to identify these residues [55]) [55]

R453A <3> (<3> kinetic as well chemical modification studies have implicated the presence of an active site arginine in enzyme, whose function is to stabilize coenzyme binding, conserved arginines are converted to identify these residues [55]) [55]

R458A <3> (<3> kinetic as well chemical modification studies have implicated the presence of an active site arginine in enzyme, whose function is to stabilize coenzyme binding, conserved arginines are converted to identify these residues [55]) [55]

R463A <3> (<3> kinetic as well chemical modification studies have implicated the presence of an active site arginine in enzyme, whose function is to stabilize coenzyme binding, conserved arginines are converted to identify these residues [55]) [55]

S440A <1> (<1> the wild-type enzyme is distributed predominantly in cytoplasm (88%), with the remainder (12%) bound to cellular membranes, mutation S440A results in localization of the enzyme entirely in cytoplasm [54]) [54]

V459T/D460E/N461T <3> (<3> is described as ChAT-TET [52]) [52]

6 Stability

Temperature stability

37 <18, 19> (<18,19> 60% sucrose, 1 mg/ml bovine serum albumin, 2 h stable [22]) [22]

Additional information <1> (<1> increasing ionic strength increases susceptibility to thermal degradation [27]) [27]

General stability information

<1>, acetyl-CoA stabilizes [14]

<1>, choline chloride stabilizes [14]

<1>, increasing ionic strength increases susceptibility to proteolysis and thermal degradation [27]

<3>, ethylene glycol stabilizes [21]

<18, 19>, EDTA stabilizes [22]

<18, 19>, sucrose stabilizes [22]

<2, 3, 18, 19>, bovine serum albumin stabilizes [22, 25]

Storage stability

<1>, -20°C, partially purified preparation, several months stable, less stable in purified state [24]

<1>, -20°C, several months [37]

<1>, 4°C, 24 h, less than 10% loss of activity [18]

References

[1] McGeer, P.L.; McGeer, E.G.; Peng, J.H.: Choline acetyltransferase: purification and immunohistochemical localization. Life Sci., **34**, 2319-2336 (1984)
[2] Chao, L.P.: Choline acetyltransferase: purification and characterization. J. Neurosci. Res., **5**, 85-115 (1980)
[3] Haubrich, D.R.: Choline acetyltransferase and its inhibitors. Biol. of Cholinergic Funct. (Goldberg, A.M., Hanin, I., eds.), **1976**, 239-266 (1976)
[4] Patel, P.J.; Wohlfeil, E.R.; Stahl, S.S.; McLaughlin, K.A.; Hudson, R.A.: Redox reactive reagents inhibiting and inactivating choline acetyltransferase. Biochem. Biophys. Res. Commun., **175**, 407-413 (1991)
[5] Ostermann, C.; Dickmann, U.; Muley, T.; Mäder, M.: Large-scale purification of choline acetyltransferase and production of highly specific antisera. Eur. J. Biochem., **192**, 215-218 (1990)
[6] Raeber, A.J.; Riggio, G.; Waser, P.G.: Purification and isolation of choline acetyltransferase from the electric organ of Torpedo marmorata by affinity chromatography. Eur. J. Biochem., **186**, 487-492 (1989)
[7] Cozzari, C.; Hartman, B.K.: Choline acetyltransferase. Purification procedure and factors affecting chromatographic properties and enzyme stability. J. Biol. Chem., **258**, 10013-10019 (1983)
[8] Slemmon, J.R.; Salvaterra, P.M.; Crawford, G.D.; Roberts, E.: Purification of choline acetyltransferase from Drosophila melanogaster. J. Biol. Chem., **257**, 3847-3852 (1982)
[9] Habert, E.; Birman, S.; Mallet, J.: High-level synthesis and fate of acetylcholine in baculovirus-infected cells: characterization and purification of recombinant rat choline acetyltransferase. J. Neurochem., **58**, 1447-1453 (1992)
[10] Errampalli, D.D.; Knowles, C.O.: Properties of choline acetyltransferase from the bulb mite Rhizoglyphus echinopus. Exp. Appl. Acarol., **12**, 119-128 (1991)
[11] Carbini, L.; Rodriguez, G.; Hersh, L.B.: Kinetic and inactivation studies of recombinant Drosophila choline acetyltransferase. Brain Res. Bull., **24**, 119-124 (1990)
[12] Rylett, R.J.: Synaptosomal membrane-bound choline acetyltransferase is most sensitive to inhibition by choline mustard. J. Neurochem., **52**, 869-875 (1989)
[13] Lutz, E.M.; Lloyd, S.J.; Tyrer, N.M.: Purification of choline acetyltransferase from the locust Schistocerca gregaria and production of serum antibodies to this enzyme. J. Neurochem., **50**, 82-89 (1988)
[14] Peng, J.H.; McGeer, P.L.; McGeer, E.G.: Membrane-bound choline acetyltransferase from human brain: purification and properties. Neurochem. Res., **11**, 959-971 (1986)
[15] Rand, J.B.; Russell, R.L.: Properties and partial purification of choline acetyltransferase from the nematode Caenorhabditis elegans. J. Neurochem., **44**, 189-200 (1985)

[16] Slemmon, J.R.; Salvaterra, P.M.; Roberts, E.: Molecular characterization of choline acetyltransferase from Drosophila melanogaster. Neurochem. Int., **6**, 519-525 (1984)
[17] Peng, J.H.; Ma, K.; Sung, S.C.: Purification of chicken brain choline acetyltransferase. Neurochem. Int., **3**, 377-383 (1981)
[18] Froissart, C.; Massarelli, R.: Placental choline acetyltransferase purification and properties. Adv. Behav. Biol., **25**, 91-95 (1981)
[19] Ryan, R.L.; McClure, W.O.: Inhibition by styrylpyridines of purified rat and bovine brain choline acetyltransferase. Neurochem. Res., **6**, 163-173 (1981)
[20] Su, Y.Y.T.; Wu, J.Y.; Lam, D.M.K.: Purification and some properties of choline acetyltransferase from catfish brain. J. Neurochem., **34**, 438-445 (1980)
[21] Dietz, G.W.; Salvaterra, P.M.: Purification and peptide mapping of rat brain choline acetyltransferase. J. Biol. Chem., **255**, 10612-10617 (1980)
[22] Polsky, R.; Shuster, L.: Preparation and characterization of two isozymes of choline acetyltransferase from squid head ganglia. I. Purification and properties. Biochim. Biophys. Acta, **445**, 25-42 (1976)
[23] Polsky, R.; Shuster, L.: Preparation and characterization of two isozymes of choline acetyltransferase from squid head ganglia. II. Self-association, molecular weight determinations, and studies with inactivating antisera. Biochim. Biophys. Acta, **445**, 43-66 (1976)
[24] Roskoski, R.; Lim, C.T.; Roskoski, L.M.: Human brain and placental choline acetyltransferase: purification and properties. Biochemistry, **14**, 5105-5110 (1975)
[25] Ryan, R.L.; McClure, W.O.: Purification of choline acetyltransferase from rat and cow brain. Biochemistry, **18**, 5357-5365 (1979)
[26] Hersh L.B.; Peet, M.: Re-evaluation of the kinetic mechanism of the choline acetyltransferase reaction. J. Biol. Chem., **252**, 4796-4802 (1977)
[27] Hersh, L.B.; Peet, M.: Effect of salts on the physical and kinetic properties of human placental choline acetyltransferase. J. Neurochem., **30**, 1087-1093 (1978)
[28] Malthe-Sorensson, D.; Lea, T.; Fonnum, F.; Eskeland, T.: Molecular characterization of choline acetyltransferase from bovine brain caudate nucleus and some immunological properties of the highly purified enzyme. J. Neurochem., **30**, 35-46 (1978)
[29] Rowell, P.P.; Chiou, C.Y.: Inhibition of choline acetyltransferase by tertiary alkylaminoethyl esters. Biochem. Pharmacol., **25**, 1093-1099 (1976)
[30] Malthe-Sorenssen, D.: Molecular properties of choline acetyltransferase from different species investigated by isoelectric focusing and ion exchange adsorption. J. Neurochem., **26**, 861-865 (1976)
[31] Chao, L.P.; Wolfgram, F.: Activation, inhibition and aggregation of choline acetyltransferase (EC 2-3-1-6). J. Neurochem., **23**, 697-701 (1974)
[32] Emson, P.C.; Malthe-Soerenssen, D.; Fonnum, F.: Purification and properties of choline acetyltransferase from the nervous system of different invertebrates. J. Neurochem., **22**, 1089-1098 (1974)
[33] Roskoski, R.: Choline acetyltransferase. Inhibition by thiol reagents. J. Biol. Chem., **249**, 2156-2159 (1974)

[34] Rossier, J.; Bauman, A.; Benda, P.: Improved purification of rat brain choline acetyltransferase by using an immunoabsorbent. FEBS Lett., **32**, 231-234 (1973)
[35] Husain, S.S.; Mautner, H.G.: The purification of choline acetyltransferase of squid-head ganglia. Proc. Natl. Acad. Sci. USA, **70**, 3749-3753 (1973)
[36] Hebb, C.; Stephens, T.C.; Smith, M.W.: Effect of environmental temperature on the kinetic properties of goldfish brain choline acetyltransferase. Biochem. J., **129**, 1013-1021 (1972)
[37] Schuberth, J.: Choline acetyltransferase. Purification and effect of salts on the mechanism of the enzyme-catalyzed reaction. Biochim. Biophys. Acta, **122**, 470-481 (1966)
[38] Berman, R.; Wilson, I.B.; Nachmansohn, D.: Choline acetylase specificity in relation to biological function. Biochim. Biophys. Acta, **12**, 315-324 (1953)
[39] Potempska, A.; Loo, Y.H.; Wisniewski, M.M.: On the possible mechanism of phenylacetate neurotoxicity: inhibition of choline acetyltransferase by phenylacetyl-CoA. J. Neurochem., **42**, 1499-1501 (1984)
[40] Khan, W.A.; Dechkovskaia, A.M.; Herrick, E.A.; Jones, K.H.; Abou-Donia, M.B.: Acute sarin exposure causes differential regulation of choline acetyltransferase, acetylcholinesterase, and acetylcholine receptors in the central nervous system of the rat. Toxicol. Sci., **57**, 112-120 (2000)
[41] Habert, E.; Birman, S.; Mallet, J.: High-level synthesis and fate of acetylcholine in baculovirus-infected cells: characterization and purification of recombinant rat choline acetyltransferase. J. Neurochem., **58**, 1447-1453 (1992)
[42] Carbini, L.A.; Hersh, L.B.: Functional analysis of conserved histidines in choline acetyltransferase by site-directed mutagenesis. J. Neurochem., **61**, 247-253 (1993)
[43] Loureiro-Dos-Santos, N.E.; Reis, R.A.M.; Kubrusly, R.C.C.; De Almeida, O.M.M.S.; Gardino, P.F.; De Mello, M.C.F.; De Mello, F.G.: Inhibition of choline acetyltransferase by excitatory amino acids as a possible mechanism for cholinergic dysfunction in the central nervous system. J. Neurochem., **77**, 1136-1144 (2001)
[44] Quirin-Stricker, C.; Mauvais, C.; Schmitt, M.: Transcriptional activation of human choline acetyltransferase by AP2- and NGF-induced factors. Mol. Brain Res., **49**, 165-174 (1997)
[45] Pendrak, K.; Lin, T.; Stone, R.A.: Ciliary ganglion choline acetyltransferase activity in avian macrophthalmos. Exp. Eye Res., **60**, 237-243 (1995)
[46] Cooke, L.J.; Rylett, R.J.: Inhibitors of serine/threonine phosphatases increase membrane-bound choline acetyltransferase activity and enhance acetylcholine synthesis. Brain Res., **751**, 232-238 (1997)
[47] Wattanathorn, J.; Kotchabhakdi, N.; Casalotti, S.O.; Baldwin, B.A.; Govitrapong, P.: Ethanol alone or with dexamethasone alters the kinetics of choline acetyltransferase. Eur. J. Pharmacol., **313**, 69-72 (1996)
[48] Zambrzycka, A.; Alberghina, M.; Strosznajder, J.B.: Effects of aging and amyloid-b peptides on choline acetyltransferase activity in rat brain. Neurochem. Res., **27**, 277-281 (2002)

[49] Faivre-Bauman, A.; Loudes, C.; Neveu, I.; Naveilhan, P.; Vantini, G.; Epelbaum, J.; Onteniente, B.: Decreased choline acetyltransferase activity in nerve growth factor-transgenic mice during brain development. Neurosci., **62**, 333-336 (1994)
[50] Haugaard, N.; Levin, R.M.: Regulation of the activity of choline acetyl transferase by lipoic acid. Mol. Cell. Biochem., **213**, 61-63 (2000)
[51] Tavares, N.N.; Hasson-Voloch, A.: Choline acetyltransferase from the electric organ of Electrophorus electricus (L.)-physicochemical characterization and immunochemical identification. Z. Naturforsch. C, **53**, 407-415 (1998)
[52] Cronin, C.N.: Redesign of choline acetyltransferase specificity by protein engineering. J. Biol. Chem., **273**, 24465-24469 (1998)
[53] Loennerberg, P.; Lendahl, U.; Funakoshi, H.; Aerhlund-Richter, L.; Persson, H.; Ibanez, C.F.: Regulatory region in choline acetyltransferase gene directs developmental and tissue-specific expression in transgenic mice. Proc. Natl. Acad. Sci. USA, **92**, 4046-4050 (1995)
[54] Dobransky, T.; Davis, W.L.; Rylett, R.J.: Functional characterization of phosphorylation of 69-kDa human choline acetyltransferase at serine 440 by protein kinase C. J. Biol. Chem., **276**, 22244-22250 (2001)
[55] Wu, D.; Hersh, L.B.: Identification of an active site arginine in rat choline acetyltransferase by alanine scanning mutagenesis. J. Biol. Chem., **270**, 29111-29116 (1995)

Carnitine O-acetyltransferase 2.3.1.7

1 Nomenclature

EC number
2.3.1.7

Systematic name
acetyl-CoA:carnitine O-acetyltransferase

Recommended name
carnitine O-acetyltransferase

Synonyms
CARAT <10, 14> [32]
CAT <1, 3-7, 9-11, 14, 17> [6, 13, 18, 19, 23, 25-29, 31, 33, 35, 37]
CATC
CRAT <14> [37]
CT-CAT <6> [30]
CarAc <15> [3]
H-CAT <10> [30]
P-CAT <9> [30]
S-CAT1 <18> [30]
S-CAT2 <18> [30]
acetyl-CoA-carnitine O-acetyltransferase
acetylcarnitine transferase
carnitine acetyl coenzyme A transferase
carnitine acetylase
carnitine acetyltransferase
carnitine-acetyl-CoA transferase

CAS registry number
9029-90-7

2 Source Organism

<1> *Aspergillus niger* (A60 MZKIBK (NRLL 2270) [28]) [28]
<2> *Bos taurus* (beef [8, 11, 13, 27]; bovine [11, 22, 28, 33]; ox [4]; Holstein bulls [11]) [4, 8, 11, 13, 22, 27, 28, 33]
<3> *Camelus dromedarius* (arabian camel [25-27]) [25-27]
<4> *Candida albicans* (dimorphic yeast, C316 [31]) [31]
<5> *Candida curvata* [28]

<6> *Candida tropicalis* (yeast [6, 9, 10, 14, 15, 17, 19, 22, 28-31]; ATCC 32113 [9, 19]; (Castellani) Berkhout strain pK233(ATCC 20336) [6, 10, 15, 30, 31]) [6, 9, 10, 14, 15, 17, 19, 22, 28-31]
<7> *Candida utilis* [28]
<8> *Cavia porcellus* (guinea pig [4, 8]) [4, 8]
<9> *Columba sp.* (pigeon [1, 3-5, 8, 10-12, 14, 16, 18, 22, 23, 26, 28, 30, 32, 35, 36]; commercial preparation, Boehringer Mannheim [5]; commercial preparation, Sigma [16, 18, 23, 36]) [1, 3-5, 7-12, 14, 16, 18, 22, 23, 25-28, 30, 32, 35, 36]
<10> *Homo sapiens* (human [3, 20, 22, 25-28, 30-32, 34, 35, 37]) [3, 20, 22, 25-28, 30-32, 34, 35, 37]
<11> *Lipomyces starkeyi* [28]
<12> *Mus musculus* (mouse, Uni Prot accession number P47934 [37]) [37]
<13> *Mus musculus* (mouse, cDNA clone MMRNACAR [32]) [32]
<14> *Mus musculus* (mouse [13, 17, 24-28, 32, 37]; swiss mice [13]; S3T3(ATCC CCL-92) [32]; NIH3T3 (ATCC CRL-1658) [32]; Balb/c [24]) [13, 17, 24-28, 32, 37]
<15> *Oryctolagus cuniculus* (rabbit [3, 36]; New Zealand white [3]) [3, 36]
<16> *Ovis aries* (sheep [1, 4, 11]; ram [11]) [1, 4, 11]
<17> *Rattus norvegicus* (rat [7, 8, 10-14, 18, 20-22, 25-28, 33, 34, 36]; albino [7]; Fischer 344 [36]; Sprague-Dawley [13, 33]) [7, 8, 10-14, 18, 20-22, 25-28, 33, 34, 36]
<18> *Saccharomyces cerevisiae* (baker's yeast [22, 25, 28-30, 32]) [22, 25, 28-30, 32]
<19> *Sus scrofa* (pig [1, 2, 8, 10, 20, 28]; swine [7, 26]; porcine [32]) [1, 2, 7, 8, 10, 20, 26, 28, 32]
<20> *Torulopsis bovina* [13, 22]

3 Reaction and Specificity

Catalyzed reaction

acetyl-CoA + carnitine = CoA + O-acetylcarnitine (also acts on propanoyl-CoA and butanoyl-CoA cf. EC 2.3.1.21 carnitine O-palmitoyltransferase and EC 2.3.1.137 carnitine O-octanoyltransferase)

Reaction type

acyl group transfer

Natural substrates and products

S O-acetylcarnitine + CoASH <1-19> (<17> key mitochondrial enzyme for fuel utilization, essential functions are to regenerate CoA, which allows peroxisomal β-oxidation to proceed, and to facilitate transport of acetyl moieties to mitochondria for oxidation [36]; <2> mitochondrial fatty acid transport [11]; <18> acetate metabolism, enzyme transports activated acetyl group from the cytosol into the mitochondrial matrix [22]; <4> key enzyme for the translocation of acetyl units between intracellular compartments [31]) (Reversibility: r <1-19> [1-20, 22-37]) [1-20, 22-37]

- **P** acetyl-CoA + carnitine
- **S** acetyl-CoA + carnitine <1-19> (Reversibility: r <1-19> [1-20, 22-37]) [1-20, 22-37]
- **P** O-acetylcarnitine + CoASH
- **S** acyl-CoA + carnitine <1-20> (Reversibility: r <1-20> [1-37]) [1-37]
- **P** acylcarnitine + CoASH
- **S** acylcarnitine + CoASH <1-20> (Reversibility: r <1-20> [1-37]) [1-37]
- **P** acyl-CoA + carnitine

Substrates and products

- **S** 3-keto-4-pentenoyl-CoA + L-(-)carnitine <9> (Reversibility: ? <9> [16]) [16]
- **P** 3-keto-4-pentenoylcarnitine + CoASH
- **S** O-acetoacetyl(-)carnitine + CoASH <17> (Reversibility: r <17> [8]) [8]
- **P** acetoacetyl-CoA + L-(-)carnitine
- **S** O-acetyl(-)carnitine + CoASH <1-20> (Reversibility: r <1-20> [1-20, 22-37]) [1-20, 22-37]
- **P** acetyl-CoA + L-(-)carnitine
- **S** acetoacetyl-CoA + L-(-)carnitine <17> (Reversibility: r <17> [8]) [8]
- **P** O-acetoacetyl(-)carnitine + CoASH
- **S** acetyl-CoA + (-)-norcarnitine <19> (<19> β-hydroxy-γ-dimethylaminobutyrate [2]) (Reversibility: r <19> [2]) [2]
- **P** acetylnorcarnitine + CoASH
- **S** acetyl-CoA + (trishydroxymethyl)glycine <9> (Reversibility: ? <9> [5]) [5]
- **P** acetyl(trishydroxymethyl)glycine + CoASH
- **S** acetyl-CoA + DL-β-hydroxy-γ-aminobutyrate <19> (<19> non-methylated derivative of carnitine [2]) (Reversibility: r <19> [2]) [2]
- **P** β-hydroxybutyrate + CoASH
- **S** acetyl-CoA + DL-norcarnitine <19> (<19> β-hydroxy-γ-dimethylaminobutyrate [2]) (Reversibility: r <19> [2]) [2]
- **P** acetylnorcarnitine + CoASH
- **S** acetyl-CoA + L-(-)carnitine <1-20> (Reversibility: r <1-20> [1-20, 22-37]) [1-20, 22-37]
- **P** O-acetyl(-)carnitine + CoASH
- **S** acetyl-CoA + choline <9, 15> (Reversibility: ? <9, 15> [3]) [3]
- **P** acetylcholine + CoASH
- **S** acetyl-CoA + tris(hydroxymethyl)aminoethansulfonic acid <9> (Reversibility: ? <9> [5]) [5]
- **P** acetyl-tris(hydroxymethyl)aminoethansulfonic acid + CoASH
- **S** acetyl-CoA + tris(hydroxymethyl)aminomethane <9> (Reversibility: ? <9> [5]) [5]
- **P** acetyl-tris(hydroxymethyl)aminomethane + CoASH
- **S** bromoacetyl-CoA + L-(-)carnitine <9> (Reversibility: ? <9> [16]) [16]
- **P** bromoacetylcarnitine + CoASH

S decanoyl-CoA + carnitine <9, 14, 17, 19> (Reversibility: r <9, 14, 17, 19> [10, 12, 13, 20]) [10, 12, 13, 20]
P decanoylcarnitine + CoASH
S decanoylcarnitine + CoASH <9, 10, 14, 17, 19> (Reversibility: r <9, 10, 14, 17, 19> [10, 12, 13, 20]) [10, 12, 13, 20]
P decanoyl-CoA + carnitine
S hexanoyl-CoA + carnitine <9, 14, 17, 19> (Reversibility: r <9, 14, 17, 19> [10, 12, 13]) [10, 12, 13]
P hexanoylcarnitine + CoASH
S hexanoylcarnitine + CoASH <9, 14, 17, 19> (Reversibility: r <9, 14, 17, 19> [10, 12, 13]) [10, 12, 13]
P hexanoyl-CoA + carnitine
S isobutyryl-CoA + carnitine <2> (Reversibility: r <2> [11]) [11]
P isobutyrylcarnitine + CoASH
S isobutyrylcarnitine + CoASH <2> (Reversibility: r <2> [11]) [11]
P isobutyryl-CoA + carnitine
S isovaleryl-CoA + carnitine <2> (Reversibility: r <2> [11]) [11]
P isovalerylcarnitine + CoASH
S isovalerylcarnitine + CoASH <2> (Reversibility: r <2> [11]) [11]
P isovaleryl-CoA + carnitine
S malonyl-CoA + L-(-)carnitine <6> (Reversibility: r <6> [9]) [9]
P malonylcarnitine + CoASH
S malonylcarnitine + CoASH <6> (<6> slight transfer of the acyl group to carnitine [9]) (Reversibility: r <6> [9]) [9]
P malonyl-CoA + L-(-)carnitine
S n-butyryl-CoA + carnitine <2-4, 6, 10, 14, 17> (Reversibility: r <2-4, 6, 10, 14, 17> [9, 11-13, 20, 26, 27, 31]) [9, 11-13, 20, 26, 27, 31]
P n-butyrylcarnitine + CoASH
S n-butyrylcarnitine + CoASH <2-4, 6, 10, 14, 17> (Reversibility: r <2-4, 6, 10, 14, 17> [9, 11-13, 20, 26, 27, 31]) [9, 11-13, 20, 26, 27, 31]
P n-butyryl-CoA + carnitine
S n-valeryl-CoA + carnitine <2> (Reversibility: r <2> [11]) [11]
P n-valerylcarnitine + CoASH
S n-valerylcarnitine + CoASH <2> (Reversibility: r <2> [11]) [11]
P n-valeryl-CoA + carnitine
S octanoyl-CoA + carnitine <2, 10, 14, 17> (Reversibility: r <2, 10, 14, 17> [11-13, 20, 31]) [11-13, 20, 31]
P octanoylcarnitine + CoASH
S octanoylcarnitine + CoASH <2, 10, 14, 17> (Reversibility: r <2, 10, 14, 17> [11-13, 20, 31]) [11-13, 20, 31]
P octanoyl-CoA + carnitine
S propionyl-CoA + L-(-)carnitine <2-4, 6, 10, 14, 17> (Reversibility: r <2-4, 6, 10, 14, 17> [8-11, 13, 14, 20, 21, 26, 27, 31]) [8-11, 13, 14, 20, 21, 26, 27, 31]
P propionylcarnitine + CoASH
S propionylcarnitine + CoASH <2-4, 6, 10, 14, 17> (Reversibility: r <2-4, 6, 10, 14, 17> [8-11, 13, 14, 20, 21, 26, 27, 31]) [8-11, 13, 14, 20, 21, 26, 27, 31]

P propionyl-CoA + L-(-)carnitine

S Additional information <4, 6, 9, 17, 19, 20> (<9> hydroxyethylpiperazine-ethansulfonic acid is no substrate [5]; <9> 3-hydroxy-5,5-dimethylhexanoic acid, 3-acetoxy-5,5-dimethylhexanoic acid, R-(+)-3-hydroxy-5,5-dimethylhexanoic acid and S-(-)-3-hydroxy-5,5-dimethylhexanoic acid are not substrates [23]; <19> only (-) optical isomers are substrates, unnatural (+) isomers are competitive inhibitors [2]; <17> O-acetyl-(+)-carnitine and reduced glutathione will not serve as substrates, succinyl-CoA or β-hydroxy-β-methylglutaryl-CoA are no substrates [8]; <2,9> D-carnitine is no substrate [11,23]; <6> crotonyl-CoA is no substrate [9]; <6> pentanoyl-CoA, hexanoyl-CoA, decanoyl-CoA and palmitoyl-CoA are no substrates [10]; <4,6> octanoyl-CoA and palmitoyl-CoA are no substrates [31]; <17> palmityl-CoA does not act as a substrate for the liver enzyme [7]; <3,10> palmitoyl-CoA is no substrate [20,27]; <20> no carnitine acyltransferase activity with acyl CoAs of carbon chain lengths greater than 3 [13]) [2, 5, 7-10, 13, 23, 27, 31]

P ?

Inhibitors

(+)-acetylcarnitine <19> (<19> competitive inhibition [2]) [2]
(+)-carnitine <19> [2]
(-)-acetylcarnitine <19> (<19> noncompetitive inhibition [2]) [2]
(2R,6S:2S,6R)-6-carboxymethyl-2-hydroxy-2,4,4-trimethylmorpholinium chloride <9, 17> (<9,17> hemiacetylcarnitinium [18]) [18]
3-acetoxy-5,5-dimethylhexanoic acid <9> [23]
3-hydroxy-5,5-dimethylhexanoic acid <9> [23]
3-keto-4-pentenoyl-CoA <9> (<9> inhibitor only in presence of L-carnitine [16]) [16]
4-hydroxynonenal <17> [12]
5,5'-dithio-bis(2-nitrobenzoic acid) <2, 3, 14, 17, 19> [2, 10, 11, 17, 25]
Ca^{2+} <1, 2, 6, 10, 18> [10, 11, 20, 25, 28]
Cu^{2+} <1, 3> [25, 28]
$CuSO_4$ <3> [25]
D-carnitine <2, 9, 14> [11, 13]
DL-acetyl-β-methylcholine <19> [2]
DL-acetylcarnitine <19> (<19> competitive inhibition [2]) [2]
DL-carnitine <17> (<17> inhibitor for the reverse reaction [7]) [7]
Fe^{2+} <17> [36]
$FeSO_4$ <17> [36]
Hg^{2+} <2, 3, 19> [25, 28]
$HgCl_2$ <3, 19> [2, 25]
Mg^{2+} <1, 6, 18> [10, 25, 28]
Mn^{2+} <1> [28]
N-ethylmaleimide <2, 17, 19> [2, 7, 11]
Zn^{2+} <2, 3, 14, 19> (<14> inhibition only in reverse direction [13]) [11, 13, 25, 28]
$ZnSO_4$ <3> [25]

acetyl-D-carnitine <9> [11]
acetylcarnitine <2, 17, 19> (<17> competitive inhibition, inhibitor for the forward reaction [7]) [2, 7, 11]
acetylcholine <9> (<9> competitively inhibits formation of acetylcarnitine [3]) [3]
bromoacetyl-CoA <9> [16]
bromoacetylcarnitine <9, 17> [25]
carnitine nitrile <19> [2]
choline <9, 19> (<9> competitively inhibits formation of acetylcarnitine [3]) [2, 3]
deoxycarnitine <19> (<19> competitive inhibition [2]) [2]
γ-aminobutyrate <19> [2]
iodoacetamide <3, 17, 19> [2, 7, 25]
iodoacetic acid <17> [12, 25]
malondialdehyde <17> [36]
malonyl-CoA <17> [13]
methoxycarbonyl-CoA disulfide <3> [25]
p-chloromercuribenzoic acid <6> [10]
p-chloromercuriphenyl sulfonic acid <17> [12]
p-hydroxymercuribenzoate <17, 19> [2, 7]
palmitoyl-CoA <2, 9, 17> [7, 11]
prostigmine <19> [2]
tensilon <19> [2]

Activating compounds

clofibrate <17> (<17> mitochondrial enzyme increases in specific activity [8]; <17> enzyme is induced by [18]) [7, 8, 18, 30]
diethylhexylphthalate <17> (<17> enzyme is induced by [18]) [18]
humic acid <14> [24]
streptozotocin <17> (<17> increases enzyme in liver [8]) [8]

Specific activity (U/mg)

0.002 <4> (<4> carbon source glucose, peroxisomal enzyme [31]) [31]
0.005 <17> (<17> liver [8]) [8]
0.021 <6> [9]
0.04 <4> (<4> carbon source serum, peroxisomal enzyme [31]) [31]
0.06 <4> (<4> carbon source glucose, mitochondrial enzyme [31]) [31]
0.25 <17> (<17> skeletal muscle [8]) [8]
0.27 <4> (<4> carbon source serum, mitochondrial enzyme [31]) [31]
4.5 <4> (<4> carbon source triolein, mitochondrial enzyme [31]) [31]
8.8 <4> (<4> carbon source triolein, peroxisomal enzyme [31]) [31]
10.65 <1> [28]
71 <2> (<2> source heart [11]) [11]
78.75 <10> [20]
80 <6> (<6> cytoplasmic CAT [19]) [19]
82.83 <17> [7]
85 <2, 9, 17> (<2> source spermatozoon [11]) [7, 11]
88 <6> (<6> peroxisomal CAT [19]) [19]

88.5 <6> (<6> peroxisomal enzyme [10]) [10]
94 <3> [26, 27]
113 <6> (<6> mitochondrial CAT [19]) [10, 19]
119 <9> [1]
120 <17> (<17> purified from postmitochondrial fraction [12]) [12]
200 <18> [22]
220 <17> (<17> purified from liver homogenate [12]) [12]
240 <17> (<17> purified from mitochondrial fraction [12]) [12]

K_m-Value (mM)

0.009 <2> (CoASH) [11]
0.012 <2> (acetyl-CoA, <2> sperm enzyme [11]) [11]
0.014 <4> (acetyl-CoA, <4> peroxisomal enzyme [31]) [31]
0.014 <2> (propionyl-CoA) [11]
0.015 <2> (acetyl-CoA) [11]
0.0153 <14> (CoASH) [13, 27]
0.0166 <14> (butyryl-CoA) [27]
0.0173 <3> (propionyl-CoA) [27]
0.0177 <18> (acetyl-CoA) [22]
0.018 <17> (CoASH, <17> supplementation with R-α-lipoic acid [36]) [36]
0.019 <17> (acetylcarnitine, <17> supplementation with R-α-lipoic acid [36]) [36]
0.0191 <14> (octanoyl-CoA) [13]
0.02 <2> (acetyl-CoA, <2> heart enzyme [11]) [11]
0.02 <14> (hexanoyl-CoA) [13]
0.021 <10> (acetyl-CoA) [31]
0.0213 <10> (acetyl-CoA) [20, 27]
0.022 <17> (CoASH, <17> substrate hexanoylcarnitine, enzyme from liver homogenate [12]) [12]
0.024 <17> (CoASH, <17> substrate decanoylcarnitine, enzyme from liver homogenate [12]) [12]
0.027 <17> (CoASH, <17> substrate decanoylcarnitine, enzyme from mitochondria [12]; <17> substrate butyrylcarnitine and octanoylcarnitine, enzyme from liver homogenate [12]) [12]
0.0277 <3> (CoASH) [27]
0.028 <17> (CoASH, <17> substrate acetylcarnitine, enzyme from liver homogenate [12]) [12]
0.028 <10> (propionyl-CoA) [20, 27, 31]
0.0287 <14> (decanoyl-CoA) [13]
0.029 <3> (butyryl-CoA) [27]
0.03 <17> (CoASH) [7]
0.032 <17> (CoASH, <17> substrate acetylcarnitine, enzyme from mitochondria [12]) [12, 27]
0.034 <9> (CoASH) [8, 27]
0.034 <4> (propionyl-CoA, <4> peroxisomal enzyme [31]) [31]
0.036 <6> (acetyl-CoA, <6> mitochondrial enzyme [10,31]) [10, 31]
0.037 <9> (CoASH) [7]

0.037 <9> (propionyl-CoA) [27]
0.038 <17> (CoASH, <17> substrate butyrylcarnitine, enzyme from mitochondria [12]) [12]
0.038 <4> (acetyl-CoA, <4> mitochondrial enzyme [31]) [31]
0.038 <17> (butyryl-CoA) [7]
0.039 <3> (L-(-)-carnitine) [26]
0.041 <19> (CoASH) [2]
0.041 <2> (n-butyryl-CoA) [11]
0.042 <6> (acetyl-CoA, <6> peroxisomal enzyme [10,31]) [10, 31]
0.0435 <10> (butyryl-CoA) [20, 27]
0.044 <17> (CoASH, <17> substrate hexanoylcarnitine, enzyme from mitochondria [12]) [12]
0.044 <3> (acetyl-CoA) [26]
0.044 <9> (butyryl-CoA) [27]
0.049 <2> (isobutyryl-CoA) [11]
0.0503 <10> (octanoyl-CoA) [20]
0.054 <17> (CoASH, <17> substrate octanoylcarnitine, enzyme from mitochondria [12]) [12]
0.0549 <10> (hexanoyl-CoA) [20]
0.064 <10> (decanoyl-CoA) [20]
0.069 <17> (CoASH) [8]
0.071 <2> (n-valeryl-CoA) [11]
0.076 <18> (acetyl-CoA) [22]
0.086 <14> (carnitine) [13]
0.086 <10> (carnitine, <10> cosubstrate propionyl-CoA [20]) [20]
0.09 <1> (CoASH, <1> mitochondrial enzyme [28]) [28]
0.097 <10> (carnitine, <10> cosubstrate acetyl-CoA [20]) [20]
0.1 <1> (CoASH) [28]
0.1 <6> (propionyl-CoA, <6> peroxisomal enzyme [10,31]) [10, 31]
0.11 <1> (CoASH, <1> cytosolic enzyme [28]) [28]
0.12 <2, 9> (L-(-)-carnitine) [7, 8, 11]
0.12 <10> (carnitine, <10> cosubstrate hexanoyl-CoA [20]) [20]
0.121 <14> (L-carnitine, <14> cosubstrate butanoyl-CoA [13]) [13]
0.13 <6> (acetyl-CoA) [9]
0.148 <10> (carnitine, <10> cosubstrate octanoyl-CoA [20]) [20]
0.15 <17> (L-(-)-carnitine) [8]
0.152 <10> (carnitine, <10> cosubstrate butyryl-CoA [20]) [20]
0.162 <6> (propionyl-CoA, <6> mitochondrial enzyme [10,31]) [10, 31]
0.17 <18> (L-(-)-carnitine) [22]
0.185 <4> (propionyl-CoA, <4> mitochondrial enzyme [31]) [31]
0.23 <18> (L-(-)-carnitine) [22]
0.244 <3> (L-(-)-carnitine) [26]
0.257 <6> (CoASH, <6> mitochondrial enzyme [10]) [10]
0.28 <17> (acetylcarnitine) [7]
0.3 <2> (acetyl-L-(-)-carnitine, <2> heart enzyme [11]) [11]
0.304 <6> (CoASH, <6> peroxisomal enzyme [10]) [10]
0.307 <3> (acetyl-DL-carnitine) [26]

0.31 <19> ((-)carnitine) [2]
0.344 <14> (L-carnitine, <14> cosubstrate hexanoyl-CoA [13]) [13]
0.36 <2> (acetyl-L-(-)-carnitine, <2> sperm enzyme [11]) [11]
0.39 <17> (acetylcarnitine, <17> enzyme from liver homogenate [12]) [12]
0.395 <14> (L-carnitine, <14> cosubstrate octanoyl-CoA [13]) [13]
0.4 <9> (carnitine) [3]
0.417 <6> (acetylcarnitine, <6> peroxisomal enzyme [10]) [10]
0.42 <10> (acetylcarnitine) [20]
0.48 <17> (butyrylcarnitine, <17> enzyme from liver homogenate [12]) [12]
0.519 <14> (L-carnitine, <14> cosubstrate decanoyl-CoA [13]) [13]
0.58 <10> (carnitine, <10> cosubstrate decanoyl-CoA [20]) [20]
0.6 <17> (acetylcarnitine, <17> enzyme from mitochondria [12]) [12]
0.622 <6> (L-carnitine, <6> mitochondrial enzyme, second substrate acetyl-CoA [10]) [10]
0.639 <6> (acetylcarnitine, <6> mitochondrial enzyme [10]) [10]
0.64 <17> (decanoylcarnitine, <17> enzyme from mitochondria [12]) [12]
0.65 <10> (propionylcarnitine) [20]
0.655 <6> (L-carnitine, <6> peroxisomal enzyme, second substrate propionyl-CoA [10]) [10]
0.66 <17> (butyrylcarnitine, <17> enzyme from mitochondria [12]) [12]
0.66 <17> (decanoylcarnitine, <17> enzyme from liver homogenate [12]) [12]
0.719 <6> (L-carnitine, <6> peroxisomal enzyme, second substrate acetyl-CoA [10]) [10]
0.72 <17> (DL-carnitine) [7]
0.774 <6> (L-carnitine, <6> mitochondrial enzyme, second substrate propionyl-CoA [10]) [10]
1 <1> (carnitine, <1> mitochondrial enzyme [28]) [28]
1.1 <1> (carnitine, <1> cytosolic enzyme [28]) [28]
1.1 <15> (choline, <15> brain enzyme [3]) [3]
1.27 <17> (octanoylcarnitine, <17> enzyme from mitochondria [12]) [12]
1.39 <10> (octanoylcarnitine) [20]
1.6 <19> ((-)-norcarnitine) [2]
1.62 <17> (octanoylcarnitine, <17> enzyme from liver homogenate [12]) [12]
2 <17> (CoASH) [36]
2.28 <17> (hexanoylcarnitine, <17> enzyme from liver homogenate [12]) [12]
2.51 <17> (hexanoylcarnitine, <17> enzyme from mitochondria [12]) [12]
3.2 <19> (DL-norcarnitine) [2]
3.3 <6> (DL-carnitine) [9]
5 <15> (choline, <15> heart enzyme [3]) [3]
8.7 <9> (choline) [3]
40 <19> (DL-β-hydroxy-γ-aminobutyrate) [2]
50 <9> ((trishydroxymethyl)glycine) [5]
Additional information <17> (<17> K_m acetylcarnitine 1.25 mg/ml [36]) [36]

K_i-Value (mM)

0.0046 <1> (Cu^{2+}) [28]
0.015 <2> (CoASH, <2> sperm enzyme [11]) [11]
0.016 <2> (palmitoyl-CoA) [11]
0.021 <2> (CoASH, <2> heart enzyme [11]) [11]
0.047 <2> (acetyl-CoA, <2> heart enzyme [11]) [11]
0.052 <2> (acetyl-CoA, <2> sperm enzyme [11]) [11]
0.2 <17> (DL-carnitine) [7]
0.25 <19> ((+)-acetylcarnitine) [2]
0.25 <2> (L-(-)-carnitine, <2> heart enzyme [11]) [11]
0.31 <2> (L-(-)-carnitine, <2> sperm enzyme [11]) [11]
0.39 <19> (DL-acetylcarnitine) [2]
0.62 <19> (DL-carnitine) [2]
0.81 <2> (acetyl-L-(-)-carnitine, <2> heart enzyme [11]) [11]
0.89 <9> ((2R,6S:2S,6R)-6-carboxymethyl-2-hydroxy-2,4,4-trimethylmorpholinium chloride) [18]
0.98 <2> (acetyl-L-(-)-carnitine, <2> sperm enzyme [11]) [11]
1 <14> (D-carnitine) [13]
1.1 <2> (D-carnitine) [11]
1.2 <19> (deoxynorcarnitine) [2]
1.5 <1> (Mg^{2+}) [28]
2.1 <19> ((+)carnitine) [2]
2.3 <19> (deoxycarnitine) [2]
2.6 <1> (Ca^{2+}) [28]
4.1 <9> (3-acetoxy-5,5-dimethylhexanoic acid, <9> reverse direction [23]) [23]
4.72 <17> ((2R,6S:2S,6R)-6-carboxymethyl-2-hydroxy-2,4,4-trimethylmorpholinium chloride) [18]
5 <1> (Mn^{2+}) [28]
7.5 <9> (S-(-)-3-hydroxy-5,5-dimethylhexanoic acid) [23]
8 <17> (palmityl-CoA, <17> substrate carnitine [7]) [7]
8.3 <9> (3-hydroxy-5,5-dimethlhexanoic acid) [23]
12 <19> (prostigmine) [2]
15 <19> (tensilon) [2]
18.6 <17> (palmityl-CoA, <17> substrate acetyl-CoA [7]) [7]
20.3 <9> (R-(+)-3-hydroxy-5,5-dimethylhexanoic acid) [23]
27 <19> (choline) [2]
44 <19> (γ-aminobutyrate) [2]
71 <19> (DL-acetyl-β-methylcholine) [2]
920 <17> (acetylcarnitine, <17> substrate DL-carnitine [7]) [7]

pH-Optimum

7.2 <6> [28]
7.2-8 <3> [26]
7.3 <1> [28]
7.5-7.8 <4, 6> (<4,6> peroxisomal enzyme [31]) [31]
7.5-8 <18> [22]

8 <4, 6, 17> (<4> mitochondrial enzyme [31]) [10, 12, 31]
8.5 <14> [13]
8.7 <10, 19> [20, 28]

pH-Range
4-9 <2> [4]
4.5-8.5 <16> [4]
5.5-10 <14> [13]
6-9 <3> (<3> activity declines rapidly on either side of the optimum pH [26]) [26]
6-10 <10> [20]
7-9 <18> [22]

Temperature optimum (°C)
37 <17> [7]

Temperature range (°C)
20-40 <17> [7]

4 Enzyme Structure

Molecular weight
51000 <9> [8]
55000 <9> (<9> gel filtration [1,7]) [1, 7]
56000 <17> (<17> purified from mitochondrial fraction of liver, rats fed on a diet containing ethyl *p*-chlorophenoxyisobutyrate [12]; <17> gel filtration [7]) [7, 12]
58000 <9> [16, 27]
59000 <3, 17> (<17> microsomal and peroxisomal enzyme [8]; <3> gel filtration, native and SDS-PAGE [27]) [8, 27]
60000 <6, 14> (<6> SDS-PAGE [9]; <14> gel filtration [13]; <14> SDS-PAGE [24]) [9, 13, 24, 27]
60500 <10> (<10> gel filtration [20]) [20, 27, 28]
61000 <2> (<2> heart enzyme, gel filtration, SDS-PAGE [11]) [11]
61900 <17> (<17> purified from mitochondrial fraction, gel filtration [12]) [12]
62000 <2, 17> (<2> sperm enzyme, gel filtration, SDS-PAGE [11]) [8, 11, 21, 28]
64000 <6> (<6> SDS-PAGE [19,30]) [19, 30]
66000 <6> (<6> mitochondrial CAT [29]) [29]
66100 <17> (<17> purified from liver homogenate, gel filtration [12]) [12]
66450 <6> (<6> deduced amino acid sequence, mitochondrial enzyme [30]) [30]
67000 <17> [27]
67500 <17> (<17> purified from liver homogenate, SDS-PAGE [12,28]) [12, 28]
68000 <6> (<6> peroxisomal and cytosolic CAT [29]) [29]

68190 <6> (<6> deduced amino acid sequence, peroxisomal enzyme [30]) [30]
69000 <17> (<17> post-mitochondrial fraction, 2 polypeptides 67500 + 69000 cannot be separated from each other [12]; <17> translation product of the enzyme by the reticulocyte lysate protein-synthesizing system [12]) [12]
70710 <6> (<6> nucleotide sequence analysis [30]) [30]
75000 <9> (<9> SDS-PAGE [7]) [7]
77000 <1> (<1> gel filtration [28]) [28]
400000 <18> [22]
415000 <6> (<6> peroxisomal enzyme, meniscus depletion technique [14]) [14]
420000 <6> (<6> gel filtration, analytical ultracentrifugation [14,19]) [14, 19]
426000 <6> (<6> mitochondrial enzyme, meniscus depletion technique [14]) [14]
540000 <6> (<6> gel filtration [19]) [19]

Subunits

dimer <1, 6, 17> (<17> 1 * 36500 + 1 * 27000, purified from mitochondrial fraction, monomeric form converted into dimeric by proteolytic modification after disruption of mitochondria, SDS-PAGE [12,28]; <17> 1 * 34000 + 1 * 25000, purified from mitochondrial fraction of liver, rats fed on a diet containing ethyl *p*-chlorophenoxyisobutyrate, SDS-PAGE [7,8,12,13]; <1> 2 * 39000, SDS-PAGE [28]) [7, 12, 13, 27, 28]
hexamer <18> (<18> 6 * 65000, SDS-PAGE [22]) [22]
monomer <2, 3, 9, 10, 14, 17> (<9> 1 * 75000, SDS-PAGE [7]; <17> 1 * 62000, monomeric in presence of catalytic amounts of substrate, SDS-PAGE [21]; <2> 1 * 61000, heart enzyme, SDS-PAGE [11]; <2> 1 * 62000, sperm enzyme, SDS-PAGE [11]; <14> 1 * 60000, SDS-PAGE [13]; <10> 1 * 60500, SDS-PAGE [20]; <3> 1 * 59000, SDS-PAGE [27]) [7, 11, 13, 20, 21, 27, 37]
octamer <6> (<6> 8 * 64000, SDS-PAGE, immunoblotting [19]; <6> 4 * 64000 + 4 * 57000, peroxisomal enzyme, heterooligomer, SDS-PAGE [14,15,19]; <6> 4 * 64000 + 4 * 52000, mitochondrial enzyme, heterooligomer, SDS-PAGE [14,15,19]) [14, 15, 19]

5 Isolation/Preparation/Mutation/Application

Source/tissue

brain <10, 15, 17> [3, 11, 20, 36]
breast muscle <9> [1, 4, 7-10, 14, 16, 18, 22, 23, 26, 36]
brown adipose tissue <17> [20, 22]
diaphragm <17> [36]
fibroblast <10, 14> [20, 32]
heart <2, 9, 10, 14, 15, 17, 19> [1-4, 7, 8, 11, 13, 20-22, 25-28, 32-34]
kidney <10, 17, 19> [8, 20]
liver <8-10, 14, 16, 17> [1, 4, 7, 8, 10-14, 17, 18, 20-22, 24-26, 28, 30-34]

muscle <3, 9, 16, 17> [1, 4, 7-12, 14, 16, 18, 22, 23, 25-27, 35, 36]
mycelium <1> [28]
myocyte <17> [33]
nervous system <17> [20]
retina <17> [25]
semen <10> [22]
skeletal muscle <3, 16, 17> [8, 11, 25-27]
small intestine <8> [8]
soleus <17> [36]
spermatozoon <2, 16, 17> [11]
testis <14, 17> [20, 32]

Localization

cytoplasm <6, 14> [6, 19, 24]
cytosol <1, 6> [28, 29]
endoplasmic reticulum <9, 10> (<9> smooth and rough [8]) [8, 35]
microsome <3, 6, 9, 14, 17> (<3,17> membrane bound [7,8,26]) [7, 8, 10, 11, 14, 17, 19, 25-27]
mitochondrial matrix <6, 17, 18, 20> [4, 6, 22, 30, 36]
mitochondrion <1, 3, 4, 6, 8, 10, 14, 17, 18, 20> (<3,9,10,17> membrane-bound [4,20,25-27,33]) [4, 7, 8, 10-15, 17-22, 25-36]
peroxisome <3, 4, 6, 10, 14, 17> [8, 10-15, 17, 19-21, 25-31, 33-36]
soluble <17> [8]

Purification

<1> (mitochondrial and cytosolic enzyme [28]) [28]
<2> [8, 11]
<3> [25-27]
<4> (partially [31]) [31]
<6> (mitochondrial and peroxisomal enzyme [10]) [9, 10, 14, 15, 19, 28, 30, 31]
<9> [1, 4, 7, 8, 18, 25, 27, 35]
<10> [3, 20, 25, 27, 31, 35]
<14> [13, 17, 25-27]
<15> [3]
<17> (purified from mitochondrial fraction and from homogenate of frozen liver [12]; partial [8]) [7, 8, 12, 20, 21, 25-27, 30, 36]
<18> [22, 25]
<19> (partially [1,2,7,8]) [1, 2, 7, 8]

Crystallization

<9> [1, 7, 8, 18]
<10> (crystals grown by hanging-drop vapor-diffusion belong to the orthorhombic space group P2(1)2(1)2(1) with unit-cell parameters a: 137.65 A, b: 84.76 A, c: 57.65 A [35]) [35]
<14> (sitting drop vapor diffusion method, space group C2, cell dimensions for the free enzyme crystal a: 158.9 A, b: 89.6 A, c: 119.4 A, β: 127.5° [37]) [37]

Cloning

<6> (both enzymes encoded by one gene, CT-CAT, cloned and expressed in Saccharomyces cerevisiae [29]; genomic DNA clone isolated from the yeast lambda EMBL library sequenced and expressed in Saccharomyces cerevisiae [30]) [29, 30]
<9> (cloning, sequencing and heterologous expression of cDNA [30]) [30]
<10> (single-gene product with multiple targeting signals [34]; cDNA cloned, mapping of the corresponding gene to chromosome 9q34.1 [30]; hCAT, cloned and expressed in Escherichia coli [35]) [30, 34, 35, 37]
<14> (sequence analysis of full-length cDNA clone [32]) [32, 37]
<18> (cDNA cloned and sequenced [30]) [30]

Application

analysis <6, 9> (<9> enzyme in pure form facilitates assay of carnitine and its acyl derivatives in tissues [1]; <9> in medicine and biochemical practice, the enzyme isolated from breast muscle is used analytically [9]; <6> specific activity is higher than in commercial preparations from other sources, enzyme prepared from yeast may be used for determination of carnitine in biological materials, this preparation is free of acetyl-CoA-deacetylase activity [9]) [1, 9]
medicine <10> (<10> pharmacological applications, patients suffering from CARAT deficiency, serious neurological, motor, repiratory and heart problems observed [32]; <10> CAT is clinically used in evaluating L-CA and its esters in body fluids and tissues [25]; <10> dysregulation of the enzyme can lead to serious diseases in humans, inherited deficiency in CRAT activity can lead to neurological and heart problems, patients suffering from Alzheimer 's disease also have reduced CRAT activity, promising target for therapeutic development against several human diseases like diabetes [37]) [20, 25, 32, 37]

6 Stability

pH-Stability

6-7.5 <17> (<17> microsomal transferase is more stable at pH 7.5 than at pH 6.0 [8]) [8]
6-9 <18> (<18> below pH 6.0 enzyme undergoes irreversible inactivation [22]) [22]

Temperature stability

4 <9> (<9> precipitated enzyme is stable at [1]) [1]
4-25 <17> (<17> stored in a refrigerator for 2 days or incubated at 25°C for 5 h, monomeric form remains unchanged [12]) [12]
37-50 <17> (<17> enzyme retains total activity when kept at 37°C for 30 min, rapid loss of activity above 37°C, loses half of the activity at 40°C for 30 min, but is completely inactivated when kept at 44°C for 30 min or 50°C for 5 min, 75% loss of activity at 37°C for 8 h, only 15% loss when kept at 25°C for 8 h [7]) [7]

General stability information

<6>, inactivated under the conditions of freeze/thaw [10]
<9>, enzyme is abolished by detergents, hypotonic media or freeze-thawing [8]
<14>, proteolytic degradation can occur during purification from liver [13]
<17>, enzyme is stable throughout purification procedure [12]
<17>, retains activity indefinitely when kept frozen, pure enzyme retains activity when kept frozen for long periods of time, inactivation of pure preparations occurs during dialysis [7]
<17>, very unstable [8]

Storage stability

<2>, 4°C, retains 60-80% activity over a period of several months [11]
<9>, 0-4°C, loses little activity when stored for months [7]
<9>, 4°C, enzyme is abolished by prolonged storage [8]
<10>, -20°C, when stored in 50% glycerol, maintains its activity unaltered for several months [20]
<15>, -20°C, 0.05% bovine albumin, repeated thawing and freezing does not significantly affect enzyme activity over a period of several months [3]
<17>, -20°C, 50 mM potassium phosphate buffer, pH 7.5, purified enzyme can be stored for several months without loss of activity [12]
<17>, -20°C, frozen mitochondrial fraction can be stored for several months without loss of activity [12]
<17>, -20°C, peroxisomal and microsomal fractions from isopycnic sucrose density gradients, activity is stable during storage for at least 3 months [8]
<17>, -20°C, peroxisomes and microsomes stored in 0.4 M KCl, 0.02% sodium azide, 150 mM Tris-HCl, pH, retains at least 95% of their activity when stored for 3 months [8]
<17>, 0-4°C, loses little activity when stored for months [7]
<17>, 4°C, peroxisomal and microsomal fractions from isopycnic sucrose density gradients, activity is stable during storage for 10 days [8]
<17>, 4°C, peroxisomes and microsomes stored in 0.4 M KCl, 0.02% sodium azide, 150 mM Tris-HCl, pH, retains at least 95% of their activity when stored for 18 days [8]

References

[1] Chase, J.F.A.; Pearson, D.J.; Tubbs, P.K.: The preparation of crystalline carnitine acetyltransferase. Biochim. Biophys. Acta, **96**, 162-165 (1965)
[2] Fritz, I.B.; Schultz, S.K.: Carnitine acetyltransferase. II. Inhibition by carnitine analogues and by sulfhydryl reagents. J. Biol. Chem., **240**, 2188-2192 (1965)
[3] White, H.L.; Wu, J.C.: Choline and carnitine acetyltransferases of heart. Biochemistry, **12**, 841-846 (1973)

[4] Edwards, Y.H.; Chase, J.F.A.; Edwards, M.R.; Tubbs, P.K.: Carnitine acetyltransferase: the question of multiple forms. Eur. J. Biochem., **46**, 209-215 (1974)

[5] Christiansen, R.Z.; Bremer, J.: Acetylation of tris(hydroxymethyl)aminomethane (Tris) and Tris derivatives by carnitine acetyltransferase. FEBS Lett., **86**, 99-102 (1978)

[6] Kawamoto, S.; Ueda, M.; Nozaki, C.; Yamamura, M.; Tanaka, A.; Fukui, S.: Localization of carnitine acetyltransferase in peroxisomes and in mitochondria of n-alkane-grown Candida tropicalis. FEBS Lett., **96**, 37-40 (1978)

[7] Mittal, B.; Kurup, C.K.R.: Purification of clofibrate-induced carnitine acetyltransferase from rat liver mitochondria. Biochim. Biophys. Acta, **619**, 90-97 (1980)

[8] Bieber, L.L.; Markwell, M.A.K.: Peroxisomal and microsomal carnitine acetyltransferases. Methods Enzymol., **71**, 351-358 (1981)

[9] Claus, R.; Kaeppeli, O.; Fiechter, A.: Purification of carnitine acetyltransferase from Candida tropicalis. Anal. Biochem., **127**, 376-379 (1982)

[10] Ueda, M.; Tanaka, A.; Fukui, S.: Peroxisomal and mitochondrial carnitine acetyltransferases in alkane-grown yeast Candida tropicalis. Eur. J. Biochem., **124**, 205-210 (1982)

[11] Huckle, W.R.; Tamblyn, T.M.: Purification and properties of carnitine acetyltransferases from bovine spermatozoa and heart. Arch. Biochem. Biophys., **226**, 94-110 (1983)

[12] Miyazawa, S.; Ozasa, H.; Furuta, S.; Osumi, T.; Hashimoto, T.: Purification and properties of carnitine acetyltransferase from rat liver. J. Biochem., **93**, 439-451 (1983)

[13] Farrell, S.O.; Fiol, C.J.; Reddy, J.K.; Bieber, L.L.: Properties of purified carnitine acyltransferases of mouse liver peroxisomes. J. Biol. Chem., **259**, 13089-13095 (1984)

[14] Ueda, M.; Tanaka, A.; Fukui, S.: Characterization of peroxisomal and mitochondrial carnitine acetyltransferases purified from alkane-grown Candida tropicalis. Eur. J. Biochem., **138**, 445-449 (1984)

[15] Ueda, M.; Tanaka, A.; Horikawa, S.; Numa, S.; Fukui, S.: Synthesis in vitro of precursor-type carnitine acetyltransferase with messenger RNA from Candida tropicalis. Eur. J. Biochem., **138**, 451-457 (1984)

[16] Zhong, J.; Fong, J.C.; Schulz, H.: Inhibition of carnitine acetyltransferase by metabolites of 4-pentenoic acid. Arch. Biochem. Biophys., **240**, 524-529 (1985)

[17] Bieber, L.L.; Fiol, C.: Purification and assay of carnitine acyltransferases. Methods Enzymol., **123**, 276-284 (1986)

[18] Gandour, R.D.; Colucci, W.J.; Stelly, T.C.; Brady, P.S.; Brady, L.J.: Active-site probes of carnitineacyltransferases. Inhibition of carnitine acetyltransferase by hemiacetyl carnitinium, a reaction intermediate analogue. Biochem. Biophys. Res. Commun., **138**, 735-741 (1986)

[19] Kozulic, B.; Käppeli, O.; Meussdoerffer, F.; Fiechter, A.: Characterization of a soluble carnitine acetyltransferase from Candida tropicalis. Eur. J. Biochem., **168**, 245-250 (1987)

[20] Bloisi, W.; Colombo, I.; Garavaglia, B.; Giardini, R.; Finocchiaro, G.; Didonato, S.: Purification and properties of carnitine acetyltransferase from human liver. Eur. J. Biochem., **189**, 539-546 (1990)
[21] Chung, C.; Chung, C.D.; Bieber, L.L.: Purification of heart and liver mitochondrial carnitine acetyltransferase. Protein Expr. Purif., **2**, 426-431 (1991)
[22] Kispal, G.; Cseko, J.; Alkonyi, I.; Sandor, A.: Isolation and characterization of carnitine acetyltransferase from S. cerevisiae. Biochim. Biophys. Acta, **1085**, 217-222 (1991)
[23] Saeed, A.; McMillin, J.B.; Wolkowicz, P.E.; Brouillette, W.J.: Carnitine acyltransferase enzymic catalysis requires a positive charge on the carnitine cofactor. Arch. Biochem. Biophys., **305**, 307-312 (1993)
[24] Lu, F.J.; Huang, T.S.; Chen, Y.S.: Effects of humic acid-metal complexes on hepatic carnitine palmitoyltransferase, carnitine acetyltransferase and catalase activities. Environ. Toxicol. Chem., **13**, 435-441 (1994)
[25] Alhomida, A.S.: Inhibition studies of the carnitine acetyltransferase from skeletal muscle of the camel (Camelus dromedarius) by sulfhydryl reagents and metal ions. Biochem. Mol. Biol. Int., **39**, 923-931 (1996)
[26] Alhomida, A.S.; Al-Jafari, A.A.; Duhaiman, A.S.; Rabbani, N.; Junaid, M.A.: Kinetic properties of purified carnitine acetyltransferase from the skeletal muscle of Arabian camel (Camelus dromedarius). Biochimie, **78**, 204-208 (1996)
[27] Alhomida, A.S.; Duhaiman, A.S.; Al-Jafari, A.A.; Junaid, M.A.: Purification of carnitine acetyltransferase from skeletal muscle of the camel (Camelus dromedarius). Mol. Cell. Biochem., **165**, 95-101 (1996)
[28] Jernejc, K.; Legisa, M.: Purification and properties of carnitine acetyltransferase from citric acid producing Aspergillus niger. Appl. Biochem. Biotechnol., **60**, 151-158 (1996)
[29] Kawachi, H.; Atomi, H.; Ueda, M.; Hashimoto, N.; Kobayashi, K.; Yoshida, T.; Kamasawa, N.; Osumi, M.; Tanaka, A.: Individual expression of Candida tropicalis peroxisomal and mitochondrial carnitine acetyltransferase-encoding genes and subcellular localization of the products in Saccharomyces cerevisiae. J. Biochem., **120**, 731-735 (1996)
[30] Kawachi, H.; Atomi, H.; Ueda, M.; Tanaka, A.: Peroxisomal and mitochondrial carnitine acetyltransferases of the n-alkane-assimilating yeast Candida tropicalis. Analysis of gene structure and translation products. Eur. J. Biochem., **238**, 845-852 (1996)
[31] Sheridan, R.; Ratledge, C.: Changes in cell morphology and carnitine acetyltransferase activity in Candida albicans following growth on lipids and serum and after in vivo incubation in mice. Microbiology, **142**, 3171-3180 (1996)
[32] Brunner, S.; Kramar, K.; Denhardt, D.T.; Hofbauer, R.: Cloning and characterization of murine carnitine acetyltransferase: evidence for a requirement during cell cycle progression. Biochem. J., **322**, 403-410 (1997)
[33] Abbas, A.S.; Wu, G.; Schulz, H.: Carnitine acetyltransferase is not a cytosolic enzyme in rat heart and therefore cannot function in the energy-linked regulation of cardiac fatty acid oxidation. J. Mol. Cell. Cardiol., **30**, 1305-1309 (1998)

[34] Ramsay, R.R.: The carnitine acyltransferases: modulators of acyl-CoA-dependent reactions. Biochem. Soc. Trans., **28**, 182-186 (2000)
[35] Lian, W.; Govindasamy, L.; Gu, Y.; Kukar, T.; Agbandje-McKenna, M.; McKenna, R.; Wu, D.: Crystallization and preliminary X-ray crystallographic studies on recombinant human carnitine acetyltransferase. Acta Crystallogr. Sect. D, **58**, 1193-1194 (2002)
[36] Liu, J.; Killilea, D.W.; Ames, B.N.: Age-associated mitochondrial oxidative decay: improvement of carnitine acetyltransferase substrate-binding affinity and activity in brain by feeding old rats acetyl-L-carnitine and/or R-α -lipoic acid. Proc. Natl. Acad. Sci. USA, **99**, 1876-1881 (2002)
[37] Jogl, G.; Tong, L.: Crystal structure of carnitine acetyltransferase and implications for the catalytic mechanism and fatty acid transport. Cell, **112**, 113-122 (2003)

Phosphate acetyltransferase 2.3.1.8

1 Nomenclature

EC number
2.3.1.8

Systematic name
acetyl-CoA:phosphate acetyltransferase

Recommended name
phosphate acetyltransferase

Synonyms
PTA
acetyltransferase, phosphate
phosphoacylase
phosphotransacetylase

CAS registry number
9029-91-8

2 Source Organism

<1> *Rhodopseudomonas palustris* [13]
<2> *Methanosarcina thermophila* (strain TM-1 [29]) [14, 27, 29, 30]
<3> *Clostridium kluyveri* [1-4, 6-9, 18, 22, 23]
<4> *Chlorogonium elongatum* (green alga [5]) [5]
<5> *Veillonella alcalescens* [10, 11]
<6> *Clostridium thermoaceticum* [12]
<7> *Bacillus subtilis* (wild-type and mutant deficient in enzyme activity [31]) [15, 31]
<8> *Bradyrhizobium japonicum* [16]
<9> *Azotobacter vinelandii* [24]
<10> *Clostridium acidiurici* [17]
<11> *Escherichia coli* [19, 20]
<12> *Lactobacillus fermentum* (strain 36, IFO No. 3071 [21]) [21]
<13> *Thermotoga maritima* (hyperthermophilic eubacterium [25]) [25]
<14> *Clostridium acetobutylicum* [26]
<15> *Lactobacillus sanfranciscensis* [28]
<16> *Sinurhizobium meliloti* (wild-type and mutant strains [32]) [32]

3 Reaction and Specificity

Catalyzed reaction

acetyl-CoA + phosphate = CoA + acetyl phosphate (<4,5> rapid equilibrium random bi-bi reaction mechanism [5,10]; <3> evidence against an acyl-enzyme intermediate [6]; <3> reaction mechanism [6,9,22,23])

Reaction type

acyl group transfer

Natural substrates and products

S acetyl-CoA + phosphate <1-16> (Reversibility: r <2-5, 7, 12, 13, 15> [2, 3, 5, 6, 7, 8, 9, 10, 14, 15, 21, 22, 23, 25, 27, 28, 30, 31]; ? <1, 6, 8, 9, 10, 11, 14, 16> [1, 4, 11, 12, 13, 16, 17, 18, 19, 20, 24, 26, 29, 32]) [1-28, 30, 31, 32]
P CoA + acetyl phosphate <1-15> [1-24, 25, 26, 27, 28, 30, 31]
S Additional information <4> (<4> physiological function in anaerobic metabolism of eukaryotic green algae rather than in aerobic acetate activation [5]) [5]
P ?

Substrates and products

S 3'-dephospho-CoA + acetyl phosphate <12> (<12> rate is about one tenth of the activity with CoA [21]) (Reversibility: ? <12> [21]) [21]
P acetyl CoA
S acetyl-CoA + phosphate <1-12, 13, 14, 15, 16> (<2> rate of acetyl-CoA synthesis is 10times greater than rate of acetyl phosphate synthesis [14]; <7> formation of acetyl-CoA is favored [15]; <5> rate of acetyl-CoA synthesis is 6.5times greater than rate of acetyl phosphate synthesis [10]; <7> formation of acetyl-CoA is favored [15]; <3> in reverse reaction specific for CoA [1,2]) (Reversibility: r <2-5, 7, 12, 13, 15> [2, 3, 5, 6, 7, 8, 9, 10, 14, 15, 21, 22, 23, 25, 27, 28, 30, 31]; ? <1, 6, 8, 9, 10, 11, 14, 16> [1, 4, 11, 12, 13, 16, 17, 18, 19, 20, 24, 26, 29, 32]) [1-24, 25, 26, 27, 28, 30, 31, 32]
P CoA + acetyl phosphate <1-15>
S arsenate + acetyl-CoA <3> (Reversibility: ? <3> [1, 3, 8, 23]) [1, 3, 8, 23]
P acetyl arsenate + CoA <3> [8, 23]
S butyryl-CoA + phosphate <3, 7, 13> (<3> at a rate 0.01 as rapid as acetyl-CoA [1]) (Reversibility: ? <3, 7, 13> [1, 15, 25]) [1, 15, 25]
P CoA + butyryl phosphate
S propionyl-CoA + phosphate <3, 7, 13> (<3> at a rate 0.1 to 0.5 as rapid as acetyl-CoA [1]) (Reversibility: ? <3, 7, 13> [1, 15, 25]) [1, 15, 25]
P CoA + propionyl phosphate
S Additional information <3> (<3> arsenolysis [1,3]) [1, 3]
P ?

Inhibitors

$(NH_4)_2SO_4$ <1> (<1> activation at low concentration, inhibition at high concentration [13]) [13]
$(NH_4)_2SO_4$ <2> (<2> above 10 mM [14]) [14]

2,2'-dipyridyl <3> [2]

2,3-butanedione <2> (<2> almost complete loss of wild-type enzyme activity after 10 min at 10 mM [27]) [27]

5,5'-dithiobis(2-nitrobenzoic acid) <3> [6]

ADP <3, 5, 7> (<4> no effect [5]; <5> 50% inhibition at 6 mM, $MgCl_2$ reverses inhibition [10]; <7> inhibitory at 1 mM [15]) [1, 10, 15]

AMP <7> (<7> inhibitory at 1 mM [15]) [15]

ATP <3, 5, 7> (<4> no effect [5]; <5> 50% inhibition at 1.5 mM, $MgCl_2$ reverses inhibition [10]; <7> inhibitory at 1 mM [15]) [1, 10, 11, 15]

Ba^{2+} <7, 12> (<7> 15% inhibition at 1 mM, 76% inhibition at 10 mM [15]) [15, 21]

Ca^{2+} <7, 12> (<7> 50% inhibition at 1 mM [15]) [15, 21]

Cs^+ <12> [21]

$CuSO_4$ <3> [2]

K^+ <3> [2]

KCN <3> (<3> weak [2]) [2]

Li^+ <3, 12> (<3> 23% inhibition [3]) [1, 3, 21]

LiCl <1> (<1> slight [13]) [13]

Mg^{2+} <3, 12> [2, 21]

$MgCl_2$ <1> (<1> activation at low concentration, inhibition at high concentration [13]) [13]

Mn^{2+} <6, 7> (<7> 50% inhibition at 1 mM [15]) [12, 15]

$MnCl_2$ <3> (<3> weak [2]) [2]

N-ethylmaleimide <2, 3, 12> (<12> above 0.1 mM [21]) [6, 21, 30]

Na^+ <2, 3, 8> (<3> 85% inhibition [3]; <3> acts as inhibitor in the presence of NH_4^+ or K^+, competitive inhibition [7]) [1-3, 7, 14, 16]

NaCl <1> (<1> slight [13]) [13]

S-dimethylarsino-CoA <3, 4> (<3> irreversible inhibition, phosphate protects [4]) [4, 5]

SO_4^{2-} <1, 2> [13, 14]

Tris(hydroxymethyl)aminomethane <3> (<3> weak [2]) [2]

Tris-citrate <10> [17]

acetyl phosphate <3> [23]

acetyl-CoA <3, 4, 5> (<4> competitive [5]; <3> competitive [6]; <5> competitive [10]) [5, 6, 10]

arsenate <2, 3, 8, 10> (<10> 50% inhibition at 10 mM [17]) [6, 14, 16, 17]

citrate buffer <3> (<3> 0.1 M, pH 8.0 [1,3]) [1, 3]

coenzyme A <1, 3, 5> (<1> strong substrate inhibition [13]; <3> competitive [6]; <5> competitive [10]; <3> competitive with respect to acetyl-CoA, non competitive with respect to phosphate [23]) [6, 8, 13, 10, 23]

desulfo-CoA <2, 3, 11> (<3> competitive [4]; <3> competitive [6]; <2> strong competitive inhibitor [27]) [4, 6, 19, 20, 27]

diethylbarbiturate <3, 10> (<3> potassium diethylbarbiturate buffer, 0.1 M, pH 8.0 [1,3]) [1, 3, 17]

diphosphate <3, 10> (<3> non competitive [6]; <10> 50% inhibition at 12 mM [17]) [6, 17]

iodoacetamide <4> (<4> 70% inactivation after 1 min at 5 mM [5]) [5]

iodoacetic acid <4> (<4> 76% inactivation after 4 min at 5 mM [5]) [5]
p-chloromercuribenzoate <3, 12> (<12> above 0.1 mM [21]) [2, 9, 21]
palmitoyl-CoA <7> (<7> competitive [15]) [15]
phenylglyoxal <2> (<2> 33% inhibition after preincubation with phenylglyoxal [30]) [30]
phosphate <2-4, 10> (<3> inhibition of arsenolysis [3]; <4> competitive [5]; <3> competitive with respect to acyl phosphate, non competitive with respect to CoA [6]; <2> end product inhibitor [14]) [3, 5, 6, 14, 17, 23, 27]
potassium diphosphate <3> (<3> 0.1 M, pH 8.0 [1,3]) [1, 3]
potassium phosphate <2> (<2> above 10 mM [14]) [14]
Additional information <3, 10> (<3> overview: product inhibition [23]; <10> inhibition by various buffer systems [17]) [23, 17]

Activating compounds

1,3-dimercaptopropanol <3> (<3> activates at 0.01 M [3]) [3]
cysteine <3> (<3> activates at 0.01 M [3]) [3]
hydrogen sulfide <3> (<3> activates at 0.01 M [3]) [3]
pyruvate <11> [20]
thiolglycolic acid <3> (<3> activates at 0.01 M [3]) [3]

Metals, ions

Fe^{2+} <10> (<10> required, Fe^{2+} or Mn^{2+}, Mn^{2+} is 50-75% as effective as Fe^{2+} [17]) [17]
K^+ <1-3, 4, 7, 8, 12> (<1> KCl activates [13]; <2> K^+ or NH_4^+ at concentration above 10 mM required for maximum activity [14]; <6> no effect [12]; <3> enzyme is inactive in absence of K^+ or NH_4^+, maximal activation at about 0.02 M [1]; <1,3,8,12> activates [3,13,16,21]; <3> rate of arsenolysis increases with increasing potassium or ammonium salt concentration and reaches a maximum rate at 0.02 to 0.03 M salt concentration [3]; <3> enzyme is activated by low concentrations of NH_4^+, K^+ and Na^+, sequence of effectiveness: NH_4^+, K^+, Na^+ [7]; <7> 3.4fold stimulation at 10 mM [15]) [1, 3, 5, 7, 9, 13-16, 21]
Mn^{2+} <10> (<10> required, Fe^{2+} or Mn^{2+}, Mn^{2+} is 50-75% as effective as Fe^{2+} [17]) [17]
NH_4^+ <1-3, 4, 7, 8, 12> (<1,12> activates [13,21]; <1> NH_4Cl activates [13]; <3> K^+ or NH_4^+ at concentration above 10 mM required for maximum activity, enzyme is inactive in absence of K^+ or NH_4^+, maximal activation at about 0.02 M [1]; <3> rate of arsenolysis increases with increasing potassium or ammonium salt concentration and reaches a maximum rate at 0.02 to 0.03 M salt concentration [3]; <3> enzyme is activated by low concentrations of NH_4^+, K^+ and Na^+, sequence of effectiveness: NH_4^+, K^+, Na^+ [7]; <3> required for maximum activity, optimum at 7 mM $(NH_4)_2SO_4$ [2]; <6> no effect [12]; <2> above 10 mM [14]; <7> 5.2fold stimulation at 10 mM [15]) [1-3, 5, 7, 13-16, 21]
Na^+ <3> (<3> less effective than NH_4^+ and K^+ [7]) [7]
Rb^+ <12> (<12> stimulates [21]) [21]
Tris <4> (<4> maximum activity with Tris buffer, with HEPES and MES catalysis is lowered by ca. 15% [5]) [5]

phosphate <3> (<3> 1 mM, weak activation [2]) [2]
Additional information <3, 10> (<3> mechanism of activation by univalent cations [7]; <10> no activity with: Fe^{3+}, Zn^{2+}, Mg^{2+}, Co^{2+}, Ni^{2+}, Sn^{2+}, Cu^{2+}, Mo^{6+}, K^+, NH_4^+ [17]) [7, 17]

Turnover number (min^{-1})

0.28 <2> (desulfo-CoA, <2> R133E mutant [27]) [27]
0.75 <2> (desulfo-CoA, <2> R133A mutant [27]) [27]
8 <2> (desulfo-CoA, <2> R133K mutant [27]) [27]
17 <2> (CoA, <2> R87E mutant [27]) [27]
23 <2> (desulfo-CoA, <2> R87E mutant [27]) [27]
28 <2> (desulfo-CoA, <2> R133Q mutant [27]) [27]
41 <2> (desulfo-CoA, <2> R87K mutant [27]) [27]
44 <2> (CoA, <2> R133E mutant [27]) [27]
79 <2> (acetyl phosphate, <2> C159A mutant [30]) [30]
85 <2> (CoA, <2> R310Q mutant [30]) [30]
89 <2> (CoA, <2> C159A mutant [30]) [30]
98 <2> (acetyl phosphate, <2> R310Q mutant [30]) [30]
98 <2> (desulfo-CoA, <2> R87Q mutant [27]) [27]
182 <2> (acetyl phosphate, <2> C159A/C277A/C312A/C325A mutant [30]) [30]
184 <2> (desulfo-CoA, <2> R87A mutant [27]) [27]
198 <2> (CoA, <2> C159A/C277A/C312A/C325A mutant [30]) [30]
272 <2> (desulfo-CoA, <2> wild-type [27]) [27]
296 <2> (acetyl phosphate, <2> R87Q mutant [30]) [30]
327 <2> (acetyl phosphate, <2> R287Q mutant [30]) [30]
357 <2> (CoA, <2> R287Q mutant [30]) [30]
373 <2> (CoA, <2> R87K mutant [27]) [27]
418 <2> (CoA, <2> R133K mutant [27]) [27]
631 <2> (CoA, <2> R133A mutant [27]) [27]
749 <2> (acetyl phosphate, <2> R133Q mutant [30]) [30]
1146 <2> (CoA, <2> R133Q mutant [27]) [27]
1225 <2> (CoA, <2> R87A mutant [27]) [27]
1390 <2> (CoA, <2> R87Q mutant [27]) [27]
1448 <2> (acetyl phosphate, <2> R28Q mutant [30]) [30]
1474 <2> (CoA, <2> R87Q mutant [30]) [30]
1559 <2> (CoA, <2> R28Q mutant [30]) [30]
2434 <2> (acetyl phosphate, <2> wild tpye [30]) [30]
2669 <2> (acetyl phosphate, <2> C312A mutant [30]) [30]
2736 <2> (CoA, <2> R133Q mutant [30]) [30]
3095 <2> (acetyl phosphate, <2> C277A/C312A/C325A mutant [30]) [30]
3217 <2> (CoA, <2> C277A/C312A/C325A mutant [30]) [30]
3226 <2> (CoA, <2> wild tpye [30]) [30]
3324 <2> (CoA, <2> C312A mutant [30]) [30]
3453 <2> (acetyl phosphate, <2> C325A mutant [30]) [30]
3595 <2> (CoA, <2> C325A mutant [30]) [30]
3980 <2> (acetyl phosphate, <2> C159S mutant [30]) [30]

4110 <2> (acetyl phosphate, <2> wild-type after expression in E. coli [30]) [30]
4412 <2> (CoA, <2> C159S mutant [30]) [30]
4880 <2> (CoA, <2> wild-type [27]) [27]
5252 <2> (CoA, <2> wild-type after expression in E. coli [30]) [30]
5694 <2> (acetyl phosphate, <2> C277A mutant [30]) [30]
6004 <2> (CoA, <2> C277A mutant [30]) [30]

Specific activity (U/mg)

12.8 <6> [12]
31.6 <1> [13]
225 <13> [25]
396 <5> [11]
950 <3> [3]
1124 <12> [21]
1150 <7> [31]
1371 <7> [15]
1940 <15> [28]
2469 <2> [14]
7100 <10> [17]
7140 <3> [18]
9006 <2> [29]
9100 <3> (<3> after crystallization [2]) [2]
Additional information <3-5, 14> (<3,4> assay methods [1,5]; <14> specific activities of recombinant plasmids expressed in Escherichia coli and Clostridium acetobutylicum [26]) [1, 5, 10, 26]

K_m-Value (mM)

0.0086 <5> (acetyl-CoA) [10]
0.0231 <13> (acetyl-CoA) [25]
0.024 <13> (acetyl phosphate) [25]
0.03 <13> (CoA) [25]
0.034 <2> (CoA, <2> R287Q mutant [30]) [30]
0.037 <2> (CoA, <2> R310Q mutant [30]) [30]
0.043 <2> (acetyl phosphate, <2> R87Q mutant [30]) [30]
0.058 <7> (butyryl-CoA) [15]
0.06 <7> (acetyl-CoA) [15]
0.07 <2> (CoA, <2> wild-type [27]) [27]
0.071 <2> (CoA, <2> C159S mutant [30]) [30]
0.073 <2> (CoA, <2> C159A/C277A/C312A/C325A mutant [30]) [30]
0.073 <2> (acetyl phosphate, <2> R133Q mutant [30]) [30]
0.078 <2> (CoA, <2> C325A mutant [30]) [30]
0.078 <4> (acetyl-CoA) [5]
0.086 <2> (CoA, <2> C277A/C312A/C325A mutant [30]) [30]
0.087 <12> (CoA) [21]
0.089 <2> (CoA, <2> wild-type after expression in E. coli [30]) [30]
0.09 <2> (CoA, <2> wild tpye [30]) [30]
0.09 <4> (CoA, <4> arsenolysis [5]) [5]

0.092 <2> (CoA, <2> C312A mutant [30]) [30]
0.093 <4> (CoA) [5]
0.093 <2> (CoA, <2> C159A mutant [30]) [30]
0.093 <2> (CoA, <2> C277A mutant [30]) [30]
0.1 <15> (CoA) [28]
0.11 <2> (CoA, <2> R133K mutant [27]) [27]
0.111 <13> (phosphate) [25]
0.15 <1> (CoA) [13]
0.15 <7> (propionyl-CoA) [15]
0.166 <2> (acetyl phosphate, <2> R287Q mutant [30]) [30]
0.17 <2> (acetyl phosphate, <2> wild tpye [30]) [30]
0.18 <2> (acetyl phosphate, <2> wild-type after expression in E. coli [30]) [30]
0.187 <2> (acetyl phosphate, <2> C277A mutant [30]) [30]
0.191 <2> (acetyl phosphate, <2> C159S mutant [30]) [30]
0.198 <2> (acetyl phosphate, <2> C312A mutant [30]) [30]
0.206 <2> (acetyl phosphate, <2> C159A mutant [30]) [30]
0.222 <2> (acetyl phosphate, <2> C159A/C277A/C312A/C325A mutant [30]) [30]
0.23 <2> (acetyl phosphate, <2> C325A mutant [30]) [30]
0.253 <2> (CoA, <2> R28Q mutant [30]) [30]
0.254 <2> (acetyl phosphate, <2> C277A/C312A/C325A mutant [30]) [30]
0.27 <2> (CoA, <2> R87E mutant [27]) [27]
0.311 <4> (acetyl phosphate) [5]
0.33 <5> (CoA) [10]
0.36 <2> (CoA, <2> R133A mutant [27]) [27]
0.56 <3> (CoA) [2]
0.59 <5> (acetyl phosphate) [10]
0.6 <15> (acetyl-CoA) [28]
0.66 <3, 4> (acetyl phosphate) [2, 5]
0.7 <2> (CoA, <2> R133E mutant [27]) [27]
0.7 <2> (CoA, <2> R133Q mutant [27]) [27]
0.747 <2> (CoA, <2> R87Q mutant [30]) [30]
0.75 <2> (CoA, <2> R87K mutant [27]) [27]
0.775 <2> (acetyl phosphate, <2> R310Q mutant [30]) [30]
0.8 <2> (desulfo-CoA, <2> R133K mutant [27]) [27]
0.8 <2> (desulfo-CoA, <2> R87E mutant [27]) [27]
0.9 <2> (CoA, <2> R87A mutant [27]) [27]
1 <2> (desulfo-CoA, <2> R133E mutant [27]) [27]
1.1 <2> (CoA, <2> R87Q mutant [27]) [27]
1.3 <15> (acetyl phosphate) [28]
1.3 <2> (desulfo-CoA, <2> R133A mutant [27]) [27]
1.3 <2> (desulfo-CoA, <2> R87Q mutant [27]) [27]
1.32 <2> (acetyl phosphate, <2> R28Q mutant [30]) [30]
1.4 <2> (desulfo-CoA, <2> wild-type [27]) [27]
1.694 <2> (CoA, <2> R133Q mutant [30]) [30]

3.9 <2> (desulfo-CoA, <2> R87K mutant [27]) [27]
4 <2> (desulfo-CoA, <2> R133Q mutant [27]) [27]
4.7 <1> (acetyl phosphate) [13]
5.44 <4> (phosphate) [5]
6 <2> (desulfo-CoA, <2> R87A mutant [27]) [27]
6.7 <15> (phosphate) [28]
9.3 <5> (phosphate) [10]
Additional information <3> (<3> kinetic studies [9,22,23]; <3> effects of monovalent kations [7]) [7, 9, 14, 22, 23]

K_i-Value (mM)
0.001 <2> (desulfo-CoA) [27]
0.004 <11> (desulfo-CoA) [20]
0.02 <3> (desulfo-CoA) [4]
0.32 <2> (phosphate, <2> R133Q mutant [27]) [27]
0.52 <2> (phosphate, <2> R87Q mutant [27]) [27]
0.81 <2> (phosphate, <2> wild-type enzyme [27]) [27]

pH-Optimum
7 <2> [14]
7.4 <3> [2]
7.4-8.2 <3> [3]
7.4-8.4 <3> [1]
7.6 <4, 6, 7, 12> (<4> both directions [5]; <12> Tris buffer [21]) [5, 12, 15, 21]
7.8 <1> [13]
8-8.1 <10> [17]
8.1-9.1 <15> [28]
Additional information <4> (<4> maximum activity with Tris buffer, with HEPES and MES catalysis is lowered by ca. 15% [5]) [5]

pH-Range
6.6-8.2 <3> (<3> about 50% of activity maximum at pH 6.6 pH and 8.2 [2]) [2]
6.6-9 <3> [1]
6.8-8.6 <3> (<3> pH 6.8: about 25% of activity maximum, pH 8.6: about 90% of activity maximum [3]) [3]
7-8.7 <10> (<10> pH 7.0: about 10% of activity maximum, pH 8.7: about 20% of activity maximum [17]) [17]
Additional information <3> (<3> pH effects on enzymatic reaction [8,22]) [8, 22]

Temperature optimum (°C)
28 <4> (<4> both directions [5]) [5]
35-45 <2> [14]
38 <1> [13]
49-58 <15> [28]
75 <6> [12]
90 <13> [25]

Temperature range (°C)
22-40 <3> (<3> only small effect of temperature in this range [2]) [2]

4 Enzyme Structure

Molecular weight
35200 <2> (<2> calculated from amino acid sequence [29]) [29]
35500 <15> (<15> calculated from amino acid sequence [28]) [28]
36200 <14> (<14> calculated from amino acid sequence [26]) [26]
38000-41000 <3> (<3> ultracentrifugation [2]) [2]
40000 <15> (<15> SDS-PAGE [28]) [28]
52000 <2> (<2> gel filtration [14]) [14]
54500 <1> (<1> gel filtration [13]) [13]
63000-75000 <10> (<10> gel filtration [17]) [17]
68000 <12> (<12> gel filtration [21]) [21]
75000-80000 <5> (<5> gel filtration [11]) [11]
76000 <7> (<7> gel filtration [31]) [31]
88000 <6> (<6> gel filtration [12]) [12]
90000 <7> (<7> gel filtration [15]) [15]
170000 <13> (<13> gel filtration [25]) [25]
280000 <4> (<4> gel filtration, SDS-PAGE [5]) [5]

Subunits
? <3> (<3> x * 70000, SDS-PAGE [6]) [6]
dimer <5, 7> (<5> 2 * 35000-40000, SDS-PAGE [11]; <7> 2 * 36000, SDS-PAGE [31]) [11, 31]
hexamer <4> (<4> 6 * 44000, SDS-PAGE [5]) [5]
monomer <1, 2> (<1> 1 * 52500, SDS-PAGE [13]; <2> 1 * 43000, SDS-PAGE [14]) [13, 14]
tetramer <6, 13> (<6> 4 * 20000, SDS-PAGE [12]; <13> α_4, homotetramer, 4 * 34000, SDS-PAGE [25]) [12, 25]

5 Isolation/Preparation/Mutation/Application

Source/tissue
commercial preparation <3> [9]

Localization
soluble <2> [14]

Purification
<1> [13]
<2> (homogeneity [29]) [14, 29, 30]
<3> (affinity chromatography, 73% pure [18]) [1-3, 18]
<4> [5]
<5> (partial [10]) [10, 11]

<6> [12]
<7> (600fold [15]; fusion protein with β-galactosidase [31]) [15, 31]
<9> (enzyme is associated with the pyruvate dehydrogenase complex [24]) [24]
<10> [17]
<12> [21]
<13> (apparent homogeneity [25]) [25]
<15> (homogeneity, biotinylated fusion protein [28]) [28]

Crystallization
<3> [2]

Cloning
<2> [29, 30]
<7> [31]
<14> [26]
<15> [28]
<16> [32]

Engineering
C159A <2> (<2> K_m similar to wild-type enzyme [30]) [30]
C159A/C277A/C312A/C325A <2> (<2> K_m similar to wild-type enzyme [30]) [30]
C159S <2> (<2> K_m similar to wild-type enzyme [30]) [30]
C277A <2> (<2> K_m similar to wild-type enzyme [30]) [30]
C277A/C312A/C325A <2> (<2> K_m similar to wild-type enzyme [30]) [30]
C312A <2> (<2> K_m similar to wild-type enzyme [30]) [30]
C325A <2> (<2> K_m similar to wild-type enzyme [30]) [30]
R133A <2> (<2> altered kinetic properties, increased K_m for CoA [27]) [27]
R133E <2> (<2> altered kinetic properties, increased K_m for CoA [27]) [27]
R133K <2> (<2> altered kinetic properties, increased K_m for CoA [27]) [27]
R133Q <2> (<2> altered kinetic properties, increased K_m for CoA [27]) [27]
R133Q <2> (<2> increased K_m for CoA, decreased K_m for acetyl phosphate [30]) [30]
R287Q <2> (<2> decreased K_m for CoA [30]) [30]
R28Q <2> (<2> increased K_m [30]) [30]
R310Q <2> (<2> decreased K_m for CoA [30]) [30]
R87A <2> (<2> altered kinetic properties, increased K_m for CoA [27]) [27]
R87E <2> (<2> altered kinetic properties, increased K_m for CoA [27]) [27]
R87K <2> (<2> altered kinetic properties, increased K_m for CoA [27]) [27]
R87Q <2> (<2> altered kinetic properties, increased K_m for CoA [27]) [27]
R87Q <2> (<2> increased K_m for CoA, decreased K_m for acetyl phosphate [30]) [30]

6 Stability

pH-Stability

6 <3> (<3> 1.5 mg/ml protein concentration, 0.15 M ammonium sulfate, 40°C, 5 min, stable [2]) [2]

7-10 <15> (<15> stable [28]) [28]

Temperature stability

37 <5> (<5> half-life: 20 min [11]) [11]

40 <3> (<3> 5 min, 1.5 mg/ml protein concentration, 0.15 M ammonium sulfate, pH 6, stable [2]) [2]

45 <3, 12> (<3> 5 min, 1.5 mg/ml protein concentration, 0.15 M ammonium sulfate, pH 6, 15% loss of activity [2]; <12> unstable, can be stabilized by ammonium sulfate [21]) [2, 21]

50 <3, 4> (<3> 5 min, 1.5 mg/ml protein concentration, 0.15 M ammonium sulfate, pH 6, 48% loss of activity [2]; <4> $t_{1/2}$ 18 min [5]) [2, 5]

60 <1, 3, 15> (<3> 5 min, 1.5 mg/ml protein concentration, 0.15 M ammonium sulfate, pH 6, 98% loss of activity [2]; <1> pH 8.0, 50 mM Tris/HCl, complete loss of activity after 1 min [13]; <15> inactivation after 15 min [28]) [2, 13, 28]

70 <2> (<2> 5 min, stable up to 70°C in absence of additional salts [14]) [14]

80 <2, 13> (<2> 5 min, complete inactivation in absence of additional salts [14]; <13> stable for 120 min [25]) [14, 25]

88 <10> (<10> 60 min, 44% loss of activity [17]) [17]

90 <13> (<13> 60% loss of activity after 120 min [25]) [25]

100 <6, 13> (<13> 60% loss of activity after 120 min [25]; <6> 2 min, complete inactivation after boiling [12]) [12, 25]

Additional information <2, 12> (<2> sulfate and phosphate partially protect against heat inactivation [14]) [14, 21]

Oxidation stability

<2>, stable to air [14]

General stability information

<1>, $(NH_4)_2SO_4$, Na_2SO_4, NaCl or KCl stabilizes [13]

<1>, $MgCl_2$ destabilizes [13]

<1>, NH_4Cl sligthly stabilizes [13]

<1>, ammonium sulfate and potassium phosphate stabilize [13]

<1>, ethylene glycol 20% v/v stabilizes [13]

<1>, labile in dilute solutions or at elevated temperatures [13]

<1>, repeated freezing and thawing inactivates [13]

<2>, sulfate and phosphate partially protect against heat inactivation [14]

<3>, activity is lost upon dialysis and cannot be restored by addition of known cofactors or crude boiled extracts [1]

<3>, dilute solutions are instable, in frozen state stable [2]

<4>, stabilized by addition of 200 mM ammonium sulfate and 2-5 mM mercaptans to the extraction buffer [5]

<5>, divalent cations, e.g. $FeSO_4$, $Fe(NH_4)_2SO_4$, $MgCl_2$, $MnCl_2$, $MgSO_4$ or ATP increase lability [11]
<5>, increased stability is obtained by adding a reducing agent and a component of the reaction [11]
<10>, 50-60% loss of activity after dialysis for 8 h against 0.05 M Tris buffer, Fe^{2+} and dithiothreitol stabilize [17]
<12>, lyophilization causes almost complete loss of enzyme activity [21]

Storage stability
<1>, -24°C, $(NH_4)_2SO_4$, 50-70% loss of activity after some months [13]
<2>, stable for at least 3 h when stored on ice [14]
<3>, -10°C, stable for several months [1]
<3>, -20°C, 0.2 M phosphate or Tris-HCl, pH 8.0, several months [3]
<3>, 0-4°C, 2.7-3.0 M ammonium sulfate, months [2]
<5>, 4°C or -20°C, protein concentration 0.5 mg/ml, several weeks [11]
<13>, -20°C, no loss of activity for several weeks [25]

References

[1] Stadtman, E.R.: Phosphotransacetylase from Clostridium kluyveri. Methods Enzymol., **1**, 596-599 (1955)
[2] Bergmeyer, H.U.; Holz, G.; Klotsch, H.; Lang, G.: Phosphotransacetylase aus Clostridium kluyveri. Biochem. Z., **338**, 114-121 (1963)
[3] Stadtman, E.R.: The purification and properties of phosphotransacetylase. J. Biol. Chem., **196**, 527-534 (1952)
[4] Duhr, E.F.; Owens, M.S.; Barden, R. E.: Irreversible inhibition of phosphotransacetylase by S-dimethylarsino-CoA. Biochim. Biophys. Acta, **749**, 84-90 (1983)
[5] Kreuzberg, K.; Umlauf, H.; Blaschkowski, H.P.: Purification and properties of phosphotransacetylase from the eucaryotic green alga Chlorogonium elongatum. Biochim. Biophys. Acta, **842**, 22-29 (1985)
[6] Henkin, J.; Abeles, R.H.: Evidence against an acyl-enzyme intermediate in the reaction catalyzed by clostridial phosphotransacetylase. Biochemistry, **15**, 3472-3479 (1976)
[7] Kyrtopoulos, S.A.; Satchell, D.P.N.: Kinetic studies with phosphotransacetylase. V. The mechanism of activation by univalent cations. Biochim. Biophys. Acta, **321**, 126-142 (1973)
[8] Kyrtopoulos, S.A.; Satchell, D.P.N.: Kinetic studies with phosphotransacetylase. II. The acetylation of arsenate by acetyl coenzyme A. Biochim. Biophys. Acta, **268**, 334-343 (1972)
[9] Hibbert, F.; Kyrtopoulos, S.A.; Satchell, D.P.N.: Kinetic studies with phosphotransacetylase. Biochim. Biophys. Acta, **242**, 39-54 (1971)
[10] Pelroy, R.A.; Whiteley, H.R.: Kinetic properties of phosphotransacetylase from Veillonella alcalescens. J. Bacteriol., **111**, 47-55 (1972)
[11] Whiteley, H.R.; Pelroy, R.A.: Purification and properties of phosphotransacetylase from Veillonella alcalescens. J. Biol. Chem., **247**, 1911-1917 (1972)

[12] Drake, H.L.; Hu, S.I.; Wood, H.G.: Purification of five components from Clostridium thermoaceticum which catalyze synthesis of acetate from pyruvate and methyltetrahydrofolate. Properties of phosphotransacetylase. J. Biol. Chem., **256**, 11137-11144 (1981)
[13] Vigenschow, H.; Schwarm, H.M.; Knobloch, K.: Purification and characterization of a phosphotransacetylase from Rhodopseudomonas palustris. Biol. Chem. Hoppe-Seyler, **367**, 957-962 (1986)
[14] Lundie, L.L.; Ferry, J.G.: Activation of acetate by Methanosarcina thermophila. Purification and characterization of phosphotransacetylase. J. Biol. Chem., **264**, 18392-18396 (1989)
[15] Rado, T.A.; Hoch, J.A.: Phosphotransacetylase from Bacillus subtilis: purification and physiological studies. Biochim. Biophys. Acta, **321**, 114-125 (1973)
[16] Preston, G.G.; Zeiher, C.; Wall, J.D.; Emerich, D.W.: Acetate-activating enzymes of Bradyrhizobium japonicum bacteroids. Appl. Environ. Microbiol., **55**, 165-170 (1989)
[17] Robinson, J.R.; Sagers, R.D.: Phosphotransacetylase from Clostridium acidiurici. J. Bacteriol., **112**, 465-473 (1972)
[18] Smith, L.T.; Kaplan, N.O.: Purification of phosphotransacetylase by affinity chromatography. Anal. Biochem., **95**, 2-7 (1979)
[19] Shimizu, M.; Suzuki, T.; Kameda, K.Y.; Abiko, Y.: Phosphotransacetylase of Escherichia coli B, purification and properties. Biochim. Biophys. Acta, **191**, 550-558 (1969)
[20] Suzuki, T.: Phosphotransacetylase of Escherichia coli B, activation by pyruvate and inhibition by NADH and certain nucleotides. Biochim. Biophys. Acta, **191**, 559-569 (1969)
[21] Nojiri, T.; Tanaka, F.; Nakayama, I.: Purification and properties of phosphotransacetylase from Lactobacillus fermenti. J. Biochem., **69**, 789-801 (1971)
[22] Kyrtopoulos, S.A.; Satchell, D.P.N.: Kinetic studies with phosphotransacetylase. 3. The acylation of phosphate ions by acetyl coenzyme A. Biochim. Biophys. Acta, **276**, 376-382 (1972)
[23] Kyrtopoulos, S.A.; Satchell, D.P. N.: Kinetic studies with phosphotransacetylase. IV. Inhibition by products. Biochim. Biophys. Acta, **276**, 383-391 (1972)
[24] Bresters, T.W.; Krul, J.; Scheepens, P.C.; Veeger, C.: Phosphotransacetylase associated with the pyruvate dehydrogenase complex from the nitrogen fixing Azotobacter vinelandii. FEBS Lett., **22**, 305-309 (1972)
[25] Bock, A.K.; Glasemacher, J.; Schmidt, R.; Schonheit, P.: Purification and characterization of two extremely thermostable enzymes, phosphate acetyltransferase and acetate kinase, from the hyperthermophilic eubacterium Thermotoga maritima. J. Bacteriol., **181**, 1861-1867 (1999)
[26] Boynton, Z.L.; Bennett, G.N.; Rudolph, F.B.: Cloning, sequencing, and expression of genes encoding phosphotransacetylase and acetate kinase from Clostridium acetobutylicum ATCC 824. Appl. Environ. Microbiol., **62**, 2758-2766 (1996)

[27] Iyer, P.P.; Ferry, J.G.: Role of arginines in coenzyme A binding and catalysis by the phosphotransacetylase from Methanosarcina thermophila. J. Bacteriol., **183**, 4244-4250 (2001)
[28] Knorr, R.; Ehrmann, M.A.; Vogel, R.F.: Cloning of the phosphotransacetylase gene from Lactobacillus sanfranciscensis and characterization of its gene product. J. Basic Microbiol., **41**, 339-349 (2001)
[29] Latimer, M.T.; Ferry, J.G.: Cloning, sequence analysis, and hyperexpression of the genes encoding phosphotransacetylase and acetate kinase from Methanosarcina thermophila. J. Bacteriol., **175**, 6822-6829 (1993)
[30] Rasche, M.E.; Smith, K.S.; Ferry, J.G.: Identification of cysteine and arginine residues essential for the phosphotransacetylase from Methanosarcina thermophila. J. Bacteriol., **179**, 7712-7717 (1997)
[31] Shin, B.S.; Choi, S.K.; Park, S.H.: Regulation of the Bacillus subtilis phosphotransacetylase gene. J. Biochem., **126**, 333-339 (1999)
[32] Summers, M.L.; Denton, M.C.; McDermott, T.R.: Genes coding for phosphotransacetylase and acetate kinase in Sinorhizobium meliloti are in an operon that is inducible by phosphate stress and controlled by PhoB. J. Bacteriol., **181**, 2217-2224 (1999)

Acetyl-CoA C-acetyltransferase 2.3.1.9

1 Nomenclature

EC number
2.3.1.9

Systematic name
acetyl-CoA:acetyl-CoA C-acetyltransferase

Recommended name
acetyl-CoA C-acetyltransferase

Synonyms
2-methylacetoacetyl-CoA thiolase
3-oxothiolase
acetoacetyl CoA thiolase
acetoacetyl-CoA thiolase
acetyl coenzyme A thiolase
acetyl-CoA acetyltransferase
acetyl-CoA:N-acetyltransferase
acetyltransferase, acetyl coenzyme A
β-acetoacetyl coenzyme A thiolase
β-ketothiolase <9> [2]
thiolase II

CAS registry number
9027-46-7

2 Source Organism

<1> *Gallus gallus* [4, 6]
<2> *Clostridium kluyveri* [2, 7, 21]
<3> *Rattus norvegicus* (2 forms: A and B [11,13]) [8, 11, 13, 23, 32, 34]
<4> *Zoogloea ramigera* [3, 15-18, 27, 33, 36]
<5> *Candida tropicalis* (ATCC 20336 [29]) [20, 29, 31]
<6> *Clostridium acetobutylicum* (ATCC 824 [19]) [19]
<7> *Escherichia coli* (wild-type and constitutive mutant [24]) [24, 25]
<8> *Bos taurus* (2 forms: A and B [10]; 3 forms: I, A, B [12]) [1, 10, 12]
<9> *Sus scrofa* [2, 5, 9, 14]
<10> *Bradyrhizobium japonicum* [22]
<11> *Caulobacter crescentus* [26]
<12> *Zoogloea ramigera* [17]

<13> *Catharanthus roseus* (L. G. Don, herbarium number LEF 920116 [28]) [28]
<14> *Rhizobium sp.* (cicer, strain CC 1192 [29]) [29, 30]
<15> *Streptomyces collinus* [35]
<16> *Enterococcus faecalis* [37]
<17> *Halobacterium sp.* (strain ZP-6, haloarchaeon [38]) [38]
<18> *Helianthus annuus* (sunflower [39]) [39]

3 Reaction and Specificity

Catalyzed reaction

2 acetyl-CoA = CoA + acetoacetyl-CoA (<3,6,8> mechanism [8,10,19]; <6,10> ping-pong mechanism [19,22]; <4> bi bi ping pong mechanism [3]; <4> proposed two-step ping pong mechanism [33]; <4> reaction involves 2 chemically distinct steps: acetyl group transfer from acetyl-CoA to Cys89, and transfer of this acetyl group to a second acetyl-CoA in the Claisen condensation step to form acetoacetyl-CoA [36])

Reaction type

acyl group transfer
condensation
thiolytic cleavage

Natural substrates and products

S CoA + acetoacetyl-CoA <1-18> (<2> enzyme of fatty acid oxidation cycle [2]; <3> liver mitochondrial isoenzyme catalyzes the first step in biosynthesis of ketone bodies [11]; <4> involved in generation of acetoacetyl-CoA for poly-β-hydroxybutyrate synthesis [15]; <5> cytosolic thiolase I is essential for the mevalonate pathway [31]; <3> peroxisomal thiolase most probably catalyzes the first reaction in peroxisomal cholesterol and dolichol synthesis [34]) (Reversibility: ? <1-18> [1-39]) [1-39]
P acetyl-CoA + acetyl-CoA <1-18> [1-39]

Substrates and products

S CoA + acetoacetyl-10-bis-demethylpantetheine 11-pivaloate <4> (Reversibility: ? <4> [15]) [15]
P acetyl-CoA + acetyl-10-bis-demethylpantetheine 11-pivaloate <4> [15]
S CoA + acetoacetyl-CoA <1-18> (<4> most active in thiolysis [15]; <10> specific for acetoacetyl-CoA, no activity with ketodecanoyl-CoA, dithiothreitol and 2-mercaptoethanol [22]; <7> thiolase II from constitutive mutant is specific for acetoacetyl-CoA [24]; <9> thiolysis is preferred, highly specific for acetoacetyl-CoA [2]; <3> thiolysis is strongly preferred [8]; <8> rate of synthesis is 0.31 and 0.08 the rate of thiolysis for isoenzyme A and B, respectively [10]; <15> enzyme has both thiolase and acetyl-CoA:acyl carrier protein transacylase activity [35]; <16> enzyme has both thiolase and 3-hydroxy-3-methylglutaryl-CoA reductase activity [37]) (Reversibility: r <1-17> [1-38]; ir <18> [39]) [1-39]
P acetyl-CoA + acetyl-CoA <1-18> [1-39]

S CoA + acetoacetyl-S-(11-methoxymethyl)pantetheine <4> (Reversibility: ? <4> [15]) [15]
P acetyl-CoA + acetyl-S-(11-methoxymethyl)pantetheine
S CoA + acetoacetyl-S-(11-t-butyldimethylsilyl)pantetheine <4> (Reversibility: ? <4> [15]) [15]
P acetyl-CoA + acetyl-S-(11-t-butyldimethylsilyl)pantetheine <4> [15]
S CoA + acetoacetyl-S-(D-pantetheine) 11-pivalate <4> (Reversibility: ? <4> [15]) [15]
P acetyl-CoA + acetyl-S-(D-pantetheine) 11-pivalate <4> [15]
S CoA + acetoacetyl-S-(L-pantetheine) 11-pivalate <4> (Reversibility: ? <4> [15]) [15]
P acetyl-CoA + acetyl-S-(L-pantetheine) 11-pivalate <4> [15]
S CoA + acetoacetyl-S-homopantetheine 12-pivalate <4> (Reversibility: ? <4> [15]) [15]
P acetyl-CoA + acetyl-S-homopantetheine 12-pivalate <4> [15]
S CoA + acetoacetyl-S-pantetheine <4> (Reversibility: ? <4> [15]) [15]
P acetyl-CoA + acetyl-S-pantetheine <4> [15]
S acetoacetyl-S-pantetheine + acetyldithio-CoA <4> (Reversibility: r <4> [27]) [27]
P acetyl-S-pantetheine + 3-ketobutyryldithio-CoA <4> [27]
S acetyl-CoA + acetyldithio-CoA <4> (Reversibility: r <4> [27]) [27]
P CoA + 3-ketobutyryldithio-CoA <4> [27]

Inhibitors

11-chloro-10-oxoundecanoic acid <9> (<9> acetyl-CoA or CoA protect [9]) [9]
11-chloro-10-oxoundecanoyl-CoA <9> (<9> acetyl-CoA or CoA protect [9]) [9]
2,3-pentadienoyl-S-pantetheine 11-pivalate <4> (<4> half-life for inactivation: 1.9 min [18]) [18]
2-butynoyl-CoA <9> (<9> 0.1 mM, 74% inactivation after 10 min [5]) [5]
2-oxo-5-(1-hydroxy-2,4,6-heptatriynyl)-1,3-dioxolone-4-heptanoic acid <3> (<3> natural product isolated from actinomycete culture L-660,631, 0.00001 mM, 505 inhibition [23]) [23]
3-butynoyl-CoA <9> (<9> 0.01 mM, 95% inactivation after 10 min, acetoacetyl-CoA or 0.8 mM CoA protect [5]) [5]
3-hydroxybutyryl-CoA <9> [14]
3-pentenoyl-S-pantetheine 11-pivalate <4> (<4> half-life for inactivation: 0.26 min [18]) [18]
3-pentynoyl-CoA <9> (<9> 0.1 mM, complete inactivation after 10 min, acetoacetyl-CoA or 0.8 mM CoA protect [5]) [5]
3-pentynoylpantetheine <9> (<9> 1 mM, complete inactivation after 10 min [5]) [5]
4-bromocrotonyl-CoA <9> (<9> 0.1 mM, complete inactivation after 10 min, acetoacetyl-CoA protects [5]) [5]
4-bromocrotonylpantetheine <9> (<9> 1 mM, complete inactivation after 10 min [5]) [5]

5,5'-dithiobis(2-nitrobenzoate) <4> (<4> 0.4 mM, 87% inhibition [3]) [3]
5-chloro-4-oxopentanoyl-CoA <9> (<9> acetyl-CoA or CoA protect [9]) [9]
7-chloro-6-oxoheptanoic acid <9> (<9> acetyl-CoA or CoA protect [9]) [9]
7-chloro-6-oxoheptanoyl-CoA <9> (<9> acetyl-CoA or CoA protect [9]) [9]
9-chloro-8-oxononanoic acid <9> (<9> acetyl-CoA or CoA protect [9]) [9]
9-chloro-8-oxononanoyl-CoA <9> (<9> acetyl-CoA or CoA protect [9]) [9]
ATP <6> (<6> 10 mM, 41% inhibition [19]) [19]
Ca^{2+} <14> (<14> 10 and 25 mM, 25 and 50% inhibition of thiolysis [30]) [30]
CoA <1, 3, 6, 8, 14, 16> (<8> both directions [10]; <1> higher than 0.03 mM, inhibition in cleavage direction [4]; <6> inhibition in condensation direction [19]; <3> competitive vs. acetoacetyl-CoA in thiolysis [8]; <14> substrate inhibition above 0.05 mM, 50% inhibition of synthesis at 0.06 mM in bacteroids [30]) [4, 8, 10, 19, 30, 37]
Mg^{2+} <3, 10, 14> (<3> inhibits the rate of acetoacetyl-CoA thiolysis but not the rate of synthesis of acetoacetyl-CoA [8]; <10> 60 mM, 45% inhibition of thiolysis, 69% inhibition of synthesis [22]; <14> 2 and 5 mM, 20 and 40% inhibition of synthesis, respectively [30]) [8, 22, 30]
N-ethylmaleimide <4, 7> (<4> 0.4 mM, 53% inhibition [3]) [3, 25]
NADH <4, 10> (<4,10> inhibition of thiolysis [3,22]) [3, 22]
NADPH <4> [3]
acetoacetyl-CoA <3, 7, 8> (<7> substrate inhibition above 0.105 mM [25]; <3> isoenzyme A: above 0.015 mM, isoenzyme B: above 0.01 mM [13]) [10, 13, 25]
acetyl-CoA <9> [14]
acryl-S-pantetheine 11-pivalate <4> [18]
bromoacetyl oxoester <4> (<4> acetyl-pantetheine 11-pivalate analog [16]) [16]
bromoacetyl thioester <4> (<4> acetyl-pantetheine 11-pivalate analog [16]) [16]
bromoacetylamide <4> (<4> acetyl-pantetheine 11-pivalate analog [16]) [16]
butyryl-CoA <6> (<6> 1 mM, 42% inhibition [19]) [19]
citraconic anhydride <9> [9]
dec-3-ynoic acid <9> (<9> irreversible inhibition [9]) [9]
dithionitrobenzoate <6> (<6> low but significant inhibition [19]) [19]
iodoacetamide <3, 4, 6, 7, 10> (<4> 0.4 mM, 36% inhibition [3]; <3> 0.005 mM, inactivation half-life: 3 min [8]; <4> 125fold molar excess, rapid inactivation [15]; <4> reacts with Cys89 [16]; <6> 0.3 mM, 93% inhbition after 30 min [19]; <10> 0.5 mM, 96% inhibition, acetoacetyl-CoA partly protects [22]; <7> 2 mM, complete inactivation after approx. 60 min, 0.3 mM acetoacetyl-CoA or 1.6 mM acetyl-CoA protect [25]) [3, 8, 15, 16, 19, 22, 25]
iodoacetate <10> [22]
p-chloromercuribenzoate <4, 7> (<4> 0.4 mM, 95% inhibition [3]) [3, 25]
sodium borohydride <7, 9> (<9> inactivation in the presence of either acetoacetyl-CoA or acetyl-CoA [5]) [5, 25]
sulfhydryl reagents <7> [25]
thiolactomycin <15> (<15> 0.36 mM, 50% inhibition [35]) [35]

Cofactors/prosthetic groups

Additional information <2> (<2> enzyme contains selenomethionine [7]) [7]

Activating compounds

KCl <17> (<17> maximal activity at 4.5 M [38]) [38]
NaCl <17> (<17> maximal activity at 4.5 M [38]) [38]
lovostatin <3> (<3> 1.9fold activity increase in rats treated with lovostatin [32]) [32]

Metals, ions

Ca^{2+} <14> (<14> can replace Mg^{2+}, 5 mM, 90% of activity with Mg^{2+}, inhibition above 10 mM [30]) [30]
Mg^{2+} <14> (<14> maximal activity at 5-10 mM, inhibition above [30]) [30]
Mn^{2+} <14> (<14> can replace Mg^{2+}, 5 mM, 90% of activity with Mg^{2+}, inhibition above [30]) [30]

Turnover number (min^{-1})

4.6 <15> (acetoacetyl-CoA, <15> C92S mutant thiolase [35]) [35]
126 <4> (acetyl-CoA) [27]
840 <4> (3-ketobutyryldithio-CoA) [27]
840 <4> (acetoacetyl-S-pantetheine) [27]
1770 <1> (CoA, presence of Mg^{2+}) [4]
2170 <15> (acetoacetyl-CoA, <15> native thiolase [35]) [35]
2404 <15> (acetoacetyl-CoA, <15> recombinant thiolase [35]) [35]
10440 <4> (acetoacetyl-S-pantetheine) [15]
10620 <4> (acetoacetyl-S-homopantetheine 12-pivalate) [15]
15360 <4> (acetoacetyl-S-(L-pantetheine) 11-pivalate) [15]
15960 <4> (acetoacetyl-10-bis-demethylpantetheine 11-pivaloate) [15]
21180 <4> (acetoacetyl-S-(11-methoxymethyl)pantetheine) [15]
24600 <4> (acetoacetyl-CoA) [27]
26040 <4> (acetoacetyl-S-(11-t-butyldimethylsilyl)pantetheine) [15]
27900 <4> (acetoacetyl-CoA) [15]
28140 <4> (acetoacetyl-S-(D-pantetheine) 11-pivalate) [15]
37200 <4> (acetoacetyl-CoA) [33]
54000 <1> (thiolytic cleavage) [4, 6]

Specific activity (U/mg)

0.05 <8> [1]
0.16 <4> (<4> acetoacetyl-CoA synthesis [15]) [15]
29 <5> [20]
33 <18> [39]
36.6 <15> [35]
37 <3> [8]
40.98 <11> [26]
58.2 <9> [9]
59.8 <5> [29]
91 <4> [3]
100-130 <3> (<3> peroxisomal thiolase, thiolysis [34]) [34]
102.9 <10> [22]

115 <2> [21]
163 <14> [30]
203 <1> [4, 6]
216 <6> [19]
412 <4> (<4> thiolysis of acetoacetyl-CoA [15]) [15]
506 <7> (<7> thiolase II [24]) [24]
833.3 <17> [38]
1078 <7> [25]
Additional information <1> (<1> assay method [4]) [4]

K_m-Value (mM)

0.0038 <15> (acetoacetyl-CoA, <15> native thiolase [35]) [35]
0.004 <3> (acetoacetyl-CoA, <3> isoenzyme B, thiolysis [13]) [13]
0.0048 <6> (CoA) [19]
0.006 <3> (acetoacetyl-CoA, <3> isoenzyme A, thiolysis [13]) [13]
0.0062 <15> (acetoacetyl-CoA, <15> C92S mutant thiolase [35]) [35]
0.0062 <3> (acetyl-CoA, <3> isoenzyme A [13]) [13]
0.0064 <1> (CoA) [4]
0.008 <3> (CoA, <3> peroxisomal thiolase, thiolysis [34]) [34]
0.0085 <4> (CoA) [3]
0.009 <3> (acetoacetyl-CoA, <3> peroxisomal thiolase, thiolysis [34]) [34]
0.0098 <15> (acetoacetyl-CoA, <15> recombinant thiolase [35]) [35]
0.01 <16> (CoA, <16> thiolysis [37]) [37]
0.01 <4, 8> (acetoacetyl-CoA, <8> isoenzyme A, thiolysis [10]) [3, 10]
0.011 <18> (acetoacetyl-CoA) [39]
0.014 <18> (CoA) [39]
0.016 <3> (acetoacetyl-CoA, <3> cytosolic thiolase, thiolysis [34]) [34]
0.019 <10> (acetoacetyl-CoA) [22]
0.02 <3> (CoA, <3> cytosolic thiolase, thiolysis [34]) [34]
0.022 <3> (CoA) [13]
0.023 <5> (acetoacatyl-CoA, <5> recombinant cytosolic thiolase I, thiolysis [29]) [29]
0.024 <4> (acetoacetyl-CoA) [15]
0.025 <8> (CoA, <8> isoenzyme A, thiolysis [10]) [10]
0.025 <5> (acetoacatyl-CoA, <5> peroxisomal thiolase I, thiolysis [29]) [20, 29]
0.03 <10> (CoA) [22]
0.03 <5> (CoA, <5> peroxisomal thiolase I, thiolysis [29]) [29]
0.03 <3> (acetyl-CoA, <3> isoenzyme A, in the presence of 3.5 mM NH_4^+ [11]) [11]
0.032 <5> (CoA, <5> recombinant peroxisomal thiolase I, thiolysis [29]) [29]
0.032 <6, 8> (acetoacetyl-CoA, <8> isoenzyme B, thiolysis [10]) [10, 19]
0.035 <3> (acetoacetyl-CoA, <3> peroxisomal thiolase, synthesis [34]) [34]
0.037 <5> (CoA, <5> cytosolic thiolase I, thiolysis [29]) [29]
0.0385 <1> (acetoacetyl-CoA) [4]
0.042 <14> (acetoacetyl-CoA, <14> thiolysis [30]) [30]

0.042 <5> (acetoacetyl-CoA, <5> cytosolic thiolase I, thiolysis [29]) [29]
0.05 <5> (CoA, <5> recombinant cytosolic thiolase I, thiolysis [29]) [29]
0.056 <14> (CoA, <14> thiolysis [30]) [30]
0.057 <5> (acetoacetyl-CoA, <5> recombinant peroxisomal thiolase I, thiolysis [29]) [29]
0.073 <4> (acetoacetyl-S-(D-pantetheine) 11-pivalate) [15]
0.074 <4> (acetoacetyl-S-(11-t-butyldimethylsilyl)pantetheine, <4> cosubstrate CoA [15]) [15]
0.08 <3> (acetoacetyl-CoA, <3> cytosolic thiolase, synthesis [34]) [34]
0.08 <8> (acetyl-CoA, <8> enzyme B [10]) [10]
0.088 <16> (acetoacetyl-CoA, <16> thiolysis [37]) [37]
0.091 <8> (acetyl-CoA, <8> isoenzyme A, synthesis [10]) [10]
0.091 <8> (acetyl-CoA, <8> isoenzyme B, synthesis [10]) [10]
0.104 <10> (acetyl-CoA) [22]
0.12 <4> (acetoacetyl-S-(11-methoxymethyl)pantetheine) [15]
0.154 <8> (CoA, <8> isoenzyme B, thiolysis [10]) [10]
0.21 <4> (acetoacetyl-10-bis-demethylpantetheine 11-pivalate) [15]
0.25 <4> (acetoacetyl-S-homopantetheine 12-pivalate) [15]
0.27 <1, 6> (acetyl-CoA) [4, 19]
0.33 <4> (acetyl-CoA) [3]
0.38 <14> (acetyl-CoA, <14> enzyme from bacteroids [30]) [30]
0.46 <4> (acetoacetyl-S-pantetheine) [15]
0.47 <7> (acetyl-CoA, <7> synthesis [25]) [25]
0.6 <16> (acetyl-CoA, <16> synthesis [37]) [37]
0.69 <5> (acetyl-CoA, <5> recombinant cytosolic thiolase I, thiolysis [29]) [29]
0.74 <5> (acetyl-CoA, <5> recombinant peroxisomal thiolase I, thiolysis [29]) [29]
0.77 <5> (acetyl-CoA, <5> cytosolic thiolase I, thiolysis [29]) [29]
1.05 <5> (acetyl-CoA, <5> peroxisomal thiolase I, thiolysis [29]) [29]
1.06 <14> (acetyl-CoA, <14> synthesis [30]) [30]

K_i-Value (mM)

0.0014 <9> (9-chloro-8-oxononanoyl-CoA) [9]
0.0014 <3> (acetoacetyl-CoA, <3> isoenzyme A [13]) [13]
0.0016 <8> (acetoacetyl-CoA, <8> isoenzyme A [10]) [10]
0.002 <9> (7-chloro-6-oxoheptanoyl-CoA) [9]
0.0025 <9> (11-chloro-10-oxoundecanoyl-CoA) [9]
0.006 <14> (CoA, <14> enzyme from bacteroids [30]) [30]
0.011 <14> (CoA) [30]
0.013 <9> (4-bromocrotonyl-CoA) [5]
0.015 <9> (5-chloro-4-oxopentanoyl-CoA) [9]
0.025 <9> (3-pentynoyl-CoA) [5]
0.028 <9> (11-chloro-10-oxoundecanoic acid) [9]
0.05 <8> (CoA) [10]
0.067 <3> (CoA) [8]
0.12 <16> (CoA) [37]

0.49 <9> (9-chloro-8-oxononanoic acid) [9]
1.25 <4> (3-pentenoyl-S-pantetheine 11-pivalate) [18]
1.54 <4> (2,3-pentadienoyl-S-pantetheine 11-pivalate) [18]
10 <9> (dec-3-ynoic acid) [9]
11.4 <9> (7-chloro-6-oxoheptanoic acid) [9]

pH-Optimum
5.5-7.9 <6> [19]
7 <10> (<10> condensation [22]) [22]
7.5 <4> (<4> condensation [3]) [3]
7.8 <10, 14> (<10> thiolysis [22]; <14> thiolysis, more than 90% of maximal activity between pH 7.5 and pH 8.3 [30]) [22, 30]
8.1 <5, 14> (<5> cytosolic and peroxisomal thiolase I, synthesis [29]; <14> synthesis, more than 90% of maximal activity between pH 7.5 and pH 8.5 [30]) [29]
8.1-8.4 <8> (<8> isoenzyme A and B, acetoacetyl-CoA synthesis [10]) [10]
8.3 <5> (<5> cytosolic and peroxisomal thiolase I [29]) [29]
8.4 <1> (<1> cleavage of acetoacetyl-CoA, 85% of maximal activity between pH 8.0 and pH 9.0 [4]) [4]
8.5 <4, 18> (<4> thiolysis [3]; <18> 505 of maximal activity at pH 7.5 and pH 9.5, no activity below pH 6.5 [39]) [3, 39]
9.5 <16> (<16> thiolysis [37]) [37]
10.5 <16> (<16> synthesis [37]) [37]

pH-Range
5.5-9 <6> (<6> 69% of maximal activity at pH 9.0 [19]) [19]
6-9 <4> (<4> approx. 80% of maximal activity at pH 6.0, approx. 40% at pH 9.0 [3]) [3]
6.5-9.2 <4> (<4> approx. 50% of maximal activity at pH 6.5 and pH 9.2, thiolysis [3]) [3]

Temperature optimum (°C)
25 <2> (<2> assay at [21]) [21]
30 <1> (<1> assay at [4,6]) [4, 6]
37 <16> (<16> acetoacetyl-CoA synthesis and thiolysis [37]) [37]

4 Enzyme Structure

Molecular weight
80000 <17> (<17> gel filtration [38]) [38]
147600 <8> (<8> isoenzyme B, sedimentation analysis [12]) [12]
150000 <8> (<8> isoenzyme A, gel filtration [12]) [12]
151000 <3> (<3> gel filtration, isoenzyme A and B [11]) [11]
152000 <4, 8> (<8> isoenzyme B, gel filtration [12]; <4> analytical ultracentrifugation [33]) [12, 33]
154600 <8> (<8> isoenzyme A, sedimentation analysis [12]) [12]

155000 <2, 8> (<2> gel filtration [7]; <8> isoenzyme I, gel filtration [12]) [7, 12]
156000 <7> (<7> sucrose density gradient sedimentation [25]) [25]
160000 <3> (<3> peroxisomal thiolase, gel filtration [34]) [34]
162000 <4> [15]
164000 <18> (<18> gel filtration [39]) [39]
166000 <7> (<7> gel filtration [25]) [25]
169000 <1> (<1> sedimentation equilibrium [4,6]) [4, 6]
170000 <3> (<3> gel filtration [8]) [8]
180000 <10, 14> (<10> gel filtration [22]; <14> gel filtration [30]) [22, 30]
185000 <15> (<15> native thiolase, gel filtration [35]) [35]
187000 <15> (<15> recombinant thiolase, gel filtration [35]) [35]
188000 <1> (<1> gel filtration [4]) [4]
190000 <4> (<4> gel filtration [3]) [3]
193000 <10> (<10> native PAGE [22]) [22]
240000 <5> (<5> peroxisomal thiolase I, gel filtration [29]) [29]
250000 <5> (<5> cytosolic thiolase I, gel filtration [29]) [29]

Subunits

? <11, 16, 18> (<11> x * 45000, SDS-PAGE [26]; <16> x * 88712, MALDI-mass spectrometry [37]; <16> x * 86496, deduced from nucleotide sequence [37]; <18> x * 63000, SDS-PAGE [39]) [26, 37, 39]
dimer <17> (<17> 2 * 43000, CTAB-PAGE [38]; <17> 2 * 60000, SDS-PAGE [38]) [38]
hexamer <5> (<5> 6 * 41000, SDS-PAGE [20]; <5> 6 * 41000, cytosolic and peroxisomal thiolase I, SDS-PAGE [29]) [20, 29]
tetramer <1-4, 6-8, 10, 14, 15> (<4> 4 * 44000, SDS-PAGE [3]; <6> 4 * 44000, SDS-PAGE [19]; <10> 4 * 44000, SDS-PAGE [22]; <1> 4 * 41000, SDS-PAGE [4,6]; <2> 4 * 38000-40000, SDS-PAGE [7]; <3> 4 * 40000, enzyme A and B, SDS-PAGE [8,11]; <8> 4 * 39000, SDS-PAGE [12]; <7> 4 * 42000, SDS-PAGE [25]; <12> 4 * 40598, deduced from nucleotide sequence [17]; <2> 4 * 41000, SDS-PAGE [21]; <14> 4 * 44000, SDS-PAGE [30]; <4> 4 * 40495, electrospray ionization mass spectroscopy [33]; <3> 4 * 42000, peroxisomal thiolase, SDS-PAGE [34]; <15> 4 * 46000-48000, recombinant thiolase, SDS-PAGE [35]; <15> 4 * 45000-47000, native thiolase, SDS-PAGE [35]) [3, 4, 6, 7, 11, 12, 17, 19, 21, 22, 25, 30, 33, 34, 35]

5 Isolation/Preparation/Mutation/Application

Source/tissue

bacteroid <10, 14> [22, 30]
cotyledon <18> [39]
free-living state <14> [30]
heart <9> [2, 5, 9, 14]
liver <1, 3, 8> [1, 4, 6, 8, 10-13, 23, 32, 34]
stem <13> [28]

Localization

cytosol <1, 3, 5> (<5> cytosolic thiolase I [29]; <5> N-terminal sequence targets thiolase I to the peroxisome [31]) [4, 6, 8, 23, 29, 31]
glyoxysome <18> [39]
peroxisome <3, 5> (<5> peroxisomal thiolase I [29]) [20, 29, 34]

Purification

<1> (cytosolic enzyme [6]; liver enzyme, ammonium sulfate, calcium phosphate gel, phosphocellulose column, DEAE-cellulose, hydroxylapatite, Sephadex G-200 [4]) [4, 6]
<2> (CM-Sephadex [21]) [21]
<2> (single-step purification [21]; DEAE-cellulose, ammonium sulfate, gel filtration, DEAE-cellulose, Matrex gel green A [7]) [7, 21]
<3> (2 forms: A and B [11]; pH 5.5, DEAE-cellulose, calcium phosphate column, celulose phosphate column [8]; peroxisomal thiolase, DEAE-cellulose, phosphocellulose column, Blue-Sepharose [34]) [8, 11, 34]
<4> (protamine sulfate, DEAE-cellulose, phosphocellulose, Sephadex G-200 [3]; ammonium sulfate, DEAE-Sepharose, hydroxyapatite, phosphocellulose, Blue Sepharose [15]) [3, 15]
<5> (DEAE-Sepharose, butyl-Toyopearl, Cellulofine [20]; cytosolic thiolase I [29]) [20, 29]
<6> (ammonium sulfate, DEAE-Sephacel, Blue Sepharose, AX 300 column [19]) [19]
<7> (constitutive mutant, thiolase II, heat treatment, phosphocellulose [24]; ammonium sulfate, Sephadex G-150, DEAE-Sephadex, QAE-Sephadex [24]) [24]
<8> (ammonium sulfate, calcium phosphate gel, ethanol, partial purification [1]; isoenzymes A and B, ammonium sulfate, cellulose phosphate column, crystallization [10]) [1, 10]
<9> (ammonium sulfate, acetone, acid treatment, CM-Sephadex, Sepharose 6B [9]) [9]
<10> (ammonium sulfate, hydroxylapatite, phosphocellulose [22]) [22]
<11> (ammonium sulfate, Sephacryl S-200, Superose 6, Mono Q [26]) [26]
<13> (Q-Sepharose, S-Sepharose, Superose 6, partially purified [28]) [28]
<14> (ammonium sulfate, Econo-Q, Superose 6, Mono Q, HA-Ultrogel, Mono Q [30]) [30]
<15> (Q-Sepharose, ammonium sulfate, Sephacryl S-200, Mono Q, FPLC S-300 [35]) [35]
<16> (recombinant His-tagged thiolase, Ni-affinity column [37]) [37]
<17> (poly ethylene glycol 6000, DEAE-cellulose 52, benzyl-Sepharose, hydroxylapatite [38]) [38]
<18> (hydroxyapatite, Superdex, Fractogel [39]) [39]

Crystallization

<4> (hangig-drop vapor diffusion at 21°C, crystal structure at 2.0 A resolution [33]; wild-type thiolase and acetylated thiolase complexed with CoA, C89A mutant thiolase complexed with acetyl-CoA and acetoacetyl-CoA, Q64A mutant thiolase [36]) [33, 36]
<8> [10]

Cloning

<4> (expression in Escherichia coli [33]) [17, 33, 36]
<5> (expression of cytosolic and peroxisomal thiolase I in Saccharomyces cerevisiae [29]) [29]
<15> (expression in Escherichia coli [35]) [35]
<16> (expression in Escherichia coli [37]) [37]

Engineering

C378G <4> (<4> mutation eliminates the ability of thiolase to catalyze proton abstraction from C_2 of acetyl-CoA [18]) [18]
C89A <4> (<4> no thiolytic activity towards acetoacetyl-CoA [36]) [36]
C92S <15> (<15> k_{cat} decreases to 0.2% of that for the recombinant thiolase [35]) [35]
Q64A <4> (<4> 30% lower k_{cat} than that of the wild-type [36]) [36]

6 Stability

Temperature stability

0 <17> (<17> 1 h, complete loss of activity [38]) [38]
60 <15> (<15> 15 min, loss of 47% activity [35]) [35]
63 <11> (<11> 5 min, stable [26]) [26]
70 <17> (<17> 10 h, 4 M KCl, 20% loss of activity [38]) [38]
85 <17> (<17> 10 h, 4 M KCl, 35% loss of activity [38]) [38]

Oxidation stability

<7>, sensitive to borohydride reduction [25]

General stability information

<4>, urea: 2.5 M, 17 h, 50% loss of activity, 5 M, 45 min, 50% loss of activity, 7 M, 1 min, 50% loss of activity [15]
<10>, dithiothreitol stabilizes [22]
<10>, glycerol stabilizes [22]
<10>, unstable in dilute solution, less than 0.5 mg/ml protein [22]
<17>, more than 10% sucrose, 2-mercaptoethanol or dithiothreitol are necessary to maintain activity [38]
<18>, labile enzyme, partially stabilized by 1-2 mM dithiothreitol and 20% ethylene glycol [39]

Storage stability

<1>, -90°C, 12 months [4, 6]
<4>, -20°C, 50 mM Tris-HCl, pH 8.0, 0.5 mM dithiothreitol, glycerol [3]
<6>, 4°C, 1 mM dithiothreitol, 10% v/v glycerol, N_2 atmosphere, 1 month, less than 10% loss of activity [19]
<7>, 0°C, 0.05 M Tris-HCl, pH 7.8, 2 mM dithiothreitol, 1 mM EDTA, 20% glycerol, 6 weeks, 50% loss of activity [25]
<8>, -10°C, less than 0.2 mg/ml protein, 50 mM Tris, pH 7.2, 0.5 mM dithiothreitol, 8 d, 60-70% loss of activity [10]

<10>, -20°C, in dilute solution, less than 0.5 mg/ml protein, 24 h, complete loss of activity [22]
<14>, -20°C, 20 mM Tris-HCl, pH 7.8, 0.5 mM 1 mM, EDTA, 15 d, loss of more than 50% activity [30]
<14>, -20°C, 20 mM Tris-HCl, pH 7.8, 0.5 mM dithiothreitol, 1 mM, EDTA, 15 d, loss of 20% activity [30]
<14>, -20°C, 20 mM Tris-HCl, pH 7.8, 0.5 mM dithiothreitol, 1 mM, EDTA, 20% glycerol, 60 d, loss of less than 20% activity [30]
<17>, 25°C, 1 week, 4.5 M KCl, no loss of activity [38]
<17>, 25°C, 72 h, 200 mM KCl, 90% loss of activity [38]

References

[1] Stern, J.R.; Drummond, G.I.; Coon, M.J.; Del Campillo, A.: Enzymes of ketone body metabolism. I. Purification of an acetoacetate-synthesizing enzyme from ox liver. J. Biol. Chem., **235**, 313-317 (1960)
[2] Lynen, F.; Ochoa, S.: Enzymes of fatty acid metabolism. Biochim. Biophys. Acta, **12**, 299-314 (1953)
[3] Nishimura, T.; Saito, T.; Tomita, K.: Purification and properties of β-ketothiolase from Zoogloea ramigera. Arch. Microbiol., **116**, 21-27 (1978)
[4] Clinkenbeard, K.D.; Sugiyama, T.; Lane, M. D.: Cytosolic acetoacetyl-CoA thiolase from chicken liver. Methods Enzymol., **35 B**, 167-173 (1975)
[5] Holland, P.C.; Clark, M.G.; Bloxham, D. P.: Inactivation of pig heart thiolase by 3-butynoyl coenzyme A, 3-pentynoyl coenzyme A, and 4-bromocrotonyl coenzyme A. Biochemistry, **12**, 3309-3315 (1973)
[6] Clinkenbeard, K.D.; Sugiyama, T.; Moss, J.; Reed, W.D.; Lane, M.D.: Molecular and catalytic properties of cytosolic acetoacetyl coenzyme A thiolase from avian liver. J. Biol. Chem., **248**, 2275-2284 (1973)
[7] Hartmanis, M.G.N.; Stadtman, T. C.: Isolation of a selenium-containing thiolase from Clostridium kluyveri: identification of the selenium moiety as selenomethionine. Proc. Natl. Acad. Sci. USA, **79**, 4912-4916 (1982)
[8] Middleton, B.: The kinetic mechanism and properties of the cytoplasmic acetoacetyl-coenzyme A thiolase from rat liver. Biochem. J., **139**, 109-121 (1974)
[9] Bloxham, D.P.; Chalkley, R.A.; Coghlin, S. J.; Salam, W.: Synthesis of chloromethyl ketone derivatives of fatty acids. Their use as specific inhibitors of acetoacetyl-coenzyme A thiolase, cholesterol biosynthesis and fatty acid synthesis. Biochem. J., **175**, 999-1011 (1978)
[10] Huth, W.; Jonas, R.; Wunderlich, I.; Seubert, W.: On the mechanism of ketogenesis and its control. Purification, kinetic mechanism and regulation of different forms of mitochondrial acetoacetyl-CoA thiolases from ox liver. Eur. J. Biochem., **59**, 475-489 (1975)
[11] Schwabe, D.; Huth, W.: Immunochemical aspects, molecular and kinetic properties of multiple forms of acetyl-CoA acetyltransferase from rat liver mitochondria. Biochim. Biophys. Acta, **575**, 112-120 (1979)

[12] Jonas, R.; Huth, W.: Acetyl-CoA acetyltransferase from bovine liver mitochondria. Molecular properties of multiple forms. Biochim. Biophys. Acta, **527**, 379-390 (1978)
[13] Huth, W.; Dierich, C.; von Oeynhausen, V.; Seubert, W.: Multiple mitochondrial forms of acetoacetyl-CoA thiolase in rat liver: possible regulatory role in ketogenesis. Biochem. Biophys. Res. Commun., **56**, 1069-1077 (1974)
[14] Olowe, Y.; Schulz, H.: Regulation of thiolases from pig heart. Control of fatty acid oxidation in heart. Eur. J. Biochem., **109**, 425-429 (1980)
[15] Davis, J.T.; Moore, R.N.; Imperiali, B.; Pratt, A.J.; Kobayashi, K.; Masamune, S.; Sinskey, A.J.; Walsh, C.T.; Fukui, T.; Tomita, K.: Biosynthetic thiolase from zoogloea ramigera. I. Preliminary characterization and analysis of proton transfer reaction. J. Biol. Chem., **262**, 82-89 (1987)
[16] Davis, J.T.; Chen, H.H.; Moore, R.; Nishitani, Y.; Masamune, S.; Sinskey, A.J.; Walsh, C.T.: Biosynthetic thiolase from Zoogloea ramigera. II. Inactivation with haloacetyl CoA analogs. J. Biol. Chem., **262**, 90-96 (1987)
[17] Peoples, O.P.; Masamune, S.; Walsh, S.T.; Sinskey, A.J.: Biosynthetic thiolase from Zoogloea ramigera. III. Isolation and characterization of the structural gene. J. Biol. Chem., **262**, 97-102 (1987)
[18] Palmer, M.A.J.; Differding, E.; Gamboni, R.; Williams, S.F.; Peoples, O.P.; Walsh, C.T.; Sinskey, A.J.; Masamune, S.: Biosynthetic thiolase from Zoogloea ramigera. Evidence for a mechanism involving Cys-378 as the active site base. J. Biol. Chem., **266**, 8369-8375 (1991)
[19] Wiesenborn, D.P.; Rudolph, F.B.; Papoutsakis, E.T.: Thiolase from Clostridium acetobutylicum ATCC 824 and its role in the synthesis of acids and solvents. Appl. Environ. Microbiol., **54**, 2717-2722 (1988)
[20] Kurihara, T.; Ueda, M.; Tanaka, A.: Peroxisomal acetoacetyl-CoA thiolase and 3-ketoacyl-CoA thiolase from an n-alkane-utilizing yeast, Candida tropicalis: purification and characterization. J. Biochem., **106**, 474-478 (1989)
[21] Sliwkowski, M.X.; Hartmanis, M.G.N.: Simultaneous single-step purification of thiolase and NADP-dependent 3-hydroxybutyryl-CoA dehydrogenase from Clostridium kluyveri. Anal. Biochem., **141**, 344-347 (1984)
[22] Suzuki, F.; Zahler, W.L.; Emerich, D. W.: Acetoacetyl-CoA thiolase of Bradyrhizobium japonicum bacteroids: purification and properties. Arch. Biochem. Biophys., **254**, 272-281 (1987)
[23] Greenspan, M.D.; Yudkovitz, J.B.; Chen, J.S.; Hanf, D.P.; Chang, M.N.; Chiang, P.Y.C.; Chabala, J.C.; Alberts, A.W.: The inhibition of cytoplasmic acetoacetyl-CoA thiolase by a triyne carbonate (L-660, 631). Biochem. Biophys. Res. Commun., **163**, 548-553 (1989)
[24] Feigenbaum, J.; Schulz, H.: Thiolases of Escherichia coli: purification and chain length specificities. J. Bacteriol., **122**, 407-411 (1975)
[25] Duncombe, G.R.; Frerman, F.E.: Molecular and catalytic properties of the acetoacetyl-coenzyme A thiolase of Escherichia coli. Arch. Biochem. Biophys., **176**, 159-170 (1976)
[26] O'Connell, M.A.; Orr, G.; Shapiro, L.: Purification and characterization of fatty acid β-oxidation enzymes from Caulobacter crescentus. J. Bacteriol., **172**, 997-1004 (1990)

[27] Anderson, V.E.; Bahnson, B.J.; Wlassics, I.D.: The reaction of acetyldithio-CoA, a readily enolized anaolg of acetyl-CoA with thiolase from Zoogloea ramigera. J. Biol.Chem., **266**, 6255-6261 (1990)
[28] Heijden, R.v.d.; Verpoorte, R.; Duine, J.A.: Biosynthesis of 3S-hydroxy-3-methylglutaryl-coenzyme A in Catharanthus roseus: Acetoacetyl-CoA thiolase and HMG-CoA synthase show similar chromatographic behavior. Plant Physiol. Biochem., **32**, 807-812 (1994)
[29] Kanayama, N.; Himeda, Y.; Atomi, H.; Ueda, M.; Tanaka, A.: Expression of acetoacetyl-CoA thiolase isoenzyme genes of n-alkane-assimilating yeast, Candida tropicalis: isoenzymes in two intracellular compartments are derived from the same genes. J. Biochem., **122**, 616-621 (1997)
[30] Kim, S.A.; Copeland, L.: Acetyl coenzyme A acetyltransferase of Rhizobium sp. (Cicer) strain CC 1192. Appl. Environ. Microbiol., **63**, 3432-3437 (1997)
[31] Kanayama, N.; Ueda, M.; Atomi, H.; Tanaka, A.: Genetic evaluation of physiological functions of thiolase isoenzymes in the n-alkane-assimilating yeast Candida tropicalis. J. Bacteriol., **180**, 690-698 (1998)
[32] Honda, A.; Salen, G.; Nguyen, L.B.; Xu, G.; Tint, G.S.; Batta, A.K.; Shefer, S.: Regulation of early cholesterol biosynthesis in rat liver: effects of sterols, bile acids, lovastatin, and BM 15.766 on 3-hydroxy-3-methylglutaryl coenzyme A synthase and acetoacetyl coenzyme A thiolase activities. Hepatology, **27**, 154-159 (1998)
[33] Modis, Y.; Wierenga, R.K.: A biosynthetic thiolase in complex with a reaction intermediate: the crystal structure provides new insights into the catalytic mechanism. Structure, **7**, 1279-1290 (1999)
[34] Antonenkov, V.D.; Croes, K.; Waelkens, E.; Van Veldhoven, P.P.; Mannaerts, G.P.: Identification, purification and characterization of an acetoacetyl-CoA thiolase from rat liver peroxisomes. Eur. J. Biochem., **267**, 2981-2990 (2000)
[35] Lobo, S.; Florova, G.; Reynolds, K.A.: A Streptomyces collinus thiolase with novel acetyl-CoA:acyl carrier protein transacylase activity. Biochemistry, **40**, 11955-11964 (2001)
[36] Kursula, P.; Ojala, J.; Lambeir, A.M.; Wierenga, R.K.: The catalytic cycle of biosynthetic thiolase: a conformational journey of an acetyl group through four binding modes and two oxyanion holes. Biochemistry, **41**, 15543-15556 (2002)
[37] Hedl, M.; Sutherlin, A.; Wilding, E.I.; Mazzulla, M.; McDevitt, D.; Lane, P.; Burgner, J.W.; Lehnbeuter, K.R.; Stauffacher, C.V.; Gwynn, M.N.; Rodwell, V.W.: Enterococcus faecalis acetoacetyl-coenzyme A thiolase/3-hydroxy-3-methylglutaryl-coenzyme A reductase, a dual-function protein of isopentenyl diphosphate biosynthesis. J. Bacteriol., **184**, 2116-2122 (2002)
[38] Liu, T.; Liu, S.J.; Xue, Y.; Ma, Y.; Zhou, P.: Purification and characterization of an extremely halophilic acetoacetyl-CoA thiolase from a newly isolated Halobacterium strain ZP-6. Extremophiles, **6**, 97-102 (2002)
[39] Oeljeklaus, S.; Fischer, K.; Gerhardt, B.: Glyoxysomal acetoacetyl-CoA thiolase and 3-oxoacyl-CoA thiolase from sunflower cotyledons. Planta, **214**, 597-607 (2002)

Hydrogen-sulfide S-acetyltransferase 2.3.1.10

1 Nomenclature

EC number
2.3.1.10

Systematic name
acetyl-CoA:hydrogen-sulfide S-acetyltransferase

Recommended name
hydrogen-sulfide S-acetyltransferase

Synonyms
acetyltransferase, hydrogen sulfide
hydrogen sulfide acetyltransferase

CAS registry number
9029-92-9

2 Source Organism

<1> *Columba sp.* (pigeon [1]) [1]
<2> *Clostridium kluyveri* [1]

3 Reaction and Specificity

Catalyzed reaction
acetyl-CoA + hydrogen sulfide = CoA + thioacetate

Reaction type
acyl group transfer

Natural substrates and products
S acetyl-CoA + hydrogen sulfide <1, 2> (Reversibility: ? <1, 2> [1]) [1]
P CoA + thioacetate <1, 2> [1]

Substrates and products
S acetyl-CoA + hydrogen sulfide <1, 2> (<1,2> probably the reverse reaction of an enzyme that catalyzes the reaction of SH-CoA with thioacetate and sulfanilamide, 2-mercaptoethanol or choline [1]) (Reversibility: ? <1, 2> [1]) [1]
P CoA + thioacetate <1, 2> [1]

pH-Optimum
7 <1, 2> (<1,2> assay at [1]) [1]

Temperature optimum (°C)
30 <1, 2> (<1,2> assay at [1]) [1]

5 Isolation/Preparation/Mutation/Application

Source/tissue
liver <1> [1]

6 Stability

General stability information
<1, 2>, acetone fractionation inactivates [1]
<1, 2>, alcohol fractionation inactivates [1]
<1, 2>, ammonium sulfate fractionation inactivates [1]

References

[1] Brady, R.O.; Stadtman, E.R.: Enzymatic thioltransacetylation. J. Biol. Chem., **211**, 621-629 (1954)

Thioethanolamine S-acetyltransferase 2.3.1.11

1 Nomenclature

EC number
2.3.1.11

Systematic name
acetyl-CoA:thioethanolamine S-acetyltransferase

Recommended name
thioethanolamine S-acetyltransferase

Synonyms
thioethanolamine acetyltransferase
thioltransacetylase B

CAS registry number
9029-93-0

2 Source Organism

<1> *Columba sp.* (pigeon [1]) [1]
<2> *Clostridium kluyveri* [1]

3 Reaction and Specificity

Catalyzed reaction
acetyl-CoA + thioethanolamine = CoA + S-acetylthioethanolamine

Reaction type
acyl group transfer

Natural substrates and products
S acetyl-CoA + thioethanolamine <1, 2> (Reversibility: ? <1, 2> [1]) [1]
P CoA + S-acetylthioethanolamine <1, 2> [1]

Substrates and products
S acetyl-CoA + thioethanolamine <1, 2> (Reversibility: ? <1, 2> [1]) [1]
P CoA + S-acetylthioethanolamine <1, 2> [1]

Specific activity (U/mg)
7.1 <1, 2> [1]

5 Isolation/Preparation/Mutation/Application

Source/tissue
liver <1> [1]

Purification
<1, 2> [1]

References

[1] Brady, R.O.; Stadtman, E.R.: Enzymatic thioltransacetylation. J. Biol. Chem., **211**, 621-629 (1954)

Dihydrolipoamide S-acetyltransferase 2.3.1.12

1 Nomenclature

EC number
2.3.1.12

Systematic name
acetyl-CoA:dihydrolipoamide S-acetyltransferase

Recommended name
dihydrolipoamide S-acetyltransferase

Synonyms
DHLTA
E2
E2p
acetyltransferase, lipoate
dihydrolipoate acetyltransferase
dihydrolipoic transacetylase
dihydrolipoyl acetyltransferase
lipoate acetyltransferase
lipoate transacetylase
lipoic acetyltransferase
lipoic acid acetyltransferase
lipoic transacetylase
lipoylacetyltransferase
thioltransacetylase A
transacetylase X

CAS registry number
9032-29-5

2 Source Organism

<1> *Escherichia coli* (<1> various wild-type and mutant strains [28]; K-12, wild-type and mutant strains [39]; K-12 [41]; wild-type and mutants [72]) [1, 16, 18, 21, 22, 26-28, 33-36, 39, 41-45, 71, 72]
<2> *Geobacillus stearothermophilus* [2, 13, 46, 53, 56, 60, 61, 65]
<3> *Pelobacter carbinolicus* [3]
<4> *Saccharomyces cerevisiae* [4, 58]
<5> *Azotobacter vinelandii* (wild-type and mutants [72]) [5-8, 12, 18, 20, 52, 63, 71, 72]

<6> *Bos taurus* [9-11, 17, 23, 29-32, 38, 40, 54, 55, 70]
<7> *Homo sapiens* [14, 64, 66, 75]
<8> *Bacillus subtilis* [19, 57]
<9> *mammalia* [46, 54]
<10> *Rattus norvegicus* [15, 24]
<11> *Salmonella typhimurium* [25]
<12> *Sus scrofa* [37, 54]
<13> *Hansenula miso* [47]
<14> *Columba sp.* (pigeon [48]) [48]
<15> *Brassica oleracea* (broccoli [49]) [49]
<16> *Brassica oleracea var. botrytis* (cauliflower [49]) [49]
<17> *Neurospora crassa* [50]
<18> *Rhodospirillum rubrum* [51]
<19> *Neisseria meningitidis* [59]
<20> *Arabidopsis thaliana* [62, 68]
<21> *Solanum tuberosum* (potato [67]) [67]
<22> *Proteobacter carbinolicus* [69]
<23> *Zea mays* (maize [73]) [73]
<24> *Mus musculus* (cell lines BML1-ME, NIH3T3, PA317 and PE 501 [74]) [74]

3 Reaction and Specificity

Catalyzed reaction

acetyl-CoA + dihydrolipoamide = CoA + S-acetyldihydrolipoamide (<2,6> reaction mechanism [38,60])

Reaction type

acyl group transfer

Natural substrates and products

S dihydrolipoamide + acetyl-CoA <1-24> (<6> acetyl group is transferred from the S-acetyldihydrolipoyl moiety bound covalently to the transacetylase to CoA [38]) (Reversibility: r <6> [38]; ? <1-24> [1-37, 39-75]) [1-75]
P S-acetyldihydrolipoamide + CoA <1-24> [1-75]
S Additional information <3> (<3> protein bound acetyldihydrolipoyl moiety + CoA [3]) [3]
P ?

Substrates and products

S dihydrolipoamide + acetyl-CoA <1-24> (Reversibility: r <6> [38]; ? <1-24> [1-37, 39-75]) [1-75]
P S-acetyldihydrolipoamide + CoA <1-24> [1-75]

Inhibitors

1,2-diselenolane-3-valeric acid <6> (<6> competitive with respect to dihydrolipoamide and non-competitive with respect to acetyl-CoA [38]) [38]
CoA <6> (<6> competitive to acetyl-CoA, non-competitive to dihydrolipoamide [38]) [38]

N-acetylimidazole <1> (<1> modifies tyrosyl residues and dissociates the enzyme into subunits [42]) [42]
S-acetyldihydrolipoamide <6> (<6> competitive to dihydrolipoamide, non-competitive to acetyl-CoA [38]) [38]
guanidine hydrochloride <7> (<7> 50% inhibition at 0.3 M, complete inhibition at 0.7-1 M [66]) [66]
iodine <1> (<1> complete inactivation at 2.5 mol iodine per mol of subunit [43]) [43]
lipoamidase <6> (<6> complete loss of activity after 3 h [38]) [38]
palmitoyl-CoA <6> (<6> competitive to acetyl-CoA, non-competitive to dihydrolipoamide [38]) [38]

Turnover number (min^{-1})

0.25 <5> (acetyl-CoA, <5> with dihydrolipoamide, H610C mutant [63]) [63]
0.86 <5> (acetyl-CoA, <5> with dihydrolipoamide, S558A mutant [63]) [63]
16.9 <5> (acetyl-CoA, <5> with dihydrolipoamide, N614D mutant [63]) [63]
144 <5> (acetyl-CoA, <5> with dihydrolipoamide, wild-type enzyme [63]) [63]

Specific activity (U/mg)

3.89 <20> (<20> fusion protein with glutathione S-transferase [68]) [68]
4.09 <20> (<20> after removal of glutathione S-transferase moiety [68]) [68]
19.4 <7> [75]
19.7 <12> [37]
47.1 <1> (<1> catalytic domain [71]) [71]
58.2 <19> [59]
64.1 <5> [72]
79.4 <5> [5]
119 <5> (<5> catalytic domain [71]) [71]
191 <23> [73]
509 <3, 22> [3, 69]
Additional information <1, 6> (<1> specific activities in various wild-type and mutant strains [28]; <6> specific activities of fragments after protease cleavage [30]; <6> specific activity of native enzyme and proteolytic fragments in various experiments [31]; <1> specific activity after growth on various carbon sources [39]) [28, 30, 31, 33, 39]

K_m-Value (mM)

0.013 <2> (acetyl-CoA, <2> catalytic domain after expression in Escherichia coli [60]) [60]
0.022 <5> (acetyl-CoA, <5> wild-type enzyme [63]) [63]
0.028 <5> (acetyl-CoA, <5> S558A mutant [63]) [63]
0.03-0.1 <6> (acetyl-CoA) [38]
0.036 <5> (acetyl-CoA, <5> H610C mutant [63]) [63]
0.052 <5> (acetyl-CoA, <5> N614D mutant [63]) [63]
0.13-0.4 <6> (dihydrolipoamide) [38]
1.08 <5> (dihydrolipoamide, <5> N614D mutant [63]) [63]

1.2 <2> (dihydrolipoamide, <2> catalytic domain after expression in Escherichia coli [60]) [60]
2.2 <3, 22> (dihydrolipoamide) [3, 69]
4.2 <1> (dihydrolipoamide) [39]
5.3 <5> (dihydrolipoamide, <5> H610C mutant [63]) [63]
5.37 <5> (dihydrolipoamide, <5> wild-type enzyme [63]) [63]
6.4 <5> (dihydrolipoamide, <5> S558A mutant [63]) [63]

pH-Optimum
6.7 <22> [69]
7.8-8.5 <12> [37]
Additional information <3> (<3> reactivity depending on buffer system [3]) [3]

Temperature optimum (°C)
25 <12> [37]

4 Enzyme Structure

Molecular weight
30000 <2> (<2> gel filtration in the presence of guanidinium hydrochloride [60]) [60]
46930 <23> (<23> calculation from nucleotide sequence [73]) [73]
50080 <20> (<20> calculation from nucleotide sequence [68]) [68]
51000 <6> (<6> gel filtration [32]) [32]
52000-53000 <23> (<23> SDS-PAGE [73]) [73]
57000 <2> (<2> SDS-PAGE [56]) [56]
58000-59000 <8> (<8> SDS-PAGE [57]) [57]
58890 <24> (<24> calculation from nucleotide sequence [74]) [74]
59000 <24> (<24> SDS-PAGE [74]) [74]
60000 <20> (<20> calculation from nucleotide sequence [62]; <20> mature protein, SDS-PAGE [68]) [62, 68]
60100-64500 <1> (<1> sedimentation equilibrium, depending on buffer [33]) [33]
63000 <5> (<5> sedimentation equilibrium [8]) [8]
64260 <2> (<2> calculation from nucleotide sequence [53]) [53]
64910 <5> (<5> calculation from nucleotide sequence [6]) [6]
65960 <1> (<1> calculation from nucleotide sequence [22]) [22]
67000 <20> (<20> plastid precursor protein, SDS-PAGE [68]) [68]
68000 <10> (<10> SDS-PAGE [24]) [24]
70000 <6, 19> (<6,19> SDS-PAGE [31,59]) [31, 59]
73000-75000 <6> (<6> SDS-PAGE [11]) [11]
74000 <6, 12> (<6> SDS-PAGE [29,32,55]; <12> SDS-PAGE and sedimentation equilibrium data [37]) [29, 32, 37, 55]
78000 <1> (<1> SDS-PAGE [33]) [33]

80000 <1, 20> (<1,20> SDS-PAGE [21,62]; <1> SDS-PAGE and calculated from amino acid sequence [41]) [21, 41, 62]
81000 <1> (<1> SDS-PAGE, gel filtration [27]) [27]
83000 <1, 5> (<5> SDS-PAGE [8]; <1> wild type enzyme, SDS-PAGE [16]) [8, 16]
87000 <11> (<11> SDS-PAGE [25]) [25]
127000 <10> (<10> fusion protein of 11000 Da fragment with β-galactosidase, SDS-PAGE [15]) [15]
150000 <1> (<1> no component of pyruvate dehydrogenase complex, sucrose density gradient centrifugation [39]) [39]
200000 <5, 7> (<5> sedimentation equilibrium centrifugation, light scattering, sedimentation and diffusion coefficients [20]; <7> at 1.8 - 2.8 M guanidine hydrochloride, complex dissociates at higher guanidine hydrochloride levels to a monomeric form with MW 82000 [66]) [20, 66]
500000 <3, 22> (<3> above, gel filtration [3]; <22> gel filtration [69]) [3, 69]
530000 <5> (<5> gel filtration [7]) [7, 8]
670000 <1, 5> (<1,5> gel filtration [71]) [71]
864000 <1> (<1> crystallization experiments [34]) [34]
Additional information <1, 2, 6, 5, 19, 21> (<1,2> molecular weight of lipolyl domains [1,2,21]; <1,5,6,8,19,21> molecular weight of proteolytic fragments [8,11,19,29,30,31,32,33,59,67,71]; <6> molecular weight of fragments after treatment with trypsin [10]; <1> molecular weight of E2 complexed with E1 or E3 [36]; <2> high molecular mass species in the absence of guanidinium hydrochloride [60]) [1, 2, 8, 10, 11, 19, 21, 29, 30, 31, 32, 36, 59, 60, 67, 71]
1000000 <1> (<1> sedimentation equilibrium centrifugation [44]) [44]
1470000 <1> (<1> sedimentation equilibrium experiments in 0.83 M acetic and and 0.005 M sodium chloride [45]) [45]
1548000 <1> (<1> native enzyme [33]) [33]
1560000 <5> (<5> fluorescence anisotropy decay experiments [12]) [12]
1800000-1970000 <12> (<12> calculation from sedimentation and diffusion constants, meniscus depletion method [37]) [37]
2700000 <23> (<23> gel filtration [73]) [73]
2800000 <6> (<6> calculation from subunit composition of pyruvate dehydrogenase complex [30]) [30]
3100000 <9> [46]

Subunits

dimer <1> (<1> multimer formation is probably lost by loss of a small segment during genetic rearrangement [39]; <1> 2 * 36000, in the presence of dilute acetic acid [45]) [39, 45]
polymer <1-6, 7, 9, 12, 21, 22> (<9> 60 * 52000 [46]; <6> domain structure of subunits: lipoyl domain MW 28000, subunit binding domain MW 26000, sedimentation equilibrium [29]; <6> gel filtration in presence of guanidine-HCl, anomalous migration on SDS [32]; <2,6> 60 subunits [30,46]; <6> 30 * 70000, SDS-PAGE, each polypeptide chains has 2 domains [31]; <5> 24 * 27500, SDS-PAGE, light scattering experiments, quarternary structure [7]; <5> 30 * 66000, SDS-PAGE, sedimentation equilibrium [20]; <2> x * 46265,

calculation from nucleotide sequence [53]; <2> primary structure of lipoyl domain [13]; <4> x * 48546, calculation from nucleotide sequence [4]; <1> 24 * 65959, calculation from nucleotide sequence [22]; <1> domain structure [1]; <1> 22 or 24 subunits [35]; <3> x * 60000, SDS-PAGE [3]; <2,9,21> stoichiometry of pyruvate dehydrogenase complexes [46,54,56,67]; <6> 60 * 52000 [23]; <1> 24 identical subunits, data from crystallographic, biochemical and electron microscopic methods [34]; <12> 24 * 74000 SDS-PAGE, gel filtration in 6 M guanidine-HCl, sedimentation equilibrium in 6 M guanidine-HCl [37]; <12> 24 * 48000 [54]; <1> 24 * 40000, linked by noncovalent bonds [42]; <1> 24 * 36000, linked by noncovalent bonds [44]; <2> 60 polypeptide chains, crystal structure [65]; <6> 60 * 52000 [54]; <22> 24 * 60000, SDS-PAGE [69]; <6> 60 subunits [70]; <1,5> 24 subunits [71]; <7> 60 * 42000, SDS-PAGE [75]) [1, 3, 4, 7, 8, 9, 13, 20, 22, 29-33, 34, 35, 37, 42, 46, 53, 54, 56, 65, 67, 69, 70, 71, 75]

trimer <7> (<7> 3 * 82000, gel filtration, trimeric form occurs in solutions with 4 M guanidine hydrochloride [66]) [66]

Additional information <6, 21> (<21> treatment with NaCl results in 2 fractions with E2 activity, MW 55000 and MW 78000, SDS-PAGE, gel filtration [67]; <6> enzyme forms stable aggregates [70]) [67, 70]

Posttranslational modification

lipoprotein <1, 4-6, 7, 9, 12, 19, 23> (<1> 3 lipoyl domains separated by regions rich in alanine and proline [16,21]; <1> 3 lipoyl domains [1]; <4> 1 lipoyl domain [4]; <5> dynamic properties of lipoyl domains [12]; <9,23> 1 lipoyl domain [46,73]; <6> 30 polypeptide chains with 2 lipoyl domains each [31]; <12> 23 mol of lipoic acid per mol of enzyme [37]; <1> 1 molecule lipoic acid per polypeptide chain [44]; <19> 2 lipoyl domains per polypeptide chain [59]; <7> 2 lipoyl domains [75]) [1, 4, 12, 16, 21, 29, 31, 33, 34, 37, 38, 44, 46, 54, 59, 61, 73, 75]

5 Isolation/Preparation/Mutation/Application

Source/tissue

flower <20> [68]
heart <6, 10, 12> [9-11, 17, 23, 24, 29, 37, 40, 54, 55, 70]
kidney <6> [9-11, 23, 30-32, 38, 40, 54, 55, 70]
leaf <20> [68]
liver <7, 10> [14, 15]
muscle <14> [48]
mycelium <17> [50]
root <20> [68]
seedling <23> [73]
silique <20> [68]

Localization

chloroplast <20> [68]
mitochondrial inner membrane <6> (<6> pyruvate dehydrogenase complex bound to inner mitochondrial membrane [17]) [17]
mitochondrion <6, 9, 15, 16, 23> (<9> pyruvate dehydrogenase complex bound to inner mitochondrial membrane [54]) [31, 40, 49, 54, 55, 73]
outer membrane <19> [59]
soluble <5> (<5> wild-type enzyme is soluble, whereas mutant enzyme with deletion of all three lipoyl domains aggregates [72]) [72]

Purification

<1> (isolation of lipoyl domains [1,21]; wild-type and mutants with deletions of lipolyl domains [16]; E1-E2 complex [26]; pyruvate dehydrogenase complex and tryptic fragments of E2 [33]; pyruvate dehydrogenase complex, stoichiometry E1:E2:E3 is 1.56:1:0.89 [35]; pure [41]; fusion protein of 6 amino acids form β-galactosidase, the apa-4 region and the catalytic domain of E2 [71]) [1, 2, 16, 18, 21, 26, 33, 35, 36, 41, 42, 43, 44, 45, 71, 72]
<2> (peptides after limited proteolysis [13]) [13, 56, 60, 61]
<3> [3]
<4> [58]
<5> (proteolytic fragments, isolation of lipoyl domain and catalytic domain [8]; catalytic domain mutants [63]; fusion protein of 6 amino acids form β-galactosidase, the papa-4 region and the catalytic domain of E2 [71]) [5, 6, 8, 52, 63, 71, 72]
<6> (preparation of E2-X subcomplex: E2 i.e. EC 2.3.1.12, X i.e. component X of mammalian pyruvate dehydrogenase complex [9]; from pyruvate dehydrogenase complex [31,40,55]; purification of tryptic fragments [29]) [9, 11, 12, 29, 31, 40, 55]
<7> (highly [75]) [14, 66, 75]
<8> (pyruvate dehydrogenase complex, composed of EC 1.2.4.1, EC 1.8.1.4, EC 2.3.1.12 [57]) [19, 57]
<10> (from pyruvate dehydrogenase complex [24]) [24]
<11> (E2-E3 subcomplex of pyruvate dehydrogenase, i.e. dihydrolipoamide acetyltransferase and dehydrogenase [25]) [25]
<12> [37]
<13> (purification of pyruvate dehydrogenase complex, composed of EC 1.2.4.1, EC 1.8.1.4, EC 2.3.1.12 [47]) [47]
<14> (purification from pyruvate dehydrogenase complex, composed of EC 1.2.4.1, EC 1.8.1.4, EC 2.3.1.12 [48]) [48]
<15> (pyruvate dehydrogenase complex, composed of EC 1.2.4.1, EC 1.8.1.4, EC 2.3.1.12 [49]) [49]
<16> (pyruvate dehydrogenase complex, composed of EC 1.2.4.1, EC 1.8.1.4, EC 2.3.1.12 [49]) [49]
<17> (pyruvate dehydrogenase complex, composed of EC 1.2.4.1, EC 1.8.1.4, EC 2.3.1.12 [50]) [50]
<18> (pyruvate dehydrogenase complex, composed of EC 1.2.4.1, EC 1.8.1.4. EC 2.3.1.12 [51]) [51]

<19> [59]
<20> (fusion protein with glutathione S-transferase [68]) [68]
<21> [67]
<22> (homogeneity [69]) [69]
<23> [73]

Renaturation

<1> (reconstitution of E1 with E2-E3 complex with a stoichiometry of E1:E2:E3 of 2:1:1) [26]
<1> (reconstitution of pyruvate dehydrogenase complex, maximum activity is produced when transacetylase accommodates 12 pyruvate dehydrogenase dimers and 6 flavoprotein dimers) [36]
<1> (reconstitution of pyruvate dehydrogenase complex, stoichiometry pyruvate dehydrogenase:transacetylase:flavoprotein is 1:0.35:0.4) [42]
<1> (treatment with dilute acetic acid solution results in dissociation into inactive subunits with MW 70000, removal of the acid results in restoration of enzymatic activity) [45]
<2> (treatment with guanidine hydrochloride and its subsequent removal results in little recovery of the core, but full recovery of X) [61]
<2> (treatment with guanidine hydrochloride results in dissociation into subunits, removal of guanidine hydrochloride results in refolding of the enzyme with 95% of the original activity) [60]
<3> (reconstitution of acetoin dehydrogenase enzyme system [3]; <11> reconstitution of pyruvate dehydrogenase complex with components from Escherichia coli and Salmonella typhimurium) [25]
<5> (reconstitution with components from Escherichia coli and Azotobacter vinelandii, optimum catalytic stoichiometry of E1:E2:E3 is 1:1:0.5) [18]
<7> (treatment with guanidine hydrochloride and its subsequent removal results in refolding) [66]
<12> (reconstitution of pyruvate dehydrogenase complex, optimum catalytic stoichiometry of E1:E2:E3 is 5.2:11.5:2) [37]
<6, 22> (reconstitution of pyruvate dehydrogenase complex) [29, 69]

Crystallization

<1> (E2-E3 subcomplex of pyruvate dehydrogenase, E2: EC 2.3.1.12, E3: EC 1.8.1.4, various crystallization conditions [34]) [34]
<2> (E3 dimer with E2 and a small binding domain E2BD, hanging drop vapor diffusion method [65]) [65]
<5> (hanging drop vapor diffusion method [63, 71]) [63, 71]
<6> (pyruvate dehydrogenase complex [40]) [40]

Cloning

<1> (wild-type and mutants [72]) [1, 22, 28, 71, 72]
<2> (expression of subgene for lipoyl domain in Escherichia coli [2]; catalytic domain, residues 173-427 [60]) [2, 53, 60]
<4, 5> (wild-type and deletion mutants, expression in Escherichia coli [5]; wild-type and mutants [72]) [4, 5, 6, 63, 71, 72]
<7> [14, 15]

<7> [64, 75]
<19> [59]
<20> [62, 68]
<23> [73]
<24> (knock-out mutants [74]) [74]

Engineering

F231A <7> (<7> ability to be post-translationally lipoylated remains [64]) [64]
G610C <5> (<5> 500fold reduced enzymatic activity [63]) [63]
I229A <7> (<7> ability to be post-translationally lipoylated remains [64]) [64]
I229A/F231A <7> (<7> ability to be post-translationally lipoylated remains, far-UV CD spectrum differs from wild-type enzyme [64]) [64]
K244Q <1> (<1> unlipoylatable domain [1]) [1]
L189A <7> (<7> ability to be post-translationally lipoylated remains [64]) [64]
M194A <7> (<7> ability to be post-translationally lipoylated remains [64]) [64]
M197A <7> (<7> ability to be post-translationally lipoylated remains [64]) [64]
N614D <5> (<5> 9fold reduced enzymatic activity [63]) [63]
S558A <5> (<5> 200fold reduced enzymatic activity [63]) [63]
V188A <7> (<7> ability to be post-translationally lipoylated remains, far-UV CD spectrum differs from wild-type enzyme [64]) [64]

6 Stability

pH-Stability

3-11 <1> (<1> isolated E. coli lipoyl domains stay soluble in this range [1]) [1]
6-8.4 <5> (<5> stable [63]) [63]

Temperature stability

60 <12> (<12> stable up to [37]) [37]
70 <1, 2> (<1> 1 h stable, isolated lipoyl domains [1]; <2> incubation converts core of the enzyme into an unidentified active molecular form X, loss of 10% enzyme activity after 5 h [61]) [1, 61]
78 <2> (<2> temperatures below result in less than 5% loss in enzyme activity [61]) [61]
80 <2> (<2> partly precipitation above [61]) [61]

General stability information

<12>, freezing and thawing does not cause loss of activity [37]
<3, 6, 22>, sensitive to proteolysis [3, 54, 69]

Storage stability

<5>, liquid nitrogen, 50 mM potassium phosphate buffer pH 7.0, 0.5 mM EDTA, 0.05 mM phenylmethylsulfonylfluoride [8]
<12>, -18°C, 0.05 mM potassium phosphate buffer, pH 7.0, 0.5 mM EDTA, more than 6 months [37]

References

[1] Ali, S.T.; Guest, J.R.: Isolation and characterization of lipoylated and unlipoylated domains of the E2p subunit of the pyruvate dehydrogenase complex of Escherichia coli. Biochem. J., **271**, 139-145 (1990)
[2] Dardel, F.; Packman, L.C.; Perham, R.N.: Expression in Escherichia coli of a sub-gene encoding the lipoyl domain of the pyruvate dehydrogenase complex of Bacillus stearothermophilus [published erratum appears in FEBS Lett 1990 Jul 30;268(1):306]. FEBS Lett., **264**, 206-210 (1990)
[3] Oppermann, F.B.; Schmidt, B.; Steinbuchel, A.: Purification and characterization of acetoin:2,6-dichlorophenolindophenol oxidoreductase, dihydrolipoamide dehydrogenase, and dihydrolipoamide acetyltransferase of the Pelobacter carbinolicus acetoin dehydrogenase enzyme system. J. Bacteriol., **173**, 757-767 (1991)
[4] Niu, X.D.; Browning, K.S.; Behal, R.H.; Reed, L.J.: Cloning and nucleotide sequence of the gene for dihydrolipoamide acetyltransferase from Saccharomyces cerevisiae. Proc. Natl. Acad. Sci. USA, **85**, 7546-7550 (1988)
[5] Hanemaaijer, R.; Westphal, A.H.; Berg, A.; van Dongen, W.; de Kok, A.; Veeger, C.: The gene encoding dihydrolipoyl transacetylase from Azotobacter vinelandii. Expression in Escherichia coli and activation and isolation of the protein. Eur. J. Biochem., **181**, 47-53 (1989)
[6] Hanemaaijer, R.; Janssen, A.; de Kok, A.; Veeger, C.: The dihydrolipoyltransacetylase component of the pyruvate dehydrogenase complex from Azotobacter vinelandii. Molecular cloning and sequence analysis. Eur. J. Biochem., **174**, 593-599 (1988)
[7] Hanemaaijer, R.; Westphal, A.H.; van der Heiden, T.; de Kok, A.; Veeger, C.: The quaternary structure of the dihydrolipoyl transacetylase component of the pyruvate dehydrogenase complex from Azotobacter vinelandii. A reconsideration. Eur. J. Biochem., **179**, 287-292 (1989)
[8] Hanemaaijer, R.; de Kok, A.; Jolles, J.; Veeger, C.: The domain structure of the dihydrolipoyl transacetylase component of the pyruvate dehydrogenase complex from Azotobacter vinelandii. Eur. J. Biochem., **169**, 245-252 (1987)
[9] Powers-Greenwood, S.L.; Rahmatullah, M.; Radke, G.A.; Roche, T.E.: Separation of protein X from the dihydrolipoyl transacetylase component of the mammalian pyruvate dehydrogenase complex and function of protein X. J. Biol. Chem., **264**, 3655-3657 (1989)
[10] Rahmatullah, M.; Gopalakrishnan, S.; Andrews, P.C.; Chang, C.L.; Radke, G.A.; Roche, T.E.: Subunit associations in the mammalian pyruvate dehydrogenase complex. Structure and role of protein X and the pyruvate dehy-

drogenase component binding domain of the dihydrolipoyl transacetylase component. J. Biol. Chem., **264**, 2221-2227 (1989)
[11] Rahmatullah, M.; Gopalakrishnan, S.; Radke, G.A.; Roche, T.E.: Domain structures of the dihydrolipoyl transacetylase and the protein X components of mammalian pyruvate dehydrogenase complex. Selective cleavage by protease Arg C. J. Biol. Chem., **264**, 1245-1251 (1989)
[12] Hanemaaijer, R.; Masurel, R.; Visser, A.J.W.G.; de Kok, A.; Veeger, C.: Time-resolved fluorescence studies on the dihydrolipoyl transacetylase (E2) component of the pyruvate dehydrogenase complex from Azotobacter vinelandii. FEBS Lett., **238**, 285-288 (1988)
[13] Packman, L.C.; Borges, A.; Perham, R.N.: Amino acid sequence analysis of the lipoyl and peripheral subunit-binding domains in the lipoate acetyltransferase component of the pyruvate dehydrogenase complex from Bacillus stearothermophilus. Biochem. J., **252**, 79-86 (1988)
[14] Thekkumkara, T.J.; Jesse, B.W.; Ho, L.; Raefsky, C.; Pepin, R.A.; Javed, A.A.; Pons, G.; Patel, M.S.: Isolation of a cDNA clone for the dihydrolipoamide acetyltransferase component of the human liver pyruvate dehydrogenase complex. Biochem. Biophys. Res. Commun., **145**, 903-907 (1987)
[15] Matuda, S.; Matuo, S.; Nakano, K.; Saheki, T.: Molecular cloning of cDNA for rat liver lipoate acetyltransferase. A component of pyruvate dehydrogenase complex. Biochem. Biophys. Res. Commun., **142**, 953-957 (1987)
[16] Guest, J.R.; Lewis, H.M.; Graham, L.D.; Packman, L.C.; Perham, R.N.: Genetic reconstruction and functional analysis of the repeating lipoyl domains in the pyruvate dehydrogenase multienzyme complex of Escherichia coli. J. Mol. Biol., **185**, 743-754 (1985)
[17] Phelps, A.; Lindsay, J.G.: Mammalian pyruvate dehydrogenase complex binds tightly to the mitochondrial inner membrane. Biochem. Soc. Trans., **14**, 893 (1986)
[18] De Kok, A.; Westphal, A.H.: Hybrid pyruvate dehydrogenase complexes reconstituted from components of the complexes from Escherichia coli and Azotobacter vinelandii. Eur. J. Biochem., **152**, 35-41 (1985)
[19] Lowe, P.N.; Hodgson, J.A.; Perham, R.N.: Limited proteolysis of the pyruvate dehydrogenase multienzyme complex of Bacillus subtilis. Biochem. J., **225**, 249-253 (1985)
[20] Bosma, H.J.; de Kok, A.; van Markwijk, B.W.; Veeger, C.: The size of the pyruvate dehydrogenase complex of Azotobacter vinelandii. Association phenomena. Eur. J. Biochem., **140**, 273-280 (1984)
[21] Packman, L.C.; Hale, G.; Perham, R.N.: Repeating functional domains in the pyruvate dehydrogenase multienzyme complex of Escherichia coli. EMBO J., **3**, 1315-1319 (1984)
[22] Guest, J.R.; Darlison, M.G.; Spencer, M.E.; Stephens, P.E.: Cloning and sequence analysis of the pyruvate and 2-oxoglutarate dehydrogenase complex genes of Escherichia coli. Biochem. Soc. Trans., **12**, 220-223 (1984)
[23] Wu, T.L.; Reed, L.J.: Subunit binding in the pyruvate dehydrogenase complex from bovine kidney and heart. Biochemistry, **23**, 221-226 (1984)
[24] Matuda, S.; Shirahama, T.; Saheki, T.; Miura, S.; Mori, M.: Purification and immunochemical studies of pyruvate dehydrogenase complex from rat

heart, and cell-free synthesis of lipoamide dehydrogenase, a component of the complex. Biochim. Biophys. Acta, **741**, 86-93 (1983)
[25] Seckler, R.; Binder, R.; Bisswanger, H.: Purification and properties of the pyruvate dehydrogenase complex from Salmonella typhimurium and formation of hybrids with the enzyme complex from Escherichia coli. Biochim. Biophys. Acta, **705**, 210-217 (1982)
[26] Graupe, K.; Abusaud, M.; Karfunkel, H.; Bisswanger, H.: Reassociation of the pyruvate dehydrogenase complex from Escherichia coli: Kinetic measurements and binding studies by resonance energy transfer. Biochemistry, **21**, 1386-1394 (1982)
[27] Danson, M.J.; Porteous, C.E.: Pyruvate dehydrogenase multienzyme complex of Escherichia coli: determination of the Mr of the lipoate acetyltransferase component. FEBS Lett., **133**, 112-114 (1981)
[28] Guest, J.R.; Stephens, P.E.: Molecular cloning of the pyruvate dehydrogenase complex genes of Escherichia coli. J. Gen. Microbiol., **121**, 277-292 (1980)
[29] Bleile, D.M.; Hackert, M.L.; Pettit, F.H.; Reed, L.J.: Subunit structure of dihydrolipoyl transacetylase component of pyruvate dehydrogenase complex from bovine heart. J. Biol. Chem., **256**, 514-519 (1981)
[30] Kresze, G.B.; Ronft, H.: Bovine kidney pyruvate dehydrogenase complex. Limited proteolysis and molecular structure of the lipoate acetyltransferase component. Eur. J. Biochem., **112**, 589-599 (1980)
[31] Machicao, F.; Wieland, O.H.: Subunit structure of dihydrolipoamide acetyltransferase component of pyruvate dehydrogenase complex from bovine kidney. Hoppe-Seyler's Z. Physiol. Chem., **361**, 1093-1106 (1980)
[32] Kresze, G.B.; Dietl, B.; Ronft, H.: Mammalian lipoate acetyltransferase: molecular weight determination by gel filtration in the presence of guanidinium chloride. FEBS Lett., **112**, 48-50 (1980)
[33] Bleile, D.M.; Munk, P.; Oliver, R.M.; Reed, L.J.: Subunit structure of dihydrolipoyl transacetylase component of pyruvate dehydrogenase complex from Escherichia coli. Proc. Natl. Acad. Sci. USA, **76**, 4385-4389 (1979)
[34] Fuller, C.C.; Reed, L.J.; Oliver, R.M.; Hackert, M.L.: Crystallization of a dihydrolipoyl transacetylase - dihydrolipoyl dehydrogenase subcomplex and its implications regarding the subunit structure of the pyruvate dehydrogenase complex from Escherichia coli. Biochem. Biophys. Res. Commun., **90**, 431-438 (1979)
[35] Danson, M.J.; Hale, G.; Johnson, P.; Perham, R.N.: Molecular weight and symmetry of the pyruvate dehydrogenase multienzyme complex of Escherichia coli. J. Mol. Biol., **129**, 603-617 (1979)
[36] Reed, L.J.; Pettit, F.H.; Eley, M.H.; Hamilton, L.; Collins, J.H.; Oliver, R.M.: Reconstitution of the Escherichia coli pyruvate dehydrogenase complex. Proc. Natl. Acad. Sci. USA, **72**, 3068-3072 (1975)
[37] Hamada, M.; Otsuka, K.I.; Tanaka, N.; Ogasahara, K.; Koike, K.; Hiraoka, T.; Koike, M.: Purification properties and subunit composition of pig heart lipoate acetyltransferase. J. Biochem., **78**, 187-197 (1975)
[38] Butterworth, P.J.; Tsai, S.; Eley, M.H.; Roche, T.E.; Reed, L.J.: A kinetic study of dihydrolipoyl transacetylase from bovine kidney. J. Biol. Chem., **250**, 1921-1925 (1975)

[39] Bisswanger, H.; Henning, U.: A new dihydrolipoamide transacetylase in Escherichia coli K12. Biochim. Biophys. Acta, **321**, 143-148 (1973)
[40] Linn, T.C.; Pelley, J.W.; Pettit, F.H.; Hucho, F.; Randall, D.D.; Reed, L.J.: α-Keto acid dehydrogenase complexes. XV. Purification and properties of the component enzymes of the pyruvate dehydrogenase complexes from bovine kidney and heart. Arch. Biochem. Biophys., **148**, 327-342 (1972)
[41] Vogel, O.; Beikirch, H.; Muller, H.; Henning, U.: The subunit structure of the Escherichia coli K-12 pyruvate dehydrogenase complex. The dihydrolipoamide transacetylase component. Eur. J. Biochem., **20**, 169-178 (1971)
[42] Schwartz, E.R.; Reed, L.J.: α-Keto acid dehydrogenase complexes. XII. Effects of acetylation on the activity and structure of the dihydrolipoyl transacetylase of Escherichia coli. J. Biol. Chem., **244**, 6074-6079 (1969)
[43] Schwartz, E.; Reed, L.J.: α-Keto acid dehydrogenase complexes. IX. Effects of iodination of the tyrosyl residues on the properties of the dihydrolipoyl transacetylase of Escherichia coli. J. Biol. Chem., **243**, 639-643 (1968)
[44] Henney, H.R.; Willms, C.R.; Muramatsu, T.; Mukherjee, B.B.; Reed, L.J.: α-Keto acid dehydrogenase complexes. VII. Isolation and partial characterization of the polypeptide chains in the dihydrolipoyl transacetylase of Escherichia coli. J. Biol. Chem., **242**, 898-901 (1967)
[45] Willms, C.R.; Oliver, R.M.; Henney, H.R.; Mukherjee, B.B.; Reed, L.J.: α-Keto acid dehydrogenase complexes. VI. Dissociation and reconstitution of the dihydrolipoyl transacetylase of Escherichia coli. J. Biol. Chem., **242**, 889-897 (1967)
[46] Reed, L.J.; Yeaman, S.J.: Pyruvate dehydrogenase. The Enzymes, 3rd Ed. (Boyer, P.D., ed.), **18**, 77-95 (1987)
[47] Harada, T.; Hirabayashi, T.: Pyruvate dehydrogenase complex from Hansenula. Methods Enzymol., **89**, 420-423 (1982)
[48] Furuta, S.; Hashimoto, T.: Pyruvate dehydrogenase complex from pigeon breast muscle. Methods Enzymol., **89**, 414-420 (1982)
[49] Randall, D.D.: Pyruvate dehydrogenase complex from broccoli and cauliflower. Methods Enzymol., **89**, 408-414 (1982)
[50] Harding, R.W.; Caroline, D.F.; Wagner, R.P.: Pyruvate dehydrogenase complex from Neurospora. Methods Enzymol., **89**, 386-391 (1982)
[51] Luderitz, R.; Klemme, J.H.: Isolation and characterization of a membrane-bound pyruvate dehydrogenase complex from the phototropic bacterium Rhodopseudomonas capsulata. Z. Naturforsch. C, **32c**, 351-361 (1977)
[52] De Graaf-Hess, A.C.; de Kok, A.: Binding and dissociation of the pyruvate dehydrogenase complex of Azotobacter vinelandii on thiol–Sepharose. FEBS Lett., **143**, 261-264 (1982)
[53] Borges, A.; Hawkins, C.F.; Packman, L.C.; Perham, R.N.: Cloning and sequence analysis of the genes encoding the dihydrolipoamide acetyltransferase and dihydrolipoamide dehydrogenase components of the pyruvate dehydrogenase multienzyme complex of Bacillus stearothermophilus. Eur. J. Biochem., **194**, 95-102 (1990)
[54] Yeaman, S.J.: The mammalian 2-oxoacid dehydrogenases: a complex family. Trends Biochem. Sci., **11**, 293-296 (1986)

[55] Pettit, F.H.; Reed, L.J.: Pyruvate dehydrogenase complex from bovine kidney and heart. Methods Enzymol., **89**, 376-386 (1982)

[56] Henderson, C.E.; Perham, R.N.: Purificaton of the pyruvate dehydrogenase multienzyme complex of Bacillus stearothermophilus and resolution of its four component polypeptides. Biochem. J., **189**, 161-172 (1980)

[57] Visser, J.; Kester, H.; Huigen, A.: Purification and some properties of the pyruvate dehydrogenase complex of Bacillus subtilis. FEMS Microbiol. Lett., **9**, 227-232 (1980)

[58] Kresze, G.B.; Ronft, H.: Pyruvate dehydrogenase complex from baker's yeast. 2. Molecular structure, dissociation, and implications for the origin of mitochondria. Eur. J. Biochem., **119**, 581-587 (1981)

[59] Ala' Aldeen, D.A.A.; Westphal, A.H.; De Kok, A.; Weston, V.; Atta, M.S.; Baldwin, T.J.; Bartley, J.; Borriello, S.P.: Cloning, sequencing, characterization and implications for vaccine design of the novel dihydrolipoyl acetyltransferase of Neisseria meningitidis. J. Med. Microbiol., **45**, 419-432 (1996)

[60] Allen, M.D.; Perham, R.N.: The catalytic domain of dihydrolipoyl acetyltransferase from the pyruvate dehydrogenase multienzyme complex of Bacillus stearothermophilus. Expression, purification and reversible denaturation. FEBS Lett., **413**, 339-343 (1997)

[61] Aso, Y.; Nakajima, A.; Meno, K.; Ishiguro, M.: Thermally induced changes of lipoate acetyltransferase inner core isolated from the Bacillus stearothermophilus pyruvate dehydrogenase complex. Biosci. Biotechnol. Biochem., **65**, 698-701 (2001)

[62] Guan, Y.; Rawsthorne, S.; Scofield, G.; Shaw, P.; Doonan, J.: Cloning and characterization of a dihydrolipoamide acetyltransferase (E2) subunit of the pyruvate dehydrogenase complex from Arabidopsis thaliana. J. Biol. Chem., **270**, 5412-5417 (1995)

[63] Hendle, J.; Mattevi, A.; Westphal, A.H.; Spee, J.; de Kok, A.; Teplyakov, A.; Hol, W.G.: Crystallographic and enzymatic investigations on the role of Ser558, His610, and Asn614 in the catalytic mechanism of Azotobacter vinelandii dihydrolipoamide acetyltransferase (E2p). Biochemistry, **34**, 4287-4298 (1995)

[64] Jackson, J.C.; Vinluan, C.C.; Dragland, C.J.; Sundararajan, V.; Yan, B.; Gounarides, J.S.; Nirmala, N.R.; Topiol, S.; Ramage, P.; Blume, J.E.; Aicher, T.D.; Bell, P.A.; Mann, W.R.: Heterologously expressed inner lipoyl domain of dihydrolipoyl acetyltransferase inhibits ATP-dependent inactivation of pyruvate dehydrogenase complex, Identification of important amino acid residues. Biochem. J., **334**, 703-711 (1998)

[65] Mande, S.S.; Sarfaty, S.; Allen, M.D.; Perham, R.N.; Hol, W.G.J.: Protein-protein interactions in the pyruvate dehydrogenase multienzyme complex: dihydrolipoamide dehydrogenase complexed with the binding domain of dihydrolipoamide acetyltransferase. Structure, **4**, 277-286 (1996)

[66] McCartney, R.G.; Sanderson, S.J.; Lindsay, J.G.: Refolding and reconstitution studies on the transacetylase-protein X (E2/X) subcomplex of the mammalian pyruvate dehydrogenase complex: Evidence for specific binding of the dihydrolipoamide dehydrogenase component to sites on reassembled E2. Biochemistry, **36**, 6819-6826 (1997)

[67] Millar, A.H.; Leaver, C.J.; Hill, S.A.: Characterization of the dihydrolipoamide acetyltransferase of the mitochondrial pyruvate dehydrogenase complex from potato and comparisons with similar enzymes in diverse plant species. Eur. J. Biochem., **264**, 973-981 (1999)
[68] Mooney, B.P.; Miernyk, J.A.; Randall, D.D.: Cloning and characterization of the dihydrolipoamide S-acetyltransferase subunit of the plastid pyruvate dehydrogenase complex (E2) from Arabidopsis. Plant Physiol., **120**, 443-452 (1999)
[69] Oppermann, F.B.; Schmidt, B.; Steinbuchel, A.: Purification and characterization of acetoin:2,6-dichlorophenolindophenol oxidoreductase, dihydrolipoamide dehydrogenase, and dihydrolipoamide acetyltransferase of the Pelobacter carbinolicus acetoin dehydrogenase enzyme system. J. Bacteriol., **173**, 757-767 (1991)
[70] Roche, T.E.; Powers-Greenwood, S.L.; Shi, W.F.; Zhang, W.B.; Ren, S.Z.; Roche, E.D.; Cox, D.J.; Sorensen, C.M.: Sizing of bovine heart and kidney pyruvate dehydrogenase complex and dihydrolipoyl transacetylase core by quasielastic light scattering. Biochemistry, **32**, 5629-5637 (1993)
[71] Schulze, E.; Westphal, A.H.; Obmolova, G.; Mattevi, A.; Hol, W.G.; de Kok, A.: The catalytic domain of the dihydrolipoyl transacetylase component of the pyruvate dehydrogenase complex from Azotobacter vinelandii and Escherichia coli. Expression, purification, properties and preliminary X-ray analysis. Eur. J. Biochem., **201**, 561-568 (1991)
[72] Schulze, E.; Westphal, A.H.; Veenhuis, M.; de Kok, A.: Purification and cellular localization of wild type and mutated dihydrolipoyltransacetylases from Azotobacter vinelandii and Escherichia coli expressed in E. coli. Biochim. Biophys. Acta, **1120**, 87-96 (1992)
[73] Thelen, J.J.; Muszynski, M.G.; David, N.R.; Luethy, M.H.; Elthon, T.E.; Miernyk, J.A.; Randall, D.D.: The dihydrolipoamide S-acetyltransferase subunit of the mitochondrial pyruvate dehydrogenase complex from maize contains a single lipoyl domain. J. Biol. Chem., **274**, 21769-21775 (1999)
[74] Wang, L.; Kaneko, S.; Kagaya, M.; Ohno, H.; Honda, M.; Kobayashi, K.: Molecular cloning, and characterization and expression of dihydrolipoamide acetyltransferase component of murine pyruvate dehydrogenase complex in bile duct cancer cells. J. Gastroenterol., **37**, 449-454 (2002)
[75] Yang, D.; Song, J.; Wagenknecht, T.; Roche, T.E.: Assembly and full functionality of recombinantly expressed dihydrolipoyl acetyltransferase component of the human pyruvate dehydrogenase complex. J. Biol. Chem., **272**, 6361-6369 (1997)

Glycine N-acyltransferase 2.3.1.13

1 Nomenclature

EC number
2.3.1.13

Systematic name
acyl-CoA:glycine N-acyltransferase

Recommended name
glycine N-acyltransferase

Synonyms
acyltransferase, glycine
glycine acyltransferase
glycine-N-acylase
Additional information (not identical with EC 2.3.1.68 or EC 2.3.1.71)

CAS registry number
9029-95-2

2 Source Organism

<1> *Homo sapiens* [2, 7, 8, 9, 10]
<2> *Bos taurus* [1, 5, 6, 10]
<3> *Macacca mulatta* [2]
<4> *Oryctolagus cuniculus* (New Zealand white rabbits [3]) [3]
<5> *Rattus norvegicus* (Sprague-Dawley rats [3]) [3, 4]
<6> *Ovis aries* (ovine [7,8]) [7, 8]

3 Reaction and Specificity

Catalyzed reaction
acyl-CoA + glycine = CoA + N-acylglycine (<2> sequential reaction mechanism, acyl-CoA substrate adds to the enzyme first, glycine adds before CoA leaves and the peptide product dissociates last [1])

Reaction type
acyl group transfer

Natural substrates and products

S acyl-CoA + glycine <5> (<5> enzyme is believed to catalyze the formation of both the straight-chain and branched-chain acylglycines from the corresponding acyl-CoA esters [4]) (Reversibility: ? <5> [4]) [4]
P CoA + N-acylglycine <5> [4]

Substrates and products

S acyl-CoA + L-asparagine <2> (Reversibility: ? <2> [1, 5]) [1, 5]
P ?
S acyl-CoA + L-glutamine <2> (Reversibility: ? <2> [1, 5]) [1, 5]
P ?
S acyl-CoA + glycine <1-6> (<1,3,5> the CoA derivatives of a number of aliphatic and aromatic acids, e.g. benzoyl-CoA and butyryl-CoA can act as acyl donor [2,4]; <4,5> phenylacetyl-CoA acts as acyl donor [3]; <1,3> salicyl-CoA acts as acyl donor [2]; <2> phenylacetyl-CoA or indoleacetyl-CoA can not act as acyl donor [1,5]) (Reversibility: ? <1-6> [1-8]) [1-8]
P CoA + N-acylglycine <1-6> [1-8]
S benzoyl-CoA + L-alanine <1, 2> (Reversibility: ? <1, 2> [10]) [10]
P ?
S benzoyl-CoA + L-asparagine <1, 2> (Reversibility: ? <1, 2> [10]) [10]
P ?
S benzoyl-CoA + L-glutamic acid <1, 2> (Reversibility: ? <1, 2> [10]) [10]
P ?
S benzoyl-CoA + L-glutamine <1, 2> (Reversibility: ? <1, 2> [10]) [10]
P ?
S benzoyl-CoA + glycine <1, 2, 6> (Reversibility: ? <1, 2, 6> [6, 8]) [6, 8]
P CoA + N-benzoylglycine <1, 2, 6> [6, 8]
S butyryl-CoA + glycine <2> (<2> 19% of activity with benzoyl-CoA [6]) (Reversibility: ? <2> [6]) [6]
P CoA + N-butyrylglycine <2> [6]
S heptanoyl-CoA + glycine <2> (<2> 3.9% of activity with benzoyl-CoA [6]) (Reversibility: ? <2> [6]) [6]
P CoA + N-heptanoylglycine <2> [6]
S isovaleryl-CoA + glycine <1, 2, 6> (<2> 8.4% of activity with benzoyl-CoA [6]) (Reversibility: ? <1, 2, 6> [6, 8]) [6, 8]
P CoA + N-isovalerylglycine <1, 2, 6> [6, 8]
S octanoyl-CoA + glycine <1, 6> (Reversibility: ? <1, 6> [8]) [8]
P CoA + N-octanoylglycine <1, 6> [8]
S salicyl-CoA + glycine <1, 2, 6> (<2> 15% of activity with benzoyl-CoA [6]) (Reversibility: ? <1, 2, 6> [6, 8]) [6, 8]
P CoA + N-salicylglycine <1, 2, 6> [6, 8]

Inhibitors

5,5'-dithiobis(2-nitrobenzoate) <2> [1]
K_3PO_4<1, 6> (<1> 100 mM, 68% inhibition [8]; <6> 100 mM, 43% inhibition [8]) [8]
KCl <1, 2, 6> (<2> 100 mM, 50% inhibition [6]; <1> 100 mM, 79% inhibition [8]; <6> 100 mM, 88% inhibition [8]) [6, 8]

Mg^{2+} <2> [1]
N-benzoyl-L-alanine <2> (<2> competitive vs. benzoyl-CoA [10]) [10]
N-benzoyl-L-asparagine <2> (<2> competitive vs. benzoyl-CoA [10]) [10]
N-benzoyl-L-glutamic acid <2> (<2> competitive vs. benzoyl-CoA [10]) [10]
N-benzoyl-L-serine <2> (<2> competitive vs. benzoyl-CoA [10]) [10]
N-benzoylglycine <2> [6]
NaCl <1, 6> (<1> 100 mM, 53% inhibition [8]; 100 mM, 57% inhibition [8]) [8]
Ni^{2+} <2> [1]
Zn^{2+} <2> [1]
benzoyl-CoA <2> [1]
benzoylglycine <2> (<2> competitive vs. benzoyl-CoA [10]) [10]
butyryl-CoA <2> [1]
p-chloromercuribenzoate <2> (<2> 1 mM, 90% inhibition after 40 min, 20% activity can be recovered by adding 20fold excess of dithiothreitol, benzoyl-CoA protects [6]) [1, 6]

Metals, ions

Cs^+ <2> (<2> K^+, Rb^+, Na^+, Li^+, Cs^+ or NH_4^+ required for activity [1]) [1]
K^+ <2> (<2> K^+, Rb^+, Na^+, Li^+, Cs^+ or NH_4^+ required for activity [1]) [1]
Li^+ <2> (<2> K^+, Rb^+, Na^+, Li^+, Cs^+ or NH_4^+ required for activity [1]) [1]
NH_4^+ <2> (<2> K^+, Rb^+, Na^+, Li^+, Cs^+ or NH_4^+ required for activity [1]) [1]
Na^+ <2> (<2> K^+, Rb^+, Na^+, Li^+, Cs^+ or NH_4^+ required for activity [1]) [1]
Rb^+ <2> (<2> K^+, Rb^+, Na^+, Li^+, Cs^+ or NH_4^+ required for activity [1]) [1]

Specific activity (U/mg)

0.105 <5> (<5> activity in 176000 g supernantant of Triton X-100 solubilized preparations [3]) [3]
0.19 <1> (<1> activity in liver 4 h after birth [9]) [9]
0.7 <1> [10]
2.51 <1> (<1> activity in liver of a 7 months old child [9]) [9]
6.38 <1> (<1> mean activity in liver of children aged 18 months to 11 years [9]) [9]
6.5 <1> (<1> activity in liver of adults aged 24 to 40 years [9]) [9]
9.7 <2> [10]
10 <2> (<2> acyl donor benzoyl-CoA [6]) [6]
13.9 <1> [8]
32 <6> [7]
500 <2> [5]

K_m-Value (mM)

0.003 <3> (glycine, <3> acyl donor salicyl-CoA [2]) [2]
0.006 <3> (glycine, <3> acyl donor benzoyl-CoA [2]) [2]
0.02 <3> (benzoyl-CoA, <3> cosubstrate glycine [2]) [2]
0.03 <3> (salicyl-CoA, <3> cosubstrate glycine [2]) [2]
0.15 <4> (phenylacetyl-CoA, <4> cosubstrate glycine [3]) [3]
0.5 <3> (glycine, <3> acyl donor butyryl-CoA [2]) [2]
0.7 <3> (butyryl-CoA, <3> cosubstrate glycine [2]) [2]

6 <2> (glycine, <2> acyl donor benzoyl-CoA [6]) [6]
6.2 <2> (glycine, <2> acyl donor benzoyl-CoA [10]) [10]
6.4 <1> (glycine, <1> acyl donor benzoyl-CoA [10]) [10]
7.6 <2> (glycinc, <2> acyl donor salicyl-CoA [6]) [6]
13 <1> (benzoyl-CoA, <1> cosubstrate glycine [10]) [10]
15.2 <1> (benzoyl-CoA, <1> cosubstrate alanine [10]) [10]
20 <4> (glycine, <4> acyl donor phenylacetyl-CoA [3]) [3]
22.5 <6> (benzoyl-CoA) [8]
35.6 <6> (salicyl-CoA) [8]
41 <2> (benzoyl-CoA, <2> cosubstrate alanine [10]) [10]
57.9 <1> (benzoyl-CoA) [8]
73.3 <6> (isovaleryl-CoA) [8]
79 <2> (glycine, <2> acyl donor butyryl-CoA [6]) [6]
83.7 <1> (salicyl-CoA) [8]
105 <2> (benzoyl-CoA, <2> cosubstrate glutamine [10]) [10]
108.6 <6> (octanoyl-CoA) [8]
124 <1> (isovaleryl-CoA) [8]
129 <2> (asparagine, <2> acyl donor benzoyl-CoA [10]) [10]
157 <2> (benzoyl-CoA, <2> cosubstrate asparagine [10]) [10]
160 <2> (benzoyl-CoA, <2> cosubstrate glycine [10]) [10]
198 <1> (octanoyl-CoA) [8]
353 <2> (glutamine, <2> acyl donor benzoyl-CoA [10]) [10]
997 <1> (alanine, <1> acyl donor benzoyl-CoA [10]) [10]
998 <2> (benzoyl-CoA, <2> cosubstrate glutamic acid [10]) [10]
1150 <2> (glutamic acid, <2> acyl donor benzoyl-CoA [10]) [10]
1573 <2> (alanine, <2> acyl donor benzoyl-CoA [10]) [10]

K_i-Value (mM)
0.075 <2> (N-benzoylglycine) [6]
0.1 <2> (N-benzoylglycine) [10]
3 <2> (N-benzoyl-L-asparagine) [10]
4.1 <2> (N-benzoyl-L-serine) [10]
7.9 <2> (N-benzoyl-L-alanine) [10]
8 <2> (N-benzoyl-L-glutamic acid) [10]

pH-Optimum
8.4-8.6 <1-3> [1, 2]
8.8-9.5 <4> (<4> kidney enzyme [3]) [3]

pH-Range
8-10 <4> (<4> approx. 35% of maximal activity at pH 8.0, approx. 40% of maximal activity at pH 10.0, kidney enzyme [3]) [3]

Temperature optimum (°C)
37 <4, 5> (<4,5> assay at [3]) [3]

4 Enzyme Structure

Molecular weight

24000 <3> (<3> gel filtration [2]) [2]
33000 <2> (<2> SDS-PAGE, sucrose density gradient centrifugation [1,5]) [1, 5]
33500 <2> (<2> gel filtration [6]) [6]

Subunits

monomer <1, 2, 6> (<2> 1 * 33000, SDS-PAGE [1]; <2> 1 * 33500, SDS-PAGE [6]; <6> 1 * 33000, SDS-PAGE [7]; <1> 1 * 30000, SDS-PAGE [8]; <1> 27000, SDS-PAGE [10]; <2> 1 * 36000, SDS-PAGE [10]) [1, 6, 7, 8, 10]

5 Isolation/Preparation/Mutation/Application

Source/tissue

kidney <4, 5> [3, 4]
liver <1-6> [1-5, 7, 9]

Localization

mitochondrion <1-5> (<4,5> kidney enzyme is located in the matrix, probably not membrane bound [3]) [1-5]

Purification

<1> (ammonium sulfate, hydroxylapatite, Biogel P 100, Blue-dextran agarose, chromatofocusing [8]) [2, 8, 10]
<2> (DE-52, gel filtration, chromatofocusing [6]) [1, 5, 6, 10]
<3> (ammonium sulfate, Bio-gel P-150 [2]) [2]
<6> (ammonium sulfate, DEAE-Sephacel, heparin-Sepharose 4B, Biogel P 100, hydroxylapatite, Blue-Dextran agarose [7]) [7]

6 Stability

General stability information

<2>, enzyme fractions from hydroxylapatite column are inactivated by dialysis or concentration by ultracentrifugation [1, 5]

Storage stability

<2>, 4°C, 1-3 weeks, about 25% loss of activity [1, 5]

References

[1] Nandi, D.L.; Lucas, S.V.; Webster, L.T.: Benzoyl-coenzyme A:glycine N-acyltransferase and phenylacetyl-coenzyme A:glycine N-acyltransferase from

bovine liver mitochondria. Purification and characterization. J. Biol. Chem., **254**, 7230-7237 (1979)
[2] Webster, L.T.; Siddiqui, U.A.; Lucas, S.V.; Strong, J.M.; Mieyal, J.J.: Identification of separate acyl CoA:glycine and acyl-CoA:L-glutamine N-acyltransferase activities in mitochondrial fractions from liver of rhesus monkey and man. J. Biol. Chem., **251**, 3352-3358 (1976)
[3] James, M.O.; Bend, J.R.: A radiochemical assay for glycine N-acyltransferase activity. Some properties of the enzyme in rat and rabbit. Biochem. J., **172**, 285-291 (1978)
[4] Koelvraa, AS.; Gregersen, N.: Acyl-CoA: glycine N-acyltransferase: organelle localization and affinity toward straight- and branched-chained acyl-CoA esters in rat liver. Biochem. Med. Metab. Biol., **36**, 98-105 (1986)
[5] Webster, L.T.: Benzoyl-CoA: amino acid and phenylacetyl-CoA: amino acid N-acyltransferases. Methods Enzymol., **77**, 301-308 (1981)
[6] Kelley, M.; Vessey, D.A.: Isolation and characterization of mitochondrial acyl-CoA: glycine N-acyltransferase from kidney. J. Biochem. Toxicol., **8**, 65-69 (1993)
[7] Mawal, Y.R.; Qureshi, I.A.: An immunodetection method for the quantitation of human acyl CoA: glycine N-acyltransferase in biological samples. Biochem. Mol. Biol. Int., **34**, 595-601 (1994)
[8] Mawal, Y.R.; Qureshi, I.A.: Purification to homogeneity of mitochondrial acyl CoA: glycine N-acyltransferase from human liver. Biochem. Biophys. Res. Commun., **205**, 1373-1379 (1994)
[9] Mawal, Y.; Paradis, K.; Qureshi, I.A.: Developmental profile of mitochondrial glycine N-acyltransferase in human liver. J. Pediat., **130**, 1003-1007 (1997)
[10] Van der Westhuizen, F.H.; Pretorius, P.J.; Erasmus, E.: The utilization of alanine, glutamic acid, and serine as amino acid substrates for glycine N-acyltransferase. J. Biochem. Mol. Toxicol., **14**, 102-109 (2000)

Glutamine N-phenylacetyltransferase 2.3.1.14

1 Nomenclature

EC number
2.3.1.14

Systematic name
phenylacetyl-CoA:L-glutamine α-N-phenylacetyltransferase

Recommended name
glutamine N-phenylacetyltransferase

Synonyms
acyl-CoA:amino acid N-acyltransferase
glutamine phenylacetyltransferase
phenylacetyl-CoA:L-glutamine N-acetyltransferase
phenylacetyltransferase, glutamine

CAS registry number
9030-00-6

2 Source Organism

<-7> no activity in *Felis sp.* (cat [1]) [1]
<-6> no activity in *Canis sp.* (dog [1]) [1]
<-5> no activity in *Equus sp.* (horse [1]) [1]
<-4> no activity in *Oryctolagus cuniculus* (rabbit [1]) [1]
<-3> no activity in *Macaca fuscata* (monkey [1]) [1]
<-2> no activity in *Rattus norvegicus* (rat [1]) [1]
<-1> no activity in *Ovis aries* (sheep [1]) [1]
<1> *Homo sapiens* [1]
<2> *Macaca fuscata* (monkey [2]) [2]

3 Reaction and Specificity

Catalyzed reaction
phenylacetyl-CoA + L-glutamine = CoA + α-N-phenylacetyl-L-glutamine

Reaction type
acyl group transfer

Natural substrates and products

S phenylacetyl-CoA + L-glutamine <1> (Reversibility: ? <1> [1]) [1]
P CoA + α-N-phenylacetyl-L-glutamine <1> [1]

Substrates and products

S benzoyl-CoA + glycine <1, 2> (Reversibility: ? <1, 2> [1, 2]) [1, 2]
P CoA + benzoylglycine <1> [1]
S phenylacetyl-CoA + L-glutamine <1> (Reversibility: ? <1> [1]) [1]
P CoA + α-N-phenylacetyl-L-glutamine <1> [1]
S phenylacetyl-CoA + glycine <1, 2> (Reversibility: ? <1, 2> [1, 2]) [1, 2]
P CoA + phenylacetylglycine <1> [1]

Specific activity (U/mg)

Additional information <2> (<2> specific activity in various organs [2]) [2]

pH-Optimum

8.8-10 <1> [1]

Temperature optimum (°C)

37.5 <1> (<1> assay at [1]) [1]

4 Enzyme Structure

Molecular weight

30000 <2> (<2> gel filtration [2]) [2]

5 Isolation/Preparation/Mutation/Application

Source/tissue

kidney <1, 2> [1, 2]
liver <1, 2> [1, 2]
Additional information <2> (<2> no activity in heart and lung [2]) [2]

Localization

mitochondrion <1, 2> (<2> inner membrane and matrix [2]) [1, 2]

Purification

<1> (partial [1]) [1]
<2> (partial [2]) [2]

6 Stability

General stability information

<1>, 50% loss of activity after lyophilization [1]

Storage stability

<1>, -15°C, stable for several months [1]

References

[1] Moldave, K.; Meister, A.: Synthesis of phenylacetylglutamine by human tissue. J. Biol. Chem., **229**, 463-476 (1957)
[2] Asaoka, K.: Enzymes that metabolize acyl-coenzyme A in the monkey-their distribution, properties and roles in an alternative pathway for the excretion of nitrogen. Int. J. Biochem., **23**, 429-434 (1991)

Glycerol-3-phosphate O-acyltransferase 2.3.1.15

1 Nomenclature

EC number
2.3.1.15

Systematic name
acyl-CoA:sn-glycerol-3-phosphate 1-O-acyltransferase

Recommended name
glycerol-3-phosphate O-acyltransferase

Synonyms
3-glycerophosphate acyltransferase
ACP:sn-glycerol-3-phosphate acyltransferase
sn-glycerol 3-phosphate acyltransferase
sn-glycerol-3-phosphate acyltransferase
acyltransferase, glycerol phosphate
α-glycerophosphate acyltransferase
glycerol 3-phosphate acyltransferase
glycerol phosphate acyltransferase
glycerol phosphate transacylase
glycerophosphate acyltransferase
glycerophosphate transacylase
Additional information <12, 16> (<16> enzyme may be identical with EC 2.3.1.42 [24]; <12> enzyme not identical with EC 2.3.1.42 [22]) [22, 24]

CAS registry number
9029-96-3

2 Source Organism

<1> *Pisum sativum* (pea, var. kleine Rheinländerin [5,7,8]; petit Provençal [12]) [1, 5, 7-9, 12, 50, 55]
<2> *Spinacia oleracea* (spinach, var. Subito [5]; var. Mona Lisa [7]; var. Vital [8]) [5, 7, 8, 50, 52, 55]
<3> *Carthamus tinctorius* (safflower, var. Gila [2]) [2]
<4> *Brassica campestris* (turnip rape, var. Bele [2]) [2]
<5> *Persea americana* (avocado, cv. Fuerte [39]) [39, 47, 50]
<6> *Euglena sp.* [3]
<7> *Euglena gracilis* (var. bacillaris, achloroplastic mutant [10]) [10]

<8> *Cucurbita moschata* (squash, isoenzymes AT1-3 [28]) [28, 50, 52, 55, 56, 61, 62, 63]
<9> *Solanum tuberosum* (potato [9]) [9, 50]
<10> *Theobroma cacao* (cocoa [11]) [11]
<11> *Tetrahymena pyriformis* (WH-14, ciliated protozoon [23]) [23]
<12> *Rattus norvegicus* (female [4,21]; male [20,22,37]; Wistar strain [17,37]; Sprague-Dawley strain [18,20,22]) [4, 6, 16-22, 37, 45, 46, 51, 54, 58]
<13> *Oryctolagus cuniculus* [13, 14]
<14> *Cavia porcellus* (Hartley strain, female [15]) [15, 51, 51]
<15> *Schistocerca gregaria* (african locust, cv. Forskäl, male adult [38]) [38]
<16> *Saccharomyces cerevisiae* (S288c [24]; OC-2 [43]) [24, 43]
<17> *Escherichia coli* (B [31,33,34,41,42]; K-12 [26,30,34-36]; K-12 derivative strains 8, wild-type genotype [27,29,35]; A324 and 205 [35]; strain 8 overproducing mutants CV15, lsA-genotype and BB 26-36, plsB-genotype [35]; overproducing strains VL3/pVL1, contains hybrid plasmid bearing plsB structural gene and VL3/plB3-4 [25,27,29,40]; strain ML 308-225, i-z-y+ [32]) [6, 25-36, 40-42, 49, 54]
<18> *Mus musculus* [44, 48, 51, 53]
<19> *Arabidopsis thaliana* [50, 55, 56, 62, 63]
<20> *Ricinus communis* (castor bean [50]) [50]
<21> *Cucumis sativus* (cucumber [50]) [50]
<22> *Phaseolus vulgaris* (kidney bean [50]) [50]
<23> *Rattus norvegicus* [54]
<24> *Mortierella ramanniana* (var. angulispora [60]) [60]
<25> *Elaeis guineesis* (oil palm [55]) [55, 62]
<26> *Cucurbita moschata* (cv. Shirogikuza, CmATS1,1 encoding oleate-selective isozyme AT1 [57]) [57]
<27> *Cucurbita moschata* (cv. Shirogikuza, CmATS1,2 encoding non-selective isoforms AT2 and AT3 [57]) [57]
<28> *Saccharomyces cerevisiae* (Gatlp [59]) [59]
<29> *Saccharomyces cerevisiae* (Gat2p [59]) [59]

3 Reaction and Specificity

Catalyzed reaction

acyl-CoA + sn-glycerol 3-phosphate = CoA + 1-acyl-sn-glycerol 3-phosphate (<17> mechanism [30]; <8> reaction proceeds via a compulsory-ordered ternary with acyl-ACP binding before glycerol 3-phosphate [63])

Reaction type

acyl group transfer

Natural substrates and products

S acyl-CoA + sn-glycerol 3-phosphate <1, 2, 10, 12, 17> (<1, 2> initial reaction of glycolipid biosynthesis [5]; <10> initial reaction of cocoa butter biosynthesis [11]; <12> initial reaction of phosphoglycerol biosynthesis

[16]; <17> initial step in membrane phospholipid biosynthesis [40]) (Reversibility: ? <1, 2, 10, 12, 17> [5, 11, 16, 26, 40]) [5, 11, 16, 26, 40]

P CoA + 1-acyl-sn-glycerol 3-phosphate <1, 2, 10, 12, 17> [5, 11, 16, 26, 40]

S acyl-[acyl-carrier-protein] + sn-glycerol 3-phosphate <17> (<17> acyl-[acyl-carrier-protein] is probably the physiological substrate [49]) (Reversibility: ? <17> [49]) [49]

P acyl-carrier-protein + 1-acyl-sn-glycerol 3-phosphate <17> [49]

Substrates and products

S acyl-CoA + sn-glycerol 3-phosphate <1-4, 6, 7, 9-17, 18, 24, 28, 29> (<1, 2, 9> specific for glycerol 3-phosphate [7-9]; <1, 2, 9, 10, 12, 13, 17> exclusively acylates glycerol 3-phosphate at the C-1-position [7-9, 11, 13, 17, 19, 25, 34-37, 40-42]; <13> pulmonary enzyme acetylates C-1 and C-2-position [14]; <1, 2, 9, 17> dihydroxyacetone phosphate is ineffective as acyl-acceptor [7-9, 36]; <17> dihydroxyacetone phosphate, ethylene glycol phosphate and 1,3-propanediol phosphate can replace glycerol 3-phosphate [40]; <3, 4> acyl-selectivity [2]; <17> temperature-dependent specificity of reconstituted enzyme [30]; <1, 12, 17> saturated acyl-CoAs preferred [6, 9, 16, 30, 35, 36]; <16> broad specificity [43]; <11, 13, 17> saturated and unsaturated acyl-CoAs are substrates [13, 23, 34, 42]; <1, 2> chloroplastic enzyme has no strict acyl-CoA specificity [8]; <12> no acyl-donors are hexanoyl-CoA, octanoyl-CoA [19]; <16> enzyme has both 3-glycerophosphate acyltransferase and dihydroxyacetonephosphate acyltransferase activity [24]; <18> no activity with hexanoyl-CoA, enzyme prefers saturated fatty acyl-CoAs [48]; <17> preference for saturated fatty acyl-CoAs [49]; <28, 29> enzyme has both 3-glycerophosphate acyltransferase and dihydroxyacetone phosphate acyltransferase activity [59]; <8> very low activity with butanoyl-CoA and hexanoyl-CoA [63]) (Reversibility: ? <1-4, 6, 7, 9-17, 18, 24, 28, 29> [1-17, 19, 23, 25, 30, 34-38, 40-43, 48, 59, 60]) [1-17, 19, 23, 24, 25, 30, 34-38, 40-43, 48, 49, 59, 60, 63]

P CoA + 1-acyl-sn-glycerol 3-phosphate <1, 2, 7, 10, 12-14, 17, 18, 24, 28, 29> [5, 7, 8, 10-13, 15, 16, 24, 25, 40, 42, 48, 49, 59, 60, 63]

S acyl-[acyl-carrier-protein] + sn-glycerol 3-phosphate <1, 2, 8, 17> (<1, 2> specific for glycerol 3-phosphate, ACP-thioesters preferred over CoA-thioester [7]) (Reversibility: ? <1, 2, 8, 17> [7, 25, 26, 36, 40, 63]) [7, 25, 26, 36, 40, 63]

P acyl-carrier-protein + 1-acyl-sn-glycerol 3-phosphate <1, 2, 8, 17> [7, 25, 26, 36, 40, 63]

S arachidonoyl-CoA + sn-glycerol 3-phosphate <12, 14> (<14> 20% of the reaction rate with palmitoyl-CoA [15]; <12> poor substrate [17, 37]) (Reversibility: ? <12, 14> [15, 17, 37]) [15, 17, 37]

P CoA + 1-arachidonoyl-sn-glycerol 3-phosphate <12, 14> [15, 17, 37]

S cis-vaccenoyl-CoA + sn-glycerol 3-phosphate <17> (Reversibility: ? <17> [25, 31, 34]) [25, 31, 34]

P CoA + cis-vaccenoyl-sn-glycerol 3-phosphate <17> [31]

S cis-vaccenoyl-[acyl-carrier-protein] + sn-glycerol 3-phosphate <17> (Reversibility: ? <17> [25, 40]) [25, 40]

P acyl-carrier-protein + 1-cis-vaccenoyl-sn-glycerol 3-phosphate <17> [25, 40]

S erucoyl-CoA + sn-glycerol 3-phosphate <3, 4, 13> (<3, 4> poor substrate [2]) (Reversibility: ? <3, 4, 13> [2, 13]) [2, 13]

P CoA + 1-erucoyl-sn-glycerol 3-phosphate <3, 4, 13> [2, 13]

S linolenoyl-CoA + sn-glycerol 3-phosphate <1, 2> (Reversibility: ? <1, 2> [8]) [8]

P CoA + 1-linolenoyl-sn-glycerol 3-phosphate <1, 2> [8]

S linoleoyl-CoA + sn-glycerol 3-phosphate <1, 2, 9, 11-14> (<14> 40% of the reaction rate with palmitoyl-CoA [15]; <12> poor substrate [17, 37]; <14> no substrate for DTNB-resistant isozyme [15]; <12> 2% of activity with palmitoyl-CoA [19]) (Reversibility: ? <1, 2, 9, 11-14> [8, 9, 13, 15, 17, 23, 37]) [8, 9, 13, 15, 17, 23, 37]

P CoA + 1-linoleoyl-sn-glycerol 3-phosphate <1, 2, 9, 11-14> [8, 9, 13, 15, 17, 23, 37]

S myristoyl-CoA + sn-glycerol 3-phosphate <11, 12, 16, 17, 24> (<11, 12, 16, 17> i.e. tetradecanoyl-CoA [19, 23-25, 42, 43]; <24> 20% of activity with oleoyl-CoA [60]) (Reversibility: ? <11, 12, 16, 17, 24> [19, 23-25, 42, 43, 60]) [19, 23-25, 42, 43, 60]

P CoA + 1-myristoyl-sn-glycerol 3-phosphate <11, 12, 16, 17, 24> [19, 23-25, 42, 43, 60]

S n-decanoyl-CoA + sn-glycerol 3-phosphate <12> (Reversibility: ? <12> [19]) [19]

P CoA + 1-n-decanoyl-sn-glycerol 3-phosphate <12> [19]

S n-dodecanoyl-CoA + sn-glycerol 3-phosphate <16, 17, 24> (<16,1 7> i.e. lauroyl-CoA, poor substrate [24, 42]; <24> very low activity [60]) (Reversibility: ? <16, 17, 24> [24, 25, 42, 60]) [24, 25, 42, 60]

P CoA + 1-n-dodecanoyl-sn-glycerol 3-phosphate <16, 17, 24> [24, 25, 42, 60]

S octanoyl-CoA + sn-glycerol 3-phosphate <24> (<24> very low activity [60]) (Reversibility: ? <24> [60]) [60]

P CoA + 1-n-octanoyl-sn-glycerol 3-phosphate <24> [60]

S oleoyl-CoA + sn-glycerol 3-phosphate <1, 2, 6, 8, 9, 11-17, 19, 24, 26, 28, 29> (<1> i.e. octadecenoyl-CoA [1]; <1> preferred substrate [8, 12]; <1,2> preferred in an equimolar mixture of palmitoyl-CoA, stearoyl-CoA and oleoyl-CoA [8]; <14> 50% of the reaction rate with palmitoyl-CoA [15]; <17> 16% of the reaction rate with palmitoyl-CoA [35]; <12> 7% of activity with palmitoyl-CoA [19]; <12> poor substrate [17, 37]; <2> slight preference over palmitoyl-CoA [52]; <19> preferred over palmitoyl-CoA with a ratio of approx. 3:1 [56]; <26> isoenzyme AT1, preferred over palmitoyl-CoA [57]; <29> low activity [59]; <24> best acyl-donor [60]; <19> preferred over palmitoyl-CoA [62]) (Reversibility: ? <1, 2, 6, 8, 9, 11-17, 19, 24, 26, 28, 29> [1, 3-5, 7-9, 12, 13, 15, 17, 19, 20, 23-25, 30, 35, 37, 38, 40-43, 52, 56, 57, 59, 60]) [1, 3-5, 7-9, 12, 13, 15, 17, 19, 20, 23-25, 30, 35, 37, 38, 40-43, 52, 56, 57, 59, 60, 62]

P CoA + 1-oleoyl-sn-glycerol 3-phosphate <1, 2, 6, 8, 9, 11-17, 19, 24, 26, 28, 29> [1, 3-5, 7-9, 12, 13, 15, 17, 19, 20, 23-25, 30, 35, 37, 38, 40-43, 52, 56, 57, 59, 60, 62]

S oleoyl-[acyl-carrier-protein] + sn-glycerol 3-phosphate <1, 2, 8, 19> (<1, 2> best substrate [7, 12]; <8> no preference over palmitoyl- acyl-carrier-protein [55]; <19> oleoyl- acyl-carrier-protein is preferred over palmitoyl-acyl-carrier-protein [55]) (Reversibility: ? <1, 2, 8, 19> [7, 12, 55]) [7, 12, 55]

P acyl-carrier-protein + 1-oleoyl-sn-glycerol 3-phosphate <1, 2, 8, 19> [7, 12, 55]

S palmitoleoyl-CoA + sn-glycerol 3-phosphate <1, 2, 17, 28, 29> (Reversibility: ? <1, 2, 17, 28, 29> [8, 25, 42, 59]) [8, 25, 42, 59]

P CoA + 1-palmitoleoyl-sn-glycerol 3-phosphate <1, 2, 17, 28, 29> [8, 25, 42, 59]

S palmitoyl-CoA + sn-glycerol 3-phosphate <1, 2, 6, 7, 8, 9-17, 18, 24, 28, 29> (<2, 12, 14, 16, 17> i.e. n-hexadecanoyl-CoA, best substrate [4, 5, 15, 17, 24, 30, 31, 35, 42]; <12> mitochondrial enzyme, preferred over oleoyl-CoA [4]; <1> microsomal enzyme, preferred over oleoyl-CoA [5]; <12> best acyl donor [17, 37]; <17> 4fold better substrate than oleoyl-CoA [35]; <18> recombinant mitochondrial acyltransferase prefers palmitoyl-CoA over oleoyl-CoA [44]; <18> best acyl-donor [48]; <8> stromal acyltransferase, preferred acyl-donor [52]; <8> wild-type acyltransferase uses palmitoyl-CoA and oleoyl-CoA at comparable rates, recombinant mutant acyltransferase prefers oleoyl-CoA [56]; <29> preferred acyl-donor [59]; <24> 20% of activity with oleoyl-CoA [60]) (Reversibility: ? <1, 2, 6, 7, 8, 9-17, 18, 24, 28, 29> [1, 3-5, 8-11, 13, 15-27, 29-38, 40-43, 44, 52, 59, 60]) [1, 3-5, 8-11, 13, 15-27, 29-38, 40-43, 44, 48, 52, 56, 59, 60]

P CoA + 1-palmitoyl-sn-glycerol 3-phosphate <1, 2, 6, 7, 8, 9-17, 18, 24, 28, 29> [1, 3-5, 8-11, 13, 15-27, 29-38, 40-43, 44, 48, 52, 56, 59, 60]

S palmitoyl-[acyl-carrier-protein] + sn-glycerol 3-phosphate <1, 2, 8, 17, 19> (Reversibility: ? <1, 2, 8, 17, 19> [7, 12, 25, 40, 55]) [7, 12, 25, 40, 55]

P acyl-carrier-protein + 1-palmitoyl-sn-glycerol 3-phosphate <1, 2, 8, 17, 19> [7, 12, 25, 40, 55]

S stearoyl-CoA + sn-glycerol 3-phosphate <1, 2, 10, 12-14, 16, 17, 18, 24, 28, 29> (<14> i.e. n-octadecanoyl-CoA, 30% of the reaction rate with palmitoyl-CoA [15]; <12> poor substrate [19]; <18> less than 20% of activity with palmitoyl-CoA [48]; <29> low activity [59]; <24> 20% of activity with oleoyl-CoA [60]) (Reversibility: ? <1, 2, 10, 12-14, 16, 17, 18, 24, 28, 29> [8, 11, 13, 15, 17, 19, 25, 37, 42, 43, 48, 59, 60]) [8, 11, 13, 15, 17, 19, 25, 37, 42, 43, 48, 59, 60]

P CoA + 1-stearoyl-sn-glycerol 3-phosphate <1, 2, 10, 12-14, 16, 17, 18, 24, 28, 29> [8, 11, 13, 15, 17, 19, 25, 37, 42, 43, 48, 59, 60]

S stearoyl-[acyl-carrier-protein] + sn-glycerol 3-phosphate <1, 2> (Reversibility: ? <1, 2> [7]) [7]

P acyl-carrier-protein + 1-stearoyl-sn-glycerol 3-phosphate <1, 2> [7]

S undecanoyl-CoA + sn-glycerol 3-phosphate <17> (Reversibility: ? <17> [25]) [25]

P CoA + 1-undecanoyl-sn-glycerol 3-phosphate <17> [25]

Inhibitors

1-monooleoylglycerol <12> (<12> mechanism [21]) [21]

1-monooleoylglycerol 3-phosphate <12> (<12> non-competitive, mitochondrial isozyme [21]) [21]

1-monooleoylglycerol amide <12> (<12> mechanism [21]) [21]

1-monooleoylglycerol ether <12> (<12> mechanism [21]) [21]

1-palmitoyl-sn-glycerol 3-phosphate <12> [19]

2,3-butanedione <12> (<12> 5 mM, 54% and 21% inhibition of microsomal and mitochondrial acyltransferase respectively [54]) [54]

2-mercaptoethanol <14> [15]

2-monooleoylglycerol <12> (<12> mechanism [21]) [21]

2-monooleoylglycerol amide <12> (<12> mechanism [21]) [21]

2-monooleoylglycerol ether <12> (<12> mechanism [21]) [21]

3-[(3-cholamidopropyl)-dimethylammonio]-1-propanesulfonic acid <12> (<12> trivial name CHAPS, complete inhibition [46]) [46]

5,5'-dithiobis(2-nitro-benzoate) <12, 14, 17, 24> (<12> inhibition of microsomal enzyme, mitochondrial enzyme is not inhibited [6]; <14> 1mM, 20-30% inhibition of activity on palmitoyl-CoA, complete inhibition of activity on other acyl-CoAs [15]; <17> 0.15mM, 90% inhibition in sonicated vesicles [32]; <24> 1 mM, 77% inhibition [60]) [6, 15, 32, 37, 60]

Ca^{2+} <12, 17> (<12> higher concentration, small amounts of phospholipids protect, [17]; <17> in the absence of NaCl [25]) [17, 25]

Ca^{2+} <24> (<24> 5 mM, 70% inhibition [60]) [60]

Cd^{2+} <12> [17]

EDTA <24> (<24> 1 mM, 79% inhibition [60]) [60]

Mg^{2+} <17, 24> (<17> in the absence of NaCl [25]; <24> 5 mM, 75% inhibition [60]) [25, 60]

$MgCl_2$ <12> (<12> inhibition of microsomal enzyme [6]) [6]

N-ethylmaleimide <6, 12, 14, 16, 17, 24> (<12> 0.5-10 mM, 80-90% inhibition of microsomal acyltransferase, mitochondrial acyltransferase is not inhibited [18]; <12> strong inhibition of microsomal isozyme, mitochondrial enzyme is not inhibited [4,6,16]; <16> 0.4 mM, 60% inhibition after 10 min [24]; <17> 10 mM, 65% inhibition in sonicated vesicles, 2 mM, 93% inhibition in vesicles treated with 0.3% toluene [32]; <12> 4 mM, approx. 50% inhibition of the solubilized acyltransferase, membrane-bound enzyme is not inhibited [45]; <24> 5 mM, 79% inhibition [60]) [3, 4, 6, 15, 16, 18, 21, 22, 24, 32, 37, 45, 60]

NaCl <1> (<1> 100 mM, 30% inhibition [12]) [12]

Triton X-100 <12, 14, 16, 17> (<16> strong, above 0.15 mg/ml, [24]; <17> phospholipids restore activity [35]) [6, 15, 20, 22, 24, 26, 35]

Zn^{2+} <12> [17]

acetanhydride <17> [32]

acetone <12> (<12> inhibition of microsomal enzyme [6]) [6]

acyl-CoA <12> (<12> strong substrate inhibition [37]) [37]

aminophenylboronate <12, 23> (<12> mixed-type inhibition of microsomal and mitochondrial acyltransferase vs. glycerol 3-phosphate [54]) [54]

antibodies to cocoa seed enzyme <10> (<10> raised in rabbit [11]) [11]

bovine serum albumin <17> (<17> above 6 mg/ml, at 0.01 mg palmitoyl-CoA depending on palmitoyl-CoA and phospholipid concentration [25]) [25]
bromelain <12> (<12> strong inhibition of microsomal enzyme [20]) [20]
cardiolipin <12> (<12> inhibition of mitochondrial enzyme [16]) [16, 19]
cholate <12, 17> (<17> phospholipids restore activity [35]; <12> complete inhibition [46]) [19, 35, 46]
cis-9,10-methylenehexadecanoic acid <17> (<17> 0.15 mM, 50% inhibition, non-competitive to glycerol 3-phosphate, synergism with oleate, palmitoleate or cis-vaccenate [33]) [33]
cis-vaccenate <17> (<17> 0.45 mM, 50% inhibition, synergism with cis-9,10-methylenehexadecanoic acid [33]) [33]
cis-vaccenoyl-CoA <17> [31]
cyclohexandione <18> (<18> 50 mM, almost complete inhibition of mitochondrial acyltransferase [53]) [53]
deoxycholate <12, 16, 17> (<16> above 0.35 mg/ml [24]; <17> phospholipids restore activity [35]; <12> complete inhibition [46]) [20, 24, 35, 46]
detergents <17> (<17> e.g. Brij 35, Brij 58, Lubrol 17A-10, phospholipids restore activity [35]) [35]
diacylglycerophosphate <12> (<12> i.e. phosphatidic acid, inhibition of mitochondrial enzyme [16]) [16, 19, 46]
diethyldicarbonate <12> (<12> 0.3 mM, approx. 67% inhibition of microsomal acyltransferase, approx. 40% inhibition of mitochondrial acyltransferase, approx. 60% inhibition of recombinant mitochondrial acyltransferase, activity is restored to approx 50% by hydroxylamine [54]) [54]
dihydroxyacetone phosphate <12, 14> (<12> microsomal isozyme [4]; <14> palmitoyl-CoA-specific isozyme [15]) [4, 15]
diisopropylfluorophosphate <17> (<17> weak inhibition [32]) [32]
dithiothreitol <14> (<14> palmitoyl-CoA-specific isozyme [15]) [15]
iodoacetamide <12> (<12> strong inhibition of microsomal enzyme, mitochondrial enzyme is not inhibited [6]) [6, 16]
monoacylglycerophosphate <12, 17> (<12> i.e. lysophosphatidic acid, inhibition of mitochondrial enzyme [16]) [16, 19, 30, 46]
oleate <1, 17> (<1,2> not oleoyl-CoA [7]; <17> 0.31 mM, 50% inhibition, synergism with cis-9,10-methylenehexadecanoic acid [33]) [12, 33]
p-chloromercuribenzoate <14, 17, 24> (<17> 10 mM, 88% inhibition in sonicated vesicles, 1 mM, 98% inhibition in vesicles treated with 0.3% toluene, 2-mercaptoethanol partially restores activity, not inhibited by *p*-chloromercuriphenylsulfonic acid [32]; <24> 1 mM, 64% inhibition [60]) [15, 32, 60]
palmitoleate <17> (<17> 0.31 mM, 50% inhibition, synergism with cis-9,10-methylenehexadecanoic acid [33]) [33]
palmitoyl-CoA <12, 14, 16, 17> (<12> microsomal enzyme, above 0.1 mM [18]; <16> 0.08 mM [24]; <14> bovine serum albumin attenuates [15]) [15, 18, 24, 25]
papain <12> (<12> strong inhibition of microsomal enzyme [20]) [20]
phenylglyoxal <12, 18, 23> (<18> 50 mM, approx. 70% inhibition of mitochondrial acyltransferase [53]; <12,23> 1 mM, 88% and 59% inhibition of microsomal and mitochondrial acyltransferase, respectively [54]) [53, 54]

phosphate <17> [49]
phosphatidylethanolamine <17> (<17> 1,2-unsaturated species [31]) [31]
phosphatidylglycerol <17> (<17> above 1.5 mM, activation below [31]) [31]
phosphatidylinositol <12> (<12> inhibition of microsomal enzyme [16,17]) [16, 17]
phosphatidylserine <12> (<12> inhibition of microsomal enzyme [16,17]) [16, 17]
polymyxin B <12> (<12> 0.5 mg/ml, approx. 75% inhibition of microsomal enzyme, slight stimulation of mitochondrial enzyme [6,20]) [6, 20]
pronase <12> (<12> strong inhibition of microsomal enzyme [20]) [20]
trypsin <12, 16, 17> (<12> 0.04 mg/ml, approx. 80% inhibition of microsomal enzyme within 3 min at 37°C, inhibition of the mitochondrial enzyme only in the presence of 0.05% deoxycholate or 0.1% Triton X-100 [20]; <16> 0.06 mg/ml, 80% inhibition after 10 min [24]) [20, 24, 26]
Additional information <12, 17, 18> (<12> microsomal isozyme, not inhibited by sn-1,2-diolein, 1,2-dioleoylglycerol ether, dioleoylphosphatidic acid [21]; <17> not inhibited by palmitate, elaidate, trans-vaccenate [33]; <12> liver microsomal enzyme is inhibited by all proteases [16]; <12> liver mitochondrial enzyme, not inhibited by trypsin and chymotrypsin [16]; <12> mitochondrial enzyme, not inhibited by N-ethylmaleimide [18]; <12> mitochondrial enzyme, not inhibited by bromelain, papain, pronase and trypsin [20]; <18> recombinant mitochondrial acyl transferase, not inhibited by N-ethylmaleimide [48]) [16, 18, 20, 21, 33, 48]

Cofactors/prosthetic groups

acyl-CoA <1> (<1> activation, above 0.02 mM, in the presence of bovine serum albumin [8]) [8]

Activating compounds

(3-morpholino)propanesulfonic acid-NaOH <1> (<1> i.e. MOPS-NaOH, activation, up to 30 mM [8]) [8]
ATP <17> (<17> activation of acyltransferase in isolated membrane preparations [49]) [49]
KF <13> (<13> stimulation of activity in microsomal preparation [14]) [14]
Lubrol PX <12> (<12> above the critical micellar concentration [46]) [46]
Triton X-100 <16> (<16> i.e. octylphenoxy polyethoxyethanol, slight stimulation at 0.1 mg/ml [24]) [24]
Zwittergent 3-14 <12> (<12> above the critical micellar concentration [46]) [46]
acetone <12> (<12> activation of mitochondrial enzyme, inhibition of microsomal enzyme [16]) [16]
asolectin <12> (<12> i.e. crude mixture of soybean phospholipids, activation of mitochondrial enzyme [16,19]) [16, 19]
bovine serum albumin <12, 14, 17> (<17> activation at 1 mg/ml [27]; <14,17> alleviates detergent effect of palmitoyl-CoA [15,25,40]; <17> alleviates detergent effect of palmitoyl-CoA and oleoyl-CoA [40]; <17> not necessary with acyl-ACP derivatives or mixed micellar samples [40]) [15, 19, 25, 27, 40]

cardiolipin <12> (<12> activation of reconstituted mitochondrial enzyme [46]) [46]
dioleoyl phosphatidic acid <24> (<24> very strong activation [60]) [60]
dioleoyl phosphatidylserine <24> (<24> best phospholipid for activation [60]) [60]
dithiothreitol <24> (<24> 1 mM, 82% activation [60]) [60]
non-site-specific proteases <12> (<12> activation, e.g. proteinase K, subtilisin, not chymotrypsin, trypsin, mitochondrial enzyme, degree of inhibition increases with decreasing ionic strength [16]) [16]
phosphatidylcholine <12> (<12> i.e. lecithin, activation of mitochondrial enzyme [16,46]) [16, 19, 37, 46]
phosphatidylethanolamine <12, 17> (<12> activation of microsomal enzyme [16]; <17> inefficient, pH-dependent [40]; <17> seems to be required for optimal activity [35]; <12> activation of reconstituted mitochondrial acyl transferase [46]) [16, 35, 40, 46]
phosphatidylglycerol <12, 17> (<17> 1.5 mM, 30-40fold activation of solubilized enzyme, inhibition above [31]; <17> required for activation and stabilization [34]; <12> highest activation of reconstituted mitochondrial acyltransferase in the presence of Lubrol PX [46]) [25, 26, 31, 34, 40, 46]
phosphatidylinositol <12> (<12> activation of mitochondrial enzyme [46]) [46]
phosphatidylserine <12> (<12> activation of mitochondrial enzyme [16,19,46]) [16, 19, 46]
phospholipids <12, 17, 24> (<17> absolutely required [35,36,49]; <17> activation, e.g. phosphatidylglycerol and cardiolipin, i.e. diphosphatidyl glycerol, ratio 2/1, protect against inhibition by palmitoyl-CoA [25]; <17> reconstituted activity of Triton X-100-solubilized enzyme, maximal reconstitution at a phospholipid/Triton X-100 ratio of 20/1 [25]; <17> bilayers required for activity: phosphatidylethanolamine/-glycerol, ratio 4/1, phosphatidylethanolamine/cardiolipin, ratio 4/1 [35,36]; <17> phosphatidylethanolamine/phosphatidylglycerol/cardiolipin, ratio 6/1/1 [40]; <17> solubilization and reconstitution in phospholipids activates latent glycerol 3-phosphate activity [29]; <24> required for activity [60]) [17, 19, 25-27, 29, 30, 35, 36, 40, 49, 60]
polymyxin B <12> (<12> activation of mitochondrial enzyme, inhibition of microsomal enzyme [6,16,20]) [6, 16, 20]
proteases <12> (<12> activation, microsomal enzyme [16]) [16]
Additional information <17> (<17> not activated by 1,2-diacyl glycerol 3-phosphate [31]) [31]

Metals, ions

Ca^{2+} <12> (<12> requirement [17]; <12> activation of microsomal and mitochondrial enzyme [16,19]; <17> inhibitory in the absence of NaCl [25]) [16, 17, 19, 37]
Co^{2+} <12> (<12> activation, can replace Ca^{2+} to some extent [17]; <12> at high concentrations [37]) [17, 19, 37]
KCl <17> (<17> activation at 200 mM [25]) [25, 42]
LiCl <17> (<17> activation at 200 mM [25]) [25]

Mg^{2+} <12, 15, 17> (<12> activation, slightly less than Ca^{2+} [37]; <17> inhibitory in the absence of NaCl [25]; <15> 1 mM, optimal activity in microsomal fraction [38]) [16, 19, 37, 38, 49]
Mn^{2+} <12, 24> (<12> activation, can replace Ca^{2+} to some extent [17]; <12> less effective than Ca^{2+} [37]; <24> 2.5 mM, 53% activation, 10 mM, 100% activation [60]) [17, 19, 37, 60]
NaCl <17> (<17> activation at 200 mM [25]) [25]
Additional information <1, 9, 17> (<1,17> not activated by divalent ions [8,40]; <9> not activated by EDTA or EGTA up to 10 mM [9]) [8, 9, 40]

Specific activity (U/mg)

0.0022 <12> [17, 37]
0.0125 <1, 17> (<17> strain 8 [29]) [12, 29]
0.0138 <12> [19]
0.015 <2, 17> (<17> reconstituted enzyme [26]) [8, 26]
0.021 <12> (<12> mitochondrial enzyme [16]) [16]
0.043 <12> (<12> mitochondrial acyltransferase reconstituted in asolectin [46]) [46]
0.105 <17> [35, 36]
0.126 <17> (<17> strain VL3/pVL1 [29]) [29]
0.19 <12> (<12> mitochondrial acyltransferase reconstituted in phosphatidylglycerol/phosphatidylethanolamine mixture [46]) [46]
0.192 <18> (<18> recombinant mitochondrial acyl transferase [48]) [48]
0.32 <17> (<17> strain VL3/plB3-4 [29]) [29]
0.66 <2> [7]
0.69 <10> [11]
1.45 <24> [60]
1.74 <1> (<1> isoform 1 [8]) [8]
1.74-3.18 <1> [7]
2.3-8 <17> [25]
3-4 <17> (<17> reconstituted enzyme [27]) [27]
3.2 <1> (<1> isoform 2 [8]) [8]
6.8 <17> [25]
10 <26> (<26> isoenzyme AT1, approx. value [57]) [57]

K_m-Value (mM)

0.0003-0.0007 <1, 2> (oleoyl-ACP) [7]
0.002 <17> (palmitoyl-CoA) [25]
0.0028 <8> (oleoyl-ACP, <8> L261F mutant acyltransferase [62]) [62]
0.003 <8> (oleoyl-ACP) [62]
0.003-0.0034 <8> (acyl-ACP, <8> recombinant N-terminal truncated acyltransferase [63]) [63]
0.0031 <12> (palmitoyl-CoA, <12> mitochondrial acyl transferase reconstituted in 1,2-dioleoylphosphatidylglycerol [46]) [46]
0.0032 <2> (palmitoyl-ACP) [7]
0.0032 <18> (palmitoyl-CoA, <18> R318A mutant mitochondrial acyltransferase [53]) [53]

0.0033 <2> (stearoyl-ACP) [7]
0.0034 <8> (palmitoyl-ACP) [62]
0.004 <10> (stearoyl-CoA) [11]
0.0046 <17> (palmitoyl-CoA, <17> strain 8, reconstituted enzyme [27]) [22, 27, 49]
0.0046 <18> (palmitoyl-CoA, <18> R318K mutant mitochondrial acyltransferase [53]) [53]
0.005 <17> (oleoyl-CoA) [25]
0.0056 <1> (palmitoyl-ACP) [7]
0.0056 <17> (palmitoyl-CoA, <17> strain 8, membrane-associated enzyme [27,31]) [27, 31]
0.006 <12> (palmitoyl-CoA, <12> mitochondrial acyl transferase reconstituted in asolectin [46]) [46]
0.006 <17> (palmitoyl-CoA, <17> strain VL3/pVL1 bearing plasmid which overproduces glycerol 3-phosphate acyltransferase, membrane-associated enzyme [27,31]) [27, 31]
0.0064 <12> (stearoyl-CoA, <12> mitochondrial acyl transferase reconstituted in 1,2-dioleoylphosphatidylglycerol [46]) [46]
0.007 <12> (palmitoyl-CoA, <12> mitochondrial acyl transferase reconstituted in 1,2-dipalmitoylphosphatidylglycerol [46]) [46]
0.0078 <17> (palmitoyl-CoA, <17> strain VL3/pLB3-4 bearing plasmid which overproduces glycerol 3-phosphate acyltransferase, membrane-associated enzyme [27,31]) [27, 31]
0.0084 <12> (oleoyl-CoA, <12> mitochondrial acyl transferase reconstituted in 1,2-dioleoylphosphatidylglycerol [46]) [46]
0.0092 <8> (palmitoyl-ACP, <8> L261F mutant acyltransferase [62]) [62]
0.011 <1> (oleoyl-CoA) [12]
0.011 <18> (palmitoyl-CoA, <18> wild-type mitochondrial acyltransferase [53]) [53]
0.011 <12> (stearoyl-CoA, <12> mitochondrial acyl transferase reconstituted in asolectin [46]) [46]
0.012 <12> (myristoyl-CoA, <12> mitochondrial acyl transferase reconstituted in 1,2-dioleoylphosphatidylglycerol [46]) [46]
0.012 <17> (palmitoyl-CoA, <17> strain VL3/pVL1 bearing plasmid which overproduces glycerol 3-phosphate acyltransferase, reconstituted enzyme [27]) [22, 27]
0.013 <12> (oleoyl-CoA, <12> mitochondrial acyl transferase reconstituted in asolectin [46]) [46]
0.013 <12> (stearoyl-CoA, <12> mitochondrial acyl transferase reconstituted in 1,2-dipalmitoylphosphatidylglycerol [46]) [46]
0.014 <6> (oleoyl-CoA) [3]
0.015 <12, 24> (oleoyl-CoA, <12> mitochondrial acyl transferase reconstituted in 1,2-dipalmitoylphosphatidylglycerol [46]) [46, 60]
0.015 <17> (palmitoyl-ACP) [25]

0.016 <12> (myristoyl-CoA, <12> mitochondrial acyl transferase reconstituted in 1,2-dipalmitoylphosphatidylglycerol [46]) [46]
0.016 <12> (myristoyl-CoA, <12> mitochondrial acyl transferase reconstituted in asolectin [46]) [46]
0.017 <17> (palmitoyl-CoA, <17> strain VL3/pLB3-4 bearing plasmid which overproduces glycerol 3-phosphate acyltransferase, reconstituted enzyme [27]) [22, 27]
0.02 <18> (oleoyl-CoA, <18> recombinant mitochondrial acyltransferase [44]) [44]
0.02 <18> (palmitoyl-CoA, <18> recombinant mitochondrial acyltransferase [44]) [44]
0.021 <6> (palmitoyl-CoA) [3]
0.025 <24> (glycerol 3-phosphate) [60]
0.025 <17> (vaccenoyl-ACP) [25]
0.03 <16> (sn-glycerol 3-phosphate) [24]
0.033 <1> (sn-glycerol 3-phosphate) [12]
0.05 <17> (sn-glycerol 3-phosphate, <17> acyl-CoA [36]) [24, 36]
0.07 <17> (acyl-ACP) [36]
0.083 <8> (oleoyl-CoA, <8> recombinant stromal acyltransferase [52]) [52]
0.1 <5> (glycerol 3-phosphate, <5> CHAPS-solubilized acyl transferase [47]) [47]
0.1 <2> (oleoyl-CoA, <2> recombinant stromal acyltransferase [52]) [52]
0.1 <13> (sn-glycerol 3-phosphate, <13> assay with microsomal preparation [14]) [14]
0.1-0.2 <12> (sn-glycerol 3-phosphate, <12> microsomal enzyme [16]) [16]
0.102 <17> (sn-glycerol 3-phosphate, <17> strain VL3/pVL1 bearing plasmid which overproduces glycerol 3-phosphate acyltransferase, membrane-associated enzyme [27,31]) [27, 31]
0.105 <17> (sn-glycerol 3-phosphate, <17> strain 8, membrane-associated enzyme [27,31]) [27, 31]
0.108 <17> (sn-glycerol 3-phosphate, <17> strain VL3/pLB3-4 bearing plasmid which overproduces glycerol 3-phosphate acyltransferase, membrane-associated enzyme [27,31]) [27, 31]
0.11 <8> (glycerol 3-phosphate, <8> L261F mutant acyltransferase, acyl-donor oleoyl-CoA [62]) [62]
0.117-0.143 <8> (glycerol 3-phosphate, <8> acyl donor palmitoyl-CoA [62]) [62]
0.12 <8, 17> (glycerol 3-phosphate, <8> acyl donor oleoyl-CoA [62]) [49, 62]
0.14 <8> (palmitoyl-CoA, <8> recombinant stromal acyltransferase [52]) [52]
0.143 <8> (glycerol 3-phosphate, <8> acyl donor palmitoyl-CoA [62]) [62]
0.15 <8> (glycerol 3-phosphate, <8> L261F mutant acyltransferase, acyl-donor palmitoyl-CoA [62]) [62]
0.15 <17> (sn-glycerol 3-phosphate, <17> acyl donor palmitoyl-CoA or palmitoyl-ACP [25]) [25]

0.156 <17> (sn-glycerol 3-phosphate, <17> membrane-bound enzyme [31]) [31]
0.2 <12> (sn-glycerol 3-phosphate) [17, 37]
0.3-0.5 <15> (sn-glycerol 3-phosphate, <15> assay with microsomal fraction [38]) [38]
0.46 <12> (glycerol 3-phosphate, <12> mitochondrial acyl transferase reconstituted in 1,2-dioleoylphosphatidylglycerol [46]) [46]
0.5 <14> (sn-glycerol 3-phosphate, <14> assay with microsomal preparation [15]) [15]
0.56 <17> (sn-glycerol 3-phosphate, <17> strain 8 [35]) [35]
0.67 <18> (glycerol 3-phosphate, <18> mitochondrial acyl transferase [48]) [48]
0.7 <1> (sn-glycerol 3-phosphate, <1> isoforms 1 and 2 [8]) [8]
0.79 <17> (sn-glycerol 3-phosphate, <17> strain 8, reconstituted enzyme [27]) [22, 27]
0.82 <12> (glycerol 3-phosphate, <12> mitochondrial acyl transferase reconstituted in asolectin [46]) [46]
0.92 <17> (sn-glycerol 3-phosphate, <17> strain VL3/pVL1 bearing plasmid which overproduces glycerol 3-phosphate acyltransferase, reconstituted enzyme [27]) [22, 27]
0.93 <12> (glycerol 3-phosphate, <12> mitochondrial acyl transferase reconstituted in 1,2-dipalmitoylphosphatidylglycerol [46]) [46]
1 <12> (sn-glycerol 3-phosphate, <12> mitochondrial enzyme [16]) [16]
1.1 <17> (sn-glycerol 3-phosphate, <17> strain VL3/pLB3-4 bearing plasmid which overproduces glycerol 3-phosphate acyltransferase, reconstituted enzyme [27]) [22, 27]
1.5 <2> (palmitoyl-CoA, <2> recombinant stromal acyltransferase [52]) [52]
2.9 <6> (sn-glycerol 3-phosphate) [3]
3 <17> (sn-glycerol 3-phosphate, <17> solubilized enzyme in the presence of phosphatidylglycerol [31]) [31]
3-3.2 <2, 17> (sn-glycerol 3-phosphate, <2> acyl donors: palmitoyl-ACP or stearoyl-ACP [7]; <17> reconstituted enzyme [31]; <17> strain BB26-36 [35]) [7, 31, 35]
4.2 <10> (sn-glycerol 3-phosphate) [11]
Additional information <12, 17> (<17> kinetic cooperativity towards glycerol 3-phosphate and palmitoyl-CoA [27]; <12> solubilization increases K_m of mitochondrial acyltransferase for glycerol 3-phosphate [45]) [27, 45]

K_i-Value (mM)

0.036 <12> (1-monooleoylglycerol ether, <12> microsomal enzyme [21]) [21]
0.05 <12> (1-monooleoylglycerol ether, <12> mitochondrial enzyme [21]) [21]
0.055 <12> (1-monooleoylglycerol 3-phosphate, <12> microsomal enzyme [21]) [21]
0.06 <12> (1-monooleoylglycerol, <12> mitochondrial enzyme [21]) [21]
0.063 <12> (sn-2-monooleoylglycerol, <12> mitochondrial enzyme [21]) [21]
0.067 <12> (1-monooleoylglycerol, <12> microsomal enzyme [21]) [21]

0.082 <12> (sn-2-monooleoylglycerol ether, <12> microsomal enzyme [21]) [21]
0.093 <12> (sn-2-monooleoylglycerol ether, <12> mitochondrial enzyme [21]) [21]
0.095 <12> (1-monooleoylglycerol amide, <12> microsomal enzyme [21]) [21]
0.096 <12> (sn-2-monooleoylglycerol, <12> microsomal enzyme [21]) [21]
0.475 <12> (sn-2-monooleoylglycerol amide, <12> microsomal enzyme [21]) [21]
0.54 <16> (dihydroxyacetone phosphate) [24]
2.9 <12> (aminophenylboronate, <12> microsomal acyltransferase [54]) [54]
8.2 <12> (aminophenylboronate, <12> mitochondrial acyltransferase [54]) [54]
10.3 <12> (aminophenylboronate, <12> recombinant mitochondrial acyltransferase [54]) [54]

pH-Optimum

6-8 <18> (<18> mitochondrial acyl transferase [48]) [48]
6.6 <1> (<1> 2 isoforms [8]; <1> pI: 6.6 [12]) [8, 11, 12]
6.6-7 <16> [24]
6.6-9 <12> (<12> both microsomal and mitochondrial enzymes exhibit a broad optimum [16,17,37]) [16, 17, 37]
7 <16, 24> (<16> 50% of maximal activity at pH 7.4 [24]) [24, 60]
7.4 <1, 5, 13> (<13> broad optimum [14]; <1> isoforms 1 and 2 [8]; <5> CHAPS-solubilized microsomal acyl transferase [47]) [8, 14, 47]
7.5 <9> [9]
8.3 <17> [33]
8.5 <15, 17> [25, 36, 38]
9.2 <17> (<17> reconstituted enzyme [26]) [26]
9.5 <17> (<17> membrane-associated enzyme [26]) [26]

pH-Range

6.6-8.6 <12> (<12> approx. half-maximal activity at pH 6.6 and 8.6 [22]) [22]
6.6-8.9 <16> (<16> maximal activity at pH 6.6, approx. 30% of maximal activity at pH 8.9 [24]) [24]
7-8.6 <17> (<17> approx. half-maximal activity at pH 7.0 and 8.6 [33]) [33]
7.5-9.2 <17> (<17> approx. half-maximal activity at pH 7.5 and 9.2, reconstituted enzyme [26]) [26]
8-9.5 <17> (<17> approx. half-maximal activity at pH 8.0 and 9.5, membrane-associated enzyme [26]) [26]

Temperature optimum (°C)

20 <17> (<17> reconstituted enzyme, 2 optima [26]) [26]
30 <17> (<17> reconstituted enzyme [30]) [30]
30-35 <17> (<17> membrane-associated enzyme [26]) [26]
35 <17> (<17> reconstituted enzyme, 2 optima [26]) [26]

Temperature range (°C)

10-50 <17> (<17> approx. half-maximal activity at 10°C and 50°C [26]) [26]
20-40 <17> (<17> approx. half-maximal activity at 20°C and about 70% of maximal activity at 40°C [26]) [26]

4 Enzyme Structure

Molecular weight

30000 <8> (isoenzyme AT1, gel filtration [28]) [28]
40000 <8> (isoenzymes AT2 and AT3, gel filtration [28]) [28]
42000 <1> (<1> 2 soluble chloroplast isoforms, gel filtration [8]) [8]
60000-85000 <12> (<12> mitochondrial enzyme, gel filtration [16]) [16]
88000 <17> (<17> gel filtration [27]) [27]
95000-105000 <12> (<12> mitochondrial acyltransferase, gel filtration [46]) [46]
200000 <10> (<10> gel filtration [11]) [11]
Additional information <17> (<17> amino acid analysis [25]; <17> peptide mapping [29]; <17> different molecular masses depending on Triton X-100 concentration [27]) [25, 27, 29]

Subunits

? <1, 17, 24, 26> (<17> x * 83000, SDS-PAGE [25]; <1> x * 41000, putative mature enzyme, deduced from nucleotide sequence [1]; <1> x * 40500, SDS-PAGE [12]; <26> x * 40500 + x * 40000 + x * 39500, isoenzyme AT1, SDS-PAGE, immunoblot [57]; <24> x * 45000, SS-PAGE [60]) [1, 12, 25, 27, 29, 40, 57, 60]
decamer <10> (<10> 10 * 20000, SDS-PAGE [11]) [11]
monomer <8, 12> (<12> 1 * 80000-85000, mitochondrial enzyme, SDS-PAGE [16]; <8> 1 * 30000, isoenzyme AT1, SDS-PAGE [28]; <8> 1 * 40000, isoenzymes AT2 and AT3, SDS-PAGE [28]; <12> 1 * 85000, mitochondrial acyltransferase, SDS-PAGE [46]) [16, 28, 46]
Additional information <17> (<17> monomer or oligomer, depending on Triton X-100 concentration [27]; <17> membranes of overproducing E. coli strain VL3/pLB3-4 contain predominantly the dimeric form [40]) [27, 40]

Posttranslational modification

lipoprotein <12> [19]
side-chain modification <12> (<12> activity of microsomal enzyme is probably regulated by phosphorylation-dephosphorylation [6]) [6]

5 Isolation/Preparation/Mutation/Application

Source/tissue

adipocyte <18> (<18> mitochondrial acyltransferase [44]) [44]
brain <12> [20]
cotyledon <8, 10, 26> [28, 11, 57]

endosperm <20> [50]
fat body <15> [38]
harderian gland <14> [15]
heart <13> [13]
leaf <1, 2, 7> [5, 7-10, 12]
liver <12, 14, 18, 23> (<18> mitochondrial acyltransferase [44]) [6, 16-20, 37, 44, 46, 51, 54, 58]
lung <13> [14]
mammary gland <12> (<12> lactating [4,6]) [4, 6]
mesocarp <5> [39, 47]
seed <10> [11]
seedling <1-4> (<1> etiolated [5]) [1, 2, 5]
tuber <9> [9]
Additional information <12, 17> (<12,17> distribution, overview [6]) [6]

Localization

chloroplast <1, 2, 8> (<1,2,8> soluble [1,8,28]) [1, 7, 8, 28]
chloroplast stroma <2, 8> [28, 52]
cytoplasm <10> [11]
membrane <24> [60]
microsome <1-7, 11-14, 16> (<1-4,12> membrane-bound [2,5,6]; <5> active site of acyl transferase is probably accesible to the cytoplasmic face of the endoplasmic reticulum, integral to membrane [47]) [2-6, 10, 13-18, 20-23, 37, 39, 43, 47]
mitochondrial inner membrane <12, 17> (<12> significant acyltransferase activity in inner membrane [45]; <17> cytoplasmic side of inner membrane [49]) [30, 40, 45, 49]
mitochondrial intermembrane space <1> [9]
mitochondrial membrane <12> (<12> membrane-bound [6,19]) [6, 19]
mitochondrial outer membrane <9, 12> (<9> cytosolic side of outer-membrane [9]; <12> spans transverse plane of outer membrane [16]; <12> N- and C-termini and the active site of acyltransferase face the cytosolic face of the outer mitochondrial membrane, proposed model for orientation [58]) [9, 16, 45, 46, 58]
mitochondrion <1, 7, 12, 13, 14, 16, 18, 20, 23> (<14> DTNB-resistant isozyme [15]; <1> intermembrane space [9]) [4, 9, 10, 14-16, 18, 20, 21, 43, 44, 48, 50, 53, 54, 58]
plasma membrane <3, 4, 17> (<17> integral cytoplasmic membrane protein [25]; <17> deeply located in the membrane [32]; <17> active site on outside surface [26]; <17> cytoplasmic side [40]) [2, 25-36, 40-42]
soluble <5> [39]
Additional information <13, 15> (<13,15> subcellular distribution [14,38]) [14, 38]

Purification

<1> (2 isoforms of different pI-values [8]; partial, Tween-80/NaCl-solubilized [5]; ammonium sulfate, Sephadex G-100, DEAE-Trisacryl, Mono Q [12]; recombinant acyltransferase, Mono Q [55]) [1, 5, 7, 8, 12, 55]

<2> (partial, Tween-80/NaCl-solubilized [5]; recombinant acyltransferase, Mono Q [55]) [5, 7, 8, 55]
<5> (partial purification of microsomal acyl transferase on glycerophosphorylethanolamine affinity column [47]) [47]
<7> (Triton X-100-solubilized [10]) [10]
<8> (isoenzyme AT1, [acyl-carrier protein]-affinity chromatography, anion-exchange column, isoenzymes AT2 and AT3, [acyl-carrier protein]-affinity chromatography, hydroxylapatite [28]; recombinant acyltransferase, Mono Q [55]) [28, 55]
<10> (ammonium sulfate, DEAE-cellulose, Sephacryl S-300 [11]) [11]
<12> (microsomal enzyme, Sepharose 2B, sucrose gradient, partial purification [16]; mitochondrial enzyme, Sepharose Q, BioGel HAT, octyl-Sepharose CL-4B, partial purification [16]; mitochondrial enzyme, cholate, ammonium sulfate, partial purification [19]; mitochondrial acyltransferase, Bio-Gel HAT, Rotofor, Sephacryl S-300 [46]) [16, 17, 19, 37, 46, 51]
<17> (Triton X-100-solubilized [6,25]; Matrex Gel Red A, octyl-Sepharose CL-4B, hydroxylapatite [25,40]; solubilized [30,31]; solubilized, by non-ionic detergents [40]; solubilized, by non-ionic detergents and reconstituted in unilamellar phospholipid vesicles [30,40]; partial purification [26,27,29,35,36]; affinity chromatography on 6-phosphogluconic acid-Sepharose 4B [34]) [6, 25-27, 29-31, 34-36, 40]
<18> (recombinant His-tagged acyl transferase, Ni^{2+}-affinity column [48]) [48]
<19> (recombinant acyltransferase, Mono Q [55]) [55]
<24> (heparin-Sepharose, Yellow 86-Sepharose, Superdex 200, hydroxylapatite [60]) [60]
<25> (recombinant acyltransferase, Mono Q [55]) [55]
<26> (isoenzyme AT1, DEAE-Toyopearl, [acyl-carrier protein]-affinity chromatography, MonoP [57]) [57]

Renaturation

<12> (reconstitution in asolection or phosphatidylglycerol/phosphatidylethanolamine mixture [46]) [46]
<17> (reconstitution in dispersed E. coli phospholipids [25]; <17> reconstitution in phospholipid bilayers [35]; <17> reconstitution in vesicles [26, 27, 36, 40]; <17> reconstitution in cardiolipin vesicles [27]; <17> reconstitution in cardiolipin/C12E8, i.e. non-ionic detergent, ratio 2:1 mixed micelles [40]; <17> reconstitution in phosphatidylcholine/phosphatidylglycerol, ratio 4:1, vesicles, 90% of active site of the reconstituted enzyme are located at the outside surface [26,30]; <17> reconstituted at a phospholipid-detergent ratio of 20:1 [36]; reconstituted in phospholipid vesicles containing phosphatidylethanolamine, phosphatidylglycerol and diphosphatidylglecerol in the approx. molar ratio of 6:1:1 [40]) [25, 26, 27, 29, 30, 35, 36, 40, 49]
<18> (reconstitution effectiveness of phospholipids in descending order: crude soybean phosphatidylcholine, phosphatidylserine, phosphatidylinositol, phosphatidylethanolamine, cardiolipin [48]) [48]

Crystallization

<8> (hanging-drop vapour diffusion, poly(ethylene glycol) 4000 as precipitant, 1.9 A resolution [55]) [55, 61]

Cloning

<1> (cloning of cDNA [1]; chloroplast acyltransferase [50]) [1, 50]
<1> (expression in Escherichia coli [55]) [55]
<2> (chloroplast acyltransferase [50]; expression of stromal acyltransferase in Escherichia coli [52]) [50, 52, 55]
<8> (expression of stromal acyltransferase in Escherichia coli [52]; expression of wild-type and PCR-mutated cDNA in Escherichia coli [56]; cloning of CmATS1,1 and CmATS1,2 [57]; expression of N-terminal truncated enzyme in Escherichia coli, overexpression of wild-type acyltransferase in tobacco [63]) [52, 55, 56, 57, 62, 63]
<12> (transient expression of epitope-tagged acyltransferase in CHO K1 cells [58]) [58]
<17> (plsB structural gene encoding glycerol 3-phosphate acyltransferase [40]) [40, 49]
<18> (expression of mitochondrial enzyme in CHO cells [44]; expression of mitochondrial acyl transferase in Sf9 insect cells [48]; cloning of cDNA [51]) [44, 48, 51]
<19> (chloroplast acyltransferase [50]; expression in Escherichia coli [55]; expression in Escherichia coli [56]; overexpression in tobacco [63]) [50, 55, 56, 62, 63]
<21> (chloroplast acyltransferase [50]) [50]
<22> (chloroplast acyltransferase [50]) [50]
<23> (expression of mitochondrial acyltransferase in Sf21 insect cells [54]) [54]
<25> (expression in Escherichia coli [55]) [55, 62]
<28> (expression in Escherichia coli [59]) [59]
<29> (expression in Escherichia coli [59]) [59]

Engineering

E142A <8> (<8> no activity [62]) [62]
E315A <18> (<18> 28% of wild-type mitochondrial acyltransferase activity [53]) [53]
E315Q <18> (<18> 47% of wild-type mitochondrial acyltransferase activity [53]) [53]
F313A <18> (<18> 17% of wild-type mitochondrial acyltransferase activity [53]) [53]
F313Y <18> (<18> 88% of wild-type mitochondrial acyltransferase activity [53]) [53]
H194S <8> (<8> 79% of wild-type activity [62]) [62]
K193S <8> (<8> no activity [62]) [62]
L261F <8> (<8> 90% of wild-type activity [62]) [62]
L261F <8> (<8> change from a non-selective acyltransferase to an oleate-selective acyltransferase [62]) [62]

L352F <25> (<25> change from a non-selective acyltransferase to an oleate-selective acyltransferase, 117% of wild-type activity [62]) [62]
R235S <8> (<8> no activity [62]) [62]
R237S <8> (<8> no activity [62]) [62]
R278H <18> (<18> 10% of wild-type mitochondrial acyltransferase activity [53]) [53]
R278K <18> (<18> 76% of wild-type mitochondrial acyltransferase activity [53]) [53]
R279A <18> (<18> 101% of wild-type mitochondrial acyltransferase activity [53]) [53]
R279H <18> (<18> 95% of wild-type mitochondrial acyltransferase activity [53]) [53]
R279K <18> (<18> 101% of wild-type mitochondrial acyltransferase activity [53]) [53]
R318H <18> (<18> 2.6% of wild-type mitochondrial acyltransferase activity [53]) [53]
R318K <18> (<18> 11% of wild-type mitochondrial acyltransferase activity [53]) [53]
S331P <8> (<8> 94% of wild-type activity [62]) [62]
S331P <8> (<8> no change in acyl-donor selectivity [62]) [62]
T141S <8> (<8> 106% of wild-type activity [62]) [62]

6 Stability

General stability information

<12>, 1 mM dithiothreitol stabilizes solubilized enzyme [45]
<12>, asolectin stabilizes cholate-containing preparations [19]
<12>, bovine serum albumin stabilizes [16, 19]
<12>, cholate inactivates during storage [19]
<12>, repeated freezing and thawing inactivates [19]
<17>, 400 mM NaCl or KCl, 5 mM 2-mercaptoethanol, 5 mM $MgCl_2$, stabilize membrane-bound or reconstituted enzyme [35, 36]
<17>, column elution during chromatography causes aggregation leading to irreversible inactivation [25]
<17>, glycerol, 20%, stabilizes [35]
<17>, ion exchange chromatography, hydrophobic interaction chromatography, affinity chromatography, gel filtration and isoelectric focusing inactivate enzyme [26]
<17>, phosphatidylglycerol stabilizes [34]
<17>, phospholipids reconstitute Triton X-100-solubilized enzyme [25]
<12, 17>, phospholipids stabilize [6, 16]

Storage stability

<1>, -20°C, 50% glycerol, at least 3 weeks, no loss of activity [12]
<1>, -20°C, 50% w/v glycerol, 3 years, no loss of activity [8]
<1>, 4°C, 3 weeks, complete loss of activity [12]

<1, 9>, -20°C, 50% v/v glycerol, soluble enzyme at least 2 weeks [9]
<2>, -20°C, 50% glycerol, at least 6 months [7]
<12>, -70°C, 2 months after the removal of cholate [19]
<12>, -70°C, 35% loss of activity after 2 days [19]
<12>, 4-6°C, microsomal enzyme, $t_{1/2}$: 4 h [20]
<12>, 4-6°C, mitochondrial enzyme, 4 h, no loss of activity [20]
<12>, 4°C, unstable, microsomal enzyme [21]
<16>, -15°C, 3-5 mg/ml protein, membrane preparation, at least 2 months [24]
<17>, -20°C, at least 6 months [36]
<17>, -20°C, at least a few weeks [34]
<17>, -40°C, solubilized enzyme, at least 1 year, no loss of activity [26]
<17>, -70°C, solubilized, indefinitely stable [40]
<17>, -80°C, at least 10 months [25]

References

[1] Weber, S.; Wolter, F.P.; Buck, F.; Frentzen, M.: Purification and cDNA sequencing of an oleate-selective acyl-ACP: sn-glycerol-3-phosphate acyltransferase from pea chloroplasts. Plant Mol. Biol., **17**, 1067-1076 (1991)

[2] Bafor, M.; Stobart, A.K.; Stymne, S.: Properties of the glycerol acylating enzymes in microsomal preparations from the developing seeds of safflower (Carthamus tinctorius) and turnip rape (Brassica campestris) and their ability to assemble cocoa-butter type fats. J. Am. Oil Chem. Soc., **67**, 217-225 (1990)

[3] Hershenson, S.; Ernst-Fonberg, M.L.: Comparison of 1-acylglycerophosphate and glycerophosphate acyltransferases from Euglena microsomes. Int. J. Biochem., **16**, 219-223 (1984)

[4] Bramley, P.M.; Grigor, M.R.: Acyl-CoA:sn-glycerol 3-phosphate acetyltransferase isozymes in rat mammary gland. Biochem. Int., **5**, 199-204 (1982)

[5] Frentzen, M.: Comparison of certain properties of membrane bound and solubilized acetyltransferase activities of plant microsomes. Plant Sci., **69**, 39-48 (1990)

[6] Bell, R.M.; Coleman, R.A.: Enzymes of triacylglycerol formation in mammals. The Enzymes, 3rd Ed. (Boyer, P.D., ed.), **16**, 87-111 (1983)

[7] Frentzen, M.; Heinz, E.; McKeon, T.A.; Stumpf, P.K.: Specificities and selectivities of glycerol-3-phosphate acyltransferase and monoacylglycerol-3-phosphate acyltransferase from pea and spinach chloroplasts. Eur. J. Biochem., **129**, 629-636 (1983)

[8] Bertrams, M.; Heinz, E.: Positional specificity and fatty acid selectivity of purified sn-glycerol 3-phosphate acyltransferases from chloroplasts. Plant Physiol., **68**, 653-657 (1981)

[9] Frentzen, M.; Neuburger, M.; Joyard, J.; Douce, R.: Intraorganelle localization and substrate specificities of the mitochondrial acyl-CoA: sn-glycerol-3-phosphate O-acyltransferase and acyl-CoA: 1-acyl-sn-glycerol-3-phos-

phate O-acyltransferase from potato tubers and pea leaves. Eur. J. Biochem., **187**, 395-402 (1990)
[10] Grobovsky, L.V.; Hershenson, S.; Ernst-Fonberg, M.L.: Isolation of sn-glycerol-3-phosphate acyltransferase from achloroplastic Euglena. FEBS Lett., **102**, 261-264 (1979)
[11] Fritz, P.K.; Kauffman, J.M.; Robertson, C.A.; Wilson, M.R.: Cocoa butter biosynthesis. Purification and characterization of a soluble sn-glycerol-3-phosphate acyltransferase from cocoa seeds. J. Biol. Chem., **261**, 194-199 (1986)
[12] Douady, D.; Dubacq, J.P.: Purification of acyl-CoA: glycerol-3-phosphate acyltransferas from pea leaves. Biochim. Biophys. Acta, **921**, 615-619 (1987)
[13] Zaror-Behrens, G.; Kako, K.J.: Positional and fatty acid specificity of monoacyl- and diacylglycerol 3-phosphate formation by rabbit heart microsomes. Biochim. Biophys. Acta, **441**, 1-13 (1976)
[14] Hendry, A.T.; Possmayer, F.: Pulmonary phospholipid biosynthesis. Properties of a stable microsomal glycerophosphate acyltransferase preparation from rabbit lung. Biochim. Biophys. Acta, **369**, 156-172 (1974)
[15] Kume, K.; Shimizu, T.; Seyama, Y.: Characterization of sn-glycerol 3-phosphate acyltransferase from guinea pig harderian gland microsomes. J. Biochem., **101**, 653-660 (1987)
[16] Haldar, D.; Vancura, A.: Glycerophosphate acyltransferase from liver. Methods Enzymol., **209**, 64-72 (1992)
[17] Yamashita, S.; Numa, S.: Partial purification and properties of glycerophosphate acyltransferase from rat liver. Formation of 1-acylglycerol 3-phosphate from sn-glycerol 3-phosphate and palmityl coenzyme A. Eur. J. Biochem., **31**, 565-573 (1972)
[18] Bates, E.J.; Saggerson, E.D.: A study of the glycerol phosphate acyltransferase and dihydroxyacetone phosphate acyltransferase activities in rat liver mitochondrial and microsomal fractions. Relative distribution in parenchymal and non-parenchymal cells, effects of N-ethylmaleimide, palmitoyl-coenzyme A concentration, starvation, adrenalectomy and anti-insulin serum treatment. Biochem. J., **182**, 751-762 (1979)
[19] Monroy, G.; Chroboczek Kelker, H.; Pullman, M.E.: Partial purification and properties of an acyl coenzyme A:sn-glycerol 3-phosphate acyltransferase from rat liver mitochondria. J. Biol. Chem., **248**, 2845-2852 (1973)
[20] Carroll, M.A.; Morris, P.E.; Grosjean, C.D.; Anzalone, T.; Haldar, D.: Further distinguishing properties of mitochondrial and microsomal glycerophosphate acyltransferase and the transmembrane location of the mitochondrial enzyme. Arch. Biochem. Biophys., **214**, 17-25 (1982)
[21] Coleman, R.A.: Hepatic sn-glycerol-3-phosphate acyltransferases: effect of monoacylglycerol analogs on mitochondrial and microsomal activities. Biochim. Biophys. Acta, **963**, 367-374 (1988)
[22] Datta, N.S.; Hajra, A.K.: Does microsomal glycerophosphate acyltransferase also catalyze the acylation of dihydroxyacetone phosphate?. FEBS Lett., **176**, 264-268 (1984)
[23] Okuyama, H.; Yamada, K.; Kameyama, Y.; Ikezawa, H.; Fukushima, H.; Nozawa, Y.: Studies on lipid metabolism in Tetrahymena pyriformis: proper-

ties of acyltransferase systems. Arch. Biochem. Biophys., **178**, 319-326 (1977)
[24] Schlossman, D.M.; Bell, R.M.: Glycerolipid biosynthesis in Saccharomyces cerevisiae: sn-glycerol-3-phosphate and dihydroxyacetone phosphate acyltransferase activities. J. Bacteriol., **133**, 1368-1376 (1978)
[25] Green, P.R.; Merrill, A.H.; Bell, R.M.: Membrane phospholipid synthesis in Escherichia coli. Purification, reconstitution, and characterization of sn-glycerol-3-phosphate acyltransferase. J. Biol. Chem., **256**, 11151-11159 (1981)
[26] Kessels, J.M.M.; Van Den Bosch, H.: Characterization of reconstituted partially purified glycerophosphate acyltansferase from Escherichia coli. Biochim. Biophys. Acta, **713**, 570-580 (1982)
[27] Scheideler, M.A.; Bell, R.M.: Characterization of active and latent forms of the membrane-associated sn-glycerol-3-phosphate acyltransferase of Escherichia coli. J. Biol. Chem., **266**, 14321-14327 (1991)
[28] Nishida, I.; Frentzen, M.; Ishizaki, O.; Murata, N.: Purification of isomeric forms of acyl-[acyl-carrier-protein]:glycerol-3-phosphate acyltransferase from greening squash cotyledons. Plant Cell Physiol., **28**, 1071-1079 (1987)
[29] Scheideler, M.A.; Bell, R.M.: Efficiency of reconstitution of the membrane-associated sn-glycerol 3-phosphate acyltransferase of Escherichia coli. J. Biol. Chem., **261**, 10990-10995 (1986)
[30] Kessels, J.M.M.; Ousen, H.; Van Den Bosch, H.: Studies on reconstituted partially purified glycerophosphate acyltransferase from Escherichia coli. Eur. J. Biochem., **138**, 543-549 (1984)
[31] Kito, M.; Ishinaga, M.; Nishihara, M.: Function of phospholipids on the regulatory properties of solubilized and membrane-bound sn-glycerol-3-phosphate acyltransferase of Escherichia coli. Biochim. Biophys. Acta, **529**, 237-249 (1978)
[32] Negrel, R.; Ailhaud, G.; Mutaftschiev, S.: Comparative inhibition studies of the phosphotransferase and glycerophosphate acylation systems in membrane vesicles of Escherichia coli. Biochim. Biophys. Acta, **291**, 635-649 (1973)
[33] Kito, M.; Aibara, S.; Hasegawa, K.; Hata, T.: Inhibition of L-glycerol 3-phosphate acyltransferase from Escherichia coli by cis-9,10-methylenehexadecanoic acid. J. Biochem., **71**, 99-105 (1972)
[34] Ishinaga, M.; Nishihara, M.; Kito, M.: Purification and positional specificity of sn-glycerol-3-phosphate acyltransferase from Escherichia coli membranes. Biochim. Biophys. Acta, **450**, 269-272 (1976)
[35] Snider, M.D.; Kennedy, E.P.: Partial purification of glycerophosphate acyltransferase from Escherichia coli. J. Bacteriol., **130**, 1072-1083 (1977)
[36] Snider, M.D.: sn-Glycero-3-phosphate acyltransferase from Escherichia coli. Methods Enzymol., **71**, 546-550 (1981)
[37] Yamashita, S.; Numa, S.: Glycerophosphate acyltransferase from rat liver. Methods Enzymol., **71**, 550-554 (1981)
[38] Marchand, C.; Lemonde, A.; Beaudoin, A.R.: Isolation and properties of a glycerophosphate acylating fraction in the fat body of Schistocerca gregaria (Forskal). Can. J. Biochem., **55**, 1166-1170 (1977)

[39] Barron, E.J.; Stumpf, P.K.: Inhibition of L-glycerol 3-phosphate acyltransferase from Escherichia coli by cis-9,10-methylenehexadecaonic acid. Biochim. Biophys. Acta, **60**, 329-337 (1962)

[40] Scheideler, M.A.; Bell, R.M.: Glycerophosphate acyltransferase from Escherichia coli. Methods Enzymol., **209**, 55-63 (1992)

[41] Okuyama, H.; Wakil, S.J.: Positional specificities of acyl coenzyme A: glycerophosphate and acyl coenzyme A: monoacylglycerophosphate acyltransferases in Escherichia coli. J. Biol. Chem., **248**, 5197-5205 (1973)

[42] Okuyama, H.; Yamada, K.; Ikezawa, H.; Wakil, S.J.: Factors affecting the acyl selectivities of acyltransferases in Escherichia coli. J. Biol. Chem., **251**, 2487-2492 (1976)

[43] Yamada, K.; Okuyama, H.; Endo, Y.; Ikezawa, H.: Acyltransferase systems involved in phospholipid metabolism in Saccharomyces cerevisiae. Arch. Biochem. Biophys., **183**, 281-289 (1977)

[44] Yet, S.F.; Lee, S.; Hahm, Y.T.; Sul, H.S.: Expression and identification of p90 as the murine mitochondrial glycerol-3-phosphate acyltransferase. Biochemistry, **32**, 9486-9491 (1993)

[45] Mitchell, J.R.D.; Saggerson, E.D.: A study of glycerolphosphate acyltransferase in rat liver mitochondria - submitochondrial localization and some properties of the solubilized enzyme. Int. J. Biochem., **26**, 181-187 (1994)

[46] Vancura, A.; Haldar, D.: Purification and characterization of glycerophosphate acyltransferase from rat liver mitochondria. J. Biol. Chem., **269**, 27209-27215 (1994)

[47] Eccleston, V.S.; Harwood, J.L.: Solubilization, partial purification and properties of acyl-CoA:glycerol-3-phosphate acyltransferase from avocado (Persea americana) fruit mesocarp. Biochim. Biophys. Acta, **1257**, 1-10 (1995)

[48] Yet, S.F.; Moon, Y.K.; Sul, H.S.: Purification and reconstitution of murine mitochondrial glycerol-3-phosphate acyltransferase. Functional expression in baculovirus-infected insect cells. Biochemistry, **34**, 7303-7310 (1995)

[49] Wilkison, W.O.; Bell, R.M.: sn-Glycerol-3-phosphate acyltransferase from Escherichia coli. Biochim. Biophys. Acta, **1348**, 3-9 (1997)

[50] Murata, N.; Tasaka, Y.: Glycerol-3-phosphate acyltransferase in plants. Biochim. Biophys. Acta, **1348**, 10-16 (1997)

[51] Dircks, L.K.; Sul, H.S.: Mammalian mitochondrial glycerol-3-phosphate acyltransferase. Biochim. Biophys. Acta, **1348**, 17-26 (1997)

[52] Ferri, S.R.; Toguri, T.: Substrate specificity modification of the stromal glycerol-3-phosphate acyltransferase. Arch. Biochem. Biophys., **337**, 202-208 (1997)

[53] Dircks, L.K.; Ke, J.; Sul, H.S.: A conserved seven amino acid stretch important for murine mitochondrial glycerol-3-phosphate acyltransferase activity. Significance of arginine 318 in catalysis. J. Biol. Chem., **274**, 34728-34734 (1999)

[54] Ganesh Bhat, B.; Wang, P.; Kim, J.H.; Black, T.M.; Lewin, T.M.; Fiedorek, F.T., Jr.; Coleman, R.A.: Rat sn-glycerol-3-phosphate acyltransferase: molecular cloning and characterization of the cDNA and expressed protein. Biochim. Biophys. Acta, **1439**, 415-423 (1999)

[55] Slabas, A.R.; Simon, W.R.; Schierer, T.; Kroon, J.; Fawcett, T.; Hayman, M.; Gilroy, J.; Nishida, I.; Murata, N.; Rafferty, J.; Turnbull, A.; Rice, D.: Plant glycerol-3-phosphate-1-acyltransferase (GPAT): structure selectivity studies. Biochem. Soc. Trans., **28**, 677-679 (2000)
[56] Hayman, M.W.; Fawcett, T.; Schierer, T.F.; Simon, J.W.; Kroon, J.T.M.; Gilroy, J.S.; Rice, D.W.; Rafferty, J.; Turnbull, A.P.; Sedelnikova, S.E.; Slabas, A.R.: Mutagenesis of squash (Cucurbita moschata) glycerol-3-phosphate acyltransferase (GPAT) to produce an enzyme with altered substrate selectivity. Biochem. Soc. Trans., **28**, 680-681 (2000)
[57] Nishida, I.; Sugiura, M.; Enju, A.; Nakamura, M.: A second gene for acyl-(acyl-carrier-protein): glycerol-3-phosphate acyltransferase in squash, Cucurbita moschata cv. Shirogikuza(*), codes for an oleate-selective isozyme: molecular cloning and protein purification studies. Plant Cell Physiol., **41**, 1381-1391 (2000)
[58] Gonzalez-Baro, M.R.; Granger, D.A.; Coleman, R.A.: Mitochondrial glycerol phosphate acyltransferase contains two transmembrane domains with the active site in the N-terminal domain facing the cytosol. J. Biol. Chem., **276**, 43182-43188 (2001)
[59] Zheng, Z.; Zou, J.: The initial step of the glycerolipid pathway: identification of glycerol 3-phosphate/dihydroxyacetone phosphate dual substrate acyltransferases in Saccharomyces cerevisiae. J. Biol. Chem., **276**, 41710-41716 (2001)
[60] Mishra, S.; Kamisaka, Y.: Purification and characterization of thiol-reagent-sensitive glycerol-3-phosphate acyltransferase from the membrane fraction of an oleaginous fungus. Biochem. J., **355**, 315-322 (2001)
[61] Turnbull, A.P.; Rafferty, J.B.; Sedelnikova, S.E.; Slabas, A.R.; Schierer, T.P.; Kroon, J.T.M.; Simon, J.W.; Fawcett, T.; Nishida, I.; Murata, N.; Rice, D.W.: Analysis of the Structure, substrate specificity, and mechanism of squash slycerol-3-Phosphate (1)-acyltransferase. Structure, **9**, 347-353 (2001)
[62] Slabas, A.R.; Kroon, J.T.M.; Scheirer, T.P.; Gilroy, J.S.; Hayman, M.; Rice, D.W.; Turnbull, A.P.; Rafferty, J.B.; Fawcett, T.; Simon, W.J.: Squash glycerol-3-phosphate (1)-acyltransferase. Alteration of substrate selectivity and identification of arginine and lysine residues important in catalytic activity. J. Biol. Chem., **277**, 43918-43923 (2002)
[63] Hayman, M.W.; Fawcett, T.; Slabas, A.R.: Kinetic mechanism and order of substrate binding for sn-glycerol-3-phosphate acyltransferase from squash (Cucurbita moschata). FEBS Lett., **514**, 281-284 (2002)

Acetyl-CoA C-acyltransferase 2.3.1.16

1 Nomenclature

EC number
2.3.1.16

Systematic name
acyl-CoA:acetyl-CoA C-acyltransferase

Recommended name
acetyl-CoA C-acyltransferase

Synonyms
3-ketoacyl CoA thiolase
3-ketoacyl coenzyme A thiolase
3-ketoacyl thiolase
3-ketoacyl-CoA thiolase
3-ketothiolase
3-oxoacyl-CoA thiolase
3-oxoacyl-coenzyme A thiolase
6-oxoacyl-CoA thiolase
KAT
SCP2/3-oxoacyl-CoA thiolase
acetoacetyl-CoA β-ketothiolase
acetyl-CoA acyltransferase
acyltransferase, acetyl coenzyme A
β-ketoacyl coenzyme A thiolase
β-ketoacyl-CoA thiolase
β-ketoadipyl coenzyme A thiolase
β-ketoadipyl-CoA thiolase
β-ketothiolase
ketoacyl-CoA acyltransferase
ketoacyl-coenzyme A thiolase
long-chain 3-oxoacyl-CoA thiolase
oxoacyl-coenzyme A thiolase
pro-3-ketoacyl-CoA thiolase
thiloase B
thiolase A
thiolase I
thiolase II
thiolase III

CAS registry number
9029-97-4

2 Source Organism

<1> *Caulobacter crescentus* [14]
<2> *Saccharomyces cerevisiae* [16, 17, 29-31]
<3> *Cucumis sativus* [18]
<4> *Ralstonia eutropha* (basonym Alcaligenes eutrophus [25]) [25]
<5> *Candida tropicalis* (representative and intermediate strains [34]) [4, 34]
<6> *Pseudomonas fragi* (B-0771 [11]) [11]
<7> *Bos taurus* [1, 15]
<8> *Rattus norvegicus* (male Wistar rats [33]) [1, 5, 7, 12, 20-22, 27, 33, 35]
<9> *Sus scrofa* [1, 2, 6, 10, 15, 19]
<10> *Ovis aries* (male [1]) [1]
<11> *Escherichia coli* (ML308-225, D-1ML, K-12-Y Mel and fatty degradation mutants: fad-5, fadB64, fadA30 [24]) [3, 8, 17, 23, 24, 26]
<12> *Homo sapiens* [9]
<13> *Pseudomonas fragi* [13]
<14> *Helianthus annuus* [28]
<15> *Brassica napus* [32]
<16> *Arabidopsis thaliana* [36]

3 Reaction and Specificity

Catalyzed reaction

acyl-CoA + acetyl-CoA = CoA + 3-oxoacyl-CoA (<8> ping-pong mechanism [21,27]; <7,9> Claisen condensation [15]; <2> reaction mechanism [31]; <4> stoichiometry [25]; <6> multifunctional enzyme: 2-enoyl-CoA hydratase, 3-ketoacyl-CoA thiolase, 3-hydroxyacyl-CoA dehydrogenase [11]; <8> multifunctional enzyme: 2-enoyl-CoA hydratase, 3-ketoacyl-CoA thiolase, 3-hydroxyacyl-CoA dehydrogenase [12]; <13> multienzyme complex: EC 4.2.1.17, EC 1.1.1.35, EC 5.3.3.3, EC 5.1.2.3, EC 2.3.1.16 [13]; <11> multienzyme complex contains enoyl-CoA hydratase, L-3-hydroxyacyl-CoA dehydrogenase, 3-ketoacyl-CoA thiolase, 3-hydroxyacyl-CoA epimerase : EC 5.1.2.3 and Δ^3-cis-Δ-trans-enoyl-CoA isomerase: EC 5.3.3.3 [3]; <11> multienzyme complex: EC 4.2.1.17, EC 1.1.1.35, EC 5.3.3.3, EC 5.1.2.3, EC 2.3.1.16 [26]; <11> multienzyme complex: 3-hydroxyacyl-coenzyme A epimerase, cis-Δ^3-trans-Δ^2-enoyl-CoA isomerase, enoyl-CoA hydratase, L-3-hydroxyacyl-CoA thiolase [23]; <12> multifunctional enzyme: 2-enoyl-CoA hydratase, 3-ketoacyl-CoA thiolase and long-chain specific 3-hydroxyacyl-CoA dehydrogenase [9]; <8> sterol carrier protein X is a 3-oxoacyl CoA thiolase with intrinsic sterol carrier and lipid transfer activity [27])

Reaction type

Claisen condensation <7,9> [15]
acyl group transfer

Natural substrates and products

S CoA + 3-oxoacyl-CoA <1, 2, 8, 9, 11-13, 15, 16> (<2,8,9,15> enzyme catalyses the last step in the β-oxidation cycle [2,29,31,32,35]; <16> metabolism in the peroxisome influences metabolism in the lipid body [36]; <2,12,16> enzyme catalyses step of fatty acid β-oxidation [9,30,34,36]; <8> enzyme plays important role in peroxisomal β-oxidation, in addition the enzyme may facilitate the intraperoxysomal movement of sterols and certain other lipids [27]) (Reversibility: r <9> [2]) [2, 3, 9, 10, 12-14, 17, 24, 27, 29, 30, 32, 34, 36]

P acyl-CoA + acetyl-CoA

Substrates and products

S CoA + 3-ketolauryl-CoA <7-9> (Reversibility: ? <7-9> [15, 22]) [15, 22]

P acetyl-CoA + decanoyl-CoA

S CoA + 3-oxoacyl-CoA <1-15> (<2,7,9> two-step reaction [15,31]) (Reversibility: r <2, 4, 7-9, 11> [1-3, 25, 31, 34]; ? <5, 6, 10, 12-15> [1, 4-24, 26-30, 32, 33, 35, 36]) [1-35]

P acyl-CoA + acetyl-CoA <1-15> [1-35]

S CoA + 3-oxodecanoyl-CoA <4, 8, 7, 9, 11> (Reversibility: r <4, 9> [2, 25]; ? <8, 9> [10, 15, 21]) [2, 10, 15, 21, 23, 25]

P acetyl-CoA + octanoyl-CoA

S CoA + 3-oxododecanoyl-CoA <5, 7-9> (Reversibility: ? <5, 7-9> [4, 15, 21]) [4, 15, 21]

P acetyl-CoA + decanoyl-CoA

S CoA + 3-oxoheptanoyl-CoA <4> (Reversibility: r <4> [25]) [25]

P acetyl-CoA + pentanoyl-CoA

S CoA + 3-oxohexadecanoyl-CoA <8> (Reversibility: ? <8> [21]) [21]

P acetyl-CoA + tetradecanoyl-CoA

S CoA + 3-oxohexanoyl-CoA <4, 5, 7-9> (<7-10> highest activity with 3-oxohexanoyl-CoA [1]) (Reversibility: r <7-9> [1, 2]; ? <4, 5, 7-9> [4, 10, 15, 21, 25]) [1, 2, 4, 10, 15, 21, 25]

P acetyl-CoA + butanoyl-CoA

S CoA + 3-oxooctanoyl-CoA <4, 5, 8, 9> (Reversibility: r <4, 9> [2, 25]; ? <5, 8, 9> [4, 10, 15, 21, 27, 33]) [2, 4, 10, 15, 21, 25, 27, 33]

P acetyl-CoA + hexanoyl-CoA

S CoA + 3-oxopalmitoyl-CoA <7-9> (Reversibility: <7-9> [1]; ? <8> [33]) [1, 33]

P acetyl-CoA + tetradecanoyl-CoA

S CoA + 3-oxopentanoyl-CoA <4> (Reversibility: r <4> [25]) [25]

P acetyl-CoA + propanoyl-CoA

S CoA + acetoacetyl-CoA <4, 5, 7-9, 11, 14> (Reversibility: r <4, 7-9, 11> [1-3, 25]; ? <5, 8, 9, 14> [4, 10, 28, 33]) [1-4, 10, 25, 28, 33]

P acetyl-CoA + acetyl-CoA

S Additional information <2, 4, 7-11, 14> (<7-10> 3-oxoacyl-CoA homologues from acetoacetyl to 3-oxopalmitoyl-CoA [1]; <9> acts on 3-oxoacyl-CoA compounds of various chain lengths [2]; <11> highest activities with medium-chain substrates [3]; <11> 3-oxoacyl-coenzyme A derivatives containing 4 to 16 carbons [8]; <9> acetoacetyl-CoA is a poor substrate [2]; <2,11> highest activity with medium- and long-chain 3-oxoacyl-CoA substrates, but also shows significant activity with acetoacetyl-CoA (20-30% of the optimum with 3-oxohexanoyl-CoA) [17]; <4> active with all 3-oxoacyl-CoAs (C_4-C_{10}) tested [25]; <9> low activity with acetyl-CoA [15]; <9,14> low activity with butyryl-CoA [15,28]; <9> low activity with hexanoyl-CoA [15]; <7,9> low activity with myristoyl-CoA [15]; <14> activity of the enzyme towards 3-oxoacy-CoAs increased with increasing chain length of the substrate [28]; <2> the active site of thiolase can also catalyse the synthesis of acetyl-Co from two molecules of acetyl-CoA [31]; <8> thiolase A and B possess virtually the same substrate specificity [33]) [1-3, 8, 15, 17, 22, 25, 28, 31, 33]

P ?

Inhibitors

3-oxooctanoyl-CoA <8> [27]
4-bromo-2-octenoic acid <8> [7]
CoA <4, 8> (<4> inhibition of condensation [25]; <8> at concentrations above 0.025 mM [33]) [25, 27, 33]
Mg^{2+} <8> (<8> 25 mM, 20% inhibition [22]) [22]
N-ethylmaleimide <4, 11> [25, 26]
N-methylmaleimide <9> [2]
NAD^+ <4> [25]
$NADP^+$ <4> [25]
Tris(hydroxymethyl)aminomethane <11> [23]
acetyl-CoA <9> [6]
anti-thiolase-antibody <2> [30]
decanoyl-CoA <9> [6]
iodoacetamide <9> [2]
long-chain 3-oxoacyl-CoA compounds <9> [2]
palmitoyl-CoA <14> (<14> up to 0.03 mM [28]) [28]
semicarbazide <4> [25]

Turnover number (min^{-1})

Additional information <14> [28]

Specific activity (U/mg)

1.9 <11> [23]
26.3 <8> [12]
42 <5> [4]
75 <8> [5]
119 <8> (<8> thiolase A, substrate: 3-oxooctanoyl-CoA [33]) [33]
123 <8> (<8> thiolase B, substrate: 3-oxooctanoyl-CoA [33]) [33]
130 <13> [13]
352 <14> (<14> substrate: acetoacetyl-CoA [28]) [28]
597 <9> [2, 10]

1988 <14> (<14> substrate: 3-oxohexadecanoyl-CoA [28]) [28]
Additional information <2, 7, 8, 9, 11> (<7,9> assay method [2,15]) [2, 3, 7, 15, 24, 26, 30]

K_m-Value (mM)
0.0018 <9> (3-oxodecanoyl-CoA) [10]
0.0021 <9> (3-oxodecanoyl-CoA) [2]
0.0024 <9> (3-oxooctanoyl-CoA) [2, 10]
0.0083 <9> (3-oxohexanoyl-CoA) [2, 10]
0.0087 <9> (CoA-SH, <9> with acetoacetyl-CoA [2]) [2]
0.0088 <5> (3-oxooctanoyl-CoA) [4]
0.0093 <5> (3-oxododecanoyl-CoA) [4]
0.01 <5, 8> (acetoacetyl-CoA, <5> 3-oxodecanoyl-CoA [4]) [1, 4]
0.016 <9> (acetoacetyl-CoA) [10]
0.017 <9> (acetoacetyl-CoA) [2]
0.018 <8> (CoA) [1]
0.027 <14> (acetoacetyl-CoA) [28]
0.028 <5> (3-oxohexanoyl-CoA) [4]
0.031 <11> (acetoacetyl-CoA) [3]
0.059 <5> (3-oxopentanoyl-CoA) [4]
0.093 <4> (CoA) [25]
0.093 <5> (acetoacetyl-CoA) [4]
0.394 <4> (acetoacetyl-CoA) [25]
Additional information <8, 9, 11, 14> (<8> overview: K_m of peroxisomal and mitochondrial enzyme with various substrates [5,21]; <14> K_m values between 0.003 and 0.007 mM with substrates having a carbon chain length from 6 to 16 carbon atoms, K_m values for CoA at high and low acetoacetyl-CoA concentration [28]) [3, 5, 10, 15, 21, 27, 28, 33]

K_i-Value (mM)
0.02 <8> (CoA, <8> sterol carrier protein X-547 [27]) [27]
0.029 <8> (3-oxooctanoyl-CoA, <8> sterol carrier protein X-547 [27]) [27]

pH-Optimum
7.6 <9> [2]
8 <3> [18]
8.5 <14> (<14> pI: 8.0 [28]) [28]
Additional information <4, 8, 13> (<13> pI: 4.9 [13]; <8> higher activity in phosphate than in Tris buffer [22]; <4> pI: 6.4 [25]; <8> thiolase A pI: 8.45, thiolase B pI: 8.7 [33]) [13, 22, 25, 33]

pH-Range
7-8 <3> (<3> pH 7.0: no significant activity below, pH 8.0: maximum activity [18]) [18]
7.5-9.5 <14> (<14> pH 7.5 and 9.5: 50% maximum activity, pH 6.5: no activity [28]) [28]
7.6-8.3 <9> (<9> pH 7.6: maximum activity, pH 8.3: about 75% of maximum activity [2]) [2]

4 Enzyme Structure

Molecular weight

75000 <14> (<14> gel filtration [28]) [28]
85000 <8> (<8> thiolase A and B, gel filtration [33]) [33]
86000 <2> (<2> gel filtration [30]) [30]
89000 <8> (<8> peroxisomal enzyme, sedimentation equilibrium [21]) [21]
90000 <3> (<3> gel filtration, rate zonal centrifugation [18]) [18]
98000 <5> (<5> gel filtration [4]) [4]
154000 <8> (<8> mitochondrial enzyme, sedimentation equilibrium [21]) [21]
168000 <4> (<4> gel filtration, sucrose density gradient centrifugation [25]) [25]
200000 <9> (<9> gel filtration [2,10]) [2, 10]
240000 <13> (<13> multienzyme complex: EC 4.2.1.17, EC 1.1.1.35, EC 5.3.3.3, EC 5.1.2.3, EC 2.3.1.16, gel filtration [13]) [13]
260000 <11> (<11> multienzyme complex: EC 4.2.1.17, EC 1.1.1.35, EC 5.3.3.3, EC 5.1.2.3, EC 2.3.1.16, activity of EC 2.3.1.16 resides in the 42000 Da subunit of the tetramer, SDS-PAGE [26]) [26]
270000 <11> (<11> multienzyme complex: enoyl-CoA hydratase, L-3-hydroxyacyl-CoA dehydrogenase, 3-ketoacyl-CoA thiolase, 3-hydroxyacyl-CoA epimerase and Δ^3-cis-Δ-trans-enoyl-CoA isomerase, PAGE [3]) [3]
460000 <8> (<8> multifunctional enzyme: 2-enoyl-CoA hydratase, 3-ketoacyl-CoA thiolase, 3-hydroxyacyl-CoA dehydrogenase, gel filtration [12]) [12]

Subunits

dimer <2, 3, 5, 8, 14> (<5> 2 * 43000, SDS-PAGE [4]; <2,3,14> 2 * 45000, SDS-PAGE [18,28,30]; <8> 2 * 40000, peroxisomal enzyme, SDS-PAGE [21]; <8> 2 * 41000, mitochondrial enzyme, SDS-PAGE [21]; <8> 2 * 41000, SDS-PAGE, thiolase B dimer dissociates into monomer with low activity during cold inactivation, thiolase A maintains its dimeric structure [33]) [4, 18, 21, 28-31, 33]
tetramer <4, 8, 9, 11, 13> (<9> 4 * 46000, SDS-PAGE [2,10]; <13> 2 * 73000 + 2 * 42000, multienzyme complex: EC 4.2.1.17, EC 1.1.1.35, EC 5.3.3.3, EC 5.1.2.3, EC 2.3.1.16, SDS-PAGE [13]; <8> mitochondrial enzyme, 4 * 41000 [21]; <4> 4 * 46000, SDS-PAGE [25]; <11> 2 * 78000 + 2 * 42000, multienzyme complex: EC 4.2.1.17, EC 1.1.1.35, EC 5.3.3.3, EC 5.1.2.3, EC 2.3.1.16, activity of EC 2.3.1.16 resides in the 42000 Da subunit, SDS-PAGE [26]) [2, 10, 13, 21, 25, 26]
Additional information <2, 6> (<6> multifunctional enzyme is composed of 2 subunits: α and β, α-subunit exhibits 3-hydroxyacyl-CoA dehydrogenase and 2-enoyl-CoA hydratase activity, β-subunit exhibits 3-oxoacyl-CoA thiolase activity, primary structure of genes for α- and β-subunit [11]; <2> the high sequence similarity between the tetrameric and dimeric thiolases suggests that the tetrameric thiolases are assemble as a dimer of two thiolase dimers [31]) [11, 31]

Posttranslational modification

side-chain modification <15> (<15> deletion of the first 35 N-terminal residues including the conserved cysteine from the cDNA of enzyme, both recombinant enzymes have comparable activity in Escherichia coli, the N-terminal targeting sequence is not essential for proper folding and function of the enzyme [32]) [32]

5 Isolation/Preparation/Mutation/Application

Source/tissue

achene <14> [28]
brain <8> [1]
culture condition:acetate-grown cell <11> [24]
culture condition:aerobically-grown cell <6> [11]
culture condition:long-chain fatty acids-grown cell <11> [3]
culture condition:oleate-grown cell <11> [24]
heart <7-9> (<9> muscle [10]) [1, 2, 6, 10, 15, 19]
kidney <8> [1]
liver <7-9, 12> [1, 5, 7, 9, 12, 15, 21, 22, 27, 33, 35]
seed <3> (<3> cotyledons [18]) [18]
seedling <16> [36]
semen <10> [1]

Localization

cytosol <2, 8> (<2,8> enzyme is synthesized in the cytosol and subsequently imported into the peroxisome [29,35]) [29, 35]
glyoxysome <3, 14, 15> (<3> matrix [18]) [18, 28, 32]
membrane <11> (<11> strain D-1ML, 83% of activity [24]) [24]
mitochondrial membrane <9, 12> [9, 19]
mitochondrion <7-10> (<9> binding to inner membrane [19]) [1, 2, 7, 9, 12, 17, 19-21]
peroxisome <2, 5, 8> (<8> thiolase A from normal rat liver peroxisomes and thiolase B from livers of clofibrate-treated rats [33]) [4, 5, 20-22, 27, 30, 33, 35]

Purification

<2> [30]
<4> [25]
<5> [4]
<7> (general method for separation, quantitation and partial purification [1]) [1]
<8> (thiolase A and B [33]) [5, 12, 22, 33]
<9> [2, 10]
<10> (general method for separation, quantitation and partial purification [1]) [1]
<11> (multienzyme complex contains enoyl-CoA hydratase, L-3-hydroxyacyl-CoA dehydrogenase, 3-ketoacyl-CoA thiolase, 3-hydroxyacyl-CoA epimerase

and Δ^3-cis-Δ-trans-enoyl-CoA isomerase [3]; multienzyme complex: EC 4.2.1.17, EC 1.1.1.35, EC 5.3.3.3, EC 5.1.2.3, EC 2.3.1.16 [26]; multienzyme complex: 3-hydroxyacyl-coenzyme A epimerase, cis-Δ^3-trans-Δ^2-enoyl-CoA isomerase, enoyl-CoA hydratase, L-3-hydroxyacyl-CoA thiolase [23]) [3, 8, 23, 26]
<13> (multienzyme complex: EC 4.2.1.17, EC 1.1.1.35, EC 5.3.3.3, EC 5.1.2.3, EC 2.3.1.16 [13]) [13]
<14> [28]
<16> (recombinant enzyme of KAT gene on chromosome 2 mutant [36]) [36]
<1, 3> (single enzyme, no part of multienzyme complex of *β*-oxidation cycle [14]) [14, 18]

Crystallization
<2> [16, 29, 31]

Cloning
<5> (expression in Saccharomyces cerevisiae [34]) [34]
<8> [20]
<16> (expression of KAT gene on chromosome 2 mutant lacking the putative N-terminal peroxisomal targeting sequence in Escherichia coli [36]) [36]
<6, 8, 15> (expression in Escherichia coli [11,27]; expression in Escherichia coli of a full-length and truncated version [32]) [11, 27, 32]

Application
medicine <2> (<2> mutation in the thiolase gene causes severe health problems, such as ketoacidic attacks at the age of six months followed by severe retardation [29]) [29]

6 Stability

pH-Stability
6.6 <11> (<11> multienzyme complex, -76°C, 25% glycerol v/v, stable for months [3]) [3]
8.1 <9> (<9> 25°C, 25% glycerol v/v, stable for days [2]) [2]
9 <3> (<3> activity destroyed after 3 days [18]) [18]

Temperature stability
25 <9> (<9> pH 8.1, 25% glycerol v/v, stable for days [2]) [2]
Additional information <1, 8> (<1> enzyme is heat-labile [14]; <8> enzyme exhibits reversible cold-lability: loss of acitvity at 4°C in the presence of NaCl, half-maximal inactivation in ice in 4 min [21]; <8> thiolase B exhibits cold-lability: loss of acitvity is more pronounced at 4°C than at 30°C, in the presence of KCl, thiolase A is stable under the same conditions, thiolase A exhibits higher stabiliy than thiolase B at 46°C, lability of the thiolase B might be compensated for by the formation of a more stable multienzyme complex [33]) [14, 21, 33]

General stability information

<9>, glycerol or 2-mercaptoethanol stabilizes [2]
<9, 11>, inactivation at Tris-HCl concentration [2, 3]

Storage stability

<2>, -20°C, 50% glycerol, stable [30]
<2>, frequent freeze and thaw results in a gradual loss of activity [30]
<7>, -20°C, glycerol, several weeks [1]
<9>, 25°C, 0.75 M Tris-HCl, pH 8.1, 25% v/v glycerol, 10 mM 2-mercaptoethanol, stable for days [2]
<11>, -76°C, 0.2 M potassium phosphate, pH 6.6, 25% v/v glycerol, 10 mM 2-mercaptoethanol, stable for months [3]

References

[1] Middleton, B.: 3-Ketoacyl-CoA thiolase of mammalian tissues. Methods Enzymol., **35 B**, 128-136 (1975)

[2] Schulz, H.; Staack, H.: 3-Ketoacyl-Co-A-thiolase with broad chain length specificity from pig heart muscle. Methods Enzymol., **71**, 389-403 (1981)

[3] Binstock, J.F.; Schulz, H.: Fatty acid oxidation complex from Escherichia coli. Methods Enzymol., **71**, 403-411 (1981)

[4] Kurihara, T.; Ueda, M.; Tanaka, A.: Peroxisomal acetoacetyl-CoA thiolase and 3-ketoacyl-CoA thiolase from an n-alkane-utilizing yeast, Candida tropicalis: purification and characterization. J. Biochem., **106**, 474-478 (1989)

[5] Miyazawa, S.; Osumi, T.; Hashimoto, T.: The presence of a new 3-oxoacyl-CoA thiolase in rat liver peroxisomes. Eur. J. Biochem., **103**, 589-596 (1980)

[6] Olowe, Y.; Schulz, H.: Regulation of thiolases from pig heart. Control of fatty acid oxidation in heart. Eur. J. Biochem., **109**, 425-429 (1980)

[7] Li, J.; Schulz, H.: 4-Bromo-2-octenoic acid specifically inactivates 3-ketoacyl-CoA thiolase and thereby fatty acid oxidation in rat liver mitochondria. Biochemistry, **27**, 5995-6000 (1988)

[8] Feigenbaum, J.; Schulz, H.: Thiolases of Escherichia coli: purification and chain length specificities. J. Bacteriol., **122**, 407-411 (1975)

[9] Carpenter, K.; Pollitt, R.J.; Middleton, B.: A unique, membrane-bound, multifunctional enzyme from human liver mitochondria catalysing three steps of fatty acid β-oxidation. Biochem. Soc. Trans., **21**, 35S (1992)

[10] Staack, H.; Binstock, J.F.; Schulz, H.: Purification and properties of a pig heart thiolase with broad chain length specificity and comparison of thiolases from pig heart and Escherichia coli. J. Biol. Chem., **253**, 1827-1831 (1978)

[11] Sato, S.; Hayashi, M.; Imamura, S.; Ozeki, Y.; Kawaguchi, A.: Primary structures of the genes, faoA and faoB, from Pseudomonas fragi B-0771 which encode the two subunits of the HDT multienzyme complex involved in fatty acid β-oxidation. J. Biochem., **111**, 8-15 (1992)

[12] Uchida, Y.; Izai, K.; Orii, T.; Hashimoto, T.: Novel fatty acid β-oxidation enzymes in rat liver mitochondria. II. Purification and properties of en-

oyl-coenzyme A (CoA) hydratase/3-hydroxyacyl-CoA dehydrogenase/3-ketoacyl-CoA thiolase trifunctional protein. J. Biol. Chem., **267**, 1034-1041 (1992)
[13] Imamura, S.; Ueda, S.; Mizugaki, M.; Kawaguchi, A.: Purification of the multienzyme complex for fatty acid oxidation from Pseudomonas fragi and reconstitution of the fatty acid oxidation system. J. Biochem., **107**, 184-189 (1990)
[14] O'Connell, M.A.; Orr, G.; Shapiro, L.: Purification and characterization of fatty acid β-oxidation enzymes from Caulobacter crescentus. J. Bacteriol., **172**, 997-1004 (1990)
[15] Wrensford, L.V.; Coppola, C.; Anderson, V.E.: An acyl-coenzyme A chain length dependent assay for 3-oxoacyl-coenzyme A thiolases employing acetyldithio-coenzyme A. Anal. Biochem., **192**, 49-54 (1991)
[16] Zeelen, J.P.; Wierenga, R.K.; Erdmann, R.; Kunau, W.H.: Crystallographic studies of 3-ketoacylCoA thiolase from yeast Saccharomyces cerevisiae. J. Mol. Biol., **215**, 211-213 (1990)
[17] Middleton, B.: 3-Oxoacyl-coenzyme A thiolases and L-3-hydroxyacyl-coenzyme A dehydrogenases. Biochem. Soc. Trans., **6**, 80-83 (1978)
[18] Frevert, J.; Kindl, H.: Purification of glyoxysomal acetyl-CoA acyltransferase. Hoppe-Seyler's Z. Physiol. Chem., **361**, 537-542 (1980)
[19] Sumegi, B.; Srere, P.A.: Binding of the enzymes of fatty acid β-oxidation and some related enzymes to pig heart inner mitochondrial membrane. J. Biol. Chem., **259**, 8748-8752 (1984)
[20] Arakawa, H.; Takiguchi, M.; Amaya, Y.; Nagata, S.; Hayashi, H.; Mori, M.: cDNA-derived amino acid sequence of rat mitochondrial 3-oxoacyl-CoA thiolase with no transient presequence: structural relationship with peroxisomal isozyme. EMBO J., **6**, 1361-1366 (1987)
[21] Miyazawa, S.; Furuta, S.; Osumi, T.; Hashimoto, T.; Ui, N.: Properties of peroxisomal 3-ketoacyl-CoA thiolase from rat liver. J. Biochem., **90**, 511-519 (1981)
[22] Krahling, J.B.; Tolbert, N.E.: Peroxisomal β-ketothiolase. Arch. Biochem. Biophys., **209**, 100-110 (1981)
[23] Pramanik, A.; Pawar, S.; Antonian, E.; Schulz, H.: Five different enzymatic activities are associated with the multienzyme complex of fatty acid oxidation from Escherichia coli. J. Bacteriol., **137**, 469-473 (1979)
[24] O'Brien, W.J.; Frereman, F.E.: Evidence for a complex of three β-oxidation enzymes in Escherichia coli: induction and localization. J. Bacteriol., **132**, 532-540 (1977)
[25] Haywood, G.H.; Anderson, A.J.; Chu, L.; Dawes, E.A.: Characterization of two 3-ketothiolases possessing differing substrate specificities in the polyhydroxyalkanoate synthesizing organism Alcaligenes eutrophus. FEMS Microbiol. Lett., **52**, 91-96 (1988)
[26] Pawar, S.; Schulz, H.: The structure of the multienzyme complex of fatty acid oxidation from Escherichia coli. J. Biol. Chem., **256**, 3894-3899 (1981)
[27] Seedorf, U.; Brysch, P.; Engel, T.; Schrage, K.; Assmann, G.: Sterol carrier protein X is peroxisomal 3-oxoacyl coenzyme A thiolase with intrinsic sterol carrier and lipid transfer activity. J. Biol. Chem., **269**, 21277-21283 (1994)

[28] Oeljeklaus, S.; Fischer, K.; Gerhardt, B.: Glyoxysomal acetoacetyl-CoA thiolase and 3-oxoacyl-CoA thiolase from sunflower cotyledons. Planta, **214**, 597-607 (2002)

[29] Mathieu, M.; Zeelen, J.P.; Pauptit, R.A.; Erdmann, R.; Kunau, W.H.; Wierenga, R.K.: The 2.8 ANG. crystal structure of peroxisomal 3-ketoacyl-CoA thiolase of Saccharomyces cerevisiae: a five-layered $\alpha,\beta,\alpha,\beta,\alpha$ structure constructed from two core domains of identical topology. Structure, **2**, 797-808 (1994)

[30] Erdmann, R.; Kunau, W.H.: Purification and immunolocalization of the peroxisomal 3-oxoacyl-CoA thiolase from Saccharomyces cerevisiae. Yeast, **10**, 1173-1182 (1994)

[31] Mathieu, M.; Modis, Y.; Zeelen, J.P.; Engel, C.K.; Abagyan, R.A.; Ahlberg, A.; Rasmussen, B.; Lamzin, V.S.; Kunau, W.H.; Wierenga, R.K.: The 1.8 A crystal structure of the dimeric peroxisomal 3-ketoacyl-CoA thiolase of Saccharomyces cerevisiae: implications for substrate binding and reaction mechanism. J. Mol. Biol., **273**, 714-728 (1997)

[32] Olesen, C.; Thomsen, K.K.; Svendsen, I.; Brandt, A.: The glyoxysomal 3-ketoacyl-CoA thiolase precursor from Brassica napus has enzymic activity when synthesized in Escherichia coli. FEBS Lett., **412**, 138-140 (1997)

[33] Antonenkov, V.D.; Van Veldhoven, P.P.; Waelkens, E.; Mannaerts, G.P.: Comparison of the stability and substrate specificity of purified peroxisomal 3-oxoacyl-CoA thiolases A and B from rat liver. Biochim. Biophys. Acta, **1437**, 136-141 (1999)

[34] Kanayama, N.; Ueda, M.; Atomi, H.; Tanaka, A.: Genetic evaluation of physiological functions of thiolase isoenzymes in the n-alkane-assimilating yeast Candida tropicalis. J. Bacteriol., **180**, 690-698 (1998)

[35] Latruffe, N.; Nicolas-Frances, V.; Dasari, V.K.; Osumi, T.: Studies on regulation of the peroxisomal β-oxidation at the 3-ketothiolase step: Dissection of the rat liver thiolase B gene promoter. Adv. Exp. Med. Biol., **466**, 253-259 (1999)

[36] Germain, V.; Rylott, E.L.; Larson, T.R.; Sherson, S.M.; Bechtold, N.; Carde, J.P.; Bryce, J.H.; Graham, I.A.; Smith, S.M.: Requirement for 3-ketoacyl-CoA thiolase-2 in peroxisome development, fatty acid β-oxidation and breakdown of triacylglycerol in lipid bodies of Arabidopsis seedlings. Plant J., **28**, 1-12 (2001)

Aspartate N-acetyltransferase 2.3.1.17

1 Nomenclature

EC number
2.3.1.17

Systematic name
acetyl-CoA:L-aspartate N-acetyltransferase

Recommended name
aspartate N-acetyltransferase

Synonyms
L-aspartate N-acetyltransferase
acetyltransferase, aspartate
aspartate acetyltransferase
aspartic acetylase

CAS registry number
9029-99-6

2 Source Organism

<1> *Rattus norvegicus* (male Sprague-Dawley rat [3]) [1, 3]
<2> *Felis catus* [2]
<3> *Homo sapiens* (normal patients and patients with Canavan disease [4]) [4]

3 Reaction and Specificity

Catalyzed reaction
acetyl-CoA + L-aspartate = CoA + N-acetyl-L-aspartate

Reaction type
acyl group transfer

Natural substrates and products
S acetyl-CoA + L-aspartate <1, 2, 3> (Reversibility: ? <1, 2, 3> [1-4]) [1-4]
P CoA + N-acetyl-L-aspartic acid <1, 2, 3> [1-4]

Substrates and products
S acetyl-CoA + L-aspartate <1, 2, 3> (<1,2> highly specific [1-2]) (Reversibility: ? <1, 2, 3> [1-4]) [1-4]
P CoA + N-acetyl-L-aspartic acid <1, 2, 3> [1-4]

S acetyl-CoA + L-glutamate <1> (<1> only 10% activity compared to L-aspartate [3]) (Reversibility: ? <1> [3]) [3]
P CoA + N-acetyl-L-glutamic acid <1> [3]
S Additional information <1> (<1> reduced CoA with acetyl-AMP can substitute for acetyl-CoA, about 50% of activity with acetyl-CoA [1]) [1]
P ?

Inhibitors
acyl-AMP derivatives <1> (<1> e.g. acetyl-AMP, butyryl-AMP, strong [1]) [1]
benzoyl-CoA <1> (<1> weak [1]) [1]
fluoroacetyl-CoA <1> (<1> weak [1]) [1]
propionyl-CoA <1> (<1> weak [1]) [1]

Metals, ions
KBr <1> (<1> stimulates [1]) [1]
KCNS <1> (<1> stimulates [1]) [1]
KCl <1> (<1> 2fold enzyme activity at 0.05 - 0.1 M [1]) [1]
$KClO_3$ <1> (<1> stimulates [1]) [1]
KI <1> (<1> stimulates [1]) [1]
KNO_3 <1> (<1> stimulates [1]) [1]
LiCl <1> (<1> stimulates [1]) [1]
NH_4Cl <1> (<1> stimulates [1]) [1]
NaCl <1> (<1> stimulates [1]) [1]
$NaNO_2$ <1> (<1> stimulates [1]) [1]

Specific activity (U/mg)
0.000063 <1> (<1> brain [3]) [3]
0.000011 <1> (<1> spinal cord [3]) [3]

K_m-Value (mM)
Additional information <1> (<1> K_m value for acetyl-CoA [1]) [1]

K_i-Value (mM)
Additional information <1> (<1> Ki values for various inhibitors [1]) [1]

pH-Optimum
6-6.8 <1> [1]

pH-Range
5.2-8 <1> (<1> about 13% of maximal activity at pH 5.2, about 50% of maximal activity at pH 8.0 [1]) [1]

Temperature optimum (°C)
37 <1, 2> (<1,2> assay at [1-3]) [1-3]

5 Isolation/Preparation/Mutation/Application

Source/tissue

brain <1, 2, 3> [1-4]
spinal cord <1> [3]
Additional information <1> (<1> highest activity in the brainstem and the spinal cord, lowest activity in the retina, no activity in heart, kidney or liver of rat [3]) [3]

Localization

membrane <1> [3]
soluble <2> [2]

Purification

<1> (partial [1,3]) [1, 3]
<2> (partial [2]) [2]

6 Stability

Temperature stability

100 <1> (<1> no enzyme activity after boiling [1]) [1]

Storage stability

<1>, 0-5°C, 8-10 mg protein/ml, stable for weeks [1]

References

[1] Goldstein, F.B.: Biosynthesis of N-acetyl-L-aspartic acid. J. Biol. Chem., **234**, 2702-2706 (1959)
[2] Knizley, H. Jr.: The enzymatic synthesis of N-acetyl-L-aspartic acid by a water-insoluble preparation of a cat brain acetone powder. J. Biol. Chem., **242**, 4619-4622 (1967)
[3] Truckenmiller, M.E.; Namboodiri, M.A.A.; Brownstein, M.J.; Neale, J.H.: N-Acetylation of L-aspartate in the nervous system: differential distribution of a specific enzyme. J. Neurochem., **45**, 1658-1662 (1985)
[4] Moreno, A.; Ross, B.D.; Bluml, S.: Direct determination of the N-acetyl-L-aspartate synthesis rate in the human brain by ^{13}C MRS and [1-^{13}C]glucose infusion. J. Neurochem., **77**, 347-350 (2001)

Galactoside O-acetyltransferase 2.3.1.18

1 Nomenclature

EC number
2.3.1.18

Systematic name
acetyl-CoA:β-D-galactoside 6-acetyltransferase

Recommended name
galactoside O-acetyltransferase

Synonyms
GAT
acetyltransferase, galactoside
galactoside acetyltransferase
thiogalactoside acetyltransferase
thiogalactoside transacetylase

CAS registry number
9029-94-1

2 Source Organism

<1> *Escherichia coli* (ML308 [1,2,5]; A324-5 [3]; K12 [5]) [1-8]

3 Reaction and Specificity

Catalyzed reaction
acetyl-CoA + a β-D-galactoside = CoA + a 6-acetyl-β-D-galactoside (<1> ordered bi-bi reaction mechanism with acetyl-CoA and CoA as the first substrate and final product, respectively [3])

Reaction type
acyl group transfer

Natural substrates and products
S Additional information <1> (<1> it is suggested that the enzyme plays an important role in lactose utilization [5]) [5]
P ?

Substrates and products

S acetyl-CoA + β-D-methylgalactoside <1> (Reversibility: ? <1> [2]) [2]
P CoA + 6-O-acetyl-β-D-methylgalactoside
S acetyl-CoA + β-D-phenylgalactoside <1> (Reversibility: ? <1> [2, 7]) [2, 7]
P CoA + 6-O-acetyl-β-D-phenylgalactoside
S acetyl-CoA + β-D-phenylglucoside <1> (Reversibility: ? <1> [2]) [2]
P CoA + 6-O-acetyl-β-D-phenylglucoside
S acetyl-CoA + β-D-thiobutylgalactoside <1> (Reversibility: ? <1> [2]) [2]
P CoA + 6-O-acetyl-β-D-thiobutylgalactoside
S acetyl-CoA + β-D-thiodigalactoside <1> (Reversibility: ? <1> [2]) [2]
P CoA + 6-O-acetyl-β-D-thiodigalactoside
S acetyl-CoA + β-D-thiomethylgalactoside <1> (Reversibility: ? <1> [2]) [2]
P CoA + 6-O-acetyl-β-D-thiomethylgalactoside
S acetyl-CoA + β-D-thiophenylgalactoside <1> (Reversibility: ? <1> [2]) [2]
P CoA + 6-O-acetyl-β-D-thiophenylgalactoside
S acetyl-CoA + β-D-thiophenylglucoside <1> (Reversibility: ? <1> [2]) [2]
P CoA + 6-O-acetyl-β-D-thiophenylglucoside
S acetyl-CoA + isopropyl-β-D-thiogalactoside <1> (Reversibility: ? <1> [1-3, 5]) [1-3, 5]
P CoA + isopropyl 6-O-acetyl-β-D-thiogalactoside <1> [3, 4]
S acetyl-CoA + *o*-nitrophenyl-β-D-galactopyranoside <1> (Reversibility: ? <1> [7]) [7]
P CoA + *o*-nitrophenyl 6-O-acetyl-β-D-galactoside
S acetyl-CoA + *o*-nitrophenyl-β-D-galactoside <1> (Reversibility: ? <1> [3]) [3]
P CoA + *o*-nitrophenyl 6-O-acetyl-β-D-galactoside
S acetyl-CoA + *o*-nitrophenyl-β-D-thiogalactoside <1> (Reversibility: ? <1> [3]) [3]
P CoA + *o*-nitrophenyl 6-O-acetyl-β-D-thiogalactoside
S acetyl-CoA + *p*-nitrophenyl-β-D-galactopyranoside <1> (Reversibility: ? <1> [7]) [7]
P CoA + *p*-nitrophenyl 6-O-acetyl-β-D-galactopyranoside
S acetyl-CoA + *p*-nitrophenyl-β-D-galactoside <1> (<1> twice as effective as isopropyl-β-D-thiogalactoside [3]) (Reversibility: ? <1> [3]) [3]
P CoA + *p*-nitrophenyl 6-O-acetyl-β-D-galactoside
S acetyl-CoA + *p*-nitrophenyl-β-D-glucopyranoside <1> (Reversibility: ? <1> [7]) [7]
P CoA + *p*-nitrophenyl 6-O-acetyl-β-D-glucopyranoside
S acetyl-CoA + *p*-nitrophenyl-β-D-glucoside <1> (Reversibility: ? <1> [3]) [3]
P CoA + *p*-nitrophenyl-6-O-acetyl-β-D-glucoside
S acetyl-CoA + *p*-nitrophenyl-β-D-lactopyranoside <1> (Reversibility: ? <1> [7]) [7]
P CoA + *p*-nitrophenyl 6-O-acetyl-β-D-lactopyranoside
S acetyl-CoA + *p*-nitrophenyl-β-D-lactoside <1> (Reversibility: ? <1> [3]) [3]

P CoA + *p*-nitrophenyl 6-O-acetyl-β-D-lactoside
S acetyl-CoA + *p*-nitrophenyl-β-D-thiogalactoside <1> (Reversibility: ? <1> [3]) [3]
P CoA + *p*-nitrophenyl 6-O-acetyl-β-D-thiogalactoside
S butyryl-CoA + *p*-nitrophenyl-β-D-galactopyranoside <1> (Reversibility: ? <1> [7]) [7]
P CoA + *p*-nitrophenyl 6-O-butyryl-β-D-galactopyranoside
S propionyl-CoA + *p*-nitrophenyl-β-D-galactopyranoside <1> (Reversibility: ? <1> [7]) [7]
P CoA + *p*-nitrophenyl 6-O-propionyl-β-D-galactopyranoside
S Additional information <1> (<1> butyryl-CoA is about 10% as effective as acetyl-CoA, enzyme catalyzes the transfer of acetyl from acetyl-CoA to thiogalactosides but not or at much slower rate to galactosides except when a phenyl group is attached to the oxygen atom at carbon 1, a phenyl group substitution also allows glucosides to react [2]; <1> inactive: succinyl-CoA [2]; <1> palmitoyl-CoA [2]; <1> oleoyl-CoA [3]; <1> crotonyl-CoA [3]; <1> enzyme catalyses transfer of an acetyl group from acetyl-CoA to the 6-hydroxyl group of a β-D-galactoside [7,8]) [2, 3, 7, 8]
P ?

Inhibitors

1,10-phenanthroline <1> (<1> weak [3]) [3]
Co^{2+} <1> (<1> slight [2]) [2]
CoA <1> [3]
Cu^{2+} <1> [2]
$MgSO_4$ <1> (<1> 1 mM, addition to 5,5'-dithiobis(2-nitrobenzoic acid) assay mixture lacking EDTA inhibits, preincubation of transacetylase (2 mg/ml in 0.05 M Tris, pH 7.8) with $MgSO_4$ for 1 h at 30°C and subsequent dilution of the enzyme 1000fold in 5,5'-dithiobis(2-nitrobenzoic acid) assay mixture lacking both EDTA and the metal ion stimulates [3]) [3]
Mn^{2+} <1> (<1> slight [2]) [2]
$MnSO_4$ <1> (<1> 40% inhibition [3]) [3]
Zn^{2+} <1> (<1> slight [2]) [2]
acetyl-CoA <1> [3]
iodoacetamide <1> (<1> complete inactivation [7]) [7]
p-chloromercuribenzoate <1> [2]

Metals, ions

$MgSO_4$ <1> (<1> 1 mM, addition to 5,5'-dithiobis(2-nitrobenzoic acid) assay mixture lacking EDTA inhibits, preincubation of transacetylase (2 mg/ml in 0.05 M Tris, pH 7.8) with $MgSO_4$ for 1 h at 30°C and subsequent dilution of the enzyme 1000-fold in 5,5'-dithiobis(2-nitrobenzoic acid) assay mixture lacking both EDTA and the metal ion stimulates [3]) [3]
Mn^{2+} <1> (<1> stimulates when enzyme is initially incubated with the ion in Tris buffer but has no effect if initially incubated in phosphate or if added to the standard assay medium containing 0.05 M phosphate [1]) [1]
Additional information <1> (<1> no requirement for divalent cation [2]) [2]

Turnover number (min^{-1})

3.24 <1> (*p*-nitrophenyl-*β*-D-galactopyranoside, <1> acetyl donor: butyryl-CoA [7]) [7]
85.2 <1> (*p*-nitrophenyl-*α*-D-galactopyranoside, <1> acetyl donor: acetyl-CoA [7]) [7]
110.4 <1> (*p*-nitrophenyl-*β*-D-glucopyranoside, <1> acetyl donor: acetyl-CoA [7]) [7]
110.4 <1> (*p*-nitrophenyl-*β*-D-lactopyranoside, <1> acetyl donor: acetyl-CoA [7]) [7]
348 <1> (*p*-nitrophenyl-*β*-D-galactopyranoside, <1> acetyl donor: propionyl-CoA [7]) [7]
534 <1> (*o*-nitrophenyl-*β*-D-galactopyranoside, <1> acetyl donor: acetyl-CoA [7]) [7]
780 <1> (phenyl-*β*-D-galactopyranoside, <1> acetyl donor: acetyl-CoA [7]) [7]
882 <1> (*p*-nitrophenyl-*β*-D-galactopyranoside, <1> acetyl donor: acetyl-CoA [7]) [7]

Specific activity (U/mg)

35.3 <1> (<1> isopropyl 6-O-acetyl-*β*-D-thiogalactoside produced [1]) [1]
Additional information <1> [2, 4, 5]

K$_m$-Value (mM)

0.04 <1> (acetyl-CoA) [2]
0.045 <1> (acetyl-CoA, <1> E. coli K12 [5]) [5]
0.049 <1> (acetyl-CoA, <1> E. coli ML308 [5]) [5]
0.059 <1> (acetyl-CoA, <1> phenyl-*β*-D-galactopyranoside [7]) [7]
0.065 <1> (acetyl-CoA, <1> *p*-nitrophenyl-*β*-D-galactopyranoside [7]) [7]
0.082 <1> (acetyl-CoA, <1> *o*-nitrophenyl-*β*-D-galactopyranoside [7]) [7]
0.098 <1> (propionyl-CoA, <1> *p*-nitrophenyl-*β*-D-galactopyranoside [7]) [7]
0.104 <1> (acetyl-CoA, <1> *p*-nitrophenyl-*β*-D-galactopyranoside [7]) [7]
0.18 <1> (acetyl-CoA) [3]
0.221 <1> (butyryl-CoA, <1> *p*-nitrophenyl-*β*-D-galactopyranoside [7]) [7]
0.266 <1> (acetyl-CoA, <1> *p*-nitrophenyl-*β*-D-lactopyranoside [7]) [7]
25.2 <1> (*p*-nitrophenyl-*β*-D-galactopyranoside, <1> acetyl donor: butyryl-CoA [7]) [7]
40.2 <1> (*p*-nitrophenyl-*β*-D-galactopyranoside, <1> acetyl donor: acetyl-CoA [7]) [7]
45.6 <1> (*p*-nitrophenyl-*β*-D-galactopyranoside, <1> acetyl donor: propionyl-CoA [7]) [7]
57.9 <1> (*p*-nitrophenyl-*α*-D-galactopyranoside, <1> acetyl donor: acetyl-CoA [7]) [7]
60.2 <1> (*p*-nitrophenyl-*β*-D-galactopyranoside, <1> acetyl donor: acetyl-CoA [7]) [7]
63.4 <1> (*p*-nitrophenyl-*β*-D-galactopyranoside, <1> acetyl donor: acetyl-CoA [7]) [7]

63.4 <1> (phenyl-β-D-galactopyranoside, <1> acetyl donor: acetyl-CoA [7]) [7]
67.9 <1> (*o*-nitrophenyl-β-D-galactopyranoside, <1> acetyl donor: acetyl-CoA [7]) [7]
360 <1> (isopropyl β-D-thiogalactoside) [2]
770 <1> (isopropyl β-D-thiogalactoside) [3]
1200 <1> (isopropyl β-D-thiogalactoside, <1> E. coli K12 [5]) [5]
1500 <1> (isopropyl β-D-thiogalactoside, <1> E. coli ML308 [5]) [5]

K_i-Value (mM)
0.057 <1> (acetyl-CoA) [3]
0.068 <1> (CoA) [3]

pH-Optimum
7 <1> (<1> phosphate buffer [2]) [2]
8.2 <1> (<1> Tris buffer [2]) [2]

4 Enzyme Structure

Molecular weight
45340 <1> (<1> amino acid sequence analysis [6]) [6]
47900 <1> (<1> strain K12, high-speed equilibrium centrifugation using meniscus depletion technique [5]) [5]
49900 <1> (<1> strain ML308, high-speed equilibrium centrifugation using meniscus depletion technique [5]) [5]

Subunits
dimer <1> (<1> 2 * 24800, strain K12 and ML308, SDS-PAGE [5]; <1> 2 * 22671, amino acid sequence analysis [6]) [5, 6]

5 Isolation/Preparation/Mutation/Application

Localization
cytoplasm <1> [6]
Additional information <1> (<1> activity present in supernatant solution [2]) [2]

Purification
<1> (strain ML308 [1,2,5]; affinity chromatography [4]; K12 strain H30000 [5]; wild-type and mutant enzyme [7]) [1, 2, 4, 5, 7]

Crystallization
<1> [1, 8]

Cloning
<1> (expressed in Escherichia coli [7,8]) [7, 8]

Engineering

H115A <1> (<1> k_{cat} highly decreased, histidyl residue has an important catalytic role [7]) [7]
S162A <1> (<1> k_{cat} increased, Ser162 not catalytically important [7]) [7]
W139F <1> (<1> k_{cat} 10fold decreased [7]) [7]

6 Stability

Temperature stability

56 <1> (<1> 30 min, about 25% loss of activity [1]) [1]

Storage stability

<1>, frozen, in salt solution of low ionic strength, stable [1]
<1>, in ammonium sulfate suspension, stable for months [1]

References

[1] Zabin, I.: Crystalline thiogalactoside transacetylase. J. Biol. Chem., **238**, 3300-3306 (1963)
[2] Zabin, I.; Kepes, A.; Monod, J.: Thiogalactoside transacetylase. J. Biol. Chem., **237**, 253-257 (1962)
[3] Musso, R.E.; Zabin, I.: Substrate specificity and kinetic studies on thiogalactoside transacetylase. Biochemistry, **12**, 553-557 (1973)
[4] Zabin, I.; Fowler, A.V.: Purification of thiogalactoside transacetylase by affinity chromatography. Anal. Biochem., **136**, 493-496 (1984)
[5] Fried, V.A.: lac Thiogalactoside transacetylase of Escherichia coli K-12 and ML. J. Bacteriol., **143**, 506-509 (1980)
[6] Fowler, A.V.; Hediger, M.A.; Musso, R.E.; Zabin, I.: The amino acid sequence of thiogalactoside transacetylase of Escherichia coli. Biochimie, **67**, 101-108 (1985)
[7] Lewendon, A.; Ellis, J.; Shaw, W.V.: Structural and mechanistic studies of galactoside acetyltransferase, the Escherichia coli LacA gene product. J. Biol. Chem., **270**, 26326-26331 (1995)
[8] Wang, X.G.; Roderick, S.L.: Expression, purification, crystallization and preliminary X-ray data of Escherichia coli galactoside acetyltransferase. Acta Crystallogr. Sect. D, **D55**, 1955-1957 (1999)

Phosphate butyryltransferase 2.3.1.19

1 Nomenclature

EC number
2.3.1.19

Systematic name
butanoyl-CoA:phosphate butanoyltransferase

Recommended name
phosphate butyryltransferase

Synonyms
butyryltransferase, phosphate
phosphotransbutyrylase

CAS registry number
9030-01-7

2 Source Organism

<1> *Clostridium butyricum* [1]
<2> *Clostridium beijerinckii* (strain VPI 13437, identical with Clostridium butylicum NRRL B593 [2]) [2]
<3> *Clostridium acetobutylicum* (strain ATCC 824 [3,7]; strain NCIMB 8052 [10]; recombinant enzyme [7,8]) [3, 4, 7, 8, 10]
<4> *Dasytricha ruminantium* [5]
<5> *Butyrivibrio fibrisolvens* [6]
<6> *Clostridium saccharoacetobutylicum* [1]
<7> *Clostridium kluyveri* [1]
<8> *Clostridium sporogenes* [1]
<9> *Bacillus megaterium* [9]

3 Reaction and Specificity

Catalyzed reaction
butanoyl-CoA + phosphate = CoA + butanoylphosphate

Reaction type
acyl group transfer

Natural substrates and products

S butanoyl-CoA + phosphate <2, 5> (<2> role in energy metabolism in butyric acid bacteria [2]; <5> role in metabolism [6]) (Reversibility: ? <2, 5> [2, 6]) [2, 6]
P butanoylphosphate + CoA

Substrates and products

S 2-methyl-propionyl-CoA + phosphate <9> (Reversibility: ? <9> [9]) [9]
P 2-methylpropionylphosphate + CoA
S 3-hydroxybutanoyl-CoA + phosphate <3> (Reversibility: r <3> [8]) [8]
P 3-hydroxybutanoylphosphate + CoA
S 4-hydroxybutanoyl-CoA + phosphate <3> (Reversibility: r <3> [8]) [8]
P 4-hydroxybutanoylphosphate + CoA
S 4-hydroxypentanoyl-CoA + phosphate <3> (Reversibility: r <3> [8]) [8]
P 4-hydroxypentanoylphosphate + CoA
S acetoacetyl-CoA + phosphate <2> (Reversibility: r <2> [2]) [2]
P acetoacetylphosphate + CoA <2> [2]
S acetyl-CoA + phosphate <3> (<3> 1.3% of butanoylphosphate formation [3]) (Reversibility: ? <3> [3]) [3]
P acetylphosphate + CoA
S butanoyl-CoA + phosphate <1-9> (Reversibility: ? <1-9> [1-7, 9]; r <3> [8]) [1-9]
P butanoylphosphate + CoA <1-3, 6-8> [1-3]
S isobutanoyl-CoA + phosphate <3> (<3> 30% of butanoylphosphate formation [3]) (Reversibility: ? <3> [3]) [3]
P isobutanoylphosphate + CoA
S isopentanoyl-CoA + phosphate <3> (<3> 95% of butanoylphosphate formation [3]) (Reversibility: ? <3> [3]) [3]
P isopentanoylphosphate + CoA
S n-pentanoyl-CoA + phosphate <3> (<3> 78% of butanoylphosphate formation [3]) (Reversibility: ? <3> [3]) [3]
P n-pentanoylphosphate + CoA
S n-propionyl-CoA + phosphate <3> (<3> 23% of butanoylphosphate formation [3]) (Reversibility: ? <3> [3]) [3]
P n-propionylphosphate + CoA

Inhibitors

ADP <3> [3]
AMP <3> [3]
ATP <3> [3]
$NADP^+$ <3> [3]
NADPH <3> [3]

Metals, ions

Mn^{2+} <1, 6-8> (<1,6-8> requirement [1]) [1]

Specific activity (U/mg)

0.7 <5> [6]
1380 <3> [3]
Additional information <3> (<3> study on gene dosage effects [7]) [7]

K_m-Value (mM)

0.04 <2> (butanoyl-CoA) [2]
0.077 <3> (CoA) [3]
0.11 <3> (butanoyl-CoA) [3]
0.14 <3> (phosphate) [3]
0.26 <3> (butanoylphosphate) [3]
1.1 <2> (acetoacetyl-CoA) [2]
3.33 <2> (acetyl-CoA) [2]
6.5 <2> (phosphate, <2> cosubstrate butyryl-CoA [2]) [2]
12.9 <2> (phosphate, <2> cosubstrate acetoacetyl-CoA [2]) [2]

pH-Optimum

7.5 <2> [2]
8 <3> (<3> butanoylphosphate formation [3]) [3]
8.8 <3> (<3> butanoyl-CoA formation [3]) [3]

pH-Range

6.2-9.5 <3> (<3> butanoylphosphate formation, less than 50% of maximal activity above and below [3]) [3]
6.5-8 <2> (<2> butanoylphosphate formation, 50% of maximal activity at pH 6.5, 80% of maximal activity at pH 8.0 [2]) [2]
6.8-9.5 <3> (<3> butanoyl-CoA formation, less than 50% of maximal activity above and below [3]) [3]

4 Enzyme Structure

Molecular weight

205000 <2> (<2> gel filtration [2]) [2]
264000 <3> (<3> gel filtration [3]) [3]

Subunits

? <2> (<2> x * 33000, SDS-PAGE [2]) [2]
? <3> (<3> x * 31000, SDS-PAGE [4]) [4]
octamer <3> (<3> 8 * 31000, SDS-PAGE [3]) [3]

5 Isolation/Preparation/Mutation/Application

Localization

hydrogenosome <4> [5]

Purification

<1> [1]
<2> [2]
<3> (recombinant enzyme [8]) [3, 8]

Cloning

<3> [4, 10]

Application

biotechnology <3> (<3> use of enzyme for in vitro biosynthesis of poly(hydroxyalkanoic acid) [8]) [8]

6 Stability

Storage stability

<2>, -20°C, less purified preparation, 50% v/v glycerol, 50% loss of activity in 5 d [2]
<2>, liquid N_2, purified preparation, 50 mM sodium phosphate buffer, pH 7.0, various NaCl concentrations, 92-100% loss of activity [2]

References

[1] Valentine, R.C.; Wolfe, R.S.: Purification and role of phosphotransbutylase. J. Biol. Chem., **235**, 1948-1952 (1960)
[2] Thompson, D.K.; Chen, J.C.: Purification and properties of an acetoacetyl coenzyme A-reacting phosphotransbutyrylase from Clostridium beijerinckii (Clostridium butylicum) NRRL B593. Appl. Environ. Microbiol., **56**, 607-613 (1990)
[3] Wiesenborn, D.P.; Rudolph, F.B.; Papoutsakis, E.T.: Phosphotransbutyrylase from Clostridium acetobutylicum ATCC 824 and its role in acidogenesis. Appl. Environ. Microbiol., **55**, 317-322 (1989)
[4] Cary, J.W.; Petersen, D.J.; Papoutsakis, E.T.; Bennett, G.N.: Cloning and expression of Clostridium acetobutylicum phosphotransbutyrylase and butyrate kinase genes in Escherichia coli. J. Bacteriol., **170**, 4613-4618 (1988)
[5] Yarlett, N.; Lloyd, D.; Williams, A.G.: Butyrate formation from glucose by the rumen protozoon Dasytricha ruminantium. Biochem. J., **228**, 187-192 (1985)
[6] Miller, T.L.; Jenesel, S.E.: Enzymology of butyrate formation by Butyrivibrio fibrisolvens. J. Bacteriol., **138**, 99-104 (1979)
[7] Walter, K.A.; Mermelstein, L.D.; Papoutsakis, E.T.: Studies of recombinant Clostridium acetobutylicum with increased dosages of butyrate formation genes. Ann. N.Y. Acad. Sci., **721**, 69-72 (1994)
[8] Liu, S.J.; Steinbuchel, A.: Exploitation of butyrate kinase and phosphotransbutyrylase from Clostridium acetobutylicum for the in vitro biosynthesis of poly(hydroxyalkanoic acid). Appl. Microbiol. Biotechnol., **53**, 545-552 (2000)

[9] Vazquez, G.J.; Pettinari, M.J.; Mendez, B.S.: Phosphotransbutyrylase expression in Bacillus megaterium. Curr. Microbiol., **42**, 345-349 (2001)
[10] Oultram, J.D.; Burr, I.D.; Elmore, M.J.; Minton, N.P.: Cloning and sequence analysis of the genes encoding phosphotransbutyrylase and butyrate kinase from Clostridium acetobutylicum NCIMB 8052. Gene, **131**, 107-112 (1993)

Diacylglycerol O-acyltransferase 2.3.1.20

1 Nomenclature

EC number
2.3.1.20

Systematic name
acyl-CoA:1,2-diacylglycerol O-acyltransferase

Recommended name
diacylglycerol O-acyltransferase

Synonyms
1,2-diacylglycerol acyltransferase
DGAT
DGAT1
DGAT1A
DGAT2
DGAT2A
EC 2.3.1.124 (formerly)
acyl-CoA:1,2-dioleoyl-sn-glycerol acyltransferase
acyltransferase, diacylglycerol
diacylglycerol acyltransferase
diglyceride O-acyltransferase
diglyceride acyltransferase
palmitoyl-CoA-sn-1,2-diacylglycerol acyltransferase

CAS registry number
9029-98-5

2 Source Organism

<1> *Rattus norvegicus* (female Charles River CD [7, 17]; Sprague-Dawley [11, 12, 24]; male Wistar [43]) [1, 3, 6-9, 11-14, 17, 23, 24, 43]
<2> *Homo sapiens* [2, 26, 27, 28]
<3> *Gallus gallus* (chick [35]) [5, 36]
<4> *Bos taurus* (cow [9,15]) [9, 15, 34]
<5> *Capra hircus* [9]
<6> *Oryctolagus cuniculus* [9]
<7> *Glycine max* (soy bean, Merr. cv. Dare [18,19]) [18, 19]
<8> *Spinacia oleracea* (spinach [4]) [4]
<9> *Cuphea carthagenensis* [10]

<10> *Zea mays* (maize, Mo17 [10]) [10, 40]
<11> *Brassica napus* (cv. Reston [20]; var. Dwarf Essex, rapeseed, var. Tower, canola [10]; cv. Topas [29,42]; cv. Reston and Jet Neuf [33]; oilseed rape, cv. Jet Neuf [37,41,42]) [10, 20, 29, 33, 37, 41, 42]
<12> *Carthamus tinctorius* (safflower [22]) [22, 42]
<13> *Saccharomyces cerevisiae* [16]
<14> *Mycobacterium smegmatis* [21]
<15> *Caenorhabditis elegans* [25]
<16> *Nicotiana tabacum* [25]
<17> *Arabidopsis thaliana* [25, 30, 38]
<18> *Mus musculus* [26, 27, 35]
<19> *Mortierella ramanniana* (var angulispora, fungus [31]) [31, 32]
<20> *Helianthus annuus* (sunflower, var. Albena [39]) [39]

3 Reaction and Specificity

Catalyzed reaction
acyl-CoA + 1,2-diacylglycerol = CoA + triacylglycerol

Reaction type
acyl group transfer

Natural substrates and products
S acyl-CoA + 1,2-diacylglycerol <1, 2, 11, 18> (<1> last reaction in triacylglycerol synthesis [1,6,7,17]; <2> last step of triacylglycerol synthesis [28]; <11> enzyme catalyzing the final reaction in the sn-glycerol-3-phosphate pathway leading to TAG [29]; <18> DGAT1 is more involved in fat absorption in the intestine and DGAT2 plays important role in assembly of de novo synthesized fatty acids into VLDL particles in liver [35]) (Reversibility: ? <1, 2, 11, 18> [1, 6, 7, 17, 28, 29, 35]) [1, 6, 7, 17, 28, 29, 35]
P CoA + triacylglycerol

Substrates and products
S acyl-CoA + 1,2-diacylglycerol <1, 2, 3, 4, 8, 10-20> (<1> 2,3-diacylglycerol is acylated at 20% the rate of 1,2-isomer [6]; <1> endogenously and exogenously synthesized diacylglycerol [14]; <1> acyl-donors: broad specificity, saturated, mono-, di- and tetraenoic thioesters [8]; <1> saturated fatty acyl-CoAs from C-8 to C-12 [6]; <13> endogenous 1,2-diacylglycerol [16]; <18> acceptors: overview [26]; <20> acceptors: 1,2-diacylglycerol containing oleic acid and 1,2-diacylglycerol containing capric acid [39]; <4> broad range of donors from C-12 to C-22, broad range of acceptors from C_{12} to C_{22} [34]; <4> saturated fatty acyl-CoAs from C_4 to C_{18} [9]) (Reversibility: ? <1, 2, 3, 4, 8, 10-20> [2, 4-9, 10, 12, 14, 15-17, 20, 21-23, 25-28, 30, 31, 33, 34, 36, 37, 39, 42, 43]) [2, 4-9, 10, 12, 14, 15, 16, 17, 20, 21-23, 25-28, 30, 31, 33, 34, 36, 37, 39, 42, 43]
P CoA + triacylglycerol <1, 3> [5-7, 8, 12, 14, 17, 28, 29, 30, 31-34, 36, 37, 39, 40, 43]

S acyl-CoA + 1,2-didecanoylglycerol <12> (Reversibility: ? <12> [22]) [22]
P CoA + 3-acyl-1,2-didecanoylglycerol
S acyl-CoA + 1,2-dihexanoylglycerol <12> (Reversibility: ? <12> [22]) [22]
P CoA + 3-acyl-,2-dihexanoylglycerol
S acyl-CoA + 1,2-dilauroylglycerol <12> (Reversibility: ? <12> [22]) [22]
P CoA + 3-acyl-1,2-dilauroylglycerol
S acyl-CoA + 1,2-dimyristoylglycerol <12> (Reversibility: ? <12> [22]) [22]
P CoA + 3-acyl-1,2-dimyristoylglycerol
S acyl-CoA + 1,2-dioctanoylglycerol <12> (Reversibility: ? <12> [22]) [22]
P CoA + 3-acyl-1,2-dioctanoylglycerol
S acyl-CoA + 1,2-dioleoylglycerol <1, 8, 12> (Reversibility: ? <1, 8, 12> [4, 7, 22]) [4, 7, 22]
P CoA + 3-acyl-1,2-dioleoylglycerol
S acyl-CoA + 1,2-dipalmitin <8> (<8> no acceptor [4]) (Reversibility: ? <8> [4]) [4]
P CoA + 3-acyl-1,2-dipalmitoylglycerol
S acyl-CoA + 1,2-dipalmitoylglycerol <4> (<4> membrane bound 1,2-dipalmitoylglycerol [15]) (Reversibility: ? <4> [15]) [15]
P CoA + 3-acyl-1,2-dipalmitoylglycerol
S acyl-CoA + 1,3-dioleoylglycerol <12, 19> (<12> no acceptor [22]) (Reversibility: ? <12, 19> [22, 31]) [22, 31]
P CoA + 2-acyl-1,3-dioleoylglycerol
S acyl-CoA + 2,3-dioleoyl-sn-glycerol <12> (Reversibility: ? <12> [22]) [22]
P CoA + 1-acyl-2,3-dioleoyl-sn-glycerol
S acyl-CoA + 2-oleoylglycerol <12> (Reversibility: ? <12> [22]) [22]
P CoA + 3-acyl-2-oleoylglyerol
S acyl-CoA + glycerol <3> (Reversibility: ? <3> [5]) [5]
P CoA + acylglycerol
S acyl-CoA + lysophosphatidylcholine <19> (Reversibility: ? <19> [31]) [31]
P CoA + acyllysophosphatidylcholine
S acyl-CoA + rac-1,2-diacetylglycerol <12> (<12> no acceptor [22]) (Reversibility: ? <12> [22]) [22]
P CoA + 3-acyl-rac-1,2-diacetylglycerol
S acyl-CoA + rac-1,2-dibutyrylglycerol <12> (<12> no acceptor [22]) (Reversibility: ? <12> [22]) [22]
P CoA + 3-acyl-rac-1,2-dibutyrylglycerol
S acyl-CoA + rac-1,2-dioleoylglycerol <12, 19> (Reversibility: ? <12, 19> [22, 31]) [22, 31]
P CoA + 3-acyl-rac-1,2-dioleoylglycerol
S acyl-CoA + sn-1,2-diolein <1, 19> (Reversibility: ? <1, 19> [14, 31]) [14, 31]
P CoA + acyl-sn-1,2-diolein
S acyl-CoA + sn-2-monoolein <1, 3, 19> (Reversibility: ? <1, 3, 19> [5, 14, 31]) [5, 14, 31]
P CoA + ?

S arachidonoyl-CoA + 1,2-diacylglycerol <1, 18> (Reversibility: ? <1, 18> [7, 23, 27]) [7, 23, 27]

P CoA + 3-arachidonoyl-1,2-diacylglycerol

S butyryl-CoA + 1,2-diacylglycerol <1, 4-6> (Reversibility: ? <1, 4-6> [9, 15]) [9, 15]

P CoA + 1,2-diacyl-3-butyrylglycerol

S decanoyl-CoA + 1,2-diacylglycerol <1> (<1> saturated fatty acyl-CoAs from C-8 to C-18 with decanoyl-CoA as the best [7]) (Reversibility: ? <1> [7]) [7]

P CoA + 1,2-diacyl-3-decanoylglycerol

S erucoyl-CoA + 1,2-diacylglycerol <9, 10, 11> (<11> Jet Neuf enzyme displays an enhanced specificity for erucoyl-CoA over oleoyl-CoA [42]) (Reversibility: ? <9, 10, 11> [10]) [10, 42]

P CoA + 1,2-diacyl-3-erucoylglycerol

S erucoyl-CoA + 1,2-dioleoyl-sn-glycerol <11> (Reversibility: ? <11> [20]) [20]

P CoA + 3-erucoyl-1,2-dioleoyl-sn-glycerol

S lauroyl-CoA + 1,2-diacylglycerol <9, 10, 11, 19> (Reversibility: ? <10, 11, 9, 19> [10, 31]) [10, 31]

P CoA + 1,2-diacyl-3-lauroylglycerol

S linoleoyl-CoA + 1,2-diacylglycerol <1, 2, 18> (<1> poor donor [7]) (Reversibility: ? <1, 2, 18> [2, 7, 27]) [2, 7, 23, 27]

P CoA + 1,2-diacyl-3-linoleoylglycerol

S linoleoyl-CoA + 1,2-dioleoyl-sn-glycerol <8> (<8> i.e. diolein [4]) (Reversibility: ? <8> [4]) [4]

P CoA + 3-linoleoyl-1,2-dioleoylglycerol <8> [4]

S myristoyl-CoA + 1,2-diacylglycerol <1, 2> (Reversibility: ? <1, 2> [2, 23]) [2, 23]

P CoA + 1,2-diacyl-3-myristoylglycerol

S oleoyl-CoA + 1,2-diacylglycerol <1, 9, 10, 11, 12, 17, 18, 19, 20> (Reversibility: ? <1, 9, 10, 11, 12, 17, 18, 19, 20> [10, 22, 23, 26, 27, 29, 30, 31, 33, 39, 40, 42]) [10, 22, 23, 26, 27, 29, 30, 31, 33, 39, 40, 42]

P CoA + 1,2-diacyl-3-oleoylglycerol

S oleoyl-CoA + 1,2-dioleoyl-sn-glycerol <1, 7, 8, 11, 15, 16, 17> (<1> i.e. diolein, at 34% the rate of the acylation with stearoyl-CoA [7]) (Reversibility: ? <1, 7, 8, 11, 15, 16, 17> [4, 11, 19, 20, 25, 37]) [4, 11, 19, 20, 25, 37]

P CoA + trioleoylglycerol

S palmitoyl-CoA + 1,2-diacylglycerol <1, 3, 4, 11, 14, 17> (<1> diacylglycerol diluted in ethanol [17]) (Reversibility: ? <1, 3, 4, 11, 14, 17> [5, 6, 7, 9, 14, 15, 17, 21, 23, 30, 33, 36]) [5, 6, 7, 9, 14, 15, 17, 21, 23, 30, 33, 36]

P CoA + 1,2-diacyl-3-palmitoylglycerol

S palmitoyl-CoA + 1,2-dioleoyl-sn-glycerol <1, 8, 14> (<8> i.e. diolein [4]) (Reversibility: ? <1, 8, 14> [4, 13, 21, 43]) [4, 13, 21, 43]

P CoA + 1,2-dioleoyl-3-palmitoylglycerol

S palmitoyl-CoA + 1,2-dipalmitin <14> (Reversibility: ? <14> [21]) [21]

P CoA + tripalmitin

S palmitoyl-CoA + 1,2-distearin <14> (Reversibility: ? <14> [21]) [21]

P CoA + 3-palmitoyl-1,2-distearoylglycerol
S palmitoyl-CoA + 1,3-diolein <14> (Reversibility: ? <14> [21]) [21]
P CoA + 1,3-dioleoyl-2-palmitoylglycerol
S palmitoyl-CoA + 1,3-dipalmitin <14> (Reversibility: ? <14> [21]) [21]
P CoA + tripalmitin
S palmitoyl-CoA + 1-monoolein <14> (Reversibility: ? <14> [21]) [21]
P CoA + ?
S palmitoyl-CoA + 1-monopalmitin <14> (Reversibility: ? <14> [21]) [21]
P CoA + ?
S palmitoyl-CoA + 1-monostearin <14> (Reversibility: ? <14> [21]) [21]
P CoA + ?
S palmitoyl-CoA + didecanoylglycerol <1> (<1> i.e. dicaprin [7]) (Reversibility: ? <1> [7]) [7]
P CoA + 1,2-didecanoyl-3-palmitoylglycerol
S palmitoyl-CoA + dihexanoylglycerol <1> (<1> i.e. dicaproin [7]) (Reversibility: ? <1> [7]) [7]
P CoA + dihexanoyl-palmitoylglycerol
S palmitoyl-CoA + dipalmitoylglycerol <4> (<4> membrane-bound dipalmitoylglycerol [15]) (Reversibility: ? <4> [15]) [15]
P CoA + tripalmitoylglycerol
S stearoyl-CoA + 1,2-diacylglycerol <1, 11> (<1,11> weak donor [23,33]) (Reversibility: ? <1, 11> [23, 33]) [23, 33]
P CoA + 1,2-diacyl-3-stearoylglycerol
S stearoyl-CoA + 1,2-dioleoyl-sn-glycerol <1, 7, 8> (<8> i.e. diolein [4]) (Reversibility: ? <1, 7, 8> [4, 11, 19]) [4, 7, 19]
P CoA + 1,2-dioleoyl-3-stearoylglycerol

Inhibitors

1,2-diacyl-sn-glycerol <1> (<1> dicaprin, dimyristin, dipalmitin, high concentration [7]) [7]
1-O-hexadecyl-2-oleoyl-sn-glycerol <2> (<2> competitive [2]) [2]
1-O-hexadecylpropanediol-3-phosphocholine <1> (<1> activation, low concentration, inhibitory at high concentration [12]) [12]
1-acyl-sn-glycero-3-phosphocholine <1> (<1> activation, at low concentrations, inhibition above 0.2 mM [12]) [12]
1-hexanol <1> [7]
2-bromooctanoate <11> (<11> 50% inhibition at 1.5 mM [20]) [20]
2-butanol <1> [7]
5,5'-dithio-bis(2-nitrobenzoic acid) <1, 14> (<14> i.e. DTNB, complete inhibition, DTT protects [21]; <1> 11% inhibition at 1 mM [23]) [21, 23]
ATP <1> (<1> ATP-dependent activity reduces activity in vitro by 30-40% [11]) [11]
CHAPS <11, 20> (<20> 16.2 mM [39]) [20, 39]
$CaCl_2$ <19> (<19> 2 mM, inhibits purified enzyme but not activity in lipid body fraction [31]) [31]
DTT <1, 2> (<2> above 2 mM [2]; <1> microsomal preparation [14]) [2, 14]
EDTA <1> [17]

$HgCl_2$ <1> (<1> 90% at 0.03-0.05 mM, reversible by DTT [14]) [14]
I^- <11> (<11> substantial inhibition [33]) [33]
KCl <19> (<19> 500 mM, inhibits purified enzyme [31]) [31]
$MgCl_2$ <2, 18, 19> (<2> 20 mM [2]; <18> 100 mM [27]; <19> 5 mM, inhibits purified enzyme but not activity in lipid body fraction [31]) [2, 27, 31]
$MnCl_2$ <1> (<1> 2 mM [7]) [7]
N-ethylmaleimide <4, 14> (<14> inhibits solubilized enzyme, but not membrane-bound enzyme [21]; <4> 95% loss of activity after preincubation with 40 mM [34]) [21, 34]
NaF <3> [5]
Tergitol NP-40 <1> (<1> microsomal preparation [14]) [14]
Triton WR1339 <14> [21]
Triton X-100 <1, 11, 14, 20> (<1> above 0.05 mM [13]; <1,11> microsomal preparation [14,20]; <20> 50% inhibition at 0.1-0.2% [39]) [13, 14, 20, 21, 39]
Tween 20 <1, 11, 14> (<1> 99% inhibition at 5 mg/ml, 89% inhibition at 1 mg/ml, 49% inhibition at 0.5 mg/ml [17]) [7, 17, 20, 21]
$ZnSO_4$ <17> (<17> 3 mM [30]) [30]
acyl-CoA <1> (<1> C_{10}-CoA to C_{16}-CoA, high concentration [7]) [7]
bovine serum albumin <4, 20> (<4> binds palmitoyl-CoA and decreases activity towards palmitoyl-CoA, but not butyryl-CoA [15]; <20> activation at 0.1%, inhibition at 2.5% [39]) [15, 39]
butyryl-CoA <4> [15]
cetyltrimethylammonium bromide <11> [20]
cholate <7, 14> (<1> not up to 4 mM [13]) [19, 21]
deoxycholate <7, 11, 14> [19-21]
dioxane <1> [7]
ethanol <1> (<1> 43% inhibition when 0.04 ml are added to reaction mixture [17]) [17]
gum arabic <1> [7]
iodoacetamide <1> (<1> 15% inhibition at 1 mM, DTT protects [23]) [23]
lysophosphatidylcholine <1> (<1> activation at low concentrations, optimum at 0.075 mM, inhibitory above 0.2 mM [12]) [12]
n-octyl-β-D-glucopyranoside <11> (<11> weak [20]) [20]
niacin <17> (<17> 30% inhibition [30]) [30]
p-chloromercuribenzene sulfonate <14> (<14> complete inhibition, DTT protects [21]) [21]
p-chloromercuribenzoate <14> (<14> complete inhibition, DTT protects [21]) [21]
palmitoyl-CoA <1> (<1> high concentration [13]) [13]
sodium taurocholate <1> [12]
sodiumdodecyl sulfate <11, 14, 19> (<19> irreversible denaturation [31]) [20, 21, 31]
sphingosine <19> [31]
trifluoperazine <19> (<19> 0.5 mM, in the presence of suboptimal phosphatidic acid concentration [31]) [31]

Additional information <1, 3, 14> (<3> no inhibition by Ca^{2+} [5]; <1> 10% v/v acetone, sn-2-monoolein [14]; <14> diethyl-*p*-nitrophenylphosphate, diisopropyl fluorophosphate [21]) [5, 14, 21]

Activating compounds

1-O-hexadecylpropanediol-3-phosphocholine <1> (<1> activation, low concentration, inhibitory at high concentration [12]) [12]
1-acyl-sn-glycero-3-phosphocholine <1> (<1> activation, at low concentrations, inhibition above 0.2 mM [12]) [12]
1-acyl-sn-glycero-3-phosphoethanolamine <1> (<1> activation, to a lesser extent [12]) [12]
DTT <1> (<1> activation, microsomal preparation [23]) [23]
bovine serum albumin <1, 11, 20> (<1> activation, 1.0-1.5 mg/ml [7]; <1> 65% activation at 1 mg/ml, microsomal preparation [17]; <11> 4-5fold by 3 to 4 mg/ml [33]; <20> activation at 0.1%, inhibition at 2.5% [39]) [7, 17, 33, 39]
ethanol <1> (<1> highest activity when diacylglycerol is in ethanol [7]) [7]
glucose <18> (<18> treatment of adipocytes with glucose increases DGAT1 mRNA [35]) [35]
insulin <18> (<18> treatment of adipocytes with insulin increases DGAT2 mRNA [35]) [35]
lysophosphatidylcholine <1> (<1> activation at low concentrations, optimum at 0.075 mM, inhibitory above 0.2 mM [12]) [12]
phosphatidic acid <19> [31]
phosphatidylcholine <1> (<1> from egg yolk [43]) [43]
phosphocholine <1> (<1> from soy bean, activation [13]) [13]
phospholipid <1> (<1> microsomal [13]) [13, 14]
sucrose <11> (<11> treatment of cells with elevated sucrose levels increases DGAT activity [37,41]) [37, 41]

Metals, ions

Ca^{2+} <8> (<8> activation, to some extent [4]) [4]
Co^{2+} <8> (<8> activation to some extent [4]) [4]
Mg^{2+} <1, 2, 3, 8, 17> (<2> 1.4fold stimulation [2]; <1> 2-10 mM [7]; <8> 2fold stimulation at 20 mM [4]; <3> stimulation but no absolute requirement [5]; <1> omission results in 50% activity [7]; <1> strongly dependent [13]; <1> 2.4fold at 8 mM [17]; <17> stimulation at 3 and 10 mM [30]) [2, 4, 5, 7, 17, 30]
Mn^{2+} <3, 8> (<8> 2fold stimulation at 20 mM [4]; <3> stimulation but no absolute requirement [5]) [4, 5]
NaCl <11> (<11> 1.4fold stimulation at 0.5 M [33]) [33]

Specific activity (U/mg)

0.0003 <18> (<18> membrane fraction after insulin-starvation [35]) [35]
0.00036 <18> (<18> membrane fraction after carbohydrate-starvation [35]) [35]
0.0012 <1> [11]
0.0018 <13> [16]

0.0024-0.038 <1> [13]
0.00265 <18> (<18> membrane fraction [35]) [35]
0.00292 <7> (<7> after KCl wash [19]) [19]
0.0035-0.0134 <2> (<2> transfected cells [28]) [28]
0.00361 <3> [36]
0.0038 <1> (<1> submandibular gland [43]) [43]
0.00391 <1> (<1> parotid gland [43]) [43]
0.0073 <1> (<1> after stimulation with cytosolic fraction [11]) [11]
0.0089 <1> [8]
0.0098 <3> (<3> microsomal fraction of embryo liver after 12 days of development [36]) [36]
0.0225 <3> (<3> microsomal fraction of embryo liver after 19 days of development [36]) [36]
0.0264 <3> (<3> microsomal fraction of embryo liver after 22 days of development [36]) [36]
0.05 <1> [17]
0.05 <1> (<1> above [17]) [17]
0.119 <10> [40]
0.169-0.354 <18> [27]
0.83 <1> [24]
2.63 <19> [32]
3.3 <19> [31]
Additional information <1, 7, 10, 11, 20> (<7> microsomal preparations with detergents [19]; <1> specific activity in the nanomolar range in microsomal preparations, varies considerably between different tissues [6]; <11> specific activity in homogenates in the picomolar range [42]) [6, 19, 39, 40, 42]

K_m-Value (mM)

0.0057-0.0068 <4> (hexanoyl-CoA) [9]
0.006 <14> (palmitoyl-CoA) [21]
0.006 <7> (sn-1,2-diolein, <7> with stearoyl-CoA [19]) [19]
0.014 <14> (1,2-diolein) [21]
0.016-0.025 <4> (butyryl-CoA) [9]
0.06 <20> (oleoyl-CoA) [39]
0.113 <7> (sn-1,2-diolein, <7> with oleoyl-CoA [19]) [19]
Additional information <7> [19]

K_i-Value (mM)

0.032 <2> (1-O-hexadecyl-2-oleoyl-sn-glycerol) [2]

pH-Optimum

6.1 <2> (<2> broad [2]) [2]
6.6 <4> [34]
6.75-7.25 <17> [30]
6.8 <19> [32]
7 <11, 19> [31, 33]
7-9 <3> (<3> broad [5]) [5]
7.5-8 <1> [7]

8 <1, 8> (<1> microsomal preparation [17]) [4, 17]
8-9 <14> [21]

Temperature optimum (°C)
23 <1> (<1> assay at [7]) [7]
25 <1> (<1> assay at [6,17]) [6, 17]
27 <1, 13> (<1,13> assay at [8,16]) [8, 16]
30-40 <11> [20]
37 <19> (<19> DGAT2A [32]) [32]
Additional information <1-6, 9-12> (<1-6,9-12> assay at various temperatures from 24°C to 40°C [2,5-12,14,17,22,23]) [2, 5, 9, 10, 11, 12, 14, 22, 23]

Temperature range (°C)
20-50 <11> (<11> about 70% of maximal activity at 20°C, about half-maximal activity at 50°C [20]) [20]

4 Enzyme Structure

Molecular weight
13280 <11> (<11> N-terminal fragment, calculated DNA sequence [41]) [41]
36000-36500 <19> (<19> SDS-PAGE [32]) [32]
39500 <11> (<11> DGAT2, calculated from amino acid sequence [37]) [37]
39600 <19> (<19> DGAT2A, calculated from amino acid sequence [32]) [32]
40600 <19> (<19> DGAT1A, calculated from amino acid sequence [32]) [32]
44500 <18> (<18> SDS-PAGE [27]) [27]
47000 <18> (<18> SDS-PAGE [26]) [26]
53000 <19> (<19> SDS-PAGE [31]) [31]
56900 <11> (<11> DGAT1, calculated from amino acid sequence [37]) [37]
60000 <1> (<1> SDS-PAGE [24]) [24]
100000 <10> (<10> gel filtration [40]) [40]
150000 <2> (<2> gel filtration [28]) [28]
Additional information <1, 7> (<1> MW 350000-370000, triacylglycerol synthetase complex, gel filtration [3]; <7> amino acid composition [18]) [3, 18]
1539000 <7> (<7> delipidated enzyme, free of phospholipids, gel filtration [18]) [18]
1843000 <7> (<7> native enzyme monomer, gel filtration [18]) [18]
2000000 <11> (<11> gel filtration [33]) [33]

Subunits
? <7> (<7> x * 40800 + 2 * 28700 + 2 * 24500, monomer, SDS-PAGE, amino acid analysis [18]) [18]
tetramer <2> (<2> 4 * 50000, SDS-PAGE [28]) [28]
Additional information <7> (<7> the native enzyme is presumably composed of 10 identical monomers each consisting of 3 different subunits with a ratio of 1:2:2 [18]) [18]

5 Isolation/Preparation/Mutation/Application

Source/tissue

3T3-L1 cell <1, 18> (<18> 3T3-L1 fibroblasts differentiated into adipocytes [35]) [6, 35]
HL-60 cell <2> (<2> leukemia, HL-60 cell line [2]) [2]
adipocyte <1, 2, 4, 18> (<1> isolated from parametrial adipose [6]) [1, 6, 7, 9, 11, 17, 27, 28, 34, 35]
adipose tissue <1> [11]
brain <1, 18> [27, 43]
cotyledon <7, 9, 11> [10, 18, 19]
embryo <3, 10, 11> (<11> microspore-derived [20]) [20, 36, 40]
epididymis <1> [11]
heart <18> [27]
intestinal mucosa <1> (<1> small intestinal epithelium, decrease of specific activity from proximal to distal small intestine [14]; <1> villus cells [3]; <1> high activity [6]) [1, 3, 6, 7, 14, 24, 43]
leaf <8> [4]
leukocyte <2> [27]
liver <1, 2, 3-6, 18> [1, 5-9, 11, 12, 13, 15, 23, 27, 35, 36]
mammary gland <1, 2, 4-6> (<1> high activity [6]) [1, 6, 9, 15, 27]
muscle <4, 18> [27, 34]
parotid gland <1> [43]
peripheral blood <2> [27]
salivary gland <1> [43]
scutellum <10> [10]
seed <9-12, 17, 20> (<9-12> maturating [10,22,33,42]) [10, 22, 33, 38, 39, 42]
spleen <1> [43]
testis <2> [27]
Additional information <1> (<1> distribution, overview [1,7]) [1, 7]

Localization

lipid particle <13> [16]
lipid storage body <10, 19> [31, 40]
membrane <3, 11, 14, 18> (<14> firmly bound [21]; <3> yolk sac membrane [36]) [21, 33, 35, 36]
microsome <1-7, 11, 10, 12, 15-17, 20> (<1> intrinsic membrane protein [6]; <1> active site on cytoplasmic surface [1,6]) [1-3, 5-9, 11-15, 17-20, 22, 23, 24, 25, 34, 39, 40]
mitochondrion <3> [5]
Additional information <1> (<1> subcellular distribution [7]) [7]

Purification

<1> (partial [6,8,11,13]; with Triton X-100, deoxycholate or cholate [13]; solubilized, as part of triacyl synthetase complex [3]; partial, with 1-acylglycerolphosphorylcholine acyltransferase [23]; near homogeneity [24]) [3, 6, 8, 11, 13, 23, 24]

<7> (homogeneity [19]) [18, 19]
<8> (partial [4]) [4]
<10> [40]
<14> (solubilized by aqueous acetone [21]) [21]
<19> (4802fold [31]; MrDGAT2A and MrDGAT2B [32]) [31, 32]

Cloning

<2> (DGATsv, splice variant, deletion of 101 residues from the C-terminus, enzymatically inactive [28]) [26, 27, 28]
<11> (wild-type and N-terminal fragment [41]) [41]
<15> [25]
<16> [25]
<17> (mutants AS11 and ABX45, ABX45 germination is delayed [38]) [25, 30, 38]
<18> (DGAT2 [27]) [27, 26]
<19> (MrDGAT2A and MrDGAT2B [32]) [32]

6 Stability

Temperature stability

23 <1> (<1> at least 15 min stable [7]) [7]
25 <2> (<2> labile above, gradual inactivation [2]) [2]
28 <1> (<1> stable below, about 30% loss of activity after 15 min [7]) [7]
30 <11> (<11> rapid loss of activity [29]) [29]
37 <1> (<1> $t_{1/2}$: 5 min [7]; <1> $t_{1/2}$: 8 min [23]) [7, 23]
40 <2, 11, 19> (<11> stable below [20]; <2> 20% loss of activity after 15 min [2]; <19> complete loss of activity after 30 min [31]) [2, 20, 31]
42 <1> (<1> $t_{1/2}$: 1.5 min [7]) [7]
45 <11> (<11> $t_{1/2}$: 10 min [20]) [20]
48 <1> (<1> $t_{1/2}$: 1 min [7]) [7]
50 <2> (<2> $t_{1/2}$: 15 min [2]) [2]
65 <2> (<2> inactivation after 15 min [2]) [2]

Organic solvent stability

acetone <10, 14> (<14> aqueous, stable to [21]; <10> stable [40]) [21, 40]
benzene <10> (<10> stable [40]) [40]
diethyl ether <10> (<10> stable [40]) [40]
hexane <10> (<10> stable [40]) [40]
Additional information <10> (<10> stable in diisobutyl ketone and carbon tetrachloride [40]) [40]

General stability information

<1>, $MgCl_2$ accelerates loss of activity during storage at 4°C, dilution increases stability, cholate-solubilized activity is more stable than deoxycholate-solubilized activity [13]
<4>, freezing causes loss of 40% of the activity [34]
<14>, antichaotropic salts stabilize solubilized enzyme [21]

<14>, asolectin, GSH, DTT do not stabilize [21]
<14>, phosphate buffer, high concentration, stabilizes [21]

Storage stability

<1>, -15°C, several months, microsomal preparation [7]
<1>, -20°C, $t_{1/2}$: 2 weeks in the presence of PMSF [3]
<1>, -80°C, at least 24 h, deoxycholate solubilized enzyme [13]
<1>, -80°C, undiluted, several weeks, cholate solubilized enzyme [13]
<1>, -85°C, negligible loss of activity in microsomes after 3 months [43]
<1>, 4°C, 20-30% loss of activity after 24 h, cholate solubilized enzyme, dilution restores activity [13]
<1>, 4°C, about 65-85% loss of activity after 24 h, deoxycholate solubilized enzyme, partially restorable by dialysis against buffer lacking detergent [13]
<1>, 4°C, inactivation after 24 h in the presence of 8 mM $MgCl_2$, deoxycholate solubilized enzyme [13]
<2>, -20°C, 30 days, microsomal preparation [2]
<7>, -20°C, 3 months [19]
<11>, 4°C, 20% loss of activity after 120 h [29]
<14>, 0°C, inactivation after 4 days in Tris-HCl buffer [21]
<14>, 0°C, more than 50% of activity retained after 4 days in 0.25 M phosphate buffer [21]

References

[1] Bell, R.M.; Coleman, R.A.: Enzymes of triacylglycerol formation in mammals. The Enzymes, 3rd Ed. (Boyer, P.D., ed.), **16**, 87-111 (1983)
[2] Kawasaki, T.; Snyder, F.: Synthesis of a novel acetylated neutral lipid related to platelet-activating factor by acyl-CoA:1-O-alkyl-2-acetyl-sn-glycerol acyltransferase in HL-60 cells [published erratum appears in J Biol Chem 1988 Jul 25;263(21):10539]. J. Biol. Chem., **263**, 2593-2596 (1988)
[3] Manganaro, F.; Kuksis, A.: Rapid isolation of a triacylglycerol synthetase complex from rat intestinal mucosa. Can. J. Biochem. Cell Biol., **63**, 107-114 (1985)
[4] Martin, B.A.; Wilson, R.F.: Properties of diacylglycerol acyltransferase from spinach leaves. Lipids, **18**, 1-6 (1983)
[5] Weiss, S.B.; Kennedy, E.P.; Kiyasu, J.Y.: The enzymatic synthesis of triglycerides. J. Biol. Chem., **235**, 40-44 (1960)
[6] Coleman, R.A.: Diacylglycerol acyltransferase and monoacylglycerol acyltransferase from liver and intestine. Methods Enzymol., **209**, 98-104 (1992)
[7] Coleman, R.A.; Bell, R.M.: Triacylglycerol synthesis in isolated fat cells. Studies on the microsomal diacylglycerol acyltransferase activity using ethanol-dispersed diacylglycerols. J. Biol. Chem., **251**, 4537-4543 (1976)
[8] Yamashita, S.; Hosaka, K.; Miki, Y.; Numa, S.: Glycerolipid acyltransferases from rat liver: 1-acylglycerophosphate acyltransferase, 1-acylglycerophosphorylcholine acyltransferase, and diacylglycerol acyltransferase. Methods Enzymol., **71**, 528-536 (1981)

[9] Marshall, M.O.; Knudsen, J.: Specificity of diacylglycerol acyltransferase from bovine mammary gland, liver and adipose tissue towards acyl-CoA esters. Eur. J. Biochem., **94**, 93-98 (1979)
[10] Cao, Y.Z.; Huang, A.H.C.: Diacylglycerol acyltransferase in maturing oil seeds of maize and other species. Plant Physiol., **82**, 813-820 (1986)
[11] Rodriguez, M.A.; Dias, C.; Lau, T.E.: Reversible ATP-dependent inactivation of adipose diacylglycerol acyltransferase. Lipids, **27**, 577-581 (1992)
[12] Parthasarathy, S.; Murari, R.; Crilly, K.S.; Baumann, W.J.: Modulation of diacylglycerol acyltransferase by lysophosphatidylcholine and related monochain phospholipids. Biochim. Biophys. Acta, **664**, 249-254 (1981)
[13] Polokoff, M.A.; Bell, R.M.: Solubilization, partial purification and characterization of rat liver microsomal diacylglycerol acyltransferase. Biochim. Biophys. Acta, **618**, 129-142 (1980)
[14] Grigor, M.R.; Bell, R.M.: Separate monoacylglycerol and diacylglycerol acyltransferases function in intestinal triacylglycerol synthesis. Biochim. Biophys. Acta, **712**, 464-472 (1982)
[15] Marshall, M.O.; Knudsen, J.: Factors influencing the in vitro activity of diacylglycerol acyltransferase from bovine mammary gland and liver towards butyryl-CoA and palmitoyl-CoA. Biochim. Biophys. Acta, **617**, 393-397 (1980)
[16] Christiansen, K.: Utilization of endogenous diacylglycerol for the synthesis of triacylglycerol, phosphatidylcholine and phosphatidylethanolamine by lipid particles from bakers yeast (Saccharomyces cerevisiae). Biochim. Biophys. Acta, **574**, 448-460 (1979)
[17] Bell, R.M.; Miller, Y.: A highly sensitive assay for microsomal diacyglycerol acyltransferase activity from isolated fat cells. Anal. Biochem., **71**, 436-443 (1976)
[18] Kwanyuen, P.; Wilson, R.W.: Subunit and amino acid composition of diacylglycerol acyltransferase from germinating soybean cotyledons. Biochim. Biophys. Acta, **1039**, 67-72 (1990)
[19] Kwanyuen, P.; Wilson, R.W.: Isolation and purification of diacylglycerol acyltransferase from germinating soybean dotyledons. Biochim. Biophys. Acta, **877**, 238-245 (1986)
[20] Weselake, R.J.; Taylor, D.C.; Pomeroy, D.C.; Lawson, S.L.; Underhill, E.W.: Properties of diacylglycerol acyltransferase from microspore-derived embryos of Brassica napus. Phytochemistry, **30**, 3533-3538 (1991)
[21] Akao, T.; Kusaka, T.: Solubilization of diglyceride acyltransferase from the membrane of Mycobacterium smegmatis. J. Biochem., **80**, 723-728 (1976)
[22] Ichihara, K.; Noda, M.: Some properties of diacylglycerol acyltransferase in a pariculate fraction from maturing safflower seeds. Phytochemistry, **21**, 1895-1901 (1982)
[23] Husaka, K.; Schiele, U.; Numa, S.: Diacylglycerol acyltransferase from rat liver microsomes. Separation and acyl-donor specificity. Eur. J. Biochem., **76**, 113-118 (1977)
[24] Andersson, M.; Wettesten, M.; Boren, J.; Magnusson, A.; Sjoberg, A.; Rustaeus, S.; Olofsson, S.O.: Purification of diacylglycerol:acyltransferase from rat liver to near homogeneity. J. Lipid Res., **35**, 535-545 (1994)

[25] Bouvier-Nave, P.; Benveniste, P.; Noiriel, A.; Schaller, H.: Expression in yeast of an acyl-CoA:diacylglycerol acyltransferase cDNA from Caenorhabditis elegans. Biochem. Soc. Trans., **28**, 692-695 (2000)
[26] Cases, S.; Smith, S.J.; Zheng, Y.W.; Myers, H.M.; Lear, S.R.; Sande, E.; Novak, S.; Collins, C.; Welch, C.B.; Lusis, A.J.; Erickson, S.K.; Farese, R.V., Jr.: Identification of a gene encoding an acyl CoA:diacylglycerol acyltransferase, a key enzyme in triacylglycerol synthesis. Proc. Natl. Acad. Sci. USA, **95**, 13018-13023 (1998)
[27] Cases, S.; Stone, S.J.; Zhou, P.; Yen, E.; Tow, B.; Lardizabal, K.D.; Voelker, T.; Farese, R.V., Jr.: Cloning of DGAT2, a second mammalian diacylglycerol acyltransferase, and related family members. J. Biol. Chem., **276**, 38870-38876 (2001)
[28] Cheng, D.; Meegalla, R.L.; He, B.; Cromley, D.A.; Billheimer, J.T.; Young, P.R.: Human acyl-CoA:diacylglycerol acyltransferase is a tetrameric protein. Biochem. J., **359**, 707-714 (2001)
[29] Furukawa-Stoffer, T.L.; Byers, S.D.; Hodges, D.M.; Laroche, A.; Weselake, R.J.: Identification of N-ethylmaleimide-sensitive and -insensitive phosphatidate phosphatase activity in microspore derived cultures of oilseed rape. Plant Sci., **131**, 139-147 (1998)
[30] Hobbs, D.H.; Hills, M.J.: Expression and characterization of diacylglycerol acyltransferase from Arabidopsis thaliana in insect cell cultures. Biochem. Soc. Trans., **28**, 687-689 (2000)
[31] Kamisaka, Y.; Mishra, S.; Nakahara, T.: Purification and characterization of diacylglycerol acyltransferase from the lipid body fraction of an oleaginous fungus. J. Biochem., **121**, 1107-1114 (1997)
[32] Lardizabal, K.D.; Mai, J.T.; Wagner, N.W.; Wyrick, A.; Voelker, T.; Hawkins, D.J.: DGAT2 is a new diacylglycerol acyltransferase gene family: purification, cloning, and expression in insect cells of two polypeptides from Mortierella ramanniana with diacylglycerol acyltransferase activity. J. Biol. Chem., **276**, 38862-38869 (2001)
[33] Little, D.; Weselake, R.; Pomeroy, K.; Furukawa-Stoffer, T.; Bagu, J.: Solubilization and characterization of diacylglycerol acyltransferase from microspore-derived cultures of oilseed rape. Biochem. J., **304**, 951-958 (1994)
[34] Lozeman, F.J.; Middleton, C.K.; Deng, J.; Kazala, E.C.; Verhaege, C.; Mir, P.S.; Laroche, A.; Bailey, D.R.C.; Weselake, R.J.: Characterization of microsomal diacylglycerol acyltransferase activity from bovine adipose and muscle tissue. Comp. Biochem. Physiol. B, **130B**, 105-115 (2001)
[35] Meegalla, R.L.; Billheimer, J.T.; Cheng, D.: Concerted elevation of acyl-coenzyme A:diacylglycerol acyltransferase (DGAT) activity through independent stimulation of mRNA expression of DGAT1 and DGAT2 by carbohydrate and insulin. Biochem. Biophys. Res. Commun., **298**, 317-323 (2002)
[36] Murray, A.M.B.; Denis, R.; Speake, B.K.: Acyltransferase activities in the yolk sac membrane of the chick embryo. Lipids, **34**, 929-935 (1999)
[37] Nykiforuk, C.L.; Furukawa-Stoffer, T.L.; Huff, P.W.; Sarna, M.; Laroche, A.; Moloney, M.M.; Weselake, R.J.: Characterization of cDNAs encoding diacylglycerol acyltransferase from cultures of Brassica napus and sucrose-

mediated induction of enzyme biosynthesis. Biochim. Biophys. Acta, **1580**, 95-109 (2002)

[38] Routaboul, J.M.; Benning, C.; Bechtold, N.; Caboche, M.; Lepiniec, L.: The TAG1 locus of Arabidopsis encodes for a diacylglycerol acyltransferase. Plant Physiol. Biochem., **37**, 831-840 (1999)

[39] Triki, S.; Ben Hamida, J.; Mazliak, P.: Diacylglycerol acyltransferase in maturing sunflower seeds. Biochem. Soc. Trans., **28**, 689-692 (2000)

[40] Valencia-Turcotte, L.; Rodriguez-Sotres, R.: The treatment of purified maize oil bodies with organic solvents and exogenous diacylglycerol allows the detection and solubilization of diacylglycerol acyltransferase. Biochim. Biophys. Acta, **1534**, 14-26 (2001)

[41] Weselake, R.J.; Nykiforuk, C.L.; Laroche, A.; Patterson, N.A.; Wiehler, W.B.; Szarka, S.J.; Moloney, M.M.; Tari, L.W.; Derekh, U.: Expression and properties of diacylglycerol acyltransferase from cell-suspension cultures of oilseed rape. Biochem. Soc. Trans., **28**, 684-686 (2000)

[42] Weselake, R.J.; Pomeroy, M.K.; Furukawa, T.L.; Golden, J.L.; Little, D.B.; Laroche, A.: Developmental profile of diacylglycerol acyltransferase in maturing seeds of oilseed rape and safflower and microspore-derived cultures of oilseed rape. Plant Physiol., **102**, 565-571 (1993)

[43] Yashiro, K.; Kameyama, Y.; Mizuno-Kamiya, M.; Shin, S.O.; Fujita, A.: Microsomal diacylglycerol acyltransferase in rat parotid and submandibular glands: acylation of 1,2-dioleoyl-sn-glycerol dispersed with phospholipids. Int. J. Biochem. Cell Biol., **28**, 895-903 (1996)

Carnitine O-palmitoyltransferase 2.3.1.21

1 Nomenclature

EC number
2.3.1.21

Systematic name
palmitoyl-CoA:L-carnitine O-palmitoyltransferase

Recommended name
carnitine O-palmitoyltransferase

Synonyms
CPT
CPT I <3> (<3> outer membrane carnitine palmitoyl transferase [18]) [18]
CPT-A (enzyme from outer side of inner mitochondrial membrane)
CPT-B (enzyme from inner side of inner mitochondrial membrane)
CPTi <3> (<3> inner form, tightly membrane bound [43]; <3> inner, initially latent carnitine transferase [14]) [14, 43]
CPTo (outer form, easily released and solubilized by digitonin, <3> [43]; <3> outer malonyl-CoA inhibitable carnitine palmitoyltransferase [14]) [14]
L-carnitine palmitoyltransferase
acylcarnitine transferase
carnitine palmitoyltransferase
carnitine palmitoyltransferase I (located on the outer face of the inner mitochondrial membrane)
carnitine palmitoyltransferase II
carnitine palmitoyltransferase-A
palmitoylcarnitine transferase
palmitoyltransferase, carnitine
Additional information (cf. EC 2.3.1.7 and EC 2.3.1.137; <3> M-CPT I and L-CPT I, i.e. CPT Iβ-2 and CPT Iβ-3, are 2 different splicing variants, schematic overview [63])

CAS registry number
9068-41-1

2 Source Organism

<-1> no activity in *Pichia pastoris* [55]
<1> *Bos taurus* (CPT II [58]; enzyme forms CPT I and CPT II [36]; ox [33]; enzyme forms CPTi and CPTo [22]; CPT A, enzyme form on outer surface

of the inner mitochondrial membrane, CPT B on the inner surface [17]; 2 enzyme forms: inner membrane enzyme in mitochondria and soluble enzyme from peroxisomes [6]; calf [16]) [1-3, 5-7, 9, 16, 17, 22, 33, 36, 58]

<2> *Valanga nigricornis* (Burm., grasshopper [10]) [10]

<3> *Rattus norvegicus* (Walker 256 tumour rat [68]; recombinant liver CPT I [65]; L-CPT I [59]; 2 isoforms L-CPT I and M-CPT I [56,61,63]; 2 isoforms of CPT I in heart mitochondria, liver L-CPT I, skeletal muscle M-CPT I [50,63]; enzyme form CPT II [20,23,26,53,55]; enzyme form CPT I [18,23,25,37,39,50,51,55,66,68,70]; enzyme forms CPTo and CPTi from outer and inner mitochondrial membrane [14,32,43]) [4, 8, 14-16, 18, 20, 21, 23-27, 29-32, 34-43, 45-53, 55, 56, 59, 61, 63, 65, 66, 68-70]

<4> *Ovis aries* (CPT I [11]) [11, 31]

<5> *Homo sapiens* (CPT II deficient mutant [67]; CPT I [60]; CPT II [60]; 2 isoforms M-CPT I and L-CPT I [56]; M-CPT I [54]) [12, 19, 44, 46, 54, 56, 60, 63, 67]

<6> *Gallus gallus* (embryo [13,28]) [13, 28]

<7> *monkey* [46]

<8> *Cavia porcellus* [31]

<9> *Mus musculus* (2 isoforms L-CPT I and M-CPT I [56]) [46, 56]

<10> *Mesocricetus auratus* (2 isoforms L-CPT I and M-CPT I [56]) [56]

<11> *Sus scrofa* (2 isoforms of CPT I in liver mitochondria, liver L-CPT I, skeletal muscle M-CPT I [64]; CPT I [57]) [57, 64, 69]

<12> *Pisum sativum* (2 isoforms CPTo (I) and CPTi (II) [62]) [62]

3 Reaction and Specificity

Catalyzed reaction

palmitoyl-CoA + L-carnitine = CoA + L-palmitoylcarnitine (broad sepcificity to acyl group, over the range C_8 to C_{18}; optimal activity with palmitoyl-CoA; <3> active site three-dimensional structure model [66]; <3> CPT I acts at 2 distinct sites, a regulatory and an active site [51]; <3> 2 independent inhibitory sites [51]; <1> sigmoidal kinetics with both substrates [1]; <1> flexible active site [9]; <1> rapid-equilibrium random-order mechanism [33]; <1> random sequential mechanism [2])

Reaction type

acyl group transfer

Natural substrates and products

S acyl-CoA + L-carnitine <1-12> (<3> enzyme is involved in apoptosis induction [68]; <3,5> CPT I activity is involved in cardiac mitochondrial β-oxidation flux control [63]; <3> CPT I is the rate-limiting enzyme in β-oxidation of long-chain fatty acids [55]; <3,9> CPT I, involved in regulation of fatty acid oxidation [51,56,70]; <3,5,12> enzyme has a key function in regulation of fatty acid β-oxidation [27,44,62,67,70]) (Reversibility: r <1, 3, 5, 6, 12> [5, 8, 58, 60, 62, 63]; ? <1-11> [1-4, 6, 7, 9-57, 59, 61, 64-70]) [1-70]

P CoA + L-acylcarnitine <1-12> [1-70]

S palmitoyl-CoA + L-carnitine <3> (<3> forward reaction by CPT I and reverse reaction by CPT II [55]) (Reversibility: r <3> [55]) [55]

P L-palmitoylcarnitine + CoA <3> [55]

S Additional information <1, 3> (<3> γ-linolenic acid diet leads to several physiological effects during cancer therapy, e.g. reduction of mitochondrial metabolic activity, modification of mitochondrial outer membrane composition, increase in lipid peroxidation, cytoplasmic accumulation of acyl-CoA, reduction of CPT I activity, and malonyl-CoA sensitivity [68]; <1> responsible for transferase activity on the outer side of the inner mitochondrial membrane [2]; <1,3> participates in fatty acyl group transport into mitochondria [16,17]) [2, 16, 17, 68]

P ?

Substrates and products

S DL-palmitoyl-CoA + DL-carnitine <12> (Reversibility: r <12> [62]) [62]

P CoA + DL-palmitoylcarnitine <12> [62]

S acyl-CoA + L-carnitine <1-12> (<3> activity with different substrates of wild-type CPT I with and without γ-linoleic acid treatment [68]; <12> preference for desaturated long-chain acyl CoAs [62]; <11> substrate specificity at early developmental stages in skeletal muscle [57]; <1, 3, 5, 6, 11, 12> substrate specificity [8, 13, 19, 24, 28, 33, 57, 62]; <3> isozyme CPT II catalyzes the reaction only in the direction towards long-chain fatty acyl-CoA formation, CPT I catalyzes reaction in both directions [48]; <6,12> C_4 to C_{18} acyl-CoAs [28,62]; <3> CPT I unlike CPT II requires membrane integrity for catalytic function [47]; <3> CPT II requires preincubation with CoA-SH for manifestation of catalytic activity [48]; <3> CPT II shows greater chain-length substrate specificity for transfer of long chain fatty acyl groups from (-)-acylcarnitine derivatives to CoA-SH with greatest activity being obtained in the transfer of stearoyl and palmitoyl groups [48]; <1> effect of octylglucoside and Triton X-100 on specificity [9]; <3> overview: activity towards long-chain polyunsaturated fatty acids and their CoA esters [24]) (Reversibility: r <1, 3, 5, 6, 12> [5, 8, 16, 27, 28, 33, 36, 44, 45, 51, 58, 60, 62, 63]; ? <1-11> [1-4, 6, 7, 9-15, 17-26, 29-32, 34, 35, 37-43, 46-50, 52-57, 59, 61, 64-70]) [1-70]

P CoA + L-acylcarnitine <1-12> (<6> reverse reaction: acyl-CoA substrates are C_4 to C_{18} acylcarnitines [28]; <1> reverse reaction: myristoylcarnitine [16]; <1> reverse reaction with palmitoylcarnitine [8,16]) [1-70]

S arachidoyl-CoA + L-carnitine <5> (Reversibility: ? <5> [19]) [19]

P CoA + L-arachidoylcarnitine

S butyryl-CoA + L-carnitine <6> (Reversibility: ? <6> [13]) [13]

P CoA + L-butyrylcarnitine

S decanoyl-CoA + L-carnitine <1, 3, 5> (<1,5> best substrate [5,19]) (Reversibility: r <1, 5> [5, 44, 58]; ? <1, 3, 5> [19, 21, 22]) [5, 19, 21, 22, 44, 58]

P CoA + L-decanoylcarnitine <1, 3, 5> [5, 19, 21, 22, 44, 58]

S docosanhexaenoyl-CoA + L-carnitine <3> (Reversibility: ? <3> [65]) [65]

P CoA + L-docosanhexaenoylcarnitine
S erucoyl-CoA + L-carnitine <5> (Reversibility: ? <5> [19]) [19]
P CoA + L-erucoylcarnitine
S hexanoyl-CoA + L-carnitine <1, 5> (Reversibility: r <1, 5> [5, 44]) [5, 44]
P CoA + L-hexanoylcarnitine <1, 5> [5, 44]
S lauroyl-CoA + L-carnitine <1, 5, 6> (<1> best substrate [5]) (Reversibility: r <1, 5> [5, 44]; ? <6> [13]) [5, 13, 44]
P CoA + L-lauroylcarnitine <1, 5> [5, 44]
S linoleoyl-CoA + L-carnitine <5, 12> (<12> preference for [62]) (Reversibility: r <12> [62]; ? <5> [19]) [19, 62]
P CoA + L-linoleoylcarnitine <12> [62]
S myristoyl-CoA + L-carnitine <1, 5> (<1> best substrate [33]) (Reversibility: r <5> [44]; ? <1> [5, 16, 33]) [5, 16, 33, 44]
P CoA + L-myristoylcarnitine <1, 5> [5, 16, 44]
S octanoyl-CoA + L-carnitine <1, 5, 6> (Reversibility: r <1, 5> [5, 44]; ? <6> [13]) [5, 13, 44]
P CoA + L-octanoylcarnitine <1, 5> [5, 44]
S oleoyl-CoA + L-carnitine <3> (Reversibility: ? <3> [65]) [65]
P CoA + L-oleoylcarnitine
S palmitoyl-CoA + L-carnitine <1-3, 5, 6, 11> (<3,11> chimeric L-CPT I with exchanged C-terminals between pig and rat enzymes [69]; <1> best substrate [33]) (Reversibility: r <1, 3, 5, 6> [5, 8, 16, 28, 29, 33, 36, 38, 44, 45, 48, 51, 55, 60]; ? <1-3, 5, 6, 11> [1, 13, 10, 12, 19, 30, 37, 39, 41, 47-49, 52, 53, 59, 61, 64, 66, 68-70]) [1, 5, 8, 10, 12, 13, 16, 19, 28-30, 33, 36-39, 41, 44, 45, 47-49, 51-53, 55, 59-61, 64, 66, 68-70]
P CoA + L-palmitoylcarnitine <1, 3, 5, 6, 11>
S stearoyl-CoA + L-carnitine <3, 5> (Reversibility: r <5> [44]; ? <3, 5> [19, 52]) [19, 44, 52]
P CoA + L-stearoylcarnitine <3> [52]
S Additional information <3> (<3> acyl-CoA binding protein ACBP forms binary complexes with acyl-CoA and possibly transfers acyl-CoA to CPT I, CPT I induces conformational changes in ACBP leading to release of acyl-CoA [65]; <3> the branched chain fatty acid CoA-thioesters pristanoyl-CoA and phytanoyl-CoA, i.e. 2,6,10,14-tetramethylhexadecanoyl-CoA and 3,7,11,15-tetramethylhexadecanoyl-CoA, are no substrates in vivo, since they cannot be transported into mitochondria of liver [52]) [52, 65]
P ?

Inhibitors

11-trimethylamino-undecanoyl-DL-carnitine <1> [33]
2-[5(4-chlorophenyl)pentyl]-oxirane-2-carboxyl-CoA <1> [33]
2-bromopalmitoyl-CoA <1, 3, 5> (<1,3> irreversible [2,45]; <3,5> with addition of carnitine [19,45]; <3> no inhibition [8]) [2, 19, 32, 33, 45]
4-hydroxy phenylglyoxylate <3> (<3> CPT I, greatly reduced inhibition of protease treated enzyme [51]) [51]

CoA <3, 5> (<3> CPT I, inhibition is not affected by proteinase treatment [51]) [12, 51]
CoA esters of certain oxirane carboxylic acids <3, 5, 7, 9> (<3,5,7,9> irreversible, CPT I but not CPT II [46]) [46]
L-aminocarnitine <3, 5> (<5> inhibition of CPT II not CPT I [60]; <3> ability to act as substrate or inhibitor of CPT is dependent on the nature of CPT and on the chain length of the acyl-CoA cosubstrate [27]) [27, 60]
L-palmitoylcarnitine <3> (<3> competitive product inhibition [29]) [29]
L-sulfocarnitine <3> (<3> ability to act as substrate or inhibitor of CPT is dependent on the nature of CPT and on the chain length of the acyl-CoA cosubstrate [27]) [27]
Nagarse <3> (<3> mitochondria, malonyl-CoA protects [30]) [30]
Ro 25-0187 <3> (<3> malonyl-CoA analogue [51]; <3> CPT I, strong, inhibitory effect is drastically reduced by protease treatment of outer mitochondrial membrane [51]) [51]
S-(4-bromo-2,3-dioxobutyl)-CoA <3> (<3> inhibition of malonyl-CoA sensitive enzyme, malonyl-CoA insensitive enzyme is not inhibited [47]) [47]
Triton X-100 <3, 5, 7, 9> (<5> recombinant CPT I, complete inactivation at 5%, at least partly reversible [54]; <3> only CPTo [14]; <5> mutant CPT [12]; <3, 5, 7, 9> complete loss of CPT I activity, no loss of CPT II activity [46]; <5> erythrocyte enzyme is not affected [19]) [12, 14, 46, 54]
Tween 20 <1, 3> (<3> only CPTo at 2% and above [14]) [14, 33]
acetyl-CoA <3, 5> (<3> CPT I, inhibition is not affected by proteinase treatment [51]) [12, 42, 51]
bile acids <3> (<3> e.g. in cholestatic rats [4]) [4]
bromoacetyl-CoA <1> [33]
cardiolipin <3> (<3> inhibits conversion of palmitoylcarnitine to palmitoyl-CoA, stimulates palmitoylcarnitine formation [38]) [38]
carnitine <3> (<3> CPT I, slightly [51]) [51]
chenodeoxycholic acid <3> (<3> competitive to carnitine [4]) [4]
cholate <3> (<3> only CPTo [14]) [14]
deoxycarnitine <1> [2]
diethyl dicabonate <1> (<1> CPT II, strong, linear pseudo-first order kinetic, modification of a histidine residue, reversible by hydroxylamine, decanoyl-CoA and L-carnitine [58]) [58]
digitonin <3> (<3> CPTo and slightly CPTi [14]) [14]
dinitrophenyl analogue of etomoxir <3> (<3> i.e. 2[6-(4-chlorophenoxy)hexyl]oxirane-2-carboxylic acid, DNP-Et, specific inhibitor for liver L-CPT I, identical with the small isoform of heart CPT I, complete inhibition of L-CPT I, but not M-CPT I [50]) [50]
etomoxir-CoA <3> (<3> CPTo [15]) [15]
γ-linolenic acid <3> (<3> inhibits CPT I in vivo and reduces malonyl-CoA sensitivity, decreases affinity for 16:0 acyl-CoA substrate [68]) [68]
hemipalmitoylcarnitinium bromide <3> (<3> CPT I, strong, active-site-directed, inhibition not altered by protease treatment [51]; <3> competitve to palmitoyl-carnitine in the reverse reaction [29]; <3> i.e. HPC, (2S,6R:2R,6S)-

6-carboxymethyl-2-hydroxy-2-pentadecyl-4,4-dimethylmorpholinium bromide [29,51]) [29, 51]
malonyl-CoA <1, 3-5, 7-9, 11, 12> (<3> membrane or micelle composition and properties influence the sensitivity of CPT I to inhibition [68,70]; <3,11> sensitivity to inhibition of deletion and chimeric L-CPT I mutants [69]; <3> CPT I, sensitivity is reduced in vivo during γ-linolenic acid treatment [68]; <3> similar inhibition of wild-type CPT I and mutants A381D and H473A [66]; <11> L-CPT I [64]; <3,5,12> inhibits CPT I not CPT II [62,63]; <3> overview: recombinant chimeric proteins of L-CPT I and M-CPT I [61]; <11> CPT I, inhibition depends on age [57]; <3> L-CPT I, recombinant enzyme from Saccharomyces cerevisiae is inhibited only in intact mitochondria, not as solubilized enzyme [59]; <3,5,9> isoforms L-CPT I and M-CPT I [56,63]; <3> CPT I catalytic activity and malonyl-CoA sensitivity is located on 1 single polypeptide [55]; <3> recombinant CPT I is highly sensitive, recombinant CPT II not [55]; <5> CPT I, native and reconstituted in phospholipid micelles [54]; <3> partial proteolysis of CPT I greatly diminishes the inhibitory effect [51]; <3> inhibition of isoform L-CPT I is increased by 2[6-(4-chlorophenoxy)hexyl]oxirane-2-carboxylic acid [50]; <3-5,11,12> regulatory role in vivo [25,31,57,60,62,70]; <1,3> mitochondrial isozyme CPT I, sensitivity against malonyl-CoA is mediated by a 86 kDa malonyl-CoA binding protein complexed with CPT and other proteins of the β-oxidation, detergent and salt sensitive [21-23, 25]; <3> inhibition is pH-dependent [52]; <3, 11> inhibition is temperature dependent [18, 52, 57]; <5> Triton X-100 protects [12]; <1, 3, 5> no inhibition [1, 23, 44]; <1, 3, 5, 7, 9> CPT II not inhibited [22, 46, 54, 55, 60]; <3> CPTo [15]; <3-5, 7-9, 11> CPT I [11, 31, 37, 40, 46, 55, 56, 57, 60, 70]; <3> a malonyl-CoA insensitive enzyme and a malonyl-CoA sensitive enzyme [23, 47]; <3> benzyl alcohol, isoamyl alcohol and 2-(2-methoxyethoxy)ethyl-8-(cis-2-n-octylpropyl)octanoate decrease the ability of malonyl-CoA to inhibit CPT I [18]; <3> inhibition is very pH-dependent: malonyl-CoA concentrations causing 50% inhibition at pH 6.0, 6.5, 7.0, 7.5 and 8.0 are 0.00004, 0.001, 0.009, 0.04 and 0.2 mM, respectively [40]; <3> differences in sensitivity of hepatic and heart enzyme [35,63]; <1> inhibition of soluble peroxisomal form, no inhibition of enzyme from inner mitochondrial membrane [6]; <3> enzyme from liver mitochondrial inner membrane becomes inhibitable by malonyl-CoA if reconstituted with outer membrane malonyl-CoA binding protein [23]; <3, 5, 7, 9> reversible inhibition [45, 46]) [6, 8, 11, 12, 14, 15, 18, 19, 21-23, 25, 31, 33, 35, 37, 39, 40, 45-47, 50-52, 54-57, 59-64, 66, 68-70]
methylmalonyl-CoA <3, 4, 8> (<4> CPT I [11]; <3> partial proteolysis of CPT I slightly diminishes the inhibitory effect [51]; <3,4,8> inhibits sheep liver CPT I, no inhibition of rat liver and guinea-pig liver mitochondrial CPT I [31]) [11, 31, 51]
octyl glucoside <1, 3, 5, 7, 9> (<3> only CPTo [14]; <1> with octanoyl-CoA as substrate, competitive [9]; <3,5,7,9> complete loss of CPT I activity, no loss of CPT II activity [46]) [9, 14, 46]
palmitoyl-CoA <3> (<3> CPT I, slightly [51]; <3> product inhibition, reverse reaction [38]; <3> substrate inhibition [41,49]) [38, 41, 49, 51]

palmitoylcholine <3> (<3> competitive in the forward reaction to both substrates [29]) [29]
phosphatidylcholine <3> [38]
propionyl-CoA <3> (<3> CPT I, inhibition is not affected by proteinase treatment [51]) [51]
short chain-length fatty acylcarnitine derivatives <3> [48]
succinyl-CoA <3> (<3> partial proteolysis of CPT I slightly diminishes the inhibitory effect [51]) [51]
tetradecylglycidyl-CoA <3> (<3> and analogues, irreversible [45]; <3> CPTo [15]) [15, 45]
thiolcarnitine <3> (<3> ability to act as substrate or inhibitor of CPT is dependent on the nature of CPT and on chain length of the acyl-CoA cosubstrate [27]) [27]
trypsin <3, 6> (<3> mitochondria, malonyl-CoA protects [30]; <6> peroxisomal and mitochondrial isozymes [28]) [28, 30]
Additional information <3, 5, 11> (<11> L-CPT I behaves like a natural chimera of M-CPT I and L-CPT I due to malonyl-CoA sensitivity and K_m values compared to the enzymes of other organisms [64]; <3,5> physiological role of malonyl-CoA in the heart [63]; <3> CPT I, 2 independent binding sites for acyl-CoA inhibitors and regulation of fatty acid oxidation, the malonyl-CoA site is located on the cytoplasmic face of the outer mitochondrial membrane, the site for monocarboxylic acids and free CoA is located in the mitochondrial intermembrane space [51]; <5> different sites of inhibition of carnitine palmitoyltransferase by malonyl-CoA, acetyl-CoA and CoA [12]; <3> no inhibition by choline, hemiacetylcarnitinium [29]) [12, 29, 51, 63, 64]

Activating compounds

Triton X-100 <3> (<3> only CPTi [14]) [14]
Tween 20 <3, 6> (<3> activates only CPTi [14]; <6> activates peroxisomal enzyme [13]) [13, 14]
cardiolipin <3> (<3> inhibits conversion of palmitoylcarnitine to palmitoyl-CoA, stimulates palmitoylcarnitine formation [38]) [38]
cholate <3> (<3> activates only CPTi [14]) [14]
di(2-ethyl-hexyl)phtalate <3> [8]
octyl glucoside <3> (<3> only CPTi [14]) [14]
palmitoylcarnitine <3> (<3> high above micellar concentrations increase conversion of palmitoylcarnitine to palmitoyl-CoA by facilitating the removal of palmitoyl-CoA from the enzyme surface [38]) [38]
phosphatidylcholine liposomes <3> (<3> only CPTi [14]; <3> high above micellar concentrations increase conversion of palmitoylcarnitine to palmitoyl-CoA by facilitating the removal of palmitoyl-CoA from the enzyme surface [38]) [14, 38]
phospholipids <3> (<3> e.g. phosphatidylcholine, cardiolipin, stimulate [49]) [49]
proteins <3> (<3> e.g. albumin, fatty acid-binding protein, lambda-globulin, stimulate [49]) [49]
salicylic acid <3> (<3> activates best at 20 mM, pH changes [41]) [41]

Metals, ions

KCl <3, 6> (<6> activates peroxisomal enzyme [13]; <3> high above micellar concentrations increase conversion of palmitoylcarnitine to palmitoyl-CoA by facilitating the removal of palmitoyl-CoA from the enzyme surface [38]) [13, 38]

Specific activity (U/mg)

0.001 <3> [41]
0.0013 <3, 9> (<3> native L-CPT I in liver mitochondria [59]; <9> cell line 3T3-L1, CPT I, 6 days after differentiation [56]) [56, 59]
0.0014 <3> (<3> CPT I after γ-linoleic acid treatment [68]) [68]
0.0032 <3> (<3> mitochondrial fraction of recombinant Saccharomyces cerevisiae expressing L-CPT I [59]; <3> mitochondrial fraction of recombinant Pichia pastoris expressing CPT I [55]) [55, 59]
0.0036 <3> (<3> CPT I [68]) [68]
0.0052 <11> (<11> recombinant wild-type L-CPT I [69]) [69]
0.0062 <9> (<9> cell line 3T3-L1, CPT II, 6 days after differentiation [56]) [56]
0.0069 <3> (<3> mitochondrial fraction of recombinant Pichia pastoris expressing CPT II [55]) [55]
0.0073 <12> (<12> solubilized CPT [62]) [62]
0.0078 <3> (<3> recombinant wild-type L-CPT I [69]) [69]
0.009 <3> (<3> CPT I [51]) [51]
0.43 <3> (<3> recombinant CPT II from Sf 9 insect cells [53]) [53]
0.473 <3> (<3> purified recombinant L-CPT I, reconstituted in liposomes [70]) [70]
4.65 <6> (<6> purified mitochondrial enzyme [28]) [28]
13.2 <6> (<6> purified peroxisomal enzyme [28]) [28]
24 <3> (<3> purified enzyme [38]) [38]
29.6 <3> (<3> purified enzyme [8]) [8]
40 <1> (<1> purified enzyme [3]) [3]
42 <1> (<1> purified enzyme [1]) [1]
47.5 <1> (<1> purified enzyme [5]) [5]
Additional information <1, 3, 5, 6, 11> (<3,11> activities of deletion and chimeric L-CPT I mutants [69]; <5> CPT I and II activities in muscle biopsis [60]; <11> activity at early developmental stages in skeletal muscle [57]) [28, 33, 41, 43, 44, 57, 60, 69]

K_m-Value (mM)

0.0008 <1> (palmitoyl-CoA, <1> with C_8-CoA [33]) [2, 33]
0.0009 <1> (decanoyl-CoA) [33]
0.0012 <1> (myristoyl-CoA) [33]
0.0017 <1> (stearoyl-CoA) [33]
0.0018 <1> (hexanoyl-CoA) [33]
0.004 <1> (dodecanoyl-CoA) [33]
0.0049 <3> (palmitoyl-CoA, <3> recombinant wild-type CPT I [66]) [66]
0.0051 <1> (butyryl-CoA) [33]
0.0052 <6> (palmitoyl-CoA) [13]

0.0055 <1> (CoA) [16]
0.0085 <6> (octanoyl-CoA) [13]
0.011 <5> (lauroyl-CoA) [44]
0.011 <3> (palmitoyl-CoA, <3> CPT I [68]) [68]
0.0122 <5> (palmitoyl-CoA) [44]
0.013 <6> (lauroyl-CoA) [13]
0.017 <5> (decanoyl-CoA) [44]
0.017 <3> (palmitoyl-CoA, <3> CPT I after γ-linoleic acid treatment [68]) [68]
0.018 <1> (palmitoyl-CoA) [16]
0.018 <5> (stearoyl-CoA) [44]
0.022 <1> (acetyl-CoA) [33]
0.023 <5> (octanoyl-CoA) [44]
0.0242 <1> (palmitoyl-CoA) [1]
0.028 <3> (palmitoyl-CoA, <3> purified recombinant L-CPT I, reconstituted in liposomes [70]) [70]
0.03 <6> (butyryl-CoA) [13]
0.031 <5> (myristoyl-CoA) [44]
0.033 <3> (palmitoyl-CoA, <3> mutant A381D [66]) [66]
0.034 <3> (palmitoyl-CoA, <3> recombinant CPT II [53]) [53]
0.035 <11> (palmitoyl-CoA, <11> recombinant wild-type L-CPT I [64,69]) [64, 69]
0.039 <3> (L-carnitine, <3> purified recombinant L-CPT I, reconstituted in liposomes [70]) [70]
0.039 <1> (palmitoyl-CoA) [2]
0.042 <3> (L-carnitine) [52]
0.043 <3> (palmitoyl-CoA, <3> recombinant wild-type L-CPT I [69]) [69]
0.047 <3> (palmitoyl-CoA) [41]
0.055 <3> (palmitoyl-CoA) [52]
0.092 <3> (L-carnitine, <3> CPT I [68]) [68]
0.093 <3> (L-carnitine, <3> mutant A381D [66]) [66]
0.098 <3> (L-carnitine, <3> CPT I after γ-linoleic acid treatment [68]) [68]
0.1 <3> (L-carnitine, <3> recombinant wild-type L-CPT I [69]) [69]
0.105 <3> (L-carnitine, <3> recombinant CPT II [55]) [55]
0.108 <1> (L-carnitine, <1> with C_8-CoA [33]) [33]
0.12 <1> (L-carnitine, <1> with C_{12}-CoA [33]) [33]
0.123 <5> (L-palmitoylcarnitine) [44]
0.126 <11> (L-carnitine, <11> recombinant wild-type L-CPT I [64,69]) [64, 69]
0.129 <3> (L-carnitine, <3> recombinant wild-type CPT I [66]) [66]
0.14 <1> (L-palmitoylcarnitine) [16]
0.152 <5> (hexanoyl-CoA) [44]
0.17 <1> (L-carnitine, <1> peroxisomal soluble enzyme [6]) [6]
0.18 <1> (L-carnitine) [2]
0.185 <1> (L-carnitine, <1> with C_{16}-CoA [33]) [33]
0.2 <3> (L-carnitine, <3> isoform L-CPT I [50]) [50]
0.45 <1> (L-carnitine) [16]

0.5 <3> (L-carnitine, <3> isoform L-CPT I, with 2[6-(4-chlorophenoxy)hexyl]oxirane-2-carboxylic acid [50]) [50]
0.95 <3> (L-carnitine, <3> recombinant CPT II [53]) [53]
1.139 <5> (L-octanoylcarnitine) [44]
2.8 <1> (L-carnitine, <1> inner mitochondrial membrane [6]) [6]
Additional information <1, 3, 5, 6, 11> (<3,11> K_m values of deletion and chimeric L-CPT I mutants [69]; <11> L-CPT I behaves like a natural chimera of M-CPT I and L-CPT I due to malonyl-CoA sensitivity and K_m values compared to the enzymes of other organisms [64]; <3> overview: K_m values of recombinant chimeric proteins of L-CPT I and M-CPT I [61]; <3> kinetics [52]; <5> K_m values for carnitine with different acyl-CoA substrates [44]; <6> overview: K_m values for several acyl-CoA and acyl-carnitine substrates, both reaction directions, peroxisomal and mitochondrial isozyme [28]; <1> influence of detergent on K_m value [9]; <1> effects of malonyl-CoA on kinetics [1]; <3> alterations in long-chain acyl-CoA binding to albumin in the CPT assay appreciablly affect the reaction at commonly used substrate concentrations [41]) [1, 8, 9, 17, 18, 22, 24, 28, 33, 41, 43, 44, 49, 52, 61, 64, 69]

K_i-Value (mM)

0.00022 <5> (malonyl-CoA) [12]
0.00066 <1> (2-bromopalmitoyl-CoA) [33]
0.0016 <3> (hemipalmitoylcarnitinium bromide, <3> with palmitoyl-carnitine, reverse reaction [29]) [29]
0.0035 <5> (L-aminocarnitine, <5> CPT II [60]) [60]
0.0051 <3> (hemipalmitoylcarnitinium bromide, <3> with carnitine, forward reaction [29]) [29]
0.009 <3> (palmitoylcarnitine, <3> versus palmitoyl-CoA [29]) [29]
0.0104 <3> (palmitoylcholine, <3> versus palmitoyl-CoA [29]) [29]
0.014 <3> (palmitoylcarnitine, <3> versus carnitine [29]) [29]
0.019 <3> (palmitoylcholine, <3> versus carnitine [29]) [29]
0.0215 <3> (hemipalmitoylcarnitinium bromide, <3> with palmitoyl-CoA, forward reaction [29]) [29]
0.045 <5> (CoA) [12]
0.045 <5> (acetyl-CoA) [12]
0.17 <1> (malonyl-CoA) [33]
0.2 <1> (11-trimethylamino-undecanoyl-DL-carnitine) [33]
0.6 <3> (chenodeoxycholic acid) [4]
0.9 <1> (2-bromoacetyl-CoA) [33]
15 <1> (octyl-glucoside, <1> with octanoyl-CoA as substrate [9]) [9]
Additional information <3, 5> (<3> Ki values for malonyl-CoA in mitochondria of fasted and fed rat [35]; <5> addition of acetyl-CoA and CoA results in 3fold increase in K_i for acetyl-CoA [12]) [12, 35]

pH-Optimum

6-8.5 <1> [33]
6.5-8 <6> (<6> mitochondrial [28]) [28]
6.8 <3> (<3> assay at [23]) [23]
7 <3, 11> (<3,11> assay at [69]) [69]

7-8.2 <1> [16]
7.2 <3> (<3> assay at [70]) [70]
7.4 <1, 3, 5> (<1,3,5> assay at [19,38,58]) [19, 38, 58]
7.5 <12> (<12> assay at [62]) [62]
7.6 <5> (<5> assay at [60]) [60]
8 <1> (<1> assay at [5]) [5]
8.5 <5, 6> (<6> peroxisomal [28]) [28, 44]
Additional information <1, 3> (<1> pI: 6.0 [33]; <1> pI: 8.5 [5]) [5, 33, 70]

pH-Range
5.6-8.8 <3> (<3> L-CPT I [70]) [70]
6-8 <1, 3> [1, 47]

Temperature optimum (°C)
22 <3> (<3> assay at [70]) [70]
25 <1, 3, 5> (<1,3,5> assay at [1,5,23,44]) [1, 5, 23, 44]
27 <12> (<12> assay at [62]) [62]
30 <3, 5, 11> (<3,5,11> assay at [19,60,64,69]) [19, 60, 64, 69]
37 <3> (<3> assay at [38,68]) [38, 68]

4 Enzyme Structure

Molecular weight
63000 <1> (<1> gel filtration [2]) [2]
63500 <1> (<1> gel filtration, sucrose density gradient sedimentation [7,33]) [7, 33]
66000 <6> (<6> peroxisomes, gel filtration [28]) [28]
67000 <1> (<1> gel filtration [1]) [1]
150000 <3> (<3> CPT I and II, gel filtration [48]) [48]
260000 <6> (<6> mitochondria, gel filtration [28]) [28]
274000 <5> (<5> gel filtration [44]) [44]
280000-320000 <3> (<3> gel filtration [8]) [8]
430000 <3> (<3> CPTo and CPTi, gel filtration [43]) [43]
660000 <1> (<1> about, gel filtration [9]) [9]
Additional information <1, 3, 5> (<3> 3-dimensional structural model of CPT I active site based on X-ray diffraction analysis data from similar rat enoyl-CoA hydratase [66]; <3> CPT I is composed of different structural segments: 1. a cytosolic N-terminal domain of 46 amino acid residues, 2. 2 transmembranal parts TM1 and TM2, 3. a 27-residue loop between TM1 and 2, 4. a cytosolic catalytic C-terminal domain, amino acid sequences [61]; <1> amino acid composition [33]; <3> mitochondrial enzyme, 86 kDa malonyl-CoA binding protein complexed with CPT and other proteins of the β-oxidation [21,22]; <5> amino acid sequence of peptides [44]; <3> amino acid sequence, alignment [20,66]; <1> enzyme migrates as part of a detergent micelle with MW of 510000 Da on gel filtration [5]) [5, 20-22, 33, 44, 61, 66]

Subunits

? <1, 3, 11, 12> (<11> x * 82000, L-CPT I, SDS-PAGE [64]; <3> x * 82000, M-CPT I, SDS-PAGE [63]; <12> x * 20000, CPT I and II, SDS-PAGE [62]; <3> x * 70000, recombinant CPT II, SDS-PAGE [55]; <3> x * 88000, recombinant CPT I, SDS-PAGE [55]; <1> x * 68000, SDS-PAGE [22]; <1> x * 67000, SDS-PAGE [5,6]; <3> x * 69200, SDS-PAGE [8]; <3> x * 68000, liver, CTP II [26]) [5, 6, 8, 22, 26, 55, 62-64]

monomer <1, 3, 6> (<6> 1 * 64000, peroxisomes, SDS-PAGE [28]; <1> 1 * 62500, SDS-PAGE [2]; <1> 1 * 63500, SDS-PAGE [7,33]; <3> 1 * 86000, muscle, SDS-PAGE [45]; <3> 1 * 94000, liver, SDS-PAGE [45]) [2, 7, 28, 33, 45]

tetramer <5, 6> (<6> 4 * 69000, mitochondria [28]; <5> 4 * 66000, SDS-PAGE [44]) [28, 44]

Additional information <1, 3> (<3> CPTo and CPTi are distinct proteins, 90000 MW subunit of liver and 86000 MW subunit of muscle constitutes a component of their respective CPTo system, the 66000 MW subunit of CPTi does not constitute a part of the CPTo system [15]; <3> 86000 MW malonyl-CoA binding protein is a regulatory subunit of the carnitine palmitoyltransferase complex [21]; <1> CPTo and CPTi are associated with a complex that contains β-oxidation enzymes [22]; <3> rat liver mitochondria carnitine palmitoyltransferase may be composed of an easily dissociable catalytic unit and a malonyl-CoA sensitivity conferring regulatory component [25]) [15, 21, 22, 25]

Posttranslational modification

lipoprotein <1> (<1> 18.9 mol of bound phospholipid per mol of enzyme, cardiolipin phosphatidylethanolamine and phosphatidylcholine [1]) [1]

Additional information <3> (<3> it is suggested that the enzyme contains hydrophobic sites which require phospholipid to prevent spurious binding of palmitoyl-CoA and which normally anchor the enzyme to the mitochondrial membrane [49]) [49]

5 Isolation/Preparation/Mutation/Application

Source/tissue

3T3-L1 cell <9> (<9> CPT I and CPT II [56]) [56]

Walker carcinoma 256 cell <3> (<3> CPT I [68]) [68]

adipocyte <3, 5, 9, 10> [39, 56]

brain <3> (<3> low content L-CPT I and CPT II [56]) [56]

brown adipose tissue <3, 9, 10> (<10> CPT II, M-CPT I [56]; <3,9> CPT II, M-CPT I, L-CPT I [56]; <3> interscapular [39]) [39, 56]

duodenum <11> (<11> L-CPT I [64]) [64]

erythrocyte <3> [14, 19]

flight muscle <2> [10]

heart <1, 3, 5, 9, 11> (<3,5> CPT II, M-CPT I and low content of L-CPT I [56]; <11> L-CPT I [64]; <5,11> M-CPT I [54,64]; <3> 2 isoforms of CPT I [50]) [1, 3, 5, 6, 9, 16, 21, 22, 35, 37, 38, 46, 50, 54, 56, 63, 64]

intestine <3> (<3> small and large, L-CPT I and CPT II [56]) [56]
kidney <3, 11> (<11> L-CPT I [64]; <3> cortex and medulla, L-CPT I and CPT II [56]) [56, 64]
leaf <12> (<12> CPT I and CPT II [62]) [62]
liver <1, 3-9, 11> (<1,3> CPT II [53,55,58]; <3,11> CPT I [51,55,64,70]; <3,11> L-CPT I [50,56,59,61,64,70]; <3,11> M-CPT I [61,64,70]) [2, 4, 6, 8, 11, 13-18, 20, 23-33, 35, 36, 38, 40-53, 55, 56, 58, 59, 61, 63-65, 70]
lung <3> (<3> low content L-CPT I and CPT II [56]) [56]
ovary <3> (<3> low content of M-CPT I, and higher content of CPT II ad L-CPT I [56]) [56]
pancreatic islet <3> (<3> low content of CPT II, no CPT I [56]) [56]
skeletal muscle <3, 5, 9, 11> (<5> Musculus vastus lateralis [60];<11> intermyofibrillar and subsarcolemmal [57]; <5,11> CPT I [57,60]; <3,5> CPT II [56,60]; <3,11> M-CPT I [50,56,64]) [12, 45, 46, 50, 56, 57, 60, 63, 64]
testis <3, 11> (<11> L-CPT I, not M-CPT I [64]; <3> L-CPT I and high content of CPT II and M-CPT I [56]) [56, 64]
white adipose tissue <3, 5, 9, 10> (<3,10> CPT II, M-CPT I, L-CPT I [56]; <5> CPT II, M-CPT I [56]; <9> CPT II, L-CPT I [56]) [56]
Additional information <3> (<3> not in spleen [56]; <3> liver L-CPT I is identical with the small isoform CPT I from heart, the large isoform of CPT I from heart resembles M-CPT I from skeletal muscle [50]) [50]

Localization

chloroplast <12> (<12> inner envelope membrane with isoform CPT I on the outer, cytosolic side and isoform CPT II on the inner, stromal side of the membrane [62]) [62]
cytosol <1> (<1> peroxisomal, liver [6]) [2, 6]
membrane <1, 3, 5> (<1> membrane-bound [1]) [1, 8, 19]
mitochondrial inner membrane <3> (<3> enzyme from outer and from inner membrane have different properties [14,15]; <3,4,8> inner face of inner membrane [17,31,32,34,37,40,45]; <1,3> outer face of inner membrane [2,17,32,34,35]; <1,3> enzyme is present on both sides of the inner membrane, not on the outer membrane [16,36]) [2, 6, 14-17, 20, 23, 31, 32, 34-37, 40, 45]
mitochondrial outer membrane <3> (<3> enzyme from outer and from inner membrane have different properties [14,15]; <3,9,11> CPT I is located in the outer membrane [55,56,64,68]; <3> L-CPT I in outer membrane, native and recombinant in Saccharomyces cerevisiae [59]) [14, 15, 18, 23, 55, 56, 59, 61, 64, 68]
mitochondrion <1-5, 7-11> (<11> recombinant L-CPT I and M-CPT I in Pichia pastoris [64]; <3,5,9-11> 2 isoforms of CPT I, L-CPT I and M-CPT I [50,56,61,64,70]; <3,11> CPT I [39,51,55,57]; <3> CPT II [20,55]) [1-3, 5, 6, 8-11, 14-18, 20-25, 27, 30-43, 45-47, 49-52, 56, 57, 59, 61, 64, 68, 70]
nucleus <3> [8]
peroxisome <1, 3, 6> [4, 6, 13, 28]
Additional information <1, 3, 5> (<3> composition of the outer mitochondrial membrane is important for functionality and activity of CPT I [68];

<3,5> forward reaction by CPT I at outer mitochondrial membrane, transport of acylcarnitine through mitochondrial membrane, reverse reaction at inner face of inner membrane by CPT II, overview [55,63]; <1,3> transport mechanism for acylcaritine and carnitine through the mitochondrial inner membrane [16,17]) [16, 17, 55, 63, 68]

Purification

<1> (CPT II [58]; CPTo [22]; calf liver [16]) [1-3, 5, 16, 22, 33, 58]
<3> (recombinant His-tagged L-CPT I from Pichia pastoris [70]; partial, recombinant CPT I and II from Pichia pastoris [55]; recombinant CPT II from Sf 9 insect cells [53]; CPT II, partial [48]; liver mitochondria [38]; CPT II [26]; native CPT II from liver [20]; CPTo and CPTi, partial [43,49]) [8, 20, 26, 38, 43, 48, 49, 53, 55, 70]
<5> (recombinant M-CPT I from Pichia pastoris [54]) [44, 54]
<6> (peroxisomal [28]) [28]
<12> (partial, CPT I and II [62]) [62]

Renaturation

<3, 5> (reconstitution of recombinant L-CPT I from Pichia pastoris in phospholipids after purification with detergent extraction [70]; reconstitution of detergent-inactivated recombinant M-CPT I, purified from Pichia pastoris, by removal of detergents in presence of phospholipids [54,55]) [54, 55, 70]

Cloning

<3> (expression of His-tagged L-CPT I in Pichia pastoris [70]; expression of CPT I in Saccharomyces cerevisiae [66]; construction of 6 chimeric proteins with exchanges structure segments of L-CPT I and M-CPT I, expression of chimeric mutants and wild-type isoforms in Pichia pastoris [61]; expression of L-CPT I in Saccharomyces cerevisiae, recombinant and native enzyme show the same biochemical properties [59]; clonig of CPT I and II from liver, expression in Pichia pastoris [55]; cloning and expression of CPT II in Escherichia coli, overexpression of CPT II in Spodoptera frugiperda Sf 9 cells via baculavirus infection, DNA sequence analysis [53]; cloning of liver mitochondrial CPT II from genetic library, in vitro transcription and translation, DNA sequence analysis, expression in COS cells [20]) [20, 53, 55, 59, 61, 66, 70]
<5> (overexpression of M-CPT I in Pichia pastoris, containing about 24 copies of the expression vector [54]) [54]
<11> (in vitro transcription and translation of L-CPT I [64]; 2 isoforms of CPT I, liver L-CPT I, skeletal muscle M-CPT I, cloned from liver, expression in Pichia pastoris GS115, DNA sequence determination [64]) [64]
<3, 11> (construction of chimeric L-CPT I with deletions and exchanged C-terminal sequences between rat and pig enzymes, expression in Pichia pastoris [69]) [69]

Engineering

A381D <3> (<3> site-directed mutagenesis, activity is reduced by 86%, K_m for acyl-CoA is 6-8fold increased [66]) [66]
D454G <3> (<3> site-directed mutagenesis, loss of activity [66]) [66]

H473A <3> (<3> site-directed mutagenesis, active site mutant, no remaining activity [66]) [66]
L484P <3> (<3> site-directed mutagenesis, loss of activity [66]) [66]
P479L <3> (<3> site-directed mutagenesis, loss of activity [66]) [66]
R451A <3> (<3> site-directed mutagenesis, loss of activity [66]) [66]
S113L <5> (<5> natural missense mutation of CPT II, enzyme deficiency leads to myopathic syndroms, metabolic characterization, e.g. insulin resistance, increased content of muscle lipidsreduced lipolysis [67]) [67]
W391A <3> (<3> site-directed mutagenesis, loss of activity [66]) [66]
W452A <3> (<3> site-directed mutagenesis, loss of activity [66]) [66]
Additional information <3> (<3> CPT I mutant with deletion of Arg395 show no activity [66]) [66]

Application

pharmacology <3> (<3> CPT I, model enzyme for studies of fatty acid-induced apoptosis in cancer therapy [68]) [68]

6 Stability

pH-Stability

7-10 <3> (<3> stable for 24 h at 4°C, recombinant CPT II [53]) [53]

General stability information

<1>, dithiothreitol stabilizes [2]
<1>, glycerol, 20%, stabilizes [2]
<3>, fractionation causes rapid loss of CPTo activity, CPTi is relatively stable [15]
<3>, no loss in activity after 6 freeze/thaw cycles at -70°C or room temeprature, recombinant CPT II [53]
<3>, quite stable at all steps of purification [8]
<3>, solubilization causes rapid loss of CPTo activity, CPTi is relatively stable [15]
<5>, Triton X-100: erythrocyte plasma membrane enzyme is stable, enzyme from microsomes and mitochondrial outer membrane not [19]

Storage stability

<3>, -20°C, several months [8]
<3>, 4°C, pH 7.0-10.0, 24 h stable, recombinant CPT II [53]
<3>, room temperature, 6.5 h, no loss in activity, recombinant CPT II [53]

References

[1] Fiol, C.J.; Bieber, L.L.: Sigmoid kinetics of purified beef heart mitochondrial carnitine palmitoyltransferase. Effect of pH and malonyl-CoA. J. Biol. Chem., **259**, 13084-13088 (1984)

[2] Derrick, J.P.; Tubbs, P.K.; Ramsay, R.R.: Purification and properties of an easily solubilized L-carnitine palmitoyltransferase from beef liver mitochondria. Biochem. Soc. Trans., **14**, 698 (1986)

[3] Bieber, L.L.; Fiol, C.: Purification and assay of carnitine acyltransferases. Methods Enzymol., **123**, 276-284 (1986)

[4] Sekas, G.; Paul, H.S.: Inhibition of carnitine acyltransferase activities by bile acids in rat liver peroxisomes. Biochim. Biophys. Acta, **1125**, 396-402 (1992)

[5] Clarke, P.R.H.; Bieber, L.L.: Isolation and purification of mitochondrial carnitine octanoyltransferase activities from beef heart. J. Biol. Chem., **256**, 9861-9868 (1981)

[6] Ramsay, R.R.: The soluble carnitine palmitoyltransferase from bovine liver. A comparison with the enzymes from peroxisomes and from the mitochondrial inner membrane. Biochem. J., **249**, 239-245 (1988)

[7] Ramsay, R.R.; Derrick, J.P.; Friend, A.S.; Tubbs, P.K.: Purification and properties of the soluble carnitine palmitoyltransferase from bovine liver mitochondria. Biochem. J., **244**, 271-278 (1987)

[8] Miyazawa, S.; Ozaka, H.; Osumi, T.; Hashiomoto, T.: Purification and properties of carnitine octanoyltransferase and carnitine palmitoyltransferase from rat liver. J. Biochem., **94**, 529-542 (1983)

[9] Fiol, C.J.; Bieber, L.L.: Effects of octylglucoside and Triton X-100 on the kinetics and specificity of carnitine palmitoyltransferase. Lipids, **23**, 120-125 (1988)

[10] Lee, C.W.; Tan, C.H.: Hexokinase and carnitine palmitoyltransferase activities in flight muscles of grashopper Valanga nigricornis (Burm.). Comp. Biochem. Physiol. B, **74**, 791-792 (1983)

[11] Brindle, N.P.J.; Zammit, V.A.; Pogson, C.I.: Inhibition of sheep liver carnitine palmitoyltransferase by methylmalonyl-CoA. Biochem. Soc. Trans., **13**, 880-881 (1985)

[12] Zierz, S.; Engel, A.G.: Different sites of inhibition of carnitine palmitoyltransferase by malonyl-CoA, and by acetyl-CoA and CoA, in human skeletal muscle. Biochem. J., **245**, 205-209 (1987)

[13] Ishii, S.; Ishii, H.; Suga, T.: The presence of peroxisomal carnitine palmitoyltransferase in chick embryo liver. J. Biochem., **98**, 747-755 (1985)

[14] Murthy, M.S.R.; Pande, S.V.: Some differences in the properties of carnitine palmitoyltransferase activities of the mitochondrial outer and inner membranes. Biochem. J., **248**, 727-733 (1987)

[15] Murthy, M.S.R.; Pande, S.V.: Characterization of a solubilized malonyl-CoA-sensitive carnitine palmitoyltransferase from the mitochondrial outer membrane as a protein distinct from the malonyl-CoA-insensitive carnitine palmitoyltransferase of the inner membrane. Biochem. J., **268**, 599-604 (1990)

[16] Halperin, M.L.; Pande, S.V.: Fatty acyl group transport into mitochondria: carnitine palmitoyl transferases EC 2.3.1.23 and the carnitine-acylcarnitine translocase. Methods Enzymol., **56**, 368-378 (1979)

[17] Hoppel, C.L.: Carnitine and carnitine palmitoyltransferase in fatty acid oxidation and ketosis. Fed. Proc., **41**, 2853-2857 (1982)

[18] Kolodziej, M.P.; Zammit, V.A.: Sensitivity of inhibition of rat liver mitochondrial outer-membrane carnitine palmitoyltransferase by malonyl-CoA to chemical- and temperature-induced changes in membrane fluidity. Biochem. J., **272**, 421-425 (1990)
[19] Ramsay, R.R.; Mancinelli, G.; Arduini, A.: Carnitine palmitoyltransferase in human erythrocyte membrane. Properties and malonyl-CoA sensitivity. Biochem. J., **275**, 685-688 (1991)
[20] Woeltje, K.F.; Esser, V.; Weis, B.C.; Sen, A.; Cox, W.F.; McPaul, M.J.; Slaughter, C.A.; Foster, D.W.; McGarry, J.D.: Cloning, sequencing, and expression of a cDNA encoding rat liver mitochondrial carnitine palmitoyltransferase II. J. Biol. Chem., **265**, 10720-10725 (1990)
[21] Chung, C.H.; Woldegiorgis, G.; Dai, G.; Shrago, E.; Bieber, L.L.: Conferral of malonyl coenzyme A sensitivity to purified rat heart mitochondrial carnitine palmitoyltransferase. Biochemistry, **31**, 9777-9783 (1992)
[22] Kerner, J.; Bieber, L.: Isolation of a malonyl-CoA-sensitive CPT/β-oxidation enzyme complex from heart mitochondria. Biochemistry, **29**, 4326-4334 (1990)
[23] Ghadiminejad, I.; Saggerson, E.D.: Carnitine palmitoyltransferase (CPT2) from liver mitochondrial inner membrane becomes inhibitable by malonyl-CoA if reconstituted with outer membrane malonyl-CoA binding protein. FEBS Lett., **269**, 406-408 (1990)
[24] Gavino, G.R.; Gavino, V.C.: Rat liver outer mitochondrial carnitine palmitoyltransferase activity towards long-chain polyunsaturated fatty acids and their CoA esters. Lipids, **26**, 266-270 (1991)
[25] Woldegiorgis, G.; Fibich, B.; Contreras, L.; Shrago, E.: Restoration of malonyl-CoA sensitivity of soluble rat liver mitochondria carnitine palmitoyltransferase by reconstitution with a partially purified malonyl-CoA binding protein. Arch. Biochem. Biophys., **295**, 348-351 (1992)
[26] Ghadiminejad, I.; Saggerson, D.: A simple method for the purification of carnitine palmitoyl transferase 2 from rat liver. Biochem. Soc. Trans., **17**, 348-349 (1989)
[27] Murthy, M.S.R.; Ramsay, R.R.; Pande, S.V.: Carnitine analogues and carnitine palmitoyltransferases. Biochem. Soc. Trans., **18**, 604-605 (1990)
[28] Ishii, S.; Ishii, H.; Watanabe, T.; Suga, T.: Purification and properties of peroxisomal carnitine palmitoyltransferase in chick embryo liver. Biochim. Biophys. Acta, **924**, 530-542 (1987)
[29] Gandour, R.D.; Colucci, W.J.; Stelly, T.C.; Brady, P.S.; Brady, L.J.: Hemipalmitoylcarnitinium, a strong competitive inhibitor of purified hepatic carnitine palmitoyltransferase. Arch. Biochem. Biophys., **267**, 515-520 (1988)
[30] Kashfi, K.; Cook, G.A.: Malonyl-CoA inhibits proteolysis of carnitine palmitoyltransferase. Biochem. Biophys. Res. Commun., **178**, 600-605 (1991)
[31] Brindle, N.P.J.; Zammit, V.A.; Pogson, C.I.: Regulation of carnitine palmitoyltransferase activity by malonyl-CoA in mitochondria from sheep liver, a tissue with a low capacity for fatty acid synthesis. Biochem. J., **232**, 177-182 (1985)

[32] Brady, P.S.; Dunker, A.K.; Brady, L.J.: Characterization of hepatic carnitine palmitoyltransferase. Use of bromoacyl derivatives and antibodies. Biochem. J., **241**, 751-757 (1987)
[33] Ramsay, R.R.; Derrick, J.P.; Friend, A.S.; Tubbs, P.K.: Purification and properties of the soluble carnitine palmitoyltransferase from bovine liver mitochondria. Biochem. J., **244**, 271-278 (1987)
[34] Bartlett, K.; Sherratt, H.S.A.; Turnbull, D.M.: Inhibition of hepatic and skeletal muscle carnitine palmitoyltransferase I by 2[5(4-chlorophenyl)pentyl]-oxirane-2-carbonyl-CoA. Biochem. Soc. Trans., **12**, 688-689 (1984)
[35] Cook, G.A.: Differences in the sensitivity of carnitine palmitoyltransferase to inhibition by malonyl-CoA are due to differences in Ki values. J. Biol. Chem., **259**, 12030-12033 (1984)
[36] Brosnan, J.T.; Kopec, B.; Fritz, I.B.: The localization of carnitine palmitoyltransferase on the inner membrane of bovine liver mitochondria. J. Biol. Chem., **248**, 4075-4082 (1973)
[37] Paulson, D.J.; Ward, K.M.; Shug, A.L.: Malonyl CoA inhibition of carnitine palmityltransferase in rat heart mitochondria. FEBS Lett., **176**, 381-384 (1984)
[38] Pande, S.V.; Murthy, M.S.R.; Noel, H.: Differential effects of phosphatidylcholine and cardiolipin on carnitine palmitoyltransferase activity. Biochim. Biophys. Acta, **877**, 223-230 (1986)
[39] Saggerson, E.D.; Carpenter, C.A.: Sensitivity of brown-adipose-tissue carnitine palmitoyltransferase to inhibition by malonyl-CoA. Biochem. J., **204**, 373-375 (1982)
[40] Stephens, T.W.; Cook, G.A.; Harris, R.A.: Effect of pH on malonyl-CoA inhibition of carnitine palmitoyltransferase I. Biochem. J., **212**, 521-524 (1983)
[41] McCormick, K.; Notar-Francesco, V.J.: Importance of albumin binding in the assay for carnitine palmitoyltransferase. Biochem. J., **216**, 495-498 (1983)
[42] McCormick, K.; Notar-Francesco, V.J.; Sriwatanakul, K.: Inhibition by acetyl-CoA of hepatic carnitine acyltransferase and fatty acid oxidation. Biochem. J., **216**, 499-502 (1983)
[43] Bergstrom, J.D.; Reitz, R.C.: Studies on carnitine palmitoyl transferase: the similar nature of CPTi (inner form) and CPTo (outer form). Arch. Biochem. Biophys., **204**, 71-79 (1980)
[44] Finocchiaro, G.; Colombo, I.; DiDonato, S.: Purification, characterization and partial amino acid sequences of carnitine palmitoyl-transferase from human liver. FEBS Lett., **274**, 163-166 (1990)
[45] Declercq, P.E.; Falck, J.R.; Kuwajima, M.; Tyminski, H.; Foster, D.W.; McGarry, J.D.: Characterization of the mitochondrial carnitine palmitoyltransferase enzyme system. I. Use of inhibitors. J. Biol. Chem., **262**, 9812-9821 (1987)
[46] Woeltje, K.F.; Esser, V.; Weis, B.C.; Cox, W.F.; Schroeder, J.G.; Liao, S.L.; Foster, D.W.; McGarry, J.D.: Inter-tissue and inter-species characteristics of the mitochondrial carnitine palmitoyltransferase enzyme system. J. Biol. Chem., **265**, 10714-10719 (1990)

[47] Lund, H.: Carnitine palmitoyltransferase: characterization of a labile detergent-extracted malonyl-CoA-sensitive enzyme from rat liver mitochondria. Biochim. Biophys. Acta, **918**, 67-75 (1987)
[48] Kopec, B.; Fritz, I.B.: Comparison of properties of carnitine palmitoyltransferase I with those of carnitine palmitoyltransferase II, and preparation of antibodies to carnitine palmitoyltransferases. J. Biol. Chem., **248**, 4069-4074 (1973)
[49] Woldegiorgis, G.; Bremer, J.; Shrago, E.: Substrate inhibition of carnitine palmitoyltransferase by palmitoyl-CoA and activation by phospholipids and proteins. Biochim. Biophys. Acta, **837**, 135-140 (1985)
[50] Weis, B.C.; Cowan, A.T.; Brown, N.; Foster, D.W.; McGarry, J.D.: Use of a selective inhibitor of liver carnitine palmitoyltransferase I (CPT I) allows quantification of its contribution to total CPT I activity in rat heart. Evidence that the dominant cardiac CPT I isoform is identical to the skeletal muscle enzyme. J. Biol. Chem., **269**, 26443-26448 (1994)
[51] Kashfi, K.; Mynatt, R.L.; Cook, G.A.: Hepatic carnitine palmitoyltransferase-I has two independent inhibitory binding sites for regulation of fatty acid oxidation. Biochim. Biophys. Acta, **1212**, 245-252 (1994)
[52] Singh, H.; Poulos, A.: Substrate specificity of rat liver mitochondrial carnitine palmitoyl transferase I: evidence against α-oxidation of phytanic acid in rat liver mitochondria. FEBS Lett., **359**, 179-183 (1995)
[53] Johnson, T.M.; Mann, W.R.; Dragland, C.J.; Anderson, R.C.; Nemecek, G.M.; Bell, P.A.: Over-expression and characterization of active recombinant rat liver carnitine palmitoyltransferase II using baculovirus. Biochem. J., **309**, 689-693 (1995)
[54] Zhu, H.; Shi, J.; Cregg, J.M.; Woldegiorgis, G.: Reconstitution of highly expressed human heart muscle carnitine palmitoyltransferase I. Biochem. Biophys. Res. Commun., **239**, 498-502 (1997)
[55] De Vries, Y.; Arvidson, D.N.; Waterham, H.R.; Cregg, J.M.; Woldegiorgis, G.: Functional characterization of mitochondrial carnitine palmitoyltransferases I and II expressed in the yeast Pichia pastoris. Biochemistry, **36**, 5285-5292 (1997)
[56] Brown, N.F.; Hill, J.K.; Esser, V.; Kirkland, J.L.; Corkey, B.E.; Foster, D.W.; McGarry, J.D.: Mouse white adipocytes and 3T3-L1 cells display an anomalous pattern of carnitine palmitoyltransferase (CPT) I isoform expression during differentiation. Inter-tissue and inter-species expression of CPT I and CPT II enzymes. Biochem. J., **327**, 225-231 (1997)
[57] Schmidt, I.; Herpin, P.: Carnitine palmitoyltransferase I (CPT I) activity and its regulation by malonyl-CoA are modulated by age and cold exposure in skeletal muscle mitochondria from newborn pigs. J. Nutr., **128**, 886-893 (1998)
[58] Nic a'Bhaird, N.; Yankovskaya, V.; Ramsay, R.R.: Active sites residues of beef liver carnitine octanoyltransferase (COT) and carnitine palmitoyltransferase (CPT-II). Biochem. J., **330**, 1029-1036 (1998)
[59] Prip-Buus, C.; Cohen, I.; Kohl, C.; Esser, V.; McGarry, J.D.; Girard, J.: Topological and functional analysis of the rat liver carnitine palmitoyltransfer-

ase 1 expressed in Saccharomyces cerevisiae. FEBS Lett., **429**, 173-178 (1998)
[60] Hertel, K.; Gellerich, F.N.; Hein, W.; Zierz, S.: Kinetic investigation of carnitine palmitoyltransferases in homogenates of human skeletal muscle using L-amino-carnitine and malonyl-CoA. Adv. Exp. Med. Biol., **466**, 87-93 (1999)
[61] Jackson, V.N.; Cameron, J.M.; Fraser, F.; Zammit, V.A.; Price, N.T.: Use of six chimeric proteins to investigate the role of intramolecular interactions in determining the kinetics of carnitine palmitoyltransferase I isoforms. J. Biol. Chem., **275**, 19560-19566 (2000)
[62] Masterson, C.; Wood, C.: Carnitine palmitoyltransferases in pea leaf chloroplasts: partial purification, location, and properties. Can. J. Bot., **78**, 328-335 (2000)
[63] Eaton, S.; Fukumoto, K.; Paladio Duran, N.; Pierro, A.; Spitz, L.; Quant, P.A.; Bartlett, K.: Carnitine palmitoyl transferase I and the control of myocardial β-oxidation flux. Biochem. Soc. Trans., **29**, 245-250 (2001)
[64] Nicot, C.; Hegardt, F.G.; Woldegiorgis, G.; Haro, D.; Marrero, P.F.: Pig liver carnitine palmitoyltransferase I, with low K_m for carnitine and high sensitivity to malonyl-CoA inhibition, is a natural chimera of rat liver and muscle enzymes. Biochemistry, **40**, 2260-2266 (2001)
[65] Abo-Hashema, K.A.H.; Cake, M.H.; Lukas, M.A.; Knudsen, J.: The interaction of acyl-CoA with acyl-CoA binding protein and carnitine palmitoyltransferase I. Int. J. Biochem. Cell Biol., **33**, 807-815 (2001)
[66] Morillas, M.; Gomez-Puertas, P.; Roca, R.; Serra, D.; Asins, G.; Valencia, A.; Hegardt, F.G.: Structural model of the catalytic core of carnitine palmitoyltransferase I and carnitine octanoyltransferase (COT). Mutation of CPT 1 histidine 473 and alanine 381 and COT alanine 238 impairs the catalytic activity. J. Biol. Chem., **276**, 45001-45008 (2001)
[67] Haap, M.; Thamer, C.; Machann, J.; Tschritter, O.; Loblein, K.; Kellerer, M.; Schick, F.; Jacob, S.; Haring, H.U.; Stumvoll, M.: Metabolic characterization of a woman homozygous for the Ser113Leu missense mutation in carnitine palmitoyl transferase II. J. Clin. Endocrinol. Metab., **87**, 2139-2143 (2002)
[68] Colquhoun, A.: γ-linolenic acid alters the composition of mitochondrial membrane subfractions, decreases outer mitochondrial membrane binding of hexokinase and alters carnitine palmitoyltransferase I properties in the Walker 256 rat tumour. Biochim. Biophys. Acta, **1583**, 74-84 (2002)
[69] Nicot, C.; Relat, J.; Woldegiorgis, G.; Haro, D.; Marrero, P.F.: Pig liver carnitine palmitoyltransferase: chimera studies show that both the N- and C-terminal regions of the enzyme are important for the unusual high malonyl-CoA sensitivity. J. Biol. Chem., **277**, 10044-10049 (2002)
[70] McGarry, J.D.; Brown, N.F.: Reconstitution of purified, active and malonyl-CoA-sensitive rat liver carnitine palmitoyltransferase I: relationship between membrane environment and malonyl-CoA sensitivity. Biochem. J., **349**, 179-187 (2000)

2-Acylglycerol O-acyltransferase 2.3.1.22

1 Nomenclature

EC number
2.3.1.22

Systematic name
acyl-CoA:2-acylglycerol O-acyltransferase

Recommended name
2-acylglycerol O-acyltransferase

Synonyms
MGAT
acyl coenzyme A-monoglyceride acyltransferase
acyl-CoA:monoacylglycerol transferase
acylglycerol palmitoyltransferase
monoacylglycerol acyltransferase
monoglyceride acyltransferase
palmitoyltransferase, acylglycerol

CAS registry number
9055-17-8

2 Source Organism

<1> *Rattus norvegicus* (Wistar strain, male [10]; Zivic Miller SD-strain [12]) [2, 3, 4, 7, 8, 10, 12, 13, 16, 17, 19]
<2> *Cavia porcellus* (male [11]) [11]
<3> *Homo sapiens* [1, 5, 20]
<4> *Locusta migratoria* [9]
<5> *Mus musculus* [6, 14, 20]
<6> *Arachis hypogaea* [15]
<7> *Zonotrichia albicollis* [18]

3 Reaction and Specificity

Catalyzed reaction
acyl-CoA + 2-acylglycerol = CoA + diacylglycerol

Reaction type
acyl group transfer

Natural substrates and products

S acyl-CoA + sn-2-monoacylglycerol <1, 3, 5, 6, 7> (<1>, involved in synthesis of triacylglycerols [12]; <3>, representative enzyme of the neutral lipid pathway [1]; <5>, the enzyme plays a predominant role in dietary fat absorption in the small intestine, where it catalyzes the first step of triacylglycerol resynthesis in enterocytes for chylomicron formation and secretion [6]; <5>, MGAT2 may play an important role in dietary fat absorption [6]; <5>, in the intestine the enzyme plays a major role in the absorption of dietary fat because resynthesis of triacylglycerol is required for the assembly of lipoproteins that transport absorbed fat to other tissues [14]; <6>, diacylglycerol biosynthetic activity [15]; <1>, the enzyme is developmentally expressed, associated with physiological periods characterized by high rates of lipolysis and depends on fatty acids for energy production. During these periods the enzyme may help to retain essential fatty acids selectively [16]; <1>, the enzyme may be regulated physiologically by specific intermediates of glycerolipid metabolism. In neonatal liver, signal transduction may be linked to the synthesis of complex lipids via the monoacylglycerol pathway [17]; <7>, the specificity of the enzyme for sn-2-monoacylglycerol and the probable enhanced affinity for sn-2-monoacylglycerols of specific acyl chains may allow selected $\omega 6$ and $\omega 3$ fatty acids to be retained within the adipocyte, while nonessential fatty acids are released for β-oxidation in flight muscles [18]; <3>, the enzyme plays a major role in the absorption of dietary fat by catalyzing the resynthesis of triacylglycerol in enterocytes [20]) [1, 6, 12, 14, 15, 16, 17, 18, 20]

P CoA + diacylglycerol

Substrates and products

S acyl-CoA + sn-2-monolinolenoylglycerol <1> (<1>, in preference to sn-2-monooleoylglycerol [7]) (Reversibility: ? <1> [7]) [7]

P CoA + 1-acyl-2-linolenoylglycerol

S acyl-CoA + sn-2-monolinoleoylglycerol <1> (<1>, in preference to sn-2-monooleoylglycerol [7]) (Reversibility: ? <1> [7]) [7]

P CoA + 1-acyl-2-linoleoylglycerol

S arachidonoyl-CoA + sn-2-monooleoylglycerol <5> (Reversibility: ? <5> [14]) [14]

P CoA + 1-arachidonoyl-2-oleoylglycerol

S arachidoyl-CoA + sn-2-monooleoylglycerol <5> (Reversibility: ? <5> [14]) [14]

P CoA + 1-arachidoyl-2-oleoylglycerol

S atearoyl-CoA + sn-2-monooleoylglycerol <5> (Reversibility: ? <5> [14]) [14]

P CoA + 1-atearoyl-2-oleolyglycerol

S fatty acyl-CoA + sn-2-monoacylglycerol <1, 3> (Reversibility: ? <1, 3> [2, 4, 5, 7, 8, 12]) [2, 4, 5, 7, 8, 12]

P CoA + 1,2-diacylglycerol <1> (<1>, 88% 1,2-diacylglycerol and 12% 1,3-diacylglycerol [4]) [2, 4, 7, 12]

S lauroyl-CoA + sn-2-monooleoylglycerol <5> (Reversibility: ? <5> [14]) [14]
P CoA + 1-lauroyl-2-oleoylglycerol
S linoleoyl-CoA + sn-2-monooleoylglycerol <5> (Reversibility: ? <5> [14]) [14]
P CoA + 1-linoleoyl-2-oleoylglycerol
S n-octanoyl-CoA + sn-2-monooleoylglycerol <5> (Reversibility: ? <5> [14]) [14]
P CoA + 1-octanoyl-2-oleoylglycerol
S oleoly-CoA + sn-3-monostearoylglycerol <5> (Reversibility: ? <5> [14]) [14]
P CoA + 1-oleoyl-3-stearoylglycerol <5> [14]
S oleoyl-CoA + 1-palmitoyl-sn-glycerol <1, 6> (Reversibility: ? <1, 6> [13, 15]) [13, 15]
P CoA + 1-palmitoyl-3-oleoyl-sn-glycerol <1> (<1>, 1-palmitoyl-3-oleoyl-sn-glycerol is the predominant reaction product [13]) [13]
S oleoyl-CoA + rac-1-monooleoylglycerol <3> (Reversibility: ? <3> [20]) [20]
P CoA + dioleoylglycerol
S oleoyl-CoA + sn-1-monooleoylglycerol <5> (Reversibility: ? <5> [14]) [14]
P CoA + sn-1,3-dioleoylglycerol <5> [14]
S oleoyl-CoA + sn-2-monooleoylglycerol <1, 3, 5> (Reversibility: ? <1, 3, 5> [6, 13, 14, 20]) [6, 13, 14, 20]
P CoA + sn-1,2(2,3)-dioleoylglycerol <1, 5> (<1>, preferentially esterified at sn-1-position, 10% esterification at sn-3-position [13]) [6, 13, 14]
S palmitoyl-CoA + 1-palmitoyl-sn-glycerol <6> (Reversibility: ? <6> [15]) [15]
P CoA + ?
S palmitoyl-CoA + 2-hexadecylglycerol <2> (Reversibility: ? <2> [11]) [11]
P CoA + 2-hexadecyl-1-palmitoylglycerol
S palmitoyl-CoA + 2-palmitoyl-sn-glycerol <2, 4> (Reversibility: ? <2, 4> [9, 11]) [9, 11]
P CoA + 1,2-dipalmitoylglycerol
S palmitoyl-CoA + rac-1-monooleoylglycerol <1> (<1>, intestine, at 78% the rate of sn-2-monooleoylglycerol acylation [12]) (Reversibility: ? <1> [12]) [12]
P CoA + ?
S palmitoyl-CoA + rac-1-monooleoylglycerol ether <1> (<1>, intestine, at 45% the rate of sn-2-monooleoylglycerol acylation [12]) (Reversibility: ? <1> [12]) [12]
P CoA + ?
S palmitoyl-CoA + rac-1-monopalmitoylglycerol <2> (Reversibility: ? <2> [11]) [11]
P CoA + ?

S palmitoyl-CoA + sn-1-monooleoylglycerol <6, 7> (<7>, at 7.5% of the activity with sn-2-monooleoylglycerol [18]) (Reversibility: ? <6, 7> [15, 18]) [15, 18]

P CoA + ?

S palmitoyl-CoA + sn-1-monooleoylglycerol ether <7> (<7>, at 5.7% of the activity with sn-2-monooleoylglycerol [18]) (Reversibility: ? <7> [18]) [18]

P CoA + ?

S palmitoyl-CoA + sn-2-monooleoylglycerol <1, 2, 3, 4, 5, 7> (<1>, liver enzyme highly specific for sn-2-monooleoylglycerol [12]; <1>, intestine, at 52% the rate of sn-2-monooleoylglycerol acylation [12]) (Reversibility: ? <1, 2, 3, 4, 5, 7> [4, 5, 7, 9, 11, 12, 14, 16, 18]) [4, 5, 7, 9, 11, 12, 14, 16, 18]

P CoA + 1(3)-palmitoyl-2-oleoylglycerol

S palmitoyl-CoA + sn-2-monooleylglycerol ether <7> (<7>, at 5.7% of the activity with sn-2-monooleoylglycerol [18]) (Reversibility: ? <7> [18]) [18]

P CoA + 2-oleoyl-1-palmitoylglycerol ether

S stearoyl-CoA + 1-palmitoyl-sn-glycerol <6> (Reversibility: ? <6> [15]) [15]

P CoA + ?

S Additional information <3, 5, 7> (<3>, broad acyl-donor pattern with maximal activities for C10:0 to C16:0, the highest activity with C16:0. Preferential esterification for 2-monoacylglycerols as compared with the 1-isomers [5]; <5>, the enzyme catalyzes the acylation of rac-1-monoacylglycerol, sn-2-monoacylglycerol, and sn-3-monoacylglycerols, the enzyme prefers monoacylglycerols containing unsaturated fatty acyls [6]) [5, 6, 18]

P ?

Inhibitors

2,4,6-trinitrobenzene sulfonic acid <1> (<1>, 3 mM, 80% inactivation of liver enzyme, 45% inactivation of intestinal enzyme, palmitate partly protects, enzyme from intestine is less sensitive than enzyme from liver [12]) [12, 16]

2-monooleoylglycerol <1> (<1>, high concentration [12]) [12]

CHAPS <1, 6> (<6>, 20 mM, complete loss of activity [15]; <1>, 1.5% [7]) [7, 15]

$CaCl_2$ <1> (<1>, 12.5 mM, 20-30% inhibition [16]) [16]

Co^{2+} <7> [18]

$CuCl_2$ <7> (<7>, 10 mM [18]) [18]

EDTA <1> (<1>, weak [12]) [12]

Fe^{2+} <7> [18]

$HgCl_2$ <1> (<1>, weak [12]; <1>, 0.5 mM, complete inhibition [16]) [1, 16]

KI <1> (<1>, 300 mM, 30% inhibition [16]) [16]

$MnCl_2$ <1> (<1>, 10 mM, 18-25% [16]; <1>, 2.5 mM, 86% inhibition of the intestinal enzyme, 40% inhibition of the enzyme from suckling liver [12]; <1>, 2.5-5 mM, 40% inhibition of enzymne from neonatal liver [16]) [12, 16]

NEM <1> (<1>, 2 mM, 12% inhibition [16]) [16]
NaF <6> [15]
PCMB <1> [12]
Triton X 100 <1, 6> (<1>, no inhibition [7]; <6>, 42% loss of activity) [12, 15]
Zn^{2+} <7> [18]
acetone <1> (<1>, 10%, liver [12]) [12]
β-octylglucoside <1, 6> (<1>, 2.5% [7]; <6>, 40 mM, 50% loss of activity [15]) [7, 15]
bovine serum albumin <1> (<1>, activation below 1 mg/ml [12]) [12]
deoxycholate <1> (<1>, inhibition of liver enzyme [12]) [12]
diethyldicarbonate <1> (<1>, 0.04 mM, 50% inhibition, 0.05 mM palmitate partly protects, completely reversible by 50 mM hydroxylamine, enzyme from intestine is less sensitive than enzyme from liver [12]) [12]
oleic acid <1> [19]
palmitoyl-CoA <1, 3, 4> (<1>, high concentration [12]; <3>, bovine serum albumin protects [5]; <4>, above 0.25 mM [9]) [5, 9, 12]
phytosphingosine <1> [17]
sphinganine <1> [17]
sphingosine <1> (<1>, IC50 is 9 mol%, inhibits activity in Triton X-100 mixed micelles and isolated hepatocytes [17]) [17]
stearylamine <1> [17]
taurocholate <1> (<1>, 2% [7]) [7, 12]

Activating compounds

1,2-monooleoylglycerol ether <1> (<1>, 12.6 mol%, alone has no effect on activity, becomes an activator in presence of 10 mol% oleic acid [19]) [19]
Tween 20 <4> (<4>, maximal stimulation in presence of 1.4 mg/ml [9]) [9]
bovine serum albumin <1> (<1>, activation, liver, below 1 mg/ml, inhibitory above, not intestine [12]; <1>, stimulates [16]) [12, 16]
deoxycholate <1> (<1>, stimulates enzyme from intestinal mucosa [12]) [12]
oleic acid <6> (<6>, stimulates below 0.015 mM [15]) [15]
palmitic acid <6> (<6>, 0.02-0.05 mM, stimulates [15]) [15]
phosphatidic acid <6> (<6>, stimulates [15]) [15]
phosphatidylcholine <6> (<6>, stimulates [15]) [15]
phospholipid <1> (<1>, phosphatidylcholine/phosphatidylserine at 0.025 mg/ml stimulates 80% [7]) [7]
sn-1,2-dioleoylglycerol <1> (<1>, 0.3 mM, stimulates 44% [7]; <1>, cooperative activator, reverses the 50% inhibition caused by 10 mol% [19]) [7, 19]
sphingomyelin <1, 6> (<6>, 0.0025-0.0075 mM, activates [15]; <1>, stimulates [17]) [15, 17]
sphingosine <6> (<6>, 0.0025-0.0075 mM, activates [15]) [15]

Metals, ions

Ba^{2+} <7> (<7>, stimulates [18]) [18]
Ca^{2+} <7> (<7>, stimulates [18]) [18]
$CuCl_2$ <7> (<7>, 2.5 mM, stimulation [18]) [18]

Mn^{2+} <1, 7> (<1>, 2.5-5 mM, stimulates 160%, enzyme from adult liver [16]; <7>, stimulates [18]) [16, 18]
Na^{+} <7> (<7>, minimal stimulation [18]) [18]

Specific activity (U/mg)
0.0159 <6> [15]
1.219 <1> [7]
Additional information <1, 2> (<1>, radioassay of the stereospecificity [10]; <2>, assay procedure [11]) [10, 11]

K_m-Value (mM)
0.00565 <6> (1-oleoylglycerol) [15]
0.0065 <1> (palmitoyl-CoA) [7]
0.0071 <1> (sn-2-monolinolenoylglycerol) [7]
0.00935 <6> (oleoyl-CoA) [15]
0.011 <1> (sn-2-monolinoleoylglycerol) [7]
0.01639 <6> (1-palmitoylglycerol) [15]
0.01754 <6> (palmitoyl-CoA) [15]
0.021 <1> (sn-2-monooleoylglycerol) [7]
0.02566 <6> (stearoyl-CoA) [15]
0.047 <7> (linoleoyl-CoA) [18]
0.16 <4> (2-oleoyl-sn-glycerol) [9]

pH-Optimum
7 <1, 4, 6> [9, 15, 16]
7-9 <1> [12]
7.4 <3> [5]
8-9 <1> [2]

4 Enzyme Structure

Molecular weight
43000 <6> (<6>, gel filtration [15]) [15]
Additional information <1> (<1>, molecular weight for triacylglycerol synthetase complex of EC 2.3.1.20/EC 2.3.1.22 is 350000-370000 Da, gel filtration [3]) [3]

Subunits
? <1, 5> (<5>, x * 33000, SDS-PAGE [14]; <1>, x * 37000, SDS-PAGE [2]) [2, 14]
monomer <6> (<6>, 1 * 43000, SDS-PAGE [15]) [15]

5 Isolation/Preparation/Mutation/Application

Source/tissue
adipose tissue <5, 7> [6, 17, 18]
brown adipose tissue <5> (<5>, MGAT1 [14]) [14]

cotyledon <6> [15]
fat body <4> [9]
gut <3> (<3>, high levels of activity in developing gut [1]) [1]
kidney <5> (<5>, MGAT2 [6]; <5>, highest expression of MGAT1 mRNA [6]) [6, 14]
liver <1, 5, 7> (<1>, neonatal [4,7,12]; <5>, low expression of MGAT2 [6]; <5>, MGAT1 [14]; <1>, adult and neonatal, enhanced expression in diabetes [16]) [2, 6, 7, 8, 12, 14, 16, 17, 18, 19]
skeletal muscle <5> (<5>, low expression of MGAT2 [6]) [6]
small intestine <5> [6, 20]
small intestine mucosa <1, 2, 3> (<1>, villus cells [2,10]; <3>, fetal [1]) [1, 2, 3, 4, 5, 10, 11, 12, 13]
stomach <5> (<5>, MGAT2 [6]; <5>, highest expression of MGAT1 mRNA [6]) [6, 14]
white adipose tissue <5> (<5>, MGAT1 [14]) [14]
Additional information <1> (<1>, tissue specific isozymes [12]) [12]

Localization

microsome <1, 2, 3, 4> [3, 4, 5, 7, 9, 11, 12, 13, 16, 19]
microvillus <1> [2, 3, 10]
plasma membrane <1> [8]

Purification

<1> (partial [7,17]; as part of triacylglycerol synthetase complex [3]) [2, 3, 7, 13, 17]
<6> [15]

Cloning

<3> (expression in COS-7 cells [20]) [20]
<5> (expression of MGAT2 in AV-12, COS-7 and CACO-2 cell [6]; expression of MGAT1 cDNA in Spodoptera frugiperda cells [14]) [6, 14]

6 Stability

Temperature stability

48 <1> (<1>, 12 min, microsomes from liver retain 90% of the initial activity [12]) [12]
51 <1> (<1>, 12 min, microsomes from liver lose 23% of the initial activity, microsomes from intestinal mucosa lose 88.5% of control activity [12]; <1>, after microsomes from adult liver are incubated for 13 min, they retain 44% of the control activity. After 13.5 min of heating, neonatal liver microsomes retain 90% of the control activity [16]) [12, 16]
56 <1> (<1>, after microsomes from adult liver are incubated for 13 min, they retain 7% of the control activity. After 13.5 min of heating, neonatal liver microsomes retain 57% of the control activity [16]) [16]

Storage stability

<1>, 20°C, the purified enzyme complex EC 2.3.1.20/2.3.1.22 loses 50% of its activity after 2 weeks [3]

References

[1] Loirdighi, N.; Menard, D.; Delvin, E.; Levy, E.: Ontogeny and location of HMG-CoA reductase, ACAT, and MGAT in human small intestine. Am. J. Physiol., **273**, G62-G67 (1997)

[2] Manganaro, F.; Kuksis, A.: Purification and preliminary characterization of 2-monoacylglycerol acyltransferase from rat intestinal villus cells. Can. J. Biochem. Cell Biol., **63**, 341-347 (1985)

[3] Manganaro, F.; Kuksis, A.: Rapid isolation of a triacylglycerol synthetase complex from rat intestinal mucosa. Can. J. Biochem. Cell Biol., **63**, 107-114 (1985)

[4] Coleman, R.A.; Walsh, J.P.; Millington, D.S.; Maltby, D.A.: Stereospecificity of monoacylglycerol acyltransferase activity from rat intestine and suckling rat liver. J. Lipid Res., **27**, 158-165 (1986)

[5] Bierbach, H.: Triacylglycerol biosynthesis in human small intestinal mucosa. Acyl-CoA: monoglyceride acyltransferase. Digestion, **28**, 138-147 (1983)

[6] Cao, J.; Lockwood, J.; Burn, P.; Shi, Y.: Cloning and functional characterization of a mouse intestinal acyl-CoA: monoacylglycerol acyltransferase, MGAT2. J. Biol. Chem., **278**, 13860-13866 (2003)

[7] Bhat, B.G.; Bardes, E.S.G.; Coleman, R.A.: Solubilization and partial purification of neonatally expressed rat hepatic microsomal monoacylglycerol acyltransferase. Arch. Biochem. Biophys., **300**, 663-669 (1993)

[8] Waite, M.; Sisson, P.: Utilization of serum lipoprotein lipids by the monoacylglycerol acyltransferase. Biochim. Biophys. Acta, **450**, 301-310 (1976)

[9] Tietz, A.; Weintraub, H.; Peled, Y.: Utilization of 2-acyl-sn-glycerol by locust fat body microsomes. Specificity of the acyltransferase system. Biochim. Biophys. Acta, **388**, 165-170 (1975)

[10] Manganaro, F.; Kuksis, A.; Myher, J.J.: Radioassay of the stereospecificity of 2-monoacylglycerol acyltransferase. Anal. Biochem., **119**, 4-11 (1982)

[11] Short, V.; Brindley, D.N.; Dils, R.: A new assay procedure for monoglyceride acyltransferase. Biochem. J., **141**, 407-411 (1974)

[12] Coleman, R.A.; Haynes, E.B.: Monoacylglycerol acyltransferase. Evidence that the activities from rat intestine and suckling liver are tissue-specific isoenzymes. J. Biol. Chem., **261**, 224-228 (1986)

[13] Lehner, R.; Kuksis, A.; Itabashi, Y.: Stereospecificity of monoacylglycerol and diacylglycerol acyltransferases from rat intestine as determined by chiral phase high-performance liquid chromatography. Lipids, **28**, 29-34 (1993)

[14] Yen, C.L.E.; Stone, S.J.; Cases, S.; Zhou, P.; Farese, R.V., Jr.: Identification of a gene encoding MGAT1, a monoacylglycerol acyltransferase. Proc. Natl. Acad. Sci. USA, **99**, 8512-8517 (2002)

[15] Tumaney, A.W.; Shekar, S.; Rajasekharan, R.: Identification, purification, and characterization of monoacylglycerol acyltransferase from developing peanut cotyledons. J. Biol. Chem., **276**, 10847-10852 (2001)
[16] Mostafa, N.; Bhat, B.G.; Coleman, R.A.: Increased hepatic monoacylglycerol acyltransferase activity in streptozotocin-induced diabetes: characterization and comparison with activities from adult and neonatal rat liver. Biochim. Biophys. Acta, **1169**, 189-195 (1993)
[17] Bhat, B.G.; Wang, P.; Coleman, R.A.: Sphingosine inhibits rat hepatic monoacylglycerol acyltransferase in Triton X-100 mixed micelles and isolated hepatocytes. Biochemistry, **34**, 11237-11244 (1995)
[18] Mostafa, N.; Bhat, B.G.; Coleman, R.A.: Adipose monoacylglycerol:acylcoenzyme A acyltransferase activity in the white-throated sparrow (Zonotrichia albicollis): characterization and function in an migratory bird. Lipids, **29**, 785-791 (1994)
[19] Coleman, R.A.; Wang, P.; Bhat, B.G.: Diradylglycerols alter fatty acid inhibition of monoacylglycerol acyltransferase activity in Triton X-100 mixed micelles. Biochemistry, **37**, 5916-5922 (1998)
[20] Yen, C.L.; Farese, R.V., Jr.: MGAT2, a monoacylglycerol acyltransferase expressed in the small intestine. J. Biol. Chem., **5**, 1-24 (2003)

1-Acylglycerophosphocholine O-acyltransferase 2.3.1.23

1 Nomenclature

EC number
2.3.1.23

Systematic name
acyl-CoA:1-acyl-sn-glycero-3-phosphocholine O-acyltransferase

Recommended name
1-acylglycerophosphocholine O-acyltransferase

Synonyms
1-acyl-sn-glycero-3-phosphocholine acyltransferase
acyl coenzyme A-monoacylphosphatidylcholine acyltransferase
acyl-CoA:1-acyl-glycero-3-phosphocholine transacylase
acyltransferase, lysolecithin
lysolecithin acyltransferase
lysophosphatide acyltransferase
lysophosphatidylcholine acyltransferase

CAS registry number
9027-64-9

2 Source Organism

<1> *Rattus norvegicus* (male Wistar [20,26,29,30]) [1, 5, 7, 10, 17, 19-30, 32, 34, 37]
<2> *Mus musculus* (C57BL/10J [9]; female BALb/c [28]) [6, 9, 28]
<3> *Mesocricetus auratus* [7]
<4> *Sus scrofa* [7, 17, 33, 35, 42]
<5> *Oryctolagus cuniculus* [3, 4, 10-12, 16]
<6> *Cavia porcellus* [7, 10, 17-19]
<7> *Bos taurus* (calf [8]) [7, 8, 10, 13-15, 17, 38, 39]
<8> *Salmo gairdneri* (rainbow trout [2]) [2]
<9> *Saccharomyces cerevisiae* (OC-2 [31]) [31, 40]
<10> *mammalia* [10]
<11> *Solanum tuberosum* (potato [36]) [36]
<12> *Homo sapiens* [41, 45]
<13> *Allium porrum* (leek [46]) [43, 46]
<14> *Helianthus annuus* (sunflower [44]) [44, 46]

3 Reaction and Specificity

Catalyzed reaction

acyl-CoA + 1-acyl-sn-glycero-3-phosphocholine = CoA + 1,2-diacyl-sn-glycero-3-phosphocholine (<5> lung enzyme displays an iso-ping-pong mechanism [10,12])

Reaction type

acyl group transfer

Natural substrates and products

S acyl-CoA + 1-acyl-sn-glycero-3-phosphocholine <1, 7> (<1,7> participates in deacylation/reacylation cycle of cell membranes [1,13]; <7> determines composition of phosphocholine in myocardium [14]) (Reversibility: ? <1, 7> [1, 13]) [1, 13, 14]

P CoA + 1,2-diacyl-sn-glycero-3-phosphocholine <1, 7> [1, 13, 14]

Substrates and products

S acyl-CoA + 1-O-alkyl-2-acyl-sn-3-glycerophosphocholine <9> (Reversibility: ? <9> [40]) [40]

P CoA + 1-O-alkyl-2-acyl-sn-3-glycerophosphocholine <9> [40]

S acyl-CoA + 1-acyl-sn-glycero-3-phosphocholine <1-14> (<5> 2 isoenzymes differing in their selectivity toward saturated and unsaturated fatty acids [3]; <1> specificity for acyl-CoA with unsaturated acyl groups, 1-palmitoyl-sn-glycero-3-phosphocholine is a better acceptor than 1-stearyl-sn-glycero-3-phosphocholine [10]; <6> microsomal enzyme is more active toward unsaturated acyl-CoAs, the mitochondrial enzyme has a very high specificity toward linoleoyl-CoA [10]; <1,6> highest activity with polyunsaturated acyl-CoAs e.g. arachidonoyl-CoA, eicosatrienoyl-CoA and linoleoyl-CoA [19]; <1> selectivity for polyunsaturated fatty acids, preferred acyl donors in descending order: arachidonyl-CoA, eicosapentaenoyl-CoA, docosahexaenoyl-CoA, in platelets, 1-acyl-sn-glycero-3-phosphocholine is a better acceptor than 1-acyl-sn-glycero-3-phosphorylethanolamine [29]) (Reversibility: r <1-14> [1-46]) [1-46]

P CoA + 1,2-diacyl-sn-glycero-3-phosphocholine <1-14> [1-46]

S acyl-CoA + 1-acyl-sn-glycero-3-phosphoryldimethylethanolamine <1> (<1> preferred acyl acceptors in descending order: 1-acyl-sn-glycero-3-phosphocholine, 1-acyl-sn-glycero-3-phosphoryldimethylethanolamine, 1-acyl-sn-glycero-3-phosphorylmonomethylethanolamine, 1-acyl-sn-glycero-3-phosphorylethanolamine [24]) (Reversibility: ? <1> [24]) [24]

P CoA + 1,2-diacyl-sn-glycero-3-phosphoryldimethylethanolamine <1> [24]

S acyl-CoA + 1-acyl-sn-glycero-3-phosphorylethanolamine <1> (Reversibility: ? <1> [24]) [24]

P CoA + 1,2-diacyl-sn-glycero-3-phosphorylethanolamine <1> [24]

S acyl-CoA + 1-acyl-sn-glycero-3-phosphorylmonomethylethanolamine <1> (Reversibility: ? <1> [24]) [24]

P CoA + 1,2-diacyl-sn-glycero-3-phosphorylmonomethylethanolamine <1> [24]

S acyl-CoA + 1-decanoyl-sn-glycero-3-phosphocholine <4> (Reversibility: ? <4> [42]) [42]

P CoA + 1-decanoyl-2-acyl-sn-glycero-3-phosphocholine <4> [42]

S acyl-CoA + 1-lauroyl-sn-glycero-3-phosphocholine <4> (Reversibility: ? <4> [42]) [42]

P CoA + 1-lauroyl-2-acyl-sn-glycero-3-phosphocholine <4> [42]

S acyl-CoA + 1-myristoyl-sn-glycero-3-phosphocholine <1> (<1> preferred acyl acceptors: 1-myristyl-sn-glycero-3-phosphocholine, 1-stearyl-sn-glycero-3-phosphocholine, 1-oleoyl-sn-glycero-3-phosphocholine and 1-linoleoyl-sn-glycero-3-phosphocholine, very little activity with 1-decanoyl-sn-glycero-3-phosphocholine [22]) (Reversibility: ? <1> [22]) [22]

P CoA + 1-myristoyl-2-acyl-sn-glycero-3-phosphocholine <1> [22]

S acyl-CoA + 1-oleoyl-sn-glycero-3-phosphocholine <4, 9> (<9> preferred acyl acceptors in descending order: 1-oleoyl-sn-glycero-3-phosphocholine, 1-myristoyl-sn-glycero-3-phosphocholine, 1-lauroyl-sn-glycero-3-phosphocholine, 1-stearoyl-sn-glycero-3-phosphocholine, 1-palmitoyl-sn-glycero-3-phosphocholine, 1-decanoyl-sn-glycero-3-phosphocholine and 1-arachidonoyl-sn-glycero-3-phosphocholine [40]; <4> integral membrane enzyme, preferred acyl acceptors in descending order: 1-oleoyl-sn-glycero-3-phosphocholine, 1-stearoyl-sn-glycero-3-phosphocholine, 1-palmitoyl-sn-glycero-3-phosphocholine, 1-myristoyl-sn-glycero-3-phosphocholine, 1-lauroyl-sn-glycero-3-phosphocholine and 1-decanoyl-sn-glycero-3-phosphocholine [42]) (Reversibility: ? <4, 9> [40, 42]) [40, 42]

P CoA + 1-oleoyl-2-acyl-sn-glycero-3-phosphocholine <4, 9> [40, 42]

S acyl-CoA + 1-palmitoyl-sn-glycero-3-phosphocholine <1, 7> (<7> no activity with 1-myristoyl-sn-glycero-3-phosphocholine [13]; <1> at 0.16 mM 1-palmitoyl-sn-glycero-3-phosphocholine is preferred 3.5fold over 1-stearoyl-sn-glycero-3-phosphocholine [25]) (Reversibility: r <1, 7> [13, 25]) [13, 25]

P CoA + 1-palmitoyl-2-acyl-sn-glycero-3-phosphocholine <1, 7> [13, 25]

S acyl-CoA + 1-stearoyl-sn-glycero-3-phosphocholine <7> (<7> no activity with 1-myristoyl-sn-glycero-3-phosphocholine [13]) (Reversibility: r <7> [13]) [13]

P CoA + 1-stearoyl-2-acyl-sn-glycero-3-phosphocholine <7> [13]

S arachidonoyl-CoA + 1-acyl-sn-glycero-3-phosphocholine <1, 4, 5, 7> (<7> preferred acyl donor [13]; <1> preferred acyl donor [23]; <5> arachidonoyl-CoA is the preferred acyl donor at 0.1 mM lysophosphatidylcholine concentrations in neuronal nuclei, 1-palmitoyl-sn-glycero-3-phosphocholine is preferred over 1-oleoyl-sn-glycero-3-phosphocholine as acyl acceptor [16]; <1> preferred acyl donors in descending order: arachidonoyl-CoA, linoleoyl-CoA, oleoyl-CoA, palmitoyl-CoA, stearyl-CoA, 1-palmitoyl-sn-glycero-3-phosphocholine is preferred over 1-stearoyl-sn-glycero-3-phosphocholine as acceptor [21]; <7> preferred acyl donors in descending order: arachidonoyl-CoA, oleoyl-CoA and palmitoyl-CoA [39]; <1> preferred acyl donors in descending order: arachidonoyl-CoA, linoleoyl-CoA, oleoyl-CoA, very little activity with palmitoyl-CoA, lauroyl-CoA, myristoyl-CoA, palmitolenoyl-CoA and stearoyl-CoA [23]; <4> peripheral and integral acyltransferase, specific for arachidonoyl-CoA and oleoyl-CoA [35]) (Reversibility: r <1, 4, 5, 7> [13, 16, 23, 35]) [13, 16, 21, 23, 35, 39]

P CoA + 1-acyl-2-arachidonoyl-sn-glycero-3-phosphocholine <1, 5, 7> [13, 16, 21, 23, 35, 39]

S arachidonoyl-CoA + 1-acyl-sn-glycero-3-phosphoethanolamine <1> (<1> 57% and 6% of activity with 1-acyl-sn-glycero-3-phosphocholine in microsomes and phosphatidylcholine vesicles respectively [37]) (Reversibility: ? <1> [26]) [26, 37]

P CoA + 1-acyl-2-arachidonoyl-sn-glycero-3-phosphoethanolamine <1> [26, 37]

S arachidonoyl-CoA + 1-acyl-sn-glycero-3-phosphoinositol <1, 7> (<7> preferred acyl donor arachidonoyl-CoA, other acyl donors: cis,cis-11,14-eicosadienoyl-CoA, cis-11-eicosaenoyl-CoA, linolenoyl-CoA, linoleoyl-CoA, oleoyl-CoA, stearoyl-CoA, palmitoleoyl-CoA, palmitoyl-CoA, dodecanoyl-CoA and decanoyl-CoA, no activity with 1-acyl-sn-glycero-3-phosphocholine [15]; <1> 35% and 8% of activity with 1-acyl-sn-glycero-3-phosphocholine in microsomes and phosphatidylcholine vesicles respectively [37]) (Reversibility: ? <1, 7> [15, 26]) [15, 26, 37]

P CoA + 1-acyl-2-arachidonoyl-sn-glycero-3-phosphoinositol <1, 7> [15, 26, 37]

S arachidonoyl-CoA + 1-acyl-sn-glycero-3-phosphoserine <1> (<1> 99% and 22% of activity with 1-acyl-sn-glycero-3-phosphocholine in microsomes and phosphatidylcholine vesicles respectively [37]) (Reversibility: ? <1> [37]) [26, 37]

P CoA + 1-acyl-2-arachidonoyl-sn-glycero-3-phosphoserine <1> [37]

S cis-11-eicosaenoyl-CoA + 1-acyl-sn-glycero-3-phosphocholine <7> (<7> 13% of activity with arachidonoyl-CoA [13]) (Reversibility: r <7> [13]) [13]

P CoA + 1-acyl-2-cis-11-eicosaenoyl-sn-glycero-3-phosphocholine <7> [13]

S decanoyl-CoA + 1-acyl-sn-glycero-3-phosphocholine <7> (<7> 58% of activity with arachidonoyl-CoA [13]) (Reversibility: r <7> [13]) [13]

P CoA + 1-acyl-2-decanoyl-sn-glycero-3-phosphocholine <7> [13]

S decanoyl-CoA + 1-palmitoyl-sn-glycero-3-phosphocholine <7> (<7> highest activity with 1-palmitoyl-sn-glycero-3-phosphocholine as acceptor, preferred acyl donors in descending order: decanoyl-CoA, oleoyl-CoA, arachidonoyl-CoA, stearoyl-CoA, linolenoyl-CoA, dodecanoyl-CoA, palmitoyl-CoA [14]) (Reversibility: ? <7> [14]) [14]

P CoA + 1-palmitoyl-2-decanoyl-sn-glycero-3-phosphocholine <7> [14]

S eicosatrienoyl-CoA + 1-acyl-sn-glycero-3-phosphocholine <7> (<7> 20% of activity with arachidonoyl-CoA [13]) (Reversibility: r <7> [13]) [13]

P CoA + 1-acyl-2-eicosatrienoyl-sn-glycero-3-phosphocholine <7> [13]

S linolenoyl-CoA + 1-acyl-sn-glycero-3-phosphocholine <7> (<7> 8% of activity with arachidonoyl-CoA [13]) (Reversibility: r <7> [13]) [13]

P CoA + 1-acyl-2-linolenoyl-sn-glycero-3-phosphocholine <7> [13]

S linoleoyl-CoA + 1-acyl-sn-glycero-3-phosphocholine <7> (Reversibility: r <7> [13]) [13]
P CoA + 1-acyl-2-linoleoyl-sn-glycero-3-phosphocholine <7> [13]
S linoleoyl-CoA + 1-acyl-sn-glycero-3-phosphocholine <1, 4, 6, 7> (<6> only acyl donor of the mitochondrial enzyme, 1-acyl-sn-glycero-3-phosphocholine and 1-alkenyl-sn-glycero-3-phosphocholine can act as acyl acceptors [18]) (Reversibility: r <1, 4, 6, 7> [7, 18]) [7, 18]
P CoA + 1-acyl-2-linoleoyl-sn-glycero-3-phosphocholine <1, 4, 6, 7> [7, 18]
S myristoyl-CoA + 1-acyl-sn-glycero-3-phosphocholine <1, 4, 6, 7> (<7> 41% of activity with arachidonoyl-CoA [13]) (Reversibility: r <1, 4, 6, 7> [7, 13]) [7, 13]
P CoA + 1-acyl-2-myristoyl-sn-glycero-3-phosphocholine <1, 4, 6, 7> [7, 13]
S oleoyl-CoA + 1-acyl-sn-glycero-3-phosphocholine <1, 2, 4, 6, 7> (<7> best acyl donor [14]) (Reversibility: r <1, 2, 4, 6, 7> [6, 7, 9, 13, 14]) [6, 7, 9, 13, 14]
P CoA + 1-acyl-2-oleoyl-sn-glycero-3-phosphocholine <1, 2, 4, 6, 7> [6, 7, 9, 13, 14]
S oleoyl-CoA + 1-acyl-sn-glycero-3-phosphoinositol <2, 7> (<7> preferred acyl donors in descending order: oleoyl-CoA, palmitoyl-CoA and arachidonoyl-CoA [39]) (Reversibility: ? <2, 7> [9, 39]) [9, 39]
P CoA + 1-acyl-2-oleoyl-sn-glycero-3-phosphoinositol <2, 7> [9, 39]
S oleoyl-CoA + 1-azido-oleoyl-sn-lysophosphatidyl-(N-methyl)-choline <14> (Reversibility: ? <14> [44]) [44]
P CoA + 1-azido-oleoyl-2-oleoyl-sn-lysophosphatidyl-(N-methyl)-choline <14> [44]
S palmitoyl-CoA + 1-acyl-sn-glycero-3-phosphocholine <1, 2, 4, 6, 7> (Reversibility: r <1, 2, 4, 6, 7> [6, 7]) [6, 7]
P CoA + 1-acyl-2-palmitoyl-sn-glycero-3-phosphocholine <1, 2, 4, 6, 7> [6, 7]
S stearoyl-CoA + 1-acyl-sn-glycero-3-phosphocholine <1, 2, 4, 6, 7> (Reversibility: r <1, 2, 4, 6, 7> [6, 7]) [6, 7]
P CoA + 1-acyl-2-stearoyl-sn-glycero-3-phosphocholine <2> [6]

Inhibitors

1-acyl-sn-glycero-3-phosphocholine <1, 5> (<1> high concentration, lung enzyme [10]) [10, 12]
1-alkenylglycerophosphocholine <6> [18]
1-azido-oleoyl-sn-lysophosphatidyl-N-methyl <14> (<14> 0.2 mM, 50% inactivation after photolysis [44]) [44]
1-decyl propanediol-3-phosphorylcholine <7> [8]
1-dodecyl propanediol-3-phosphorylcholine <7> [8]
1-hexadecyl propanediol-3-phosphorylcholine <7> [8]
1-octadecyl propanediol-3-phosphorylcholine <7> [8]
1-tetradecyl propanediol-3-phosphorylcholine <7> [8]

18-(4'-azido-2-2'-hydroxybenzoylamino)-oleic acid <12> (<12> 0.4 mM; 73% inhibition, reversible inhibition in the dark, irreversible inactivation after photolysis, used as photoaffinity label [41]) [41]
ADP <7> [15]
AMP <7> [15]
Ag^+ <13> (<13> 0.000005 mM, almost complete inhibition, noncompetitive, irreversible, probably due to formation of a thiolate [46]) [46]
Berol-043 <7> (<7> 0.25%, 90% inhibition [8]) [8]
Brij 35 <1> (<1> 0.1% [26]) [26]
Brij 98 <1> (<1> 0.1% [26]) [26]
Ca^{2+} <10> (<6> 10 mM, weak inhibition [18]) [10, 18]
CoA <5> [12]
Cu^{2+} <13> (<13> 0.00001 mM, 50% inhibition, noncompetitive, irreversible, probably due to formation of a thiolate [46]) [46]
Hg^{2+} <13> (<13> 0.00001 mM, 50% inhibition, noncompetitive, irreversible, probably due to formation of a thiolate [46]) [46]
Lubrol WX <1> (<1> 0.1% [26]) [26]
Mg^{2+} <10> [10]
N-ethylmaleimide <1> (<1> liver enzyme [10]; <1> 1 mM, 30% inhibition after 5 min, 95% inhibition after 45 min [26]) [10, 26]
Nonidet P-40 <7> (<7> 0.25%, 90% inhibition [8]) [8]
Pb^{2+} <13> (<13> 0.00002 mM, 50% inhibition, noncompetitive, irreversible, probably due to formation of a thiolate [46]) [46]
Triton X-100 <1, 7> (<1> 0.1% [26]; <7> 0.05% [8]; <1> 0.5 mM, 40% inhibition, 2 mM, 92% inhibition [23]) [8, 23, 26]
Tween 20 <1> (<1> 0.1% [26]) [26]
Tween 80 <1> (<1> 0.1% [26]) [26]
WY-14643 <1> (<1> inhibition of mitochondrial enzyme [10]) [10]
Zn^{2+} <9> (<9> 0.1 mM, 67% inhibition, 1 mM, 93% inhibition [40]) [40]
acyl-CoA <1> [21]
arachidonic acid <4> (<4> 0.045 mM, 50% inhibition [42]) [42]
arachidonoyl-CoA <7> (<7> high concentration [13]) [13]
cholate <7> (<7> 1%, 90% inhibition [8]) [8]
clofibric acid <1> (<1> inhibition of mitochondrial enzyme [10]) [10]
cyclic nucleotide phosphodiesterase inhibitors <1, 5> (<1> liver and heart enzyme [10]; <5> gastric mucosa microsomes [10]) [10]
deoxycholate <7> (<7> 0.25%, 90% inhibition [8]) [8]
dibucaine <7> [15]
dipalmitoylphosphatidylcholine <5> [12]
ethanol <4> (<4> integral membrane enzyme, 10%, 20% inhibition [42]) [42]
glycocholate <7> (<7> 0.2%, 90% inhibition [8]) [8]
iodoacetate <1, 5-7, 10> (<1> liver enzyme [10]) [10]
lysolecithin <5> [12]
lysolecithin analogs <7> (<7> above 0.1 mM, C_{10}-C_{16}, competitive inhibition [8]) [8]
methanol <1> (<1> oleoyl-CoA as substrate [26]) [26]

methyllidocaine <10> (<10> inhibition of heart enzyme [10]) [10]
o-phenanthroline <9> (<9> 2 mM, 82% inhibition, 1 mM, 73% inhibition [40]) [40]
oleic acid <4> (<4> 0.08 mM, 50% inhibition [42]) [42]
p-chloromercuribenzoate <1> (<1> partially restorable by dithiothreitol [26]) [26]
p-chloromercuriphenyl sulfonic acid <1> (<1> liver enzyme [10]; <1> partially restorable by dithitothreitol [26]) [10, 26]
palmitic acid <4> (<4> 0.15 mM, 50% inhibition [42]) [42]
palmitoyl-CoA <4> (<4> 0.2 mM, 50% inhibition [42]) [42]
palmitoyltrifluoromethylketone <4> (<4> 0.15 mM, 50% inhibition [42]) [42]
phosphorylhomocholine <1, 3> [7]
photooxidized phospholipids <1> (<1> 0.0001-0.01 mM photoxoidized phospholipids preincubated with 0.1 mg microsomal membranes, approx. 45% loss of initial activity [34]) [34]
procaine <7> [15]
pronase <1, 2> [28]
tetracaine <7> [15]
trypsin <1, 2, 7> (<7> inclusion in liposomes protects [13]) [13, 28]
Additional information <1, 4-7> (<1, 4, 6, 7> not inhibited by diisopropylfluorophosphate [17]; <1> not inhibited by IAA and DTNB [26]; <5> not inhibited by 0.08-0.8 mM palmitoyl-CoA [12]; <7> not inhibited by ATP [15]; <1> activity is reduced in rigid membranes obtained by incorporation of sphingomyelin and dipalmitoylphosphatidylcholine [1]) [1, 12, 15, 17, 26]

Activating compounds

ATP <1> [1]
Ca^{2+}-ionophore A23187 <4> (<4> 0.3 units/ml, approx. 5fold activation in endothelial cells in the presence of low-density lipoprotein [33]) [33]
$MgCl_2$ <1> [1]
cardiolipin <9> (<9> 236% activation [40]) [40]
chloroquine <7> (<7> activation [15]) [15]
dioleoylphosphatidylcholine <1> (<1> incorporation in membrane elevates acyltransferase activity [30]) [30]
egg yolk phosphatidylcholine <1> (<1> incorporation in membrane elevates acyltransferase activity [30]) [30]
ethanol <4> (<4> integral membrane enzyme, 2% (v/v), 30% increase in activity [42]) [42]
iodoacetate <7> (<7> activation [15]) [15]
low-density lipoprotein <4> (<4> 1 mg/ml, approx. 10fold activation in endothelial cells, 7fold activation in the presence of both ionomycin and 1-oleoyl-2-acetyl glycerol [33]) [33]
methanol <1> (<1> activation, arachidonoyl-CoA as substrate [26]) [26]
phosphatidic acid <9> (<9> 182% activation [40]) [40]
phosphatidylethanolamine <1> (<1> incorporation in membrane elevates acyltransferase activity [30]) [30]

phosphatidylinositol <1, 9> (<1> incorporation in membrane elevates acyltransferase activity [30]; <9> 212% activation [40]) [30, 40]
phosphatidylserine <1> (<1> incorporation in membrane elevates acyltransferase activity [30]) [30]
thimerosal <7> (<7> i.e. sodium ethylmercurithiosalicylate, activation [15]) [15]
thrombin <4> (<4> 0.3 units/ml, approx. 5fold activation in endothelial cells in the presence of low-density lipoprotein [33]) [33]

Specific activity (U/mg)

0.0001 <12> (<12> activity in freshly prepared erythrocyte membranes with palmitoyl-CoA as acyl donor [45]) [45]
0.0006 <12> (<12> activity in stored erythrocyte membranes with palmitoyl-CoA as acyl donor [45]) [45]
0.0007 <12> (<12> activity in freshly prepared erythrocyte membranes with oleoyl-CoA as acyl donor [45]) [45]
0.0008 <12> (<12> activity in stored erythrocyte membranes with oleoyl-CoA as acyl donor [45]) [45]
0.002 <8> (<8> at 5°C [2]) [2]
0.0028 <1> (<1> crude extract, TLC assay [20]) [20]
0.0029 <1> (<1> crude extract, silic acid microcolumn assay [20]) [20]
0.0038 <8> (<8> at 20°C [2]) [2]
0.01 <1> (<1> partially purified enzyme, 1-palmitoyl-glycero-3-phosphocholine as substrate [37]) [37]
0.013 <1> (<1> acyl donor oleoyl-CoA, heart enzyme [10]) [10]
0.028 <7> [8]
0.042 <4> (<4> integral membrane enzyme, 1-palmitoyl-glycero-3-phosphocholine as substrate [42]) [42]
0.058 <1> (<1> acyl donor oleoyl-CoA, liver enzyme [10]) [10]
0.06 <14> (<14> 1-azido-oleoyl-sn-lysophosphatidyl-N-methyl as substrate [44]) [44]
0.079 <7> (<7> acyl donor oleoyl-CoA [13]) [13]
0.134 <7> (<7> acyl donor linoleoyl-CoA [13]) [13]
0.393 <1> [32]
0.61 <7> (<7> acyl donor arachidonoyl-CoA [13]) [13]
0.68 <1> (<1> acyl donor arachidonoyl-CoA [27]) [27]
2.3 <1> (<1> acyl donor arachidonoyl-CoA [26]) [26]
4.125 <7> (<7> 1-acylglycerophosphoinositol O-acyltransferase [15]) [15]
9-10 <1> [5]
13.33 <7> (<7> acyl donor oleoyl-CoA [14]) [14]
Additional information <1> (<1> activity is different in either its microsomal environment or vesicles and is affected by the phospholipid molecular species environment [37]) [37]

K_m-Value (mM)

0.00105-0.0057 <10> (acyl-CoA, <10> saturated and unsaturated acyl donors, human platelets [10]) [10]
0.0012 <8> (oleoyl-CoA, <8> at 20°C [2]) [2]

0.0017 <11> (1-acyl-sn-glycero-3-phosphocholine) [36]
0.0027 <1> (palmitoyl-CoA) [26]
0.0028 <1> (oleoyl-CoA) [26]
0.0032 <1> (arachidonoyl-CoA) [26]
0.0035 <1> (linoleoyl-CoA) [26]
0.004 <2> (oleoyl-CoA, <2> acyl acceptor 1-acyl-sn-glycero-3-phosphoinositol [9]) [9]
0.0065 <7> (palmitoyl-glycero-3-phosphocholine) [8]
0.0085 <5> (palmitoyl-CoA) [10, 12]
0.0091 <8> (oleoyl-CoA, <8> at 5°C [2]) [2]
0.011 <2> (oleoyl-CoA, <2> acyl acceptor 1-acyl-sn-glycero-3-phosphocholine [9]) [9]
0.012 <8> (1-acyl-glycero-3-phosphocholine, <8> at 20°C [2]) [2]
0.0124 <6> (1-acyl-sn-glycero-3-phosphocholine) [18]
0.0124 <6> (1-acyl-sn-glycero-3-phosphocholine, <6> enzyme from heart mitochondria [18]) [18]
0.0138 <2> (1-acyl-sn-glycero-3-phosphoinositol) [9]
0.026 <13> (oleoyl-CoA) [46]
0.027 <7> (1-acyl-sn-glycero-3-phosphocholine) [39]
0.032 <13> (1-acyl-sn-glycero-3-phosphocholine) [46]
0.058 <8> (1-acyl-glycero-3-phosphocholine, <8> at 5°C [2]) [2]
0.061 <5> (1-acyl-sn-glycero-3-phosphocholine) [10, 12]
0.069 <9> (1-acyl-sn-glycero-3-phosphocholine) [40]
0.092 <6> (1-alkenyl-sn-glycero-3-phosphocholine, <6> enzyme from heart mitochondria [18]) [18]
0.107 <2> (1-acyl-sn-glycero-3-phosphocholine) [9]
0.108 <7> (oleoyl-CoA) [39]
0.152 <9> (oleoyl-CoA) [40]
Additional information <1, 4-7> (<7> kinetic data of soluble and liposome-bound enzyme [13]; <7> kinetic study [14]) [12-14, 17]

K_i-Value (mM)
0.08 <7> (1-hexadecyl propanediol-3-phosphorylcholine) [8]
0.27 <7> (1-octadecyl propanediol-3-phosphorylcholine) [8]
0.275 <7> (1-tetradecyl propanediol-3-phosphorylcholine) [8]
0.375 <7> (1-dodecyl propanediol-3-phosphorylcholine) [8]
0.93 <5> (lysolecithin) [12]
1.4 <7> (1-decyl propanediol-3-phosphorylcholine) [8]

pH-Optimum
7 <1, 6, 7> (<1> transfer of oleoyl-CoA [10,22]; <1> transfer of arachidonoyl-CoA is more rapid at a higher pH, liver enzyme [10]; <6> 1-acyl-sn-glycero-3-phosphocholine and 1-alkenyl-sn-glycero-3-phosphocholine acyltransferase activities [18]) [10, 18, 22, 39]
7-8 <1, 4-7, 10> (<10> value depends on acyl donor [10]) [10, 17]
7-9 <9> [31]
8 <1> [26]

pH-Range

5-10 <6> (<6> approx. half-maximal activity at pH 5.0 and 10.0 [18]) [18]
5.5-8.5 <1> (<1> approx. half-maximal activity at pH 5.5 and 8.5 [22]) [22]
6-9 <1, 7> (<1> approx. half-maximal activity at pH 6.0 and 9.6 [26]; <7> almost no activity at pH 6.0 and pH 9.0 [39]) [26, 39]

Temperature optimum (°C)

5 <8> (<8> assay at [2]) [2]
20 <5, 8> (<5,8> assay at [2,12]) [2, 12]
25 <1, 5-7, 10> (<1,5-7,10> assay at [10,18,26]) [10, 18, 26]
30 <1> (<1> assay at [24]) [24, 26, 30]
37 <1, 2, 7> [8, 9, 13-15, 21, 22, 25, 27-29]

4 Enzyme Structure

Molecular weight

225000 <1> (<1> gel filtration [26]) [26]

Subunits

? <1, 5-7, 10> (<7> x * 43000, brain enzyme, SDS-PAGE [10,13,38]; <7> x * 64000, heart enzyme, SDS-PAGE [10,14,38]) [10, 13, 14]
Additional information <1, 14> (<1> one major protein of 21000 Da after purification of the enzyme from vesicles [37]; <14> 2 proteins of 54000 Da and 61000 Da after photoaffinity labeling [44]) [37, 44]

5 Isolation/Preparation/Mutation/Application

Source/tissue

blood <12> (<12> erythrocyte, probably 2 distinct enzymes [45]) [45]
brain <1, 2, 5, 7> (<5> cortex [16]) [9, 10, 13, 16, 34]
carcinoma cell <2> (<2> Ehrlich ascites [6]) [6]
cell culture <4> (<4> endothelial cells [33]) [33]
gastric mucosa <5> [10]
heart <1, 4, 5, 6, 7> (<7> myocardium [14]) [11, 14, 15, 17, 18, 20]
intestine <1> [7]
leaf <13> [43]
liver <1, 2, 4-8, 10> [1, 2, 5, 10, 17, 19, 22-30, 32, 37]
lung <1, 2, 5> (<2> adenoma [28]) [3, 4, 10, 12, 21, 28]
placenta <12> [41]
plasma <1> [10, 30]
platelet <1> [29]
retina <7> (<7> rod outer segment [39]) [39]
smooth muscle <10> [10]
spleen <4> [35, 42]
thymocyte <7> [8]
tuber <11> [36]

Localization

cytosol <5> [11]
endoplasmic reticulum <4> [33]
membrane <1, 2, 4, 5-7, 10, 12> (<1> plasma membrane [10,30]; <4> 2 distinct enzymes: a peripheral and a CHAPS insoluble integral membrane enzyme [35]) [6, 8, 10, 30, 42, 45]
microsome <1-10, 14> (<1,2> membrane-bound, at least a sizable portion of the active center of the enzyme is embedded in the membrane interior [28]) [1-5, 7, 9, 10, 13-17, 19, 21-29, 31, 32, 34, 37, 39, 44]
mitochondrion <6, 10, 11> (<11> probably associated with the outer membrane [36]) [10, 18, 36]
nucleus <5> (<5> membrane [16]) [16]
plastid <13> (<13> plastid membrane [43]) [43]

Purification

<1> (partially purified [24,26,27,32]; solubilized by detergent effect of substrates lysolecithin and oleoyl-CoA, ratio 1:1 [5]; solubilized with Triton X-100 [23,27,32]; Ultrogel AcA 34 [26]; <1> partially purified from vesicles [37]) [5, 23, 24, 26, 27, 32, 37]
<4> (partial purification with palmitoyl-CoA-agarose affinity chromatography [42]) [42]
<5> (partially purified [3,12]) [3, 12]
<7> (solubilization and selective extraction with lysolecithin analogs at low concentration [8]; solubilized with lysolecithin, reactive Blue-2-agarose, DEAE cellulose, Matrex Green A [13]; solubilized with CHAPS [14,15]; Cibacron blue agarose, DEAE cellulose, Matrex green A, 1-acylglycerophosphoinositol O-acyltransferase [15]; reactive blue 2-agarose, DEAE-Cellulose, Matrex green A [14]; Spharose 6B, Mono Q, severe loss of activity is partially avoided by incorporation of 20% glycerol in the elution buffer [38]) [8, 13-15, 38]
<12> (partially purified [41]) [41]
<13> (solubilization with CHAPS [43]) [43]
<14> (gel filtration in the presence of 7.25 M urea, trypsin treatment [44]) [44]

6 Stability

Temperature stability

37 <1> (<1> oleoyl-CoA specific activity loses 50% of activity after 90 min, arachidonoyl-CoA specific activity loses 45% of activity after 90 min [26]) [26]
43 <1> (<1> half-life: 12 min [23]) [23]
45 <1> (<1> 10% loss of activity after 1 min [26]) [26]
48 <1> (<1> half-life: 1 min, oleoyl-CoA specific activity [26]) [26]
50 <1> (<1> 50% loss of activity after 1 min [26]) [26]
52 <1> (<1> half-life: 1 min, arachidonoyl-CoA specific activity [26]) [26]

55 <6> (<6> 70% loss of activity after 3 min, $t_{1/2}$ about 2 min [18]) [18]
Additional information <5> (<5> thermal stability of microsomal preparation [4]) [4]

General stability information

<1>, glycerol, 50%, stabilizes [26]
<1>, low ionic strength stabilizes during purification [26]

Storage stability

<1>, -20°C, multiple freeze-thawing cycles, 6 weeks, 20% loss of activity [26]
<1>, -20°C, partially purified enzyme, several months [26]
<1>, -20°C, purified enzyme, at least 3 weeks [26]
<5>, -60°C, 2 months [12]
<7>, -20°C, at least 2 weeks [8]
<7>, -20°C, liposome-bound enzyme, at least 10 h [13]
<7>, -20°C, soluble enzyme, less than 10 h [13]
<7>, -70°C, several weeks, no loss of activity [38]

References

[1] Koshlukova, S.; Momchilova-Pankova, A.; Markovska, T.; Kumanov, K.: Phospholipid dependence of rat liver microsomal acyl:CoA synthetase and acyl-CoA:1-acyl-sn-glycero-3-phosphocholine O-acyltransferase. J. Membr. Biol., **127**, 113-119 (1992)

[2] Livermore, R.C.; Hazel, J.R.: Properties of microsomal acyl-CoA:lysophosphatidylcholine acyltransferase from liver of thermally acclimated rainbow trout, Salmo gairdneri. J. Comp. Physiol. B, **158**, 363-368 (1988)

[3] Estrada, P.; Acebal, C.; Arche, R.: Substrate selectivity of acyl-CoA:lysolecithin acyltransferase from rabbit lung. Mol. Cell. Biochem., **69**, 49-54 (1985)

[4] Estrada, P.; Acebal, C.; Bauluz, C.; Casals, C.; Arche, R.: Influence of temperature on stability and activity of lysolecithin acyltransferase and acyl-CoA hydrolase from rabbit lung. Biochem. Int., **8**, 339-346 (1984)

[5] Gavino, V.C.; Deamer, D.W.: Purification of acyl CoA:1-acyl-sn-glycero-3-phosphorylcholine acyltransferase. J. Bioenerg. Biomembr., **14**, 513-526 (1982)

[6] Waku, K.; Nakazawa, Y.: Regulation of the fatty acid composition of alkyl ether phospholipid in Ehrlich ascites tumor cells. The substrate specificities of 1-O-alkylglycerol 3-phosphate and 1-O-alkylglycero-3-phosphocholine acyltransferases. J. Biochem., **82**, 1779-1784 (1977)

[7] Lands, W.E.M.; Hart, P.: Metabolism of glycerolipids. VI. Specificities of acyl-CoA: phospholipid acyltransferases. J. Biol. Chem., **240**, 1905-1911 (1965)

[8] Weltzien, H.U.; Richter, G.; Ferber, E.: Detergent properties of water-soluble choline phosphatides. Selective solubilization of acyl-CoA:lysolecithin acyltransferase from thymocyte plasma membranes. J. Biol. Chem., **254**, 3652-3657 (1979)

[9] Wise, R.W.; Sun, G.Y.; MacQuarrie, R.: The kinetic properties of oleoyl-CoA:1-acyl-sn-glycero-3-phosphocholine O-acyltransferase from mouse-brain microsomes. Eur. J. Biochem., **109**, 201-206 (1980)
[10] Choy, P.C.; Tardi, P.G.; Mukherjee, J.J.: Lysophosphatidylcholine acyltransferase. Methods Enzymol., **209**, 80-86 (1992)
[11] Needleman, P.; Wyche, A.; Sprecher, H.; Elliott, W.J.; Evers, A.: A unique cardiac cytosolic acyltransferase with preferential selectivity for fatty acids that form cyclooxygenase/lipoxygenase metabolites and reverse essential fatty acid deficiency. Biochim. Biophys. Acta, **836**, 267-273 (1985)
[12] Arche, R.; Estrada, P.; Acebal, C.: The kinetic mechanism of acyl-CoA:lysolecithin acyltransferase from rabbit lung. Arch. Biochem. Biophys., **257**, 131-139 (1987)
[13] Deka, N.; Sun, G.Y.; MacQuarrie, R.: Purification and properties of acyl-CoA:1-acyl-sn-glycero-3-phosphocholine-O-acyltransferase from bovine brain microsomes. Arch. Biochem. Biophys., **246**, 554-563 (1986)
[14] Sanjanwala, M.; Sun, G.Y.; Cutrera, M.A.; MacQuarrie, R.: Acylation of lysophosphatidylcholine in bovine heart muscle microsomes: purification and kinetic properties of acyl-CoA: 1-acyl-sn-glycero-3-phosphocholine O-acyltransferase. Arch. Biochem. Biophys., **265**, 476-483 (1988)
[15] Sanjanwala, M.; Sun, G.Y.; MacQuarrie, R.: Purification and kinetic properties of lysophosphatidylinositol acyltransferase from bovine heart muscle microsomes and comparison with lysophosphatidylcholine acyltransferase. Arch. Biochem. Biophys., **271**, 407-413 (1989)
[16] Baker, R.R.; Chang, H.Y.: A comparison of lysophosphatidylcholine acyltransferase activities in neuronal nuclei and microsomes isolated from immature rabbit cerebral cortex. Biochim. Biophys. Acta, **666**, 223-229 (1981)
[17] Lands, W.E.M.; Hart, P.: Metabolism of glycerolipids. VI. Specificities of acyl-CoA: phospholipid acyltransferases. J. Biol. Chem., **240**, 1905-1911 (1965)
[18] Arthur, G.; Page, L.L.; Zaborniak, C.L.; Choy, P.C.: The acylation of lysophosphoradylglycerocholines in guinea-pig heart mitochondria. Biochem. J., **242**, 171-175 (1987)
[19] Hill, E.E.; Lands, W.E.M.: Incorporation of long-chain and polyunsaturated acids into phosphatidate and phosphatidylcholine. Biochim. Biophys. Acta, **152**, 645-648 (1968)
[20] Bouroudian, M.; Chautan, M.; Termine, E.; Lafont, H.; Nalbone, G.: Use of a silicic acid microcolumn to assay acyl-CoA: lysophosphatidylcholine acyltransferase. Biochim. Biophys. Acta, **960**, 253-256 (1988)
[21] Holub, B.J.; Piekarski, J.; Possmayer, F.: Relative suitability of 1-palmitoyl and 1-stearoyl homologues of 1-acyl-sn-glycerylphosphorylcholine and different acyl donors for phosphatidylcholine synthesis via acyl-CoA:1-acyl-sn-glycero-3-phosphorylcholine acyltransferase in rat lung microsomes. Can. J. Biochem., **58**, 434-439 (1980)
[22] Van den Bosch, H.; Van Golde, L.M.G.; Eibl, H.; Van Deenen, L.L.M.: The acylation of 1-acylglycero-3-phosphorylcholines by rat-liver microsomes. Biochim. Biophys. Acta, **144**, 613-623 (1967)

[23] Yamashita, S.; Nakaya, N.; Miki, Y.; Numa, S.: Separation of 1-acylglycerolphosphate acyltransferase and 1-acylglycerolphosphorylcholine acyltransferase of rat liver microsomes. Proc. Natl. Acad. Sci. USA, **72**, 600-603 (1975)
[24] Miki, Y.; Hosaka, K.; Yamashita, S.; Handa, H.; Numa, S.: Acyl-acceptor specificities of 1-acylglycerolphosphate acyltransferase and 1-acylglycerophosphorylcholine acyltransferase resolved from rat liver microsomes. Eur. J. Biochem., **81**, 433-441 (1977)
[25] Holub, B.J.; Macnaughton, J.A.; Piekarski, J.: Synthesis of 1-palmitoyl and 1-stearoyl phosphatidylcholines from mixtures of acyl acceptors via acyl-CoA:1-acyl-sn-glycero-3-phosphorylcholine acyltransferase in liver microsomes. Biochim. Biophys. Acta, **572**, 413-422 (1979)
[26] Hasegawa-Sasaki, H.; Ohno, K.: Extraction and partial purification of acyl-CoA:1-acyl-sn-glycero-3-phosphocholine acyltransferase from rat liver microsomes. Biochim. Biophys. Acta, **617**, 205-217 (1980)
[27] Yamashita, S.; Hosaka, K.; Miki, Y.; Numa, S.: Glycerolipid acyltransferases from rat liver: 1-acylglycerophosphate acyltransferase, 1-acylglycerophosphorylcholine acyltransferase, and diacylglycerol acyltransferase. Methods Enzymol., **71**, 528-536 (1981)
[28] Renooij, W.; Snyder, F.: Transmembrane orientation of palmitoyl-CoA: lysophosphatidylcholine acyltransferase in microsomes isolated from an alveolar type II cell adenoma and rat liver. Biochim. Biophys. Acta, **666**, 468-474 (1981)
[29] Iritani, N.; Ikeda, Y.; Kajitani, H.: Selectivities of 1-acylglycerophosphorylcholine acyltransferase and acyl-CoA synthetase for n-3 polyunsaturated fatty acids in platelets and liver microsomes. Biochim. Biophys. Acta, **793**, 416-422 (1984)
[30] Momchilova, A.B.; Markovska, T.T.; Koshlukova, S.E.; Koumanov, K.S.; Pankov, R.G.: Phospholipid modifications influence acyl-CoA:1-acyl-glycero-3-phosphocholine O-acyltransferase in rat liver plasma membranes. Biochem. Cell Biol., **69**, 643-648 (1991)
[31] Yamada, K.; Okuyama, H.; Endo, Y.; Ikezawa, H.: Acyltransferase systems involved in phospholipid metabolism in Saccharomyces cerevisiae. Arch. Biochem. Biophys., **183**, 281-289 (1977)
[32] Yamashita, S.; Hosaka, K.; Numa, S.: Acyl-donor specificities of partially purified 1-acylglycerophosphate acyltransferase, 2-acylglycerophosphate acyltransferase and 1-acylglycerophosphorylcholine acyltransferase from rat-liver microsomes. Eur. J. Biochem., **38**, 25-31 (1973)
[33] Ezaki, M.; Inoue, Y.; Tomita, I.; Tomita, T.: Enhancement of lysolecithin acyltransferase activity by LDL in thrombin-stimulated porcine-cultured endothelial cells. Biochim. Biophys. Acta, **1213**, 75-81 (1994)
[34] Alberghina, M.; Lupo, G.; Anfuso, C.D.; el Ghonemy, S.H.: Lipid peroxidation inhibits acyl-CoA:-1-acyl-sn-glycero-3-phosphocholine O-acyltransferase but not CTP: phosphocholine cytidylyltransferase activity in rat brain membranes. Neurochem. Int., **26**, 477-487 (1995)
[35] Kerkhoff, C.; Gehring, L.; Habben, K.; Resch, K.; Kaever, V.: Identification of two different lysophosphatidylcholine:acyl-CoA acyltransferases (LAT)

in pig spleen with putative distinct topological localization. Biochim. Biophys. Acta, **1302**, 249-256 (1996)
[36] Testet, E.; Bessoule, J.J.; Mongrand, S.; Guillot-Salomon, T.; Cantrel, C.; Cassagne, C.: Occurrence of an acyl-CoA:1-acylglycerophosphorylcholine acyltransferase in plant mitochondria. FEBS Lett., **399**, 87-91 (1996)
[37] Fyrst, H.; Pham, D.V.; Lubin, B.H.; Kuypers, F.A.: Formation of vesicles by the action of acyl-CoA:1-acyllsophosphatidylcholine acyltransferase from rat liver microsomes: optimal solubilization conditions and analysis of lipid composition and enzyme activity. Biochemistry, **35**, 2644-2650 (1996)
[38] Choy, P.C.; Skrzypczak, M.; Lee, D.; Jay, F.T.: Acyl-GPC and alkenyl/alkyl-GPC:acyl-CoA acyltransferases. Biochim. Biophys. Acta, **1348**, 124-133 (1997)
[39] Castagnet, P.I.; Giusto, N.M.: Acyl-CoA:lysophosphatidylcholine acyltransferase activity in bovine retina rod outer segments. Arch. Biochem. Biophys., **340**, 124-134 (1997)
[40] Richard, M.G.; McMaster, C.R.: Lysophosphatidylcholine acyltransferase activity in Saccharomyces cerevisiae: regulation by a high-affinity Zn^{2+} binding site. Lipids, **33**, 1229-1234 (1998)
[41] Gehring, L.; Haase, D.; Habben, K.; Kerkhoff, C.; Meyer, H.H.; Kaever, V.: Synthesis of an unsaturated fatty acid analog (18-(4'-azido-2'-hydroxybenzoylamino)-oleic acid) and its interaction with lysophosphatidylcholine: acyl-CoA-O-acyltransferase. J. Lipid Res., **39**, 1118-1126 (1998)
[42] Kerkhoff, C.; Habben, K.; Gehring, L.; Resch, K.; Kaever, V.: Substrate specificity of acyl-CoA:lysophospholipid acyltransferase (LAT) from pig spleen. Arch. Biochem. Biophys., **351**, 220-226 (1998)
[43] Akermoum, M.; Testet, E.; Cassagne, C.; Bessoule, J.J.: Complex lipid biosynthesis: phospholipid synthesis. Biochem. Soc. Trans., **28**, 713-715 (2000)
[44] Fraser, T.; Stobart, K.: Partial purification and photoaffinity labelling of sunflower acyl-CoA:lysophosphatidylcholine acyltransferase. Biochem. Soc. Trans., **28**, 715-718 (2000)
[45] Rusnak, A.; Coghlan, G.; Zelinski, T.; Hatch, G.M.: Incorporation of fatty acids into phosphatidylcholine is reduced during storage of human erythrocytes: evidence for distinct lysophosphatidylcholine acyltransferases. Mol. Cell. Biochem., **213**, 137-143 (2000)
[46] Akermoun, M.; Testet, E.; Cassagne, C.; Bessoule, J.J.: Inhibition of the plastidial phosphatidylcholine synthesis by silver, copper, lead and mercury induced by formation of mercaptides with the lyso-PC acyltransferase. Biochim. Biophys. Acta, **1581**, 21-28 (2002)

Sphingosine N-acyltransferase 2.3.1.24

1 Nomenclature

EC number
2.3.1.24

Systematic name
acyl-CoA:sphingosine N-acyltransferase

Recommended name
sphingosine N-acyltransferase

Synonyms
acyltransferase, sphingosine
ceramide synthase <2> [4]
ceramide synthetase
sphingosine acyltransferase

CAS registry number
37257-09-3

2 Source Organism

<1> *Gallus gallus* [1]
<2> *Rattus norvegicus* [1-4, 6, 7]
<3> *Cavia porcellus* [1]
<4> *Mus musculus* [5, 8, 9]
<5> *Homo sapiens* [9, 12]
<6> *Bos taurus* [10, 11]

3 Reaction and Specificity

Catalyzed reaction
acyl-CoA + sphingosine = CoA + N-acylsphingosine

Reaction type
acyl group transfer

Natural substrates and products
S acyl-CoA + sphingosine <1-4> (<1-3> enzyme catalyzes the biosynthesis of ceramides (N-acylsphingosines) [1,2]; <4> sphingolipid biosynthesis [5]) (Reversibility: ? <1-4> [1, 2, 5]) [1, 2, 5]
P CoA + N-acylsphingosine <1-4> [1, 2, 5]

Substrates and products

S acyl-CoA + sphinganine <2, 6> (<2> i.e. dihydrosphingosine [2]) (Reversibility: ? <2, 6> [2, 11]) [2, 11]

P CoA + N-acyl-sphinganine <2, 6> [2, 11]

S acyl-CoA + sphingosine <1-6> (<1-3> threo and erythro sphingosine [1,6]; <2> (2S,3R), (2R,3S), (2R,3R), (2S,3S)-sphingosine [6]; <1> palmitoyl-CoA [1]; <2> stearoyl-CoA [2,7]; <2> lignoceroyl-CoA [7]) (Reversibility: ? <1-6> [1-10]) [1-11]

P CoA + N-acylsphingosine <1-6> [1-11]

S behenoyl-CoA + sphingosine <6> (Reversibility: ? <6> [11]) [11]

P CoA + behenoylsphingosine <6> [11]

S hexanoyl-CoA + sphingosine <6> (Reversibility: ? <6> [11]) [11]

P CoA + hexanoylsphingosine <6> [11]

S lauroyl-CoA + sphingosine <6> (Reversibility: ? <6> [11]) [11]

P CoA + lauroylsphingosine <6> [11]

S oleoyl-CoA + sphingosine <6> (Reversibility: ? <6> [11]) [11]

P CoA + oleoylsphingosine <6> [11]

S palmitoyl-CoA + sphinganine <2> (Reversibility: ? <2> [3]) [3]

P CoA + dihydroceramide <2> [3]

S palmitoyl-CoA + sphingosine <6> (Reversibility: ? <6> [11]) [11]

P CoA + palmitoylsphingosine <6> [11]

S stearoyl-CoA + sphingosine <6> (Reversibility: ? <6> [11]) [11]

P CoA + stearoylsphingosine <6> [11]

Inhibitors

4,4'-diisothiocyanostilbene-2,2'-disulfonic acid <4> [5]

Ca^{2+} <1-3> [1]

fumonisin B1 <2, 4> (<2> mycotoxin produced by Fusarium moniliforme [2]; <2> 0.0001 mM, 50% inhibition [4]; <4> 0.000075 mM, more than 50% inhibition [8]) [2, 4, 8]

fumonisin B2 <2> (<2> mycotoxin produced by Fusarium moniliforme [2]) [2]

Activating compounds

L-α-phosphatidyl-L-serine <6> (<6> 6.7fold activation of solubilized enzyme [10]) [10]

L-α-phosphatidylcholine <6> (<6> 4.3fold activation of solubilized enzyme [10]) [10]

L-α-phosphatidylinositol <6> (<6> 5.3fold activation of solubilized enzyme [10]) [10]

N-(4-hydroxyphenyl)retinamide <5> (<5> induces enzyme activity in neuroblastoma cell line possibly via a posttranslational mechanism [12]) [12]

daunorubicin <4, 5> (<4,5> anthracycline used in the treatment of leukemia [9]; <4> 0.01 mM, treatment of cell cultures for 6 h, 70% increase of activity in microsomes [9]) [9]

Metals, ions

Mg^{2+} <1-3> (<1-3> stimulates [1]; <2> 1 mM, 1.5fold stimulation [3]) [1, 3]
Additional information <1-3> (<1-3> no absolute requirement for metal ions [1]) [1]

Specific activity (U/mg)

0.000049 <4> (<4> activity in endoplasmic reticulum vesicles from liver [5]) [5]
0.0148 <6> [11]

K_m-Value (mM)

0.003 <4> (stearoyl-CoA) [8]
0.035 <4> (sphinganine) [8]
0.141 <6> (palmitoyl-CoA) [11]
0.144 <6> (sphinganine) [11]
0.146 <6> (stearoyl-CoA) [11]
0.17 <6> (D-sphingosine) [10]
0.171 <6> (sphingosine) [11]
0.18 <6> (oleoyl-CoA) [11]
0.299 <6> (behenoyl-CoA) [11]
0.358 <6> (lauroyl-CoA) [11]
0.522 <6> (hexanoyl-CoA) [11]

pH-Optimum

7-7.8 <6> [10]
7.5 <1-3, 6> [1, 11]

pH-Range

5.5-9 <6> (<6> more than 70% of maximal activity between pH 7.0 and pH 8.0 [11]) [11]
6.5-8.5 <1-3> (<1-3> 40% of maximal activity at pH 6.5, 30% of maximal activity at pH 8.5 [1]) [1]

Temperature optimum (°C)

37 <2> (<2> assay at [2,7]) [2, 7]

4 Enzyme Structure

Molecular weight

240000-260000 <6> (<6> gel filtration [11]) [11]

Subunits

? <6> (<6> x * 62000 + x * 72000, SDS-PAGE [11]) [11]

5 Isolation/Preparation/Mutation/Application

Source/tissue

brain <2, 3, 4, 6> [1, 7, 8, 11]
cell culture <4, 5> (<4> neuron cell culture [8]; <4> P388 leukemia cells [9]; <5> U937 monoblastic leukemia cells [9]; <5> neuroblastoma cell line CHLA-90 [12]) [8, 9, 12]
hepatocyte <2> [2]
liver <1, 3, 4, 6> [1-7, 11]

Localization

endoplasmic reticulum cisterna <2, 4> (<2> localized at the cytosolic surface [3]) [3, 5]
microsome <1-6> [1, 2, 4, 7, 9, 10]
mitochondrion <6> [10]

Purification

<2> (liver and brain enzyme, NaSCN, Sepharose-CL-4B [7]) [7]
<6> (solubilization with 2% Triton X-100, ammonium sulfate [10]; n-octyl β-D-thioglucoside, DE-32, sphingosine-celite, Sepharose CL-6B [11]) [10, 11]

Renaturation

<6> (reconstitution of solubilized enzyme in membrane lipid liposomes [10]) [10]

6 Stability

General stability information

<2>, high concentrations of glycerol are essential in preventing deactivation of the enzyme [7]
<2>, information about stability of solubilized enzyme [7]

Storage stability

<2>, -20°C, several months [7]
<2>, 4°C, 48 h, liver enzyme loses 70% of activity, brain enzyme loses 41% of activity [7]

References

[1] Sribney, M.: Enzymatic synthesis of ceramide. Biochim. Biophys. Acta, **125**, 542-547 (1966)
[2] Merrill, A.H.; Wang, E.: Enzymes of ceramide biosynthesis. Methods Enzymol., **209**, 427-437 (1992)
[3] Hirschberg, K.; Rodger, J.; Futerman, A.H.: The long-chain sphingoid base of sphingolipids is acylated at the cytosolic surface of the endoplasmic reticulum in rat liver. Biochem. J., **290**, 751-757 (1993)

[4] Wang, E.; Norred, W.P.; Bacon, C.W.; Riley, R.T.; Merrill, A.H.: Inhibition of sphingolipid biosynthesis by fumonisins. Implications for diseases associated with Fusarium moniliforme. J. Biol. Chem., **266**, 14486-14490 (1991)
[5] Mandon, E.C.; Ehses, I.; Rother, J.; van Echten, G.; Sandhoff, K.: Subcellular localization and membrane topology of serine palmitoyltransferase, 3-dehydrosphinganine reductase, and sphinganine N-acyltransferase in mouse liver. J. Biol. Chem., **267**, 11144-11148 (1992)
[6] Stoffel, W.; Bister, K.: Stereospecificities in the metabolic reactions of the four isomeric sphinganines (dihydrosphingosines) in rat liver. Hoppe-Seyler's Z. Physiol. Chem., **354**, 169-181 (1973)
[7] Narimatsu, S.; Soeda, S.; Tanaka, T.; Kishimoto, Y.: Solubilization and partial characterization of fatty acyl-CoA: sphingosine acyltransferase (ceramide synthetase) from rat liver and brain. Biochim. Biophys. Acta, **877**, 334-341 (1986)
[8] Merrill, A.H., Jr.; van Echten, G.; Wang, E.; Sandhoff, K.: Fumonisin B1 inhibits sphingosine (sphinganine) N-acyltransferase and de novo sphingolipid biosynthesis in cultured neurons in situ. J. Biol. Chem., **268**, 27299-27306 (1993)
[9] Bose, R.; Verheji, M.; Haimovitz-Friedman, A.; Scotto, K.; Fuks, Z.; Kolesnick, R.: Ceramide synthase mediates daunorubicin-induced apoptosis: an alternative mechanism for generating death signals. Cell, **82**, 405-414 (1995)
[10] Shimeno, H.; Soeda, S.; Yasukouchi, M.; Okamura, N.; Nagamatsu, A.: Fatty acyl-CoA:sphingosine acyltransferase in bovine brain mitochondria: its solubilization and reconstitution onto the membrane lipid liposomes. Biol. Pharm. Bull., **18**, 1335-1339 (1995)
[11] Shimeno, H.; Soeda, S.; Sakamoto, M.; Kouchi, T.; Kowakame, T.; Kihara, T.: Partial purification and characterization of sphingosine N-acyltransferase (ceramide synthase) from bovine liver mitochondrion-rich fraction. Lipids, **33**, 601-605 (1998)
[12] Wang, H.; Maurer, B.J.; Reynolds, C.P.; Cabot, M.C.: N-(4-Hydroxyphenyl)-retinamide elevates ceramide in neuroblastoma cell lines by coordinate activation of serine palmitoyltransferase and ceramide synthase. Cancer Res., **61**, 5102-5105 (2001)

Plasmalogen synthase 2.3.1.25

1 Nomenclature

EC number
2.3.1.25

Systematic name
acyl-CoA:1-O-alk-1-enyl-glycero-3-phosphocholine 2-O-acyltransferase

Recommended name
plasmalogen synthase

Synonyms
1-alkenyl-GPC acyltransferase
1-alkenyl-glycero-3-phosphorylcholine:acyl-CoA acyltransferase
O-1-alkenylglycero-3-phosphorylcholine acyltransferase
lyso-PmeC ATase
lysophosphatidylcholine acyltransferase
lysoplasmenylcholine acyltransferase
synthase, plasmalogen

CAS registry number
37257-10-6

2 Source Organism

<1> *Homo sapiens* [1]
<2> *Oryctolagus cuniculus* [1]
<3> *Rattus norvegicus* (male Wistar [2]; Sprague Dawley [4]) [2, 4]
<4> *Cavia porcellus* (guinea pig [3]) [3]

3 Reaction and Specificity

Catalyzed reaction
acyl-CoA + 1-O-alk-1-enyl-glycero-3-phosphocholine = CoA + plasmenylcholine

Reaction type
acyl group transfer

Natural substrates and products

S arachidonoyl-CoA + 1-alkenyl-glycero-3-phosphorylcholine <1, 2, 3> (Reversibility: r <1, 2, 3> [1, 2, 4]) [1, 2, 4]
P CoA + 1-alkenyl-2-arachidonoyl-glycero-3-phosphorylcholine
S linoleoyl-CoA + 1-alkenyl-glycero-3-phosphorylcholine <1, 2, 3, 4> (Reversibility: r <1, 2, 3, 4> [1, 2, 3]) [1, 2, 3]
P CoA + 1-alkenyl-2-linoleoyl-glycero-3-phosphorylcholine <1, 2> [1]
S oleoyl-CoA + 1-alkenyl-glycero-3-phosphorylcholine <1, 2, 3> (Reversibility: r <1, 2, 3> [1, 2, 4]) [1, 2, 4]
P CoA + 1-alkenyl-2-oleoyl-glycero-3-phosphorylcholine

Substrates and products

S arachidonoyl-CoA + 1-alkenyl-glycero-3-phosphorylcholine <1, 2, 3> (Reversibility: r <1, 2, 3> [1, 2, 4]) [1, 2, 4]
P CoA + 1-alkenyl-2-arachidonoyl-glycero-3-phosphorylcholine
S linoleoyl-CoA + 1-alkenyl-glycero-3-phosphorylcholine <1, 2, 3, 4> (<4> heart enzyme is highly specific for linoleoyl-CoA [3]) (Reversibility: r <1, 2, 3, 4> [1, 2, 3]) [1, 2, 3]
P CoA + 1-alkenyl-2-linoleoyl-glycero-3-phosphorylcholine <1, 2> [1]
S oleoyl-CoA + 1-alkenyl-glycero-3-phosphorylcholine <1, 2, 3> (Reversibility: r <1, 2, 3> [1, 2, 4]) [1, 2, 4]
P CoA + 1-alkenyl-2-oleoyl-glycero-3-phosphorylcholine

Inhibitors

1-acyl-glycero-3-phosphorylcholine <4> (<4> non competitive [3]) [3]
Ca^{2+} <4> (<4> mitochondrial enzyme [3]) [3]

Activating compounds

Ca^{2+} <4> (<4> enhances activity of microsomal enzyme [3]) [3]
Mg^{2+} <4> (<4> enhances activity of microsomal enzyme [3]) [3]

Specific activity (U/mg)

Additional information <3> (<3> specific activity for various acyl-CoAs [2,4]) [2, 4]

K_m-Value (mM)

0.00077 <3> (oleoyl-CoA, <3> submandibular gland [4]) [4]
0.0008-0.0009 <3> (oleoyl-CoA) [4]
0.00086 <3> (oleoyl-CoA, <3> sublingual gland [4]) [4]
0.00091 <3> (oleoyl-CoA) [2]
0.0265 <3> (1-alkenyl-glycero-3-phosphorylcholine, <3> submandibular gland [4]) [4]
0.0267 <3> (1-alkenyl-glycero-3-phosphorylcholine, <3> sublingual gland [4]) [4]
0.0297 <3> (1-alkenyl-glycero-3-phosphorylcholine) [2]

pH-Optimum

6-7 <4> [3]
7.4 <1, 2> (<1,2> assay at [1]) [1]

Temperature optimum (°C)
20 <1, 2> (<1,2> assay at [1]) [1]

5 Isolation/Preparation/Mutation/Application

Source/tissue
erythrocyte <1> [1]
heart <4> [3]
muscle <2> [1]
parotid gland <3> [2]
sublingual gland <3> [4]
submandibular gland <3> [4]
testis <2> [1]
Additional information <2> (<2> no activity in rabbit brain and heart [1]) [1]

Localization
microsome <3, 4> [2, 3, 4]
mitochondrion <4> [3]
sarcoplasmic reticulum <2> (<2> of muscle [1]) [1]

Purification
<1> (partial [1]) [1]
<2> (partial [1]) [1]

6 Stability

Temperature stability
20 <1> (<1> 60 min stable [1]) [1]
55 <4> (<4> 90% loss of activity after 1 min [3]) [3]

References

[1] Waku, K.; Lands, W.E.M.: Acyl coenzyme A:1-alkenyl-glycero-3-phosphorylcholine acyltransferase action in plasmalogen biosynthesis. J. Biol. Chem., **243**, 2654-2659 (1968)
[2] Kameyama, Y.; Shinkai, A.: Characterization of 1-alkenyl-sn-glycero-3-phosphorylcholine acyltransferase activity in the microsomes from rat parotid gland. Shika Kiso Igakkai Zasshi, **42**, 319-325 (2000)
[3] Choy, P.C.; Skrzypczak, M.; Lee, D.; Jay, F.T.: Acyl-GPC and alkenyl/alkyl-GPC:acyl-CoA acyltransferases. Biochim. Biophys. Acta, **1348**, 124-133 (1997)
[4] Okada, A.: Composition of ether-linked glycerophospholipids and properties of microsomal lysoplasmenylcholine acyltransferase in rat submandibular and sublingual glands. Shika Kiso Igakkai Zasshi, **33**, 428-445 (1991)

Sterol O-acyltransferase 2.3.1.26

1 Nomenclature

EC number
2.3.1.26

Systematic name
acyl-CoA:cholesterol O-acyltransferase

Recommended name
sterol O-acyltransferase

Synonyms
ACAT
acyl coenzyme A-cholesterol-O-acyltransferase
acyl-CoA:cholesterol acyltransferase
acylcoenzyme A:cholesterol O-acyltransferase
acyltransferase, cholesterol
cholesterol acyltransferase
cholesterol ester synthase
cholesterol ester synthetase
cholesteryl ester synthetase
sterol-ester synthase
sterol-ester synthetase

CAS registry number
9027-63-8

2 Source Organism

<1> *Cavia porcellus* [4, 30]
<2> *Sus scrofa* [3, 4, 9, 23, 30]
<3> *Rattus norvegicus* (male Sprague-Dawley strain [7,8,12]; male Wistar strain [13,28]) [1, 3-5, 7, 8, 12, 13, 15, 16, 21, 28, 30, 31, 33]
<4> *Saccharomyces cerevisiae* [2, 3]
<5> *Homo sapiens* [3, 4, 6, 11, 19, 20, 22-24, 27, 29, 30, 36, 37]
<6> *Mus musculus* [4, 18, 25, 26, 30, 32]
<7> *Mesocricetus auratus* [4, 17, 20, 30]
<8> *monkey* [4, 30, 33, 35]
<9> *Oryctolagus cuniculus* [4, 5, 10, 15, 30]
<10> *Columba sp.* (pigeon [4,30]) [4, 30]
<11> *Heliothis zea* [3]

<12> *Gallus gallus* [14]
<13> *Bos taurus* [16]
<14> *Saccharomyces cerevisiae* [27]
<15> *Alligator mississippiensis* [34]

3 Reaction and Specificity

Catalyzed reaction

acyl-CoA + cholesterol = CoA + cholesterol ester (<9> mechanism, a histidine is implicated in the catalytic mechanism of the enzyme [10])

Reaction type

acyl group transfer

Natural substrates and products

S long-chain fatty acyl-CoA + cholesterol <1-3, 5-10> (<1-3,5-10> responsible for cellular synthesis of cholesterol esters in various cell types [4,5]; <1-3,5-10> may play an important role in regulation of the accumulation of cholesterol esters within smooth muscle cells of the artery wall during atherogenesis and in synthesis of cholesterol esters during hepatic very low-density lipoprotein synthesis and secretion [4]; <3> role in lipoprotein metabolism and atherogenesis [5]; <5> the high activity of intestinal enzyme renders it possible that the enzyme plays a role in cholesterol absorption [11]; <3> the enzyme plays an important role in maintenance of hepatic cholesterol homeostasis [12]; <5> the enzyme is relevant for cellular cholesterol esterification in vivo, the regulation in human mononuclear phagocytes indicates that the enzyme is also involved in foam cell formation during early atherogenesis [23]) (Reversibility: ? <1-3, 5-10> [4, 5, 11, 12, 23, 30, 33, 35]) [4, 5, 11, 12, 23, 30, 33, 35]

P CoA + cholesterol ester <1-3, 5-10> [4, 5, 11, 12, 23, 30, 33, 35]

S Additional information <8> (<8> two closely related enzymes identified that catalyse the esterification of cholesterol using acyl-CoA substrates, namely ACAT1 and ACAT2, the orientation of ACAT2 in the endoplasmic reticulum membrane, in addition to its expression only in liver and intestine, suggests that this enzyme may have as a primary function, the secretion of cholesteryl esters into apolipoprotein B-containing lipoproteins [33,35]) [33, 35]

P ?

Substrates and products

S acetyl-CoA + cholesterol <4> (<4> 16% of the activity with oleoyl-CoA [2]) (Reversibility: ? <4> [2]) [2]

P CoA + cholesteryl acetate <4> [2]

S acyl-CoA + cholesterol <1-13> (<3> the animal enzyme is highly specific for transfer of acyl groups with a single cis-double-bond 9 carbon atoms distant from the carboxyl group [1]; <3,4> acyl-CoA derivatives can be replaced by ATP, Mg^{2+}, CoA and fatty acids, in-situ production of acyl-CoA [1,2]) (Reversibility: ? <1-13> [1-18]) [1-18]

P CoA + cholesterol ester <1-13> (<3> esterification of 3β-hydroxy group [8]) [1-18]

S elaidoyl-CoA + cholesterol <3> (Reversibility: ? <3> [1]) [1]

P CoA + cholesteryl elaidate <3> [1]

S lauroyl-CoA + cholesterol <4> (<4> 62% of the activity with oleoyl-CoA [2]) (Reversibility: ? <4> [2]) [2]

P CoA + cholesteryl laurate <4> [2]

S linelaidoyl-CoA + cholesterol <3> (Reversibility: ? <3> [1]) [1]

P CoA + cholesteryl linelaidate <3> [1]

S linoleoyl-CoA + cholesterol <3> (Reversibility: ? <3> [1, 3, 4]) [1, 3, 4]

P CoA + cholesteryl linoleate <3> [1, 3, 4]

S myristoyl-CoA + cholesterol <4> (<4> 37% of the activity with oleoyl-CoA [2]) (Reversibility: ? <4> [2]) [2]

P CoA + cholesteryl myristate <4> [2]

S oleoyl-CoA + 4α-methylcholest-7-en-3β-ol <3> (Reversibility: ? <3> [4, 8]) [4, 8]

P CoA + 4α-methylcholest-7-en 3-oleate <3> [4, 8]

S oleoyl-CoA + 4α-methylcholest-7-en-3β-ol <3> (Reversibility: ? <3> [8]) [8]

P CoA + oleate ester of 4α-methylcholest-7-en-3β-ol <3> [8]

S oleoyl-CoA + 5α-cholest-8(14)-en-3β-ol-15-one <3> (Reversibility: ? <3> [7]) [7]

P CoA + cholestanyl oleate <3> [7]

S oleoyl-CoA + 5α-cholest-8(14)-en-3β-ol-15-one <3> (Reversibility: ? <3> [7]) [7]

P CoA + oleate ester of 5α-cholest-8(14)-en-3β-ol-15-one <3> [7]

S oleoyl-CoA + campesterol <3> (<3> 20% of the activity with cholesterol [8]; <3> campesterol is a poor substrate [4]) (Reversibility: ? <3> [4, 8]) [4, 8]

P CoA + campesteryl oleate <3> [4, 8]

S oleoyl-CoA + cholest-4-en-3β-ol <3> (<3> low activity [4,8]) (Reversibility: ? <3> [4, 8]) [4, 8]

P CoA + cholest-4-en 3-oleate <3> [4, 8]

S oleoyl-CoA + cholestanol <3> (<3> cholestanol is esterified by the enzyme at almost the same rate as cholesterol [4]) (Reversibility: ? <3> [4, 8]) [4, 8]

P CoA + cholestanyl oleate <3> [4, 8]

S oleoyl-CoA + cholesterol <2-6, 8, 9, 14, 15> (<3,4> highest activity [1-4,8]; <5> recombinant enzyme [27]; <5> the dimeric enzyme responds to cholesterol in essentially the same manner as the tetrameric enzyme [37]) (Reversibility: ? <2-6, 8, 9, 14, 15> [1-3, 6-8, 10-13, 15, 23-25, 27-29, 31, 34-37]) [1-4, 6-8, 10-13, 15, 23-25, 27-29, 31, 34-37]

P CoA + cholesteryl oleate <2-6, 8, 9, 14, 15> [1-4, 6-8, 10-13, 15, 23-25, 27-29, 31, 34-37]

S oleoyl-CoA + desmosterol <3> (<3> 62% of the activity with cholesterol [8]) (Reversibility: ? <3> [4, 8]) [4, 8]
P CoA + desmosteryl oleate <3> [4, 8]
S oleoyl-CoA + episterol <4> (<4> 62% of the activity with cholesterol [2]) (Reversibility: ? <4> [2]) [2]
P CoA + episteryl oleate <4> [2]
S oleoyl-CoA + ergosterol <4, 5, 14> (<4> 147% of the activity with cholesterol [2]; <14> endogenous ergosterol as substrate or exogenous ergosterol in the form of ergosterol/phosphatidylcholine vesicles as substrate [27]; <5> recombinant enzyme: restores ergosteryl oleate formation in vivo to only approximately 8% of that catalysed by yeast enzyme in wild-type cells [27]) (Reversibility: ? <4, 5, 14> [2, 3, 27]) [2, 3, 27]
P CoA + ergosteryl oleate <4, 14> [2, 3, 27]
S oleoyl-CoA + fecosterol <4> (<4> 48% of the activity with cholesterol [2]) (Reversibility: ? <4> [2]) [2]
P CoA + fecosteryl oleate <4> [2]
S oleoyl-CoA + lanosterol <4> (<4> 65% of the activity with cholesterol [2]) (Reversibility: ? <4> [2]) [2]
P CoA + lanosteryl oleate <4> [2]
S oleoyl-CoA + lathosterol <3> (<3> 41% of the activity with cholesterol [8]) (Reversibility: ? <3> [4, 8]) [4, 8]
P CoA + lathosteryl oleate <3> [4, 8]
S oleoyl-CoA + zymosterol <4> (<4> 93% of the activity with cholesterol [2]) (Reversibility: ? <4> [2]) [2]
P CoA + zymosteryl oleate <4> [2]
S palmitoleoyl-CoA + cholesterol <4> (<4> 91% of the activity with oleoyl-CoA [2]) (Reversibility: ? <4> [2]) [2]
P CoA + cholesteryl palmitoleate <4> [2]
S palmitoyl-CoA + cholesterol <3, 4, 6, 12, 13> (<4> 47% of the activity with oleoyl-CoA [2]; <3> approximately 50% of the activity with oleoyl-CoA [4]) (Reversibility: ? <3, 4, 6, 12, 13> [1-4, 14, 16, 18]) [1-4, 14, 16, 18]
P CoA + cholesteryl palmitate <3, 4, 6, 12, 13> [1-4, 14, 16, 18]
S stearoyl-CoA + cholesterol <3> (Reversibility: ? <3> [3, 4]) [3, 4]
P CoA + cholesteryl stearate <3> [3, 4]
S stearyl-CoA + cholesterol <3> (Reversibility: ? <3> [4]) [4]
P CoA + cholesteryl stearate <3> [4]
S Additional information <3-5> (<4> the enzyme has a broad substrate specificity for sterols and acyl-CoAs [2]; <3> a part of the substrate selectivity may be due to the presence of acyl-CoA hydrolase in the microsomal preparations. For this enzyme, the preferred substrate is palmitoyl-CoA which is hydrolysed about 8times faster than oleoyl-CoA [3]; <3> investigations into the sterol specificity of the enzyme [3]; <3> structural requirements of sterol substrate [8]; <3> 3-epicholesterol is not a substrate, indicating a requirement for a 3β-hydroxyl group [4,8]; <3> sitosterol, stigmasterol, ergosterol, lanosterol, cycloartenol are not substrates [4,8]; <3> campesterol is a poor substrate, indicating the alkyl side chain

plays an important role in the enzyme reaction [4,8]; <5> oleic acid ist not incorporated into cholesterol esters by placental microsomes even in the presence of ATP, CoA and $MgCl_2$ [6]; <3> maximal ester formation is obtained when the longest chain on C-20 has five carbons and either an increase or decrease in the number of carbons reduce the amount of ester formed [8]; <5> enzyme activity is regulated by the acyl-CoA binding protein ACBP [29]) [2-4, 6, 8, 29]

P ?

Inhibitors

17α-hydroxyprogesterone <5> [6]
20-methylcholesterol <3> (<3> inhibits esterification of endogenous cholesterol [3]) [3]
20α-hydroxypregn-4-en-3-one <9> [4]
5,5'-dithiobis-(2-nitrobenzoate) <5> (<5> 50% inhibition of ACAT1 at 0.01 mM [36]) [36]
5α-cholest-8(14)-en-3β-ol-15-one <3> (<3> inhibits the esterification of cholesterol [7]) [7]
5α-dihydroprogesterone <5> (<5> inhibits esterification of endogenous cholesterol [3]) [3, 6]
ADP <3> (<3> inactivates [21]) [21]
ATP <3> (<3> inactivates [21]) [21]
ATP/$MgCl_2$ <3, 13> (<13> inhibition of activity [16]) [16, 21]
CI-976 <5, 9, 14> (<5> inhibits [24]; <14> cell-free extract of wild-type cells: approximately 90% inhibition at 0.1 mM and more than 95% inhibition at 0.2 mM, intact wild-type cells: approximately 75-80% inhibition at 0.2 mM [27]; <5> cell-free extract of sat1 and sat2 cells: more than 95% inhibition at 0.2 mM, intact sat1 and sat2 cells: approximately 75-80% inhibition at 0.2 mM [27]) [24, 27, 30]
CP113818 <8> (<8> recombinant enzyme, 80-90% inhibition at 0.0002 mM [33]) [33]
Dup 128 <5, 14> (<5> inhibits [24]; <14> approximately 50% inhibition at 0.03-0.04 mM [27]; <5> 50% inhibition at 0.00013 mM for HisACAT1, 50% inhibition at 0.0001 mM for the mutants HisACAT1/p64 and HisACAT1/1-65 [37]) [24, 27, 37]
N-ethylmaleimide <5> (<5> 50% inhibition of ACAT1 at 0.5 mM [36]) [36]
PD138412 <8> (<8> recombinant enzyme, 80-90% inhibition at 0.003 mM [33]) [33]
SC-31769 <3, 5> (<3,5> 20-oxa analog of 7-ketosterol [3,4,12]) [3, 4, 12]
Triton WR1339 <15> (<15> 90% inhibition at 0.3% wt./vol. [34]) [34]
(Z)-17(20)-dehydrocholesterol <3> (<3> inhibits esterification of endogenous cholesterol [3]) [3]
acetic anhydride <9> (<9> 2 distinct tissue types with difference in reactivity, the inhibitor appears to modify a histidine residue [10,30]) [10, 30]
acetone <3, 4> (<4> about 40% inhibition at a final concentration of 1% v/v [2,3]; <3> 30-50% inhibition at a final concentration of 1% v/v, inhibition prevented by the addition of cholesterol to the solvent [12]) [2, 3, 12]

acyl-CoA binding protein <5> (<5> 3fold decrease in activity [29]) [29]

acyl-CoA binding protein/oleoyl-CoA complexes <5> (<5> strongly inhibit [29]) [29]

androstenedione <5> [6]

androsterone <3> (<3> inhibits esterification of endogenous cholesterol [3]) [3]

bovine serum albumin <9, 12> (<9,12> inhibits at high concentrations [14,15]) [14, 15]

butanol <3, 4> (<4> about 40% inhibition at a final concentration of 1% v/v [2,3]; <3> 30-50% inhibition at a final concentration of 1% v/v, inhibition prevented by the addition of cholesterol to the solvent [12]) [2, 3, 12]

cycloartenol <3> (<3> inhibits esterification of exogenous cholesterol [3]) [3]

dehydroisoandrosterone <5> [6]

deoxycorticosterone <3, 5> (<5> inhibits esterification of endogenous cholesterol [3]) [3, 6]

detergents <1-3, 5-10> (<1-3,5-10> ionic and non-ionic [4]; <3> above their critical micelle concentration, e.g. cholate, deoxycholate, cetyltriethylammonium bromide, Triton X-100, Tween 20, Tween 80, Nonidet P40, [3,12]) [3, 4, 12]

diazepam <5> (<5> competes with oleoyl-CoA for its binding to microsomal membranes [29]) [29]

diethyl dicarbonate <9> (<9> 2 distinct tissue types with difference in reactivity, DEP-sensitive subtype typified by aortic ACTAT, DEP-resistant subtype typified by liver ACAT, irreversible inhibition, the inhibitor appear to modify a histidine residue [10,30]) [10, 30]

digitonin <3> [3, 12, 13]

dihydrotestosterone <3> (<3> inhibits esterification of endogenous cholesterol [3]) [3]

ergosterol <3> (<3> inhibits esterification of exogenous cholesterol [3]) [3]

estradiol-17β <5> [6]

ethanol <3, 4> (<4> about 40% inhibition at a final concentration of 1% v/v [2,3]; <3> 30-50% inhibition at a final concentration of 1% v/v, inhibition prevented by the addition of cholesterol to the solvent [12]) [2, 3, 12]

filipin <3> [3, 12]

inhibitor from rabbit liver <3> (<3> fatty acid ester, mostly stearate, of a pentacyclic triterpene acid, irreversible inhibition, which is prevented by incubation with bovine serum albumin, half-maximal inhibition occurs at an inhibitor concentration of 0.02 mM [5]) [5]

lanosterol <3, 11> (<3,11> inhibits esterification of exogenous cholesterol [3]) [3]

lidocaine <3, 9> [4]

octimibate <5> (<5> inhibits [24]) [24]

oleoyl-CoA <3> (<3> inhibits at high concentrations, the inhibition is prevented by serum albumin [3,4,12]) [3, 4, 12]

organic solvents <3> (<3> inhibit by perturbing the membrane structure [3]) [3, 12]

p-chloromercuribenzene sulfonic acid <5> (<5> 50% inhibition of ACAT1 at 0.01 mM, Cys467 is one of the major targets for inhibition [36]) [36]
p-mercuribenzoate <9> (<9> 100% inhibition of liver and aortic enzyme at 1 mM [10]) [10]
phenylglyoxal <9> (<9> concentration-dependent inhibition of liver enzyme with 65% inhibition at 5 mM [10]) [10]
phosphatidylinositol <2> [4, 9]
phosphatidylserine <2> (<3> not inhibitory [4]) [4, 9]
polyoxyethylated cholesterol <5> [4]
pregn-5-en-3β-ol <3> (<3> inhibits esterification of endogenous cholesterol [3]) [3]
pregnenolone <5> (<5> inhibits esterification of endogenous cholesterol [3]) [3, 6]
progesterone <3, 5, 9> (<3,5> inhibits esterification of endogenous cholesterol reversibly [3]; <3> half-maximal inhibitory concentration of 0.05 mM [5]; <5> 50% inhibition at 0.02 mM [6]) [3, 4-6, 12]
sitosterol <3> (<3> inhibits esterification of exogenous cholesterol [3]) [3]
sphingomyelin <6> [4]
stigmasterol <3> (<3> inhibits esterification of exogenous cholesterol [3]) [3]
testosterone <3, 5> (<5> inhibits esterification of endogenous cholesterol [3]) [3, 6]
tetracaine <3, 9> [4]
tetranitromethane <9> (<9> 100% inhibition of liver and aortic enzyme at 1 mM [10]) [10]
Additional information <1-3, 5-10> (<5> corticosterone, cortisol and estriol have little effect, enzyme activity in placenta may be regulated by endogenously synthesized steroid hormones [6]; <3> enhancement of inactivation of the enzyme by ATP/Mg, NaF and a heat-labile cytosolic factor is consistent with a protein kinase-catalyzed phosphorilation being involved in the short term regulation of the enzyme [21]; <9> the inhibitor CI-976 given to cholesterol-fed rabbits prevents the accumulation of monocyte-macrophages within a pre-established iliac-femoral lesion and significantly reduces foam cell area [30]; <1-3,5-10> regulation by non-sterol agents [30]) [6, 21, 30]

Cofactors/prosthetic groups

CoA <3-5> (<3,4> required by the in-situ acyl-CoA production [1,2,13,30]) [1, 2, 6, 13, 30]

Activating compounds

25-hydroxycholesterol <3, 5> (<3,5> activates [12,16,24]; <13> not stimulatory [16]; <3> stimulates enzyme activity in liver microsomes in vitro with half-maximal stimulation at 0.0168 mM oxysterol. A major part of the activation of microsomal enzyme is not ascribable to increased substrate availability for the enzyme [28]) [12, 16, 24, 28, 30]
6-ketocholestanol <5> (<5> added at high concentration causes modest increase in enzyme activity [24]) [24]

7-ketocholesterol <5> (<5> added at high concentration causes modest increase in enzyme activity [24]) [24]
ATP <5> (<5> slight stimulation [6]) [6]
albumin <3, 5, 9, 12> (<5> stimulates, effect is dependent on oleoyl-CoA concentration [11]; <3> required in the assay medium to prevent the high concentration of oleoly-CoA from inhibiting the enzyme presumably by disrupting the microsomal membrane [3]; <9,12> stimulates [14,15]) [3, 11, 14, 15]
cholate <3> (<3> increase of activity in cholate-fed animals [12]) [12]
cholesterol <3, 5, 7, 8, 15> (<8> exogenous, added as phosphatidylcholine liposome or in acetone solution, stimulates [4,30]; <3> exogenous, delivered as a Triton WR-1339 detergent dispersion, increases activity [4]; <3> increase of activity in cholesterol-fed animals [12]; <3> exogenous, 25% increase of adrenal microsome activity and 2fold increase of activity in liver [15]; <3> exogenous, in liposomes or in organic solvent, stimulates [16]; <5> addition of cholesterol to frozen microsomes prepared from unfrozen liver tissue increases the enzyme activity [19]; <2,5,9,13> exogenous, not stimulatory [4,6,15,16]; <5> addition of cholesterol to microsomes prepared from frozen liver tissue does not further increase the enzyme activity [19]; <5,7> exogenous cholesterol in the liposomes is absolutely necessary for activity of the reconstituted enzyme [20]; <5> serves as an enzyme activator in vitro, in addition to its role as an enzyme substrate [24]; <15> day 40 samples, exogenous, added as phosphatidylcholine liposomes, concentration-dependent increase in activity [34]; <15> day 50 and 60 samples: not stimulatory [34]) [4, 12, 15, 19, 20, 24, 30, 34]
liver fatty acid binding protein/bovine serum albumin <5> (<5> small increase in activity [29]) [29]
low density lipoprotein <6, 7> (<7> increase of activity [17,25]) [17, 25]
mevalonolactone <3> (<3> intragastric administration, increasing sterol biosynthesis and increasing enzyme activity [12]) [12]
phosphatidylethanolamine <2> (<2> weak stimulatory effect on activity [9]) [9]
Additional information <1-3, 5-10> (<1-3,5-10> regulation by sterol [4,30]) [4, 30]

Metals, ions

Ca^{2+} <13> (<13> stimulates [16]) [16]
Mg^{2+} <3, 13> (<3> required for the acylation of free cholesterol [1]; <13> stimulates [16]; <3> required [21]) [1, 16, 21]

Specific activity (U/mg)

0.0000053-0.000008 <5> [29]
0.0000178 <12> (<12> microsomal preparation [14]) [14]
0.0000222 <6> (<6> microsomes [18]) [18]
0.00005 <5> (<5> whole homogenate of biopsies from distal duodenum-proximal jejunum [11]) [11]
0.000053 <5> (<5> microsomal preparations from ileum, mean value [11]) [11]

0.0000582 <3> (<3> microsomal fraction without 25-hydroycholesterol [28]) [28]
0.00006 <3> (<3> cholesterol added in Tween 80 [13]) [13]
0.000062 <5> (<5> microsomal preparations from duodenum, mean value [11]) [11]
0.000063 <3> (<3> cholesterol added in acetone [13]) [13]
0.00008 <2> (<2> solubilized enzyme [4,9]) [4, 9]
0.000089 <13> (<13> microsomes + 5 mM ATP + 50 mM NaF + 5 mM $MgCl_2$ + 0.005 mM cAMP + 0.005 mg protein kinase [16]) [16]
0.000093 <13> (<13> microsomes + 5 mM ATP [16]) [16]
0.0000982 <3> (<3> microsomal fraction with 25-hydroycholesterol [28]) [28]
0.000099 <13> (<13> microsomes + 5 mM ATP + 50 mM NaF + 5 mM $MgCl_2$ [16]) [16]
0.0001042 <9> (<9> aortic microsomal preparation [15]) [15]
0.000117 <13> (<13> control microsomes [16]) [16]
0.000126 <13> (<13> microsomes + 5 mM $MgCl_2$ + 50 mM NaF [16]) [16]
0.000132 <13> (<13> microsomes + 50 mM NaF [16]) [16]
0.00015 <8> (<8> kidney microsomes [35]) [35]
0.000169 <13> (<13> microsomes + 5 mM $MgCl_2$ [16]) [16]
0.00021 <5> (<5> microsomal preparations from jejunum, mean value [11]) [11]
0.000368 <15> (<15> day 30 of development, endogenous cholesterol as substrate [34]) [34]
0.000808 <15> (<15> microsomal preparation from day 40 of development measured at 33°C [34]) [34]
0.000824 <15> (<15> microsomal preparation from day 40 of development measured at 37°C [34]) [34]
0.00087 <8> (<8> small intestine microsomes [35]) [35]
0.0009 <5> (<5> recombinant enzyme [27]) [27]
0.0012 <8> (<8> recombinant enzyme ACAT1, vector: pcDNA3+, cell type: AC29 [33]) [33]
0.00135 <8> (<8> liver microsomes [35]) [35]
0.001585 <15> (<15> day 50 of development, endogenous cholesterol as substrate [34]) [34]
0.0017 <3> (<3> liver microsomes [33]) [33]
0.0019 <8> (<8> recombinant enzyme ACAT2, vector: pCMV5, cell type: AC29 [33]) [33]
0.0022 <8> (<8> recombinant enzyme ACAT2, vector: pcDNA3+, cell type: AC29 [33]) [33]
0.0037 <8> (<8> adrenal microsomes [35]) [35]
0.0038 <8> (<8> recombinant enzyme ACAT2, vector: pcDNA3+, cell type: COS [33]) [33]
0.015 <2> (<2> partially purified enzyme [4,9]) [4, 9]
3.5-4 <5> (<5> purified recombinant enzyme [36]) [36]
Additional information <1-12, 15> (<1-3,5-11> assay method [3,4]; <9> overview: normal and cholesterol-fed organs [10]; <3> overview: effect of fasting, feeding with cholesterol or cholate, atherogenic diet administration

of Triton WR 1339 and administration of mevanolactone, evidence for a circadian rhythm [12]; <12> overview: specific activities in microsomes stored at -40°C for up to 6 weeeks [14]; <9> enhancement of activity following 8-11 weeks of cholesterol feeding [15,30]; <5,7> comparison of enzyme activities of cells grown in fetal calf serum medium and in delipidated serum medium [17,20]; <6> overview: enzyme activity under various conditions [18]; <5> activation of the enzyme by adding 25-hydroxycholesterol to intact cells or to cell-free extracts [24]; <5> when yeast-expressed enzyme activity is measured in the presence of ergosterol liposomes, specific activity is reduced more than 80%, recombinant enzyme is active but unable to esterify ergosterol efficiently [27]; <15> overview: enzyme activity at various days of development with and without exogenous cholesterol [34]) [2-4, 10, 12, 14, 15, 17-20, 24, 27, 30, 34]

K_m-Value (mM)

0.0022 <3> (elaidoyl-CoA) [1]
0.0028 <3> (linoleoyl-CoA) [1]
0.004 <2, 3> (oleoyl-CoA) [1, 4]
0.0071 <5> (oleoyl-CoA, <5> wild-type enzyme [36]) [36]
0.0072 <5> (oleoyl-CoA, <5> HisACAT1, mutant HisACAT1/P64 [37]) [37]
0.0154 <5> (oleoyl-CoA, <5> mutant HisACAT1/1-65 [37]) [37]
0.025 <5> (oleoyl-CoA, <5> bound to acyl-CoA binding protein [29]) [29]
0.03 <5> (oleoyl-CoA, <5> in the absence of acyl-CoA binding protein [29]) [29]
0.038 <5> (oleoyl-CoA) [6]
0.038 <6> (palmitoyl-CoA) [18]
0.055 <3> (oleoyl-CoA) [3]
0.06 <4> (cholesterol, <4> with oleoyl-CoA [2]) [2]
0.069 <4> (oleoyl-CoA) [2]
Additional information <3, 5> (<3> use of kinetic parameters such as K_m is of limited value due to the physical properties of oleoyl-CoA in solution and the addition of serum albumin to the assay [3]; <5> for the various Cys-deficient mutant ACAT1s, the K_m value for oleoyl-coenzyme A is approximately 2fold higher than the value for the wild-type enzyme, ranging between 0.012 und 0.015 mM [36]) [3, 11, 36]

K_i-Value (mM)

0.04 <9> (diethyl dicarbonate, <9> aortic enzyme [10,30]) [10, 30]
0.05 <9> (20α-hydroxypregn-4-en-3-one) [4]
0.05 <9> (progesterone) [4]
0.5 <9> (acetic anhydride, <9> less than, aortic enzyme [10,30]) [10, 30]
1.5 <9> (diethyl dicarbonate, <9> liver enzyme [10,30]) [10, 30]
5 <9> (acetic anhydride, <9> greater than, liver enzyme [10,30]) [10, 30]

pH-Range

5.6-7.4 <12> (<12> 2fold higher activity at pH 5.6 than at pH 7.4 [14]) [14]
6.5-8.4 <5> (<5> pH 6.5: 55% of activity maximum, pH 8.4: 75% of activity maximum [11]) [11]

Temperature optimum (°C)

30 <2> (<2> assay at [9]) [9]
30-37 <1-3, 5-10> (<1-3,5-10> assay at [4]) [4]
33 <15> (<15> assay at, the reaction rate is not significantly different from that at 37°C [34]) [34]
37 <2, 3, 5, 6, 8, 9, 11, 12, 14, 15> (<2, 3, 5, 6, 8, 9, 11, 12, 14, 15> assay at [1, 3, 8, 11-15, 24, 25, 27, 29, 31, 34, 35]) [1, 3, 8, 11-15, 24, 25, 27, 29, 31, 34, 35]

4 Enzyme Structure

Molecular weight

47000 <8> (<8> immunoblotting, ACAT2 enzyme [35]) [35]
50000 <5> (<5> immunoblotting, ACAT1 enzyme and Cys-mutants [36]) [36]
52000 <5> (<5> recombinant enzyme [27]) [27]
53000 <8> (<8> immunoblotting, ACAT1 enzyme [35]) [35]
63700 <6> (<6> deduced from amino acid sequence data [26]) [26]
64000 <3, 6> (<3,6> deduced from amino acid sequence data [25,31]) [25, 31]
64810 <5> (<5> deduced from the ORF analysis of enzyme cDNA [22]) [22, 30]
71600 <14> (<14> deduced from amino acid sequence data [27]) [27]
213000 <3> (<3> approximately, liver membranes, radiation inactivation analysis [30]) [30]
Additional information <5> (<5> expected monomer molecular masses of different fusion proteins range between 33000 and 41000 Da, sucrose density gradient velocity ultracentrifugation shows molecular masses of different fusion proteins ranging from 70000 to 160000 Da, thus some fusion proteins behave as tetramers in solution and other behave as dimers in solution [37]) [37]

Subunits

? <2> (<2> x * 58000, SDS-PAGE [23]) [23]
? <5> (<5> x * 45000-50000, SDS-PAGE [30]) [30]
? <6> (<6> x * 64000, denaturing electrophoresis conditions [26]) [26]
homotetramer <5> (<5> use of chemical cross-linking agent disuccinimidyl suberate causes the formation of material two to four times the size of the monomeric enzyme when added to intact cells expressing either wild-type ACAT1 or mutant ACAT1 C92/333/345/365/387/467/516/528/546A [36]; <5> HisACAT1 [37]) [36, 37]
Additional information <5> (<5> SDS-PAGE: detection of two protein bands with apparent sizes at 50 and 56 kDa, immunoblots: existence of 100 and 200 kDa bands, clarification of these observations needed [24]; <5> deleting a dimer-forming motif from the full-length ACAT1 converts the enzyme from a homotetramer to a homodimer that is more catalytically active than the native homotetramer [37]) [24, 37]

Posttranslational modification

glycoprotein <3, 6, 8, 14> (<3,6> two potential N-glycosylation sites close to the C-terminal end of the polypeptide identified [25,31]; <14> there are six potential N-linked glycosylation sites in sat1, two of which are conserved in sat2 [27]; <8> two potential N-glycosylation motifs at N329 and N387, only the second of which is present in the human enzyme [33]) [25, 27, 31, 33]

5 Isolation/Preparation/Mutation/Application

Source/tissue

CHO cell <6, 1, 3, 6, 7, 9> (<7> CHO cells, cultured cells [4,17]) [4, 26, 32, 3, 4, 17, 20, 25, 30]
Ehrlich ascites carcinoma cell <6> [4, 18]
Kupffer cell <8> (<8> ACAT1 present [35]) [35]
adipose tissue <6, 8, 9> (<6> low expression [25]; <8> ACAT1 enzyme, high mRNA signal [33]) [10, 25, 33]
adrenal gland <3, 5, 6, 8, 9, 13> (<9> 1.7fold induction of activity after 60 days of cholesterol feeding [10]; <6> highly expressed [25]; <3,6,9> relative high level of mRNA [30]; <3> highest activity and highest mRNA level, significant immunohistochemical reactivity [31]; <8> ACAT1 enzyme, highest mRNA signal [33]; <8> ACAT1: strongly present in the cells of the cortex [35]) [4, 10, 15, 16, 25, 30, 31, 33, 35]
aorta <3, 6, 9> (<9> 41fold induction of activity after 60 days of cholesterol feeding [10]; <9> atheroschlerotic tissue [15,30]; <3> extremely low activity [31]) [10, 15, 25, 30, 31]
artery <8, 9, 10> [4]
brain <5, 6, 9, 12> (<12> neonatal [14]; <9> 1.6fold induction of activity after 60 days of cholesterol feeding [10]; <6> low expression [25]) [10, 14, 25, 30]
duodenum <5> [11]
fibroblast <5> [4, 20]
glomerulus <8> (<8> ACAT1, distal tubules and podocytes [35]) [35]
goblet cell <8> (<8> ACAT1 [35]) [35]
heart <5, 6, 8, 9> (<9> 6fold induction of activity after 60 days of cholesterol feeding [10]; <6> low expression [25]; <8> ACAT1 enzyme, high mRNA signal [33]) [10, 25, 30]
hepatocyte <3, 8> (<8> two different forms of the enzyme are present, ACAT1 and ACAT2 [33,35]) [4, 33, 35]
hepatoma cell <3> [4]
ileum <5> [11]
intestine <3, 5, 6, 8, 9> (<9> 23fold induction of activity after 60 days of cholesterol feeding [10]; <3> high activity, high mRNA level, significant immunohistochemical reactivity [31]; <8> expression of ACAT2 mRNA [33]; <8> ACAT2: strongly present in the apical third of the mucosal cells, ACAT1: strong signal in goblet cells, Paneth cells and villus macrophages [35]) [4, 7, 10, 25, 31, 33, 35]

jejunum <3, 5, 8> (<5> microsomal preparation, highest activity [11]) [7, 11, 35]
kidney <5, 8, 9> (<8> ACAT1 enzyme, high mRNA signal [33]) [10, 30, 33, 35]
liver <1-3, 5, 6, 8-10> (<9> 27fold induction of activity after 60 days of cholesterol feeding [10]; <6> enzyme levels are doubled in animals fed an atherogenic diet [25]; <3> the enzyme activity is very low in fetal livers but is largely induced shortly after birth [30]; <9> nonparenchymal cells contain 30fold more enzyme message than parenchymal cells [30]; <3> high activity, extremely low mRNA level and negligible immunohistochemical reactivity [31]; <8> expression of ACAT2 mRNA. ACAT1 enzyme: high mRNA signal [33]) [1, 3-5, 7-10, 12, 13, 15, 16, 18, 19, 21, 23, 25, 28, 30, 31, 33, 35]
lung <9> (<9> 3fold induction of activity after 60 days of cholesterol feeding [10]) [10]
macrophage <5, 6, 9> (<6,9> relative high level of mRNA [30]) [23, 30]
macrophage <8> (<8> ACAT1, villus [35]) [35]
monocyte <5> [23]
mononuclear phagocyte <5> [29]
pancreas <8, 9> (<9> 25fold induction of activity after 60 days of cholesterol feeding [10]; <8> ACAT1 enzyme, high mRNA signal [33]) [10, 33]
paneth cell <8> (<8> ACAT1 [35]) [35]
peritoneal macrophage <6, 1, 3, 6, 7, 9> (<6> highly expressed [25]; <3,6,9> messages relative abundant [30]) [4, 26, 32, 3, 4, 17, 20, 25, 30]
placenta <5> [6]
preputial gland <6> (<6> highly expressed [25,30]) [25, 30]
skeletal muscle <5, 6> (<6> low expression [25]) [25, 30]
small intestine mucosa <5> [11]
spleen <9> (<9> 7fold induction of activity after 60 days of cholesterol feeding [10]) [10]
stomach <9> [10]
testis <3, 6, 9> (<9> 1.8fold induction of activity after 60 days of cholesterol feeding [10]; <3,6,9> relative high level of mRNA [30]) [10, 25, 30]
thymus <6> [25]
trachea <8> (<8> ACAT1 enzyme, high mRNA signal [33]) [33]
yolk sac <15> (<15> embryo, membrane, at various stages of development [34]) [34]
Additional information <6, 9> (<9> overview: enzyme activity in normal and cholesterol-fed organs, identification of two distinct tissue types of the enzyme [10]; <6> effect of an atherogenic diet upon hepatic and intestinal enzyme mRNA expression and on enzyme mRNA expression in aorta and peritoneal macrophages. Enzyme activity in the liver is regulated at least in part by enzyme mRNA abundance [25]) [10, 25]

Localization

cell surface <6> (<6> approximately 10-15% of the enzyme in freshly harvested, non-attached macrophages is exposed to the extracellular space [32]) [32]

cytoplasm <6> (<6> perinuclear [32]) [32]
cytoplasmic membrane <6> (<6> reticular/nuclear [32]) [32]
cytosol <3> (<3> little activity [12]) [12]
endoplasmic reticulum <3, 5, 7, 8> (<3> rough, membranes [4,12,29,30]; <8> ACAT2: rough [33,35]) [3, 4, 12, 17, 22, 24, 29, 30, 33, 35]
membrane <3, 6> [5, 25, 30]
microsome <2-6, 9, 12, 13> (<2> cytoplasmic surface of [4]; <3> intrinsic membrane enzyme, highest activity [12]; <3,9> submicrosomal distribution [13,15]) [1-16, 18, 19, 23, 25, 28, 30]
mitochondrion <3> (<3> little activity [12]) [12]
spheroplast <4> [2]
Additional information <5, 7, 8> (<5,7> cholesterol flux into and out of the endoplasmic reticulum [30]) [30]

Purification

<2> (partial, using ammonium acetate fractionation and Sepharose 4B column chromatography [4,9]; based on solubilization in CHAPS, preparative free-flow isotachophoresis, and further separation of the protein showing enzyme activity by preparative free-flow IEF, followed by removal of nonglycosylated proteins by ConA-affinity chromatography and preparative SDS-PAGE [23]) [4, 9, 23, 30]
<5> (of the recombinant enzyme using solubilization with CHAPS und KCl, chromatography on cobalt column and chromatography on monoclonal antibody affinity column [36]; of the glutathione S-transferase-ACAT1 peptide fusion protein using chromatography on glutathione-Sepharose 4B column and of various recombinant HisACAT1s expressed in infected H5 cells using a Talon Superflow resin [37]) [36, 37]

Renaturation

<2> (reconstitution with liposomes [23]) [23]
<3> (the enzyme must be reconstituted into an artificial membrane for optimal activity [3]) [3]
<6> (solubilization of enzyme activity from Ehrlich ascites cells, the solubilized activity can be reconstituted as a liposome complex after the detergent is removed [4,18]) [4, 18]
<2, 7> (solubilization of enzyme activity by deoxycholate, reconstitution of the solubilized activity into liposome of known cholesterol and phospholipid content by employing a cholate dialysis procedure [4,9,17,30]) [4, 9, 17, 30]
<5, 7> (solubilization of the cell extracts with deoxycholate followed by dilution into preformed liposomes [20,24]) [20, 24]

Cloning

<3> (expression in human embryonic kidney A293 cells and expression in Chinese hamster ovary cells [31]) [31]
<5> (expression in enzyme-deficient mutant chinese hamster ovary cells [22,23,30]; expression of an antisense cDNA in HepG2 cells [23]; sequence analysis and human cDNA clones indicate homology to liver carboxylesterase [23]; expression in insect Sf9 cells [24,30]; expression in sat1 sat2 mutant cells

of Saccharomyces cerevisiae [27]; expression of ACAT1 in infected insect High Five cells [36]; production of glutathione S-transferase-ACAT1 fusion proteins, expression of fusion proteins in Escherichia coli and expression of various recombinant HisACAT1s in infected H5 cells [37]) [22-24, 27, 30, 36, 37]
<6> (expression in enzyme-deficient mutant chinese hamster ovary cells [25]; expression in Xenopus oocytes [26,30]) [25, 26, 30]
<7> (cDNA cloned [30]) [30]
<8> (expression in AC29 cells [35]) [35]
<8> (presence of two gene products: ACAT1 and ACAT2, the gene product ACAT2 cloned from an African green monkey liver cDNA library. Transfection of the cloned ACAT2 cDNA into two different mammalian cell types [33]) [33]
<9> (cDNA cloned [30]) [30]
<14> (two genes, SAT1 and SAT2 encode isozymes of the enzyme [27,30]) [27, 30]

Engineering

C333A <5> (<5> expressed in H5 cells, the mutant retains a significant amount of activity, Cys-333 is not essential for ACAT1 catalysis, the mutant is expressed as a single, undegraded 50-kDA band [36]) [36]
C333A/C345A/C365A <5> (<5> expressed in H5 cells, the mutant contains essentially the same activity as the wild-type ACAT1. Sensitive to *p*-chloromercuribenzene sulfonic acid inhibition [36]) [36]
C345A <5> (<5> expressed in H5 cells, the mutant retains a significant amount of activity, Cys-345 is not essential for ACAT1 catalysis, the mutant is expressed as a single, undegraded 50-kDA band [36]) [36]
C345A/C467A <5> (<5> insensitive to *p*-chloromercuribenzene sulfonic acid [36]) [36]
C365A <5> (<5> expressed in H5 cells, the mutant retains a significant amount of activity, Cys-365 is not essential for ACAT1 catalysis, the mutant is expressed as a single, undegraded 50-kDA band [36]) [36]
C387A <5> (<5> expressed in H5 cells, the mutant retains a significant amount of activity, Cys-387 is not essential for ACAT1 catalysis, the mutant is expressed as a single, undegraded 50-kDA band [36]) [36]
C467A <5> (<5> expressed in H5 cells, the mutant retains a significant amount of activity, Cys-467 is not essential for ACAT1 catalysis, the mutant is expressed as a single, undegraded 50-kDA band [36]) [36]
C516A <5> (<5> expressed in H5 cells, the mutant retains a significant amount of activity, Cys-516 is not essential for ACAT1 catalysis, the mutant is expressed as a single, undegraded 50-kDA band [36]) [36]
C528A <5> (<5> expressed in H5 cells, the mutant retains a significant amount of activity, Cys-528 is not essential for ACAT1 catalysis, the mutant is expressed as a single, undegraded 50-kDA band [36]) [36]
C546A <5> (<5> expressed in H5 cells, the mutant retains a significant amount of activity , Cys-546 is not essential for ACAT1 catalysis, the mutant is expressed as a single, undegraded 50-kDA band [36]) [36]

C92A <5> (<5> expressed in H5 cells, the mutant retains a significant amount of activity, Cys-92 is not essential for ACAT1 catalysis, the mutant is expressed as a single, undegraded 50-kDA band [36]) [36]
C92A/C333A/C345A/C365A/C387A/C467A/C516A <5> (<5> resistant to *p*-chloromercuribenzene sulfonic acid inhibition [36]) [36]
C92A/C333A/C345A/C365A/C387A/C467A/C516A/C528A/C54 <5> (<5> expressed in H5 cells or in ACAT-deficient Chinese hamster ovary cells: both mutants retain partial enzyme activity [36]; <5> this mutant is localized in endoplasmic reticulum in Chinese hamster ovary cells, homotetramer. Insensitive to *p*-chloromercuribenzene sulfonic acid [36]) [36]
C92A/C333A/C345A/C365A/C387A/C467A/C516A/C546A <5> (<5> expressed in H5 cells or in ACAT-deficient Chinese hamster ovary cells: both mutants retain partial enzyme activity. Resistant to *p*-chloromercuribenzene sulfonic acid inhibition [36]; <5> this mutant is localized in endoplasmic reticulum in Chinese hamster ovary cells [36]) [36]
C92A/C333A/C345A/C365A/C387A/C516A/C528A/C546A <5> (<5> sensitive to *p*-chloromercuribenzene sulfonic acid inhibition [36]) [36]
C92A/C333A/C345A/C365A/C467A/C516A/C546A <5> (<5> expressed in H5 cells or in ACAT-deficient Chinese hamster ovary cells: both mutants retain partial enzyme activity. Resistant to *p*-chloromercuribenzene sulfonic acid inhibition [36]) [36]
C92A/C333A/C365A/C387A/C467A/C516A/C528A/C546A <5> (<5> insensitive to *p*-chloromercuribenzene sulfonic acid [36]) [36]
C92A/C467A/C516A/C546A <5> (<5> expressed in H5 cells or in ACAT-deficient Chinese hamster ovary cells: both mutants retain partial enzyme activity [36]; <5> this mutant is localized in endoplasmic reticulum in Chinese hamster ovary cells [36]) [36]
Additional information <5> (<5> production of double or single proline mutants derived from glutathione S-transferase-ACAT1 fusion proteins shows that the dimer-forming domain is within ACAT1 amino acids 30 to 94. Production of mutants derived from the recombinant plasmid pHisACAT1 to test the effect of mutagenizing or deleting the N-terminal peptide on the oligomerization state of ACAT1 [37]) [37]

Application

medicine <5> (<5> identified as a major target for inhibition, based on the hypothesis that enzyme inhibitors may have a direct antiatherosclerotic effect within the arterial wall and a direct effect in blocking cholesterol absorption in the small intestine, and in reducing very-low-density lipoprotein secretion in the hepatocytes [30]) [30]

6 Stability

Temperature stability

50 <5> (<5> apparent $t_{1/2}$ of activity loss: approximately 12 min for HisACAT1 and 20 min for the mutant HisACAT1/1-65, for the mutant HisACAT1/

P64 the heat inactivation curve falls between the curves for HisACAT1 and HisACAT/1-65 and its shape seems to be biphasic [37]) [37]
80 <3> (<3> 10 min, complete loss of activity [12]) [12]
100 <5> (<5> liver enzyme, 10 min, activity destroyed [4]) [4]
Additional information <5> (<5> the enzyme activity in Chinese hamster cells is more thermolabile than the enzyme activity in human cells [22]; <5> enzyme activity expressed in Sf9 insect cells exhibits heat stability comparable of that of human enzyme, its heat stability is significantly greater than that of hamster enzyme [24]) [22, 24]

Organic solvent stability

Additional information <3> (<3> the use of organic solvents or strong detergents for the suspension of cholesterol must be avoided because of their affect on the enzyme [3,12]) [3, 12]

Storage stability

<2>, 4°C, Tris buffer, 1 mM EDTA, pH 7.7, 6.0 mg/ml protein, 5.0 mg/ml cholate, 5.5 mg/ml deoxycholate, 0.5 M potassium chloride, solubilized enzyme, stable for at least 18 h [9]
<3>, frozen, stable for up to 2 weeks [12]
<3>, ovary, labile to freezing [3]
<5>, -70°C, biopsies, potassium phosphate buffer, stable for at least 3 years [11]
<5>, -70°C, microsomal preparations, stable for up to 1.5 years [11]
<12>, -40°C, stable for several weeks [14]
<2, 3, 8>, -70°C, liver microsomal enzyme, stable for several months [4]
<3, 4, 11>, -80°C, stable for several weeks [3]

References

[1] Sgoutas, D.S.: Effect of geometry and position of ethylenic bond upon acyl coenzyme A: cholesterol-O-acyltransferase. Biochemistry, **9**, 1826-1833 (1970)

[2] Taketani, S.; Nishino, T.; Katsuki, H.: Characterization of sterol-ester synthetase in Saccharomyces cerevisiae. Biochim. Biophys. Acta, **575**, 148-155 (1979)

[3] Billheimer, J.T.: Cholesterol acyltransferase. Methods Enzymol., **111**, 286-293 (1985)

[4] Chang, T.Y.; Doolittle, G.M.: Acyl coenzyme A:cholesterol O-acyltransferase. The Enzymes, 3rd Ed. (Boyer, P.D., ed.), **16**, 523-539 (1983)

[5] Tabas, I.; Chen, L.L.; Clader, J.W.; McPhail, A.T.; Burnett, D.A.; Bartner, P.; Das, P.R.; Pramanik, B.N.; Puar, M.S.; Feinmark, S.J.; Zipkin, R.E.; Boykow, G.; Vita, G.; Tall, A.R.: Rabbit and human liver contain a novel pentacyclic triterpene ester with acyl-CoA:cholesterol acyltransferase inhibitory activity. J. Biol. Chem., **265**, 8042-8051 (1990)

[6] Simpson, E.R.; Burkhart, M.F.: Acyl CoA:cholesterol acyl transferase activity in human placental microsomes: inhibition by progesterone. Arch. Biochem. Biophys., **200**, 79-85 (1980)

[7] Miller, L.R.; Needleman, D.H.; Brabson, J.S.; Wang, K.S.; Schroepfer, G.J.: 5 α-Cholest-8(14)-en-3 β-ol-15-one. A competitive substrate for acyl coenzyme A: cholesterol acyl transferase. Biochem. Biophys. Res. Commun., **148**, 934-940 (1987)

[8] Tavani, D.M.; Nes, W.R.; Billheimer, J.T.: The sterol substrate specificity of acyl CoA:cholesterol acyltransferase from rat liver. J. Lipid Res., **23**, 774-781 (1982)

[9] Doolittle, G.M.; Chang, T.Y.: Solubilization, partial purification, and reconstitution in phosphatidylcholine-cholesterol liposomes of acyl-CoA:cholesterol acyltransferase. Biochemistry, **21**, 674-679 (1982)

[10] Kinnunen, P.M.; DeMichele, A.; Lange, L.G.: Chemical modification of acyl-CoA:cholesterol O-acyltransferase. 1. Identification of acyl-CoA:cholesterol O-acyltransferase subtypes by differential diethyl pyrocarbonate sensitivity. Biochemistry, **27**, 7344-7350 (1988)

[11] Helgerud, P.; Saarem, K.; Norum, K.R.: Acyl-CoA:cholesterol acyltransferase in human small intestine: its activity and some properties of the enzymic reaction. J. Lipid Res., **22**, 271-277 (1981)

[12] Erickson, S.K.; Shrewsbury, M.A.; Brooks, C.; Meyer, D.J.: Rat liver acyl-coenzyme A:cholesterol acyltransferase: its regulation in vivo and some of its properties in vitro. J. Lipid Res., **21**, 930-941 (1980)

[13] Balasubramaniam, S.; Venkatesan, S.; Mitropoulos, K.A.; Peters, T.J.: The submicrosomal localization of acyl-coenzyme A-cholesterol acyltransferase and its substrate, and of cholesteryl esters in rat liver. Biochem. J., **174**, 863-872 (1978)

[14] Marco, C.; Morillas, L.G.; Garcia-Peregrin, E.: Acyl coenzyme A:cholesterol acyltransferase in neonatal chick brain. Biochim. Biophys. Acta, **875**, 599-604 (1986)

[15] Brecher, P.; Chan, C.T.: Properties of acyl-CoA:cholesterol O-acyltransferase in aortic microsomes from atherosclerotic rabbits. Biochim. Biophys. Acta, **617**, 458-471 (1980)

[16] Suckling, K.E.; Tocher, D.R.; Smellie, C.G.; Boyd, G.S.: In vitro regulation of bovine adrenal cortical acyl-CoA: cholesterol acyltransferase and comparison with the rat liver enzyme. Biochim. Biophys. Acta, **753**, 422-429 (1983)

[17] Doolittle, G.M.; Chang, T.Y.: Acyl-CoA:cholesterol acyltransferase in Chinese hamster ovary cells. Enzyme activity determined after reconstitution in phospholipid/cholesterol liposomes. Biochim. Biophys. Acta, **713**, 529-537 (1982)

[18] Kaduce, T.L.; Schmidt, R.W.; Spector, A.A.: Acylcoenzyme A:cholesterol acyltransferase activity: solubilization and reconstitution in liposomes. Biochem. Biophys. Res. Commun., **81**, 462-468 (1978)

[19] Einarsson, K.; Benthin, L.; Ewerth, S.; Heller, G.; Staehlberg, D.; Angelin, B.: Studies on acyl-coenzyme A:cholesterol acyltransferase activity in human liver microsomes. J. Lipid Res., **30**, 739-746 (1989)

[20] Cadigan, K.M.; Chang, T.Y.: A simple method for reconstitution of CHO cell and human fibroblast acyl coenzyme A:cholesterol acyltransferase activity into liposomes. J. Lipid Res., **29**, 1683-1692 (1988)
[21] Kwok, C.T.; Smith, J.L.; Pillay, S.P.; Hardie, I.R.: Inactivation of acyl-CoA: cholesterol acyltransferase by magnesium-ATP in rat liver microsomes. Biochem. Int., **8**, 271-282 (1984)
[22] Chang, C.C.Y.; Huh, J.Y.; Cadigan, K.M.; Chang, T.Y.: Molecular cloning and functional expression of human acyl-coenzyme A:cholesterol acyltransferase cDNA in mutant Chinese hamster ovary cells. J. Biol. Chem., **268**, 20747-20755 (1993)
[23] Becker, A.; Böttcher, A.; Lackner, K.J.; Fehringer, P.; Notka, F.; Aslanidis, C.; Schmitz, G.: Purification, cloning, and expression of a human enzyme with acyl coenzyme A:cholesterol acyltransferase activity, which is identical to liver carboxylesterase. Arterioscler. Thromb., **14**, 1346-1355 (1994)
[24] Cheng, D.; Chang C.C.Y.; Qu, X.; Chang, T.: Activation of acyl-coenzyme A:cholesterol acyltransferase by cholesterol or by oxysterol in a cell-free system. J. Biol. Chem., **270**, 685-695 (1995)
[25] Uelmen, P.J.; Oka, K.; Sullivan, M.; Chang, C.C.Y.; Chang, T.Y.; Chan, L.: Tissue-specific expression and cholesterol regulation of acylcoenzyme A:cholesterol acyltransferase (ACAT) in mice. J. Biol. Chem., **270**, 26192-26201 (1995)
[26] Green, S.; Steinberg, D.; Quehenberger, O.: Cloning and expression in Xenopus Oocytes of a mouse homologue of the human acylcoenzyme A:cholesterol acyltransferase and its potential role in metabolism of oxidized LDL. Biochem. Biophys. Res. Commun., **218**, 924-929 (1996)
[27] Yu, C.; Kennedy, N.J.; Chang, C.C.Y.; Rothblatt, J.A.: Molecular cloning and characterization of two isoforms of Saccharomyces cerevisiae acyl-CoA: sterol acyltransferase. J. Biol. Chem., **271**, 24157-24163 (1996)
[28] Bhuvaneswaran, C.; Synouri-Vrettakou, S.; Mitropoulos, K.A.: Activation of acyl-CoA:cholesterol acyltransferase in rat liver microsomes by 25-hydroxycholesterol. Biochem. Pharmacol., **53**, 27-34 (1997)
[29] Kerkhoff, C.; Beuck, M.; Threige-Rasmussen, J.; Spener, F.; Knudsen, J.; Schmitz, G.: Acyl-CoA binding protein (ACBP) regulates acyl-CoA:cholesterol acyltransferase (ACAT) in human mononuclear phagocytes. Biochim. Biophys. Acta, **1346**, 163-172 (1997)
[30] Chang, T.Y.; Chang, C.C.Y.; Cheng, D.: Acyl-coenzyme A:cholesterol acyltransferase. Annu. Rev. Biochem., **66**, 613-638 (1997)
[31] Matsuda, H.; Hakamata, H.; Kawasaki, T.; Sakashita, N.; Miyazaki, A.; Takahashi, K.; Shichiri, M.; Horiuchi, S.: Molecular cloning, functional expression and tissue distribution of rat acyl-coenzyme A:cholesterol acyltransferase. Biochim. Biophys. Acta, **1391**, 193-203 (1998)
[32] Khelef, N.; Buton, X.; Beatini, N.; Wang, H.; Meiner, V.; Chang, T.; Farese, R.V.; Maxfield, F.R.; Tabas, I.: Immunolocalization of acyl-coenzyme A:cholesterol O-acyltransferase in macrophages. J. Biol. Chem., **273**, 11218-11224 (1998)
[33] Anderson, R.A.; Joyce, C.; Davis, M.; Reagan, J.W.; Clark, M.; Shelness, G.S.; Rudel, L.L.: Identification of a form of acyl-CoA:cholesterol acyltransferase

specific to liver and intestine in nonhuman primates. J. Biol. Chem., **273**, 26747-26754 (1998)
[34] Vajda, K.; Ferguson, M.W.J.; Shand, J.H.; Noble, R.C.; Speake, B.K.: Activities of acyl-CoA:cholesterol acyltransferase and cholesteryl ester hydrolase in the yolk sac membrane of the alligator (Alligator mississippiensis) embryo. Comp. Biochem. Physiol. B, **122**, 301-307 (1999)
[35] Lee, R.G.; Willingham, M.C.; Davis, M.A.; Skinner, K.A.; Rudel, L.L.: Differential expression of ACAT1 and ACAT2 among cells within liver, intestine, kidney, and adrenal of nonhuman primates. J. Lipid Res., **41**, 1991-2001 (2000)
[36] Lu, X.; Lin, S.; Chang, C.C.Y.; Chang, T.: Mutant acyl-coenzyme A: cholesterol acyltransferase 1 devoid of cysteine residues remains catalytically active. J. Biol. Chem., **277**, 711-718 (2002)
[37] Yu, C.; Zhang, Y.; Lu, X.; Chen, J.; Chang, C.C.Y.; Chang, T.: Role of the N-terminal hydrophilic domain of acyl-coenzyme A:cholesterol acyltransferase 1 on the enzymes's quaternary structure and catalytic efficiency. Biochemistry, **41**, 3762-3769 (2002)

Cortisol O-acetyltransferase 2.3.1.27

1 Nomenclature

EC number
2.3.1.27

Systematic name
acetyl-CoA:cortisol O-acetyltransferase

Recommended name
cortisol O-acetyltransferase

Synonyms
acetyltransferase, corticosteroid
acetyltransferase, cortisol
corticosteroid-21-O-acetyltransferase

CAS registry number
9076-48-6

2 Source Organism

<1> *Papio sp.* (baboon) [1]

3 Reaction and Specificity

Catalyzed reaction
acetyl-CoA + cortisol = CoA + cortisol 21-acetate

Reaction type
transfer of acetyl group

Natural substrates and products
S acetyl-CoA + cortisol <1> (Reversibility: ? <1> [1]) [1]
P CoA + cortisol 21-acetate <1> [1]

Substrates and products
S acetyl-CoA + cortisol <1> (Reversibility: ? <1> [1]) [1]
P CoA + cortisol 21-acetate <1> [1]

Temperature optimum (°C)
37 <1> (assay at [1]) [1]

5 Isolation/Preparation/Mutation/Application

Source/tissue

brain <1> (<1> highest activity in brain stem, other activity in brain cortex, cerebellum and thalamus-hypothalamus [1]) [1]

Purification

<1> [1]

References

[1] Thomas, P.J.: Cortisol acetyltransferase from baboon brain. Biochem. J., **109**, 695-696 (1968)

Chloramphenicol O-acetyltransferase 2.3.1.28

1 Nomenclature

EC number
2.3.1.28

Systematic name
acetyl-CoA:chloramphenicol 3-O-acetyltransferase

Recommended name
chloramphenicol O-acetyltransferase

Synonyms
CAT
CAT I (most frequently encountered natural variant in gram-negative bacteria, type I enzyme is associated with transposon Tn9 and a variety of F-like R plasmids)
CAT II <1, 7> (<1,7>, enzyme encoded by the incW transmissible plasmid [8]) [8]
CAT III
Pacat <16> [16]
acetyltransferase, chloramphenicol
cat-86 <17> [17]
chloramphenicol acetylase
chloramphenicol acetyltransferase
chloramphenicol transacetylase

CAS registry number
9040-07-7

2 Source Organism

<1> *Escherichia coli* (chloramphenicol-resistant strains which carry the episomal resistance transfer factor R [1]; R-factor-bearing strain W677/HJR 66 [6]; CATII is encoded by the IncW transmissible plasmid pSA [8]; strain W677/HJR66 [14]; strain J53(R387) [15]; enzyme form CAT III [19]; enzyme is encoded ny tramsposon Tn2424 [22]) [1, 2, 3, 6, 8, 9, 12, 14, 15, 19, 20, 22, 25, 26, 27]
<2> *Staphylococcus aureus* (strain C22.1 [2]; encoded by plasmids pSCS6 or pSCS7 [4]; enzyme variant C and D [13]) [2, 4, 13]
<3> *Streptococcus faecalis* [3]
<4> *Diplococcus pneumoniae* [3, 10]

<5> *Agrobacterium tumefaciens* [3, 10]
<6> *Staphylococcus sp.* (type A, type B and type C [10]; enzyme type A [13]) [3, 10, 13]
<7> *Haemophilus influenzae* (enzymes encoded by plasmid pRI234, pMR375 and pMR385 [5]; CATII [8]) [5, 8, 10]
<8> *Vibrio anguillarum* (enzyme is encoded by plasmid Rms418 [7]) [7]
<9> *Proteus mirabilis* [10]
<10> *Haemophilus parainfluenzae* [10]
<11> *Streptomyces acrimycine* [10]
<12> *Streptococcus agalactiae* [10]
<13> *Clostridium perfringens* [10]
<14> *Flavobacterium sp.* (CB60 [11]) [11]
<15> *Staphylococcus epidermidis* (enzyme variant B [13]) [13]
<16> *Pseudomonas aeruginosa* [16, 24]
<17> *Bacillus pumilus* [17]
<18> *Bacillus subtilis* [18]
<19> *Culex pipiens* (infected with either TE/3'2J/CAT or rep5/CAT/26S virus [21]) [21]
<20> *Morganella morganii* [23]

3 Reaction and Specificity

Catalyzed reaction

acetyl-CoA + chloramphenicol = CoA + chloramphenicol 3-acetate (<1>, ternary complex mechanism with a rapid equilibrium and essentially random order of addition of substrates [9])

Reaction type

acyl group transfer

Natural substrates and products

S acetyl-CoA + chloramphenicol <2, 3, 4, 5, 6> (<1>, enzymatic inactivation of chloramphenicol [6]; <2,3,4,5,6>, inducible enzyme [2,3,4]; <1,2>, inactivates chloramphenicol [2]; <1>, catabolite repression of CAT synthesis is mediated by a mechanism involving cyclic adenosine 5'-monophosphate [3]; <1>, all known R factors carrying the CAT gene in enteric bacteria mediate constitutive synthesis of the enzyme [3]) (Reversibility: ? <2, 3, 4, 5, 6> [2, 3, 4]) [2, 3, 4]

P CoA + chloramphenicol 3-acetate

Substrates and products

S acetyl-CoA + D-threo-1-*p*-nitrophenyl-2-bromoacetamido-1,3-propanediol <1, 2> (Reversibility: ? <1, 2> [2]) [2]

P CoA + ?

S acetyl-CoA + D-threo-1-phenyl-2-dichloroacetamido-1,3-propanediol <1, 2> (Reversibility: ? <1, 2> [2]) [2]

P CoA + ?

S acetyl-CoA + chloramphenicol <2> (Reversibility: ? <2> [2]) [2]
P CoA + chloramphenicol 1,3-diacetate
S acetyl-CoA + chloramphenicol <1-20> (<2>, acetylates only the biologically active D-threo stereoisomer [2]; <1>, in forward reaction formation of a ternary complex by a rapid-equilibrium mechanism, in reverse reaction rapid-equilibrium mechanism with random addition of substrates [15]) (Reversibility: r <1> [15]; ? <1-20> [1-14, 16-27]) [1-27]
P CoA + chloramphenicol 3-acetate <1-20> (<2>, + chloramphenicol 1,3-diacetate [2]) [1-27]
S acetyl-CoA + chloramphenicol 1-acetate <1> (<1>, the enzyme acetylates specifically at the 3-hydroxy position. The diacetylation is possible only because of non-enzymatic interconversion of chloramphenical 3-acetate to chloramphenicol 1-acetate at higher pH values [6]; <14>, no activity [11]) (Reversibility: ? <1> [6]) [6]
P CoA + chloramphenicol 1,3-diacetate <1> [6]

Inhibitors

1,5-difluoro-2,4-dinitrobenzene <14> [11]
2,2'-dithiobis(pyridine) <7, 9, 11, 12, 13> [10]
4,4'-dithiodipyridine <1> [8]
5,5'-dithiobis(2-nitrobenzoic acid) <1, 5, 8, 14> (<8>, 0.025 mM, 82% inhibition [7]) [3, 7, 8, 11]
EDTA <8> (<8>, 0.1 mM, 24% inhibition [7]) [7]
Hg^{2+} <8> (<8>, 0.01 mM, complete inhibition [7]) [7]
I_2 <8> (<8>, 0.1 mM, 20% inhibition [7]) [7]
NEM <14> [11]
PCMB <8, 14> (<8>, 0.1 mM, complete inhibition [7]) [7, 11]
Zn^{2+} <8> (<8>, 0.1 mM, 15% inhibition [7]) [7]
crystal violet <1> (<1>, strong, competitive for chloramphenicol, non-competitive for acetyl-CoA [12]) [12]
ethyl violet <1> (<1>, strong, competitive for chloramphenicol, non-competitive for acetyl-CoA [12]) [12]
methyl methanethiolsulfonate <1> [8]

Metals, ions

$HgCl_2$ <2> (<2>, at 0.1 M and 0.3 mM, no influence on the activity of the pSCS6-encoded enzyme. The activity of the pSCS7-encoded enzyme decreases to 10-15%. At 1.0 mM $HgCl_2$, the pSCS6-encoded enzyme retains 50% of ist original activity, complete inactivation of the enzyme from pSCS7-encoded enzyme [4]) [4]

Turnover number (min^{-1})

Additional information <1> [8, 25]

Specific activity (U/mg)

71.4 <8> [7]
130.6 <1> [6]
251.3 <1> [14]
330 <17> [17]

367.7 <14> [11]
Additional information <1, 2, 6, 15, 20> [3, 4, 13, 23]

K_m-Value (mM)

0.00247 <2> (chloramphenicol, <2>, enzyme variant C [13]) [13]
0.0025 <2> (chloramphenicol, <2>, enzyme from plasmid pSCS6 [4]) [4, 10]
0.00258 <6> (chloramphenicol) [13]
0.0027 <2> (chloramphenicol, <2>, enzyme from plasmid pSCS7 [4]) [2, 4]
0.0027 <2> (chloramphenicol, <2>, enzyme variant D [13]) [2, 13]
0.00272 <15> (chloramphenicol) [13]
0.0037 <1> (D-threo-1-*p*-nitrophenyl-2-bromoacetamido-1,3-propanediol) [2]
0.0061 <1> (chloramphenicol) [2]
0.0093 <12> (chloramphenicol) [10]
0.01 <4> (chloramphenicol) [10]
0.0103 <1> (chloramphenicol) [8]
0.011 <1> (chloramphenicol, <1>, enzyme variant CAT I [25]) [25]
0.011 <16> (chloramphenicol, <16>, enzyme variant CATA1 [24]) [24]
0.012 <1> (chloramphenicol, <1>, enzyme variant CAT III [25]) [25]
0.0166 <11> (chloramphenicol) [10]
0.0175 <10> (chloramphenicol) [10]
0.0205 <5> (chloramphenicol) [10]
0.021 <1> (D-threo-1-phenyl-2-dichloroacetamido-1,3-propanediol) [2]
0.0215 <13> (chloramphenicol) [10]
0.022 <2> (D-threo-1-*p*-nitrophenyl-2-bromoacetamido-1,3-propanediol) [2]
0.025 <17> (chloramphenicol, <17>, wild-type enzyme [17]) [17]
0.028 <17> (acetyl-CoA, <17>, wild-type enzyme [17]) [17]
0.03 <14> (chloramphenicol) [11]
0.031 <9> (chloramphenicol) [10]
0.033 <1> (chloramphenicol) [12]
0.0345 <8> (chloramphenicol) [7]
0.0465 <2> (acetyl-CoA, <2>, enzyme variant D [13]) [13]
0.055 <1> (acetyl-CoA, <1>, enzyme variant CAT I [25]) [25]
0.056 <2> (acetyl-CoA, <2>, enzyme from plasmid pSCS7 [4]) [4]
0.0561 <15> (acetyl-CoA) [13]
0.0573 <6> (acetyl-CoA) [13]
0.059 <2> (acetyl-CoA, <2>, enzyme from plasmid pSCS6 [4]) [4]
0.0609 <2> (acetyl-CoA, <2>, enzyme variant C [13]) [13]
0.065 <1, 14> (acetyl-CoA) [8, 11]
0.068 <2> (D-threo-1-phenyl-2-dichloroacetamido-1,3-propanediol) [2]
0.083 <1> (acetyl-CoA) [12]
0.093 <1> (acetyl-CoA, <1>, enzyme variant CATIII [25]) [25]
0.094 <16> (acetyl-CoA, <16>, enzyme variant CATA1 [24]) [24]
0.136 <16> (chloramphenicol, <16>, enzyme variant CATB5 [24]) [24]
0.14 <16> (chloramphenicol, <16>, enzyme variant CATB1 [24]) [24]
0.156 <8> (acetyl-CoA) [7]
0.156 <16> (chloramphenicol, <16>, enzyme variant CATB3 [24]) [24]

0.161 <16> (acetyl-CoA, <16>, enzyme variant CATB1 [24]) [24]
0.164 <16> (acetyl-CoA, <16>, enzyme variant CATB5 [24]) [24]
0.165 <16> (acetyl-CoA, <16>, enzyme variant CATB3 [24]) [24]
0.812 <16> (chloramphenicol, <16>, enzyme variant CATB7 [24]) [24]
0.821 <16> (acetyl-CoA, <16>, enzyme variant CATB7 [24]) [24]
Additional information <20> [23]

K_i-Value (mM)

Additional information <1> [12]

pH-Optimum

7.6 <1> [6]
7.8 <1, 2, 8, 20> [2, 3, 7, 14, 23]
8 <2> [4]

pH-Range

5-9 <20> (<20>, about 50% of maximal activity at pH 5.0 and 9.0 [23]) [23]
6-8.8 <2> (<2>, pH 6.0: about 55% of maximal activity, pH 8.8: about 85% of maximal activity [2]) [2]

Temperature optimum (°C)

35 <20> [23]
Additional information <1> (<1>, enzyme expression is increased at 42°C. At 42°C the enzyme is still soluble and active [20]) [20]

Temperature range (°C)

30-45 <20> (<20>, 30°C: about 20% of maximal activity, 45°C: about 55% of maximal activity [23]) [23]

4 Enzyme Structure

Molecular weight

69000 <8> (<8>, gel filtration [7]) [7]
70000 <2> (<2>, gel filtration [4]) [4]
78000 <1, 2, 20> (<1,2>, sucrose density gradient centrifugation [2]; <20>, gel filtration [23]) [2, 23]
80000 <14> (<14>, gel filtration [11]) [11]

Subunits

? <4, 5, 6, 7, 9, 10, 11, 12, 13, 16> (<4,6,9,10,12,13>, x * 22500, SDS-PAGE [10]; <16>, x * 23351, electrospray injection mass spectrometry [16]; <7>, x * 24150, enzyme encoded by plasmid pRI234 [5]; <7>, x * 24400, enzyme encoded by plasmid pMR375 and plasmid pMR385 [5]) [5, 10, 16]
tetramer <20> (<20>, 1 * 19500 + 1 * 19000 + 1 * 18000 + 1 * 17500, SDS-PAGE [23]) [23]
trimer <2, 8, 14> (<8>, 3 * 22000, SDS-PAGE [7]; <2>, 3 * 23000, SDS-PAGE [4]; <14>, 3 * 26600, SDS-PAGE [11]) [4, 7, 11]

5 Isolation/Preparation/Mutation/Application

Source/tissue

saliva <19> (<19>, Culex pipiens infected with either TE/3'2J/CAT or rep5/CAT/26S virus [21]) [21]

Localization

cytoplasm <18> [18]

Purification

<1> [3, 6, 14, 25]
<2> (enzyme variant C and D [13]) [13]
<6> (enzyme variant A [13]) [3, 4, 13]
<7> [5]
<8> (plasmid Rms418 is transferred from Vibrio anguillarum to E. coli K-12 [7]) [7]
<14> [11]
<15> (enzyme variant B [13]) [13]
<16> [16]
<17> [17]
<20> [23]

Crystallization

<1> [3, 22]
<16> [16]

Cloning

<1> [8, 19]
<8> (<8>, plasmid Rms418 is transferred from Vibrio anguillarum to Escherichia coli K-12 [7]) [7]
<16> (overexpression in Escherichia coli harboring the plasmid pYT1 [16]; cloning of enzyme form CATB7 [24]) [16, 24]
<17> (cloning of mutant cat-86 in pTB361 and transformation of Escherichia coli JM109 [17]) [17]

Engineering

A203G <17> (<17>, mutant enzyme is less stable than wild-type enzyme [17]) [17]
A203I <17> (<17>, mutant enzyme is more thermostable than wild-type [17]) [17]
C214A <1> (<1>, 95% of the in vivo produced mutant polypeptide is soluble compared to 90% for the wild-type enzyme. Mutant enzyme loses 31% of activity after 30 min at 65°C compared to 15% for the wild-type enzyme [26]) [26]
C214D <1> (<1>, 50% of the in vivo produced mutant polypeptide is soluble compared to 90% for the wild-type enzyme. Mutant enzyme loses 85% of activity after 30 min at 65°C compared to 15% for the wild-type enzyme [26]) [26]

C214E <1> (<1>, 75% of the in vivo produced mutant polypeptide is soluble compared to 90% for the wild-type enzyme. Mutant enzyme loses 84% of activity after 30 min at 65°C compared to 15% for the wild-type enzyme [26]) [26]
C214F/G219S <1> (<1>, 95% of the in vivo produced mutant polypeptide is soluble compared to 90% for the wild-type enzyme. Mutant enzyme loses 81% of activity after 30 min at 65°C compared to 15% for the wild-type enzyme [26]) [26]
C214G <1> (<1>, 80% of the in vivo produced mutant polypeptide is soluble compared to 90% for the wild-type enzyme. Mutant enzyme loses 44% of activity after 30 min at 65°C compared to 15% for the wild-type enzyme [26]) [26]
C214L <1> (<1>, 100% of the in vivo produced mutant polypeptide is soluble compared to 90% for the wild-type enzyme. Mutant enzyme loses 33% of activity after 30 min at 65°C compared to 15% for the wild-type enzyme [26]) [26]
C214P <1> (<1>, 95% of the in vivo produced mutant polypeptide is soluble compared to 90% for the wild-type enzyme. Mutant enzyme loses 88% of activity after 30 min at 65°C compared to 15% for the wild-type enzyme [26]) [26]
C214Q <1> (<1>, 95% of the in vivo produced mutant polypeptide is soluble compared to 90% for the wild-type enzyme. Mutant enzyme loses 73% of activity after 30 min at 65°C compared to 15% for the wild-type enzyme [26]) [26]
C214R <1> (<1>, 55% of the in vivo produced mutant polypeptide is soluble compared to 90% for the wild-type enzyme. Mutant enzyme loses 84% of activity after 30 min at 65°C compared to 15% for the wild-type enzyme [26]) [26]
C214S <1> (<1>, 95% of the in vivo produced mutant polypeptide is soluble compared to 90% for the wild-type enzyme. Mutant enzyme loses 32% of activity after 30 min at 65°C compared to 15% for the wild-type enzyme [26]) [26]
C214T <1> (<1>, 90% of the in vivo produced mutant polypeptide is soluble compared to 90% for the wild-type enzyme. Mutant enzyme loses 59% of activity after 30 min at 65°C compared to 15% for the wild-type enzyme [26]) [26]
C214V <1> (<1>, 95% of the in vivo produced mutant polypeptide is soluble compared to 90% for the wild-type enzyme. Mutant enzyme loses 45% of activity after 30 min at 65°C compared to 15% for the wild-type enzyme [26]) [26]
C214W <1> (<1>, 50% of the in vivo produced mutant polypeptide is soluble compared to 90% for the wild-type enzyme. Mutant enzyme loses 70% of activity after 30 min at 65°C compared to 15% for the wild-type enzyme [26]) [26]
C214Y <1> (<1>, 90% of the in vivo produced mutant polypeptide is soluble compared to 90% for the wild-type enzyme. Mutant enzyme loses 81% of

activity after 30 min at 65°C compared to 15% for the wild-type enzyme [26]) [26]
CATIII (F24A/Y25F/L29A) <1> (<1>, K_m-value for acetyl-CoA is 0.095 mM compared to 0.093 mM for wild-type CATIII, K_m-value for chloramphenicol is 0.023 mM compared to 0.012 mM for the wild-type CATIII, turnover number is 30% of the wild-type enzyme CAT III [25]) [25]
CATIII(K14E/H195A/K217A) <1> (<1>, no activity [25]) [25]
CATIII(Q92C/N146F/Y169F/I172V) <1> (<1>, K_m-value for acetyl-CoA is 0.165 mM compared to 0.093 mM for wild-type CATIII, K_m-value for chloramphenicol is 0.02 mM compared to 0.012 mM for the wild-type CATIII, turnover number is 60% of the wild-type enzyme CAT III [25]) [25]
I191V <17> (<17>, mutant enzyme is less stable than wild-type enzyme [17]) [17]
K14/K217E <1> (<1>, K_m-value for acetyl-CoA is 0.166 mM compared to 0.093 mM for wild-type CATIII, K_m-value for chloramphenicol is 0.017 mM compared to 0.012 mM for the wild-type CATIII, turnover number is 87% of the wild-type enzyme CAT III [25]) [25]
L145F <1> (<1>, folding of chloramphenicol acetyltransferase is hampered by deletion of the carboxy-terminal tail including the last residue of the carboxy-terminal α-helix. Such truncated CAT polypeptides quantitatively aggregate into cytoplasmic inclusion bodies, which results in absence of chloramphenicol-resistant phenotype for the producing host. Introduction of Phe at amino acid position 145 improves the ability of the protein to fold into a soluble, enzymatically active conformation [19]) [19]
Y33F/A203V <17> (<17>, mutant enzyme is more thermostable than wild-type [17]) [17]
[CATI (H195A)]2[CATIII(K14E/K217E)] <1> (<1>, hybrid trimer, K_m-value for acetyl-CoA is 0.072 mM compared to 0.093 mM for wild-type CATIII, K_m-value for chloramphenicol is 0.018 mM compared to 0.012 mM for the wild-type CATIII, turnover number is 14% of the wild-type enzyme CAT III [25]) [25]
[CATIII]2[CATIII(K14E/H195A/K217A)] <1> (<1>, K_m-value for acetyl-CoA is 0.143 mM compared to 0.093 mM for wild-type CATIII, K_m-value for chloramphenicol is 0.016 mM compared to 0.012 mM for the wild-type CATIII, turnover number is 80% of the wild-type enzyme CAT III [25]) [25]
[CATIII][CATIII(K14E/H195A/K217A)]2 <1> (<1>, K_m-value for acetyl-CoA is 0.198 mM compared to 0.093 mM for wild-type CATIII, K_m-value for chloramphenicol is 0.02 mM compared to 0.012 mM for the wild-type CATIII, turnover number is 82% of the wild-type enzyme CAT III [25]) [25]
[CATI][CATIII(K14E/H195A/K217E)]2 <1> (<1>, hybrid trimer, K_m-value for acetyl-CoA is 0.107 mM compared to 0.093 mM for wild-type CATIII, K_m-value for chloramphenicol is 0.02 mM compared to 0.012 mM for the wild-type CATIII, turnover number is 50% of the wild-type enzyme CAT III [25]) [25]
Additional information <1, 18> (<18>, Bacillus subtilis cells expressing a hybrid protein (LvsSS-Cat) consisting of the Bacillus amyloliquefaciens levansucrose signal peptide fused to Bacillus pumilus chloramphenicol acetyltrans-

ferase are unable to export CAT protein into the growth medium. A series of tripartite protein fusion is constructed by inserting various length of the CAT sequences between the levansucrase signal peptide and staphylococcal protein A or E. coli alkaline phosphatase. Biochemical characterization of the various CAT protein fusion reveales that multiple regions in the CAT protein are causing the export defect [18]; <1>, in soluble CATI(1-211)(X3) mutants nearly all amino acid residues are tolerated at position 212 and 213. This reflects the relative lack of importance of these residues in the folding and/or stabilization of CAT. Substitutions at position 214 do not dramatically alter the biological activity of wild-type CATI [26]) [18, 26]

Application

analysis <1> (<1>, valuable tool in studies of eukaryotic gene expression, quantitative aspects of the use of bacterial chloramphenicol acetyltransferase as a reporter system in the yeast Saccharomyces cerevisiae [27]) [27]

6 Stability

Temperature stability

53 <8> (<8>, 0.1 mM chloramphenicol and 0.1 mM acetyl-CoA, 50 mM Tris-HCl buffer, pH 7.8, half-life: 100 min [7]) [7]
55 <17> (<17>,30 min, 82% loss of activity, wild-type enzyme [17]) [17]
60 <14> (<14>, nearly complete inactivation within 10 min [11]) [11]
70 <2> (<2>, 15 min, 40% loss of activity of enzyme variant from plasmid pSCS6, 45% loss of activity of enzyme variant pSCS7 [4]) [4]
75 <1, 2, 5> (<1>, rapid inactivation [2]; <2>, remarkably resistant [2]; <5>, stable [3]) [2, 3]

Oxidation stability

<2>, photooxidation of enzyme type C in presence of methylene blue, progressive loss of activity with increase in pH, inclusion of 1 mM chloramphenicol affords complete protection against loss of activity [13]

General stability information

<14>, loss of activity by freezing and thawing can be prevented by adding 10% glycerol [11]

Storage stability

<1>, -20 C, activity in elution buffer is stable for at least 1 year [3]
<8>, -80 C, 20% glycerol, stable for more than 3 months [7]
<14>, -20 C, stable for several months [11]

References

[1] Shaw, W.V.: The enzymatic acetylation of chloramphenicol by extracts of R factor-resistant Escherichia coli. J. Biol. Chem., **242**, 687-693 (1967)

[2] Shaw, W.V.; Brodsky, R.F.: Characterization of chloramphenicol acetyltransferase from chloramphenicol-resistant Staphylococcus aureus. J. Bacteriol., **95**, 28-36 (1968)
[3] Shaw, W.V.: Chloramphenicol acetyltransferase from chloramphenicol-resistant bacteria. Methods Enzymol., **43**, 737-755 (1975)
[4] Cardoso, M.; Schwarz, S.: Characterization of the chloramphenicol acetyltransferase variants encoded by the plasmids pSCS6 and pSCS7 from Staphylococcus aureus. J. Gen. Microbiol., **138**, 275-281 (1992)
[5] Roberts, M.; Corney, A.; Shaw, W.V.: Molecular characterization of three chloramphenicol acetyltransferases isolated from Haemophilus influenzae. J. Bacteriol., **151**, 737-741 (1982)
[6] Thibault, G.; Guitard, M.; Daigneault, R.: A study of the enzymatic inactivation of chloramphenicol by highly purified chloramphenicol acetyltransferase. Biochim. Biophys. Acta, **614**, 339-349 (1980)
[7] Masuyoshi, S.; Okubo, T.; Inoue, M.; Mitsuhashi, S.: Purification and some properties of a chloramphenicol acetyltransferase mediated by plasmids from Vibrio anguillarum. J. Biochem., **104**, 131-135 (1988)
[8] Murray, I.A.; Martinez-Suarez, J.V.; Close, T.J.; Shaw, W.V.: Nucleotide sequences of genes encoding the type II chloramphenicol acetyltransferases of Escherichia coli and Haemophilus influenzae, which are sensitive to inhibition by thiol-reactive reagents. Biochem. J., **272**, 505-510 (1990)
[9] Ellis, J.; Bagshaw, C.R.; Shaw, W.V.: Substrate binding to chloramphenicol acetyltransferase: evidence for negative cooperativity from equilibrium and kinetic constants for binary and ternary complexes. Biochemistry, **30**, 10806-10813 (1991)
[10] Zaidenzaig, Y.; Fitton, J.E.; Packman, L.C.; Shaw, W.V.: Characterization and comparison of chloramphenicol acetyltransferase variants. Eur. J. Biochem., **100**, 609-618 (1979)
[11] Nolte, G.; Sussmuth, R.: Purification and characterization of chloramphenicol acetyltransferase from Flavobacterium CB60. J. Gen. Microbiol., **133**, 2115-2122 (1987)
[12] Tanaka, H.; Izaki, K.; Takahashi, H.: Some properties of chloramphenicol acetyltransferase, with particular reference to the mechanism of inhibition by basic triphenylmethane dyes. J. Biochem., **76**, 1009-1019 (1974)
[13] Fitton, J.E.; Shaw, W.V.: Comparison of chloramphenicol acetyltransferase variants in staphylococci. Purification, inhibitor studies and N-terminal sequences. Biochem. J., **177**, 575-582 (1979)
[14] Guitard, M.; Daigneault, R.: Purification of Escherichia coli chloramphenicol acetyltransferase by affinity chromatography. Can. J. Biochem., **52**, 1087-1090 (1974)
[15] Kleanthous, C.; Shaw, W.V.: Analysis of the mechanism of chloramphenicol acetyltransferase by steady-state kinetics. Evidence for a ternary-complex mechanism. Biochem. J., **223**, 211-220 (1984)
[16] Tian, Y.; Beaman, T.W.; Roderick, S.L.: Purification and crystallization of Pseudomonas aeruginosa chloramphenicol acetyltransferase. Proteins, **28**, 298-300 (1997)

[17] Chirakkal, H.; Ford, G.C.; Moir, A.: Analysis of a conserved hydrophobic pocket important for the thermostability of Bacillus pumilus chloramphenicol acetyltransferase (CAT-86). Protein Eng., **14**, 161-166 (2001)
[18] Chen, M.W.; Nagarajan, V.: Chloramphenicol acetyltransferase, a cytoplasmic protein is incompatible for export from Bacillus subtilis. J. Bacteriol., **175**, 5697-5700 (1993)
[19] Van der Schueren, J.; Robben, J.; Volckaert, G.: Misfolding of chloramphenicol acetyltransferase due to carboxy-terminal truncation can be corrected by second-site mutations. Protein Eng., **11**, 1211-1217 (1998)
[20] Kim, S.J.; Jeon, H.Y.; Kim, H.B.: Chloramphenicol acetyltransferase expression of Escherichia coli is increased at 42 DegC. Biotechnol. Tech., **11**, 435-438 (1997)
[21] Kamrud, K.I.; Olson, K.E.; Higgs, S.; Powers, A.M.; Carlson, J.O.; Beaty, B.J.: Detection of expressed chloramphenicol acetyltransferase in the saliva of Culex pipiens mosquitoes. Insect Biochem. Mol. Biol., **27**, 423-429 (1997)
[22] Zhou, M.; Lu, M.L.; Qiu, W.; Campbell, R.L.; Nahoum, V.; Lapointe, J.; Roy, P.H.; Lin, S.X.: Crystallization and preliminary X-ray diffraction analysis of the chloramphenicol acetyltransferase from Tn2424. Acta Crystallogr. Sect. D, **57**, 281-283 (2001)
[23] El-Gamal, B.; Temsah, S.; Olama, Z.; Mohamed, A.; El-Sayed, M.: Purification and characterization of chloramphenicol acetyltransferase from Morganella morganii. J. Biochem. Mol. Biol., **34**, 415-420 (2001)
[24] White, P.A.; Stokes, H.W.; Bunny, K.L.; Hall, R.M.: Characterization of a chloramphenicol acetyltransferase determinant found in the chromosome of Pseudomonas aeruginosa. FEMS Microbiol. Lett., **175**, 27-35 (1999)
[25] Day, P.J.; Murray, I.A.; Shaw, W.V.: Properties of hybrid active sites in oligomeric proteins: kinetic and ligand binding studies with chloramphenicol acetyltransferase trimers. Biochemistry, **34**, 6416-6422 (1995)
[26] Van der Schueren, J.; Robben, J.; Goossens, K.; Heremans, K.; Volckaert, G.: Identification of local carboxy-terminal hydrophobic interactions essential for folding or stability of chloramphenicol acetyltransferase. J. Mol. Biol., **256**, 878-888 (1996)
[27] Alipour, H.; Eriksson, P.; Norbeck, J.; Blomberg, A.: Quantitative aspects of the use of bacterial chloramphenicol acetyltransferase as a reporter system in the yeast Saccharomyces cerevisiae. Anal. Biochem., **270**, 153-158 (1999)

Glycine C-acetyltransferase 2.3.1.29

1 Nomenclature

EC number
2.3.1.29

Systematic name
acetyl-CoA:glycine C-acetyltransferase

Recommended name
glycine C-acetyltransferase

Synonyms
2-amino-3-ketobutyrate CoA ligase
2-amino-3-ketobutyrate coenzyme A ligase
2-amino-3-ketobutyrate-CoA ligase
acetyltransferase, glycine
α-amino-β-oxobutyrate CoA-ligase
aminoacetone synthase
aminoacetone synthetase
glycine acetyltransferase

CAS registry number
37257-11-7

2 Source Organism

<1> *Arthrobacter sp.* (grown on L-threonine [1]) [1]
<2> *Escherichia coli* (K-12-mutant SBD-76, grown on L-threonine [2,3]) [2-4]
<3> *Bos taurus* [6-9]
<4> *Escherichia coli* [5, 10]
<5> *Homo sapiens* [9]
<6> *Mus musculus* [9]
<7> *Sus scrofa* [11]

3 Reaction and Specificity

Catalyzed reaction
acetyl-CoA + glycine = CoA + 2-amino-3-oxobutanoate (<4,7> reaction mechanism [10,11]; <7> physical interaction between threonine dehydrogenase and aminoacetone synthetase demonstrated, the two enzymes form a complex [11])

Reaction type

acyl group transfer

Natural substrates and products

S acetyl-CoA + glycine <1-3, 5-7> (<1-3,7> involved in L-threonine catabolism, inducible [1,2,6-8,11]; <3-6> enzyme catalyzes the second reaction step on the main metabolic degradation pathway for threonine [9,10]) [1, 2, 6-11]

P CoA + 2-amino-3-oxobutanoate <1, 3> [1, 2, 6, 7, 11]

Substrates and products

S acetyl-CoA + aminomalonate <3> [6]

P CoA + CO_2 + 2-amino-3-oxobutanoate <3> (<3> same product as in reaction with glycine, aminomalonate is decarboxylated to glycine by the enzyme, in the absence of acetyl-CoA: glycine is the product [6]) [6]

S acetyl-CoA + glycine <1-7> (<2,3> specific for glycine [2,3,6]; <2,3> no substrate: glycine methylester [2,6]; <2> no substrates: glycinamide, 2-aminoethanol, aminomethylphosphonic acid, aminomalonate [2]; <2,3> no substrate: L-alanine [2,6]; <2> no substrates: L-serine, L-threonine, L-valine, L-leucine [2]; <3> acetyl-CoA cannot be replaced by acetyl phosphate [6]; <2,3> succinyl-CoA [2,6]; <2> not: valeryl-CoA, glutaryl-CoA [2]; <3> not: acetyl phosphate [6]) (Reversibility: r <2, 3, 5-7> [2, 6, 7, 9, 11]) [1-4, 6-11]

P CoA + 2-amino-3-oxobutanoate <1, 3-6> (<1,3-6> spontaneous decarboxylation to aminoacetone [1,6,7,9,10]) [1, 6-11]

S n-butyryl-CoA + glycine <2> (<2> reaction at 16% the rate of acetyl-CoA [2]) [2]

P CoA + 2-amino-3-oxohaxanoate

S n-propionyl-CoA + glycine <2> (<2> reaction at 127% the rate of acetyl-CoA [2]) [2]

P CoA + 2-amino-3-oxopentanoate

Inhibitors

1,2-cyclohexanedione <2> (<2> inactivation [3]) [3]
2,3-butanedione <2> (<2> inactivation [3]) [3]
4-(oxoacetyl)phenoxyacetic acid <2> [3]
Ba^{2+} <2> [2]
Ca^{2+} <2> [2]
Cd^{2+} <2> (<2> strong [2]) [2]
CoA <1, 2> (<2> strong [2]) [1, 2]
Cu^{2+} <2> (<2> strong [2]) [2]
DTT <2> [2]
Hg^{2+} <2> (<2> strong [2]) [2]
iodoacetic acid <2, 3> [2, 6]
L-cysteine <1-3> (<3> enzyme reacts reversibly with [8]) [1, 2, 8]
Mn^{2+} <2> [2]
$NaBH_4$ <2> [2]
Ni^{2+} <2> [2]

Sr^{2+} <2> [2]
Zn^{2+} <2> [2]
aminomalonic acid <2> [2]
aminomethylphosphonic acid <2> [2]
aminooxyacetate <2> [2]
glutathione <1, 2> [1, 2]
glycine methyl ester <3> (<3> inactivation [6]) [6]
hydroxylamine <2> (<2> pyridoxal 5'-phosphate restores [2]) [2]
phenylglyoxal <2> (<2> inactivation, 50% protection by pyridoxal 5'-phosphate or CoA (glycine or threonine to some extent), kinetics [3]) [3]
phenylhydrazine <2> [2]
semicarbazide <2> [2]
Additional information <1-3> (<1-3> no inhibiton by DTNB [1,2,6]; <1> not: glycine [1]; <2> not: PCMB, glycine methylester, glycinamide [2]) [1, 2, 6]

Cofactors/prosthetic groups

pyridoxal 5'-phosphate <2, 3, 4> (<2,3> requirement, stoichiometry: 1 mol per mol subunit [4,6,8]; <2> active-site mapping [3]; <2> location of pyridoxyllysine peptide in primary structure of enzyme [4]; <3> pyridoxal 5'-phosphate is covalently bound to an active site lysine, formation of a Schiff base [8]) [2-4, 6, 7, 8, 10]

Activating compounds

5,5'-dithiobis(2-nitrobenzoic acid) <2> [2]
N-ethylmaleimide <2> [2]
iodoacetamide <1> (<1> increases yield of aminoacetone by 50% [1]; <2> not [2]) [1, 2]
p-mercuribenzoate <2> [2]

Specific activity (U/mg)

2.76 <2> [2]
16.4 <3> [6]
18 <3> [7]
Additional information <1> [1]

K_m-Value (mM)

0.053 <3> (acetyl-CoA) [6]
0.059 <2> (acetyl-CoA) [2]
0.08 <2> (propionyl-CoA) [2]
12 <2> (butyryl-CoA) [2]
12 <2> (glycine) [3]
22 <3> (glycine) [6]

K_i-Value (mM)

0.14 <2> (aminomalonic acid) [2]
3.6 <2> (aminomethylphosphonic acid) [2]

pH-Optimum

7.5 <2, 3> [2, 4, 6]
Additional information <3> (<3> pI: 5.2 [6]) [6]

pH-Range

5.1-9.4 <3> (<3> absorbtion spectrum does not change significantly [8]) [8]
6-9 <2> (<2> about 30% of maximal activity at pH 6.0 and 43% at 9.0 [2]) [2]

Temperature optimum (°C)

25 <2> (<2> assay at [2,3]) [2, 3]
30 <3> (<3> assay at [6]) [6]

4 Enzyme Structure

Molecular weight

43000 <3> [8]
44000 <3> [7]
56000 <3, 7> (<3,7> gel filtration [6,11]) [6, 11]
58330 <3> (<3> calculated from amino acid composition [6]) [6]
80000 <4> (<4> dynamic ligth scattering measurement [10]) [10]
84190 <2> (<2> calculated from amino acid composition [2]) [2, 4]
85000 <2> (<2> gel filtration [2]) [2]
90480 <2> (<2> sucrose density gradient centrifugation [2]) [2]
Additional information <4> (<4> nucleotide sequence [5]; <7> when loaded together threonine dehydrogease and aminoacetone synthetase co-eludes at a molecular weight of 150000 Da [11]) [5, 11]

Subunits

dimer <2, 3, 4> (<3> 2 * 28000, SDS-PAGE, 2 * 29165, calculated from amino acid composition [6]; <2> 2 * 41690, gel filtration, denaturating conditions, 2 * 41930, SDS-PAGE, 2 * 42093, calculated from amino acid composition [2]; <4> 2 * 43000, SDS-PAGE [10]) [2, 4, 6, 10]
monomer <3> (<3> 1 * 44000, SDS-PAGE [7] ; <3> 1 * 43000, SDS-PAGE [8]) [7, 8]
Additional information <7> (<7> the threonine dehydrogenase and the aminoacetone synthetase form a complex with an apparent stoichiometry of two dimers of aminoacetone synthetase to one threonine dehydrogenase tetramer [11]) [11]

Posttranslational modification

side-chain modification <2> (<2> the phenylglyoxal-inactivated enzyme is associated with the loss of 1.5 arginine residue per ligase subunit [3]) [3]
Additional information <2> (<2> no carbohydrates [2]) [2]

5 Isolation/Preparation/Mutation/Application

Source/tissue

brain <5> [9]
heart <5> [9]

liver <3> [6-9, 11]
lung <5, 6> [9]
pancreas <5> [9]

Localization

cytoplasm <5> (<5> enzyme synthesized in cytoplasm [9]) [9]
mitochondrion <3, 5, 6> (<5> enzyme imported into mitochondrion from cytoplasma [9]) [6-9]

Purification

<2> [2, 3, 5]
<3, 4> [6, 7, 10]

Crystallization

<4> (external aldimine in the crystal structure, interaction between aldimine and the side chains in the substrate binding site explain the specificity to the substrate [10]) [10]

Cloning

<3-6> (bovine partial [9]; expressed in Escherichia coli [10]) [9, 10]

6 Stability

pH-Stability

6.5-8 <3> (<3> stable [6]) [6]

General stability information

<2>, EDTA, 1 mM, stabilizes [4]
<2>, ethylene glycol, 20%, stabilizes [4]
<3>, glycine stabilizes [6]
<2, 3>, glycerol, 10% v/v, stabilizes [2, 6]

References

[1] McGilvray, D.; Morris, J.G.: Utilization of L-threonine by a species of Arthrobacter. A novel catabolic role for aminoacetone synthase. Biochem. J., **112**, 657-671 (1969)

[2] Mukherjee, J.J.; Dekker, E.E.: Purification, properties, and N-terminal amino acid sequence of homogeneous Escherichia coli 2-amino-3-ketobutyrate CoA ligase, a pyridoxal phosphate-dependent enzyme. J. Biol. Chem., **262**, 14441-14447 (1987)

[3] Mukherjee, J.J.; Dekker, E.E.: Inactivation of Escherichia coli 2-amino-3-ketobutyrate CoA ligase by phenylglyoxal and identification of an active-site arginine peptide. Arch. Biochem. Biophys., **299**, 147-153 (1992)

[4] Mukherjee, J.J.; Dekker, E.E.: 2-Amino-3-ketobutyrate CoA ligase of Escherichia coli: stoichiometry of pyridoxal phosphate binding and location of the

pyridoxyllysine peptide in the primary structure of the enzyme. Biochim. Biophys. Acta, **1037**, 24-29 (1990)
[5] Aronson, B.D.; Ravnikar, P.D.; Somerville, R.L.: Nucleotide sequence of the 2-amino-3-ketobutyrate coenzyme A ligase (kbl) gene of E. coli. Nucleic Acids Res., **16**, 3586 (1988)
[6] Fubara, B.; Eckenrode, F.; Tressel, T.; Davis, L.: Purification and properties of aminoacetone synthetase from beef liver mitochondria. J. Biol. Chem., **261**, 12189-12196 (1986)
[7] Tong, H.; Davis, L.: 2-Amino-3-ketobutyrate-CoA ligase from beef liver mitochondria. Purification and partial sequence. J. Biol. Chem., **269**, 4057-4064 (1994)
[8] Tong, H.; Davis, L.: 2-Amino-3-ketobutyrate-CoA ligase from beef liver mitochondria: An NMR spectroscopic study of low-barrier hydrogen bonds of a pyridoxal 5'-phosphate-dependent enzyme. Biochemistry, **34**, 3362-3367 (1995)
[9] Edgar, A.J.; Polak, J.M.: Molecular cloning of the human and murine 2-amino-3-ketobutyrate coenzyme A ligase cDNAs. Eur. J. Biochem., **267**, 1805-1812 (2000)
[10] Edgar, A.J.; Polak, J.M.: Molecular cloning of the human and murine 2-amino-3-ketobutyrate coenzyme A ligase cDNAs. Eur. J. Biochem., **267**, 1805-1812 (2000)
[11] Tressel, T.; Thompson, R.; Zieske, L.R.; Menendez, M.I.T.S.; Davis, L.: Interaction between L-threonine dehydrogenase and aminoacetone synthetase and mechanism of aminoacetone production. J. Biol. Chem., **261**, 16428-16437 (1986)

Serine O-acetyltransferase 2.3.1.30

1 Nomenclature

EC number

2.3.1.30

Systematic name

acetyl-CoA:L-serine O-acetyltransferase

Recommended name

serine O-acetyltransferase

Synonyms

L-serine acetyltransferase
SAT <2, 10, 16-21, 24-27> [20, 22, 23, 25-27]
SATase
acetyltransferase, serine
serine acetyltransferase
serine transacetylase
Additional information <3, 13> (<10, 14, 18-20> serine acetyltransferase and O-acetylserine (thiol) lyase form the cysteine synthase complex [16,23]; <13> serine acetyltransferase and O-acetylserine (thiol) lyase form the cysteine synthase complex, consisting of 2 of each enzymes as a tetramer [15]; <3> physical association of serine acetyltransferase with about 5% of the total cellular O-acetylserine sulfhydrylase to form a multienzyme complex given the trivial name cysteine synthethase [7]) [7, 15, 16, 23]

CAS registry number

9023-16-9

2 Source Organism

<1> *Brassica chinensis* [4]
<2> *Escherichia coli* (CysE gene [27]; strain C600, CysE gene [25]; EC1801, gene CysE [23]; B [1,6]) [1, 5, 6, 19, 23, 25-27]
<3> *Salmonella typhimurium* (strain Cys- DW 18.1 [10]; physical association of serine acetyltransferase with about 5% of the total cellular O-acetylserine sulfhydrylase to form a multienzyme complex given the trivial name cysteine synthethase [7]) [1, 5, 7, 10]
<4> *Phaseolus vulgaris* (kidney bean [2]) [2, 3]
<5> *Paracoccus denitrificans* (strain 8944 [8]) [8]
<6> *Spinacia oleracea* [9, 21, 24]

<7> *Citrullus vulgaris* (watermelon [11]) [11, 17, 24]
<8> *Arabidopsis thaliana* (gene SAT1 [12]) [12]
<9> *Escherichia coli* [12]
<10> *Arabidopsis thaliana* (gene SAT1-6, mitochondrial isozyme [12,23]) [12, 23, 25]
<11> *Arabidopsis thaliana* (gene SAT5 [12]) [12]
<12> *Pisum sativum* (3 isozymes specific for subcellular compartment [13]) [13]
<13> *Arabidopsis thaliana* (3 isozymes specific for subcellular compartment: SAT-c, SAT-m, SAT-p [14,24]) [14, 15, 17, 24]
<14> *Cyanidioschyzon merolae* (chloroplast isoform, constitutive gene cmSAT [16]; red alga [16]) [16]
<15> *Aspergillus nidulans* (allelic gene cysA1 and cysA103 [18]) [18]
<16> *Solanum tuberosum* [20]
<17> *Trypanosoma cruzi* [22]
<18> *Arabidopsis thaliana* (gene SAT-1, isoform A [23]) [23]
<19> *Arabidopsis thaliana* (gene SAT5, isoform B [23]) [23, 25]
<20> *Arabidopsis thaliana* (gene SAT52, isoform C [23]) [23, 25]
<21> *Salmonella typhimurium* (gene CysE [23]) [23]
<22> *Allium tuberosum* (Chinese chive [24]) [24]
<23> *Nicotiana tabacum* (cv. SNN [25]) [25]
<24> *Arabidopsis thaliana* (gene SAT A, mitochondrial isozyme [25]) [25]
<25> *Nicotiana tabacum* (gene SAT1 [25]) [25]
<26> *Nicotiana tabacum* (gene SAT4 [25]) [25]
<27> *Nicotiana tabacum* (gene SAT7 [25]) [25]

3 Reaction and Specificity

Catalyzed reaction

acetyl-CoA + L-serine = CoA + O-acetyl-L-serine (<2> steady-state random-order mechanism [27]; <2> binding motif for O-acetylserine (thiol) lyase is located within 10 amino acid residues at the C-terminal end [26]; <10,18-20> a C-terminal hexapeptide-repeat domain is common for Arabidopsis thaliana isoforms and several other enzyme DNA sequences, it has a catalytic bifunctionality as serine acetyltransferase and in interaction with O-acetylserine (thiol) lyase and is involved in cysteine biosynthesis regulation, computational modeling [23]; <3> ping pong Bi Bi mechanism [10])

Reaction type

acyl group transfer

Natural substrates and products

S acetyl-CoA + L-serine <2-8, 10-27> (<18-20> rate-limiting step of cysteine biosynthesis [23]; <6, 7, 13, 17, 22> key role in the sulfur assimilatory pathway [22, 24]; <6-8, 13> regulatory role in cysteine biosynthesis [11, 12, 14, 15, 21]; <2-4, 6> first enzyme in L-cysteine synthesis pathway [1, 3, 5, 9]; <5> enzyme in metabolic pathway of sulfur from sulfide to methionine [8]) (Reversibility: r <3> [7, 10]; ir <2-7, 13, 15, 22> [1, 3, 5, 8, 9, 11, 15, 18, 24]; ? <6, 8-14, 16-21, 23-27> [12-14, 16, 17, 19-23, 25, 26]) [1, 3, 5, 8-26]

P CoA + O-acetyl-L-serine <2-8, 10-27> [1, 3, 5, 7-26]

S Additional information <7, 13, 15> (<15> enzyme performs L-serine acetylation, but has evolutionary evolved from serine O-acetyltransferase, high DNA sequence homology with the latter [18]; <13> cysteine synthase is a bienzyme complex build of serine acetyltransferase and O-acetylserine (thiol) lyase, no channeling of substrate within the complex [15]; <7> multienzyme complex is formed between serine acetyltransferase and cysteine synthase possibly leading to metabolic channeling [11]) [11, 15, 18]

P ?

Substrates and products

S acetyl-CoA + L-serine <1-27> (<3> equilibrium constant in direction of L-serine acetylation [10]; <1> dependent on L-serine and acetyl-CoA [4]; <4,5> specific for L-serine [2,8]) (Reversibility: r <2, 3> [7, 10, 27]; ir <2-7, 13, 15, 22> [1, 3, 5, 8, 9, 15, 18, 24]; ? <1, 2, 4, 6-14, 16-21, 23-27> [2, 4, 6, 11-14, 16, 17, 19-23, 25, 26]) [1-27]

P CoA + O-acetyl-L-serine <1-8, 10-27> [1-27]

S acetyl-CoA + L-threonine <2> (<2> 0.5% of the activity with L-serine [1]; <4> no activity [2]) (Reversibility: ? <2> [1]) [1]

P CoA + O-acetyl-L-threonine

S Additional information <4> (<4> no activity: O-acetylserine, O-acetylhomoserine [2]) [2]

P ?

Inhibitors

CoA <2, 4, 5> (<5> weak [8]) [2, 8, 27]

D-cysteine <4, 5> (<17> no inhibition [22]) [2, 8]

KCN <5> [8]

L-alanine <2> (<2> noncompetitive against acetyl-CoA [27]) [27]

L-cysteic acid <2> [1]

L-cysteine <2-7, 13, 14, 17, 22> (<22> weak [24]; <6,7,22> feed-back inhibition [24]; <7> recombinant wild-type and mutants, not mutant M280I, inhibitor binding motif [17]; <13> feed-back inhibition of cytosolic isozyme SAT-c, not of mitochondrial and plastidic isozymes SAT-m and SAT-p [14,24]; <7,17> allosteric inhibition [11,22]) [1, 2, 5, 6, 8, 11, 14, 16, 17, 21, 22, 24, 26]

L-cystine <17> (<17> allosteric inhibition [22]) [22]

L-homocysteine <4, 5, 17> (<4,5,17> weak [2,8,22]) [2, 8, 22]

L-homoserine <5, 17> (<5,17> weak [8,22]) [8, 22]

L-methionine <4, 5> (<4> weak [2]) [2, 8]

L-serine <2, 3> (<3> product inhibition, competitive against CoA and noncompetitive against O-acetyl-L-serine [10]; <2> above 3 mM, weak [1]) [1, 10]

N-acetyl-L-cysteine <2, 17> (<17> weak [22]) [1, 22]

N-ethylmaleimide <5> [8]

O-acetyl-L-serine <3> (<3> product inhibition, competitive against acetyl-CoA and noncompetitive against O-acetyl-L-serine [10]) [10]
S-methyl-L-cysteine <3> (<3> competitive against O-acetyl-L-serine, noncompetitive against CoA [10]) [10]
glutathione <17> (<17> weak, allosteric inhibition [22]) [22]
glycine <2, 3> (<2,3> competitive against L-serine, noncompetitive against acetyl-CoA [10,27]) [10, 27]
hydroxylamine <4, 5> [2, 8]
iodoacetamide <5> [8]
p-chloromercuribenzoate <4, 5> [2, 8]
Additional information <2, 7> (<2> no inhibition by N-acetylserine [27]; <7> β-(pyrazol-1-yl)-L-alanine has no inhibitory effect [11]) [11, 27]

Cofactors/prosthetic groups
pyridoxal 5'-phosphate <13> (<13> 4 mol per mol of cysteine synthase complex [15]) [15]

Metals, ions
Additional information <2> (<2> salt concentrations less than 0.02 M or greater than 0.2 M appreciably decrease reaction rate [1]) [1]

Turnover number (min^{-1})
4.08 <2> (acetyl-CoA, <2> acetyl-CoA hydrolysis [27]) [27]
12000 <3> (acetyl-CoA, <3> cosubstrate L-serine [7]) [7]

Specific activity (U/mg)
0.00036 <18> (<18> purified recombinant mitochondrial isozyme A, mutant H327A [23]) [23]
0.00134 <18> (<18> purified recombinant mitochondrial isozyme A, mutant V353E [23]) [23]
0.0025 <2> (<2> wild-type strain C600 [25]) [25]
0.00513 <18> (<18> purified recombinant mitochondrial isozyme A, mutant G354A [23]) [23]
0.032 <18> (<18> purified recombinant mitochondrial isozyme A, wild-type [23]) [23]
0.89 <4> [2]
4.4 <13> (<13> recombinant protein, crude extract from E. coli [15]) [15]
14.7 <7> (<7> recombinant protein [11]) [11]
46 <14> (<14> recombinant enzyme, crude extract [16]) [16]
70 <13> (<13> purified recombinant enzyme [15]) [15]
71.6 <2> [6]
114 <3> (<3> purified enzyme [5]) [5]
205 <6> (<6> purified enzyme, pH 7.5 [9]) [9]
397 <3> (<3> purified nezyme [7]) [7]
Additional information <1-4, 6> (<2> at acetyl-CoA concentration 15fold higher than L-serine concentration, a nonproductive ternary complex of enzyme-CoA-L-serine is formed, kinetics [27]; <1,6> coupled assay system [4,9]) [3-5, 9, 10, 27]

K_m-Value (mM)

0.02 <13> (acetyl-CoA, <13> mitochondrial isozyme SAT-m [14]) [14]
0.1 <3, 5> (acetyl-CoA) [7, 8]
0.11 <2> (L-serine) [1]
0.12 <3> (acetyl-CoA) [5]
0.13 <7, 13> (acetyl-CoA) [11, 17, 24]
0.145 <18> (acetyl-CoA, <18> recombinant isozyme A, wild-type [23]) [23]
0.16 <13> (acetyl-CoA, <13> chloroplastidic isozyme SAT-p [14]) [14]
0.176 <18> (acetyl-CoA, <18> recombinant isozyme A, mutant G354A [23]) [23]
0.2 <4> (acetyl-CoA) [2]
0.24 <17> (L-serine) [22]
0.28 <13> (acetyl-CoA, <13> cytosolic isozyme SAT-c [14]) [14]
0.35 <6> (acetyl-CoA) [9]
0.4 <5> (L-serine) [8]
0.56 <2> (acetyl-CoA) [1]
0.59 <7, 13> (L-serine) [11, 17, 24]
0.6 <4> (L-serine) [2]
0.7 <3> (L-serine) [5]
0.77 <3> (L-serine) [7]
0.91 <13> (acetyl-CoA, <13> free recombinant enzyme [15]) [15]
0.92 <17> (acetyl-CoA) [22]
1.64 <13> (L-serine, <13> chloroplastidic isozyme SAT-p [14]) [14]
1.68 <13> (L-serine, <13> mitochondrial isozyme SAT-m [14]) [14]
2.29 <6> (L-serine) [9]
2.4 <18> (L-serine, <18> recombinant isozyme A, wild-type [23]) [23]
2.71 <13> (L-serine, <13> cytosolic isozyme SAT-c [14]) [14]
2.84 <13> (L-serine, <13> free recombinant enzyme [15]) [15]
25.55 <18> (L-serine, <18> recombinant isozyme A, mutant G354A [23]) [23]
Additional information <7, 13> (<7> K_m for L-serine and acetyl-CoA of recombinant L-cysteine-binding motif mutants, overview [24]; <7> K_m of mutant enzymes [17]; <13> cysteine synthase bienzyme complex [15]) [15, 17, 24]

K_i-Value (mM)

0.0037 <7> (L-cysteine, <7> versus L-serine [11]) [11]
0.0056 <7> (L-cysteine, <7> versus acetyl-CoA [11]) [11]
0.0074 <13> (L-cysteine, <13> versus acetyl-CoA, isozyme SAT-c [14]) [14]
0.0108 <13> (L-cysteine, <13> versus L-serine, isozyme SAT-c [14]) [14]
0.012 <17> (L-cysteine, <17> versus L-serine [22]) [22]
0.064 <7> (CoA, <7> versus acetyl-CoA [11]) [11]
0.14 <17> (L-cysteine, <17> versus acetyl-CoA [22]) [22]
Additional information <3> (<3> overview [10]) [10]

pH-Optimum

7.5 <2, 10> (<2,10> assay at [23,26]) [23, 26]
7.5-8 <5> [8]
7.6 <8> (<8> assay at [12]) [12]

7.6-7.8 <2> [1]
7.6-8 <7> (<7> Tris-buffer [11]) [11]
8-8.5 <6> [9]
8.3-8.5 <4> [2]
8.5 <4> (<4> assay at [3]) [3]

Temperature optimum (°C)
22 <8> (<8> assay at [12]) [12]
25 <2, 6, 10> (<2,6,10> assay at [9,23,26]) [9, 23, 26]
30 <4, 5> (<4,5> assay at [2,3,8]) [2, 3, 8]

Temperature range (°C)
15-70 <6> [9]

4 Enzyme Structure

Molecular weight
125000 <2> (<2> ultracentrifugation experiments [6]) [6]
145000 <2> (<2> gel filtration [26]) [26]
160000 <2> (<2> gel filtration [6]) [6]
163000 <3> (<3> gel filtration [7]) [7]
300000 <13> (<13> cysteine synthase bienzyme complex, gel filtration [15]) [15]
350000 <1> (<1> gel filtration [4]) [4]
450000-500000 <7> (<7> multienzyme complex of L-serine acetyltransferase and L-cysteine synthase, gel filtration [11]) [11]
Additional information <7, 10, 12, 18-21> (<10,18-21> amino acid sequence alignment [23]; <12> amino acid sequence comparison [13]; <7> multi-enzyme complex is formed between serine acetyltransferase and cysteine synthase possibly leading to metabolic channeling [11]) [11, 13, 23]
2000000 <13> (<13> enzyme aggregate, gel filtration [15]) [15]

Subunits
? <3, 6, 8, 9, 12-14, 17> (<17> x * 38400, amino acid sequence determination [22]; <14> x * 43737, native enzyme, amino acid sequence determination [16]; <14> x * 31000, recombinant enzyme, SDS-PAGE [16]; <13> x * 33000-36500, 3 recombinant isozymes, SDS-PAGE [14]; <12> x * 34330, cytosolic isozyme, amino acid sequence determination [13]; <8,9> x * 32000-35000, SDS-PAGE [12]; <6> x * 33000, SDS-PAGE [9]; <3> x * 30100, SDS-PAGE [7]) [7, 9, 12-14, 16, 22]
tetramer <2, 13> (<13> 2 * 34000, serine acetyltransferase + 2 * 36000, O-acetylserine (thiol) lyase, forming the cysteine synthase complex, SDS-PAGE [15]; <2> 4 * 29261, ultracentrifugation, gel filtration, crystallographic data [6]) [6, 15]
Additional information <1> (<1> enzyme exists in a high-molecular weight enzyme complex with cysteine synthase [4]) [4]

5 Isolation/Preparation/Mutation/Application

Source/tissue

hypocotyl <7> [11]
leaf <1, 4, 6, 8, 10, 12, 13, 16> (<13> SAT-c, SAT-m, SAT-p [14]; <10> SAT1-6 gene, major part [12]; <8> SAT1 gene [12]) [3, 4, 9, 12, 13, 14, 20]
root <8, 10> (<10> SAT1-6 gene, minor part [12]; <8> SAT1 gene [12]) [12]
seedling <4, 6, 7> (<6> green and etiolated [21]; <7> hypocotyl of etiolated seedlings [11]) [2, 3, 11, 21]
Additional information <7> (<7> accumulation in etiolated plants [11]) [11]

Localization

chloroplast <6, 7, 13, 14, 16, 22> (<14> chloroplast isozyme, gene cmSAT [16,24]; <13> isozyme SAT-p, stroma [14]; <12> 10% of total activity [13]; <6> stroma [9]) [9, 13-16, 20, 21, 24]
cytosol <6, 7, 12-14, 22> (<14> small amount of enzyme [16]; <13> isozyme SAT-c [14,24]; <12> 14% of total activity [13]) [13, 14, 16, 21, 24]
mitochondrion <4, 7, 12, 13, 22> (<13> isozyme SAT-m, matrix [14,24]; <12> 76% of total activity [13]; <4> between one-third and two-thirds of the activity [3]) [3, 13, 14, 24]
protoplast <12> [13]

Purification

<1> [4]
<2> (recombinant wild-type and mutants from E. coli [26]; recombinant from E. coli [6,27]) [1, 6, 26, 27]
<3> (recombinant [10]) [5, 7, 10]
<4> (partially [2]) [2]
<5> (partially [8]) [8]
<6> [9]
<12> (partially, 3 isozymes [13]) [13]
<13> (recombinant from E. coli [15]; partially, recombinant isozymes SAT-p, SAT-m, and SAT-c from E. coli [14]) [14, 15]
<14> (recombinant His-tagged protein from E. coli [16]) [16]
<18> (recombinant wild-type and mutants as S-tagged protein from E. coli [23]) [23]

Crystallization

<2> (hanging drop vapour diffusion method, 8-16% polyethylene glycol 1000, 10 mM Tris-HCl, pH 7.5, 2-4 weeks, room temperature, structure analysis [6]) [6]

Cloning

<2> (overexpression [27]; C-terminal deletion mutants, expression in Escherichia coli strain JM70 [26]; insertional inactivation of wild-type CysE gene from Escherichia coli C600 [25]; construction of an inactivated mutant by targeted disruption of CysE gene for usage in complementation assays [23]; construction of cysE gene mutants by random PCR mutagenesis, expression

in Escherichia coli cysteine auxotrophic strain JM39-8 [19]; overexpression in Escherichia coli, construction of an expression vector which increases the expression to 17% of the soluble cell protein [6]) [6, 19, 23, 25, 26]
<3> (expression in Salmonella typhimurium Cys$^-$ strain DW 18.1 from plasmid [10]) [10]
<6> (expression in Cys$^-$ Escherichia coli mutant and functional complementation, transient expression as GFP-fusion and β-glucuronidase-fusion protein in plants [21]) [21]
<7> (expression of wild-type and mutants in Escherichia coli [17,24]; isolation of cDNA clone from Citrullus vulgaris genetic library by complementation of a Cys$^-$ Escherichia coli strain JM 39/5, single copy gene, DNA and amino acid sequence determination [11]) [11, 17, 24]
<8> (expression of SAT1 gene as glutathione-S-transferase fusion protein in Escherichia coli DH5α, complementation of serine acetyltransferase mutant Escherichia coli strain JM39, single copy gene, amino acid sequence determination [12]; SAT1 is closely linked to SAT5, high DNA sequence homology [12]) [12]
<10> (expression in Escherichia coli SAT inactivated mutant [23]) [23]
<12> (cloning and DNA sequencing of 3 isozymes, complementation of Cys$^-$ mutant Escherichia coli strain JM39 with the 3 isozymes, sequence analysis of cytosolic isozyme [13]) [13]
<13> (expression of wild-type and mutants in Escherichia coli [17]; overexpression in Escherichia coli [15]; expression of 3 isozymes SAT-m, SAT-p, SAT-c as GFP-fusion proteins in Arabidopsis thaliana [14]; overexpression of 3 isozymes SAT-m, SAT-p, SAT-c in Escherichia coli BL21 (DE3) [14]) [14, 15, 17]
<14> (expression of His-tagged protein encoded on gene cmSAT in Escherichia coli, functional complementation of SAT-deficient Escherichia coli strain JM15, DNA and amino acid sequence determination [16]) [16]
<15> (expression in Escherichia coli, functional complementation of Escherichia coli cysA$^-$ mutant, DNA and amino acid sequence detemination [18]) [18]
<17> (expression as glutathione-S-transferase fusion protein, amino acid sequence [22]) [22]
<18> (expression of wild-type and mutants in Escherichia coli SAT inactivated mutant as S-tagged protein [23]) [23]
<24> (cloned in expression plasmid with and without mitochondrial transit peptide for transformation of Nicotiana tabacum plants [25]) [25]
<25> (expression in inactivated Escherichia coli mutant, unaltered L-cysteine sensitivity compared to wild-type [25]) [25]
<26> (expression in inactivated Escherichia coli mutant, insensitive against inhibition by L-cysteine [25]) [25]
<27> (expression in inactivated Escherichia coli mutant, unaltered L-cysteine sensitivity compared to wild-type [25]) [25]

Engineering

E166G/M201V <2> (<2> random PCR mutagenesis, cysE mutant, more insensitive to inhibition by L-cysteine than the wild-type, reduced enzyme activity [19]) [19]
E278L <7> (<7> site-directed mutagenesis, about 1.7fold more sensitive to inhibition by L-cysteine [17,24]) [17, 24]
E282A <18> (<18> site-directed mutagenesis in C-terminal hexapeptide-repeat domain, enhanced activity, complementation of inactivated E. coli mutant [23]) [23]
F285Y <7> (<7> site-directed mutagenesis, about 1.7fold less sensitive to inhibition by L-cysteine [17,24]) [17, 24]
G277C <7> (<7> site-directed mutagenesis, about 27fold less sensitive to inhibition by L-cysteine [17,24]) [17, 24]
G354A <18> (<18> site-directed mutagenesis in C-terminal hexapeptide-repeat domain, reduced activity, partial complementation of inactivated E. coli mutant [23]) [23]
H282Q <7> (<7> site-directed mutagenesis, about 3.5fold less sensitive to inhibition by L-cysteine [17,24]) [17, 24]
H309A <18> (<18> site-directed mutagenesis in C-terminal hexapeptide-repeat domain, strongly reduced activity [23]) [23]
H327A <18> (<18> site-directed mutagenesis in C-terminal hexapeptide-repeat domain, nearly no activity [23]) [23]
M201R <2> (<2> random PCR mutagenesis, cysE mutant, more insensitive to inhibition by L-cysteine than the wild-type, reduced enzyme activity [19]) [19]
M201T <2> (<2> random PCR mutagenesis, cysE mutant, more insensitive to inhibition by L-cysteine than the wild-type, reduced enzyme activity [19]) [19]
M256A <2> (<2> random PCR mutagenesis, cysE mutant, more insensitive to inhibition by L-cysteine than the wild-type, reduced enzyme activity [19]) [19]
M280I <7> (<7> site-directed mutagenesis, insensitive to inhibition by L-cysteine [17]) [17]
N51K/R91H/H233Y <2> (<2> random PCR mutagenesis, cysE mutant, more insensitive to inhibition by L-cysteine than the wild-type, reduced enzyme activity [19]) [19]
P252R <2> (<2> random PCR mutagenesis, cysE mutant, more insensitive to inhibition by L-cysteine than the wild-type, reduced enzyme activity [19]) [19]
S253L <2> (<2> random PCR mutagenesis, cysE mutant, more insensitive to inhibition by L-cysteine than the wild-type, reduced enzyme activity [19]) [19]
S279T <7> (<7> site-directed mutagenesis, as sensitive to inhibition by L-cysteine as the wild-type [17,24]) [17, 24]
T167K <2> (<2> random PCR mutagenesis, cysE mutant, more insensitive to inhibition by L-cysteine than the wild-type, reduced enzyme activity [19]) [19]

V353E <18, 21> (<21> strain BB1 [23]; <18,21> mutant enzyme does not bind O-acetylserine (thiol) lyase [23]; <18> site-directed mutagenesis in C-terminal hexapeptide-repeat domain, nearly no activity [23]) [23]
Additional information <2, 3, 7, 13, 16, 23, 24> (<2> construction of deletion mutants lacking 10 to 25 amino acid residues at the C-terminal end, deletions alter the sensitivity against L-cysteine inhibition and the interaction with O-acetylserine (thiol) lyase, overview, ΔC30 mutant is inactive [26]; <23,24> transformation via Agrobacterium tumefaciens, transgenic Nicotiana tabacum plants overexpressing recombinant Arabidopsis thaliana SAT A isozyme, expression with and without mitochondrial transit peptide [25]; <7,13> transgenic Arabidopsis plants expressing recombinant wild-type SATase are inhibited by L-cysteine [24]; <7,13> SAT of transgenic Arabidopsis plants overexpressing the point-mutated watermelon gene are not inhibited by L-cysteine, regardless if the recombinant enzyme is targeted to the chloroplast or expressed in the cytosol [24]; <16> construction of transgenic plants by introduction of E. coli cysE gene, targeted into the chloroplasts by fusing a signal sequence of rubisco from Arabidopsis thaliana, crude Solanum tuberosum leaf extract show 20fold higher enzyme activity [20]; e<2> cysteine and cystine production level of mutants [19]; <3> CysE mutants show strongly reduced enzyme activity [1]) [1, 19, 20, 24, 25]

Application

agriculture <16> (<16> metabolic engineering, i.e. construction of transgenic plants, of Solanum tuberosum to enhance the endogenous content of sulfur-containing compounds [20]) [20]

6 Stability

Temperature stability

0 <2> (<2> SAT ΔC10 mutant, lacking 10 amino acid residues at the C-terminal end is inactivated after 12 h [26]) [26]
50 <4> (<4> half-life: 25 min [2]) [2]
60 <4> (<4> half-life: 2 min [2]) [2]
80 <6> (<6> 5 min, 50% loss of activity [9]) [9]
90 <6> (<6> 5 min, complete loss of activity [9]) [9]

Storage stability

<3>, -20°C, stable for months [5]
<3>, 4°C, unstable [5]
<4>, -20°C, 1 week, stable [2]
<4>, 2°C, 50% loss of activity after 1 week [2]
<6>, 20°C, 0.4 M NaCl, 2 weeks, stable [9]

References

[1] Kredich, N.M.; Tomkins, G.M.: The enzymic synthesis of L-cysteine in Escherichia coli and Salmonella typhimurium. J. Biol. Chem., **241**, 4955-4965 (1966)

[2] Smith, I.K.; Thompson, J.F.: Purification and characterization of L-serine transacetylase and O-acetyl-L-serine sulfhydrylase from kidney bean seedlings (Phaseolus vulgaris). Biochim. Biophys. Acta, **227**, 288-295 (1971)

[3] Smith, I.K.: Studies of L-cysteine biosynthetic enzymes in Phaseolus vulgaris L.. Plant Physiol., **50**, 477-479 (1972)

[4] Nakamura, K.; Hayama, A.; Masada, M.; Fukushima, K.; Tamura, G.: Purification and some properties of plant serine acetyltransferase. Plant Cell Physiol., **29**, 689-693 (1988)

[5] Kredich, N.M.; Becker, M.A.: Cysteine biosynthesis: serine transacetylase and O-acetylserine sulfhydrylase (Salmonella typhimurium). Methods Enzymol., **17**, 459-470 (1971)

[6] Wigley, D.B.; Derrick, J.P.; Shaw, W.V.: The serine acetyltransferase from Escherichia coli. Over-expression, purification and preliminary crystallographic analysis. FEBS Lett., **277**, 267-271 (1990)

[7] Baecker, P.A.; Wedding, R.T.: Purification of serine acetyltransferase, a component of a multienzyme complex, by immunoadsorption and selective dissociation of the complex. Anal. Biochem., **102**, 16-21 (1980)

[8] Burnell, J.N.; Whatley, F.R.: Sulphur metabolism in Paracoccus denitrificans. Purification, properties and regulation of serine transacetylase, O-acetylserine sulphydrolase and β-cystathionase. Biochim. Biophys. Acta, **481**, 246-265 (1977)

[9] Ruffet, M.L.; Droux, M.; Douce, R.: Purification and kinetic properties of serine acetyltransferase free of O-acetylserine(thiol)lyase from spinach chloroplasts. Plant Physiol., **104**, 597-604 (1994)

[10] Leu, L.S.; Cook, P.F.: Kinetic mechanism of serine transacetylase from Salmonella typhimurium. Biochemistry, **33**, 2667-2671 (1994)

[11] Saito, K.; Yokoyama, H.; Noji, M.; Murakoshi, I.: Molecular cloning and characterization of a plant serine acetyltransferase playing a regulatory role in cysteine biosynthesis from watermelon. J. Biol. Chem., **270**, 16321-16326 (1995)

[12] Murillo, M.; Fogilia, R.; Diller, A.; Lee, S.; Leustek, T.: Serine acetyltransferase from Arabidopsis thaliana can functionally complement the cysteine requirement of a cysE mutant strain of Escherichia coli. Cell. Mol. Biol. Res., **41**, 425-433 (1996)

[13] Ruffet, M.L.; Lebrun, M.; Droux, M.; Douce, R.: Subcellular distribution of serine acetyltransferase from Pisum sativum and characterization of an Arabidopsis thaliana putative cytosolic isoform. Eur. J. Biochem., **227**, 500-509 (1995)

[14] Noji, M.; Inoue, K.; Kimura, N.; Gouda, A.; Saito, K.: Isoform-dependent differences in feedback regulation and subcellular localization of serine

acetyltransferase involved in cysteine biosynthesis from Arabidopsis thaliana. J. Biol. Chem., **273**, 32739-32745 (1998)
[15] Droux, M.; Ruffet, M.L.; Douce, R.; Job, D.: Interactions between serine acetyltransferase and O-acetylserine (thiol) lyase in higher plants. Structural and kinetic properties of the free and bound enzymes. Eur. J. Biochem., **255**, 235-245 (1998)
[16] Toda, K.; Takano, H.; Miyagishima, S.y.; Kuroiwa, H.; Kuroiwa, T.: Characterization of a chloroplast isoform of serine acetyltransferase from the thermo-acidiphilic red alga Cyanidioschyzon merolae. Biochim. Biophys. Acta, **1403**, 72-84 (1998)
[17] Inoue, K.; Noji, M.; Saito, K.: Determination of the sites required for the allosteric inhibition of serine acetyltransferase by L-cysteine in plants. Eur. J. Biochem., **266**, 220-227 (1999)
[18] Grynberg, M.; Topczewski, J.; Godzik, A.; Paszewski, A.: The Aspergillus nidulans cysA gene encodes a novel type of serine O-acetyltransferase which is homologous to homoserine O-acetyltransferases. Microbiology, **146**, 2695-2703 (2000)
[19] Takagi, H.; Kobayashi, C.; Kobayashi, S.I.; Nakamori, S.: PCR random mutagenesis into Escherichia coli serine acetyltransferase: isolation of the mutant enzymes that cause overproduction of L-cysteine and L-cystine due to the desensitization to feedback inhibition. FEBS Lett., **452**, 323-327 (1999)
[20] Harms, K.; Von Ballmoos, P.; Brunold, C.; Hofgen, R.; Hesse, H.: Expression of a bacterial serine acetyltransferase in transgenic potato plants leads to increased levels of cysteine and glutathione. Plant J., **22**, 335-343 (2000)
[21] Noji, M.; Takagi, Y.; Kimura, N.; Inoue, K.; Saito, M.; Horikoshi, M.; Saito, F.; Takahashi, H.; Saito, K.: Serine acetyltransferase involved in cysteine biosynthesis from spinach: molecular cloning, characterization and expression analysis of cDNA encoding a plastidic isoform. Plant Cell Physiol., **42**, 627-634 (2001)
[22] Nozaki, T.; Shigeta, Y.; Saito-Nakano, Y.; Imada, M.; Kruger, W.D.: Characterization of transsulfuration and cysteine biosynthetic pathways in the protozoan hemoflagellate, Trypanosoma cruzi. Isolation and molecular characterization of cystathionine β-synthase and serine acetyltransferase from Trypanosoma. J. Biol. Chem., **276**, 6516-6523 (2001)
[23] Wirtz, M.; Berkowitz, O.; Droux, M.; Hell, R.: The cysteine synthase complex from plants: mitochondrial serine acetyltransferase from Arabidopsis thaliana carries a bifunctional domain for catalysis and protein-protein interaction. Eur. J. Biochem., **268**, 686-693 (2001)
[24] Noji, M.; Saito, K.: Molecular and biochemical analysis of serine acetyltransferase and cysteine synthase towards sulfur metabolic engineering in plants. Amino Acids, **22**, 231-243 (2002)
[25] Wirtz, M.; Hell, R.: Production of cysteine for bacterial and plant biotechnology: Application of cysteine feedback-insensitive isoforms of serine acetyltransferase. Amino Acids, **24**, 195-203 (2003)
[26] Mino, K.; Hiraoka, K.; Imamura, K.; Sakiyama, T.; Eisaki, N.; Matsuyama, A.; Nakanishi, K.: Characteristics of serine acetyltransferase from Escheri-

chia coli deleting different lengths of amino acid residues from the C-terminus. Biosci. Biotechnol. Biochem., **64**, 1874-1880 (2000)
[27] Hindson, V.J.; Shaw, W.V.: Random-order ternary complex reaction mechanism of serine acetyltransferase from Escherichia coli. Biochemistry, **42**, 3113-3119 (2003)

Homoserine O-acetyltransferase 2.3.1.31

1 Nomenclature

EC number
2.3.1.31

Systematic name
acetyl-CoA:L-homoserine O-acetyltransferase

Recommended name
homoserine O-acetyltransferase

Synonyms
L-homoserine O-acetyltransferase
acetyltransferase, homoserine
homoserine acetyltransferase
homoserine transacetylase
homoserine-O-transacetylase

CAS registry number
9030-72-2

2 Source Organism

<1> *Saccharomyces cerevisiae* (MET2 gene [5]; haploid wild-type strain (a-mating type) and met-auxotroph mutant strains 6,8,12,13,15, and 17 [6]) [1, 5, 6]
<2> *Ascobolus immersus* [7]
<3> *Neurospora sp.* [2, 8]
<4> *Brevibacterium flavum* (mutant strain M-116 derived from wild-type strain 2247 [3]) [1, 3]
<5> *Bacillus polymyxa* (met-auxotroph strain M4 [4]) [1, 4]
<6> *Leptospira meyeri* (metX gene [9]) [9]
<7> *Haemophilus influenzae* (met2 gene [10]) [10]
<8> *Corynebacterium glutamicum* (metA gene [11]) [11]

3 Reaction and Specificity

Catalyzed reaction
acetyl-CoA + L-homoserine = CoA + O-acetyl-L-homoserine (<1,4,5,7> ping-pong reaction mechanism [1,3,10])

Reaction type

acyl group transfer

Natural substrates and products

- **S** acetyl-CoA + L-homoserine <3, 4, 6-8> (<7> control point in the pathway [10]; <4,6-8> first reaction in L-methionine biosynthesis [3,9-11]) (Reversibility: r <7> [10]; ? <3, 4, 6, 8> [3, 8, 9, 11]) [3, 8-11]
- **P** CoA + O-acetyl-L-homoserine <3, 6-8> (<3> intermediate in methionine biosynthesis [8]) [8-11]

Substrates and products

- **S** acetyl-CoA + 3-amino-1-propanol <7> (<7> low activity [10]) (Reversibility: r <7> [10]) [10]
- **P** ?
- **S** acetyl-CoA + 4-nitrophenyl acetate <7> (<7> low activity [10]) (Reversibility: r <7> [10]) [10]
- **P** ?
- **S** acetyl-CoA + D-homoserine <7> (Reversibility: r <7> [10]) [10]
- **P** CoA + O-acetyl-D-homoserine
- **S** acetyl-CoA + L-homoserine <1, 3-8> (<7,8> preference for acetyl-CoA [10,11]; <1> not D-homoserine or any other hydroxy-L-amino acid [6]; <3> acetyl phosphate, acetyl carnitine, acetyl-[acyl-carrier-protein] from E. coli [2,8]; <3> acetyl imidazole cannot replace acetyl-CoA [8]; <1,4> succinyl-CoA cannot replace acetyl-CoA [3,6]) (Reversibility: r <7> [10]; ir <1, 3> [2, 6]; ? <1, 3-6, 8> [1, 3-5, 8, 9, 11]) [1-6, 8-11]
- **P** CoA + O-acetyl-L-homoserine <1, 3-8> [2-4, 6, 8-11]
- **S** acetyl-CoA + γ-hydroxybutyric acid <7> (Reversibility: r <7> [10]) [10]
- **P** ?
- **S** butyryl-CoA + L-homoserine <7> (Reversibility: ? <7> [10]) [10]
- **P** CoA + O-butyryl-L-homoserine
- **S** crotonyl-CoA + L-homoserine <7> (<7> low activity [10]) (Reversibility: r <7> [10]) [10]
- **P** CoA + O-crotonyl-L-homoserine
- **S** glutaryl-CoA + L-homoserine <7> (<7> very low activity [10]) (Reversibility: r <7> [10]) [10]
- **P** CoA + O-glutaryl-L-homoserine
- **S** propionyl-CoA + L-homoserine <7> (Reversibility: ? <7> [10]) [10]
- **P** CoA + O-propionyl-L-homoserine
- **S** propionyl-CoA + L-homoserine <7> (Reversibility: ? <7> [10]) [10]
- **P** CoA + O-propionyl-L-homoserine
- **S** succinyl-CoA + L-homoserine <7> (<7> very low activity [10]) (Reversibility: r <7> [10]) [10]
- **P** CoA + O-succinyl-L-homoserine
- **S** Additional information <1, 3, 5> (<1,3,5> catalyzes reversible L-homoserine/O-acetyl-L-homoserine exchange in the absence of acetyl-CoA [2,4,6,8]; <1> deacylates acetyl-CoA in the absence of L-homoserine [6]) [2, 4, 6, 8]
- **P** ?

Inhibitors

DL-penicillamine <1> (<1> 3.3 mM, complete inhibition [6]) [6]
L-cysteine <1> (<1> weak [6]) [6]
L-homocysteine <1> (<4> not [3]) [6]
L-methionine <1, 3, 5> (<1> repression of enzyme expression [6]; <5> addition of S-adenosyl-L-methionine enhances the inhibitory effect [4]; <1,4> no inhibition [3,6]; <3,5> multivalent feed-back inhibition [2,4]) [2, 4, 6]
O-acetyl-L-homoserine <1, 4> [3, 6]
O-acetyl-L-serine <1> (<1> weak [6]) [6]
O-phospho-L-homoserine <1> (<1> weak [6]) [6]
O-succinyl-L-homoserine <1> (<1> weak [6]) [6]
S-adenosyl-L-homocysteine <1> (<1> weak [6]) [6]
S-adenosyl-L-methionine <1, 3-5> (<5> addition of L-methionine enhances the inhibitory effect [4]; <3,5> multivalent feed-back inhibition [2,4]; <1,4> weak [3,6]) [2-4, 6]
Zn^{2+} <5> (<5> reversible [4]) [4]
acetylcarnitine <3> [8]
ethionine <5> [4]
Additional information <1, 4, 5> (<1> no inhibition by $MgCl_2$, $MnCl_2$, $ZnCl_2$, hydroxylamine, phenylhydrazine, D-homoserine, methionine, L-tyrosine, L-serine and nucleotides [6]; <5> no inhibition by Cu^{2+}, Cd^{2+}, Co^{2+}, Fe^{2+}, Hg^{2+}, Mg^{2+}, Mn^{2+}, Ni^{2+}, CoA [4]; <4> effects of several amino acids with and without addition of L-methionine, overview [3]) [3, 4, 6]

Turnover number (min^{-1})

300 <7> (3-amino-1-propanol) [10]
468 <7> (4-nitrophenyl acetate) [10]
900 <7> (crotonyl-CoA) [10]
1620 <7> (butyryl-CoA) [10]
1800 <7> (propionyl-CoA) [10]
1860 <7> (O-acetyl-L-homoserine) [10]
3000 <7> (4-hydroxybutyric acid) [10]
4680 <7> (D-homoserine) [10]
5520 <7> (L-homoserine) [10]
5520 <7> (acetyl-CoA) [10]

Specific activity (U/mg)

0.0036 <3> [8]
0.143 <4> [3]
1.17 <5> [4]
3.6 <3> [2]
67.4 <1> [6]
Additional information <3> [8]

K_m-Value (mM)

0.012 <1> (acetyl-CoA, <1> deacylation [6]) [6]
0.027 <1> (acetyl-CoA) [6]
0.05 <4> (acetyl-CoA) [1, 3]

0.09 <7> (propionyl-CoA) [10]
0.1 <1> (acetyl-CoA) [1]
0.13 <7> (L-homoserine) [10]
0.13 <7> (crotonyl-CoA) [10]
0.14 <7> (acetyl-CoA) [10]
0.2 <5> (L-homoserine) [4]
0.2 <5> (acetyl-CoA) [1, 4]
0.21 <7> (butyryl-CoA) [10]
0.28 <7> (glutaryl-CoA) [10]
0.36 <7> (succinyl-CoA) [10]
0.62 <7> (CoA) [10]
0.85 <7> (O-acetyl-L-homoserine) [10]
1 <1> (L-homoserine) [6]
1.3 <1> (L-homoserine) [1]
1.4 <7> (4-nitrophenyl acetate) [10]
2.8 <4> (L-homoserine) [3]
4.7 <7> (D-homoserine) [10]
10 <5> (L-homoserine, <5> exchange reaction [4]) [4]
19 <7> (4-hydroxybutyric acid) [10]
71 <7> (3-amino-1-propanol) [10]

K_i-Value (mM)
0.003 <5> (Zn^{2+}) [4]

pH-Optimum
7.3-7.5 <3> (<3> assay at [8]) [8]
7.5 <1, 7> (<7> assay at [10]) [6, 10]
7.5-8 <5> [4]
8 <4> [3]
Additional information <1> (<1> pI: 4.0 [6]) [6]

pH-Range
5.5-9 <7> [10]
6.6-9 <5> (<5> more than half-maximal activity over this range [4]) [4]

Temperature optimum (°C)
8 <5> (<5> assay at [4]) [4]
25 <7> (<7> assay at [10]) [10]
30 <1, 3, 4> (<1,3,4> assay at [2,3,6,8]) [2, 3, 6, 8]

4 Enzyme Structure

Molecular weight
40000 <5> (<5> sucrose density gradient centrifugation, PAGE [4]) [4]
41000 <8> (<8> amino acid sequence determination [11]) [11]
73000 <7> (<7> gel filtration [10]) [10]

101000-104000 <1> (<1> sucrose density gradient centrifugation, gel filtration [6]) [6]
Additional information <8> (<8> amino acid sequence comparison [11]) [11]

Subunits
dimer <7> (<7> 2 * 35000-40000, SDS-PAGE and electrospray ionization-mass spectrometry [10]) [10]
monomer <5> (<5> 1 * 37000, SDS-PAGE [4]) [4]

5 Isolation/Preparation/Mutation/Application

Purification
<1> (partial, mutant strain 6 [6]) [1, 6]
<3> (partial [2,8]) [2, 8]
<4> (partial [3]) [3]
<5> (partial [4]) [4]
<7> (recombinant from E. coli [10]) [10]

Cloning
<1> (met2-gene, expression in Escherichia coli DH1, DNA and amino acid sequence determination [5]) [5]
<2> (MET2-gene, expression in Escherichia coli DH1, DNA and amino acid sequence determination [7]) [7]
<6> (overexpression of metX in Escherichia coli BL21 (DE3), complementation of Escherichia coli metA deficient mutant, amino acid sequence determination [9]) [9]
<7> (met2 gene, expression in Escherichia coli BL21 (DE3) [10]) [10]
<8> (metA gene, complementation of Escherichia coli metA deficient mutant, DNA sequence determination [11]) [11]

Engineering
Additional information <1, 3, 4> (<3> methionine deficient strains do not perform the L-homoserine/O-acetyl-L-homoserine exchange reaction [8]; <1,4> methionine auxotroph strains show no activity [3,6]) [3, 6, 8]

6 Stability

pH-Stability
7.8 <1> (<1> most stable [6]) [6]

Temperature stability
0 <5> (<5> $t_{1/2}$: 15 min [4]) [4]
8 <5> (<5> $t_{1/2}$: 10 min [4]) [4]
25 <5> (<5> most rapid and irreversible inactivation [4]) [4]

Storage stability

<5>, -10°C, $t_{1/2}$: 3 months [4]

References

[1] Yamagata, S.: Acetyl-CoA: L-homoserine O-acetyltransferase of the yeast Saccharomyces: substrate kinetics. Kyoyobu Kenkyu Hokoku, **24**, 27-28 (1988)

[2] Nagai, S.; Kerr, D.: Homoserine transacetylase (Neurospora). Methods Enzymol., **17B**, 442-445 (1971)

[3] Miyajima, R.; Shiio, I.: Regulation of aspartate family amino acid biosynthesis in Brevibacterium flavum. VII. Properities of homoserine O-transacetylase. J. Biochem., **73**, 1061-1068 (1973)

[4] Wyman, A.; Paulus, H.: Purification and properties of homoserine transacetylase from Bacillus polymyxa. J. Biol. Chem., **250**, 3897-3903 (1975)

[5] Langin, T.; Faugeron, G.; Goyon, C.; Nicolas, A.; Rossignol, J.L.: The MET2 gene of Saccharomyces cerevisiae: molecular cloning and nucleotide sequence. Gene, **49**, 283-293 (1986)

[6] Yamagata, S.: Partial purification and some properties of homoserine O-acetyltransferase of a methionine auxotroph of Saccharomyces cerevisiae. J. Bacteriol., **169**, 3458-3463 (1987)

[7] Goyon, C.; Faugeron, G.; Rossignol, J.L.: Molecular cloning and characterization of the met2 gene from Ascobolus immersus. Gene, **63**, 297-308 (1988)

[8] Nagai, S.; Flavin, M.: Acetylhomoserine. An intermediate in the fungal biosynthesis of methionine. J. Biol. Chem., **242**, 3884-3895 (1967)

[9] Bourhy, P.; Martel, A.; Margarita, D.; Saint Girons, I.; Belfaiza, J.: Homoserine O-acetyltransferase, involved in the Leptospira meyeri methionine biosynthetic pathway, is not feedback inhibited. J. Bacteriol., **179**, 4396-4398 (1997)

[10] Born, T.L.; Franklin, M.; Blanchard, J.S.: Enzyme-catalyzed acylation of homoserine: mechanistic characterization of the Haemophilus influenzae met2-encoded homoserine transacetylase. Biochemistry, **39**, 8556-8564 (2000)

[11] Park, S.D.; Lee, J.Y.; Kim, Y.; Kim, J.H.; Lee, H.S.: Isolation and analysis of metA, a methionine biosynthetic gene encoding homoserine acetyltransferase in Corynebacterium glutamicum. Mol. Cell., **8**, 286-294 (1998)

Lysine N-acetyltransferase 2.3.1.32

1 Nomenclature

EC number
2.3.1.32

Systematic name
acetyl-phosphate:L-lysine N^6-acetyltransferase

Recommended name
lysine N-acetyltransferase

Synonyms
acetyltransferase, lysine
lysine acetyltransferase

CAS registry number
37257-12-8

2 Source Organism

<1> *Bos taurus* [1]
<2> *Mus musculus* [2]

3 Reaction and Specificity

Catalyzed reaction
acetyl phosphate + L-lysine = phosphate + N^6-acetyl-L-lysine

Reaction type
acyl group transfer

Natural substrates and products
S acetyl phosphate + L-lysine <1> (Reversibility: ? <1> [1]) [1]
P phosphate + N^6-acetyl-L-lysine <1> [1]
S acetyl phosphate + L-ornithine <1> (<1> best substrate [1]) (Reversibility: ? <1> [1]) [1]
P phosphate + N^5-acetyl-L-ornithine <1> [1]

Substrates and products
S acetyl phosphate + L-lysine <1> (Reversibility: ? <1> [1]) [1]
P phosphate + N^6-acetyl-L-lysine <1> [1]

S acetyl phosphate + L-ornithine <1> (<1> best substrate [1]) (Reversibility: ? <1> [1]) [1]
P phosphate + N^5-acetyl-L-ornithine <1> [1]

Inhibitors
phosphate <1> (<1> above 0.1 mM [1]) [1]
Additional information <1> (<1> no inhibition by cysteine, glutathione, ascorbic acid, Zn^{2+}, Mn^{2+}, Co^{2+}, Ni^{2+}, Fe^{3+} or acetate [1]) [1]

Metals, ions
Ca^{2+} <1> (<1> slight stimulation [1]) [1]
Mg^{2+} <1> (<1> slight stimulation [1]) [1]
Additional information <1> (<1> no activation by Zn^{2+}, Mn^{2+}, Co^{2+}, Ni^{2+}, Fe^{3+} or acetate [1]) [1]

Specific activity (U/mg)
1.43 <1> [1]

K_m-Value (mM)
1.32 <1> (L-ornithine) [1]
1.35 <1> (acetyl phosphate, <1> cosubstrate L-ornithine [1]) [1]
4.24 <1> (L-lysine) [1]
12.4 <1> (acetyl phosphate, <1> cosubstrate L-lysine [1]) [1]

pH-Optimum
8.6 <1> (<1> various buffer systems [1]) [1]

pH-Range
8.3-9.2 <1> (<1> about half-maximal activity at pH 8.3, about 70% of maximal activity at pH 9.2 [1]) [1]

Temperature optimum (°C)
38 <1> (<1> assay at [1]) [1]

5 Isolation/Preparation/Mutation/Application

Source/tissue
insular cortex <2> (<2> novel taste learning elicited biphasic (acute and long-tasting) activation of two distinct lysine acetyltransferase activities along with the EPK/MAPK cascade in insular cortex [2]) [2]
liver <1> [1]

Purification
<1> [1]

6 Stability

Storage stability
<1>, 0°C, 5 h [1]

References

[1] Paik, W.K.; Kim, S.: Enzyme synthesis of ε-N-acetyl-L-lysine. Arch. Biochem. Biophys., **108**, 221-229 (1964)
[2] Swank, M.W.; Sweatt, J.D.: Increased histone acetyltransferase and lysine acetyltransferase activity and biphasic activation of the ERK/RSK cascade in insular cortex during novel taste learning. J. Neurosci., **21**, 3383-3391 (2001)

Histidine N-acetyltransferase 2.3.1.33

1 Nomenclature

EC number
2.3.1.33

Systematic name
acetyl-CoA:L-histidine N-acetyltransferase

Recommended name
histidine N-acetyltransferase

Synonyms
HISAT <2> [2]
acetylhistidine synthetase
acetyltransferase, histidine
histidine acetyltransferase

CAS registry number
9027-59-2

2 Source Organism

<1> *Fundulus heteroclitus* (killifish [1]) [1]
<2> *Oreochromis niloticus* (Nile tilapia [2,3]) [2, 3]

3 Reaction and Specificity

Catalyzed reaction
acetyl-CoA + L-histidine = CoA + N-acetyl-L-histidine

Reaction type
acyl group transfer

Natural substrates and products
S acetyl-CoA + L-histidine <1, 2> (Reversibility: ? <1, 2> [1-3]) [1-3]
P CoA + N-acetyl-L-histidine <1, 2> [1-3]

Substrates and products
S acetyl-CoA + 1-methyl-L-histidine <2> (<2> 23% activity compared to that with L-histidine [2]) (Reversibility: ? <2> [2]) [2]
P CoA + 1-methyl-N-acetyl-L-histidine <2> [2]

S acetyl-CoA + 3-methyl-L-histidine <2> (<2> 71% activity compared to that with L-histidine [2]) (Reversibility: ? <2> [2]) [2]
P CoA + 3-methyl-N-acetyl-L-histidine <2> [2]
S acetyl-CoA + L-histidine <1, 2> (<2> strictly specific for the substrates [2]; <1> with D-histidine 12-20% of the activity with L-histidine [1]) (Reversibility: ? <1, 2> [1-3]) [1-3]
P CoA + N-acetyl-L-histidine <1, 2> [1-3]

Inhibitors

2-mercaptoethanol <2> (<2> slightly inhibitory [2]) [2]
Co^{2+} <2> [2]
Cu^{2+} <2> (<2> complete inhibition [2]) [2]
Hg^{2+} <2> (<2> strong inhibition [2]) [2]
N-ethylmaleimide <2> [2]
N-ethylmorpholine-HCl buffer <2> (<2> slightly inhibitory [2]) [2]
Ni^{2+} <2> (<2> strong inhibition [2]) [2]
Tris-HCl buffer <2> (<2> slightly inhibitory [2]) [2]
Zn^{2+} <2> [2]
coenzyme A <2> [2]
dithiothreitol <2> [2]
hydroxylamine <2> (<2> slightly inhibitory [2]) [2]
o-phenanthroline <2> (<2> slightly inhibitory [2]) [2]
p-chloromercuribenzoate <2> (<2> strong inhibition [2]) [2]
Additional information <2> (<2> no inhibition by EDTA, Cd^{2+}, Mg^{2+}, Mn^{2+}, Sr^{2+}, Ca^{2+}, Pb^{2+} [2]) [2]

Activating compounds

ATP <1> (<1> 2.6fold stimulation [1]) [1]
acetyl-CoA <1> (<1> 15fold stimulation [1]) [1]
acetylcholine <1> (<1> 2.5fold stimulation [1]) [1]
glycine <1> (<1> 1.9fold stimulation [1]) [1]
sodium acetate <1> (<1> 3.1fold stimulation [1]) [1]

Metals, ions

Fe^{2+} <2> (<2> stimulates [2]) [2]
Additional information <2> (<2> no divalent metal ions required [2]) [2]

Specific activity (U/mg)

4.3 <2> (<2> partially purified enzyme [2]) [2]
Additional information <2> (<2> enzyme activity during peroids of starvation and feeding in brain and lens [3]; <2> tissue distribution [2]) [2, 3]

K_m-Value (mM)

0.027 <2> (acetylcoenzyme A) [2]
0.45 <2> (L-histidine) [2]

pH-Optimum

7-9.5 <2> (<2> broad optimum, assay at pH 7.8 [2]) [2]
7.7 <1> [1]

pH-Range

7-7.7 <1> (<1> pH 7.0: 63% of activity maximum, pH 7.7: activity maximum [1]) [1]

Temperature optimum (°C)

25 <1> (<1> assay at [1]) [1]
37 <2> [2]

Temperature range (°C)

37-50 <2> (<2> 37°C: optimum, 50°C: less than 10% activity [2]) [2]

4 Enzyme Structure

Molecular weight

39000 <2> (<2> gel filtration [2]) [2]

Subunits

monomer <2> (<2> 1 * 38500, SDS-PAGE [2]) [2]

5 Isolation/Preparation/Mutation/Application

Source/tissue

brain <1, 2> (<2> the amount of enzyme activity is related to starvation and feeding of the fish [3]) [1-3]
lens <2> (<2> the amount of enzyme activity is related to starvation and feeding of the fish [3]) [2, 3]
skeletal muscle <2> (<2> the amount of enzyme activity is related to starvation and feeding of the fish [3]) [3]

Purification

<1> (partial [1]) [1]
<2> (partial [2]) [2]

References

[1] Baslow, M.H.: N-Acetyl-L-histidine synthetase activity from the brain of the killifish. Brain Res., **3**, 210-213 (1966)
[2] Yamada, S.; Tanaka, Y.; Furuichi, M.: Partial purification and characterization of histidine acetyltransferase in brain of Nile tilapia (Oreochromis niloticus). Biochim. Biophys. Acta, **1245**, 239-247 (1995)
[3] Yamada, S.; Tanaka, Y.; Sameshima, M.; Furuichi, M.: Effects of starvation and feeding on tissue Na-acetylhistidine levels in Nile tilapia Oreochromis niloticus. Comp. Biochem. Physiol. A, **109A**, 277-283 (1994)

D-Tryptophan N-acetyltransferase 2.3.1.34

1 Nomenclature

EC number
2.3.1.34

Systematic name
acetyl-CoA:D-tryptophan N-acetyltransferase

Recommended name
D-tryptophan N-acetyltransferase

Synonyms
D-tryptophan acetyltransferase
acetyl-CoA-D-tryptophan-α-N-acetyltransferase
acetyltransferase, D-tryptophan

CAS registry number
37257-13-9

2 Source Organism

<1> *Saccharomyces cerevisiae* [1]

3 Reaction and Specificity

Catalyzed reaction
acetyl-CoA + D-tryptophan = CoA + N-acetyl-D-tryptophan

Reaction type
acyl group transfer

Natural substrates and products
S acetyl-CoA + D-tryptophan <1> (Reversibility: ? <1> [1]) [1]
P CoA + acetyl-D-tryptophan

Substrates and products
S acetyl-CoA + D-tryptophan <1> (<1> L-isomer is not accepted [1]) (Reversibility: ? <1> [1]) [1]
P CoA + acetyl-D-tryptophan <1> [1]

pH-Optimum
8.1 <1> (<1> assay at [1]) [1]

Temperature optimum (°C)
30 <1> (<1> assay at [1]) [1]

5 Isolation/Preparation/Mutation/Application

Purification
<1> (partial [1]) [1]

References

[1] Zenk, M.H.; Schmitt, J.: Enzymatische Acetylierung von D-Tryptophan. Naturwissenschaften, **51**, 510-511 (1964)

Glutamate N-acetyltransferase 2.3.1.35

1 Nomenclature

EC number
2.3.1.35

Systematic name
N^2-acetyl-L-ornithine:L-glutamate N-acetyltransferase

Recommended name
glutamate N-acetyltransferase

Synonyms
N-acetyl-L-glutamate synthetase
N-acetylglutamate synthase
N-acetylglutamate synthetase
acetylglutamate synthetase
acetylglutamate-acetylornithine transacetylase
acetylglutamic synthetase
acetylglutamic-acetylornithine transacetylase
acetylornithinase
acetylornithine glutamate acetyltransferase
acetyltransferase, glutamate
α-N-acetyl-L-ornithine:L-glutamate N-acetyltransferase
glutamate acetyltransferase
ornithine acetyltransferase
ornithine transacetylase

CAS registry number
37257-14-0

2 Source Organism

<1> *Methanobrevibacter arboriphilus* [3]
<2> *Chlorella vulgaris* [4]
<3> *Chlamydomonas reinhardtii* [1]
<4> *Pisum sativum* [2]
<5> *Methanobacterium thermoautotrophicum* [3]
<6> *Methanococcus vannielii* [3]
<7> *Streptomyces clavuligerus* [5]
<8> *Saccharomyces cerivisiae* [6, 8, 11]
<9> *Methanococcus jannaschii* [7]

<10> *Thermotoga neapolitana* [7]
<11> *Geobacillus stearothermophilus* [7, 9, 10]

3 Reaction and Specificity

Catalyzed reaction

N^2-acetyl-L-ornithine + L-glutamate = L-ornithine + N-acetyl-L-glutamate

Reaction type

acyl group transfer

Natural substrates and products

S N^2-acetyl-L-ornithine + L-glutamate <1-11> (<3,7-11> first step in arginine biosynthesis [1,5-7,9,10]; <10,11> bifunctional enzyme catalyzes also the fifth step of arginine biosynthesis [7,9,10]) (Reversibility: r <3, 8> [1, 6]; ? <1, 2, 4-7, 9-11> [2-5, 7-11]) [1-11]
P L-ornithine + N-acetyl-L-glutamate

Substrates and products

S N^2-acetyl-L-ornithine + H_2O <3, 11> (<3> hydrolytic activity is 1% of the transferase activity [1]) (Reversibility: ? <3, 11> [1, 9]) [1, 9]
P L-ornithine + acetic acid
S N^2-acetyl-L-ornithine + L-glutamate <1-11> (<3> specific for L-glutamate and L-ornithine [1]; <7> specific for L-glutamate [5]) (Reversibility: r <3, 8> [1, 6]; ? <1, 2, 4-7, 9-11> [2-5, 7-11]) [1-11]
P L-ornithine + N-acetyl-L-glutamate
S N^2-butyryl-L-ornithine + H_2O <3> (Reversibility: ? <3> [1]) [1]
P L-ornithine + butyric acid
S N^2-propionyl-L-ornithine + glutamate <3> (Reversibility: ? <3> [1]) [1]
P L-ornithine + N-propionylglutamate
S acetyl-CoA + L-glutamate <10, 11> (Reversibility: ? <10, 11> [7]) [7]
P CoA + N-acetyl-L-glutamate

Inhibitors

2-methylornithine <4> [2]
L-ornithine <8-11> [7-9]
N-bromoacetylornithine <4> [2]
p-chloromercuribenzoate <3> [1]
Additional information <3> (<3> not: iodoacetate [1]) [1]

Cofactors/prosthetic groups

Additional information <3> (<3> no specific cofactor [1]) [1]

Specific activity (U/mg)

0.087 <7> [5]
17.7 <3> [1]
22 <8> [6]
27 <11> [7]

175 <9> [7]
290 <10> [7]

K_m-Value (mM)
0.13 <3> (L-glutamate) [1]
0.2 <2> (α-N-acetyl-L-ornithine) [4]
0.4 <11> (α-N-acetyl-L-ornithine) [9]
1 <8> (α-N-acetyl-L-ornithine) [6]
1.5 <8> (L-ornithine) [6]
2.7 <11> (α-N-acetyl-L-ornithine) [7]
3.6 <7> (α-N-acetyl-L-ornithine) [5]
5.5 <3> (α-N-acetyl-L-ornithine) [1]
6.25 <4> (α-N-acetyl-L-ornithine) [2]
7.2 <8> (L-glutamate) [6]
8.1 <10> (α-N-acetyl-L-ornithine) [7]
9.6 <9> (α-N-acetyl-L-ornithine) [7]
11.3 <9> (L-glutamate) [7]
14.65 <4> (L-glutamate) [2]
17.1 <8> (N-acetylglutamate) [6]
19.2 <11> (L-glutamate) [7]
27.9 <10> (L-glutamate) [7]

pH-Optimum
7.5 <8> [6]
7.5-8 <7> [5]
7.5-9 <3> [1]
8 <9-11> [7]
8-8.5 <2> [4]

pH-Range
7-10 <9-11> [7]

Temperature optimum (°C)
37 <3, 4> (<3,4> assay at [1,2]) [1, 2]
75 <11> [7]
90 <10> [7]
95 <9> (<9> more than 95°C [7]) [7]

Temperature range (°C)
20-65 <11> [9]

4 Enzyme Structure

Molecular weight
57000 <8> (<8> gel filtration [6]) [6]
83000 <7> (<7> gel filtration [5]) [5]
110000 <11> (<11> gel filtration [9]) [9]

Subunits

heterodimer <8> (<8> $\alpha\beta$, 1 * 31000 + 1 * 26000, SDS-PAGE [6,11]) [6, 11]
heterotetramer <7, 9-11> (<7> $\alpha_2\beta_2$, 2 * 19000 + 2 * 25000, SDS-PAGE [5]) [5, 7, 10]

5 Isolation/Preparation/Mutation/Application

Source/tissue

cotyledon <4> [2]

Localization

mitochondrial matrix <8> [6, 8, 11]

Purification

<2> [4]
<3> [1]
<7> [5]
<8> [6]

Cloning

<7> (overexpression in Escherichia coli [5]) [5]
<8> (expression in Escherichia coli and yeast arg2 [8]) [8, 11]
<9> (expression in Escherichia coli [7]) [7]
<10> (expression in Escherichia coli [7]) [7]
<11> (expression in Escherichia coli [7,9]) [7, 9]

6 Stability

Temperature stability

70 <11> (<11> inactivation after 20 min [9]) [9]

Oxidation stability

<1, 5, 6>, insensitive to O_2 [3]

General stability information

<3>, N-acetylglutamate stabilizes and protects against inactivating effect of heat and 4 M urea [1]
<3>, N^2-acetylornithine stabilizes [1]
<3>, complete inactivation after dialysis against distilled water or 10 mM phosphate buffer, pH 7.5, for 16 h at 0°C [1]
<3>, glutamate decreases stability [1]

Storage stability

<3>, 0°C, 30% loss of activity after storage overnight [1]
<8>, -20°C, 10% loss activity per week [6]
<8>, 4°C, 10% loss activity per week [6]

References

[1] Staub, M.; Denes, G.: Mechanism of arginine biosynthesis in Chlamydomonas reinhardti. I. Purification and properties of ornithine acetyltransferase. Biochim. Biophys. Acta, **128**, 82-91 (1966)

[2] Jain, J.C.; Shargool, P.D.: A modified assay system for enzymes involved in N-acetyl group transfer reactions: its use to study enzymes involved in ornithine biosynthesis in plants. Anal. Biochem., **138**, 25-29 (1984)

[3] Meile, L.; Leisinger, T.: Enzymes of arginine biosynthesis in methanogenic bacteria. Experientia, **40**, 899-900 (1984)

[4] Morris, C.J.; Thompson, J.F.: Acetyl coenzyme A-glutamate acetyltransferase and N^2-acetylornithine-glutamate acetyltransferase of Chlorella. Plant Physiol., **55**, 960-967 (1975)

[5] Kershaw, N.J.; McNaughton, H.J.; Hewitson, K.S.; Hernandez, H.; Griffin, J.; Hughes, C.; Greaves, P.; Barton, B.; Robinson, C.V.; Schofield, C.J.: ORF6 from the clavulanic acid gene cluster of Streptomyces clavuligerus has ornithine acetyltransferase activity. Eur. J. Biochem., **269**, 2052-2059 (2002)

[6] Liu, Y.; Van Heeswijck, R.; Hoj, P.; Hoogenraad, N.: Purification and characterization of ornithine acetyltransferase from Saccharomyces cerevisiae. Eur. J. Biochem., **228**, 291-296 (1995)

[7] Marc, F.; Weigel, P.; Legrain, C.; Almeras, Y.; Santrot, M.; Glansdorff, N.; Sakanyan, V.: Characterization and kinetic mechanism of mono- and bifunctional ornithine acetyltransferases from thermophilic microorganisms. Eur. J. Biochem., **267**, 5217-5226 (2000)

[8] Crabeel, M.; Abadjieva, A.; Hilven, P.; Desimpelaere, J.; Soetens, O.: Characterization of the Saccharomyces cerevisiae ARG7 gene encoding ornithine acetyltransferase, an enzyme also endowed with acetylglutamate synthase activity. Eur. J. Biochem., **250**, 232-241 (1997)

[9] Sakanyan, V.; Charlier, D.; Legrain, C.; Kochikyan, A.; Mett, I.; Pierard, A.; Glansdorff, N.: Primary structure, partial purification and regulation of key enzymes of the acetyl cycle of arginine biosynthesis in Bacillus stearothermophilus: dual function of ornithine acetyltransferase. J. Gen. Microbiol., **139**, 393-402 (1993)

[10] Marc, F.; Weigel, P.; Legrain, C.; Glansdorff, N.; Sakanyan, V.: An invariant threonine is involved in self-catalyzed cleavage of the precursor protein for ornithine acetyltransferase. J. Biol. Chem., **276**, 25404-25410 (2001)

[11] Abadjieva, A.; Hilven, P.; Pauwels, K.; Crabeel, M.: The yeast ARG7 gene product is autoproteolyzed to two subunit peptides, yielding active ornithine acetyltransferase. J. Biol. Chem., **275**, 11361-11367 (2000)

D-Amino-acid N-acetyltransferase 2.3.1.36

1 Nomenclature

EC number
2.3.1.36

Systematic name
acetyl-CoA:D-amino-acid N-acetyltransferase

Recommended name
D-amino-acid N-acetyltransferase

Synonyms
D-amino acid acetyltransferase
D-amino acid-α-N-acetyltransferase
acetyltransferase, D-amino acid

CAS registry number
37257-15-1

2 Source Organism

<1> *Saccharomyces cerevisiae* [1]

3 Reaction and Specificity

Catalyzed reaction
acetyl-CoA + a D-amino acid = CoA + an N-acetyl-D-amino acid

Reaction type
acyl group transfer

Natural substrates and products
S acetyl-CoA + D-alanine <1> (Reversibility: ir <1> [1]) [1]
P CoA + N-acetyl-D-alanine
S acetyl-CoA + D-asparagine <1> (Reversibility: ir <1> [1]) [1]
P CoA + N-acetyl-D-asparagine
S acetyl-CoA + D-cysteine <1> (Reversibility: ir <1> [1]) [1]
P CoA + N-acetyl-D-cysteine
S acetyl-CoA + D-cystine <1> (Reversibility: ir <1> [1]) [1]

P CoA + N-acetyl-D-cystine

S acetyl-CoA + D-glutamine <1> (Reversibility: ir <1> [1]) [1]

P CoA + N-acetyl-D-glutamine

S acetyl-CoA + D-histidine <1> (Reversibility: ir <1> [1]) [1]

P CoA + N-acetyl-D-histidine

S acetyl-CoA + D-isoleucine <1> (Reversibility: ir <1> [1]) [1]

P CoA + N-acetyl-D-isoleucine

S acetyl-CoA + D-leucine <1> (Reversibility: ir <1> [1]) [1]

P CoA + N-acetyl-D-leucine

S acetyl-CoA + D-methionine <1> (Reversibility: ir <1> [1]) [1]

P CoA + N-acetyl-D-methionine

S acetyl-CoA + D-norvaline <1> (Reversibility: ir <1> [1]) [1]

P CoA + N-acetyl-D-norvaline

S acetyl-CoA + D-phenylalanine <1> (Reversibility: ir <1> [1]) [1]

P CoA + N-acetyl-D-phenylalanine

S acetyl-CoA + D-proline <1> (Reversibility: ir <1> [1]) [1]

P CoA + N-acetyl-D-proline

S acetyl-CoA + D-serine <1> (Reversibility: ir <1> [1]) [1]

P CoA + N-acetyl-D-serine

S acetyl-CoA + D-threonine <1> (Reversibility: ir <1> [1]) [1]

P CoA + N-acetyl-D-threonine

S acetyl-CoA + D-tryptophan <1> (Reversibility: ir <1> [1]) [1]

P CoA + N-acetyl-D-tryptophan <1> [1]

S acetyl-CoA + D-tyrosine <1> (Reversibility: ir <1> [1]) [1]

P CoA + N-acetyl-D-tyrosine

S acetyl-CoA + D-valine <1> (Reversibility: ir <1> [1]) [1]

P CoA + N-acetyl-D-valine

S acetyl-CoA + norleucine <1> (Reversibility: ir <1> [1]) [1]

P CoA + N-acetyl-D-norleucine

S propionyl-CoA + D-alanine <1> (Reversibility: ir <1> [1]) [1]

P CoA + N-propionyl-D-alanine

S propionyl-CoA + D-allo-isoleucine <1> (Reversibility: ir <1> [1]) [1]

P CoA + N-propionyl-D-allo-isoleucine

S propionyl-CoA + D-glutamine <1> (Reversibility: ir <1> [1]) [1]

P CoA + N-propionyl-D-glutamine

S propionyl-CoA + D-leucine <1> (Reversibility: ir <1> [1]) [1]

P CoA + N-propionyl-D-leucine

S propionyl-CoA + D-serine <1> (Reversibility: ir <1> [1]) [1]

P CoA + N-propionyl-D-serine

S propionyl-CoA + D-tryptophan <1> (Reversibility: ir <1> [1]) [1]

P CoA + N^2-propionyl-D-tryptophan <1> [1]

Substrates and products

S acetyl-CoA + D-alanine <1> (Reversibility: ir <1> [1]) [1]

P CoA + N-acetyl-D-alanine

S acetyl-CoA + D-asparagine <1> (Reversibility: ir <1> [1]) [1]

P CoA + N-acetyl-D-asparagine

S acetyl-CoA + D-cysteine <1> (Reversibility: ir <1> [1]) [1]
P CoA + N-acetyl-D-cysteine
S acetyl-CoA + D-cystine <1> (Reversibility: ir <1> [1]) [1]
P CoA + N-acetyl-D-cystine
S acetyl-CoA + D-glutamine <1> (Reversibility: ir <1> [1]) [1]
P CoA + N-acetyl-D-glutamine
S acetyl-CoA + D-histidine <1> (Reversibility: ir <1> [1]) [1]
P CoA + N-acetyl-D-histidine
S acetyl-CoA + D-isoleucine <1> (Reversibility: ir <1> [1]) [1]
P CoA + N-acetyl-D-isoleucine
S acetyl-CoA + D-leucine <1> (Reversibility: ir <1> [1]) [1]
P CoA + N-acetyl-D-leucine
S acetyl-CoA + D-methionine <1> (Reversibility: ir <1> [1]) [1]
P CoA + N-acetyl-D-methionine
S acetyl-CoA + D-norvaline <1> (Reversibility: ir <1> [1]) [1]
P CoA + N-acetyl-D-norvaline
S acetyl-CoA + D-phenylalanine <1> (Reversibility: ir <1> [1]) [1]
P CoA + N-acetyl-D-phenylalanine
S acetyl-CoA + D-proline <1> (Reversibility: ir <1> [1]) [1]
P CoA + N-acetyl-D-proline
S acetyl-CoA + D-serine <1> (Reversibility: ir <1> [1]) [1]
P CoA + N-acetyl-D-serine
S acetyl-CoA + D-threonine <1> (Reversibility: ir <1> [1]) [1]
P CoA + N-acetyl-D-threonine
S acetyl-CoA + D-tryptophan <1> (Reversibility: ir <1> [1]) [1]
P CoA + N-acetyl-D-tryptophan <1> [1]
S acetyl-CoA + D-tyrosine <1> (Reversibility: ir <1> [1]) [1]
P CoA + N-acetyl-D-tyrosine
S acetyl-CoA + D-valine <1> (Reversibility: ir <1> [1]) [1]
P CoA + N-acetyl-D-valine
S acetyl-CoA + norleucine <1> (Reversibility: ir <1> [1]) [1]
P CoA + N-acetyl-D-norleucine
S propionyl-CoA + D-alanine <1> (Reversibility: ir <1> [1]) [1]
P CoA + N-propionyl-D-alanine
S propionyl-CoA + D-allo-isoleucine <1> (Reversibility: ir <1> [1]) [1]
P CoA + N-propionyl-D-allo-isoleucine
S propionyl-CoA + D-glutamine <1> (Reversibility: ir <1> [1]) [1]
P CoA + N-propionyl-D-glutamine
S propionyl-CoA + D-leucine <1> (Reversibility: ir <1> [1]) [1]
P CoA + N-propionyl-D-leucine
S propionyl-CoA + D-serine <1> (Reversibility: ir <1> [1]) [1]
P CoA + N-propionyl-D-serine
S propionyl-CoA + D-tryptophan <1> (Reversibility: ir <1> [1]) [1]
P CoA + N^2-propionyl-D-tryptophan <1> [1]

S Additional information <1> (<1> L-isomers are not accepted, glycine and D-taurine are poor substrates, propionyl-CoA can replace acetyl-CoA [1]) [1]

P ?

Inhibitors

$(NH_4)_2SO_4$ <1> (<1> strong inhibitor at 1 mM [1]) [1]
$ZnCl_2$ <1> (<1> strong inhibitor at 1 mM [1]) [1]
o-iodosobenzoate <1> (<1> strong inhibitor at 1 mM [1]) [1]
p-chloromercuribenzoate <1> (<1> strong inhibitor at 1 mM [1]) [1]

Specific activity (U/mg)

2.97 <1> (<1> D-tryptophan [1]) [1]

K_m-Value (mM)

0.227 <1> (acetyl-CoA) [1]
0.263 <1> (propionyl-CoA) [1]
5.5 <1> (D-tryptophan) [1]

pH-Optimum

7.3-7.5 <1> (<1> in phosphate buffer [1]) [1]
8.4 <1> (<1> in Tris-HCl buffer [1]) [1]

Temperature optimum (°C)

40 <1> (<1> in Tris-HCl buffer [1]) [1]

5 Isolation/Preparation/Mutation/Application

Purification

<1> (2000fold [1]) [1]

6 Stability

Storage stability

<1>, -15°C, stable for several months [1]

References

[1] Zenk, M.H.; Schmitt, J.H.: Purification and properties of acetyl-CoA: D-amino acid-α-N-acetyltransferase from yeast. Biochem. Z., **342**, 54-65 (1965)

5-Aminolevulinate synthase 2.3.1.37

1 Nomenclature

EC number
2.3.1.37

Systematic name
succinyl-CoA:glycine C-succinyltransferase (decarboxylating)

Recommended name
5-aminolevulinate synthase

Synonyms
5-aminolevulinate synthetase
5-aminolevulinic acid synthetase
ALA synthase
ALA synthetase
ALA-S <14> [47]
ALAS
α-aminolevulinic acid synthase
aminolevulinate synthase
aminolevulinate synthetase
aminolevulinic acid synthase
aminolevulinic acid synthetase
aminolevulinic synthetase
δ-ALAS <8> [51]
δ-aminolevulinate synthase
δ-aminolevulinate synthetase
δ-aminolevulinic acid synthase
δ-aminolevulinic acid synthetase
δ-aminolevulinic synthetase
synthase, aminolevulinate
synthetase, aminolevulinate

CAS registry number
9037-14-3

2 Source Organism

<-1> no activity in *Chromatium sp.* [13]
<1> *Homo sapiens* (2 isoforms ALAS-1 and ALAS-2, a houskeeping form and an erythroid-specific form [46]) [1, 13, 43, 46]

<2> *Rhodopseudomonas palustris* [8]
<3> *Salmo gairdneri* (2 enzyme forms in cytosol and mitochondria [9]) [9]
<4> *Propionibacterium shermanii* [13]
<5> *Solanum tuberosum* [2]
<6> *Micrococcus denitrificans* (i.e. *Paracoccus denitrificans* [43]; in cells grown anaerobically and in iron-containing nitrate medium, overview [3]) [3, 43]
<7> *Rhodopseudomonas sphaeroides* (2 isozymes [37]; Y [24,37,38]; enzyme exists in an inactive or low activity form which can under certain conditions become activated, spontaneous activation in presence of oxygen or an oxidizing agent shows a pH dependence with an optimum at about pH 7.0, assisted by 2-mercaptoethanol [13]) [4, 12-14, 18, 19, 21-24, 32, 35, 37, 38, 43]
<8> *Rattus norvegicus* (non-specific, housekeeping isoform ALAS-N and erythroid-specific isoform ALAS-E [41,46]; 2 isoforms in cytosol and mitochondria [36]; drug-induced [31]) [5, 6, 11, 13, 15, 16, 26, 27, 30, 31, 34, 36, 41, 43, 44, 46, 51]
<9> *Cavia porcellus* [13, 36]
<10> *Oryctolagus cuniculus* [13, 20]
<11> yeast [13]
<12> *Spinacia oleracea* [13]
<13> *Gallus gallus* (2 isoforms ALAS-1 and ALAS-2, a houskeeping form and an erythroid-specific form [46]; animals after acetylphenylhydrazine-induced anemia [7]) [7, 11, 13, 25, 29, 43, 46]
<14> *Saccharomyces cerevisiae* [28, 43, 45, 47]
<15> *Protaminobacter ruber* (2 isoenzymes: I, constitutive, and II, inducible related to bacteriochlorophyll formation and light conditions [33]) [32, 33]
<16> *Rhodospirillum rubrum* [10, 13]
<17> *Mus musculus* (2 isoforms ALAS-1 and ALAS-2, a houskeeping form and an erythroid-specific form [46]; gene hemA [45]; inducible in erythroleukemia cells [40]; hypoxia-inducible [39]; erythroid-specific isoform [17,39,42,48-50,52]; gene ALAS-E [42]; gene ALAS-2 [17,39]) [17, 39, 40, 42-46, 48-50, 52, 54]
<18> *Spirillium intersonii* [13]
<19> *Euglena gracilis* [43]
<20> *Bradyrhizobium japonicum* [43]
<21> *Rhodobacter capsulatus* [43]
<22> *Rhizobium meliloti* [43]
<23> *Aspergillus nidulans* [43]
<24> *Agrobacterium radiobacter* [43]
<25> *Plasmodium falciparum* [53]

3 Reaction and Specificity

Catalyzed reaction

succinyl-CoA + glycine = 5-aminolevulinate + CoA + CO_2 (<17> three-step kinetic process, ordered kinetic mechanism, reaction mechanism [54]; <17> Lys313 acts as a general base during formation of the quinonoid reaction intermediates [52]; <17> Tyr121, Asp279, Arg439, and Lys313 are involved in substrate and cofactor binding, mechanism, subunit localisation [42, 44, 45, 48-50]; <14, 17> the active site is located at the subunit interface and contains catalytically essential residues from the two subunits [45]; <17> pyridoxal 5'-phosphate binding site, sequence and function of glycine-rich motif [43, 44]; <17> cysteine in heme-regulatory motif [43]; <17> covalent binding of pyridoxal 5'-phosphate and glycine to active site Lys131 is required for optimal activity [42]; <1, 4, 6-14, 16, 17, 19> mechanism [6, 12-14, 19, 38, 43, 52]; <7> ordered bi-bi mechanism in which glycine binds first and 5-aminolevulinic acid dissociates last [38, 43])

Reaction type

acyl group transfer
condensation <1,4,7-13,16-18> [13,49]

Natural substrates and products

S succinyl-CoA + glycine <1, 4, 5, 7-14, 16-18, 25> (<1, 7, 8, 13, 14, 17> regulatory mechanisms in hepatic and erythroid cells [43]; <17> cellular iron status plays a regulatory role [40]; <1, 4, 7-14, 16-18> key enzyme in tetrapyrrole biosynthesis [13, 46, 47]; <1, 4, 7-14, 16, 17> rate-limiting enzyme of heme biosynthesis [1, 5, 11, 13, 39, 40, 45, 52, 54]; <17> first step in heme biosynthesis pathway [17, 52, 54]) (Reversibility: ? <1, 4, 5, 7-14, 16-18, 25> [1, 2, 5, 11, 13, 17, 39, 40, 43, 45-47, 51-54]) [1, 2, 5, 11, 13, 17, 39, 40, 43, 45-47, 51-54]

P 5-aminolevulinate + CoA + CO_2 <1, 5, 7, 8, 13, 14, 17, 25> (<5> common intermediate in the biosynthesis of chlorophyll and heme [2]) [2, 13, 39, 40, 43, 45-47, 51-54]

S Additional information <1> (<1> enzyme deficiency causes X-linked sideroblastic anemia [46]) [46]

P ?

Substrates and products

S acetyl-CoA + glycine <1, 4, 7-13, 16> (<1,4,7-13,16> at 2% the rate of the reaction with succinyl-CoA [13]; <7> low activity [12]) (Reversibility: ? <1, 4, 7-13, 16> [12, 13]) [12, 13]

P ?

S α-glutamyl-CoA + glycine <1, 4, 7-13, 16> (<1,4,7-13,16> at 9% the rate of the reaction with succinyl-CoA [13]) (Reversibility: ? <1, 4, 7-13, 16> [13]) [13]

P 6-amino-5-oxohexanoate + CoA + CO_2

S butyryl-CoA + glycine <7> (<7> low activity [12]) (Reversibility: ? <7> [12]) [12]

P ?

S propionyl-CoA + glycine <7> (<7> low activity [12]) (Reversibility: ? <7> [12]) [12]

P ?

S succinyl-CoA + glycine <1-19, 25> (<1, 6-8, 13, 14, 19> absolutely specific for glycine [43]; <2, 7, 8> absolute requirement for succinyl-CoA [4, 5, 8]; <1, 2, 4, 7-13, 16> absolute requirement for glycine [4, 5, 8, 12, 13]; <7> enzyme utilizes acetyl-CoA, propionyl-CoA and butyryl-CoA at much lower rates [12]; <7> glycine is formed by the reverse reaction under optimum conditions at 4-5% the rate of 5-aminolevulinate formed by the forward reaction under optimum conditions [14]) (Reversibility: r <7> [14]; ir <1, 4, 7-13, 16, 18> [13]; ? <1-3, 5, 6-10, 13-17, 19, 25> [1-12, 15-54]) [1-54]

P 5-aminolevulinate + CoA + CO_2 <1-19, 25> [1-54]

S succinyl-CoA-monomethyl ester + glycine <1, 4, 7-13, 16> (<1, 4, 7-13, 16> at 80% the rate of the reaction with succinyl-CoA [13]) [13]

P 5-aminolevulinic acid methyl ester + CoA + CO_2

Inhibitors

2-hydroxyethyldisulfide <7> [13]
2-mercaptoethanol <10> [13]
2-mercaptoethanol <7> (<7> slightly at high concentration, 10 mM [4]) [4, 35]
2-oxoglutarate <1, 4, 7-13, 16> [13]
ATP <7> [24, 38]
Co^{2+} <8> (<8> accumulation of Co^{2+}-protoporphyrin containing products of hemolysis [51]) [6, 51]
CoA <7> (<7> noncompetitive [38]) [38]
Cu^{2+} <7, 16> (<7> at low concentration [35]; <16> isoenzyme a [10]) [10, 12, 13, 35]
Fe^{2+} <7> [12, 13, 35]
Fe^{3+} <7, 10> (<7> slight [35]; <7> no inhibition [13]) [13, 35]
$FeCl_3$ <7> (<7> addition of imidazole at 0.2 mM prior to $FeCl_3$ prevents inhibition [24]) [24]
GTP <7> [38]
Hg^{2+} <7, 16> (<7> 2-mercaptoethanol protects partly [35]; <16> isoenzyme a [10]) [10, 35]
L-alanine <7, 13> [13]
L-cysteine <13> [13]
L-penicillamine <13> [13]
L-threonine <7, 13> [13]
Mg-protoporphyrin <7> [23, 35]
Mg^{2+} <8> (<8> inhibition of complex form, stimulation of stripped form [30]) [30]
Mn^{2+} <8> [6]
N-ethylmaleimide <7, 8, 14> (<7> 2-mercaptoethanol protect [35]; <8> glycine and pyridoxal 5'-phosphate protect [6]) [6, 28, 35]

Na^+ <8> (<8> inhibition of complex form, stimulation of stripped form [30]) [30]
NaCl <8> (<8> fetal rat liver enzyme [34]) [34]
Zn^{2+} <7> [35]
aminomalonate <7, 13, 25> (<25> weak [53]; <7> competitive versus glycine [14]) [12-14, 53]
aminomethylphosphonate <1, 4, 7-13, 16> [13, 14]
β-alanine <7, 13> [13]
bilirubin <8> [6]
borohydride <7> (<7> in presence of pyridoxal 5'-phosphate [13,14]; <7> aminomalonate protects [14]) [13, 14]
chloramphenicol <6> [3]
chlorophyllide a <5> [2]
cyanide <13> [13]
δ-aminolevuinate <2, 7> (<7> competitive against glycine, noncompetitive against succinyl-CoA [38]; <2> product inhibition [8]) [8, 38]
deuteroporphyrin <8> [6]
diethyl aminomalonate <7, 13> [13]
dithiothreitol <7> [4]
divalent cations <8, 13> (<13> Mg^{2+} or Ca^{2+}: optimal at concentration of 50 mM, inhibition above 200 mM [25]; <8> at high concentration, relative order of inhibition: Co^{2+}, Mn^{2+}, Mg^{2+}, Ca^{2+} [6]) [6, 25]
ethanolamine <7, 13, 16, 25> [13, 53]
ferroheme <1, 4, 7-13, 16> (<14> not heme [28]) [13]
glutathione <10> [13]
hemin <1, 4-17, 25> (<1,8,13,14,17,25> inhibition of the import of the enzyme into the mitochondrial matrix [43,53]; <6-8,10> feed-back inhibition [3,20,21,26,36]; <15> no inhibition [32]; <8> no inhibition of rat fetal liver enzyme [34]; <7> reversible, noncompetitive [12]; <7> albumin protects [21]) [2-4, 6, 7, 10, 12, 13, 20, 21, 23, 25, 26, 28, 32, 35, 36, 43, 53]
hemoglobin <1, 4, 7-13, 16> [13]
iodoacetamide <7> [13, 35]
iron-deuteroporphyrin <8> [6]
iron-mesoporphyrin <8> [6]
iron-protoporphyrin <8> [6]
isobutylamine <7, 13> [13]
light <5> (<5> above 30 lux intensity [2]) [2]
mesoporphyrin <8> [6]
metalloporphyrins <1, 4, 7-13, 16> (<1,4,7-13,16> containing Co^{2+}, Zn^{2+} or Mg^{2+} [13]) [6, 13]
monovalent cations <8, 13> (<13> Na^+ or Li^+: optimal at 0.5 M, not inhibitory up to 1 M [25]; <8> in Mg^{2+}-containing medium: Na^+ or Li^+ at 100 mM have no effect on activity, higher concentrations are totally inhibitory [26]) [25, 26]
myoglobin <1, 4, 7-13, 16> [13]

p-chloromercuribenzoate <7, 8, 13, 14, 16> (<7> 2-mercaptoethanol protect partly [35]; <8> pyridoxal 5'-phosphate protects [6]; <16> isoenzyme a [10]) [6, 10, 12, 25, 28, 35]
p-mercuribenzoate <7, 13> [13]
propanolol <8> [51]
protoheme <1, 8, 13, 17> (<1,8,13,17> feed-back inhibition, non-specific isoform ALAS-1 [46]) [46]
protoporphyrin <6-8> (<7> weak [35]) [3, 6, 35]
puromycin <6> [3]
pyruvate <1, 4, 7-13, 16> [13]
succinyl-CoA <8> (<8> non-specific isoform ALAS-N in absence of ATP [41]) [41]
tergitol <8> (<8> fetal enzyme [34]) [34]
Additional information <2, 5, 7, 8> (<5> pheophytin and chlorophyll are not inhibitory [2]; <2> cysteine, glutathione, cystine or 2-mercaptoethanol diminishes activity in presence of 5-aminolevulinate [8]; <7> product inhibition pattern for the forward reaction and the reverse reaction [14]; <8> presence of high salt concentration in the sample buffer and running buffer of aggregated enzyme preparation results in inactivation [26]) [2, 8, 14, 26]

Cofactors/prosthetic groups

N-methyl-pyridoxal 5'-phosphate <17> (<17> in vitro, mutant D279A [49]) [49]
pyridoxal 5'-phosphate <1-24> (<17> noncovalently bound [46]; <17> content in ALAS-E wild-type and Lys313 mutants [42]; <5> activates [2]; <7> weekly bound [12]; <7> fully active with 0.05 mM pyridoxal 5'-phosphate [12]; <7> bound to the enzyme through a Schiff base linkage [13,18]; <7> 1 or 2 molecules of cofactor are bound per subunit of MW 50000 in hemiapoenzyme or holoenzyme, required [19]; <7> contains 1 molar equivalent of pyridoxal 5'-phosphate [37]; <7> absolute requirement [4]; <1,5,6-8,13-15,17,19> required [2,25,28,32,42,43,46-50,54]) [1-50, 52-54]

Activating compounds

2-allyl-2-isopropylacetamide <8, 9, 13> (<8,9,13> strong induction of enzyme in liver [13,43]) [13, 43]
3,5-di-carbethoxy-1,4-dihydrocillidine <8, 13> (<8,13> strong induction of enzyme in liver [43]) [43]
ATP <5, 8> (<8> slight activation, erythroid-specific isoform [41]; <5> activates in assays with crude enzyme extracts [2]) [2, 41]
CoA <5> (<5> activates [2]) [2]
EDTA <7> (<7> required [35]) [35]
NaCl <8> (<8> required for maximal activity [5]) [5]
O2 <17> (<17> 1%, 3fold increase in expression level, promotor activation in transiently transfected HeLa cells, hypoxia-inducible erythroid-isoform [39]) [39]
activator protein <7, 8> (<8> from rat liver mitochondria possibly activates by formation of high molecular weight aggregates of δ-aminolevulinic acid synthetase [15]; <7> from Rhodopseudomonas sphaeroides [23]) [15, 23]

butyric acid <17> (<17> induction of enzyme expression, erythroleukemia cells [40]) [40]
dimethylsulfoxide <17> (<17> induction of enzyme expression, erythroleukemia cells [40]) [40]
dithiothreitol <7> (<7> activates [4]) [4]
hexamethylene diacetamide <17> (<17> induction of enzyme expression, erythroleukemia cells [40]) [40]
lubrol <8> (<8> activates, fetal enzyme [34]) [34]
native activator protein <7, 8> (<7> purified from cell extract [23]; <8> purified from hepatic extract [15]) [15, 23]
phosphate <16> (<16> enzyme is maximally active in presence of 0.005 mM phosphate [10]) [10]
vitamin B_{12} <7, 15> (<7,15> stimulates [32]) [32]
Additional information <2, 17> (<17> iron depletion does not increase isoform II promotor activity [39]; <2> 1-2 low molecular-weight compounds which stimulate or stabilize activity found by dialysis of crude extract [8]) [8, 39]

Metals, ions

Ca^{2+} <13> (<13> optimal at concentration of 50 mM, inhibition above 200 mM [25]) [25]
Li^+ <8, 13> (<13> optimal at 0.5 M, not inhibitory up to 1 M [25]; <8,13> in Mg^{2+}-containing medium: at 100 mM no effect on activity, higher concentrations are totally inhibitory [25,26]) [25, 26]
Mg^{2+} <5, 8, 13> (<5> activates [2]; <8> inhibition of complex form, stimulation of stripped form [30]; <13> optimal at concentration of 50 mM, inhibition above 200 mM [25]) [2, 25, 30]
Na^+ <8, 13> (<8> inhibition of complex form, stimulation of stripped form [30]; <13> optimal at 0.5 M, not inhibitory up to 1 M [25]; <8,13> in Mg^{2+}-containing medium: at 100 mM no effect on activity, higher concentrations are totally inhibitory [25,26]) [25, 26, 30]
iron <1, 8, 13, 17> (<1,8,13,17> regulation of erythroid-specific isoform ALAS-2 via 5'-iron responsive element, i.e. IRE [46]) [46]
potassium ferricyanide <7> (<7> activates at 0.5 mM [4]) [4]
Additional information <8> (<8> partially purified enzyme requires high cation concentrations, equivalent to 0.3 M NaCl, for maximum activation, monovalent cations, i.e. Na^+, K^+, Li^+, Rb^+ or NH_4^+, in concentrations from 0.15 M to 0.30 M or divalent cations, i.e. Mg^{2+}, Ca^{2+} or Mn^{2+}, at lower concentrations of 0.03 M to 0.10 M are effective [5]) [5]

Turnover number (min^{-1})

0.57 <17> (succinyl-CoA) [17, 46]
1.1 <17> (succinyl-CoA, <17> recombinant erythroid isoform mutant G142C [44]) [44]
2.13 <17> (succinyl-CoA, <17> recombinant erythroid isoform mutant Y121H [48]) [48]
3.8 <17> (succinyl-CoA, <17> recombinant erythroid isoform mutant G144T [44]) [44]

5.8 <17> (succinyl-CoA, <17> recombinant erythroid isoform mutant D279E [49]) [49]
7.2 <17> (succinyl-CoA, <17> recombinant erythroid isoform mutant G144S [44]) [44]
7.9 <17> (succinyl-CoA, <17> recombinant erythroid isoform mutant G144A [44]) [44]
10 <17> (succinyl-CoA, <17> recombinant erythroid isoform [49]) [49]
14.1 <17> (succinyl-CoA, <17> recombinant erythroid isoform mutant Y121F [48]) [48]
18.5 <17> (succinyl-CoA, <17> recombinant erythroid isoform [44]) [44]
30.5 <17> (succinyl-CoA, <17> recombinant erythroid mutant R439K [50]) [50]
39.5 <17> (succinyl-CoA, <17> recombinant erythroid isoform [48,50]) [48, 50]
50.5 <17> (succinyl-CoA, <17> recombinant erythroid mutant R433L [50]) [50]
80.9 <17> (succinyl-CoA, <17> recombinant erythroid mutant R433K [50]) [50]

Specific activity (U/mg)

0.0037 <6> (<6> crude enzyme extract [3]) [3]
0.0057 <5> (<5> maximal enzyme activity after 3 weeks of storage of tubers at 0-15°C at 30 lux [2]) [2]
0.018 <8> (<8> purified enzyme [5]) [5]
0.035 <8> (<8> purified enzyme [26]) [26]
0.091 <8> (<8> purified enzyme [30]) [30]
0.16 <8> (<8> purified enzyme [11,31]) [11, 31]
0.167 <7> (<7> isozyme II [37]) [37]
0.35 <15> (<15> purified enzyme [32]) [32]
0.375 <7> (<7> purified isozyme I [37]) [37]
0.558 <7> (<7> purified enzyme [35]) [35]
0.58 <13> (<13> purified enzyme [11,29]) [11, 29]
0.62 <6> (<6> purified enzyme [3]) [3]
0.65 <14> (<14> purified enzyme [28]) [28]
1.22 <8> (<8> purified enzyme [27]) [27]
1.34 <7> (<7> purified enzyme [32]) [32]
1.87 <8> (<8> purified erythroid-specific isoform [41]) [41]
2.17 <7> (<7> purified enzyme [4]) [4]
2.17-2.83 <7> (<7> purified enzyme [18]) [18]
2.22 <7> (<7> purified enzyme [12]) [12]
Additional information <1, 4, 5, 7, 8, 13-18> (<1,4,7,8,10,13,16,18> various tissues, overview [13]; <14> enzyme expressed in transgenic tabacco plants [47]; <7> activation of inactive enzyme form required activator protein and L-cystine [23]; <7> active and inactive forms, inactive form can be activated by addition of rat mitochondrial protein extract [22]; <5> EDTA required for the assay to inhibit 5-aminolevulinate dehydratase [2]) [2, 13, 17, 19, 22-25, 33, 34, 47]

K_m-Value (mM)

0.00064 <17> (succinyl-CoA, <17> recombinant mutant G144A [44]) [44]
0.00077 <15> (pyridoxal 5'-phosphate) [32]
0.001-0.01 <8> (pyridoxal 5'-phosphate) [36]
0.0012 <17> (succinyl-CoA, <17> recombinant erythroid isoform and mutant G144S [44]) [44]
0.0016 <17> (succinyl-CoA, <17> recombinant mutant Y121H [48]) [48]
0.0019 <17> (succinyl-CoA, <17> recombinant mutant G144T [44]) [44]
0.002 <8, 14, 17> (succinyl-CoA, <17> recombinant erythroid isoform wild-type [48-50]; <8,17> recombinant mutant G142C [44]; <8> erythroid-specific isoform [41]) [28, 41, 44, 48-50]
0.0022 <17> (succinyl-CoA, <17> recombinant erythroid isoform mutant R433K [50]) [50]
0.003 <8> (pyridoxal phosphate) [5]
0.0032 <17> (succinyl-CoA, <17> recombinant erythroid isoform mutant R433L [50]) [50]
0.01 <6> (succinyl-CoA) [3]
0.011 <8> (succinyl-CoA) [27]
0.013 <13> (succinyl-CoA) [7]
0.014 <17> (succinyl-CoA, <17> recombinant mutant Y121F [48]) [48]
0.015 <7> (pyridoxal 5'-phosphate) [32]
0.02 <8> (succinyl-CoA) [30]
0.025 <8> (succinyl-CoA) [5]
0.027 <17> (succinyl-CoA, <17> recombinant erythroid isoform mutant R439K [50]) [50]
0.035 <17> (succinyl-CoA, <17> recombinant erythroid isoform mutant D279E [49]) [49]
0.055 <17> (succinyl-CoA, <17> recombinant protein [17]) [17, 46]
0.07 <8> (succinyl-CoA) [5]
0.1 <25> (succinyl-CoA, <25> recombinant enzyme [53]) [53]
0.2 <8> (succinyl-CoA) [36]
1.05 <15> (glycine) [32]
1.52 <17> (succinyl-CoA, <17> recombinant mutant dimer K149A/K313A [45]) [45]
1.82 <17> (succinyl-CoA) [45]
2.5 <8> (glycine, <8> papain untreated enzyme [27]) [27]
3 <14> (glycine) [28]
4 <13> (glycine) [7]
5 <8> (succinyl-CoA) [26]
6.2 <17> (glycine, <17> recombinant mutant G144T [44]) [44]
6.5 <8> (glycine, <8> erythroid-specific isoform [41]) [41]
6.95 <17> (glycine, <17> recombinant mutant dimer K149A/K313A [45]) [45]
7.5 <8> (glycine, <8> papain treated enzyme [27]) [27]
8.33 <7> (glycine) [32]
8.39 <17> (glycine) [45]
9.3 <17> (glycine, <17> recombinant mutant G144A [44]) [44]

10 <6, 8> (succinyl-CoA) [3, 5]
11.7 <17> (glycine, <17> recombinant erythroid isoform [48,50]) [48, 50]
11.9 <17> (glycine, <17> recombinant mutant G144S [44]) [44]
12 <6, 8> (glycine) [3, 30]
12.5 <17> (glycine, <17> recombinant erythroid isoform [44]) [44]
14.4 <17> (glycine, <17> recombinant erythroid isoform mutant R433K [50]) [50]
18.4 <17> (glycine, <17> recombinant erythroid isoform mutant R433L [50]) [50]
19 <8> (glycine) [36]
23 <17> (glycine, <17> recombinant erythroid isoform wild-type [49]) [49]
50 <25> (glycine, <25> recombinant enzyme [53]) [53]
51 <17> (glycine, <17> recombinant protein [17]) [17, 46]
52.2 <17> (glycine, <17> recombinant mutant Y121H [48]; <17> recombinant mutant G142C [44]) [44, 48]
103 <17> (glycine, <17> recombinant erythroid isoform mutant R439K [50]) [50]
140 <17> (glycine, <17> recombinant erythroid isoform mutant D279E [49]) [49]
400 <17> (glycine, <17> recombinant mutant Y121F [48]) [48]
Additional information <1, 2, 4, 7-13> (<7,8,10,13> overview [13]) [8, 10, 13, 38]

K_i-Value (mM)

0.023 <7> (aminomalonate) [14]
0.035 <13> (hemin) [25]
0.12 <7> (CoA, <7> versus glycine [38]) [38]
0.2 <8> (metalloporphyrin) [6]
0.23 <7> (5-aminolevulinate, <7> versus glycine [38]) [38]
10 <7> (aminomethylphosphonate) [14]
Additional information <7> [38]

pH-Optimum

7 <5, 6> (<6> 0.1 M phosphate buffer [3]) [2, 3]
7-8 <13> [7]
7.1-7.5 <16> (<16> isoenzyme a [10]) [10]
7.2 <7> [35]
7.4 <14> [28]
7.4-7.6 <7, 8> [4, 5]
7.5 <2, 3, 6, 8, 17> (<6> 0.1 M Tris-HCl buffer [3]) [3, 5, 8, 9, 26, 48]
7.6 <8> (<8> erythroid-specific isoform [41]) [27, 41]
8 <7, 15> [32]
8.5 <17> (<17> erythroid-specific isoform [42]) [42]
Additional information <7, 8> (<7> pI [37]; <7> pI: 5.5 [18]; <8> pI: 7.5 [15]) [15, 18, 37]

pH-Range

5.8-8.6 <14> (<14> about 50% of activity maximum at pH 5.8 and pH 8.6 [28]) [28]
6.5-9.5 <17> [48]

Temperature optimum (°C)

15 <5> [2]
30 <8, 14> (<8,14> assay at [27,28,41]) [27, 28, 41]
37 <2, 7, 8, 13, 15> (<7,8,13,15> assay at [11,12,18,32]) [8, 11, 12, 18, 32]

4 Enzyme Structure

Molecular weight

57000 <7> (<7> gel filtration [4]) [4]
58000 <8> (<8> monomer, gel filtration [15]) [15]
61000 <7> (<7> gel filtration [35]) [35]
61000-64000 <15> (<15> isoenzyme II, gel filtration [33]) [33]
61000-65000 <2> [8]
62000-68000 <7> (<7> gel filtration [18]) [18]
64840 <8> (<8> erythroid-specific isoform precursor, DNA-sequence determination [41]) [41]
65000-70000 <16> [10]
68000 <6> (<6> gel filtration [3]) [3]
70000 <3> (<3> mitochondrial enzyme, cytosolic enzyme is a precursor of the mitochondrial enzyme [9]) [9]
76000 <8> (<8> precursor form, gel filtration [31]) [31]
77000 <8> (<8> mitochondria, gel filtration [36]) [36]
80000 <7> (<7> active form, gel filtration [22]) [22]
81000 <7> (<7> gel filtration [12]) [12]
87000 <7, 13> (<13> gel filtration [25]; <7> gel filtration [32]) [25, 32]
93000 <14> (<14> gel filtration [28]) [28]
97000 <15> (<15> gel filtration [32]) [32]
100000 <7> (<7> inactive form, gel filtration [22]; <7> enzyme I and II, gel filtration [24]) [22, 24]
105000 <7> (<7> enzyme I, gel filtration [37]) [37]
107000-110000 <8> (<8> stripped form, sucrose density gradient centrifugation, gel filtration, disc gel electrophoresis [30]) [30]
110000 <3, 7, 8> (<3> cytosolic enzyme, is a precursor of the mitochondrial enzyme [9]; <8> sucrose density gradient centrifugation, gel filtration [27]; <7> enzyme II, gel filtration [37]) [9, 27, 37]
112000 <17> (<17> gel filtration [45,46]) [45, 46]
115000 <13> (<13> gel filtration [7]) [7]
118000 <8> (<8> dimer, gel filtration [15]) [15]
120000 <8> (<8> gel filtration [26]) [26]
158000 <25> (<25> gel filtration [53]) [53]
178000 <8> (<8> cytosol, gel filtration [36]) [36]

200000 <10> (<10> gel filtration [13]) [13]
218000 <8> (<8> tetramer of enzyme and native activator protein, gel filtration [15]) [15]
295000 <8> (<8> gel filtration, presence of NaCl [5]) [5]
500000 <8> (<8> gel filtration, absence of NaCl [5]) [5]
Additional information <1, 6-8, 13, 14, 17, 20, 21, 23, 24> (<1,6-8,13,14,17,20,21,23,24> amino acid sequence alignment [43,45,48]; <8> amino acid sequence determination of the purified protein [41]; <7> amino acid composition [18]; <8> mitochondrial enzyme has a minimum MW of 70000 (SDS-PAGE) and is apparently synthesized as a larger precursor of minimum MW 76000 [31]) [13, 18, 31, 41, 43, 45, 48]

Subunits

? <7, 8, 13, 25> (<25> x * 63000, about, SDS-PAGE [53]; <8> x * 49000, erythroid-specific isoform, SDS-PAGE [41]; <13> x * 68000, possibly degraded enzyme, SDS-PAGE [29]; <7> x * 49000, SDS-PAGE [18]; <7> x * 57000, SDS-PAGE [4]) [4, 18, 29, 41, 53]
dimer <7, 8, 13, 14, 17> (<17> 2 * 56000, SDS-PAGE [45,46]; <8> 2 * 54000, SDS-PAGE [15]; <14> 2 * 53000, SDS-PAGE [28]; <13> 2 * 62200, SDS-PAGE [7]; <7,13> 2 * 49000, SDS-PAGE [12,25]; <8> 2 * 58000, mitochondria, SDS-PAGE [26]; <17> 2 * 56000, mouse enzyme expressed in E. coli, SDS-PAGE [17]; <8> 2 * 51000, cytosol, SDS-PAGE [27,30]) [7, 12, 15, 17, 25-28, 30, 45, 46]
monomer <2, 7> (<2> 1 * 61000-65000 [8]; <7> 1 * 57000, SDS-PAGE [4]) [4, 8]
Additional information <14, 17> (<14,17> the active site is located at the subunit interface and contains catalytically essential residues from the two subunits [45]) [45]

5 Isolation/Preparation/Mutation/Application

Source/tissue

blood <13> (<13> cells [7]) [7]
cell culture <2, 6> (<6> high activity in cells grown anaerobically in defined medium, low in cells grown in an iron-deficient medium and in cells grown aerobically [3]) [3, 8]
embryo <8, 13> [11, 25, 29, 34]
erythrocyte <1, 13> (<1> enzyme in erythrocytes is genetically distinct from that in other tissue [1]) [1, 13, 43]
erythroleukemia cell line <17> (<17> erythroid-specific isoform [39,40]) [39, 40]
harderian gland <8> [13]
liver <1, 3, 8-10, 13> (<8> porphyric [36]; <8,13> fetal [11,25,29,34]) [5, 6, 9, 11, 13, 15, 16, 25-28, 30, 31, 34, 36, 43]
reticulocyte <8, 10> [13, 20, 41]
tuber <5> (<5> dim light and cold-stored, green skin peeling [2]) [2]

Additional information <1, 8, 13, 14, 17, 25> (<25> import into mitochondria [53]; <1,8,13,14,17> enzyme is synthezised in the cytosol as the precursor protein and then imported into the mitochondria matrix and cleaved to the mature enzyme, the targeting information is encoded in nonoverlapping regions of the presequence [43,46]) [43, 46, 53]

Localization

cytosol <3, 8, 10> (<3> minor activity [9]) [5, 6, 9, 20, 27, 30, 36]
microsome <1, 8-10> [13]
mitochondrial inner membrane <8, 13> (<8,13> loosely bound [11]) [11]
mitochondrial matrix <8, 13> [11, 16]
mitochondrion <1, 3, 8-10, 13, 14, 25> (<3> major activity [9]) [7, 9, 11, 13, 15, 16, 20, 26, 28, 29, 31, 34, 36, 43, 53]
Additional information <1, 8-10> (<8> processing of enzyme during to targeting to subcellular compartments [36]; <1,8-10> synthesized in microsomes and then migrates to the mitochondria [13]) [13, 36]

Purification

<6> (high activity in cells grown anaerobically in defined medium, low in cells grown in an iron-deficient medium and in cells grown aerobically [3]) [3]
<7> (2 forms from low-iron medium [24]; isoforms I and II [37]; inactive form [22]; activator protein [23]) [4, 12, 13, 18, 22-24, 32, 35, 37]
<8> (erythroid-specific isoform from reticulocytes, papain digestion method [41]; partial [36]; partial, fetal enzyme [34]; 2 forms: a complex form and a stripped form that represents a catalytically active protein component of the complex form [30]; difficulties due to the tendency of the enzyme to form large MW aggregates and to its susceptibility to proteolytic degradation [11]) [5, 11, 13, 26, 27, 30, 31, 34, 36, 41]
<9> [36]
<10> [13]
<13> (partial [7]) [7, 11, 25, 29]
<14> [28]
<15> (partial purification, 2 isozymes [33]) [32, 33]
<16> (2 isoenzymes: a and b [10]) [10]
<17> (recombinant of His-tagged wild-type and mutants from E. coli [45,48]; recombinant wild-type and mutants from E. coli [44,49]; erythroid-specific isoform [41]; recombinant wild-type and Lys313 mutants from E. coli [42]; recombinant from E. coli [17,46]) [17, 41, 42, 44-46, 48, 49]

Cloning

<1> (liver and erythroid enzyme [43]) [43]
<7> [43]
<8> (liver enzyme [43]; erythroid-specific isoform ALAS-E, DNA-sequencing and sequence analysis, amino acid sequence determination of precursor protein [41]) [41, 43]
<13> (cDNA from erythroid genetic library [43]) [43]

<14> (fusion protein of enzyme and plastid-targeting small subunit of ribulose bisphosphate carboxylase for construction of transgenic plants [47]; gene hem, functional complementation of Saccharomyces cerevisiae hem1 mutant [43]) [43, 47]
<17> (overexpression of His-tagged wild-type and mutants in Escherichia coli [48,50]; expression of His-tagged wild-type, mutant K313A, mutant R149A and dimer mutant K313A/R149A, each subunit from 1 plasmid, in Escherichia coli hemA$^-$ mutant [45]; expression of wild-type and mutants in Escherichia coli [44]; expression of erythroid-specific isoform ALAS-E and mutants in Escherichia coli [42,49]; expression of erythroid-specific isoform in MEL mutant under control of metallothionin promotor [40]; transient expression of erythroid-specific isoform and promotor-mutant in HeLa cells, luciferase reporter gene [39]; overexpression of mouse erythroid enzyme in Escherichia coli [17]) [17, 39, 40, 43-45, 48-50]
<20> [43]
<21> [43]
<22> [43]
<25> (expression in Escherichia coli [53]) [53]

Engineering

C145R <14> (<14> mutant strain G205, lacks enzyme activity, but can complement mutant strain Ole3 with mutation G344C [45]) [45]
D279A <17> (<17> exchange mutant of potential cofactor binding residue Asp279, no activity, dissociation constant for pyridoxal 5'-phosphate is 19fold increased, different mode of cofactor binding, no formation of quinonoid reaction intermediate, which can be restored by addition of analogue N-methyl-pyridoxal 5'-phosphate [49]) [49]
D279E <17> (<17> exchange mutant of potential cofactor binding residue Asp279, 30fold reduced catalytic efficiency for succinyl-CoA compared to the wild-type [49]) [49]
G142C <17> (<17> glycine-rich motif mutant, 15fold increased dissociation constant value for binding of cofactor pyridoxal 5'-phosphate, 6% turnover compared to the wild-type, 4fold increase of K_m-value for glycine [44]) [44]
G144A <17> (<17> glycine-rich motif mutant, 8.5fold increased dissociation constant value for binding of cofactor pyridoxal 5'-phosphate, 43% turnover compared to the wild-type, unaltered K_m-values for the substrates [44]) [44]
G144S <17> (<17> glycine-rich motif mutant, 8fold increased dissociation constant value for binding of cofactor pyridoxal 5'-phosphate, 39% turnover compared to the wild-type, unaltered K_m-values for the substrates [44]) [44]
G144T <17> (<17> glycine-rich motif mutant, 24.5fold increased dissociation constant value for binding of cofactor pyridoxal 5'-phosphate, 21% turnover compared to the wild-type, unaltered K_m-values for the substrates [44]) [44]
K313A <17> (<17> mutation site located at the active site of 1 subunit, functional complementation of E. coli mutant strain hemA$^-$, no activity [45]; <17> site-directed mutagenesis, mutants of erythroid-specific isoform, exchange of active site lysine residue 313, binding of pyridoxal 5'-phosphate

and glycine noncovalently, reduced activity, because covalent binding is required [42]) [42, 45]
K313A/R149A <17> (<17> each mutation site located on 1 subunit, 2 plasmids, coexpression of the dimer in E. coli hemA$^-$, functional complementation, 26% activity compared to wild-type [45]) [45]
K313G <17> (<17> site-directed mutagenesis, mutants of erythroid-specific isoform, exchange of active site lysine residue 313, binding of pyridoxal 5'-phosphate and glycine noncovalently, reduced activity, because covalent binding is required [42]) [42]
K313H <17> (<17> site-directed mutagenesis, mutants of erythroid-specific isoform, exchange of active site lysine residue 313, binding of pyridoxal 5'-phosphate and glycine noncovalently, reduced activity, because covalent binding is required [42]; <17> formation of quinonoid reaction intermediates [52]) [42, 52]
K313R <17> (<17> formation of quinonoid reaction intermediates [52]) [52]
N157Y/N162S <14> (<14> mutant strain G101, lacks enzyme activity, but can complement mutant strain G220 with mutation T452R [45]) [45]
R149A <17> (<17> mutation site located at the active site of 1 subunit, functional complementation of E. coli mutant strain hemA$^-$, no activity [45]) [45]
R433K <17> (<17> active site mutant, 2fold increased activity [50]) [50]
R433L <17> (<17> active site mutant, similar to the wild-type [50]) [50]
R439K <17> (<17> active site mutant, 77% activity compared to the wild-type, 9-13fold increased K_m for both substrates, 5fold increased dissociation constant for glycine [50]) [50]
R439L <17> (<17> active site mutant, no activity, 30fold increased dissociation constant for glycine [50]) [50]
Y121F <17> (<17> exchange mutant of potential cofactor binding residue Tyr121, 5% activity compared to the wild-type, K_m for glycine is 5fold increased, lower affinity for pyridoxal 5'phosphate [48]) [48]
Y121H <17> (<17> exchange mutant of potential cofactor binding residue Tyr121, 36% activity compared to the wild-type, K_m for glycine is 34fold increased, lower affinity for pyridoxal 5'phosphate [48]) [48]
Additional information <14, 17> (<14> construction of transgenic plants of Nicotiana tabacum wild-type and mutant deficient in chlorophyll biosynthesis expressing the enzyme targeted to the plastids, functional complementation of the mutant, regulation [47]; <17> expression of erythroid-specific isoform in MEL mutant under control of metallothionin promotor results in induction of enzyme activity by addition of Zn^{2+} and Cd^{2+} in absence of DMSO [40]; <17> stable MEL mutant, no induction of enzyme expression by dimethylsulfoxide, hexamethylene diacetamide and butyric acid, decline of enzyme activity after DMSO application [40]; <17> site-directed mutagenesis of hypoxia-inducible factor-1, i.e. HIF-1, binding site in promotor sequence -328/-318, reveales HIF-1 like activation of enzyme expression during hypoxia [39]) [39, 40, 47]

Application

nutrition <14> (<14> enzyme expressed in transgenic Nicotiana tabacum plants demonstrate functional complementation in the chlorophyll biosynthesis and open strategies for producing tolerance against inhibitors of the C_5 pathway [47]) [47]
pharmacology <25> (<25> enzyme is a target for drug development because of its immunological and inhibitor specificity [53]) [53]

6 Stability

pH-Stability

6.8-8 <8> (<8> unstable below pH 6.8 and above pH 8.0, irreversible loss of activity [5]) [5]

Temperature stability

37 <7> (<7> 60 min, 20% loss of activity with addition of 2-mercaptoethanol and glycerol, 60% loss of activity without addition of 2-mercaptoethanol and glycerol [4]) [4]
Additional information <5, 7, 8, 17> (<17> thermal stability of glycine-rich motif mutants is reduced compared to the wild-type [44]; <7> temperature insensitive [13]; <5> room temperature, inactivation within 30 min [2]) [2, 13, 16, 44]

General stability information

<7>, freezing and thawing: precipitation and loss of activity [35]
<7>, stable to freezing and thawing [4]
<8>, partially purified enzyme requires high cation concentration, equivalent to 0.3 M NaCl for maximum stabilization [5]
<1, 8, 13, 17>, short half-life [46]
<7, 15>, holoenzyme is more stable than apoenzyme [32]
<7, 15>, pyridoxal 5'-phosphate stabilizes [32]
<8, 13>, enzyme is susceptible to proteolytic degradation during isolation, tendency to form aggregates [43]

Storage stability

<2>, -15°C, not stable [8]
<2>, 0-4°C, 18 h stable [8]
<5>, -30°C, complete loss of activity overnight [2]
<7>, -15°C, 0.1 mM pyridoxal 5'-phosphate, 10% glycerol, 20% loss of activity after 6 weeks [37]
<7>, -15°C, 50 mM phosphate buffer, pH 6.8, 10% glycerol, 1 mM 2-mercaptoethanol, stable for several months [4]
<7>, 4°C, 50 mM phosphate buffer, pH 6.8, 10% glycerol, 1 mM 2-mercaptoethanol, 10% loss of activity after 2 weeks [4]
<7>, 5°C, 50 mM phosphate buffer, pH 7.2, 10 mM 2-mercaptoethanol, 20% loss of activity after 1 week [35]
<8>, -15°C, loss of 25% activity within 10 days [36]

<8>, -20°C, 20% loss of activity after 1 month [26]
<8>, 4°C, loss of 40% activity within 10 days [36]
<13>, -15°C, 20 mM potassium phosphate buffer, pH 7.6, 1 mM dithioerythritol, 0.01 mM pyridoxal 5'-phosphate, 30% loss of activity after 2 weeks [25]
<13>, 4°C, 20 mM potassium phosphate buffer, pH 7.6, 1 mM dithioerythritol, 0.01 mM pyridoxal 5'-phosphate, 20% loss of activity after 2 weeks [25]
<15>, 4°C, 30 min, 50% loss of apoenzyme of Protaminobacter ruber [32]
<8, 13>, -195°C, following an initial loss of 20% of activity due to freeze-thawing the enzyme loses no further activity for at least 6 months [11]
<8, 13>, 4°C, 50 mM Tris-HCl, 100 mM glycine, 0.1 mM pyridoxal 5'-phosphate, 1 mM dithioerythritol, pH 7.6, 5 mM 5'-AMP, stable for at least 4 months [11]

References

[1] Bishop, D.F.; Henderson, A.S.; Astrin, K.H.: Human δ-aminolevulinate synthase: assignment of the housekeeping gene to 3p21 and the erythroid-specific gene to the X chromosome. Genomics, **7**, 207-214 (1990)
[2] Ramaswamy, N.K.; Madhusudanan Nair, P.: δ-Aminolevulinic acid synthetase from cold-stored potatoes. Biochim. Biophys. Acta, **293**, 269-277 (1973)
[3] Tait, G.H.: Aminolaevulinate synthetase of Micrococcus denitrificans. Purification and properties of the enzyme, and the effect of growth conditions on the enzyme activity in cells. Biochem. J., **131**, 389-403 (1973)
[4] Warnick, G.R.; Burnham, B.F.: Regulation of prophyrin biosynthesis. Purification and characterization of δ-aminolevulinic acid synthase. J. Biol. Chem., **246**, 6880-6885 (1971)
[5] Scholnick, P.L.; Hammaker, L.E.; Marver, H.S.: Souble δ-aminolevulinic acid synthetase of rat liver. I. Some properties of the partially purified enzyme. J. Biol. Chem., **247**, 4126-4131 (1972)
[6] Scholnick, P.L.; Hammaker, L.E.; Marver, H.S.: Soluble δ-aminolevulinic acid synthetase of rat liver. II. Studies related to the mechanism of enzyme action and hemin inhibition. J. Biol. Chem., **247**, 4132-4137 (1972)
[7] Ades, I.Z.; Friedland, D.M.: Properties of chicken erythrocyte δ-aminolevulinate synthase. Int. J. Biochem., **20**, 965-969 (1988)
[8] Viale, A.A.; Wider, E.A.; Battle, A.M.Del C.: Extraction and characterization of δ-aminolevulinate synthetase. Comp. Biochem. Physiol. B, **87**, 607-613 (1987)
[9] Fernandez, J.; Gonzalez, O.; Martin, M.; Amil, M.R.: Trout (Salmo gairdenri R.) liver 5-aminolevulinate synthetase. Comp. Biochem. Physiol. B, **86**, 245-249 (1987)
[10] Kreit, J.: Control of 5-aminolevulinic acid synthetase activity in phototrophic facultative bacteria. Purification and some properties of the enzyme from Rhodospirillum rubrum. Biochimie, **63**, 439-444 (1981)
[11] Borthwick, I.A.; Srivastava, G.; Pirola, B.A.; May, B.K.; Elliott, W.H.: Purification of hepatic mitochondrial 5-aminolevulinate synthase. Methods Enzymol., **123**, 395-401 (1986)

[12] Jordan, P.M.; Laghai-Newton, A.: Purification of 5-aminolevulinate synthase. Methods Enzymol., **123**, 435-443 (1986)

[13] Jordan, P.M.; Shemin, D.: δ-Aminolevulinic acid synthetase. The Enzymes, 3rd Ed. (Boyer, P.D., ed.), **7**, 339-356 (1972)

[14] Nandi, D.L.: Studies on δ-aminolevulinic acid synthase of Rhodopseudomonas spheroides. Reversibility of the reaction, kinetic, spectral, and other studies related to the mechanism of action. J. Biol. Chem., **253**, 8872-8877 (1978)

[15] Simpson, D.M.; Beattie, D.S.: Purification of a specific activator of solubilized δ-aminolevulinic acid synthetase from rat liver mitochondria. J. Biol. Chem., **255**, 1630-1634 (1980)

[16] Scotto, A.W.; Chang, L.F.L.; Beattie, D.S.: The characterization and submitochondrial localization of δ-aminolevulinic acid synthase and an associated amidase in rat liver mitochondria using an improved assay for both enzymes. J. Biol. Chem., **258**, 81-90 (1983)

[17] Ferreira, G.C.; Dailey, H.A.: Expression of mammalian 5-aminolevulinate synthase in Escherichia coli. Overproduction, purification, and characterization. J. Biol. Chem., **268**, 584-590 (1993)

[18] Davies, R.C.; Neuberger, A.: Control of 5-aminolaevulinate synthetase activity in Rhodopseudomonas spheroides. Purification and properties of the high-activity form of the enzyme. Biochem. J., **177**, 649-659 (1979)

[19] Davies, R.C.; Neuberger, A.: Control of 5-aminolaevulinate synthetase activity in Rhodopseudomonas spheroides. Binding of pyridoxal phosphate to 5-aminolaevulinate synthetase. Biochem. J., **177**, 661-671 (1979)

[20] Ibrahim, N.G.; Gruenspecht, N.R.; Freedman, M.L.: Hemin feedback inhibition at reticulocyte δ-aminolevulinic acid synthetase and δ-aminolevulinic acid dehydratase. Biochem. Biophys. Res. Commun., **80**, 722-728 (1978)

[21] Porra, R.J.; Irving, E.A.: The nature of the inhibition of δ-aminolevulinic acid synthetase by hemin. Arch. Biochem. Biophys., **148**, 37-43 (1972)

[22] Tuboi, S.; Hayasaka, S.: Control of δ-aminolevulinate synthetase activity in Rhodopseudomonas spheroides. I. Partial purification of the inactive form of fraction I. Arch. Biochem. Biophys., **146**, 282-290 (1971)

[23] Hayasaka, S.; Tuboi, S.: Control of δ-aminolevulinate synthetase activity in Rhodopseudomonas spheroides. III. Partial purification of the fraction I activating enzyme and the occurrence of two forms of fraction II. J. Biochem., **76**, 157-168 (1974)

[24] Clement-Metral, J.D.; Fanica-Gaignier, M.: 5-Aminolevulinic-acid synthetases from Rhodopseudomonas spheroides Y. Comparison of the purification and properties of enzymes extracted from bacteria grown in different iron concentrations. Eur. J. Biochem., **59**, 73-77 (1975)

[25] Whiting, M.J.; Granick, S.: δ-Aminolevulinic acid synthase from chick embryo liver mitochondria. I. Purification and some properties. J. Biol. Chem., **251**, 1340-1346 (1976)

[26] Paterniti, J.R.; Beattie, D.S.: δ-Aminolevulinic acid synthetase from rat liver mitochondria. Purification and properties. J. Biol. Chem., **254**, 6112-6118 (1979)

[27] Nakakuki, M.; Yamauchi, K.; Hayashi, N.; Kikuchi, G.: Purification and some properties of δ-aminolevulinate synthase from the rat liver cytosol fraction and immunochemical identity of the cytosolic enzyme and the mitochondrial enzyme. J. Biol. Chem., **255**, 1738-1745 (1980)
[28] Volland, C.; Felix, F.: Isolation and properties of 5-aminolevulinate synthase from the yeast Saccharomyces cerevisiae. Eur. J. Biochem., **142**, 551-557 (1984)
[29] Borthwick, I.A.; Srivastava, G.; Brooker, J.D.; May, B.K.; Elliott, W.H.: Purification of 5-aminolaevulinate synthase from liver mitochondria of chick embryo. Eur. J. Biochem., **129**, 615-620 (1983)
[30] Ohashi, A.; Kikuchi, G.: Purification and some properties of two forms of δ-aminolevulinate synthase from rat liver cytosol. J. Biochem., **85**, 239-247 (1979)
[31] Srivastava, G.; Borthwick, I.A.; Brooker, J.D.; May, B.K.; Elliott, W.H.: Purification of rat liver mitochondrial δ-aminolaevulinate synthase. Biochem. Biophys. Res. Commun., **109**, 305-312 (1982)
[32] Sato, K.; Ishida, K.; Mutsushika, O.; Shimizu, S.: Purification and some properties of δ-aminolevulinate acid synthases from Protaminobacter ruber and Rhodopseudomonas speroides. Agric. Biol. Chem., **49**, 3415-3421 (1985)
[33] Sato, K.; Ishida, K.; Shirai, M.; Shimizu, S.: Occurence and some properties of two types of δ-amino levulinic acid synthase in a facultative methylotoph, Protaminobacter ruber. Agric. Biol. Chem., **49**, 3423-3428 (1985)
[34] Murthy, V.V.; Woods, J.S.: Solubilization and partial purification of mitochondrial δ-aminolevulinate synthase from fetal rat liver. Biochim. Biophys. Acta, **350**, 240-246 (1974)
[35] Yubisui, T.; Yoneyama, Y.: δ-Aminolevulinic acid synthetase of Rhodopseudomonas spheroides: purification and properties of the enzyme. Arch. Biochem. Biophys., **150**, 77-85 (1972)
[36] Whiting, M.J.; Elliott, W.H.: Purification and properties of solubilized mitochondrial δ-aminolevulinic acid synthetase and comparison with the cytosol enzyme. J. Biol. Chem., **247**, 6818-6826 (1972)
[37] Fanica-Gaignier, M.; Clement-Metral, J.: 5-Aminolevulinic-acid synthetase of Rhodopseudomonas spheroides Y. Purification and some properties. Eur. J. Biochem., **40**, 13-18 (1973)
[38] Fanica-Gaignier, M.; Clement-Metral, J.: 5-Aminolevulinic-acid synthetase of Rhodopseudomonas spheroides Y. Kinetic mechanism and inhibition by ATP. Eur. J. Biochem., **40**, 19-24 (1973)
[39] Hofer, T.; Wenger, R.H.; Kramer, M.F.; Ferreira, G.C.; Gassmann, M.: Hypoxic up-regulation of erythroid 5-aminolevulinate synthase. Blood, **101**, 348-350 (2003)
[40] Lake-Bullock, H.; Dailey, H.A.: Biphasic ordered induction of heme synthesis in differentiating murine erythroleukemia cells: role of erythroid 5-aminolevulinate synthase. Mol. Cell. Biol., **13**, 7122-7132 (1993)
[41] Munakata, H.; Yamagami, T.; Nagai, T.; Yamamoto, M.; Hayashi, N.: Purification and structure of rat erythroid-specific δ-aminolevulinate synthase. J. Biochem., **114**, 103-111 (1993)

[42] Ferreira, G.C.; Vajapey, U.; Hafez, O.; Hunter, G.A.; Barber, M.J.: Aminolevulinate synthase: lysine 313 is not essential for binding the pyridoxal phosphate cofactor but is essential for catalysis. Protein Sci., **4**, 1001-1006 (1995)

[43] Ferreira, G.C.; Gong, J.: 5-Aminolevulinate synthase and the first step of heme biosynthesis. J. Bioenerg. Biomembr., **27**, 151-159 (1995)

[44] Gong, J.; Kay, C.J.; Barber, M.J.; Ferreira, G.C.: Mutations at a glycine loop in aminolevulinate synthase affect pyridoxal phosphate cofactor binding and catalysis. Biochemistry, **35**, 14109-14117 (1996)

[45] Tan, D.; Ferreira, G.C.: Active site of 5-aminolevulinate synthase resides at the subunit interface. Evidence from in vivo heterodimer formation. Biochemistry, **35**, 8934-8941 (1996)

[46] Dailey, H.A.; Dailey, T.A.: Expression and purification of mammalian 5-aminolevulinate synthase. Methods Enzymol., **281**, 336-340 (1997)

[47] Zavgorodnyaya, A.; Papenbrock, J.; Grimm, B.: Yeast 5-aminolevulinate synthase provides additional chlorophyll precursor in transgenic tobacco. Plant J., **12**, 169-178 (1997)

[48] Tan, D.; Barber, M.J.; Ferreira, G.C.: The role of tyrosine 121 in cofactor binding of 5-aminolevulinate synthase. Protein Sci., **7**, 1208-1213 (1998)

[49] Gong, J.; Hunter, G.A.; Ferreira, G.C.: Aspartate-279 in aminolevulinate synthase affects enzyme catalysis through enhancing the function of the pyridoxal 5'-phosphate cofactor. Biochemistry, **37**, 3509-3517 (1998)

[50] Tan, D.; Harrison, T.; Hunter, G.A.; Ferreira, G.C.: Role of arginine 439 in substrate binding of 5-aminolevulinate synthase. Biochemistry, **37**, 1478-1484 (1998)

[51] Kaliman, P.A.; Barannik, T.V.: Regulation of δ-aminolevulinate synthase activity during the development of oxidative stress. Biochemistry, **64**, 699-704 (1999)

[52] Hunter, G.A.; Ferreira, G.C.: Lysine-313 of 5-aminolevulinate synthase acts as a general base during formation of the quinonoid reaction intermediates. Biochemistry, **38**, 3711-3718 (1999)

[53] Varadharajan, S.; Dhanasekaran, S.; Bonday, Z.Q.; Rangarajan, P.N.; Padmanaban, G.: Involvement of δ-aminolevulinate synthase encoded by the parasite gene in de novo heme synthesis by Plasmodium falciparum. Biochem. J., **367**, 321-327 (2002)

[54] Zhang, J.; Ferreira, G.C.: Transient state kinetic investigation of 5-aminolevulinate synthase reaction mechanism. J. Biol. Chem., **277**, 44660-44669 (2002)

[Acyl-carrier-protein] S-acetyltransferase 2.3.1.38

1 Nomenclature

EC number
2.3.1.38

Systematic name
acetyl-CoA:[acyl-carrier-protein] S-acetyltransferase

Recommended name
[acyl-carrier-protein] S-acetyltransferase

Synonyms
ACAT <8> [11]
ACT <10> [14]
[ACP] acetyltransferase
[acyl-carrier-protein]acetyltransferase
acetyl coenzyme A-acyl-carrier-protein transacylase
acetyl coenzyme A:ACP transacylase <2> [1]
acetyl-CoA-ACP-transacylase <2> [10]
acetyl-CoA:ACP transacylase <8> [11]
acetyl-CoA:acyl carrier protein transacylase <10> [14]
acetyltransferase, [acyl-carrier-protein]
malonyl-CoA/acetyl-CoA:acyl carrier protein S-acyltransferase <9> (<9> i.e. MAT [13]) [12, 13]
Additional information <8-10> (cf. EC 2.3.1.85 and EC 2.3.1.86; <10> enzyme has both a acetyl-CoA:acyl carrier protein transacylase activity and a much higher thiolase activity [14]; <9> dual specificity of a malonyl-CoA/acetyl-CoA:acyl carrier protein S-acyltransferase [12]; <8> possibility that β-ketoacyl-[acyl-carrier-protein] synthetase III has intrinsic acetyl-CoA:[acyl-carrier-protein] transacylase activity and the separated acetyl-CoA:[acyl-carrier-protein] transacylase would be a proteolytic fragment of β-ketoacyl-[acyl-carrier-protein] synthetase III [11]) [11, 12, 14]

CAS registry number
37257-16-2

2 Source Organism

<1> *Escherichia coli* (B [7,9]; strain E26 [4]) [2-4, 7, 9]
<2> *Euglena gracilis* (v. bacillaris [10]; Z [1]) [1, 10]
<3> *Spinacia oleracea* (var. Viroflay [5]) [5, 6]

<4> *Cuphea lutea* [6]
<5> *Carthamus tinctorius* (safflower [6,8]) [6, 8]
<6> *Brassica juncea* [6]
<7> *Pisum sativum* [6]
<8> *Persea sp.* (avocado [11]) [11]
<9> *Rattus norvegicus* [12, 13]
<10> *Streptomyces collinus* (gene fadA [14]) [14]

3 Reaction and Specificity

Catalyzed reaction

acetyl-CoA + [acyl-carrier protein] = CoA + acetyl-[acyl-carrier protein] (<9> Arg-606 is responsible for malonyl-CoA specificity of the bifunctional enzyme [12]; <9> mechanism [12]; <1> Bi Bi ping pong mechanism, acetyl-enzyme intermediate [7])

Reaction type

acyl group transfer

Natural substrates and products

S acetyl-CoA + [acyl-carrier-protein] <2, 3, 5> (<2> first step of fatty acid synthesis [10]; <3> increase in enzyme level in reconsituted FAS system enhances production of medium length fatty acids [6]; <2> enzyme is component of fatty acid synthase complex II [1,10]; <3> rate-limiting step of fatty acid synthesis [5,6]; <5> part of non-associated fatty acid synthase system of plants and prokaryotes [8]) (Reversibility: ? <2, 3, 5> [1, 5, 6, 8, 10]) [1, 5, 6, 8, 10]

P CoA + acetyl-[acyl-carrier-protein] <2, 3, 5> [1, 2, 5, 6, 8, 10]

Substrates and products

S acetyl-CoA + [acyl-carrier-protein] <1-10> (<10> specific for acetyl-CoA, utilizes [acyl-carrier-protein] of diverse origins with differing affinities [14]; <9> dual specificity of a malonyl-CoA/acetyl-CoA:acyl carrier protein S-acyltransferase [12,13]; <1> equilibrium in forward direction [9]; <1> via acetyl-enzyme intermediate [2,4,7]; <1> pantetheine, not 2-mercaptoethanol can replace [acyl-carrier-protein] [2]; <1> acetyl-S-pantetheine can replace acetyl-CoA [2,3]) (Reversibility: r <1> [2, 4, 7, 9]; ? <2-10> [1, 3, 5, 6, 8, 10-14]) [1-14]

P CoA + acetyl-[acyl-carrier-protein] <1-10> [1-14]

S acetyl-CoA + pantetheine <9> (<9> octanoyl-S-pantetheine, decanoyl-S-pantetheine, and dodecanoyl-S-pantetheine can also serve as substrates for the backward reaction [12]; <9> dual specificity of a malonyl-CoA/acetyl-CoA:acyl carrier protein S-acyltransferase [12,13]) (Reversibility: r <9> [12, 13]) [12, 13]

P CoA + acetyl-S-pantetheine <9> [12, 13]

S butyryl-CoA + [acyl-carrier-protein] <1, 3> (<3> 35% activity of that with acetyl-CoA [5]; <1> 9.8% activity of that with acetyl-CoA [2,4]) (Reversibility: ? <1, 3> [2-5]) [2-5]
P CoA + butyryl-[acyl-carrier-protein]
S hexanoyl-CoA + [acyl-carrier-protein] <1, 3> (<3> 18% activity of that with acetyl-CoA [5]; <1> 4.6% activity of that with acetyl-CoA [2,4]) (Reversibility: ? <1, 3> [2-5]) [2-5]
P CoA + hexanoyl-[acyl-carrier-protein]
S octanoyl-CoA + [acyl-carrier-protein] <1, 3> (<3> poor substrate [5]; <1> 10.3% activity of that with acetyl-CoA [4]; <3> 7% activity of that with acetyl-CoA [5]) (Reversibility: ? <1, 3> [3-5]) [3-5]
P CoA + octanoyl-[acyl-carrier-protein]
S propionyl-CoA + [acyl-carrier-protein] <1> (<1> transacylation at 23.4% the rate of acetyl-CoA [2,4]) (Reversibility: ? <1> [2, 4]) [2, 4]
P CoA + propionyl-[acyl-carrier-protein]
S Additional information <1, 3, 8, 9> (<9> enzyme is part of fatty acid synthetase complex, consisting of 2 subunits [13]; <1,3,8> no substrate: malonyl-CoA [2,5,9,11]; <1> 2-mercaptoethanol does not substitute for [acyl-carrier-protein] [9]; <1> in vertebrates, yeast and Mycobacteria acetyl transferase activity is a domain of the multifunctional polypeptide chains of fatty acid synthase EC 2.3.1.85 and EC 2.3.1.86 [2,3]) [2, 3, 5, 9, 11, 13]
P ?

Inhibitors

CoA <1> (<1> competitive [7]) [7]
N-ethylmaleimide <1-3> (<2> weak [10]; <1> strong [9]; <1> acetyl-CoA protects [3,7,9]) [3-7, 9, 10]
PCMB <3> [5]
acetyl-CoA <1> (<1> at high concentration [7]) [7]
acyl-carrier-protein <1> (<1> at high concentration [7]) [7]
arsenite <3> [5, 6]
cerulenin <8> (<8> antibiotic, weak inhibition [11]) [11]
iodoacetamide <1, 10> (<2> no inhibition [10]; <1,10> strong [9,14]; <1,10> acetyl-CoA protects [3,7,9,14]) [3, 4, 7, 9, 14]
methyl methanethiosulfonate <1> (<1> weak [7]) [7]
p-chloromercuribenzoate <2> (<2> strong inhibition [10]) [10]
p-substituted mercuribenzoate <3> [6]
pantetheine <1> (<1> weak, competitive [7]) [7]
thiolactomycin <8, 10> [11, 14]
Additional information <1, 2> (<2> no inhibition by diisopropylphosphofluoridate and antibiotic cerulenin [10]; <1> no inhibition by PMSF [7]) [7, 10]

Metals, ions

Additional information <1> (<1> high ionic strength required for maximal activity [2,9]) [2, 9]

Turnover number (min^{-1})

1.35 <10> (acetyl-CoA, <10> recombinant wild-type, crude Streptomyces glaucescens [acyl-carrier-protein] [14]) [14]
3.05 <10> (acetyl-CoA, <10> native wild-type, crude E. coli [acyl-carrier-protein] [14]) [14]
114 <9> (acetyl-CoA, <9> recombinant purified wild-type enzyme [12]) [12]
204 <9> (acetyl-CoA, <9> recombinant purified mutant R606K [12]) [12]
252 <9> (acetyl-CoA, <9> recombinant purified mutant R606A [12]) [12]

Specific activity (U/mg)

0.000005 <7> [6]
0.000009 <3> (<3> crude enzyme extract [6]) [6]
0.000018 <5> [6]
0.000064 <4> [6]
0.00019 <6> [6]
0.00023 <5> (<5> crude extract [8]) [8]
0.0056 <3> (<3> partially purified enzyme [5]; purified enzyme [6]) [5, 6]
0.022 <10> (<10> purified native wild-type enzyme [14]) [14]
0.03 <2> (<2> partially purified enzyme [10]) [10]
0.059 <1> [4]
0.06 <1> [2]
0.3 <1> (<1> purified enzyme [9]) [9]
0.8 <1> (<1> approx., purified enzyme [7]) [7]
2.68 <9> (<9> recombinant purified wild-type enzyme [12]) [12]
3.22 <9> (<9> recombinant purified wild-type enzyme, carboxyterminally FLAG-tagged [13]) [13]
3.81 <9> (<9> recombinant purified wild-type enzyme, carboxyterminally His-tagged [13]) [13]
4.53 <9> (<9> recombinant purified mutant R606K [12]) [12]
17.8 <9> (<9> recombinant purified mutant R606A [12]) [12]
Additional information <2-7, 9> (<2,3> fatty acid synthase II assay [1,6]) [1, 6, 12, 13]

K_m-Value (mM)

0.0018 <9> (acetyl-CoA, <9> recombinant purified mutant R606A [12]) [12]
0.0029 <9> (acetyl-CoA, <9> recombinant purified mutant R606K [12]) [12]
0.0039 <9> (acetyl-CoA, <9> recombinant purified wild-type enzyme [12]) [12]
0.0043 <10> (acetyl-CoA, <10> native wild-type enzyme [14]) [14]
0.008 <3> (acetyl-CoA) [5, 6]
0.0083 <3> (butyryl-CoA) [5, 6]
0.0086 <3> (hexanoyl-CoA) [5, 6]
0.0087 <10> (acetyl-CoA, <10> recombinant wild-type enzyme [14]) [14]
0.035 <1> (acyl-carrier-protein) [7]
0.05 <1> (acetyl-CoA) [7]

K_i-Value (mM)

0.03 <1> (CoA) [7]
3.2 <1> (pantetheine) [7]

pH-Optimum

6 <1> [4]
6.5 <1> (<1> assay at [7]) [2, 7, 9]
6.8 <2> (<2> assay at [1]) [1]
7 <8> [11]
8 <2> (<2> assay at [10]) [10]
8.1 <3> [5, 6]

pH-Range

5-8.5 <1> (<1> about 60% of maximal activity at pH 5.0 and pH 8.5 [4]) [4]
6.2-7.5 <8> (<8> about half-maximal activity at pH 6.0 and pH 7.5 [11]) [11]
7.5-9.1 <3> (<3> about half-maximal activity at pH 7.5 and about 75% of maximal activity at pH 9.1 [5]) [5]

Temperature optimum (°C)

25 <2, 5> (<2,5> assay at [1,8]) [1, 8]
30 <1, 10> (<1,10> assay at [7,14]) [7, 9, 14]
35 <2> (<2> assay at [10]) [10]
38 <1, 3-7> (<1,3-7> assay at [4-6]) [4-6]

4 Enzyme Structure

Molecular weight

18500 <8> (<8> gel filtration [11]) [11]
48000 <3> (<3> gel filtration [5,6]) [5, 6]
60000 <2> (<2> gel filtration [10]) [10]
61000 <1> (<1> gel filtration [7]) [7]
147000 <2> (<2> gel filtration [1]) [1]
185000 <10> (<10> gel filtration [14]) [14]
Additional information <10> (<10> amino acid sequence, comparison [14]) [14]

Subunits

dimer <1> (<1> 2 * 29000, SDS-PAGE [7]) [7]
tetramer <10> (<10> 4 * 45000, SDS-PAGE [14]) [14]
Additional information <1> (<1> vertebrates, yeast and Mycobacteria fatty acid synthetases contain all individual activities on one or two multifunctional polypeptide chains, plant, E. coli and other prokaryotic fatty acid synthases are non-associated systems of individual enzymes [2,3]) [2, 3]

5 Isolation/Preparation/Mutation/Application

Source/tissue

fruit <8> [11]
leaf <3, 7> [5, 6]
seed <4-6> [6, 8]

Localization

chloroplast <2, 8> (<2> fatty acid synthase II [10]) [10, 11]
cytosol <2> (<2> fatty acid synthase II [10]) [10]
membrane <8> [11]

Purification

<1> [4, 7, 9]
<2> (partial [10]) [10]
<3> (partial [5]) [5, 6]
<8> (partial [11]) [11]
<9> (recombinant mutants from E. coli [12]) [12]
<10> (native wild-type and recombinant wild-type and mutant from E. coli [14]) [14]

Renaturation

<9> (recombinant enzyme purified from E. coli, refolding after treatment of inclusion bodies with urea [12]) [12]

Cloning

<9> (expression of wild-type and His-tagged mutant S581A as His-tagged proteins in Escherichia coli and as FLAG-tagged proteins in Sf9 insect cells via baculovirus infection [13]; expression of wild-type and mutants in Escherichia coli BL21 (DE3) [12]) [12, 13]
<10> (cloning and sequencing of fadA gene, expression in Escherichia coli of wild-type and C92S mutant [14]) [14]

Engineering

C92S <10> (<10> no remaining activity acetyl-CoA:acyl carrier protein transacylase activity, but remaining strongly reduced thiolase activity [14]) [14]
R606A <9> (oligonucleotide-directed in vitro mutagenesis, 6.6fold increase in activity about 16000fold increase in selectivity for acetyl-CoA against malonyl-CoA, increased production of medium-chain length fatty acids [12]) [12]
R606K <9> (oligonucleotide-directed in vitro mutagenesis, 1.7fold increase in activity about 16fold increase in selectivity for acetyl-CoA against malonyl-CoA [12]) [12]
S581A <9> (nearly no remaining activity, below 0.03% compared to wild-type [13]) [13]

6 Stability

Temperature stability

40 <1> (<1> 30% loss of activity after 40 min [4]) [4]
50 <1> (<1> $t_{1/2}$: 18 min [4]) [4]
60 <1> (<1> $t_{1/2}$: 9 min [4]) [4]
100 <1> (<1> inactivation after 1 min [4]) [4]
Additional information <10> [14]

General stability information

<1>, EDTA at 5 mM stabilize [4]

Storage stability

<1>, -10°C, unpredictable instability on storage, sometimes total loss of activity occured immediately after thawing [4]
<1>, -20°C, at least 1 month [9]
<1>, -20°C, less than 10% of activity retained after 1 year [9]
<1>, -20°C, several months [7]
<1>, 0°C, inactivation within several days [4]
<3>, -70°C, several months [5, 6]

References

[1] Hendren, R.W.; Bloch, K.: Fatty acid synthetases from Euglena gracilis. Separation of component activities of the ACP-dependent fatty acid synthetase and partial purification of the β-ketoacyl-ACP synthetase. J. Biol. Chem., **255**, 1504-1508 (1980)
[2] Vagelos, R.P.: Acyl group transfer (acyl carrier protein). The Enzymes, 3rd Ed. (Boyer, P.D., ed.), **8**, 155-199 (1973)
[3] Prescott, D.J.; Vagelos, P.R.: Acyl carrier protein. Adv. Enzymol. Relat. Areas Mol. Biol., **36**, 269-311 (1972)
[4] Williamson, I.P.; Wakil, S.J.: Studies on the mechanism of fatty acid synthesis. XVII. Preparation and general properties of acetyl coenzyme A and malonyl coenzyme A-acyl carrier protein transacylases. J. Biol. Chem., **241**, 2326-2332 (1966)
[5] Shimakata, T.; Stumpf, P.K.: Purification of plant acetyl-CoA: acyl carrier protein transacylase. Methods Enzymol., **122**, 53-59 (1986)
[6] Shimakata, T.; Stumpf, P.K.: The purification and function of acetyl coenzyme A:acyl carrier protein transacylase. J. Biol. Chem., **258**, 3592-3598 (1983)
[7] Lowe, P.N.; Rhodes, S.: Purification and characterization of [acyl-carrier-protein] acetyltransferase from Escherichia coli. Biochem. J., **250**, 789-796 (1988)
[8] Shimakata, T.; Stumpf, P.K.: The procaryotic nature of the fatty acid synthetase of developing Carthamus tinctorius L. (Safflower) seeds. Arch. Biochem. Biophys., **217**, 144-154 (1982)

[9] Alberts, A.W.; Majerus, P.W.; Vagelos, P.R.: Acetyl-CoA acyl carrier protein transacylase. Methods Enzymol., **14**, 50-53 (1969)
[10] Worsham, L.M.S.; Williams, S.G.; Ernst-Fonberg, M.L.: Early catalytic steps of Euglena gracilis chloroplast type II fatty acid synthase. Biochim. Biophys. Acta, **1170**, 62-71 (1993)
[11] Gulliver, B.S.; Slabas, A.R.: Acetoacyl-acyl carrier protein synthase from avocado: its purification, characterization and clear resolution from acetyl CoA: ACP transacylase. Plant Mol. Biol., **25**, 179-191 (1994)
[12] Rangan, V.S.; Smith, S.: Alteration of the substrate specificity of the malonyl-CoA/acetyl-CoA:acyl carrier protein S-acyltransferase domain of the multifunctional fatty acid synthase by mutation of a single arginine residue. J. Biol. Chem., **272**, 11975-11978 (1997)
[13] Rangan, V.S.; Joshi, A.K.; Smith, S.: Fatty acid synthase dimers containing catalytically active β-ketoacyl synthase or malonyl/acetyltransferase domains in only one subunit can support fatty acid synthesis at the acyl carrier protein domains of both subunits. J. Biol. Chem., **273**, 34949-34953 (1998)
[14] Lobo, S.; Florova, G.; Reynolds, K.A.: A Streptomyces collinus thiolase with novel acetyl-CoA:acyl carrier protein transacylase activity. Biochemistry, **40**, 11955-11964 (2001)

[Acyl-carrier-protein] S-malonyltransferase 2.3.1.39

1 Nomenclature

EC number
2.3.1.39

Systematic name
malonyl-CoA:[acyl-carrier-protein] S-malonyltransferase

Recommended name
[acyl-carrier-protein] S-malonyltransferase

Synonyms
[acyl carrier protein]malonyltransferase
malonyl coenzyme A-acyl carrier protein transacylase
malonyl transacylase
malonyl transferase
malonyl-CoA-acyl carrier protein transacylase
malonyltransferase, [acyl-carrier-protein]
Additional information <1, 14> (<1,14> in vertebrates, yeast and mycobacteria malonyl transferase activity is a domain of the multifunctional polypeptide chains of fatty acid synthase EC 2.3.1.85 and EC 2.3.1.86 [2,21,22]) [2, 21, 22]

CAS registry number
37257-17-3

2 Source Organism

<1> *Escherichia coli* [1-3, 5-10, 12, 16, 18-20, 24, 25]
<2> *Anabaena variabilis* [11]
<3> *Euglena gracilis* (Z [4]) [4]
<4> *Glycine max* (soy bean, strain AMSOY71 [12]) [12]
<5> *Spinacia oleracea* (spinach [11,14]) [11, 14]
<6> *Persea americana* (avocado [13]) [13]
<7> *Cuphea lutea* (Lythraceae [14]) [14]
<8> *Carthamus tinctorius* (safflower [14,15]) [14, 15]
<9> *Brassica juncea* (rape [14]) [14]
<10> *Pisum sativum* (pea [14]) [14]
<11> *Pseudomonas sp.* (61-3 [19]) [19]
<12> *Pseudomonas aeruginosa* [20]
<13> *Mycobacterium tuberculosis* [17]

<14> *Rattus norvegicus* (malonyl transferase domain of EC 2.3.1.85 [21,22]) [21, 22]
<15> *Streptomyces glaucescens* (produces tetracenomicin C [23]) [23]
<16> *Streptomyces coelicolor* [16, 18]

3 Reaction and Specificity

Catalyzed reaction

malonyl-CoA + [acyl-carrier protein] = CoA + malonyl-[acyl-carrier protein] (<1> mechanism, formation of a malonyl-enzyme intermediate [1,2,6,7]; <1> isolation and characterization of an enzyme-intermediate [7,8]; <2,5,16> mechanism, kinetic studies [11,16])

Reaction type

acyl group transfer

Natural substrates and products

S malonyl-CoA + acyl-carrier-protein <1, 6> (<1> initial reaction in de novo fatty acid synthesis, part of non-associated fatty acid synthase system of plants and prokaryotes [1,3]; <6> the enzyme is a component of the fatty acid synthetase system [13]) (Reversibility: ? <1, 6> [1, 13] ; r <1> [3]) [1, 3, 13]
P CoA + malonyl-acyl-carrier-protein <1, 6> [1, 3, 13]
S Additional information <15> (<15> the enzyme is responsible for charging the tetracenomycin M acyl-carrier-protein with malonate in vivo, a key step in the synthesis of the deca(polyketide) precursor of tetracenomycin C. This implies the existence of a functional connection between fatty acid and polyketide metabolism in this bacterium [23]) [23]
P ?

Substrates and products

S malonyl-CoA + N-(N-acetyl-β-alanyl)cysteamine <1> (Reversibility: ? <1> [2, 6]) [2, 6]
P CoA + N-(N-acetyl-β-alanyl)-S-malonylcysteamine
S malonyl-CoA + N-acetylcysteamine <1, 16> (<16> in the presence of high N-acetylcysteamine concentrations the forward reaction can be driven to completion [18]) (Reversibility: ? <1, 16> [2, 6, 18]) [2, 6, 18]
P CoA + N-acetyl-S-malonylcysteamine
S malonyl-CoA + acyl-carrier-protein <1-10, 12, 13, 15, 16> (<1-10> via stable covalent malonyl-enzyme intermediate [1,4,11-15]; <1,12,15> characterization of acyl-carrier protein [1,20,23]; <1> the enzyme is very specific for the malonyl group [5,6]; <1,4> acetyl-CoA cannot replace malonyl-CoA [1,2,5,6,8,12]; <1> participation of a malonyl-enzyme intermediate in the reaction [8]; <4> both isozymes may not be completely specific for malonyl-CoA as the substrate [12]; <13> holo-acyl-carrier-protein from Mycobacterium tuberculosis constitutes the preferred substrate for the enzyme, however the enzyme uses both E. coli holo-acyl-carrier-pro-

tein and Mycobacterium tuberculosis holo-acyl-carrier-protein as substrates for transacylation in vivo and in vitro [17]; <15> the recombinant enzyme is able to use E. coli acyl-carrier-protein, Streptomyces glaucescens acyl-carrier-protein as well as the tetracenomycin M acyl-carrier-protein component of the tetracenomycin type II polyketide synthase [23]) (Reversibility: ? <1-10, 12, 13, 15, 16> [1, 2, 4-9, 11-18, 20, 23]; r <1> [3]) [1-9, 11-18, 20, 23]

P CoA + malonyl-[acyl-carrier-protein] <1-10, 12, 13, 15, 16> [1-9, 11-18, 20, 23]

S malonyl-CoA + pantetheine <1, 6, 14> (<14> reaction catalysed by the malonyl transferase domain of EC 2.3.1.85 [21,22]) (Reversibility: ? <1, 6, 14> [1, 2, 5, 6, 13, 21, 22]) [1, 2, 5, 6, 13, 21, 22]

P malonyl-pantetheine + CoA

S malonyl-pantetheine + acyl-carrier-protein <1> (Reversibility: ? <1> [1, 5]) [1, 5]

P pantetheine + malonyl-[acyl-carrier-protein]

S Additional information <1> (<1> mercaptoethanol is inactive as a substitute for acyl-carrier-protein [5]) [5]

P ?

Inhibitors

CoA <1, 2, 5> (<2,5> kinetics [11]; <4> not inhibitory [12]) [6, 11]

N-ethylmaleimide <1, 4> (<1> 79% inhibition at 0.1 mM [5]; <1> pH-dependent inhibition of purified enzyme [7]; <4> 40% of isozyme MCT2 activity lost at 1mM, no effect over MCT1, both MCT1 and MCT2 inactivate at 10 mM [12]; <1> malonyl-CoA protects [7]) [5, 7, 12]

acetoacetyl-CoA <4> (<4> inhibition of both isozymes at 0.02 mM and at 0.1 mM [12]) [12]

acetyl-CoA <1> (<1> competitive to malonyl-CoA [2,6]) [2, 6]

diethyldicarbonate <4> (<4> inhibits MCT1 but not MCT2 at 1 mM [12]) [12]

iodoacetamide <1> (<1> pH-dependent [2,7]; <1> 59% inhibition at 0.1 mM [5]; <1> malonyl-CoA protects [2,7]) [2, 3, 5, 7]

iodoacetate <1> [2]

malonyl-CoA <4> (<4> inhibition of MCT1 at 0.01 mM and inhibition of both isozymes at 0.02 mM and 0.04 mM [12]) [12]

malonyl-CoA <4> (<4> inhibits both MCT1 and MCT2 [12]) [12]

methylmalonyl-CoA <4> (<4> inhibition of MCT1 at 0.1 mM [12]) [12]

p-chloromercuribenzoate <1> (<1> strong [1]; <1> pH-independent [2]) [1, 2]

p-toluenesulfonyl fluoride <1> (<1> strong inhibitor [6]) [6]

phenylmethanesulfonyl fluoride <1, 16> (<1> malonyl-CoA protects [2]; <4> not inhibitory [12]) [1, 2, 6, 7, 16]

propionyl-CoA <4> (<4> inhibition of both isozymes at 0.1 mM [12]) [12]

reduced CoA <4> (<4> inhibition of both isozymes at 0.1 mM [12]) [12]

succinyl-CoA <4> (<4> inhibition of MCT2 at 0.02 mM and inhibition of both isozymes at 0.1 mM [12]) [12]

sulfhydryl compounds <1> (<1> purified enzyme [7]) [5, 7]
Additional information <1, 5-10> (<1> no inhibition by sulfhydryl inhibitors [1]; <1> apparently homogeneous enzyme not inhibited by sulfhydryl agents, N-ethylmaleimide and iodoacetamide, highly purified enzyme stimulated by dithiothreitol [7]; <1> pH-dependent inhibition of malonyl-enzyme intermediate formation by N-ethylmaleimide [8,12]; <5> arsenite [14]; <6> inhibition of the enzyme by thiol-alkylating reagents [13]; <1> inhibition of malonyl-enzyme intermediate formation by *p*-chloromercuribenzoate, reversible by 2-mercaptoethanol, prevented by preincubation with malonyl-CoA [8]; <1> inhibition of malonyl-enzyme intermediate formation by phenylmethanesulfonyl fluoride, malonyl-CoA protects [8]) [1, 2, 7, 8, 12-14]

Activating compounds
2-mercaptoethanol <4> (<4> activation [12]) [12]
dithiothreitol <1> (<1> stimulates [1]) [1]
malonate <4> (<4> at 0.1 mM, MCT1 is doubled in activity and MCT2 is 30-40% than control [12]) [12]

Turnover number (min^{-1})
0.056 <16> (N-acetylcysteamine, <16> S97A mutant [18]) [18]
0.41 <16> (N-acetylcysteamine, <16> N-acetylcysteamine as acceptor thiol, wild-type enzyme [18]) [18]
0.6 <14> (malonyl CoA, <14> R606A mutant [22]) [22]
7.2 <14> (malonyl CoA, <14> R606K mutant [22]) [22]
72 <14> (malonyl CoA, <14> wild-type [22]) [22]
25200 <16> (malonyl CoA, <16> hexahistidine-tagged fusion protein, reaction with E. coli acyl-carrier-protein [16]) [16]
27000 <16> (malonyl CoA, <16> hexahistidine-tagged fusion protein, reaction with Streptomyces coelicolor acyl-carrier-protein [16]) [16]
94800 <1> (malonyl CoA, <1> reaction with E. coli acyl-carrier-protein [16]) [16]

Specific activity (U/mg)
0.015 <14> (<14> R606A mutant [22]) [22]
0.041 <1> (<1> strain LA2-89 [9]) [9]
0.0683 <5> (<5> purified enzyme [11]) [11]
0.154 <14> (<14> R606K mutant [22]) [22]
0.1541 <2> (<2> purified enzyme [11]) [11]
0.32 <4> (<4> isozyme MCT1 [12]) [12]
1 <4> (<4> isozyme MCT2 [12]) [12]
1.5 <1> (<1> strain JM101 [9]) [9]
1.61 <14> (<14> wild-type [22]) [22]
2.3 <1> [3]
7.04 <1> [5]
9.59 <6> [13]
780 <16> (<16> hexahistidine-tagged fusion protein, at pH 6.5 and 30°C [16]) [16]
1851 <1> (<1> purified enzyme [7]) [7]

Additional information <1, 5, 7-10, 16> (<1> in LA2-89 and JM101 transformants [9]; <16> hexahistidine-tagged fusion protein [16]) [9, 14, 16]

K_m-Value (mM)

0.0013 <14> (malonyl-CoA, <14> R606K mutant [22]) [22]
0.00134 <6> (pantetheine) [13]
0.0019 <14> (malonyl-CoA, <14> wild-type [22]) [22]
0.00326 <6> (malonyl-CoA) [13]
0.0094 <4> (malonyl CoA) [12]
0.011 <1> (malonyl-CoA, <1> wild-type and mutant [25]) [25]
0.0126 <13> (malonyl-CoA, <13> recombinant enzyme [17]) [17]
0.0141 <13> (holo-acyl-carrier protein, <13> holo-acyl-carrier protein from Mycobacterium tuberculosis [17]) [17]
0.015 <4> (malonyl CoA) [12]
0.0162 <14> (malonyl-CoA, <14> R606A mutant [22]) [22]
0.0208 <16> (malonyl-CoA, <16> wild-type enzyme [18]) [18]
0.0239 <16> (malonyl-CoA, <16> S97A mutant [18]) [18]
0.042 <6> (acyl-carrier-protein) [13]
0.06 <16> (malonyl-CoA, <16> hexahistidine-tagged fusion protein, reaction with Streptomyces coelicolor acyl-carrier-protein [16]) [16]
0.073 <16> (acyl-carrier-protein, <16> hexahistidine-tagged fusion protein, reaction with Streptomyces coelicolor acyl-carrier-protein [16]) [16]
0.25 <1> (malonyl-CoA, <1> reaction with E. coli acyl-carrier-protein [16]) [16]
0.3 <2> (malonyl-CoA, <2> sequential mechanism [11]) [11]
0.327 <16> (malonyl-CoA, <16> hexahistidine-tagged fusion protein, reaction with E. coli acyl-carrier-protein [16]) [16]
0.351 <1> (acyl-carrier-protein, <1> reaction with E. coli acyl-carrier-protein [16]) [16]
0.4 <2, 5> (acyl-carrier protein, <2,5> sequential mechanism [11]) [11]
0.4 <2> (malonyl-CoA, <2> ping-pong mechanism [11]) [11]
0.5 <5> (malonyl-CoA, <5> sequential mechanism [11]) [11]
0.6 <2> (acyl-carrier protein, <2> ping-pong mechanism [11]) [11]
0.776 <16> (acyl-carrier-protein, <16> reaction with E. coli acyl-carrier-protein [16]) [16]
1.3 <5> (acyl-carrier protein, <5> ping-pong mechanism [11]) [11]
1.4 <5> (malonyl-CoA, <5> ping-pong mechanism [11]) [11]
Additional information <16> (<16> the K_m for N-acetylcysteamine is significantly higher than for an acyl-carrier-protein [18]) [18]

K_i-Value (mM)

0.009 <2> (CoA, <2> with respect to acyl carrier protein [11]) [11]
0.015 <2> (CoA, <2> with respect to malonyl-CoA [11]) [11]
0.023 <5> (CoA, <5> with respect to malonyl-CoA [11]) [11]
0.03 <5> (CoA, <5> with respect to acyl carrier protein [11]) [11]
Additional information <2, 5> [11]

pH-Optimum

6-7 <1> [3]
6-7.5 <1> (<1> for the malonyl-enzyme intermediate formation. The enzyme accepts malonyl groups from either malonyl-CoA or malonyl-acyl-carrier-protein to form the enzyme-intermediate, the malonyl group of the intermediate can be transferred to both CoA and acyl-carrier-protein [8]) [8]
6.5 <1, 16> (<16> assay at [16]) [5, 16]
6.5-8.5 <1> [2, 7]
6.8 <13> (<13> assay at [17]) [17]
7 <6> [13]
7.3 <16> (<16> assay at [18]) [18]
7.5-8.5 <4> [12]
8 <5, 15> (<5> higher activity in Tris buffer than in potassium phosphate buffer [11]; <15> assay at [23]) [11, 23]
8.1 <1> (<1> assay at [9]) [9]
8.3 <6> [13]
8.5 <2, 4> (<2> higher activity in Tris buffer than in potassium phosphate buffer [11]) [11, 12]

pH-Range

5-10 <1> (<1> about half-maximal activity at pH 5.0, about 80% of maximal activity at pH 10.0 [7]) [7]
6.2-9.8 <6> (<6> about half-maximal activity at pH 6.2, about 65% of maximal activity at pH 9.8 [13]) [13]
7.5-8.5 <4> (<4> maximal activity [12]) [12]

Temperature optimum (°C)

20 <4> (<4> assay at [12]) [12]
25 <1-3, 5, 13, 16> (<1-3,5,13,16> assay at [4,7-9,11,17,18]) [4, 7-9, 11, 17, 18]
30 <1, 6, 16> (<1,6,16> assay at [5,13,16]) [5, 13, 16]
38 <1> [3]

4 Enzyme Structure

Molecular weight

27000 <1> (<1> polypeptide encoded by the fabD89 allele, Western immunoblot analysis with antiserum raised against wild-type E. coli enzyme [9]; polypeptide encoded by the fabD89 allele which contains an amber mutation at codon position 257, enzymatically inactive [25]) [9, 25]
30700 <13> (<13> calculated from amino acid sequence, consistent with the observed molecular mass of approximately 33000 Da for the enzyme containing the 6-His N-terminal extension [17]) [17]
31000 <5> (<5> HPLC gel filtration [11]) [11]
32000 <1> [9, 24]
32300 <1> (<1> from amino acid sequence [10]) [10]
34110 <16> (<16> electrospray mass spectrometry, hexahistidine-tagged fusion protein [16]) [16]

36000 <2> (<2> HPLC gel filtration [11]) [11]
36660 <1> (<1> carboxymethylated enzyme: sedimentation equilibrium centrifugation [2,7]) [2, 7]
37000 <1> (<1> gel filtration [7]) [7]
40500 <6> (<6> gel filtration, elution from Sephadex G-100 [13]) [13]
42000 <14> (<14> mutant proteins R606A and R606K, Western analysis [22]) [22]
43000 <4> (<4> HPLC-gel permeation chromatography, 2 isozymic forms HPLC analysis [12]) [12]

Subunits
? <1> (<1> x * 35000, SDS-PAGE [6]) [6]
monomer <1, 14> (<1> 1 * 35000, SDS-PAGE, alkylated: 1 * 35500, SDS-PAGE, native: 1 * 36500, SDS-PAGE [2,7] ; <14> malonyl transferase domain of EC 2.3.1.85, gel filtration [21]) [2, 7, 21]
Additional information <1> (<1> vertebrates, yeast and mycobacteria fatty acid synthetases contain all individual activities on one or two multifunctional polypeptide chains, plant, E. coli and other prokaryotic fatty acid synthases are non-associated systems of individual enzymes [2]) [2]

5 Isolation/Preparation/Mutation/Application

Source/tissue
leaf <4, 5> (<4> isozymes 1 and 2 [12]) [11, 12]
mesocarp <6> (<6> mesocarp [13]) [13]
seed <4, 8, 9> (<4> isozyme 1 is predominant [12]) [12, 14, 15]

Localization
cytoplasm <1> [5]
plastid <6> (<6> from the mesocarp [13]) [13]

Purification
<1> (DEAE-cellulose and DEAE-Sephadex chromatography [1]; using streptomycin sulfate, column chromatography on DEAE-cellulose, ammonium sulfate fractionation, chromatography on Sephadex G-100 I column, DEAE-Sephadex column and Sephadex G-100 II column [5]; chromatography on DEAE-cellulose, Sephadex G-100, Sephadex G-75, DEAE-Sephadex, and preparative polyacrylamide gel electrophoresis [7]) [2, 3, 5-7, 9, 10]
<2> (ammonium sulfate fractionation and DEAE-cellulose [11]) [11]
<3> (partial [4]) [4]
<4> (of 2 isozymes MCT1 and MCT2, using salt fractionation, ion-exchange chromatography, and chromatofocusing [12]) [12]
<5> (ammonium sulfate fractionation and DEAE-cellulose [11,20]) [11, 20]
<6> (ammonium sulfate fractionation and column chromatography on DEAE-cellulose [13]) [13]
<12> (purification of the enzyme as a hexahistidine fusion protein [20]) [20]
<13> (Ni^{2+}-chelate chromatography [17]) [17]

<14> (of cytosolic recombinant transacylase, resulting from the expression of DNA corresponding to residues 488-809 of EC 2.3.1.85 in Sf9 cells, purified with column chromatography on DE-53 anion exchanger, high-performance liquid chromatography on an Ultraspherogel SEC3000 and gel filtration [21]; of the mutants R606A und R606K [22]) [21, 22]
<16> (of the hexahistidine-tagged fusion protein, using Ni(II) affinity chromatography, column chromatography on Q-Sepharose, phenyl Superose and Mono-Q [16]) [16, 18]

Renaturation
<14> (malonyl transferase domain of EC 2.3.1.85: refolding in vitro of the recombinant protein expressed in E. coli. A shortened transacylase, consisting of EC 2.3.1.85 residues 488-409 can be repeatedly denatured and renatured in vitro with reproducible high recovery and no loss in specific activity [21]; of the mutant proteins R606A and R606K [22]) [21, 22]

Cloning
<1> (cloning, nucleotide sequence and expression of the fabD-gene in an appropriate Escherichia coli expression vector [9,10]; <1> effects of overexpression on the fatty acid composition of the membrane phospholipids of Escherichia coli [10]; cloning and nucleotide sequence analysis of the enzyme gene, overexpression of fabD89 gene in Escherichia coli [25]) [9, 10, 25]
<11> (cloning and sequencing of the enzyme gene using equivalent gene of Escherichia coli HB101 as a probe. The overexpression of the enzyme gene induces monomer supply for polyhydroxybutyrate production in Escherichia coli HB101 [19]) [19]
<12> (cloning and sequencing of the enzyme gene, expressed in Escherichia coli BL21 [20]) [20]
<13> (a hexa-histidine-tagged enzyme overexpressed in Escherichia coli [17]) [17]
<14> (expression in Escherichia coli of cDNAs of various lengths encoding the second domain of EC 2.3.1.85. Expression of a shortened transacylase, consisting of EC 2.3.1.85 residues 488-809 in Spodoptera frugiperda [21]; expression of mutant proteins R606A and R606K in Escherichia coli [22]) [21, 22]
<15> (expression of the enzyme gene in Escherichia coli [23]) [23]
<16> (expression of five mutants in Escherichia coli [18]; expressed in Escherichia coli [19]; expressed in Escherichia coli as a hexahistidine-tagged fusion protein [16]) [16, 18, 19]

6 Stability

pH-Stability
Additional information <1> (<1> of malonyl-enzyme intermediate [8]) [8]

Temperature stability

42 <1> (<1> strain LA2-89 carrying the fabD89 allele which contains an amber mutation at codon position 257: the enzyme activity is almost completely inactivated by pre-incubation of the extract at this temperature [25]) [25]
50 <1, 4> (<1> 1 h stable [3]; <4> isozyme MCT1: 50% loss of activity after 5 min and 60% loss of activity after 10 min, isozyme MCT2: 30% increase of activity after 10 min [12]) [3, 12]
60 <1> (<1> stable for at least 40 min [3]) [3]
80 <1, 2, 5> (<1> 20% loss of activity after 20 min [3,5]; <2,5> inactivation after 20 min [11]) [3, 5, 11]
100 <1> (<1> $t_{1/2}$: 15 min [3]) [3]
Additional information <1> (<1> the enzyme is quite stable to heat [5]) [5]

General stability information

<1>, salt concentrations above 0.1 M inactivate [2]
<4>, freezing leads to 20% loss of isozyme MCT1-activity [12]
<4>, glycerol, 20% v/v, stabilizes isozyme MCT1 [12]
<6>, EDTA stabilizes [13]
<6>, glycerol, 15% v/v, stabilizes [13]
<16>, enzyme labile during chromatography [16]
<1, 6>, dithiothreitol stabilizes [7, 13]

Storage stability

<1>, -10°C, several weeks [3]
<1>, -20°C, 2-4 months [5]
<4>, -80°C, stable in the presence of glycerol [12]
<6>, -20°C, potassium-phosphate buffer, pH: 7.0, 1 mM dithiothreitol, 0.5 mM EDTA, 15% v/v glycerol [13]
<2, 5>, -20°C, frozen in liquid nitrogen, several months [11]

References

[1] Prescott, D.J.; Vagelos, P.R.: Acyl carrier protein. Adv. Enzymol. Relat. Areas Mol. Biol., **36**, 269-311 (1972)
[2] Vagelos, R.P.: Acyl group transfer (acyl carrier protein). The Enzymes, 3rd Ed. (Boyer, P.D., ed.), **8**, 155-199 (1973)
[3] Williamson, I.P.; Wakil, S.J.: Studies on the mechanism of fatty acid synthesis. XVII. Preparation and general properties of acetyl coenzyme A and malonyl coenzyme A-acyl carrier protein transacylases. J. Biol. Chem., **241**, 2326-2332 (1966)
[4] Hendren, R.W.; Bloch, K.: Fatty acid synthetases from Euglena gracilis. Separation of component activities of the ACP-dependent fatty acid synthetase and partial purification of the β-ketoacyl-ACP synthetase. J. Biol. Chem., **255**, 1504-1508 (1980)
[5] Alberts, A.W.; Majerus, P.W.; Vagelos, P.R.: Malonyl-CoA acyl carrier protein transacylase. Methods Enzymol., **14**, 53-56 (1969)

[6] Joshi, V.C.: Mechanism of malonyl-coenzyme A-acyl carrier protein transacylase. Biochem. J., **128**, 43P-44P (1972)
[7] Ruch, F.E.; Vagelos, P.R.: The isolation and general properties of Escherichia coli malonyl coenzyme A-acyl carrier protein transacylase. J. Biol. Chem., **248**, 8086-8094 (1973)
[8] Ruch, F.E.; Vagelos, P.R.: Characterization of a malonyl-enzyme intermediate and identification of the malonyl binding site in malonyl coenzyme A-acyl carrier protein transacylase of Escherichia coli. J. Biol. Chem., **248**, 8095-8106 (1973)
[9] Verwoert, I.I.G.S.; Verbree, E.C.; VanDerLinden, K.H.; Nijkamp, H.J.J.; Stuitje, A.R.: Cloning, nucleotide sequence, and expression of the Escherichia coli fabD gene, encoding malonyl coenzyme A-acyl carrier protein transacylase. J. Bacteriol., **174**, 2851-2857 (1992)
[10] Magnuson, K.; Oh, W.; Larson, T.J.; Cronan, J.E.: Cloning and nucleotide sequence of the fabD gene encoding malonyl coenzyme A-acyl carrier protein transacylase of Escherichia coli. FEBS Lett., **299**, 262-266 (1992)
[11] Stapleton, S.R.; Jaworski, J.G.: Characterization and purification of malonyl-coenzyme A:[acyl-carrier protein] transacylases from spinach and Anabaena variabilis. Biochim. Biophys. Acta, **794**, 240-248 (1984)
[12] Guerra, D.J.; Ohlrogge, J.B.: Partial purification and characterization of two forms of malonyl-coenzyme A:acyl carrier protein transacylase from soybean leaf tissue. Arch. Biochem. Biophys., **246**, 274-285 (1986)
[13] Caughey, I.; Kekwick, R.G.O.: The characteristics of some components of the fatty acid synthetase system in the plastids from the mesocarp of avocado (Persea americana) fruit. Eur. J. Biochem., **123**, 553-561 (1982)
[14] Shimakata, T.; Stumpf, P.K.: The purification and function of acetyl coenzyme A:acyl carrier protein transacylase. J. Biol. Chem., **258**, 3592-3598 (1983)
[15] Shimakata, T.; Stumpf, P.K.: The procaryotic nature of the fatty acid synthetase of developing Carthamus tinctorius L. (Safflower) seeds. Arch. Biochem. Biophys., **217**, 144-154 (1982)
[16] Szafranska, A.E.; Hitchman, T.S.; Cox, R.J.; Crosby, J.; Simpson, T.J.: Kinetic and mechanistic analysis of the malonyl CoA:ACP transacylase from Streptomyces coelicolor indicates a single catalytically competent serine nucleophile at the active site. Biochemistry, **41**, 1421-1427 (2002)
[17] Kremer, L.; Nampoothiri, K.M.; Lesjean, S.; Dover, L.G.; Graham, S.; Betts, J.; Brennan, P.J.; Minnikin, D.E.; Locht, C.; Besra, G.S.: Biochemical characterization of acyl carrier protein (AcpM) and malonyl-CoA:AcpM transacylase (mtFabD), two major components of Mycobacterium tuberculosis fatty acid synthase II. J. Biol. Chem., **276**, 27967-27974 (2001)
[18] Dreier, J.; Li, Q.; Khosla, C.: Malonyl-CoA:ACP transacylase from Streptomyces coelicolor has two alternative catalytically active nucleophiles. Biochemistry, **40**, 12407-12411 (2001)
[19] Taguchi, K.; Aoyagi, Y.; Matsusaki, H.; Fukui, T.; Doi, Y.: Over-expression of 3-ketoacyl-ACP synthase III or malonyl-CoA-ACP transacylase gene induces monomer supply for polyhydroxybutyrate production in Escherichia coli HB101. Biotechnol. Lett., **21**, 579-584 (1999)

[20] Kutchma, A.J.; Hoang, T.T.; Schweizer, H.P.: Characterization of a Pseudomonas aeruginosa fatty acid biosynthetic gene cluster: purification of acyl carrier protein (ACP) and malonyl-coenzyme A:ACP transacylase (FabD). J. Bacteriol., **181**, 5498-5504 (1999)
[21] Rangan, V.S.; Serre, L.; Witkowska, H.E.; Bari, A.; Smith, S.: Characterization of the malonyl-/acetyltransacylase domain of the multifunctional animal fatty acid synthase by expression in Escherichia coli and refolding in vitro. Protein Eng., **10**, 561-566 (1997)
[22] Rangan, V.S.; Smith, S.: Alteration of the substrate specificity of the malonyl-CoA/acetyl-CoA:acyl carrier protein S-acyltransferase domain of the multifunctional fatty acid synthase by mutation of a single arginine residue. J. Biol. Chem., **272**, 11975-11978 (1997)
[23] Summers, R.G.; Ali, A.; Shen, B.; Wessel, W.A.; Hutchinson, C.R.: Malonyl-coenzyme A:acyl carrier protein acyltransferase of Streptomyces glaucescens: a possible link between fatty acid and polyketide biosynthesis. Biochemistry, **34**, 9389-9402 (1995)
[24] Serre, L.; Verbree, E.C.; Dauter, Z.; Stuitje, A.R.; Derewenda, Z.S.: The Escherichia coli malonyl-CoA:acyl carrier protein transacylase at 1.5 Å resolution. J. Biol. Chem., **270**, 12961-12964 (1995)
[25] Verwoert, II; Verhagen, E.F.; van der Linden, K.H.; Verbree, E.C.; Nijkamp, H.J.; Stuitje, A.R.: Molecular characterization of an Escherichia coli mutant with a temperature-sensitive malonyl coenzyme A-acyl carrier protein transacylase. FEBS Lett., **348**, 311-316 (1994)

Acyl-[acyl-carrier-protein]-phospholipid O-acyltransferase 2.3.1.40

1 Nomenclature

EC number
2.3.1.40

Systematic name
acyl-[acyl-carrier protein]:O-(2-acyl-sn-glycero-3-phospho)-ethanolamine O-acyltransferase

Recommended name
acyl-[acyl-carrier-protein]-phospholipid O-acyltransferase

Synonyms
acyl-acyl carrier protein (acyl-ACP) synthase
acyltransferase, acyl-[acyl carrier protein]-phospholipid

CAS registry number
37257-18-4

2 Source Organism

<1> *Escherichia coli* (B [1]) [1, 4]
<2> *Cucurbita moschata* (squash [2]) [2]
<3> *Escherichia coli* (K 12 derivates LCH 25, 26, 28, 55, 56 [3]) [3]

3 Reaction and Specificity

Catalyzed reaction
acyl-[acyl-carrier protein] + O-(2-acyl-sn-glycero-3-phospho)-ethanolamine = [acyl-carrier protein] + O-(1-β-acyl-2-acyl-sn-glycero-3-phospho)ethanolamine (<1,3> bifunctional protein exhibits both 2-acylglycerophosphoethanolamine acyltransferase and acyl-acyl carrier protein synhtase activities in vitro [3, 4])

Reaction type
acyl group transfer

Substrates and products
S acyl-(acyl-carrier-protein) + O-(2-acyl-sn-glycero-3-phospho) ethanolamine <1> (<1> specific for C_1-position of glycerol-phosphate backbone, the acyl-component of the acyl-(acyl-carrier-protein) may be 3-hydroxy-

decanoate, 3-hydroxylaurate, 3-hydroxymyristate and 3-hydroxyoctanoate (tentatively identified) [1]) (Reversibility: ? <1> [1]) [1]

P (acyl-carrier-protein) + O-(1-β-acyl-2-acyl-sn-glycero-3-phospho) ethanolamine <1> [1]

Activating compounds

Additional information <1> (<1> no stimulation by coenzyme A, ATP, GTP, UTP, glucosamine 1-phosphate, UDP-N-acetylglucosamine, Triton X-100, lipid A, alkaline treated lipid A or alkaline treated lipid A annealed with E. coli phospholipids [1]) [1]

Metals, ions

Mg^{2+} <1> (<1> activation, 0.2 M [1]) [1]

Specific activity (U/mg)

Additional information <3> [3]

pH-Optimum

7.2 <1> [1]

Additional information <2> (<2> pI at 4°C: 5.5 (isozyme AT3), 5.6 (isozyme AT2), 6.6 (isozyme AT1) [2]) [2]

pH-Range

5-10 <1> (<1> about half-maximal activity at pH 5.0, about 60% of maximal activity at pH 10.0 [1]) [1]

Temperature optimum (°C)

37 <1> (<1> assay at [1]) [1]

4 Enzyme Structure

Molecular weight

30000 <2> (<2> isozyme AT1, gel filtration [2]) [2]
40000 <2> (<2> isozymes AT2 and 3, gel filtration [2]) [2]
81000 <2> (<2> [2]) [2]

Subunits

monomer <2> (1 * 30000, isozyme AT1, 1 * 40000, Cucurbita moschata isozymes AT2 and 3, SDS-PAGE) [2]

5 Isolation/Preparation/Mutation/Application

Source/tissue

cotyledon <2> (<2> greening [2]) [2]

Localization

chloroplast stroma <2> [2]
membrane <1, 3> (<3> inner [3]) [3, 4]
particle-bound <1> [1]

Purification

<1> (partial, solubilization by sonic disruption) [1]
<2> (acyl-carrier-protein affinity chromatography) [2]

Cloning

<3> (expression in Escherichia coli [3]) [3]

Engineering

H36A <1> (<1> mutant lacks 2-acylglycerophosphoethanolamine acyltransferase activity, whereas it retains acyl-acyl carrier protein synthase activity [4]) [4]

6 Stability

Temperature stability

100 <1> (<1> inactivation after 2-5 min [1]) [1]

References

[1] Taylor, S.S.; Heath, E.C.: The incorporation of β-hydroxy fatty acids into a phospholipid of Escherichia coli B. J. Biol. Chem., **244**, 6605-6616 (1969)
[2] Nishida, I.; Frentzen, M.; Ishizaki, O.; Murata, N.: Purification of isomeric forms of acyl-[acyl-carrier-protein]:glycerol-3-phosphate acyltransferase from greening squash cotyledons. Plant Cell Physiol., **28**, 1071-1079 (1987)
[3] Jackowski, S.; Jackson, P.D.; Rock, C.O.: Sequence and function of the aas gene in Escherichia coli. J. Biol. Chem., **269**, 2921-2928 (1994)
[4] Heath, R.J.; Rock, C.O.: A conserved histidine is essential for glycerolipid acyltransferase catalysis. J. Bacteriol., **180**, 1425-1430 (1998)

3-Oxoacyl-[acyl-carrier-protein] synthase 2.3.1.41

1 Nomenclature

EC number

2.3.1.41

Systematic name

acyl-[acyl-carrier-protein]:malonyl-[acyl-carrier-protein] C-acyltransferase (decarboxylating)

Recommended name

3-oxoacyl-[acyl-carrier-protein] synthase

Synonyms

3-ketoacyl-ACP synthase <3> [18]
3-ketoacyl-acyl carrier protein synthase
KAS <3, 5, 12, 14, 15, 22> [18-20, 22, 23, 25, 27, 32]
acyl-malonyl(acyl-carrier-protein)-condensing enzyme
β-ketoacyl acyl carrier protein synthase
β-ketoacyl synthetase
β-ketoacyl-ACP synthetase
β-ketoacyl-[acyl carrier protein] synthase
β-ketoacyl-acyl carrier protein synthetase
β-ketoacylsynthase
condensing enzyme
fatty acid condensing enzyme
synthase, 3-oxoacyl-[acyl-carrier-protein]
Additional information <1, 2, 12, 13, 15, 22> (<22> synthase III has both acetyl coenzyme A-acyl-carrier-protein transacylase activity, EC 2.3.1.38, a β-ketoacyl-[acyl-carrier-protein]-transacylase activity EC 2.3.1.41 [32]; <15> 3 structural classes of condensing enzymes [27]; <13> gene DpsC, encoding apolyketide synthase, exhibits also β-keto-[acyl-carrier-protein] synthase III activity, which is restricted for propionyl-[acyl-carrier-protein] as the only [acyl-carrier-protein]-substrate [21]; <12> enzyme shows also acetyl coenzyme A-acyl-carrier-protein transacylase activity EC 2.3.1.38, 12% compared to β-keto-[acyl-carrier-protein] synthase activity EC 2.3.1.41 [20]; <1,2> in vertebrates, yeast and Mycobacteria β-ketoacyl synthase activity is a domain of the multifunctional polypeptide chains of fatty acid synthase EC 2.3.1.85 and EC 2.3.1.86 [13,15]) [13, 15, 20, 21, 27, 32]

CAS registry number

9077-10-5

2 Source Organism

<1> *Escherichia coli* (synthase I [30]; synthases I, II, and III [28]; synthase III, gene fabH [19,24]; synthases I and II [3,6]; B [6]; K12 strains UC1, WN1 is a fabF-mutant lacking synthase II, and various mutants, overview [3]) [1-6, 13, 15, 19, 24, 26, 28, 30]
<2> *Clostridium kluyveri* [15]
<3> *Spinacia oleracea* (synthase I and II [7,11,17]; synthase III [18]) [7, 11, 12, 17, 18]
<4> *Hordeum vulgare* (cv. Bonus [12]) [12]
<5> *Brassica napus* (cv. Westar [19]; synthase I and II [9,10]) [9, 10, 19]
<6> *Brassica juncea* [17]
<7> *Euglena gracilis* (Z [14]) [14]
<8> *Petroselinum hortense* [8]
<9> *Carthamus tinctorius* (safflower [16,17]) [16, 17]
<10> *Cuphea lutea* [17]
<11> *Pisum sativum* [17]
<12> *Streptomyces glaucescens* (synthase III, gene fabH [20,25]) [20, 25]
<13> *Streptomyces peucetius* (synthase III-like activity [21]) [21]
<14> *Allium porrum* (synthase II [22]) [22]
<15> *Synechocystis sp.* (isozyme II [27]; synthase II [23]) [23, 27]
<16> *Bacillus subtilis* (genes fabH1 and fabH2, 2 synthase III isozymes [24]) [24]
<17> *Streptococcus pneumoniae* (synthase III, gene fabH [26]) [26]
<18> *Haemophilus influenzae* [26]
<19> *Mycobacterium tuberculosis* (synthase I, i.e. KasA [29]) [29]
<20> *Mycobacterium tuberculosis* (synthase II, i.e. KasB [29]) [29]
<21> *Rattus norvegicus* [31]
<22> *Plasmodium falciparum* (from human cell culture [32]; synthase III [32]) [32]
<23> *Staphylococcus aureus* (strain ATCC 35556 [33]; synthase III, gene fabH [33]) [33]
<24> *Cuphea lanceolata* (synthase IV [34]) [34]
<25> *Coriandrum sativum* (seed plastid-specific isoform with similarities to synthase I type [35]) [35]

3 Reaction and Specificity

Catalyzed reaction

acyl-[acyl-carrier protein] + malonyl-[acyl-carrier protein] = 3-oxoacyl-[acyl-carrier protein] + CO_2 + [acyl-carrier protein] (<1> mechanism and active site structure and function [30]; <15> structure of active site, substrate binding pocket, and subunit structure in relation to substrate specificity of the 3 isozyme types [27]; <1> synthaseII/FabB: His-HIs-Cys catalytic triad [28]; <1, 17, 18, 22> synthaseIII/FabH: His-Asn-Cys catalytic triad [26, 28,

32]; <12, 15, 21, 25> active site cysteine [23, 25, 27, 31, 33, 35]; <13> acetylation of the synthase III-like polyketid synthase at Ser118 [21]; <1, 21> mechanism [13, 15, 31]; <1> structure [3])

Reaction type

acyl group transfer
condensation <1, 5, 13, 12, 15, 17, 21, 23, 24> [1, 4, 5, 9, 21, 23, 25, 26, 27, 30, 31, 33, 34]
decarboxylation <1, 12, 21, 23> [4, 15, 25, 30, 31, 33]

Natural substrates and products

S 4-hexadecenoyl-[acyl-carrier-protein] + malonyl-[acyl-carrier-protein] <25> (<25> reaction is part of petroselinic acid biosynthesis, synthase IV [35]) (Reversibility: ? <25> [35]) [35]
P 3-oxo-6-octadecenoyl-[acyl-carrier-protein] + CO_2 + [acyl-carrier-protein]
S acyl-[acyl-carrier-protein] + malonyl-[acyl-carrier-protein] <1-3, 5, 12, 16, 19, 22-24> (<24> key enzyme for regulation of medium-chain fatty acid synthesis in seeds [34]; <23> initiation of fatty acid biosynthesis [33]; <22,23> part of fatty acid synthase II [32,33]; <19> part of fatty acid synthase system I and II, involved in cell wall mycolic acids biosynthesis, overview [29]; <16> initiation of fatty acid biosynthesis with acetyl-CoA as a primer, overview substrate specificity [24]; <12> initiation of both straight- and branched-chain fatty acid biosynthesis resulting in a wide variety of produced fatty acids in vivo [20,25]; <1,5> involved in biosynthesis of unsaturated and saturated fatty acids [19]; <1,3,5> involved in biosynthesis of saturated fatty acids [1,2,4,9,11]; <1> involved in biosynthesis of mono-unsaturated long-chain fatty acids [6]; <1,2> responsible for chain elongation during de novo fatty acid synthesis [15]; <1-3> part of non-associated fatty acid synthase system of plants and prokaryotes [1,2,6,11,15]) (Reversibility: r <1, 5> [2, 4, 9]; ir <1, 23> [1, 33]; ? <1, 3, 5, 12, 16, 19, 22, 24> [6, 11, 15, 19, 20, 24, 25, 29, 32, 34]) [1, 2, 4, 6, 9, 11, 15, 19, 20, 24, 25, 29, 32-34]
P 3-oxoacyl-[acyl-carrier-protein] + CO_2 + [acyl-carrier-protein]

Substrates and products

S 4-hexadecenoyl-[acyl-carrier-protein] + malonyl-[acyl-carrier-protein] <25> (<25> best substrate, highly specific for [35]) (Reversibility: ? <25> [35]) [35]
P 3-oxo-6-octadecenoyl-[acyl-carrier-protein] + CO_2 + [acyl-carrier-protein]
S acetyl-CoA + malonyl-[acyl-carrier-protein] <1, 17, 18, 22> (<22> synthase III [32]; <17> synthase III, best substrate [26]) (Reversibility: ir <22> [32]; ? <1, 17, 18> [26]) [26, 32]
P acetoacyl-[acyl-carrier-protein] + CO_2 + CoA <1, 17, 18, 22> [26, 32]
S acetyl-[acyl-carrier-protein] + malonyl-[acyl-carrier-protein] <1, 3, 8, 12, 19, 20, 23> (<19,20> Kas A and KasB [29]; <12,23> synthase III [20,33]; <3> synthase III, recombinant enzyme from E. coli is absolutely specific

for [18]; <3> synthase I [11]) (Reversibility: r <1, 3> [2, 6, 18]; ir <1, 23> [1, 33]; ? <1, 3, 8, 12, 19, 20> [8, 11, 13, 15, 17, 20, 29]) [1, 2, 6, 8, 11, 13, 15, 17, 18, 20, 29, 33]

P acetoacyl-[acyl-carrier-protein] + CO_2 + [acyl-carrier-protein] <1, 3, 8, 23> [1, 2, 6, 8, 18, 33]

S acyl-CoA + [acyl-carrier-protein] <1> (<1> catalyzes fatty acyl transfer between CoA and [acyl-carrier-protein], but no direct transfer to the enzyme as with acyl-[acyl-carrier-protein] [4,15]; <1> tetradecanoyl-CoA and cis-9-hexadecenoyl-CoA are poor substrates [4]) (Reversibility: r <1> [4]; ? <1> [15]) [4, 15]

P acyl-[acyl-carrier-protein] + CoA <1> [4]

S butyryl-[acyl-carrier-protein] + malonyl-[acyl-carrier-protein] <1, 12, 23> (<12,23> synthase III [20,33]) (Reversibility: ir <1> [1]; ? <12, 23> [20, 33]) [1, 20, 33]

P β-ketohexanoyl-[acyl-carrier-protein] + CO_2 + [acyl-carrier-protein]

S cis-9-hexadecenoyl-CoA + malonyl-[acyl-carrier-protein] <1> (Reversibility: ? <1> [6]) [6]

P 3-oxo-hexadecenoyl-[acyl-carrier-protein] + CO_2 + CoA

S cis-9-hexadecenoyl-[acyl-carrier-protein] + malonyl-[acyl-carrier-protein] <1> (<1> best substrate, synthase II [3,6]; <1> poor substrate, synthase I [3,6]) (Reversibility: r <1> [4, 6]; ? <1> [3, 15]) [3, 4, 6, 15]

P 3-oxo-cis-11-octadecanoyl-[acyl-carrier-protein] + CO_2 + [acyl-carrier-protein]

S cis-mono-unsaturated acyl-[acyl-carrier-protein] + malonyl-[acyl-carrier-protein] <1, 2> (<1> from C-10 to C-16, e.g. cis-5-decenoyl-[acyl-carrier-protein], cis-5-dodecenoyl-[acyl-carrier-protein], cis-7-tetradecenoyl-[acyl-carrier-protein], cis-9-hexenoyl-[acyl-carrier-protein], i.e. palmitoleoyl-[acyl-carrier-protein], poor substrates [3,4,6]; <1> not cis-vaccenoyl-[acyl-carrier-protein], i.e. cis-11-octenoyl-[acyl-carrier-protein] [3,4,15]) (Reversibility: r <1> [4]; ? <1, 2> [3, 13, 15]) [3, 4, 13, 15]

P cis-mono-unsaturated 3-oxoacyl-[acyl-carrier-protein] + CO_2 + [acyl-carrier-protein]

S decanoyl-[acyl-carrier-protein] + malonyl-[acyl-carrier-protein] <1, 3, 24> (<24> synthase IV [34]; <3> synthase I [11]) (Reversibility: ? <1, 3, 24> [7, 11, 15, 34]) [7, 11, 15, 34]

P dodecanoyl-[acyl-carrier-protein] + CO_2 + [acyl-carrier-protein]

S dodecanoyl-[acyl-carrier-protein] + malonyl-[acyl-carrier-protein] <1, 3> (Reversibility: ? <1, 3> [3, 15, 17]) [3, 15, 17]

P 3-oxotetradecanoyl-[acyl-carrier-protein] + CO_2 + [acyl-carrier-protein]

S eicosanoyl-[acyl-carrier-protein] + malonyl-[acyl-carrier-protein] <19, 20> (<19,20> Kas A and KasB [29]) (Reversibility: ? <19, 20> [29]) [29]

P 3-oxodocosanoyl-[acyl-carrier-protein] + CO_2 + [acyl-carrier-protein]

S hexadecanoyl-CoA + malonyl-[acyl-carrier-protein] <19, 20> (<19,20> Kas A and KasB, preference for [29]) (Reversibility: ? <19, 20> [29]) [29]

P 3-oxooctadecanoyl-[acyl-carrier-protein] + CO_2 + CoA

S hexadecanoyl-[acyl-carrier-protein] + malonyl-[acyl-carrier-protein] <1, 3, 5, 19, 20, 25> (<19,20> KasA and KasB, preference for [29]; <1,3,5>

i.e. palmitoyl-[acyl-carrier-protein] [3,7,10,11]; <3> only synthase II, best substrate [11]; <3> preferred substrate, synthase III [17]; <3> poor substrate [7]; <1> not [13,15]) (Reversibility: ? <1, 3, 5, 19, 20, 25> [3, 7, 10, 11, 17, 29, 35]) [3, 7, 10, 11, 17, 29, 35]

P 3-oxooctadecanoyl-[acyl-carrier-protein] + CO_2 + [acyl-carrier-protein]

S hexanoyl-[acyl-carrier-protein] + malonyl-[acyl-carrier-protein] <1, 3, 5, 15> (<3> best substrate, synthase I [7,11]; <1> characterization of hexanoyl-enzyme intermediate [5]) (Reversibility: r <1> [4, 5]; ir <1> [1]; ? <3, 5, 15> [7, 10, 11, 17, 23]) [1, 4, 5, 7, 10, 11, 17, 23]

P 3-oxooctanoyl-[acyl-carrier-protein] + CO_2 + [acyl-carrier-protein]

S isobutyryl-CoA + malonyl-[acyl-carrier-protein] <12, 23> (<12> isobutyryl-CoA is a precursor of branched-chain fatty acid synthesis [25]; <12,23> synthase III [20,25,33]) (Reversibility: ? <12, 23> [20, 25, 33]) [20, 25, 33]

P ?

S malonyl-[acyl-carrier-protein] <1, 12> (<1> wild-type synthase I and mutants C163A and C163S [30]; <1> catalyzes malonyl-[acyl-carrier-protein] decarboxylation independent of fatty acyl-[acyl-carrier-protein] [4]; <1> decarboxylation with 2-3% the rate of β-ketoacyl-[acyl-carrier-protein]-synthesis [15]) (Reversibility: ? <1, 12> [4, 15, 25, 30]) [4, 15, 25, 30]

P acetyl-[acyl-carrier-protein] + CO_2 <1> [4, 15, 25, 30]

S octadecanoyl-[acyl-carrier-protein] + malonyl-[acyl-carrier-protein] <5> (<5> synthase I and II [10]) (Reversibility: ? <5> [10]) [10]

P eicosanoyl-[acyl-carrier-protein] + CO_2 + CoA

S octanoyl-[acyl-carrier-protein] + malonyl-[acyl-carrier-protein] <1, 15> (Reversibility: r <1> [5]; ir <1> [1]; ? <15> [23]) [1, 5, 23]

P 3-oxodecanoyl-[acyl-carrier-protein] + CO_2 + [acyl-carrier-protein] <1> [5]

S palmitoyl-[acyl-carrier-protein] + malonyl-CoA <14> (Reversibility: ? <14> [22]) [22]

P 3-oxostearoyl-[acyl-carrier-protein] + CO_2 + CoA

S propionyl-[acyl-carrier-protein] + malonyl-[acyl-carrier-protein] <1, 13> (<13> absolutely specific for [21]) (Reversibility: ir <1> [1]; ? <13> [21]) [1, 21]

P β-ketopentanoyl-[acyl-carrier-protein] + CO_2 + [acyl-carrier-protein]

S saturated acyl-[acyl-carrier-protein] + malonyl-[acyl-carrier-protein] <1-12, 14-24> (<24> preference for medium-chain length [acyl-carrier-protein]-substrates, synthase IV [34]; <22> substrate [acyl-carrier-protein]: malarial [acyl-carrier-protein] + 4'-phosphopantetheine constituting holo-Pf[acyl-carrier-protein] [32]; <19> KasA and KasB, preference for long-chain acyl-[acyl-carrier-protein]s with at least 16 C-atoms [29]; <12> synthase III, broad specificity [20,25]; <1,5> fatty acid composition of Brassica napus wild-type and transgenic plants overexpressing the E. coli synthase III [19]; <1,3,12,16,17-20,24> substrate specificity, overview [1,3,4,6,7,11,15,20,24-26,29,34]; <1> not hexadecanoyl-[acyl-carrier-protein] [13,15]; <3> not octodecanoyl-[acyl-carrier-protein], i.e stearoyl-[acyl-carrier-protein] [7,11]; <1> specific for acyl-[acyl-carrier-protein] thioesters, not acyl-CoA or acyl pantetheine thioesters [13,15]) (Reversibility: r <1, 3, 5, 15> [2, 4-6, 9, 18, 27]; ir <1, 22, 23> [1, 32, 33]; ? <1-12, 14-21, 24> [7, 8, 11-17, 19, 20, 22-26, 28-31, 34]) [1-20, 22-34]

P 3-oxoacyl-[acyl-carrier-protein] + CO_2 + [acyl carrier protein] <1-12, 14-24> (<14> synthase II, parenchyma and lamina epidermis: palmitoyl-[acyl-carrier-protein] is the only product, sheath epidermis: acyl-[acyl-carrier-protein] up to C_{22} chain length [22]) [1-20, 22-34]

S stearoyl-[acyl-carrier-protein] + malonyl-CoA <14> (<14> only in sheath epidermis [22]) (Reversibility: ? <14> [22]) [22]

P 3-oxoeicosanoyl-[acyl-carrier-protein] + CO_2 + CoA

S tetradecanoyl-[acyl-carrier-protein] + malonyl-[acyl-carrier-protein] <1, 3> (<1> synthase I [30]; <3> i.e. myristoyl-[acyl-carrier-protein], only synthase II [11]; <1> poor substrate [4]) (Reversibility: r <1> [4]; ? <1, 3> [3, 11, 15, 17, 30]) [3, 4, 11, 15, 17, 30]

P 3-oxohexadecanoyl-[acyl-carrier-protein] + CO_2 + [acyl-carrier-protein]

S Additional information <1, 3, 12, 16, 17> (<17> synthase III, FabH, has a preference for straight short-chain CoA primers [26]; <16> synthase III isozymes also utilize iso- and anteiso-branched-chain acyl-CoA primers as substrates [24]; <3> synthase I is specific for short-chain fatty acid prolongation while synthase II is active in C_{18} formation from C_{16} [11]; <12> malonyl-CoA-CO_2 exchange reaction [20]; <1> enzyme also catalyzes a malonyl-CoA-CO_2 exchange reaction, overview: substrates and kinetics [4]; <1> synthase I and II have nearly the same substrate specificity [6]) [4, 6, 11, 20, 24, 26]

P ?

Inhibitors

4-phenyl-5-phenylimino-[1,2,4]dithiazolidin-3-one <23> (<23> i.e. HR19 [33]; <23> strong, synthase III [33]) [33]

5-chloro-4-phenyl-[1,2]-dithiol-3-one <23> (<23> i.e. HR12 [33]; <23> complete inhibition at 0.01 nM, synthase III [33]) [33]

N-ethylmaleimide <1-3> (<1> after incubation with 2-mercaptoethanol [2]; <3> 5 mM [7,11]; <1> acyl-[acyl-carrier-protein]-derivatives protect [1]; <1,2> no inhibition of malonyl-[acyl-carrier-protein]-decarboxylation [15]) [1, 2, 7, 11, 13, 15, 18]

[acyl-carrier-protein] <1, 1> (<1> wild-type synthase I and mutants, not mutants C163S and K328A, overview [30]) [30]

acetyl-CoA <1> [1]

acyl-CoA <1> (<1> at high concentrations [1]) [1]

arsenite <3> (<3> synthase II and I [11]; <3> synthase III [18]; <3> weak at 1 mM [7]) [7, 11, 17, 18]

butyryl-CoA <12> (<12> synthase III mutants C122A and C122Q: inhibition of decarboxylation activity in presence of [25]) [25]

cerulenin <1, 3, 9, 14, 15, 17-19, 23-25> (<25> strong [35]; <24> synthase IV [34]; <23> weak, synthase III [33]; <19> Kas A and KasB [29]; <1> FabB: inhibition is irreversible [28]; <1> analysis of the binding mechanism, mimicks the condensation transition state [28]; <15> at 0.05 mM: wild-type is

completely inactivated, mutant I108F and A193M show only 10% reduced activity [23]; <15> 12-carbon mycotoxin, irreversible inhibition, binding motif [23]; <14> synthase II, moderate [22]; <5> no inhibition [10]; <3> antibiotic, protection by hexanoyl-[acyl-carrier-protein], decanoyl-[acyl-carrier-protein], and tetradecanoyl-[acyl-carrier-protein] [7]) [7, 11, 16, 22, 23, 26, 28, 29, 33-35]
compound SB418011 <1, 17, 18> (<17> strong, synthase III [26]) [26]
iodoacetamide <1, 21> (<21> kinetics [31]; <1> synthase I and II [6]; <1> after incubation with 2-mercaptethanol [2]; <1> acyl-[acyl-carrier-protein]-derivatives, e.g. acetyl-[acyl-carrier-protein], hexanoyl-CoA, octanoyl-CoA, decanoyl-CoA protect [1,4,6]; <1> no inhibition of malonyl-[acyl-carrier-protein]-decarboxylation [15]) [1, 2, 4, 6, 13, 15, 31]
isobutyryl-CoA <12> (<12> synthase III mutants C122A and C122Q: inhibition of decarboxylation activity in presence of [25]) [25]
p-chloromercuribenzoate <3> (<3> complete inhibition at 1 mM [7,11]) [7, 11]
thiolactomycin <1, 12, 17-19, 23> (<19> KasA and KasB [29]; <1> FabB: inhibition is reversible, competitive [28]; <1> analysis of the binding mechanism, mimics malonyl-[acyl-carrier-protein] [28]; <1,12,17,23> synthase III [20,26,33]; <17,23> weak [26,33]) [20, 26, 28, 29, 33]
urea <1> (<1> about 50% inhibition at 1 M, complete inactivation at 4 M [5]) [5]

Activating compounds

2-mercaptoethanol <1> (<1> required [5]; <1> required in high concentration, 0.5 M [2]; <1> activation [1,13,15]) [1, 2, 5, 13, 15]
EDTA <1> (<1> 7 mM, stimulates [2]; <1> activation [3,15]) [2, 3, 15]
dithiothreitol <1> (<1> required [3,5]; <1> can replace 2-mercaptoethanol [2]; <1> activation [2,3,15]) [2, 3, 5, 15]
phosphate <1> (<1> activation, 0.2 M [13]) [13]

Turnover number (min^{-1})

0.6 <20> (eicosanoyl-[acyl-carrier-protein], <20> KasB [29]) [29]
1.4 <20> (malonyl-[acyl-carrier-protein], <20> KasB [29]) [29]
1.6 <20> (palmitoyl-[acyl-carrier-protein], <20> KasB [29]) [29]
4.5 <19> (eisosanoyl-[acyl-carrier-protein], <19> KasA [29]) [29]
4.8 <19> (malonyl-[acyl-carrier-protein], <19> KasA [29]) [29]
5.3 <19> (palmitoyl-[acyl-carrier-protein], <19> KasA [29]) [29]

Specific activity (U/mg)

0.0000034 <20> (<20> substrate palmitoyl-CoA [29]) [29]
0.000006 <19> (<19> substrate palmitoyl-CoA [29]) [29]
0.000032 <20> (<20> substrate palmitoyl-[acyl-carrier-protein] [29]) [29]
0.000088 <19> (<19> substrate palmitoyl-[acyl-carrier-protein] [29]) [29]
0.005 <17> (<17> synthase III [26]) [26]
0.0083 <3> (<3> purified enzyme [7]) [7]
0.403 <1> (<1> purified enzyme [1]) [1]
0.8 <5> (<5> purified synthase I [9]) [9]

3.22 <7> [14]
3.81 <1> (<1> purified enzyme [2]) [2]
5.5 <1> (<1> synthase I [3]) [3]
6.3 <1> (<1> synthase II [3]) [3]
7 <3> (<3> purified native enzyme [18]) [18]
7.5 <1> (<1> purified synthase II [6]) [6]
8.04 <8> (<8> partially purified enzyme [8]) [8]
9 <1> (<1> purified synthase I [6]) [6]
9.7 <1> [4]
14 <1> [15]
Additional information <1, 3, 5, 9-11, 16, 21> (<21> mutants [31]; <1,5> wild-type and transgenic Brassica napus plants, expressing the E. coli synthase III [19]; <3> comparison of activity, recombinant and native enzyme [18]; <9-11> fatty acid synthase activity, malonyl-CoA incorporation [17]) [9-11, 17-19, 24, 31]

K_m-Value (mM)

0.00032 <23> (isobutyryl-CoA, <23> synthase III [33]) [33]
0.00041 <12> (isobutyryl-CoA, <12> synthase III [20]) [20]
0.00071 <12> (butyryl-CoA, <12> synthase III [20]) [20]
0.0014 <20> (palmitoyl-[acyl-carrier-protein], <20> Kas B [29]) [29]
0.002 <20> (eicosanoyl-[acyl-carrier-protein], <20> Kas B [29]) [29]
0.0023 <23> (butyryl-CoA, <23> synthase III [33]) [33]
0.0024 <12> (acetyl-CoA, <12> synthase III [20]) [20]
0.0025 <19> (eicosanoyl-[acyl-carrier-protein], <19> Kas A [29]) [29]
0.0031 <24> (decanoyl-[acyl-carrier-protein], <24> synthase IV [34]) [34]
0.0032 <19> (palmitoyl-[acyl-carrier-protein], <19> Kas A [29]) [29]
0.0036 <3> (palmitoyl-[acyl-carrier-protein], <3> synthase II [11]) [11]
0.0037 <12> (malonyl-[acyl-carrier-protein], <12> synthase III [20]) [20]
0.004 <3> (malonyl-CoA, <3> in presence of [acyl-carrier-protein] [7]) [7]
0.0055 <3> (malonyl-CoA) [11]
0.0062 <23> (acetyl-CoA, <23> synthase III [33]) [33]
0.0075 <3> (hexanoyl-[acyl-carrier-protein]) [7]
0.0091 <3> (tetradecanoyl-[acyl-carrier-protein]) [7]
0.012 <1> (cis-3-decenoyl-[acyl-carrier-protein], <1> synthase I [6]) [6]
0.0133 <3> (decanoyl-[acyl-carrier-protein], <3> synthase II [11]) [11]
0.0135 <19, 20> (malonyl-[acyl-carrier-protein], <19,20> Kas A and KasB [29]) [29]
0.0139 <3> (tetradecanoyl-[acyl-carrier-protein], <3> synthase II [11]) [11]
0.014 <1> (cis-3-decenoyl-[acyl-carrier-protein], <1> synthase II [6]) [6]
0.016 <3> (malonyl-CoA, <3> without [acyl-carrier-protein] [7]) [7]
0.017 <1> (cis-9-hexadecenoyl-[acyl-carrier-protein], <1> synthase II [6]) [6]
0.0186 <17> (malonyl-[acyl-carrier-protein], <17> synthase III [26]) [26]
0.019 <1> (acetyl-[acyl-carrier-protein], <1> pH 7.0 [2]) [2]
0.025 <1> (malonyl-[acyl-carrier-protein], <1> pH 7.0 [2]) [2]
0.031 <1> (acetyl-[acyl-carrier-protein], <1> synthase I [6]) [6]

0.04 <1> (acetyl-[acyl-carrier-protein], <1> synthase II [6]; <1> acyl-carrier-protein, fatty acyl transfer between tetradecanoyl-CoA and acyl-carrier-protein [4]) [4, 6]
0.0403 <17> (acetyl-CoA, <17> synthase III [26]) [26]
0.138 <1> (cis-9-hexadecenoyl-[acyl-carrier-protein], <1> synthase I [6]) [6]
Additional information <1, 3, 21> (<21> mutants [31]; <1> kinetic data of synthase I and II at different temperatures with various substrates [3]) [3, 7, 13, 15, 31]

pH-Optimum

5.5-6.1 <1> (<1> synthase II [6]) [6]
6.3 <5> [10]
6.6-7.7 <12> (<12> synthase III [20]) [20]
6.8 <1, 19> (<1> assay at [30]; <19> KasA, HEPES buffer [29]) [29, 30]
7 <1, 12, 16, 20> (<20> KasB, phosphate buffer [29]; <1,12,16> assay at [1,20,24]) [1, 20, 24, 29]
7-7.8 <1> [2, 15]
7.2 <1, 23> (<23> synthase III [33]; <1> synthase I [6]) [6, 33]
7.4 <22> (<22> assay at [32]) [32]
7.5 <25> (<25> assay at [35]) [35]
8 <14> (<14> assay at [22]) [22]
8.1-8.5 <3> [11]

pH-Range

5.2-7.6 <1> (<1> synthase II, about half-maximal activity at pH 5.2 and pH 7.6 [6]) [6]
5.6-7.2 <5> (<5> synthase II, about half-maximal activity at pH 5.6 and pH 7.2 [10]) [10]
5.8-8 <23> (<23> synthase III [33]) [33]
5.9-6.8 <5> (<5> synthase I, about half-maximal activity at pH 5.9 and pH 6.8 [10]) [10]
6-8.2 <3> [7]
6.3-8 <1> (<1> synthase I, about half-maximal activity at pH 6.3 and pH 8.0 [6]) [6]
6.5-8 <1> (<1> sharp drop in activity below pH 6.5 and above pH 8.0 [2]) [2, 15]
6.8-9.1 <3> [11]

Temperature optimum (°C)

20 <1> (<1> assay at [30]) [30]
25 <1, 3> (<1,3> assay at [1,2,7]) [1, 2, 7]
27 <1> (<1> assay at [3]) [3]
30 <1, 8, 12> (<1,8,12> assay at [8,13,20]) [8, 13, 20]
35 <7> (<7> assay at [14]) [14]
37 <1, 3, 5, 16, 19, 20, 22> (<1,3,5,16,22> assay at [3,4,7,10,24,32]) [3, 4, 7, 10, 24, 29, 32]

Temperature range (°C)

30-42 <19, 20> (<19,20> 30-50% activity at 30°C and 42°C [29]) [29]

4 Enzyme Structure

Molecular weight

38000 <13> (<13> synthase III-like polyketid synthase, gel filtration, amino acid sequence determination [21]) [21]
56000 <3> (<3> synthase I, gel filtration [7,11,17]) [7, 11, 17]
57500 <3> (<3> synthase II, gel filtration [11,17]) [11, 17]
60000 <17> (<17> synthase III, gel filtration [26]) [26]
63000 <3> (<3> synthase III, gel filtration [18]) [18]
66000 <1> (<1> synthase I, gel filtration [6]; <1> equilibrium centrifugation [15]) [6, 15]
67600 <1, 23> (<1,23> gel filtration [1,33]) [1, 33]
70000 <8> (<8> about, gel filtration [8]) [8]
72000 <12> (<12> recombinant enzyme, gel filtration [20]) [20]
76000 <1> (<1> synthase II, gel filtration [6]) [6]
80000 <1> (<1> synthase I, sedimentation equilibrium method [3]) [3]
85000 <1> (<1> synthase II, sedimentation equilibrium method [3]) [3]
86700 <5> (<5> synthase I, gel filtration [9,10]) [9, 10]
87400 <5> (<5> synthase II, gel filtration [10]) [10]
87800 <5> (<5> synthase II, gel filtration [9]) [9]
90000 <3, 4> (<3,4> gel filtration [12]) [12]
118000 <7> (<7> gel filtration [14]) [14]
Additional information <1, 16-18, 22, 23, 25> (<1> synthase I mutants [30]; <1,16-18,22,23,25> amino acid sequence alignment [24,26,32,33,35]; <1> amino acid composition, synthase I and II [3]) [3, 24, 26, 30, 32, 33, 35]

Subunits

? <16> (<16> x * 42700, FabH1, SDS-PAGE [24]; <16> x * 39800, FabH2, SDS-PAGE [24]) [24]
dimer <1, 3-5, 12, 15, 23> (<15> crystal structure analysis of subunits [27]; <12,23> 2 * 37000, recombinant synthase III, SDS-PAGE [20,33]; <1> 2 * 34000, SDS-PAGE [13]; <1> 2 * 35000, SDS-PAGE [15]; <1> 2 * 37000, gel filtration after 6 M guanidinium chloride treatment [15]; <3> 2 * 40500, synthase III, SDS-PAGE [18]; <5> 2 * 43000, SDS-PAGE [9,10]; <1> 2 * 43000-44000, synthase I, SDS-PAGE [3]; <1> 2 * 44000-45000, synthase II, SDS-PAGE [3]; <4> 2 * 45000, SDS-PAGE [12]; <3> 2 * 46000, SDS-PAGE [12]; <4> 2 * 46000, α_2, synthase I, SDS-PAGE [12]) [3, 9, 10, 12, 13, 15, 18, 20, 27, 33]
monomer <13> (<13> 1 * 38000, synthase III-like polyketid synthase, SDS-PAGE [21]) [21]
Additional information <1, 2> (<1> synthase I mutants [30]; <1,2> vertebrates, yeast and Mycobacteria fatty acid synthetases contain all individual activities on one or two multifunctional polypeptide chains, plant, E. coli and other prokaryotic fatty acid synthases are non-associated systems of individual enzymes [13,15]) [13, 15, 30]

5 Isolation/Preparation/Mutation/Application

Source/tissue

cell suspension culture <8> [8]
epidermis <14> (<14> sheath and lamina, and underlying parenchyma [22]) [22]
leaf <3, 4, 6, 9-11, 14> [7, 11, 12, 17, 18, 22]
seed <3, 5, 6, 9, 24, 25> (<25> specific isoform [35]; <24> embryo [34]; <3,5,6,9,25> germinating [9,10,16,17,35]) [9, 10, 16, 17, 34, 35]
Additional information <22> (<22> developmental stages [32]) [32]

Localization

chloroplast <3, 4, 7, 25> (<3,4> stroma [12]) [12, 14, 35]
cytosol <1, 22> [13, 32]

Purification

<1> (recombinant synthase I and mutants as His-tagged protein from E. coli [30]; recombinant synthases I, i.e. FabA, II, i.e. FabB, and III, i.e.FabH [28]; partial [2]; synthases I and II [3,6]; isolation of hexanoyl-enzyme intermediate complex [5]) [1-3, 5, 6, 13, 15, 28, 30]
<3> (synthase I [7,17]; synthases II [11,17]; synthase III, native wild-type and recombinant from E. coli E103S [18]) [7, 11, 17, 18]
<4> (3 proteins α_2, α,β and β_2 [12]) [12]
<5> (synthases I and II [9]) [9, 10]
<7> [14]
<8> (partial [8]) [8]
<12> (wild-type and mutants [25]; recombinant from E. coli as His-tagged protein [20,25]) [20, 25]
<13> (synthase III-like polyketid synthase recombinant from Streptomyces lividans [21]) [21]
<15> (synthase II, recombinant from E. coli [23]) [23]
<16> (recombinant His-tagged proteins FabH1 and FabH2 from E. coli [24]) [24]
<17> (recombinant His-tagged synthase III from E. coli [26]) [26]
<18> (recombinant His-tagged protein from E. coli [26]) [26]
<19, 20> (recombinant His-tagged protein from E. coli [29]) [29]
<22> (His-tagged [acyl-carrier-protein], maltose-binding protein tagged full length and truncated synthase III, recombinant from E. coli [32]) [32]
<23> (recombinant His-tagged synthase III from E. coli [33]) [33]
<24> (recombinant His-tagged synthase IV from E. coli [34]) [34]
<25> (recombinant His-tagged seed-specific isoform from E. coli [35]) [35]

Crystallization

<1> (structure analysis, three-dimensional model of enzyme complexed with inhibitors thiolactomycin and cerulenin [28]) [13, 15, 28]
<15> (hanging drop vapour diffusion method, protein solution: 25 mM Tris-HCl, pH 8.0, 300 mM NaCl, 75 mM imidazole, reservoir solution: 0.085 M tri-sodium citrate, 0.17 M ammonium acetate, 22% w/v polyethylene glycol 4000,

15% v/v glycerol as cryo-protectant, 15% w/v 1,2,3-heptanetriol, x-ray structure analysis, three-dimensional model [27]) [27]

Cloning

<1> (expression of synthase I and mutants as His-tagged proteins in Escherichia coli [30]; synthases I, II, and III, recombinant expression [28]; synthase III, overexpression in Escherichia coli strain JM101, construction of binary expression vectors for transformation of Brassica napus [19]) [19, 28, 30]
<4> (cDNA cloned into lambda2AB-II [12]) [12]
<12> (plasmid-based expression of synthase III wild-type and mutants in Streptomyces glaucescens and His-tagged in Escherichia coli [25]; overexpression of fabH in Escherichia coli as His-tagged protein, synthase III [20]) [20, 25]
<13> (expression of synthase III-like polyketid synthase in Streptomyces lividans [21]) [21]
<15> (expression in E.. coli [27]; expression of wild-type and mutant synthase II in Escherichia coli [23]) [23, 27]
<16> (expression of His-tagged FabH1 and FabH2 in Escherichia coli INVαF' [24]) [24]
<17> (overexpression of His-tagged FabH in Escherichia coli BL21 (DE3) [26]) [26]
<18> (expression as His-tagged protein in Escherichia coli [26]) [26]
<22> (DNA sequence analysis, expression of full length and truncated synthase III and mutants as maltose-binding protein fusion protein and His-tagged [acyl-carrier-protein] in Escherichia coli [32]) [32]
<23> (synthase III, overexpression of His-tagged FabH in Escherichia coli BL21 (DE3) [33]) [33]
<24> (expression of His-tagged synthase IV in Escherichia coli BL21 (DE3) [34]) [34]
<25> (construction of a cDNA library from developing seeds, DNA sequence analysis, expression as His-tagged enzyme in Escherichia coli [35]) [35]
<19, 20> (expression of KasA and KasB as His-tagged protein in Escherichia coli [29]) [29]

Engineering

A193M <15> (<15> site-directed mutagenesis, hydrophobic acyl-binding pocket is decreased in size by 2.55fold, 12fold reduced catalytic activity [23]) [23]
C122A <12> (<12> site-directed mutagenesis, 75% enhanced production of straight-chain fatty acids relative to branched-chain fatty acids compared to wild-type [25]) [25]
C122Q <12> (<12> site-directed mutagenesis, 500% enhanced production of straight-chain fatty acids relative to branched-chain fatty acids compared to wild-type [25]) [25]
C122S <12> (<12> site-directed mutagenesis, enhanced production of straight-chain fatty acids relative to branched-chain fatty acids compared to wild-type [25]) [25]

C159A <22> (<22> site-directed mutagenesis, active site mutant, strongly reduced activity [32]) [32]
C159S <22> (<22> site-directed mutagenesis, active site mutant, strongly reduced activity [32]) [32]
C161Q <21> (<21> site-directed mutagenesis, active site Cys mutant, no condensation activity, but 377fold increased decarboxylation activity [31]) [31]
C163A <1> (<1> synthase I, site-directed mutagenesis, active site mutant, no activity, but decarboxylates malonyl-[acyl-carrier-protein] [30]) [30]
C163S <1> (<1> synthase I, site-directed mutagenesis, active site mutant, increased activity, decarboxylates malonyl-[acyl-carrier-protein] [30]) [30]
D306A <1> (<1> synthase I, site-directed mutagenesis, active site mutant, increased activity [30]) [30]
E309A <1> (<1> synthase I, site-directed mutagenesis, active site mutant, reduced activity [30]) [30]
H298A <1> (<1> synthase I, site-directed mutagenesis, active site mutant, reduced activity [30]) [30]
H333A <1> (<1> synthase I, site-directed mutagenesis, active site mutant, increased activity [30]) [30]
I108F <15> (<15> site-directed mutagenesis, hydrophobic acyl-binding pocket is decreased in size by 1.75fold, 38fold reduced catalytic activity towards octanoyl-[acyl-carrier-protein] compared to the activity towards hexanoy-[acyl-carrier-protein] [23]) [23]
K328A <1> (<1> synthase I, site-directed mutagenesis, active site mutant, reduced activity [30]) [30]
Additional information <1, 5, 21> (<21> triple and quadruple mutants of fatty acid synthase complex including mutation sites in the β-ketoacyl-[acyl-carrier-protein] synthase. e.g. H293A, H331A, K326A, S581A, kinetics, overview [31]; <1,5> transformation of Brassica napus via Agrobacterium tumefaciens infection with binary vector overexpressing E. coli fabH gene, targeted expression in cytoplasm and plastids is seed specific, modified fatty acid profile of seed oil [19]) [19, 31]

Application

medicine <15> (<15> target for the development of drugs for the treatment of cancer and tuberculosis [23,27]) [23, 27]
nutrition <15> (<15> target for the engineering of plant seed oils [23,27]) [23, 27]
pharmacology <15, 19, 23> (<23> synthase III is a target for drug development against multi-drug resistant strains [33]; <19> involved in biosynthesis of cell wall mycolic acids, therefore target for discovery of anti-tuberculosis agents [29]; <15> target for the development of drugs for the treatment of cancer and tuberculosis, involved in biosynthesis of precursors of pharmacological agents [23,27]) [23, 27, 29, 33]

6 Stability

Temperature stability

43 <1> (<1> $t_{1/2}$ synthase I: 18 min, $t_{1/2}$ synthase II: 81 min [6]) [6]
100 <1> (<1> inactivation after 1 min [13]) [13]

Oxidation stability

<1>, oxidation by formic acid [5]

General stability information

<1>, 2-mercaptoethanol stabilizes [1]
<1>, EDTA stabilizes [1]
<1, 5>, unstable during purification [1, 9]
<3>, Triton X-100 stabilizes [18]
<3>, glycerol stabilizes [18]
<3, 5>, dithiothreitol stabilizes [10, 18]
<5>, Triton X-100 facilitates extraction from oil seeds [10]
<5>, proteinase inhibitors stabilize during purification [10]
<7>, freezing and thawing inactivates [14]

Storage stability

<1>, -20°C, 90% loss of activity within 1 year [13]
<1>, -20°C, at least 1 month [13]
<1>, -20°C, complete loss of activity in the presence of EDTA and 2-mercaptoethanol [1]
<1>, -20°C, several months [2]
<3>, -20°C, $t_{1/2}$: 1 month [7]
<3>, -70°C, at least 1 month [11]
<3>, -80°C, several weeks in the presence of 20% v/v glycerol, 0.1% Triton X-100 and dithiothreitol [18]
<5>, -70° to -20°C, 90% loss of activity within 7 days [10]

References

[1] Toomey, R.E.; Wakil, S.J.: Studies on the mechanism of fatty acid synthesis. XVI. Preparation and general properties of acyl-malonyl acyl carrier protein-condensing enzyme from Escherichia coli. J. Biol. Chem., **241**, 1159-1165 (1966)
[2] Alberts, A.W.; Majerus, P.W.; Vagelos, P.R.: β-Ketoacyl acyl carrier protein synthase. Methods Enzymol., **14**, 57-60 (1969)
[3] Garwin, J.L.; Klages, A.L.; Cronan, J.E.: Structural, enzymatic, and genetic studies of β-ketoacyl-acyl carrier protein synthases I and II of Escherichia coli. J. Biol. Chem., **255**, 11949-11956 (1980)
[4] Alberts, A.W.; Bell, R.M.; Vagelos, P.R.: Acyl carrier protein. XV. Studies of β-ketoacyl-acyl carrier protein synthetase. J. Biol. Chem., **247**, 3190-3198 (1972)

[5] D'Agnolo, G.; Rosenfeld, I.S.; Vagelos, P.R.: β-Ketoacyl-acyl carrier protein synthetase. Characterization of the acyl-enzyme intermediate. J. Biol. Chem., **250**, 5283-5288 (1975)

[6] D'Agnolo, G.; Rosenfeld, I.S.; Vagelos, P.R.: Multiple forms of β-ketoacyl-acyl carrier protein synthetase in Escherichia coli. J. Biol. Chem., **250**, 5289-5294 (1975)

[7] Shimakata, T.; Stumpf, P.L.: Purification and characterization of β-ketoacyl-[acyl-carrier-protein] synthetase I from Spinacia oleracea leaves. Arch. Biochem. Biophys., **220**, 39-45 (1983)

[8] Schuz, R.; Ebel, J.; Hahlbrock, K.: Partial purification of β-ketoacyl-acyl acrrier protein synthase from a higher plant. FEBS Lett., **140**, 207-209 (1982)

[9] MacKintosh, R.W.; Hardie, D.G.; Slabas, A.R.: β-Ketoacyl-acyl-carrier protein synthase from developing seeds of oilseed rape (Brassica napus). Biochem. Soc. Trans., **17**, 686-687 (1989)

[10] MacKintosh, R.W.; Hardie, D.G.; Slabas, A.R.: A new assay procedure to study the induction of β-ketoacyl-ACP synthase I and II, and the complete purification of β-ketoacyl-ACP synthase I from developing seeds of oilseed rape (Brassica napus). Biochim. Biophys. Acta, **1002**, 114-124 (1989)

[11] Shimakata, T.; Stumpf, P.L.: Isolation and function of spinach leaf β-ketoacyl-[acyl-carrier-protein] synthases. Proc. Natl. Acad. Sci. USA, **79**, 5808-5812 (1982)

[12] Siggaard-Andersen, M.; Kauppinen, S.; von Wettstein-Knowles, P.: Primary structure of a cerulenin-binding β-ketoacyl-[acyl carrier protein] synthase from barley chloroplasts. Proc. Natl. Acad. Sci. USA, **88**, 4114-4118 (1991)

[13] Prescott, D.J.; Vagelos, P.R.: Acyl carrier protein. Adv. Enzymol. Relat. Areas Mol. Biol., **36**, 269-311 (1972)

[14] Hendren, R.W.; Bloch, K.: Fatty acid synthetases from Euglena gracilis. Separation of component activities of the ACP-dependent fatty acid synthetase and partial purification of the β-ketoacyl-ACP synthetase. J. Biol. Chem., **255**, 1504-1508 (1980)

[15] Vagelos, R.P.: Acyl group transfer (acyl carrier protein). The Enzymes, 3rd Ed. (Boyer, P.D., ed.), **8**, 155-199 (1973)

[16] Shimakata, T.; Stumpf, P.K.: The procaryotic nature of the fatty acid synthetase of developing Carthamus tinctorius L. (safflower) seeds. Arch. Biochem. Biophys., **217**, 144-154 (1982)

[17] Shimakata, T.; Stumpf, P.K.: The purification and function of acetyl coenzyme A:acyl carrier protein transacylase. J. Biol. Chem., **258**, 3592-3598 (1983)

[18] Clough, R.; Matthis, A.L.; Barnum, S.R.; Jaworski, J.G.: Purification and characterization of 3-ketoacyl-acyl carrier protein synthase III from spinach. A condensing enzyme utilizing acetyl-coenzyme A to initiate fatty acid synthesis. J. Biol. Chem., **267**, 20992-20998 (1992)

[19] Verwoert, I.I.G.S.; van der Linden, K.H.; Walsh, M.C.; Nijkamp, H.J.J.; Stuitje, A.R.: Modification of Brassica napus seed oil by expression of the Escherichia coli fabH gene, encoding 3-ketoacyl-acyl carrier protein synthase III. Plant Mol. Biol., **27**, 875-886 (1995)

[20] Han, L.; Lobo, S.; Reynolds, K.A.: Characterization of β-ketoacyl-acyl carrier protein synthase III from Streptomyces glaucescens and its role in initiation of fatty acid biosynthesis. J. Bacteriol., **180**, 4481-4486 (1998)
[21] Bao, W.; Sheldon, P.J.; Hutchinson, C.R.: Purification and properties of the Streptomyces peucetius DpsC β-ketoacyl:acyl carrier protein synthase III that specifies the propionate-starter unit for type II polyketide biosynthesis. Biochemistry, **38**, 9752-9757 (1999)
[22] Domergue, F.; Post-Beittenmiller, D.: Biochemical characterization of 3-ketoacyl-acyl carrier protein synthase II from leek epidermis. Biochem. Soc. Trans., **28**, 610-613 (2000)
[23] Val, D.; Banu, G.; Seshadri, K.; Lindqvist, Y.; Dehesh, K.: Re-engineering ketoacyl synthase specificity. Structure, **8**, 565-566 (2000)
[24] Choi, K.H.; Heath, R.J.; Rock, C.O.: β-ketoacyl-acyl carrier protein synthase III (FabH) is a determining factor in branched-chain fatty acid biosynthesis. J. Bacteriol., **182**, 365-370 (2000)
[25] Smirnova, N.; Reynolds, K.A.: Engineered fatty acid biosynthesis in Streptomyces by altered catalytic function of β-ketoacyl-acyl carrier protein synthase III. J. Bacteriol., **183**, 2335-2342 (2001)
[26] Khandekar, S.S.; Gentry, D.R.; Van Aller, G.S.; Warren, P.; Xiang, H.; Silverman, C.; Doyle, M.L.; Chambers, P.A.; Konstantinidis, A.K.; Brandt, M.; Daines, R.A.; Lonsdale, J.T.: Identification, substrate specificity, and inhibition of the Streptococcus pneumoniae β-ketoacyl-acyl carrier protein synthase III (FabH). J. Biol. Chem., **276**, 30024-30030 (2001)
[27] Moche, M.; Dehesh, K.; Edwards, P.; Lindqvist, Y.: The crystal structure of β-ketoacyl-acyl carrier protein synthase II from Synechocystis sp. at 1.54 Å resolution and its relationship to other condensing enzymes. J. Mol. Biol., **305**, 491-503 (2001)
[28] Price, A.C.; Choi, K.H.; Heath, R.J.; Li, Z.; White, S.W.; Rock, C.O.: Inhibition of β-ketoacyl-acyl carrier protein synthases by thiolactomycin and cerulenin. Structure and mechanism. J. Biol. Chem., **276**, 6551-6559 (2001)
[29] Schaeffer, M.L.; Agnihotri, G.; Volker, C.; Kallender, H.; Brennan, P.J.; Lonsdale, J.T.: Purification and biochemical characterization of the Mycobacterium tuberculosis β-ketoacyl-acyl carrier protein synthases KasA and KasB. J. Biol. Chem., **276**, 47029-47037 (2001)
[30] McGuire, K.A.; Siggaard-Andersen, M.; Bangera, M.G.; Olsen, J.G.; von Wettstein-Knowles, P.: β-Ketoacyl-[acyl carrier protein] synthase I of Escherichia coli: aspects of the condensation mechanism revealed by analyses of mutations in the active site pocket. Biochemistry, **40**, 9836-9845 (2001)
[31] Witkowski, A.; Joshi, A.K.; Smith, S.: Mechanism of the β-ketoacyl synthase reaction catalyzed by the animal fatty acid synthase. Biochemistry, **41**, 10877-10887 (2002)
[32] Waters, N.C.; Kopydlowski, K.M.; Guszczynski, T.; Wei, L.; Sellers, P.; Ferlan, J.T.; Lee, P.J.; Li, Z.; Woodard, C.L.; Shallom, S.; Gardner, M.J.; Prigge, S.T.: Functional characterization of the acyl carrier protein (PfACP) and β-ketoacyl ACP synthase III (PfKASIII) from Plasmodium falciparum. Mol. Biochem. Parasitol., **123**, 85-94 (2002)

[33] He, X.; Reynolds, K.A.: Purification, characterization, and identification of novel inhibitors of the β-ketoacyl-acyl carrier protein synthase III (FabH) from Staphylococcus aureus. Antimicrob. Agents Chemother., **46**, 1310-1318 (2002)
[34] Schutt, B.S.; Abbadi, A.; Loddenkotter, B.; Brummel, M.; Spener, F.: β-Ketoacyl-acyl carrier protein synthase IV: a key enzyme for regulation of medium-chain fatty acid synthesis in Cuphea lanceolata seeds. Planta, **215**, 847-854 (2002)
[35] Mekhedov, S.; Cahoon, E.B.; Ohlrogge, J.: An unusual seed-specific 3-ketoacyl-ACP synthase associated with the biosynthesis of petroselinic acid in coriander. Plant Mol. Biol., **47**, 507-518 (2001)

Glycerone-phosphate O-acyltransferase 2.3.1.42

1 Nomenclature

EC number
2.3.1.42

Systematic name
acyl-CoA:glycerone-phosphate O-acyltransferase

Recommended name
glycerone-phosphate O-acyltransferase

Synonyms
DAP-AT <2, 8> [25]
DHAPAT <1-3, 5, 6> [23, 24, 26]
acyltransferase, dihydroxyacetone phosphate
dihydroxyacetone phosphate acyltransferase
Additional information (<6> Gat1 and Gat2 proteins are glycerol-3-phosphate/dihydroxyacetone phosphate dual substrate-specific sn-1 acyltransferases [22]; <1,2,6> may be identical with EC 2.3.1.15 [1,12,18]; <1,2,6> is not identical with EC 2.3.1.15 [3,11,26])

CAS registry number
37257-19-5

2 Source Organism

<-1> no activity in *Schizosaccharomyces pombe* [26]
<1> *Rattus norvegicus* (male Sprague-Dawley [11,13]; female Charles River CD-strain [12,15]; female Wistar [14,17]) [1, 7, 11-15, 17, 20, 21, 26]
<2> *Oryctolagus cuniculus* (adult male New Zealand White [3]) [1, 3, 7, 25, 26]
<3> *Cavia porcellus* (female, Hartley strain [6]; male Dunkin and Hartley strain [13]) [2, 4-9, 13, 26]
<4> *Mus musculus* (male albino [10]) [10, 27]
<5> *Homo sapiens* [16, 23, 24, 26]
<6> *Saccharomyces cerevisiae* (strain S288c [18]) [18, 22, 26]
<7> *Saccharomyces carlsbergensis* [19, 26]
<8> *Homo sapiens* [25]
<9> *Homo sapiens* [27]

3 Reaction and Specificity

Catalyzed reaction

acyl-CoA + glycerone phosphate = CoA + acylglycerone phosphate (A membrane protein. Uses CoA derivatives of palmitate, stearate and oleate, with highest activity on palmitoyl-CoA)

Reaction type

acyl group transfer

Natural substrates and products

S palmitoyl-CoA + dihydroxyacetone phosphate <1-6, 9> (<4,9> recombinant in Saccharomyces cerevisiae cells [27]; <6> no precursor for glycerol ether lipids [26]; <1-3,5,6> obligate precursor in ether lipids biosynthesis [26]; <1-3,5,6> reaction in phosphatidic acid biosynthesis [3,26]; <1,2,5> peroxisomal enzyme involved in ether lipid biosynthesis [1,3,14-16,23]; <3,6> initiates glycerolipid, especially ether-linked glycerolipid, biosynthesis in higher eukaryotes [9,18]) (Reversibility: r <3> [26]; ir <1, 2, 5, 6> [26]; ? <1-6, 9> [1, 3, 9, 10, 14-16, 18, 23-25, 27]) [1, 3, 9, 10, 14-16, 18, 23-27]

P CoA + monopalmitoyldihydroxyacetone phosphate <1-3, 5, 6> [26]

S Additional information <5> (<5> enzyme activity is reduced with disorders: neonatal adrenoleukodystrophy, infantile Refsum disease, and hyperpipecolic acidemia [26]; <5> enzyme amount is reduced by 90% in Zellweger syndrome [24,26]; <5> patients affected with rhizomelic chondrodysplasia punctata show reduced enzyme activity due to a defect in peroxisomal targeting [23,26]) [23, 24, 26]

P ?

Substrates and products

S acyl-CoA + dihydroxyacetone phosphate <1-6, 8, 9> (<3> i.e. glycerone phosphate, saturated acyl-CoAs preferred over unsaturated [2,5]; <1,6> no acyl-donors are C_4-C_{10}-acyl-CoAs [12,18]; <5> no acyl-acceptor is glycerol 3-phosphate [16]) (Reversibility: r <3> [5, 26]; ir <1, 2, 5, 6> [2, 8, 26]; ? <1-6, 8, 9> [1, 3, 4, 6, 7, 9-18, 20-25, 27]) [1-18, 20-27]

P CoA + acyldihydroxyacetone phosphate <1-6, 8> (<1-6> i.e. acylglycerone phosphate [1,2,4,5,7-18]) [1, 2, 4, 5, 7-18, 20-26]

S arachidonoyl-CoA + dihydroxyacetone phosphate <3> (<3> i.e. eicosatetraenoyl-CoA, poor substrate [2,5]) (Reversibility: r <3> [5]; ? <3> [2]) [2, 5]

P CoA + arachidonoyldihydroxyacetone phosphate

S cis-9-hexadecenoyl-CoA + dihydroxyacetone phosphate <5, 6> (Reversibility: ? <5, 6> [22, 24]) [22, 24]

P CoA + cis-9-hexadecenoyldihydroxyacetone phosphate <6> [22]

S dodecanoyl-CoA + dihydroxyacetone phosphate <1, 3, 6> (<6> i.e. lauroyl-CoA, poor substrate [18]) (Reversibility: ? <1, 3, 6> [2, 12, 18]) [2, 12, 18]

P CoA + dodecanoyldihydroxyacetone phosphate

S elaidoyl-CoA + dihydroxyacetone phosphate <3> (<3> poor substrate, mitochondrial enzyme [2]) (Reversibility: ? <3> [2]) [2]
P CoA + elaidoyldihydroxyacetone phosphate
S heptadecanoyl-CoA + dihydroxyacetone phosphate <3> (Reversibility: ? <3> [2]) [2]
P CoA + heptadecanoyldihydroxyacetone phosphate
S lignoceroyl-CoA + dihydroxyacetone phosphate <3> (<3> poor substrate [2]) (Reversibility: ? <3> [2]) [2]
P CoA + lignoceroyldihydroxyacetone phosphate
S linoleoyl-CoA + dihydroxyacetone phosphate <3> (<3> poor substrate, mitochondrial enzyme [2]) (Reversibility: r <3> [5]; ? <3> [2]) [2, 5]
P CoA + linoleoyldihydroxyacetone phosphate
S nonadecanoyl-CoA + dihydroxyacetone phosphate <3> (<3> poor substrate [2]) (Reversibility: ? <3> [2]) [2]
P CoA + nonadecanoyldihydroxyacetone phosphate
S nonanoyl-CoA + dihydroxyacetone phosphate <3> (<3> poor substrate [2]) (Reversibility: ? <3> [2]) [2]
P CoA + nonanoyldihydroxyacetone phosphate
S octadecenoyl-CoA + dihydroxyacetone phosphate <6> (Reversibility: ? <6> [22]) [22]
P CoA + heptadecenoyldihydroxyacetone phosphate <6> [22]
S oleoyl-CoA + dihydroxyacetone phosphate <1, 3, 6> (<3,6> poor substrate [2,5,18]; <3> mitochondrial enzyme [2]) (Reversibility: r <3> [5]; ? <1, 3, 6> [2, 12, 18]) [2, 5, 12, 18]
P CoA + oleoyldihydroxyacetone phosphate
S palmitoyl-CoA + dihydroxyacetone phosphate <1-6> (<3> not palmitoylcarnitine [2]; <1,3,5> best substrate [2,5,12,15,24]; <1> mitochondrial enzyme [14]) (Reversibility: r <3> [5]; ir <1-3, 5, 6> [26]; ? <1-6> [1-4, 6, 8-18, 22, 24]) [2-6, 8-18, 22, 24, 26]
P CoA + monopalmitoyldihydroxyacetone phosphate <1, 3, 6> [2, 5, 13, 22]
S pentadecanoyl-CoA + dihydroxyacetone phosphate <3> (Reversibility: ? <3> [2]) [2]
P CoA + pentadecanoyldihydroxyacetone phosphate
S stearoyl-CoA + dihydroxyacetone phosphate <1-3, 5, 6> (<3> acylation at 35% the rate of palmitoyl-CoA [5]) (Reversibility: ir <1-3, 5, 6> [26]; ? <1, 3, 6> [2, 5, 12, 18, 22]) [2, 5, 12, 18, 22, 26]
P CoA + stearoyldihydroxyacetone phosphate <6> [22]
S tetradecanoyl-CoA + dihydroxyacetone phosphate <1, 3, 5, 6> (<1,3,5,6> i.e. myristoyl-CoA [2,12,18,24]) (Reversibility: ? <1, 3, 5, 6> [2, 12, 18, 24]) [2, 12, 18, 24]
P CoA + tetradecanoyldihydroxyacetone phosphate
S Additional information <6> (<6> Gat1 and Gat2 proteins are glycerol-3-phosphate/dihydroxyacetone phosphate dual substrate-specific sn-1 acyltransferases [22]) [22]
P ?

Inhibitors

5,5'-dithio-bis(2-nitrobenzoate) <1-3, 5, 6> (<1-3,5,6> i.e. DTNB [5,8,26]; <1-3,5,6> no inhibition [26]; <3> 5 mM, moderate [8]; <3> not (10 mM) [2]) [5, 8, 26]
Ca^{2+} <2> (<2> in crude enzyme extract [3]) [3]
Cd^{2+} <3> [5]
Cu^{2+} <3> [4, 5]
DL-carnitine <3> [2]
Hg^{2+} <3> [26]
Mg^{2+} <2> (<2> in crude enzyme extract [3]) [3]
Mn^{2+} <2> (<2> in crude enzyme extract [3]) [3]
N-ethylmaleimide <1, 3> (<1> microsomal enzyme: strong [12]; <1,3> weak [9,13]; <1> mitochondrial enzyme: weak [14]; <1-3,5,6> no inhibition [11,13,26]; <1-3,6> not peroxisomal enzyme [1,3,4,8,13,14,18]) [9, 12-14]
Pb^{2+} <3> [26]
SDS <1> [14]
Tris-maleate buffer <3> (<3> pH 5.0-6.0 [4]) [4]
Triton X-100 <1, 3, 5, 6> (<1> weak [11]; <6> above 0.1 mg/ml [18]; <5> above 0.025% w/v [16]; <1> above 0.2 mg/ml [12]; <1> stimulation up to 0.2 mg/ml [11]; <5> 0.01% w/v [16]; <6> 0.1 mg/ml [18]) [2, 11, 12, 14, 16-18]
Tween 20 <3> [2]
acetaldehyde <3> (<3> above 0.01 M, microsomal enzyme, above 0.2 M, solubilized peroxisomal enzyme [6]) [6]
bovine serum albumin <3> (<3> weak [2]) [2]
cholate <1, 3> (above 2 mg/ml, <3> [5]) [2, 5, 17]
chymotrypsin <1> (<1> intact vesicles [15]) [15]
coenzyme A <3> (<3> kinetics [9]) [9]
deoxycholate <1, 3> (<1> weak, high concentration [11]; <1> above 0.2 mg/ml [12]; <1> inactivation [14]) [2, 11, 12, 14]
glycerol 3-phosphate <1, 3, 6> (<3> pH 6.0-7.0 [4]; <1> tissue distribution [12]; <1> adipocyte [1]; <1,3> weak [2,11]; <1> strong, extraperoxisomal activities [14]; <6> competitive to dihydroxyacetone phosphate [18]; <2,3> not [3,6]) [1, 2, 4, 11, 12, 14, 18]
iodoacetamide <3> (<3> 5 mM, weak [5]; <3> not [8]) [5]
mercuridextran <1> (<1> intact microsomes [15]) [15]
p-chloromercuriphenylsulfonic acid <3> (<3> 5 mM [5]; <3> moderate [8]) [5, 8]
palmitoyl-CoA <3> (<3> weak, above 0.6 mM [2]; <3> above 0.08 mM [4]; <3> bovine serum albumin protects [4,8]) [2, 4, 8]
phosphatidic acid <3> [5]
phosphatidylethanolamine <3> (<3> above 1.5 mM [5]) [5]
phosphatidylinositol <3> [5]
phosphatidylserine <3> (<3> above 1.5 mM [5]) [5]
trypsin <1, 3> (<1> intact vesicles [15]; <1> not sensitive against [11]) [4, 12, 15, 17]

Activating compounds

F^- <3> (<3> activation, membrane-bound enzyme [2,4]; <1> not [17]) [2, 4]
Triton X-100 <1, 3, 5, 6> (<1> stimulation, up to 0.2 mg/ml [11]; <5> 0.01% w/v [16]; <6> 0.1 mg/ml [18]; <1,6> slight stimulation [12,18]; <6> inhibition above 0.1 mg/ml [18]; <5> above 0.025% w/v [16]; <1> above 0.2 mg/ml [12]) [4, 11, 12, 16, 18]
acetaldehyde <3> (<3> 130-400% stimulation at 10-250 mM, whole peroxisomes [6]) [6, 26]
asolectin <3> (<3> activation, i.e. crude soybean phospholipid preparation containing phosphatidylcholine, phosphatidylethanolamine and phosphatidylinositol [4,5,8]) [4, 5, 8]
bovine serum albumin <3> (<3> activation, by inhibiting the formation of palmitoyl-CoA micelles [4,8]) [4, 8]
cholate <2, 3> (<2> 20fold activation, 4 mM [3]) [3, 4]
deoxycholate <1, 2> (<2> 20fold stimulation and solubilization of membrane-bound microsomal activity [3]; <1> at low concentration [11]; <1> slight stimulation [12]) [3, 11, 12]
detergents <3> (<3> stimulation and solubilization of membrane-bound microsomal activity, e.g. Lubrol WX, cholate and deoxycholate [4]) [4]
diphosphate <4> (<4> 30fold activation at 10 mM [10]) [10]
isopropanol <3> (<3> stimulation [26]) [26]
phospholipid <3> (<3> stimulates partially purified enzyme, not membrane-bound [26]; <3> activation, e.g. phosphatidylcholine, dilauroylphosphatidylcholine [5,9]; <3> phosphatidylglycerol [5]; <3> crude peroxisomal phospholipids [5]; <3> at low concentration: phosphatidylserine and phosphatidylethanolamine, both inhibitory at high concentration [5]) [5, 8, 9, 26]
Additional information <3> (<3> heat shock, osmotic shock and freeze-thaw cycles increase enzyme activity [26]; <3> heating membranes to 50°C prior to assay increases activity of membrane-bound enzyme [8]) [8, 26]

Metals, ions

Mg^{2+} <3> (<3> stimulation [26]) [26]
Additional information <1-3> (<1,3> no activation by Mg^{2+}, partially purified [4,5,17,26]; <3> no activation by Ca^{2+} [5]; <3> no activation by Mn^{2+}, partially purified [5,26]; <2> no activation by EDTA [3]) [3-5, 17, 26]

Specific activity (U/mg)

0.000091 <5> [23]
0.00012 <1> [15]
0.00061 <5> (<5> purified enzyme [24]) [24]
0.0018 <2> (<2> peroxisomes [25]) [25]
0.0056 <1> [17]
0.106 <3> [5]
0.212 <3> [5]
0.861 <1> (<1> pH 7.4, partially purified enzyme [20]) [20]
2.47 <3> (<3> purified enzyme [9]) [9]

3.35 <3> (<3> purified enzyme [8]) [8]
Additional information <1-3, 5, 6> (<1-3,5,6> assay methods, overview [26]) [4, 13, 16, 22, 26]

K_m-Value (mM)

0.02 <4> (dihydroxyacetone phosphate) [10]
0.04 <3> (palmitoyl-CoA, <3> pH 5.5 [4]) [4]
0.04-0.06 <3> (dihydroxyacetone phosphate, <3> low concentration of substrate [26]) [26]
0.04-0.2 <3> (dihydroxyacetone phosphate, <3> membrane-bound enzyme, pH 7.4 [8]) [8]
0.06 <3> (dihydroxyacetone phosphate, <3> purified enzyme, pH 7.4 [8]) [8]
0.061-0.078 <3> (dihydroxyacetone phosphate) [9]
0.065 <3> (palmitoyl-CoA) [5]
0.1-0.5 <3> (dihydroxyacetone phosphate, <3> in presence of asolectin [5]; <3> solubilized, pH 7.4 [8]) [8]
0.2-0.25 <3> (dihydroxyacetone phosphate, <3> high concentration of substrate [26]) [26]
0.7 <2> (dihydroxyacetone phosphate) [3]
0.79 <6> (dihydroxyacetone phosphate) [18]
Additional information <1, 3> (<1> values for enzymes from various tissues [12]) [4, 5, 8, 12, 14, 26]

K_i-Value (mM)

0.07 <6> (glycerol 3-phosphate) [18]
0.12 <1> (glycerol 3-phosphate) [12]
0.76 <3> (CoA) [9]
Additional information <1> (<1> values for enzymes from various tissues [12]) [12]

pH-Optimum

5.4 <5> (<5> assay at [23]) [23]
5.5 <1-3, 5> (<1-3> membrane-bound peroxisomal enzyme [1,4,8,20]; <3> optimum without detergent, no activity at pH 5.5 in presence of cholate [4]) [1, 4, 8, 16, 20]
5.5-8.2 <5> (<5> broad optimum [16]) [16]
5.7-6 <1> (<1> intact peroxisomes [14]) [14]
6.4 <3> (<3> in presence of asolectin [5]) [5]
6.5-8 <1> (<1> peroxisomal enzyme [17]) [17]
7 <1, 6> (<1> purified microsomes [14]) [14, 18]
7-7.4 <1> [12]
7-8 <1> (<1> mitochondrial enzyme [14]) [14]
7-8.2 <3> (<3> broad [5]) [5]
7.4 <1, 3> (<1> assay at [13]; <1,3> optimum with detergent [4,20]; <3> detergent-solubilized enzyme [8]) [4, 8, 11, 13, 20]
7.5 <1, 2> (<1,2> microsomal enzyme [1]) [1]
Additional information <1-3, 5, 6> (<1-3,5,6> overview [26]) [26]

pH-Range

5.1-8.5 <1> (<1> about half-maximal activity at pH 5.1 and pH 8.5 [14]) [14]
5.4-7.3 <5> (<5> about half-maximal activity at pH 5.4 and pH 7.3 [16]) [16]
5.5-9 <3> (<3> detectable activity in this range [4]) [4]
5.5-9.5 <1> (<1> about half-maximal activity at pH 5.5 and pH 9.5 [11]) [11]
6-9 <3> (<3> about half-maximal activity at pH 6.0 and pH 9.0, in the presence of asolectin [5]) [5]
6.5-8.5 <1> (<1> about half-maximal activity at pH 6.5 and pH 8.5 [12]) [12]
7.4 <6> (<6> about half-maximal activity above [18]) [18]

Temperature optimum (°C)

23 <1, 6> (<1,6> assay at [12,15,18]) [12, 15, 18]
30 <1, 3> (<1,3> assay at [13,14]) [13, 14]
37 <1, 3, 5> (<1,3,5> assay at [2,4-6,8,9,11,16,17,24]) [2, 4-6, 8, 9, 11, 16, 17, 24]

4 Enzyme Structure

Molecular weight

60000-80000 <5> (<5> gel filtration [24]) [24]
66000 <1> (<1> gel filtration [20]) [20]
72000 <2> (<2> gel filtration [25]) [25]
77190 <8> (<8> DNA sequence determination [25]) [25]
90000 <3> (<3> gel filtration [8,9]) [8, 9]
210000 <2> (<2> heterooligomeric complex consisting of 3 polypeptides of 76, 72 and 69 kDa, sucrose density gradient centrifugation [25]) [25]
Additional information <1, 2, 8> (<1,8> amino acid sequence determination [20,25]) [20, 25]

Subunits

? <3> (<3> x * 69000, SDS-PAGE [8,9]) [8, 9]
monomer <5> (<5> 1 * 65000, SDS-PAGE [24]) [24]
Additional information <1> (<1> immunological, SDS-PAGE and amino acid sequencing analysis of partially purified protein [20]) [20]

5 Isolation/Preparation/Mutation/Application

Source/tissue

3T3-L1 cell <1> [1]
adipocyte <1, 3> [1, 7, 26]
brain <1, 4> [2, 7, 12, 26, 27]
fibroblast <5> (<5> cell culture [23]) [23]
harderian gland <2> [3, 7, 25, 26]
heart <3, 4> [2, 27]
intestine <1> (<1> mucosa [12,26]) [12, 26]
kidney <1, 3, 4> [2, 10, 12, 26, 27]

liver <1-5> [2, 4-15, 17, 20, 21, 24, 26, 27]
lung <1, 4> [12, 27]
placenta <5> [24, 26]
skin fibroblast <5> (<5> cell culture [26]) [16, 24, 26]
spleen <1, 3, 4> (<4> low activity [27]) [26, 27]
testis <4> [27]
Additional information <1-3> (<1-3> distribution, overview [1,7]) [1, 7]

Localization

endoplasmic reticulum <1> [14]
membrane <5, 6> [16, 18]
microsome <1-3, 5> (<1> low activity [21]; <2> active site exposed to lumenal surface [3]) [1-3, 11, 12, 14, 15, 17, 21, 26]
mitochondrion <1, 3> (<1> membrane-bound, may be side-reaction of glycerol 3-phosphate acyltransferase [14]; <1> outer membrane [1]) [1, 2, 4, 21]
peroxisome <1> (<1> microperoxisome of brain, intestinal mucosa [26]) [26]
peroxisome <1-5> (<1-3,5> membrane-bound [20,26]; <1> predominantly, marker enzyme for peroxisomal membrane [14]; <1,3,5> catalytic site at inner surface of peroxisomal membrane [16,17,26]) [1, 4-6, 8-10, 14, 16, 17, 20-26]
plasma membrane <1> [21]
Additional information <1, 3> (<1,3> subcellular distribution [4,13,17]; <1> suborganellar localization [14]; <3> not: cytosol [4]) [4, 13, 14, 17]

Purification

<1> (peroxisomal, partial [20]) [20]
<2> [25]
<3> (partial, solubilized with sodium cholate-KCl [5]; with taurocholate [6]; with 3-[(3-cholamidopropyl)dimethylammonio]-1-propanesulfonate, i.e. CHAPS, plus 0.15 M NaCl [9]) [5, 6, 8, 9, 26]
<5> (peroxisomal from placenta, with 3-[(3-cholamidopropyl)dimethylammonio]-1-propanesulfonate, i.e. CHAPS, plus 1 M KCl [24]) [24]

Cloning

<4> (functional expression in Saccharomyces cerevisiae, DNA and amino acid sequence determination [27]) [27]
<6> (genes GAT1 and GAT2, expression in Escherichia coli DH5α, Gat1 and Gat2 proteins are glycerol-3-phosphate/dihydroxyacetone phosphate dual substrate-specific sn-1 acyltransferases, amino acid sequence analysis [22]) [22]
<8> (enzyme cloned from a cDNA library matching EST-clones with peptides from rabbit peroxisomal enzyme, DNA sequence determination [25]) [25]
<9> (functional expression in Saccharomyces cerevisiae [27]) [27]

Engineering

Additional information <5, 6> (<6> null mutations of GAT1 andGAT2 are lethal [22]; <5> enzyme amount is reduced by 90% in Zellweger syndrome [24]; <5> deficient cell line caused by DNA 128 bp deletion leading to premature stop, strongly reduced activity [23]; <5> patients affected with rhizomelic chondrodysplasia punctata show reduced enzyme activity due to a defect in peroxisomal targeting, point mutation at nucleotide 848 leading to a premature stop [23]) [22-24]

Application

analysis <1> (<1> peroxisomal enzyme can be used as marker enzyme for peroxisomal membrane due to insensitivity to glycerol-3-phosphate [14]) [14]

6 Stability

Temperature stability

40 <3> (<3> significant loss of activity [8]) [8]
44 <1> (<1> $t_{1/2}$: 10 min [12]) [12]
48 <3, 6> (<6> $t_{1/2}$: 4 min [18]; <3> $t_{1/2}$: 13 min [4]; <3> $t_{1/2}$: 15 min [5]) [4, 5, 18]
50 <1> (<1> intact microsomes: 25% loss of activity after 30 min [11]) [11]
51 <3> (<3> $t_{1/2}$: 15 min, with asolectin [5]) [5]
55 <1, 3> (<1> 15 min, stable [11]; <3> 15 min stable, crude peroxisomal preparation [4,26]) [4, 11, 26]
60 <3> (<3> 10 min stable [4,5]; <3> crude peroxisomal preparation [4]; <3> intact peroxisomes [5]) [4, 5]
65 <3> (<3> inactivation within 15 min, crude peroxisomal preparation [4]) [4]
Additional information <1, 3> (<1,3> membrane-bound: heat-stable [4,8,11,26]; <3> solubilized: heat-labile [4,8]) [4, 8, 11, 26]

General stability information

<1>, freezing, crude preparation, stable to [13]
<3>, DEAE-Sephacel chromatography inactivates [5]
<3>, asolectin stabilizes against inactivation during freezing and thawing [5]
<3>, freezing inactivates [5]
<3>, heat shock, osmotic shock and freeze-thaw cycles increase enzyme activity at pH 7.5, at pH 5.5 activity is decreased [26]
<3>, heating membranes to 50°C prior to assay increases activity of membrane-bound enzyme [8]
<3>, isoelectric focusing inactivates [5]
<5>, dipalmitoylphosphatidylcholine stabilizes purified enzyme [24]

Storage stability

<3>, -20°C, up to 1 month [4]
<3>, 2-4°C, 30-50% loss of activity after 1 month, DTT stabilizes during storage [5]

References

[1] Bell, R.M.; Coleman, R.A.: Enzymes of triacylglycerol formation in mammals. The Enzymes, 3rd Ed. (Boyer, P.D., ed.), **16**, 87-111 (1983)

[2] Hajra, A.: Biosynthesis of acyl dihydroxyacetone phosphate in guinea pig liver mitochondria. J. Biol. Chem., **243**, 3458-3465 (1968)

[3] Rock, C.O.; Fitzgerald, V.; Snyder, F.: Properties of dihydroxyacetone phosphate acyltransferase in the harderian gland. J. Biol. Chem., **252**, 6363-6366 (1977)

[4] Jones, C.L.; Hajra, A.K.: Properties of guinea pig liver peroxisomal dihydroxyacetone phosphate acyltransferase. J. Biol. Chem., **255**, 8289-8295 (1980)

[5] Jones, C.L.; Hajra, A.K.: Solubilization and partial purification of dihydroxyacetone-phosphate acyltransferase from guinea pig liver. Arch. Biochem. Biophys., **226**, 155-165 (1983)

[6] Dobrowsky, R.T.; Ballas, L.M.: Peroxisomal dihydroxyacetone phosphate acyltransferase. Effect of acetaldehyde on the intact and solubilized activity. J. Biol. Chem., **262**, 3136-3139 (1987)

[7] Bell, R.M.: Enzymes of glycerolipid synthesis in eukaryotes. Annu. Rev. Biochem., **49**, 459-487 (1980)

[8] Webber, K.O.; Hajra, A.K.: Dihydroxyacetone phosphate acyltransferase. Methods Enzymol., **209**, 92-98 (1992)

[9] Webber, K.O.; Hajra, A.K.: Purification of dihydroxyacetone phosphate acyltransferase from guinea pig liver peroxisomes. Arch. Biochem. Biophys., **300**, 88-97 (1993)

[10] Patel, B.N.; Mackness, M.I.; Connock, M.J.: Effect of pyrophosphate on dihydroxyacetone-phosphate acyltransferase activity in male albino mouse tissues. Biochem. Soc. Trans., **13**, 249 (1985)

[11] Datta, N.S.; Hajra, A.K.: Does microsomal glycerophosphate acyltransferase also catalyze the acylation of dihydroxyacetone phosphate. FEBS Lett., **176**, 264-268 (1984)

[12] Schlossman, D.M.; Bell, R.M.: Microsomal sn-glycerol 3-phosphate and dihydroxyacetone phosphate acyltransferase activities from liver and other tissues. Arch. Biochem. Biophys., **182**, 737-742 (1977)

[13] Bates, E.J.; Saggerson, E.D.: A study of the glycerol phosphate acyltransferase and dihydroxyacetone phosphate acyltransferase activities in rat liver mitochondrial and microsomal fractions. Relative distribution in parenchymal and non-parenchymal cells, effects of N-ethylmaleimide, palmitoylcoenzyme A concentration, starvation, adrenalectomy and anti-insulin serum treatment. Biochem. J., **182**, 751-762 (1979)

[14] Declercq, P.E.; Haagsman, H.P.; Van Veldhoven, P.; Debeer, L.J.; Van Golde, L.M.G.; Mannaerts, G.P.: Rat liver dihydroxyacetone-phosphate acyltransferases and their contribution to glycerolipid synthesis. J. Biol. Chem., **259**, 9064-9075 (1984)

[15] Ballas, L.M.; Bell, R.M.: Topography of glycerolipid synthetic enzymes. Synthesis of phosphatidylserine, phosphatidylinositol and glycerolipid in-

termediates occurs on the cytoplasmic surface of rat liver microsomal vesicles. Biochim. Biophys. Acta, **665**, 586-595 (1981)
[16] Schutgens, R.B.H.; Romeyn, G.J.; Ofman, R.; Van den Bosch, H.; Tager, J.M.; Wanders, R.J.A.: Acyl-CoA:dihydroxyacetone phosphate acyltransferase in human skin fibroblasts: study of its properties using a new assay method [published erratum appears in Biochim Biophys Acta 1987 Sep 25;921(2):415]. Biochim. Biophys. Acta, **879**, 286-291 (1986)
[17] Hardeman, D.; Van den Bosch, H.: Rat liver dihydroxyacetone-phosphate acyltransferase: enzyme characteristics and localization studies. Biochim. Biophys. Acta, **963**, 1-9 (1988)
[18] Schlossman, D.M.; Bell, R.M.: Glycerolipid biosynthesis in Saccharomyces cerevisiae: sn-glycerol-3-phosphate and dihydroxyacetone phosphate acyltransferase activities. J. Bacteriol., **133**, 1368-1376 (1978)
[19] Johnston, J.M.; Paltauf, F.: Lipid metabolism in inositol-deficient yeast, Saccharomyces carlsbergensis. II. Incorporation of labeled precursors into lipids by whole cells and activities of some enzymes involved in lipid formation. Biochim. Biophys. Acta, **218**, 431-440 (1970)
[20] Causeret, C.; Bentejac, M.; Albet, S.; Teubner, B.; Bugaut, M.: Copurification of dihydroxyacetone-phosphate acyl-transferase and other peroxisomal proteins from liver of fenofibrate-treated rats. Biochimie, **79**, 423-433 (1997)
[21] Singh, H.; Beckman, K.; Poulos, A.: Exclusive localization in peroxisomes of dihydroxyacetone phosphate acyltransferase and alkyl-dihydroxyacetone phosphate synthase in rat liver. J. Lipid Res., **34**, 467-477 (1993)
[22] Zheng, Z.; Zou, J.: The initial step of the glycerolipid pathway: identification of glycerol 3-phosphate/dihydroxyacetone phosphate dual substrate acyltransferases in Saccharomyces cerevisiae. J. Biol. Chem., **276**, 41710-41716 (2001)
[23] De Vet, E.C.J.M.; Ijlst, L.; Oostheim, W.; Dekker, C.; Moser, H.W.; Van den Bosch, H.; Wanders, R.J.A.: Ether lipid biosynthesis: alkyl-dihydroxyacetonephosphate synthase protein deficiency leads to reduced dihydroxyacetonephosphate acyltransferase activities. J. Lipid Res., **40**, 1998-2003 (1999)
[24] Ofman, R.; Wanders, R.J.A.: Purification of peroxisomal acyl-CoA:dihydroxyacetonephosphate acyltransferase from human placenta. Biochim. Biophys. Acta, **1206**, 27-34 (1994)
[25] Thai, T.P.; Heid, H.; Rackwitz, H.R.; Hunziker, A.; Gorgas, K.; Just, W.W.: Ether lipid biosynthesis: isolation and molecular characterization of human dihydroxyacetonephosphate acyltransferase. FEBS Lett., **420**, 205-211 (1997)
[26] Hajra, A.K.: Dihydroxyacetone phosphate acyltransferase. Biochim. Biophys. Acta, **1348**, 27-34 (1997)
[27] Ofman, R.; Hogenhout, E.M.; Wanders, R.J.: Identification and characterization of the mouse cDNA encoding acyl-CoA:dihydroxyacetone phosphate acyltransferase. Biochim. Biophys. Acta, **1439**, 89-94 (1999)

Phosphatidylcholine-sterol O-acyltransferase 2.3.1.43

1 Nomenclature

EC number
2.3.1.43

Systematic name
phosphatidylcholine:sterol O-acyltransferase

Recommended name
phosphatidylcholine-sterol O-acyltransferase

Synonyms
LAT (lysolecithin acyltransferase [7]) [7]
LCAT (lecithin-cholesterol acyltransferase [7]) [7]
acyltransferase, lecithin-cholesterol
lecithin-cholesterol acyltransferase
lecithin:cholesterol acyltransferase
lysolecithin acyltransferase
phospholipid-cholesterol acyltransferase

CAS registry number
9031-14-5

2 Source Organism

<1> *Rattus norvegicus* (rat [30,41]) [30, 41]
<2> *Canis familiaris* (dog [20]) [20]
<3> *Oryctolagus cuniculus* (rabbit [20]) [20]
<4> *Capra hircus* (goat [20]) [20]
<5> *Phycomyces blakesleeanus* [1]
<6> *Aeromonas salmonicida* [2]
<7> *Homo sapiens* [3-8, 10-13, 15, 17-29, 31-52]
<8> *Equus caballus* (horse [9]) [9]
<9> *Sus scrofa* (pig) [14, 16, 20]
<10> *Gallus domesticus* (chicken [53]) [53]

3 Reaction and Specificity

Catalyzed reaction

phosphatidylcholine + sterol = 1-acylglycerophosphocholine + sterol ester

Reaction type

acyl group transfer
transesterification <7> [6]

Natural substrates and products

S phosphatidylcholine + sterol <7> (<7> important in metabolism and structure of plasma lipoproteins [3]; <7> enzyme is believed to act on the surface of high-density lipoproteins, catalyzes the hydrolysis of fatty acids of the sn-2-position of phosphatidylcholine and transfers the fatty acid to free cholesterol to form cholesteryl ester, transesterification is the major source of plasma cholesteryl ester in human, enzyme has the key role in the transport of cholesterol from peripheral tissues to the liver, in the interconversion of HDL subclasses and the maintenance of lipoprotein structure [6]) (Reversibility: r <7> [34]) [3, 6, 34]

P 1-acylglycerophosphocholine + sterol ester <7> [3, 6, 34]

Substrates and products

S 1,2-bis-(1-pyrenebutanoyl)-sn-glycero-3-phosphocholine + cholesterol <7> (<7> activity assay method based on this reaction [43]) (Reversibility: ? <7> [43]) [43, 47]

P cholesteryl-1-pyrenebutyrate + lysophosphatidylcholine <7> [43, 47]

S 1-O-alkyl-2-acetyl-sn-glycerol-3-phosphocholine + 1-O-acyl-2-lyso-sn-glycerol-3-phosphocholine <7> (Reversibility: ? <7> [34]) [34]

P 1-O-alkyl-2-lyso-sn-glycerol-3-phosphocholine + 1-O-acyl-2-acetyl-sn-glycerol-3-phosphocholine <7> [34]

S 1-O-alkyl-2-acetyl-sn-glycerol-3-phosphocholine + H_2O <7> (<7> hydrolysis of human platelet-activating factor PAF to lyso-PAF [34]) (Reversibility: ? <7> [34]) [34, 35]

P 1-O-alkyl-2-lyso-sn-glycerol-3-phosphocholine + acetate <7> [34, 35]

S 1-acylglyceryl phosphorylcholine + lecithin <7> (<7> transfer of an acyl group from a lecithin molecule to another on the low-density lipoprotein surface, lysolecithin acyltransferase activity [18]) (Reversibility: r <7> [18]) [18]

P lecithin + 1-acylglyceryl phosphorylcholine <7> [18]

S 2-sn-phosphorylcholinediacylglycerol + cholesterol <7> (<7> lower activity than the natural substrate [29]) (Reversibility: ? <7> [29]) [29]

P ? <7> [29]

S cholesterol + 1-O-hexadecyl-2-oleylphosphatidylcholine <7> (Reversibility: ? <7> [19]) [19]

P cholesteryl oleate + 1-O-hexadecylglycerophosphocholine <7> [19]

S cholesterol + 1-palmitoyl-2-arachidonoylphosphatidylcholine <7> (Reversibility: ? <7> [8]) [8]

P cholesteryl arachidonate + 1-palmitoylglycerophosphocholine <7> [8]

S cholesterol + 1-palmitoyl-2-docosahexanoylphosphatidylcholine <7> (Reversibility: ? <7> [8]) [8]

P cholesteryl docosahexanoate + 1-palmitoylglycerophosphocholine <7> [8]

S cholesterol + 1-palmitoyl-2-eicosapentaenoylphosphatidylcholine <7> (Reversibility: ? <7> [8]) [8]

P cholesteryl eicosapentaenoate + 1-palmitoylglycerophosphocholine <7> [8]

S cholesterol + 1-palmitoyl-2-linoleoylphosphatidylcholine <7> (Reversibility: ? <7> [8]) [8]

P cholesteryl linoleate + 1-palmitoylglycerophosphocholine <7> [8]

S cholesterol + 1-palmitoyl-2-oleoylphosphatidylcholine <7> (Reversibility: ? <7> [8, 19, 48]) [8, 19, 48]

P cholesteryl oleate + 1-palmitoylglycerophosphocholine <7> [8, 19, 48]

S cholesterol + 1-palmitoyl-2-phytanoylphosphatidylcholine <7> (Reversibility: ? <7> [19]) [19]

P cholesteryl phytanoate + 1-palmitoylglycerophosphocholine <7> [19]

S cholesterol + 1-phytanoyl-2-palmitoylphosphatidylcholine <7> (Reversibility: ? <7> [19]) [19]

P cholesteryl palmitoate + 1-phytanylglyerophosphocholine <7> [19]

S *p*-nitrophenol butyrate + H_2O <7> (<7> esterase activity [52]) (Reversibility: ? <7> [52]) [52]

P *p*-nitrophenol + butyric acid <7> [52]

S phosphatidylcholine + sterol <1-9> (<7> wide specificity for acyl acceptor: sterols with a β-configuration at carbon-3 and trans fused A/B rings [3,7,23]; <7> cholesterol as substrate [3,6,8,12,19]; <7> water as acyl acceptor: phospholipase activity [7,23]; <7> acyl donor: high degree of specificity for phospholipids containing a basic nitrogen atom [3]; <7> phosphatidylethanolamine, dimethylphosphatidylethanolamine, lecithins containing different acyl chain lengths and degrees of saturation [7,35]; <7> predominant but incomplete specificity for reaction at position 2 of mixed lecithins [15,45]) (Reversibility: r <1-9> [1-35]) [1-52]

P 1-acylglycerophosphocholine + sterol ester <1-9> [1-52]

S Additional information <6, 7, 9> (<7> phospholipase activity [15,21,32]; <7> long-chain primary alcohols can act as acyl acceptors [3,23]; <6> phospholipase A_2 activity [2]; <7> enzyme hydrolyses ester linkage at carbon-2 position of phosphatidylcholine [23]; <7> mechanism of LAT reaction is similar to that of LCAT, probably involving an acyl-enzyme intermediate, 2-acyl isomers of lysolecithin and lysophosphatidylethanolamine can act as substrates in the LAT reaction [18]; <7,9> lecithin-cholesterol single bilayer vesicles as substrate [3,6,14,32]; <6> bacterial enzyme is not specific for phosphatidylcholine, it can use human erythrocyte membrane substrates, 2-positional specificity as acyltransferase [2]; <7> no significant activity with 1-sn-phosphorylcholinediacylglycerol [29]; <1> rat enzyme has similar substrate specificity for phosphatidylcholines than human enzyme [41]; <7,1> symmetrical diacyl phosphatidylcholines are not substrates for human or rat enzymes [41]) [2, 3, 6, 14, 15, 18, 21, 23, 29, 32, 41]

P ?

Inhibitors

1,2-diphytanoylphosphatidylcholine <7> (<7> competitive inhibitor [19]) [19]
1-O-hexadecyl-2-oleoylphosphatidylcholine <7> (<7> competitive inhibitor [19]) [19]
2-mercaptoethanol <7> (<7> at high concentration [3]) [3]
5,5'-dithiobis(2-nitrobenzoic acid) <7> [6, 15]
Ag^+ <7> [15]
Triton X-100 <7> (<7> inhibition reversed by albumin [29]) [29]
α_2-macroglobulin <7> (<7> association of this protein with LCAT inhibits the activity of LCAT, and may have a role in its catabolism [51]) [51]
apolipoprotein A-II <7> [32]
apolipoprotein C-2 <9> (<9> inhibitor in high concentration [14]) [14]
carnitine <7> [6]
castanospermine <7> (<7> reduces the maximum reaction velocity but not the K_m [13]) [13]
chlorpromazine <7> [6]
cysteine <7> [6, 10]
deoxynojirimycin <7> (<7> reduces the maximum reaction velocity but not the K_m [13]) [13]
diisopropyl fluorophosphate <7> (<7> inhibits also the hydrolysis of platelet-activating factor by LCAT [34]) [6, 10, 34]
dithiothreitol <7> (<7> at high concentration [3]) [3]
lysolecithin <7> (<7> inhibition of LCAT activity [15]) [15]
lysophosphatidylcholine <7> (<7> both enantiomers, inhibition reversed by albumin [29]) [29]
methyldeoxynojirimycin <7> (<7> reduces the maximum reaction velocity but not the K_m [13]) [13]
p-hydroxymercuribenzoate <2-4, 7, 9> [6, 20]
phenylarsenoxide derivatives <7> [17]
phenylmethylsulfonyl fluoride <2-4, 7, 9> [20]
reduced glutathione <7> [6]
serum albumin <7> (<7> inhibits LAT activity [18]) [18]
sn-2-difluoroketone phosphatidylcholine analogues <7> (<7> IC50 values between 0.5 and 0.028 mM [12]) [12]
sphingomyelin <7> [35]
tunicamycin <7> (<7> reduces the maximum reaction velocity but not the K_m [13]) [13]
Additional information <7> (<7> heparin and serum albumin inhibit lysolecithin acyltransferase activity [18]; <7> serum albumin inhibits hydrolysis of platelet-activating factor by LCAT [34]; <7> hydroperoxides of lecithin act as competitive inhibitors [35,40]; <7> LCAT may experience product inhibition [35]) [18, 34, 35, 40]

Cofactors/prosthetic groups

apolipoprotein A1 <7> [6, 7]

Additional information <7> (<7> hydrolysis of platelet-activating factor by LCAT does not require apolipoprotein as cofactor [34]) [34]

Activating compounds

albumin <6, 7> (<7> phospholipase activity is dependent on [15]; <6> stimulates [2]) [2, 15]

apolipoprotein A-I <2-4, 6, 7, 9, 10> (<9> both human and pig ApoA-1 can activate pig LCAT in cholesterol-lecithin liposomes [14]; <7> required for hydrolysis of ester linkage at carbon-2 position of phosphatidylcholine and for transesterification [23]; <7> required for transferase and phospholipase activity [21,32]; <7> absolute requirement with phosphatidylcholine-cholesterol vesicles as a substrate [3,6,32]; <7> most effective activator [7]; <7> can substitute LDL as activator of the LAT activity [36]; <7> activates recombinant enzymes [39]; <7> a disulfide-linked apoprotein dimer is less effective as activator than apo A-I [42]; <7> the naturally occurring mutant T123I is defective in activation by apo A-I [52]) [2, 3, 6, 7, 14, 20, 21, 23, 32, 35, 36, 39, 42, 52, 53]

apolipoprotein A-IV <7> [35]

apolipoprotein C-1 <7> (<7> cofactor activity, less effective than apo-A-1 [3]; <7> activates LAT activity [36]) [3, 35, 36]

apolipoprotein E <7> (<7> activates LAT activity [36]) [35, 36]

low-density lipoprotein <7> (<7> required for acylation of lysolecithin [18,36]) [18, 36]

reducing agents <7> (<7> necessary for maximal activity [4]) [4]

serum albumin <7> (<7> activates LCAT activity [18,36]) [18, 36]

Additional information <7> (<7> apo-A-II, apo-C-III and apo-D inhibit activation of enzyme by apo-A-1 [3,6,32]) [3, 6, 32]

Metals, ions

Co^{2+} <7> (<7> stimulates phospholipase reaction and cholesterol esterification, EDTA suppresses stimulation [28]) [28]

Cu^{2+} <7> (<7> stimulates phospholipase reaction and cholesterol esterification, EDTA suppresses stimulation [28]) [28]

Ni^{2+} <7> (<7> stimulates phospholipase reaction and cholesterol esterification, EDTA suppresses stimulation [28]) [28]

Zn^{2+} <7> (<7> stimulates phospholipase reaction and cholesterol esterification, EDTA suppresses stimulation [28]) [28]

Additional information <6, 7> (<7> with lecithin-cholesterol vesicles as substrates in the presence of apo-A-1 the enzyme requires an ionic strength of approximately 0.05 for maximal activity [31]; <7> Ca^{2+}-independent [15]; <6> no divalent cation requirement [2]) [2, 15, 31]

Specific activity (U/mg)

0.001 <7> (<7> recombinant enzyme [50]) [50]

0.01 <7> (<7> in cholesterol-lecithin vesicles activated with apolipoprotein A1, at 37°C and pH 7.1 [32]) [32]

0.06 <7> [27]
0.08 <7> [28]
0.14 <1> (<1> in cholesterol-lecithin vesicles activated with rat apolipoprotein A1 [30]) [30]
0.15 <9> (<9> in cholesterol-lecithin vesicles activated with porcine apolipoprotein A1 [14]) [14]
0.27 <7> (<7> in cholesterol-lecithin vesicles activated with apolipoprotein A1 [17]) [17]
0.34 <7> (<7> at 37°C in cholesterol-lecithin vesicles [15]) [15]
0.43 <7> (<7> recombinant C-terminal histidine tagged enzyme, similar to plasma enzyme [48]) [48]
0.47 <9> (<9> HDL substrate at 37°C [16]) [16]
0.54 <7> (<7> at 37°C and pH 7.4 [6]) [6]
0.55 <4> (<4> in cholesterol-lecithin vesicles activated with human apolipoprotein A1 [20]) [20]
0.56 <3> (<3> in cholesterol-lecithin vesicles activated with human apolipoprotein A1 [20]) [20]
0.57 <2, 7> (<2,7> in cholesterol-lecithin vesicles activated with apolipoprotein A1 [20]) [20]
0.59 <9> (<9> in cholesterol-lecithin vesicles activated with human apolipoprotein A1 [20]) [20]
2.7 <7> (<7> in cholesterol-lecithin vesicles [3]) [3]
3.47 <7> (<7> in cholesterol-lecithin vesicles activated with apolipoprotein A1 [23]) [23]
53 <7> (<7> in cholesterol-lecithin vesicles activated with apolipoprotein A1 [5]) [5]
137 <6> (<6> in cholesterol-lecithin vesicles [2]) [2]
Additional information <7> (<7> assay methods [4,5,18]; <7> acyltransferase and esterase activities of several truncated recombinant enzymes [38]; <7> patients with Alzheimer's disease have lower LCAT activity in cerebrospinal fluid [43]; <7> LCAT activity for several phosphatidylcholine analogs with different chain insaturation [46]; <7> recombinant enzyme is half as enzymatically active as plasma LCAT [49]; activity of the wild-type and two naturally occurring mutant enzymes [52]) [4, 5, 18, 21-23, 25, 27, 28, 30, 32, 33, 38, 43, 46, 49, 52]

K_m-Value (mM)

0.000091 <7> (apo-A-1) [32]
0.14 <7> (1,2-bis-(1-pyrenebutanoyl)-sn-glycero-3-phosphocholine, <7> phospholipase activity [47]) [47]
0.19 <7> (1,2-bis-(1-pyrenebutanoyl)-sn-gycero-3-phosphocholine, <7> phospholipase activity of recombinant enzyme expressed in baby hamster kidney cells [47]) [47]
0.55 <7> (1,2-bis-(1-pyrenebutanoyl)-sn-glycero-3-phosphocholine, <7> phospholipase activity of recombinant enzyme devoid of sialic acid, expressed in baby hamster kidney cells [47]) [47]

0.57 <7> (cholesterol, <7> recombinant enzyme expressed in baby hamster kidney cells [39]) [39]
0.83 <7> (cholesterol, <7> recombinant enzyme expressed in hepatic Mc-7777 cells [39]) [39]
7.44 <7> (1,2-bis-(1-pyrenebutanoyl)-sn-glycero-3-phosphocholine) [47]
10.1 <7> (1,2-bis-(1-pyrenebutanoyl)-sn-glycero-3-phosphocholine, <7> recombinant enzyme expressed in baby hamster kidney cells [47]) [47]
20.5 <7> (1,2-bis-(1-pyrenebutanoyl)-sn-glycero-3-phosphocholine, <7> recombinant enzyme devoid of sialic acid, expressed in baby hamster kidney cells [47]) [47]
21.4 <7> (cholesterol, <7> in cholesterol-lecithin vesicles [13]) [13]
Additional information <7> (<7> lysophosphatidylcholine: apparent K_m values for LCAT and LAT activities with different apolipoprotein activators [36]; <7> apparent K_m values for LCAT and LAT activities in presence of cholesterol [36]; <7> cholesterol: apparent K_m values for LCAT and LAT activities with different apolipoprotein activators [36]; <7> apparent K_m values for LCAT and LAT activities in presence of lysophosphatidylcholine [36]; <7> K_m for two naturally occurring mutant enzymes [52]) [36, 52]

pH-Optimum

6-9 <6> (<6> activity is independent of pH in this range [2]) [2]
7.5-8 <7> (<7> lecithin-cholesterol acyltransferase activity [15]) [15]
7.6 <2, 3> [20]
7.8 <9> [20]
8 <4, 7> (<7> transesterification of high-density lipoproteins and liposomes [6]; <7> phospholipase activity [15]) [6, 15, 20]

Temperature optimum (°C)

37 <7> (<7> assay at [4,5]) [4, 5]

4 Enzyme Structure

Molecular weight

52000 <9> (<9> after cleavage of carbohydrate moiety [16]) [16]
59000 <7> (<7> sedimentation equilibrium ultracentrifugation [3,32]) [3, 32]
60000 <7> (<7> sedimentation equilibrium ultracentrifugation [26]) [26]
63000 <7> (<7> sedimentation equilibrium [24]) [24]
64000 <8> [9]
66000 <2, 4, 9> (<2,4> HPLC gel filtration [20];<9> HPLC gel filtration [16]) [16, 20]
67000 <3, 7, 9> (<3,7,9> HPLC gel filtration [20];<7> size exclusion chromatography [6]) [6, 20]
70000 <7> (<7> gel filtration [33]) [33]
83000 <7> (<7> gradient gel electrophoresis [24]) [24]
95000 <7> (<7> PAGE in absence of SDS [10]; <7> gel filtration [24]) [10, 24]
Additional information <7> (<7> mass spectrometry analysis gives a range from 55700 to 59600 [49]) [49]

Subunits

? <2, 4, 6, 7, 9, 10> (<6> x * 23600, SDS-PAGE [2]; <7> x * 65000, SDS-PAGE [6,23,24] ; <7> x * 46000, SDS-PAGE after cleavage of the N-linked carbohydrates [13]; <10> x * 64000, SDS-PAGE [53]; <2,4,7> x * 65000, SDS-PAGE [20]; <7> x * 66000, SDS-PAGE [21]; <7> x * 68000, SDS-PAGE [33]; <7> x * 49000, denatured and deglycosylated enzyme [49]; <7> x * 73000, recombinant enzyme, SDS-PAGE [49]; <7> x * 67000, SDS-PAGE [13,26];<9> x * 69000, SDS-PAGE [14]; <3,9> x * 66000, SDS-PAGE [20]) [2, 6, 13, 14, 20, 21, 23, 24, 26, 33, 39, 49, 53]

Posttranslational modification

glycoprotein <7, 9, 10> (<7> 24% carbohydrate by weight, 13 mol of mannose, 30 mol of galactose, 17 mol of glucosamine and 13 mol of sialic acid per mol of enzyme protein [3,32]; <7> 25% carbohydrate [6,10]; <7> with a 5% sialic acid content [6,10]; <7> 13% hexoses, 6.2% hexosamines [10]; <7> almost all of the carbohydrate moiety of LCAT is N-linked with part of the high-mannose and part of the complex type [13]; <9> 21% w/w carbohydrate, including hexoses, hexosamines and sialic acids [14,16]; <7> the enzyme exists as a family of molecular species having a common core polypeptide with the number of sialic acid residues varying from 10 to 16 [24]; <7> four N-glycosydation and two O-glycosylation sites attached to glycans [35]; <7> human enzyme cloned in animal cells has different degree of glycosydation [35]; <7> removal of individual carbohydrate chains affects activity [35]; <7> recombinant enzyme expressed in hepatic Mc-7777 cells has biantennary structure oligosaccharide moieties [39]; <7> glycans structure determined by mass spectrometry and linkage analysis [47]; <7> glycans in recombinant enzyme, expressed in baby hamster kidney cells, have bi- and triantennary structures with or without core fucosylation [47]; <7> a recombinant C-terminal histidine tagged enzyme has similar carbohydrate moieties than the enzyme purified from plasma [48]; <7> glycosylation in a recombinant enzyme is more extensive than that of isolated from plasma [49]) [3, 6, 10, 13, 14, 16, 24, 32, 35, 39, 47-49, 53]

5 Isolation/Preparation/Mutation/Application

Source/tissue

plasma <1-4, 7, 9, 10> [3-7, 10, 12, 13, 14, 15, 16, 17, 18, 20, 21, 22, 23, 24, 25, 26, 27, 28, 30-41, 53]
serum <7, 8> [4, 9, 11]

Purification

<1> (precipitation from plasma with dextran sulfate-Ca^{2+} followed by DEAE-Sephadex A-50 and ion-exchange chromatography, rat enzyme is not absorbed on hydroxyapatite, which is employed for the purification of the human enzyme [30]) [30]
<6> (extraction from cell membranes with Tween 20, purification by Sephadex G-100 and DEAE-Sepharose columns [2]) [2]

<7> (precipitation from plasma and purification by phenyl-sepharose and concanavalin A-Sepharose columns [5]; precipitation from plasma and purification by phenyl-agarose, DEAE-cellulose and hydroxyapatite columns [13,25]; Sephadex G-100, hydroxyapatite and affinity column chromatography [15]; affinity chromatography using a *p*-aminophenylarsenoxide-linked phase [17]; precipitation from plasma with dextran sulfate-Mg^{2+} followed by several chromatographic steps, including DEAE-Sepharose, phenyl-Sepharose and affinity chromatography [20]; antibody affinity chromatography [27]; HDL-Sepharose, agglutinin-Sepharose chromatography and polyacrylamide electrophoresis [28]; affinity chromatography using an anti-apolipoprotein D immunoglobulin-Sepharose column [32]; <7> purification of recombinant enzymes [39,49]; purification of a recombinant C-terminal histidine tagged enzyme using a cobalt affinity resin [48]; further purification of a recombinant enzyme using ACA-44 allows to remove a proteoglycan contaminant [50]) [5, 13, 15, 17, 20-28, 30-33, 39, 47-50]
<8> [9]
<9> (phenyl-Sepharose and hydroxyapatite column chromatography [14]; several chromatographic steps, including affinity chromatography [16]) [14, 16]
<10> (partial, using phenyl-Sepharose chromatography [53]) [53]
<2-4, 9> (precipitation from plasma with dextran sulfate-Mg^{2+} followed by several chromatographic steps, including DEAE-Sepharose, phenyl-Sepharose and affinity chromatography [20]) [20]

Cloning

<7> (expression in baby hamster kidney cells [34,39,50]; expression of mutant enzymes in baby hamster kidney cells [37]; expression in mice and rabbit [35]; expression of truncated enzymes in COS-1 cells [38]; expression in hepatic Mc-7777 cells [39]; expression of mutant enzymes in COS-6 cells [44]; expression of a C-terminal histidine tagged enzyme in Chinese hamster ovary cells [48]; expression in Chinese hamster ovary cells, recombinant enzyme is structurally similar to plasma LCAT [49]; expression of the naturally occurring mutants T123I and N228K in COS-1 and CHO cells [52]) [34, 35, 37, 39, 44, 48-50, 52]

6 Stability

pH-Stability

3.5-7 <7> (<7> partially purified enzyme is stable down to pH 3.5 [22]) [22]

Temperature stability

37 <7> (<7> 0.4 mM phosphate buffer, pH 7.4, ionic strength 0.001: stable up to 6 h, 39 mM phosphate buffer, pH 7.4, ionic strength 0.1: 90% loss of activity [31]) [31]

General stability information

<1>, apo-A-1 prevents inactivation in phosphate buffer [30]
<1>, less stable than human enzyme to ionic strength [30]
<6>, unaffected by repeated freezing and thawing [2]
<7>, apo-A-1, albumin and lecithin-cholesterol vesicles stabilize [3, 31]
<7>, complete loss of activity upon lyophilization [21]
<7>, glycerol stabilizes during storage at -20°C [3]
<7>, inactivation at the air-water interface is prevented by buffered media of low ionic strength [6, 31]
<7>, lability to freezing and thawing and to high ionic strength [3]
<7>, lecithin-cholesterol vesicles stabilize against inactivation to a lesser extent than apo-A-1 [31]
<7>, remarkably stable in buffers of very low ionic strength [31]

Storage stability

<7>, -20°C, 90% of activity is retained, 4 weeks [21]
<7>, 4°C, 0.4 mM phosphate buffer, 4 mM 2-mercaptoethanol, under N_2, 4 weeks [6]
<7>, 4°C, 100 h [22]
<7>, 4°C, 20% loss of activity, 24 h [21]
<7>, 4°C, 60 days [48]
<7>, 4°C, low ionic strength buffer, under nitrogen, 4 weeks [23]
<8>, 0°C, 30 days [9]

References

[1] Bartlett, K.; Keat, M.J.; Mercer, E.I.: Biosynthesis of sterols in Phycomyces blakesleeanus. Phytochemistry, **13**, 1107-1113 (1974)

[2] Buckley, J.T.; Halasa, L.N.; MacIntyre, S.: Purification and partial characterization of a bacterial phospholipid: cholesterol acyltransferase. J. Biol. Chem., **257**, 3320-3325 (1982)

[3] Doi, Y.; Nishida, T.: Lecithin-cholesterol acyltransferase from human plasma. Methods Enzymol., **71**, 753-767 (1981)

[4] Verdery III, R.B.; Gatt, S.: Assay for lecithin: cholesterol acyltransferase. Methods Enzymol., **72**, 375-384 (1981)

[5] Ogawa, Y.; Fielding, C.J.: Assay of cholesteryl ester transfer activity and purification of a cholesteryl ester transfer protein. Methods Enzymol., **111**, 274-285 (1985)

[6] Albers, J.J.; Chen, C.H.; Lacko, A.G.: Isolation, characterization, and assay of lecithin-cholesterol acyltransferase. Methods Enzymol., **129**, 763-783 (1986)

[7] Fielding, C.J.: Mechanisms of action of lecithin-cholesterol acyltransferase. Methods Enzymol., **129**, 783-790 (1986)

[8] Parks, J.S.; Thuren, T.Y.; Schmitt, J.D.: Inhibition of lecithin:cholesterol acyltransferase activity by synthetic phosphatidylcholine species contain-

ing eicosapentaenoic acid or docosahexaenoic acid in the sn-2 position. J. Lipid Res., **33**, 879-887 (1992)
[9] Yamamoto, M.; Yamamoto, I.; Tanaka, Y.; Michihiro, S.: Purification of horse (Equus caballus) serum lecithin:cholesterol acyltransferase. Comp. Biochem. Physiol. B, **88**, 363-368 (1987)
[10] Chong, K.S.; Mehrnoosh, J.; Hara, S.; Lacko, A.G.: Characterization of lecithin-cholesterol acyltransferase from human plasma. 3. Chemical properties of the enzyme. Can. J. Biochem. Cell Biol., **61**, 875-881 (1983)
[11] Lacko, A.G.; Varma, K.G.; Rutenberg, H.L.; Soloff, L.A.: Studies on enzymatic and molecular properties of lecithin:cholesterol acyltransferase. Scand. J. Clin. Lab. Invest., **33**, 29-34 (1974)
[12] Jauhiainen, M.; Yuan, W.; Gelb, M.H.; Dolphin, P.J.: Human plasma lecithin-cholesterol acyltransferase. Inhibition of the phospholipase A2-like activity by sn-2-difluoroketone phosphatidylcholine analogues. J. Biol. Chem., **264**, 1963-1967 (1989)
[13] Collet, X.; Fielding, C.J.: Effects of inhibitors of N-linked oligosaccharide processing on the secretion, stability, and activity of lecithin:cholesterol acyltransferase. Biochemistry, **30**, 3228-3234 (1991)
[14] Knipping, G.: Isolation and properties of porcine lecithin:cholesterol acyltransferase. Eur. J. Biochem., **154**, 289-294 (1986)
[15] Aron, L.; Jones, S.; Fielding, C.J.: Human plasma lecithin-cholesterol acyltransferase. Characterization of cofactor-dependent phospholipase activity. J. Biol. Chem., **253**, 7220-7226 (1978)
[16] Park, Y.B.; Lacko, A.G.: Isolation and characterization of lecithin-cholesterol acyltransferase from hog plasma. Biochim. Biophys. Acta, **877**, 179-190 (1986)
[17] Zhou, G.; Jauhiainen, M.; Stevenson, K.; Dolphin, P.J.: Human plasma lecithin:cholesterol acyltransferase. Preparation and use of immobilized *p*-aminophenylarsenoxide as a catalytic site-directed covalent ligand in enzyme purification. J. Chromatogr., **568**, 69-83 (1991)
[18] Subbaiah, P.V.: Lysolecithin acyltransferase of human plasma: assay and characterization of enzyme activity. Methods Enzymol., **129**, 790-797 (1986)
[19] Pownall, H.J.; Pao, Q.; Brockman, H.L.; Massey, J.B.: Inhibition of lecithin-cholesterol acyltransferase by diphytanoyl phosphatidylcholine. J. Biol. Chem., **262**, 9033-9036 (1987)
[20] Chen, C.H.; Albers, J.J.: A rapid large-scale procedure for purification of lecithin-cholesterol acyltransferase from human and animal plasma. Biochim. Biophys. Acta, **834**, 188-195 (1985)
[21] Mahadevan, V.; Soloff, L.A.: A method for isolating human plasma lecithin:cholesterol acyltransferase without using anti-apolipoprotein D, and its characterization. Biochim. Biophys. Acta, **752**, 89-97 (1983)
[22] Varma, K.G.; Soloff, L.A.: A method for the purification of milligram quantities of stable human phosphatidylcholine-cholesterol acyltransferase. Biochem. J., **155**, 583-588 (1976)

[23] Kitabatake, K.; Piran, U.; Kamio, Y.; Doi, Y.; Nishida, T.: Purification of human plasma lecithin:cholesterol acyltransferase and its specificity towards the acyl acceptor. Biochim. Biophys. Acta, **573**, 145-154 (1979)
[24] Doi, Y.; Nishida, T.: Microheterogeneity and physical properties of human lecithin-cholesterol acyltransferase. J. Biol. Chem., **258**, 5840-5846 (1983)
[25] Chaves, M.E.C.; Harry, D.S.; Owen, J.; McIntyre, N.: Purification of lecithin-cholesterol acyltransferase from human plasma. Biochem. Soc. Trans., **13**, 145-146 (1985)
[26] Chong, K.S.; Hara, S.; Thompson, R.E.; Lacko, A.G.: Characterization of lecithin:cholesterol acyltransferase from human plasma: II. Physical properties of the enzyme. Arch. Biochem. Biophys., **222**, 553-560 (1983)
[27] Chong, K.S.; Davidson, L.; Huttash, R.G.; Lacko, A.G.: Characterization of lecithin: cholesterol acyltransferase from human plasma: purification of the enzyme. Arch. Biochem. Biophys., **211**, 119-124 (1981)
[28] Suzue, G.; Vezina, C.; Marcel, Y.L.: Purification of human plasma lecithin:cholesterol acyltransferase and its activation by metal ions. Can. J. Biochem., **58**, 539-541 (1980)
[29] Smith, N.B.; Kuksis, A.: Stereochemical substrate requirements of lecithin:cholesterol acyltransferase and its inhibition by enantiomeric lysolecithins. Can. J. Biochem., **58**, 1286-1291 (1980)
[30] Furukawa, Y.; Urano, T.; Itoh, H.; Takahashi, C.; Kimura, S.: Isolation and properties of rat plasma lecithin-cholesterol acyltransferase. J. Biochem., **105**, 962-967 (1989)
[31] Furukawa, Y.; Nishida, T.: Stability and properties of lecithin-cholesterol acyltransferase. J. Biol. Chem., **254**, 7213-7219 (1979)
[32] Chung, J.; Abano, D.A.; Fless, G.M.; Scanu, A.M.: Isolation, properties, and mechanism of in vitro action of lecithin: cholesterol acyltransferase from human plasma. J. Biol. Chem., **254**, 7456-7464 (1979)
[33] Albers, J.J.; Cabana, V.G.; De Barden Stahl, Y.: Purification and characterization of human plasma lecithin: cholesterol acyltransferase. Biochemistry, **15**, 1084-1087 (1976)
[34] Liu, M.; Subbaiah, P.V.: Hydrolysis and transesterification of platelet-activating factor by lecithin-cholesterol acyltransferase. Proc. Natl. Acad. Sci. USA, **91**, 6035-6039 (1994)
[35] Jonas, A.: Lecithin cholesterol acyltransferase. Biochim. Biophys. Acta, **1529**, 245-256 (2000)
[36] Liu, M.; Subbaiah, P.V.: Activation of plasma lysolecithin acyltransferase reaction by apolipoproteins A-I, C-I and E. Biochim. Biophys. Acta, **1168**, 144-152 (1993)
[37] O, K.; Hill, J.S.; Wang, X.; McLeod, R.; Pritchard, P.H.: Lecithin:cholesterol acyltransferase: Role of N-linked glycosylation in enzyme function. Biochem. J., **294**, 879-884 (1993)
[38] Lee, Y.P.; Adimoolam, S.; Liu, M.; Subbaiah, P.V.; Glenn, K.; Jonas, A.: Analysis of human lecithin-cholesterol acyltransferase activity by carboxyl-terminal truncation. Biochim. Biophys. Acta, **1344**, 250-261 (1997)
[39] Ayyobi, A.F.; Lacko, A.G.; Murray, K.; Nair, M.; Li, M.; Molhuizen, H.O.F.; Pritchard, P.H.: Biochemical and compositional analyses of recombinant le-

cithin:cholesterol acyltransferase (LCAT) obtained from a hepatic source. Biochim. Biophys. Acta, **1484**, 1-13 (2000)
[40] Davit-Spraul, A.; Therond, P.; Leroy, A.; Palmade-Rieunier, F.; Rousset, C.; Moatti, N.; Legrand, A.: Inhibition of lecithin cholesterol acyltransferase by phosphatidylcholine hydroperoxides. FEBS Lett., **447**, 106-110 (1999)
[41] Hida, Y.; Furukawa, Y.; Urano, T.; Kim, H.J.; Kimura, S.: Substrate specificity of rat plasma lecithin-cholesterol acyltransferase towards a molecular species of phosphatidylcholine. Biosci. Biotechnol. Biochem., **57**, 1111-1114 (1993)
[42] Calabresi, L.; Franceschini, G.; Burkybile, A.; Jonas, A.: Activation of lecithin cholesterol acyltransferase by a disulfide-linked apolipoprotein A-I dimer. Biochem. Biophys. Res. Commun., **232**, 345-349 (1997)
[43] Demeester, N.; Castro, G.; Desrumaux, C.; De Geitere, C.; Fruchart, J.C.; Santens, P.; Mulleners, E.; Engelborghs, S.; De Deyn, P.P.; Vandekerckhove, J.; Rosseneu, M.; Labeur, C.: Characterization and functional studies of lipoproteins, lipid transfer proteins, and lecithin:cholesterol acyltransferase in CSF of normal individuals and patients with Alzheimer's disease. J. Lipid Res., **41**, 963-974 (2000)
[44] Qu, S.J.; Fan, H.Z.; Blanco-Vaca, F.; Pownall, H.J.: Effects of site-directed mutagenesis on the N-glycosylation sites of human lecithin:cholesterol acyltransferase. Biochemistry, **32**, 8732-8736 (1993)
[45] Subbaiah, P.V.; Liu, M.; Paltauf, F.: Role of sn-2 acyl group of phosphatidyl choline in determining the positional specificity of lecithin-cholesterol acyltransferase. Biochemistry, **33**, 13259-13266 (1994)
[46] Parks, J.S.; Huggins, K.W.; Gebre, A.K.; Burleson, E.R.: Phosphatidylcholine fluidity and structure affect lecithin:cholesterol acyltransferase activity. J. Lipid Res., **41**, 546-553 (2000)
[47] Lacko, A.G.; Reason, A.J.; Nuckolls, C.; Kudchodkar, B.J.; Nair, M.P.; Sundarrajan, G.; Pritchard, P.H.; Morris, H.R.; Dell, A.: Characterization of recombinant human plasma lecithin: cholesterol acyltransferase (LCAT): N-linked carbohydrate structures and catalytic properties. J. Lipid Res., **39**, 807-820 (1998)
[48] Chisholm, J.W.; Gebre, A.K.; Parks, J.S.: Characterization of C-terminal histidine-tagged human recombinant lecithin:cholesterol acyltransferase. J. Lipid Res., **40**, 1512-1519 (1999)
[49] Jin, L.; Lee, Y.P.; Jonas, A.: Biochemical and biophysical characterization of human recombinant lecithin: cholesterol acyltransferase. J. Lipid Res., **38**, 1085-1093 (1997)
[50] Nair, M.P.; Kudchodkar, B.J.; Pritchard, P.H.; Lacko, A.G.: Purification of recombinant lecithin: cholesterol acyltransferase. Protein Expr. Purif., **10**, 38-41 (1997)
[51] Krimbou, L.; Marcil, M.; Davignon, J.; Genest, J., Jr.: Interaction of lecithin:cholesterol acyltransferase (LCAT)α_2-macroglobulin complex with low density lipoprotein receptor-related protein (LRP). Evidence for an α_2-macroglobulin/LRP receptor-mediated system participating in LCAT clearance. J. Biol. Chem., **276**, 33241-33248 (2001)

[52] Adimoolam, S.; Jin, L.; Grabbe, E.; Shieh, J.J.; Jonas, A.: Structural and functional properties of two mutants of lecithin-cholesterol acyltransferase (T123I and N228K). J. Biol. Chem., **273**, 32561-32567 (1998)
[53] Hengstschlaeger-Ottnad, E.; Kuchler, K.; Schneider, W.J.: Chicken lecithin-cholesterol acyltransferase. Molecular characterization reveals unusual structure and expression pattern. J. Biol. Chem., **270**, 26139-26145 (1995)

N-Acetylneuraminate 4-O-acetyltransferase 2.3.1.44

1 Nomenclature

EC number
2.3.1.44

Systematic name
acetyl-CoA:N-acetylneuraminate 4-O-acetyltransferase

Recommended name
N-acetylneuraminate 4-O-acetyltransferase

Synonyms
acetyltransferase, N-acetylneuraminate 4-O-
sialate O-acetyltransferase

CAS registry number
51004-25-2

2 Source Organism

<1> *Tachyglossus aculeatus* (australian monotreme echidna [2]) [2]
<2> *Equus caballus* [1, 2, 3]
<3> *Equus asinus* [2]
<4> *Cavia porcellus* [4]

3 Reaction and Specificity

Catalyzed reaction
acetyl-CoA + N-acetylneuraminate = CoA + N-acetyl-4-O-acetylneuraminate

Reaction type
acyl group transfer

Natural substrates and products
S acetyl-CoA + N-acetylneuraminate <1-4> (<1-3> the 4-O-acetyl-group seems to be stable, as no conversion of this sialic acid to other O-acetylated species is observed [2]; <1-3> the 4-O-acetyltransferase is specific for the hydroxyl group at C-4 of the pyranose ring of neuraminic acid [2]) (Reversibility: r <1-3> [2]; ? <4> [4]) [1-4]
P CoA + N-acetyl-4-O-acetylneuraminate <1-4> [1-4]

Substrates and products

S acetyl-CoA + N-acetylneuraminate <1-4> (<1-3> the 4-O-acetyl-group seems to be stable, as no conversion of this sialic acid to other O-acetylated species is observed [2]; <1-3> the 4-O-acetyltransferase is specific for the hydroxyl group at C-4 of the pyranose ring of neuraminic acid [2]) (Reversibility: r <1-3> [2]; ? <4> [4]) [1-4]

P CoA + N-acetyl-4-O-acetylneuraminate <1-4> [1-4]

Inhibitors

3-dephospho-CoA <2> [3]
4,4-diisothiocyanatostilbene-2,2-disulfonate <2> [3]
ADP <2> [3]
IDP <2> [3]
CoA <2, 4> (<2> competitive [3,4]) [3, 4]
$MnCl_2$ <4> [4]
Triton-X <4> [4]
β-NAD$^+$ <2> [3]
p-chloromercuribenzoate <2, 4> [3, 4]
saponin <4> [4]

Specific activity (U/mg)

Additional information <1-3> (<1-3> assay method [2]) [2]

K_m-Value (mM)

0.00061 <4> (acetyl-CoA) [4]
0.0321 <2> (acetyl-CoA) [3]

K_i-Value (mM)

0.0042 <4> (CoA) [4]
0.24 <2> (CoA) [3]

pH-Optimum

6.6 <2> [3]
6.7 <4> (<4> in 70 mM potassium phosphate buffer and the presence of 90 mM KCl [4]) [4]
7.5 <1-3> (<1-3> assay at [2]) [2]

pH-Range

5-8.5 <4> [4]
5.2-8.5 <2> [3]

Temperature optimum (°C)

30 <4> [4]
37 <1-3> (<1-3> assay at [2,3]) [2, 3]

Temperature range (°C)

0-50 <2> [3]
0-55 <4> [4]

5 Isolation/Preparation/Mutation/Application

Source/tissue

liver <4> [4]
serum <4> [4]
submandibular gland <2> [3]
submaxillary gland <2> [1]

Localization

microsome <2> [3]

Purification

<2> [3]

References

[1] Schauer, R.: Biosynthesis of N-acetyl-O-acetylneuraminic acids. I. Incorporation of (^{14}C) acetate into sections of the submaxillary salivary gland of ox and horse. Hoppe-Seyler's Z. Physiol. Chem., **351**, 595-602 (1970)
[2] Schauer, R.: Sialic acids: metabolism of O-acetyl groups. Methods Enzymol., **138**, 611-626 (1987)
[3] Tiralongo, J.; Schmid, H.; Thun, R.; Iwersen, M.; Schauer, R.: Characterisation of the enzymatic 4-O-acetylation of sialic acids in microsomes from equine submandibular glands. Glycoconjugate J., **17**, 849-858 (2001)
[4] Iwersen, M.; Vandamme-Feldhaus, V.; Schauer, R.: Enzymic 4-O-acetylation of N-acetylneuraminic acid in guinea-pig liver. Glycoconjugate J., **15**, 895-904 (1998)

N-Acetylneuraminate 7-O(or 9-O)-acetyltransferase 2.3.1.45

1 Nomenclature

EC number

2.3.1.45

Systematic name

acetyl-CoA:N-acetylneuraminate 7-O(or 9-O)-acetyltransferase

Recommended name

N-acetylneuraminate 7-O(or 9-O)-acetyltransferase

Synonyms

N-acetylneuraminate 7(8)-O-acetyltransferase
N-acetylneuraminate 7,8-O-acetyltransferase
N-acetylneuraminate 9(7)-O-acetyltransferase
N-acetylneuraminate O7-(or O9-)acetyltransferase
acetyl-CoA:N-acetylneuraminate-7- and/or 8-O-acetyltransferase
acetyl-CoA:N-acetylneuraminate-7- or 8-O-acetyltransferase
acetyl-CoA:N-acetylneuraminate-9(7)-O-acetyltransferase
acetyl-CoA:N-acetylneuraminate-9(or7)-O-acetyltransferase
acetyltransferase, N-acetylneuraminate 9(7)-O-
glycoprotein 7(9)-O-acetyltransferase
sialate O-acetyltransferase
sialate-7(9)-O-acetyltransferase <7> [8]
Additional information <1, 2> (<1> in early days of investigation of sialate O-acetyltransferase it was not possible to discriminate clearly between 7- or 9-O-acetylated sialic acids enzymatically formed on a micro scale in order to denominate the enzyme correctly, there also exists the possibility of the existence of more than one O-acetyltransferases incorporating acetyl-groups into different positions of the sialic acid side chain, in contrast a hypothesis is presented for the requirement of only one transferase specific for the 7-OH group, after insertion of the O-acetyl group at O-7 the ester residue may migrate to O-9 afterwards a second O-acetyl group can be transferred by the same enzyme to O-7, which may migrate to O-8 and open the way for linking a third acetyl group to O-7 [1,2]) [1, 2, 4]

CAS registry number

9054-50-6

2 Source Organism

<1> *Bos taurus* (ox [1]) [1, 2, 4-6, 10]
<2> *Equus caballus* [4, 5]
<3> *Rattus norvegicus* [3, 4, 9]
<4> *Sus scrofa* [4]
<5> *Oryctolagus cuniculus* [4]
<6> *starfish* [4]
<7> *Homo sapiens* [7, 8]

3 Reaction and Specificity

Catalyzed reaction
acetyl-CoA + N-acetylneuraminate = CoA + N-acetyl-7-O(or 9-O)-acetylneuraminate (<1-6> 7-O-acetyl groups can migrate to the 9-position [3,4,9,10])

Reaction type
acyl group transfer

Natural substrates and products
S acetyl-CoA + sialic acid <1-7> (<1-6> sialic acid distribution, overview [4]; <1-6> pathway, overview [3,4]) (Reversibility: ? <1-7> [1-10]) [1-10]
P CoA + N-acetyl-7-O(or 9-O-)-acetylneuraminate <1-7> [1-10]

Substrates and products
S acetyl-CoA + CMP-N-acetyl-neuraminate <7> (Reversibility: ? <7> [8]) [8]
P CoA + CMP-9-O-acetyl-N-acetylneuraminate <7> [8]
S acetyl-CoA + N-acetylneuraminate <1-7> (<1,3> free acetylneuraminate [2,3,6]; <1,3> glycosidically bound N-acetylneuraminate [3,6,10]; <3> glycosidically bound N-acetylneuraminate, α2-6 linked [9]; <3> acetylneuraminate bound to sialic acids [3]; <1> the membrane-bound enzyme acts on endogenous, glycoprotein-bound N-acetyl- and N-glycolylneuraminic acids, the soluble enzyme acts on exogenous non-glycosidically bound N-acetyl- and N-glycolylneuraminic acids [5]) (Reversibility: ? <1-7> [1-10]) [1-10]
P CoA + N-acetyl-7-O(or 9-O-)-acetylneuraminate <1-7> (<1> O-acetylation at position 7, 9-O-acetylated product appears after migration of the acetyl group from position 7 to 9 [10]; <1> few N-acetyl-O-acetylneuraminate bound to cytosolic glycoproteins, some bound to microsomes [2]; <1> N-acetyl-7(8)-mono-O-acetyl- and N-acetyl-7,8-di-O-acetylneuraminate [6]) [1-10]
S acetyl-CoA + α-2,3-sialyllactose <7> (Reversibility: ? <7> [8]) [8]
P ?
S acetyl-CoA + α-2,6-sialyllactose <7> (Reversibility: ? <7> [8]) [8]
P ?

S Additional information <1> (<1> reaction affords the transport of the acetyl group by a translocase across the Golgi membrane, followed by the enzymatic transfer of the acetyl group onto sialic acid in the Golgi lumen [10]) [10]

P ?

Inhibitors

CoA <1, 3, 7> (<1> concentration-dependent [10]; <1> inhibition of membrane-bound enzyme [5]; <3> inhibition of transfer to endogenous acceptors, no inhibition of transfer from acetyl intermediate to free acetylneuraminate [3]) [3, 5, 8, 10]
Cu^{2+} <1> [10]
N-acetylimidazole <1> [10]
N-bromosuccinimide <1, 3> (<1> slightly [10]) [1, 3, 10]
Triton X-100 <1, 3> (<1> no inhibition [5]; <1,3> above 0.02% [1,3,10]) [1, 3, 10]
Zn^{2+} <1> [10]
diethyl dicarbonate <1, 3> (<1> inhibits intramolecular migration of the acetyl group from position 7 to 9 [10]) [1, 3, 10]
iodoacetate <1> [5]
methylamine <1> (<1> slightly [10]) [10]
p-chloromercuribenzoate <1> [5, 10]
Additional information <1> (<1> no inhibition by $MgCl_2$, $CaCl_2$, $MnCl_2$ [5]) [5]

Activating compounds

2-mercaptoethanol <1> (<1> slightly activates [10]) [10]
acetic acid anhydride <1> (<1> activates [10]) [10]
dithiothreitol <1> (<1> slightly activates [10]) [10]

Specific activity (U/mg)

Additional information <1, 2> (<1,2> assay method [4]; <1> subcellular distribution [10]) [4, 10]

K_m-Value (mM)

0.00163 <1> (acetyl-coA) [10]
0.0028 <3> (acetyl-CoA, <3> + glycosidically bound N-acetylneuraminate [3]) [3]
0.0061 <7> (acetyl-CoA) [8]
0.01 <1> (acetyl-CoA, <1> cytosolic enzyme [5]) [5]
0.059 <7> (CMP-5-N-acetyl-neuraminate) [8]
0.12 <1> (acetyl-CoA, <1> membrane-bound enzyme [5]) [5]
0.39 <1> (N-glycolylneuraminate, <1> cytosolic enzyme [5]) [5]
0.5 <1> (N-acetylneuraminate, <1> cytosolic enzyme [5]) [5]

K_i-Value (mM)

0.00074 <3> (CoA) [3]
0.0112 <1> (CoA) [10]
0.0119 <7> (CoA) [8]
0.56 <1> (CoA) [5]

pH-Optimum

6.5 <1, 3> (<1,3> assay at [3,10]) [3, 10]
7 <1, 7> [5, 8]

Temperature optimum (°C)

22 <1> (<1> assay at [10]) [10]
37 <1-3, 7> (<1-3> assay at [4]) [4, 5, 8]

5 Isolation/Preparation/Mutation/Application

Source/tissue

colon <7> (<7> mucosa [8]) [8]
liver <3> [3, 4, 9]
melanoma cell <7> [7]
submandibular gland <1, 2> [4-6, 10]
submaxillary gland <1> [1, 2]

Localization

Golgi vesicle <1, 3, 7> (<3> 7-O- and 9-O-acetylation [9]; <3> equally distributed between membrane and cytosolic fraction [3]) [3, 8-10]
cytosol <1, 3> (<7> not [8]; <3> equally distributed between membrane and cytosolic fraction [3]) [2, 3]
intracellular <1> [6]
lysosomal membrane <3> (<3> 7-O-acetylation [9]) [9]
membrane <3> [9]
microsome <1, 2, 7> (<1> low activity [2]) [2, 5, 8, 10]
plasma membrane <3> (<3> 9-O-acetylation [9]) [9]

Cloning

<7> (cloning of cDNAs from melanoma genetic library that induce 9-O-acetylation of sialic acids, expression in COS cells [7]) [7]

6 Stability

Temperature stability

22 <1> (<1> 50% loss of activity after 20 min at room temperature [1]) [1]
37 <1> (<1> activity is very labile [1]) [1]

General stability information

<1>, activity is stable to the first cycle of freezing, cannot be refrozen after thawing [1]
<1>, lyophilisation causes loss of 20-30% activity, storage in lyophilized stage leads to additional loss of 60% activity [5]
<3>, enzyme is labile and difficult to solubilize [9]

Storage stability

<1>, 0°C, 16 days, 55% loss of activity, membrane-bound enzyme [5]
<1>, frozen enzyme is stable for at least 6 months [5]
<3>, -70°C, prelabeled Golgi vesicles, 4 days [3]
<3>, frozen Golgi vesicles are stable for 3-6 months [3]

References

[1] Schauer, R.: Biosynthesis of N-acetyl-O-acetylneuraminic acids. I. Incorporation of (^{14}C) acetate into sections of the submaxillary salivary gland of ox and horse. Hoppe-Seyler's Z. Physiol. Chem., **351**, 595-602 (1970)
[2] Schauer, R.: Biosynthesis of N-acetyl-O-acetylneuraminic acids. II. Substrate and intracellular localization of bovine acetyl-coenzyme A: N-acetylneuraminate-7- and 8-O-acetyltransferase. Hoppe-Seyler's Z. Physiol. Chem., **351**, 749-758 (1970)
[3] Diaz, S.; Higa, H.H.; Varki, A.: Glycoprotein sialate 7(9)-O-acetyltransferase from rat liver Golgi vesicles. Methods Enzymol., **179**, 416-422 (1989)
[4] Schauer, R.: Sialic acids: metabolism of O-acetyl groups. Methods Enzymol., **138**, 611-626 (1987)
[5] Corfield, A.P.; Ferreira do Amarral, C.; Wember, M.; Schauer, R.: The metabolism of O-acyl-N-acylneuraminic acids. Biosynthesis of O-acylated sialic acids in bovine and equine submandibular glands. Eur. J. Biochem., **68**, 597-610 (1976)
[6] Schauer, R.M.; Wember, M.: Hydroxylation and O-acetylation of N-acetylneuraminic acid bound to glycoproteins of isolated subcellular membranes from porcine and bovine submaxillary glands. Hoppe-Seyler's Z. Physiol. Chem., **352**, 1282-1290 (1971)
[7] Shi, W.X.; Chammas, R.; Varki, A.: Induction of sialic acid 9-O-acetylation by diverse gene products: implications for the expression cloning of sialic acid O-acetyltransferases. Glycobiology, **8**, 199-205 (1998)
[8] Shen, Y.; Tiralongo, J.; Iwersen, M.; Sipos, B.; Kalthoff, H.; Schauer, R.: Characterization of the sialate-7(9)-O-acetyltransferase from the microsomes of human colonic mucosa. Biol. Chem., **383**, 307-317 (2002)
[9] Butor, C.; Diaz, S.; Varki, A.: High level O-acetylation of sialic acids on N-linked oligosaccharides of rat liver membranes. Differential subcellular distribution of 7- and 9-O-acetyl groups and of enzymes involved in their regulation. J. Biol. Chem., **268**, 10197-10206 (1993)
[10] Vandamme-Feldhaus, V.; Schauer, R.: Characterization of the enzymic 7-O-acetylation of sialic acids and evidence for enzymic O-acetyl migration from C-7 to C-9 in bovine submandibular gland. J. Biochem., **124**, 111-121 (1998)

Homoserine O-succinyltransferase 2.3.1.46

1 Nomenclature

EC number
2.3.1.46

Systematic name
succinyl-CoA:L-homoserine O-succinyltransferase

Recommended name
homoserine O-succinyltransferase

Synonyms
homoserine O-transsuccinylase
homoserine succinyltransferase
succinyltransferase, homoserine

CAS registry number
62213-51-8

2 Source Organism

<1> *Escherichia coli* (metA gene [3,4]; strain K-12 [2,3]; auxotrophic mutants derived from K-12 [3]; auxotrophic strains [1]) [1-4]
<2> *Salmonella typhimurium* (metA gene [3]) [3]

3 Reaction and Specificity

Catalyzed reaction
succinyl-CoA + L-homoserine = CoA + O-succinyl-L-homoserine (<1> ping-pong mechanism, enzyme contains a cysteine in the active site, member of the acyltransferase family [4])

Reaction type
acyl group transfer

Natural substrates and products
S succinyl-CoA + L-homoserine <1, 2> (<1> involved in cystathionine and methionine biosynthesis [1,3,4]) (Reversibility: ? <1, 2> [1, 3, 4]) [1, 3, 4]
P CoA + O-succinyl-L-homoserine <1, 2> [1, 3, 4]

Substrates and products

S glutaryl-CoA + L-homoserine <1> (Reversibility: r <1> [4]) [4]
P CoA + O-glutaryl-L-homoserine
S succinyl-CoA + D-homoserine <1> (Reversibility: r <1> [4]) [4]
P CoA + O-succinyl-D-homoserine
S succinyl-CoA + L-homoserine <1, 2> (Reversibility: r <1> [4]; ? <1, 2> [1-3]) [1-4]
P CoA + O-succinyl-L-homoserine <1, 2> [1-4]
S Additional information <1> (<1> no activity with acetyl-CoA, propionyl-CoA, butyryl-CoA, crotonyl-CoA, malonyl-CoA, and γ-hydroxybutyric acid, and γ-aminobutyric acid [4]; <1> 3-amino-1-propanol, L-serine, and L-threonine are poor substrates [4]) [4]
P ?

Inhibitors

D-methionine <1, 2> (<1,2> slight feed-back inhibition [3]) [3]
L-methionine <1, 2> (<1,2> feed-back inhibition [2,3]) [2, 3]
S-adenosyl-L-methionine <1> (<1> feed-back inhibition [2]) [2]
hydroxyethylclavam <1> (<1> oxygen analogue β-lactam antibiotic [2]) [2]
iodoacetamide <1> (<1> inhibition is pH-dependent [4]) [4]
valclavam <1> (<1> oxygen analogue β-lactam antibiotic, complete inhibition at 0.2 mM [2]) [2]
Additional information <1> (<1> overview, competitive antagonism of peptides against enzyme inhibition by valclavam [2]; <1> overview, inhibition of several fungi and procaryotes by valclavam [2]) [2]

Turnover number (min^{-1})

96 <1> (glutaryl-CoA, <1> recombinant protein [4]) [4]
313.8 <1> (O-succinyl-L-homoserine, <1> recombinant protein [4]) [4]
313.8 <1> (coenzyme A, <1> recombinant protein [4]) [4]
720 <1> (D-homoserine, <1> recombinant protein [4]) [4]
1440 <1> (L-homoserine, <1> recombinant protein [4]) [4]
1440 <1> (succinyl-CoA, <1> recombinant protein [4]) [4]

Specific activity (U/mg)

Additional information <1> (<1> cystathionine formation [1]) [1]

K_m-Value (mM)

0.17 <1> (succinyl-CoA, <1> recombinant protein [4]) [4]
0.18 <1> (glutaryl-CoA, <1> recombinant protein [4]) [4]
0.64 <1> (coenzyme A, <1> recombinant protein [4]) [4]
1.6 <1> (L-homoserine, <1> recombinant protein [4]) [4]
3.5 <1> (O-succinylhomoserine, <1> recombinant protein [4]) [4]
10 <1> (D-homoserine, <1> recombinant protein [4]) [4]

K_i-Value (mM)

0.00025 <1> (hydroxyethylvalclavam, <1> dissociation constant of enzyme-substrate-inhibitor complex [2]) [2]
0.00083 <1> (valclavam) [2]

0.00089 <1> (valclavam, <1> dissociation constant of enzyme-substrate-inhibitor complex [2]) [2]
0.00092 <1> (hydroxyethylvalclavam) [2]

pH-Optimum
7.5 <1> (<1> crude extract [1]) [1]
Additional information <1> (<1> broad maximum [4]) [4]

pH-Range
6-8.7 <1> [4]
6.9-8 <1> (<1> crude extract, about 40% of maximal activity at pH 6.9, about 80% of maximal activity at pH 8.0 [1]) [1]
Additional information <1> (<1> pK_s of 6.6 and abo ut 7.9 [4]) [4]

4 Enzyme Structure

Molecular weight
86000 <1> (<1> gel filtration [4]) [4]

Subunits
dimer <1> (<1> 2 * 35600, SDS-PAGE [4]) [4]

5 Isolation/Preparation/Mutation/Application

Purification
<1> (recombinant from E. coli BL21 (DE3) [4]) [2, 4]

Cloning
<1> (overexpression of metA gene in Escherichia coli BL21 (DE3) [4]; metA gene, expression in Escherichia coli auxotrophic strains as lacZ fusion proteins [3]) [3, 4]
<2> (metA gene, expression in Escherichia coli auxotrophic strains as lacZ fusion proteins [3]) [3]

6 Stability

General stability information
<1>, glycerol, 30% w/v, stabilizes during purification [2]

References

[1] Rowbury, R.J.; Woods, D.D.: O-Succinyl-homoserine as an intermediate in the synthesis of cystathionine by Escherichia coli. J. Gen. Microbiol., **36**, 341-358 (1964)

[2] Röhl, F.; Rabenhorst, J.; Zähner, H.: Biological properties and mode of action of clavams. Arch. Microbiol., **147**, 315-320 (1987)
[3] Mares, R.; Urbanowski, M.L.; Stauffer, G.V.: Regulation of the Salmonella typhimurium metA gene by the metR protein and homocysteine. J. Bacteriol., **174**, 390-397 (1992)
[4] Born, T.L.; Blanchard, J.S.: Enzyme-catalyzed acylation of homoserine: mechanistic characterization of the Escherichia coli metA-encoded homoserine transsuccinylase. Biochemistry, **38**, 14416-14423 (1999)

8-Amino-7-oxononanoate synthase 2.3.1.47

1 Nomenclature

EC number
2.3.1.47

Systematic name
6-carboxyhexanoyl-CoA:L-alanine C-carboxyhexanoyltransferase (decarboxylating)

Recommended name
8-amino-7-oxononanoate synthase

Synonyms
7-KAP synthetase <1> [1, 3]
7-keto-8-aminopelargonate synthetase
7-keto-8-aminopelargonic acid synthetase
7-keto-8-aminopelargonic synthetase
7-oxo-8-aminononanoate synthase
8-amino-7-oxopelargonate synthase
AONS <1> [10-12]
AOP synthase <3> [2]
KAPA synthetase <1-10> [5, 6]
synthase, 7-oxo-8-aminononanoate
synthetase, 7-keto-8-aminopelargonate

CAS registry number
9075-61-0

2 Source Organism

<1> *Escherichia coli* (gene bioF [3,10]; biotin-auxotroph mutants, derived from K-12 parent strain Y10-1 [1]) [1, 3, 5, 10-12]
<2> *Escherichia freundii* [5]
<3> *Bacillus sphaericus* (strain IFO 3525 [4]; gene bioF [2]) [2, 4-10]
<4> *Bacillus subtilis* [5]
<5> *Micrococcus roseus* [5]
<6> *Pseudomonas fluorescens* [5]
<7> *Aerobacter aerogenes* [5]
<8> *Aerobacter cloacae* [5]
<9> *Erwinia aroideae* [5]
<10> *Alcaligenes faecalis* [5]

<11> *Lipomyces starkeyi* [4]
<12> *Sporobolomyces salmonicolor* [4]
<13> *Sporobolomyces coprophilus* [4]

3 Reaction and Specificity

Catalyzed reaction

6-carboxyhexanoyl-CoA + L-alanine = 8-amino-7-oxononanoate + CoA + CO_2 (A pyridoxal-phosphate protein; <1,3> binding study of L-alanine and D-alanine to the enzyme [8,12]; <3> binding of pimeloyl-CoA causes a conformational change in the active site [8]; <3> Schiff base formation between D- or L-alanine and pyridoxal-5'-phophate in the active site [8]; <1,3> reaction mechanism [7,10-12]; <3> reaction mechanism could be shared by all α-oxoamine synthases [7]; <3> reaction mechanism, formation of the pyridoxal-phosphate-stabilized carbanion by abstraction of the C2 proton of the alanine-pyridoxal-phosphate aldimine, followed by acylation and decarboxylation [7,10])

Reaction type

acyl group transfer
condensation <1,3> [6,7,10-12]
decarboxylation <1,3> [6,7,10-12]

Natural substrates and products

S 6-carboxyhexanoyl-CoA + L-alanine <1-13> (<1-13> i.e. pimeloyl-CoA [1-8,10-12]; <3> second step in biotin biosynthesis [7,8]; <1-13> reaction in biotin biosynthesis [1-8,10-12]) (Reversibility: ir <1, 3> [2, 3, 7, 8, 12]; ? <1-13> [1, 4-6, 10, 11]) [1-8, 10-12]

P 8-amino-7-oxononanoate + CoA + CO_2 <1-13> (<1-13> i.e. 7-oxo-8-amino-pelargonic acid [1-8,10-12]) [1-8, 10-12]

Substrates and products

S 6-carboxyhexanoyl-CoA + L-alanine <1-13> (<1-13> i.e. pimeloyl-CoA [1-8,10-12]; <3,5,6> L-cysteine cannot replace L-alanine [5]; <3> absolutely specific for L-alanine [6]) (Reversibility: ir <1, 3> [2, 3, 7, 8, 12]; ? <1-13> [1, 4-6, 10, 11]) [1-8, 10-12]

P 8-amino-7-oxononanoate + CoA + CO_2 <1-13> (<1-13> i.e. 7-oxo-8-amino-pelargonic acid [1-8,10-12]) [1-8, 10-12]

S 6-carboxyhexanoyl-CoA + L-serine <1> (<1> low activity [1]; <2> not [5,6]) (Reversibility: ? <1> [1]) [1]

P 9-hydroxy-8-amino-7-oxononanoate + CoA <1> [1]

Inhibitors

1,10-phenanthroline <3> (<3> to some extent [6]) [6]
2-amino-3-hydroxy-2-methylnonadioic acid <3> (<3> competitive, transition state analogue [8]) [8]
4-carboxybutyl(1-amino-1-carboxyethyl)phosphonate <3> (<3> competitive, transition state analogue [8]) [8]

8-amino-7-oxo-8-phosphonononanoic acid <3> (<3> strong inhibition [8]) [8]
8-amino-7-oxononanoate <1> (<1> product inhibition in vivo [1]) [1]
Ag^+ <3> [6]
Cd^{2+} <3> (<3> weak [6]) [6]
Co^{2+} <3> [6]
Cu^{2+} <3> [6]
D-alanine <3, 6> (<3> Schiff base formation between D- or L-alanine and pyridoxal-5'-phophate in the active site [8]; <3> competitive [8]) [5, 6, 8]
D-cycloserine <3> [6]
D-penicillamine <3> (<3> L-enantiomer not as effective [6]) [6]
D/L-histidine <1-10> [5, 6]
L-cysteine <1, 3, 6> (<3> strong [6]) [1, 5, 6]
L-serine <3, 6> [5, 6]
Ni^{2+} <3> [6]
Zn^{2+} <3> (<3> weak [6]) [6]
biotin <3, 4> [5]
citrate <3> (<3> to some extent [6]) [6]
diisopropyl fluorophosphate <3> (<3> to some extent [6]) [6]
glycine <3, 6> [5, 6]
hydroxylamine <3> [6]
isoniazide <3> (<3> to some extent [6]) [6]
phenylhydrazine <3> [6]
pyridoxal 5'-phosphate <3> (<3> high concentrations [5]) [5]
pyruvic acid <6> [5]
semicarbazide <3> [6]
Additional information <1, 3> (<1> concentration above 10 mM of Ca^{2+}, Mg^{2+}, and Mn^{2+} cause slow precipitation of the enzyme [12]; <3> no inhibition by 8-amino-7-oxo-8-phosphonononanoic acid [8]; <3> inhibition mechanism [8]; <3> no inhibition by $HgCl_2$, EDTA, 2,2'-dipyridyl, oxalate, KCN, various amino acids [6]) [6, 12]

Cofactors/prosthetic groups

pyridoxal 5'-phosphate <1-10> (<1> pyridoxal-5'-phosphate binding site [12]; <3> Schiff base formation between D- or L-alanine and pyridoxal-5'-phophate in the active site [8]; <3> incubation of enzyme with L-alanine but without pyridoxal-5'-phosphate lead to abortive transamination [7]; <3> dependent on [2,5,7]; <3> requirement, pyridoxal-5'-phosphate protein, not replaceable by pyridoxal, pyridoxine, pyridoxine 5-phosphate, pyridoxamine, pyridoxamine 5-phosphate [6]) [1-3, 5-12]

Activating compounds

biotin <5, 6> (<5,6> stimulates [5]) [5]

Metals, ions

Additional information <3> (<3> effects of several metal ions on biotin biosynthesis [4]; <3> no metal ion requirement [6]) [4, 6]

Turnover number (min^{-1})
0.05 <1> (pimeloyl-CoA) [10]
0.06 <1> (pimeloyl-CoA) [12]
0.3 <3> (pimeloyl-CoA) [7]
30 <3> (pimeloyl-CoA) [2]

Specific activity (U/mg)
0.00016 <6> [5]
0.00225 <3> (<3> partially purified enzyme [6]) [6]
0.67 <3> [2]
Additional information <1-13> (<1> activity in biotin mutants [1]; <1-13> crude cell-extracts from various organism [4,5]) [1, 4, 5]

K_m-Value (mM)
0.001 <3> (pimeloyl-CoA) [2]
0.0015 <3> (pimeloyl-CoA) [7]
0.002 <3> (L-alanine) [7]
0.025 <1> (pimeloyl-CoA) [12]
0.07 <1> (pimeloyl-CoA) [10]
0.5 <1> (L-alanine) [12]
0.9 <1> (L-alanine) [10]
3 <3> (L-alanine) [2]

K_i-Value (mM)
0.007 <3> (8-amino-7-oxo-8-phosphonononanoic acid) [8]
0.068 <3> (4-carboxybutyl(1-amino-1-carboxyethyl)phosphonate) [8]
0.08 <3> (2-amino-3-hydroxy-2-methylnonadioic acid) [8]
0.59 <3> (D-alanine) [8]
0.99 <3> (L-cysteine) [5]
1.24 <3> (D-alanine) [5]
1.5 <3> (L-serine) [5]
3.33 <3> (glycine) [5]
3.75 <3> (D-histidine) [5]
5.6 <3> (L-histidine) [5]

pH-Optimum
7 <1, 3> (<1,3> assay at [1,2,7,8]) [1, 2, 6-8]
7.5 <1> (<1> assay at [12]) [12]
Additional information <1, 3> (<1> pI: 7.0 [3]; <3> pI: 5.2 [2]) [2, 3]

pH-Range
6.3-7.6 <1> [3]
6.5-7.5 <3> (<3> about half-maximal activity at pH 6.5 and pH 7.5 [6]) [6]

Temperature optimum (°C)
30 <1> (<1> assay at [12]) [12]
37 <1, 3> (<1,3> assay at [1,2,6]) [1, 2, 6]
60 <3> [6]

Temperature range (°C)

45-65 <3> (<3> about half-maximal activity at 45°C and 65°C [6]) [6]

4 Enzyme Structure

Molecular weight

41000 <1> [10]
41600 <1> (<1> calculated from nucleotide sequence [3]) [3]
41740 <1> (<1> electrospray mass spectrometry, holoenzyme [12]) [12]
46000 <3> (<3> gel filtration [2]) [2]
Additional information <1, 3> (<1> amino acid sequence [3,11]; <3> amino acid composition [2]) [2, 3, 11]

Subunits

monomer <3> (<3> 1 * 41000, SDS-PAGE [2]) [2]
Additional information <1> (<1> dimer, active site is formed at the interface between the monomers [11]; <1> homodimer with each monomer consisting of 3 domains [10]) [10, 11]

5 Isolation/Preparation/Mutation/Application

Purification

<1> (recombinant from E. coli [11,12]; recombinant from overexpressing E. coli strain HMS174 (DE3) [10]) [10-12]
<3> (recombinant from E. coli overproducing strain [2,6-9]; partial [6]) [2, 6-9]
<6> (partial [5]) [5]

Crystallization

<1> (Mg^{2+} added for structure analysis of the holoenzyme [12]; pH 8.5-8.9, ammonium sulfate precipitation, apoenzyme, X-ray diffraction structure analysis with and without bound pyridoxyl-5'-phosphate, model, active site residues [11,12]; pH 8.5 [10]) [10-12]
<3> (pyridoxal-5'-phosphate-bound form, pH 7.5, precipitation with polyethylene glycol, 0.2 M $(NH_4)_2SO_4$, X-ray diffraction structure analysis [9]) [9]

Cloning

<1> (bioF gene overexpression in Escherichia coli [11,12]; bioF gene, expression in strain HMS174 (DE3) [10]) [10-12]
<3> (improved overexpression in Escherichia coli bioF$^-$ strain R874 [7]; expressed in overproducing Escherichia coli strain C268/pTG3401 [2]; expression in Escherichia coli, nucleotide sequence and structure of bioF$^-$gene [3]) [2, 3]

Engineering

Additional information <1> (<1> natural mutants with altered enzyme activity [1]) [1]

Application

agriculture <3> (<3> possible target for design of inhibitors as herbicides and antibacterial agents [8]) [8]

6 Stability

Temperature stability

60 <3> (<3> and below, stable [6]) [6]
70 <3> (<3> $t_{1/2}$: 10 min [6]) [6]
80 <3> (<3> and above, inactivation [6]) [6]

General stability information

<3>, EDTA stabilizes during purification [2]
<3>, freeze-thawing inactivates [2]
<3>, proteinase inhibitors, e.g. PMSF, pepstatin A, stabilize during purification [2]

Storage stability

<3>, -20°C, 0.01 M phosphate buffer, pH 7.0, 5 mM 2-mercaptoethanol, 5 months [6]
<3>, 0°C, several months [2]

References

[1] Eidenberg, M.A.; Star, C.: Synthesis of 7-oxo-8-aminopelargonic acid, a biotin vitamer, in cell-free extracts of Escherichia coli biotin auxotrophs. J. Bacteriol., **96**, 1291-1297 (1968)
[2] Ploux, O.; Marquet, A.: The 8-amino-7-oxopelargonate synthase from Bacillus sphaericus. Purification and preliminary characterization of the cloned enzyme overproduced in Escherichia coli. Biochem. J., **283**, 327-331 (1992)
[3] Otsuka, A.J.; Buoncristiani, M.R.; Howard, P.K.; Flamm, J.; Johnson, C.; Yamamoto, R.; Uchida, K.; Cook, C.; Ruppert, J.; Matsuzaki, J.: The Escherichia coli biotin biosynthetic enzyme sequences predicted from the nucleotide sequence of the bio operon. J. Biol. Chem., **263**, 19577-19585 (1988)
[4] Izumi, Y.; Kano, Y.; Inagaki, K.; Kawase, N.; Tani, Y.; Yamada, H.: Characterization of biotin biosynthetic enzymes of Bacillus sphaericus: a dethiobiotin producing bacterium. Agric. Biol. Chem., **45**, 1983-1989 (1981)
[5] Izumi, Y.; Sato, K.; Tani, Y.; Ogata, K.: Distribution of 7-keto-8-aminopelargonic acid synthetase in bacteria and the control mechanism of the enzyme activity. Agric. Biol. Chem., **37**, 1335-1340 (1973)
[6] Izumi, Y.; Morita, H.; Tani, Y.; Ogata, K.: Partial purification and some properties of a 7-keto-8-aminopelargonic acid synthetase, an enzyme involved in biotin biosynthesis. Agric. Biol. Chem., **37**, 1327-1333 (1973)
[7] Ploux, O.; Marquet, A.: Mechanistic studies on the 8-amino-7-oxopelargonate synthase, a pyridoxal-5'-phosphate-dependent enzyme involved in biotin biosynthesis. Eur. J. Biochem., **236**, 301-308 (1996)

[8] Ploux, O.; Breyne, O.; Carillon, S.; Marquet, A.: Slow-binding and competitive inhibition of 8-amino-7-oxopelargonate synthase, a pyridoxal-5'-phosphate-dependent enzyme involved in biotin biosynthesis, by substrate and intermediate analogs: kinetic and binding studies. Eur. J. Biochem., **259**, 63-70 (1999)
[9] Spinelli, S.; Ploux, O.; Marquet, A.; Anguille, C.; Jelsch, C.; Cambillau, C.; Martinez, C.: Crystallization and preliminary x-ray study of the 8-amino-7-oxopelargonate synthase from Bacillus sphaericus. Acta Crystallogr. Sect. D, **D52**, 866-868 (1996)
[10] Webster, S.P.; Campopiano, D.J.; Alexeev, D.; Alexeeva, M.; Watt, R.M.; Sawyer, L.; Baxter, R.L.: Characterization of 8-amino-7-oxononanoate synthase: a bacterial PLP-dependent, acyl CoA condensing enzyme. Biochem. Soc. Trans., **26**, S268 (1998)
[11] Alexeev, D.; Alexeeva, M.; Baxter, R.L.; Campopiano, D.J.; Webster, S.P.; Sawyer, L.: The crystal structure of 8-amino-7-oxononanoate synthase: a bacterial PLP-dependent, acyl-CoA-condensing enzyme. J. Mol. Biol., **284**, 401-419 (1998)
[12] Webster, S.P.; Alexeev, D.; Campopiano, D.J.; Watt, R.M.; Alexeeva, M.; Sawyer, L.; Baxter, R.L.: Mechanism of 8-amino-7-oxononanoate synthase: spectroscopic, kinetic, and crystallographic studies. Biochemistry, **39**, 516-528 (2000)

Histone acetyltransferase 2.3.1.48

1 Nomenclature

EC number
2.3.1.48

Systematic name
acetyl-CoA:histone acetyltransferase

Recommended name
histone acetyltransferase

Synonyms
FAT <4, 8, 13, 17, 18> (<4, 8, 12, 13, 17, 18> when acetylating nonhistone transcription factor proteins, factor specific, overview [43]) [43]
HAT <4, 8, 11, 13, 15-18> [31-45]
acetyltransferase, histone
factor acetyltransferase <4, 8, 12, 13, 17, 18> (<4,8,12,13,17,18> enzyme form A is also able to acetylate nonhistone proteins, mostly transcription factors, overview [43]) [43]
histone acetokinase
histone acetylase
histone transacetylase
nucleosome-histone acetyltransferase
Additional information <4, 8, 13, 15-18> (<4,8,13,17,18> type A enzymes can be grouped into different families based on amino acid sequence comparison and binding motifs, overview [43]; <4,13,15,16> enzymes with histone acetyltransferase and transcription cofactor activity belong to the MYST protein family [31-34]; <4,8,13,17,18> type A enzymes are localized in the nuclei and acetylate nucleosomal histones as well as nonhistone proteins, type B enzymes can be found in the cytoplasmic fraction and are responsible for acetylation of newly synthesized histones before their translocation into the nucleus for chromatin assembly during DNA replication [43]) [31-34, 43]

CAS registry number
9054-51-7

2 Source Organism

<1> *Rattus norvegicus* (3 enzyme forms: A, B1, B2 [1]; 2 enzyme forms: A and B [6,21]; 2 enzyme forms NI and NII from nucleus and 2 enzyme forms CI and CII from cytoplasm [11]) [1, 3, 6, 11, 14, 19, 21]

<2> *Pisum sativum* (cv. Lincoln [28]; enzyme form B [28]; 2 enzyme forms: A and B [4]; enzyme form B [5]) [4, 5, 28]
<3> *Bos taurus* (calf [2,13,15-17,22,23]; 2 enzyme forms: A and B [2]; 3 enzyme forms: A, B, C [23]; enzyme forms A, B and DB [25]) [2, 13, 15-17, 22, 23, 25]
<4> *Saccharomyces cerevisiae* (genes/proteins Sas2, Sas4, Sas5 forming the SAS complex [45]; nucleosomal H2A/H4 acetyltransferase, i.e. NuA4 protein [35]; Esal gene [34]; Sas3 gene [33]; gene Gcn5 [31,39-41,43,44]; enzyme form A and B [7,10]) [7, 10, 31, 33, 34, 35, 39-41, 43-45]
<5> *Physarum polycephalum* (3 enzyme forms: A1, A2, B [8]) [8]
<6> *Sus scrofa* (enzyme form type I and type II [9]) [9, 12]
<7> *Simian virus 40* (i.e. SV40 [18]; from infected african green monkey kidney cell line CV-1 [18]) [18]
<8> *Drosophila melanogaster* (gene Gcn5 [43]) [20, 43]
<9> *Artemia nauplii* (3 enzyme forms: I, II and III [24]) [24]
<10> *Artemia salina* [26]
<11> *Zea mays* (strain Cuzco [27]; enzyme form B [27,36]) [27, 36]
<12> *Mus musculus* (gene Gcn5 [43]) [43]
<13> *Homo sapiens* (PCAF protein [37,38,40,43,44]; MORF protein, i.e. monocytic leukemia zinc finger protein-related factor [32]; recombinant enzyme hPCAF, i.e. human p300/CBP associated factor [29,31]; GCN5 [29,31,40,42-44]) [29, 31, 32, 37, 38, 40, 42-44]
<14> *Human Immunodeficiency Virus* (i.e. HIV [30]; recombinant from COS cell line and from E. coli [30]) [30]
<15> *Tetrahymena sp.* (protein p55 [31]) [31]
<16> *Homo sapiens* (fragment EST clone of MORF protein, i.e. monocytic leukemia zinc finger protein-related factor [32]) [32]
<17> *Tetrahymena thermophila* (Gcn5 [40,44]; protein p55 [43]; yeast-NuA4-like enzyme form [35]; transcription-associated enzyme form A, containing protein p55 as catalytic subunit [35]) [35, 40, 42-44]
<18> *Arabidopsis thaliana* (gene Gcn5 [43]) [43]

3 Reaction and Specificity

Catalyzed reaction

acetyl-CoA + histone = CoA + acetylhistone (a group of enzymes with differing specificities towards histone acceptors; <4, 13, 17> binding motifs [44]; <13, 16> MORF protein is also active as transcription factor [32]; <4, 13, 15> enzymes are also active as transcription coactivators and corepressors of transcription factors in gene regulation, acetylation of transcription factors, overview [31]; <4, 13> ternary complexed mechanism for PCAF and Gnc5 [38, 39]; <4, 13> PCAF and Gcn5 protein, fully ordered Bi-Bi kinetic mechanism with acetyl-CoA binding before histone H3 [37-39]; <4, 13, 17> mechanisms [31, 32, 44]; <2, 4, 13, 17> acetylation of lysine 5, 8, 12, and 16 of free histone H4 [28, 32, 35]; <1> rapid equilibrium ordered bireactant mechanism [14]; <3> ping-pong mechanism [16]; <1> ping-pong mechanism, enzyme form NII [11])

Reaction type

acyl group transfer

Natural substrates and products

S acetyl-CoA + histone <1, 4, 5, 8, 12, 13, 15, 17, 18> (<4,8,12,13,18> histone H3 is the preferred substrate [43]; <4,13> involved in chromatin remodeling and DNA repair [43]; <4> Esa1 protein is involved in cell cycle regulation [34]; <4,13,15> neutralization of positively charged lysine residues by acetylation lowering the affinity of histone octamers for the negatively charged DNA [31]; <4,13,15> the acetyl groups function as signals for interaction of histones with other regulatory proteins, chromatin remodeling [31,38]; <13> histone H1 is not acetylated in vivo [29]; <5> involved in dynamic equilibrium of core histone acetylation [8]; <8> most likely involved in acetylation of newly synthesized histones in cytoplasm prior to chromatin assembly [20]) (Reversibility: r <4, 8, 12, 13, 17, 18> [43]; ? <1, 4, 5, 8, 13, 15> [3, 8, 20, 31, 34, 38]) [3, 8, 20, 31, 34, 38, 43]

P CoA + acetylhistone <1, 4, 5, 8, 12, 13, 15, 17, 18> [3, 8, 20, 31, 34, 38, 43]

S Additional information <4, 13> (<4,13> enzyme activity is regulated by phosphorylation and interaction with other regulating protein factors [31]) [31]

P ?

Substrates and products

S acetyl-CoA + 1,4-butanediamine <1> (<1> i.e. putrescine [6]; <1> enzyme form A and B, low activity [6]) (Reversibility: ? <1> [6]) [6]

P ?

S acetyl-CoA + 1,5-pentanediamine <1> (<1> i.e. cadaverine [6]; <1> enzyme form A and B, low activity [6]) (Reversibility: ? <1> [6]) [6]

P ?

S acetyl-CoA + 1,6-hexanediamine <1> (<1> enzyme form A shows low activity , B not [6]) (Reversibility: ? <1> [6]) [6]

P ?

S acetyl-CoA + histone <1-15, 17, 18> (<13> calf thymus and HeLa cell histones [32]; <14> acetylates histones H2A, H3, and H4, but not histone 2B [30]; <2> acetylation of lysine 5, 8, 12, and 16 of free histone H4 with increasing preference [28]; <3> or chymotryptic peptides of histone [16]; <1,3-6,9,13> a group of enzymes with differing specificity towards histone acceptors, specificity of different enzyme forms [2,8-12,21,24,29,31,32]; <2> free pea and chicken histones H4, enzyme form B [28]; <2> pea histones, enzyme form A and B [4]; <2,4> chicken erythrocyte histones, enzyme form A and B [4,7]; <1> histone f1, enzyme form A [1]; <1> histone f2a1, enzyme form B1 and B2 [1]; <1> free histones [3,21]; <1> nucleosome-histones, enzyme form A, not enzyme form B [21]; <1> nucleosome-histones, enzyme form A and B [3]; <3> enzyme form B nearly exclusively acetylates histones H4 and H2a [23]; <5> histone H2A is a substrate for enzyme form A1 [8]; <5> histone H2B: poor substrate for enzyme form A2 [8]; <1> histone H1 poor substrate [6]; <10> histone H1 is a better substrate than H3 or H4 [26]; <6> chymostatic peptides of histones [12]) (Reversibility: r <4, 8, 12, 13, 17, 18> [10, 39, 43]; ir <13> [40]; ? <1-11, 13-15, 17> [1-9, 11-38, 41, 42, 44]) [1-44]

P CoA + acetylhistone <1-15, 17, 18> (<3> formation of N^6-acetyllysine as the only acetylation product [23]) [1-43]

S acetyl-CoA + histone H1 <1, 10, 13> (<13> acetylation of histone H1 only in vitro [29]; <1,13> histone H1 poor substrate [6,29]; <10> histone H1 is a better substrate than H3 or H4 [26]) (Reversibility: ? <1, 10, 13> [6, 26, 29]) [6, 26, 29]

P CoA + acetylhistone H1

S acetyl-CoA + histone H2A <3-5, 13, 14, 17> (<4,17> NuA4-like protein acetylates histone H4 and H2A [35]; <5> histone H2A is a substrate for enzyme form A1 [8]; <3> enzyme form B nearly exclusively acetylates histones H4 and H2A [23]) (Reversibility: ? <3, 5, 13, 14, 17> [8, 23, 30, 31, 33-35]) [8, 23, 30, 31, 33-35]

P CoA + acetylhistone H2A <4, 13, 14, 17> [30, 31, 33-35]

S acetyl-CoA + histone H2B <4, 5, 13> (<5> histone H2B: poor substrate for enzyme form A_2 [8]; <4> histone H2B: preferred substrate of enzyme form A [7]; <14> no substrate [30]) (Reversibility: ? <4, 5, 13> [7, 8, 31]) [7, 8, 31]

P CoA + acetylhistone H2B <4, 5, 13> [7, 8, 31]

S acetyl-CoA + histone H3 <1, 4-6, 8, 10-14, 18> (<4> recombinant and native SAS complex acetylates Lys14 [45]; <4,8,12,13,18> Gcn5 protein: preferred substrate, acetylation at Lys14 [43]; <4> Gcn5 protein: specific for Lys14 of histone H3 [41]; <13> PCAF protein [40]; <13> Lys14 of histone H3 and a peptide containing Lys14 thereof are the preferred substrates for Gnc5 and PCAF protein, as well as Gnc5 and PCAF catalytic domain [37-39]; <10> histone H1 is a better substrate than H3 or H4 [26]; <5> histone H3 is the preferred substrate of enzyme form A1 and A_2 [8]; <4> histone H3 preferred substrate of enzyme form A [7]; <1,4,6> histone H3: preferred substrate [6,12,19,31]; <11> enzyme form B, very low activity [27]) (Reversibility: r <4, 8, 12, 13, 18> [43]; ? <1, 4-6, 10, 11, 13, 14> [6-8, 12, 19, 26, 27, 30-34, 37-41, 45]) [6-8, 12, 19, 26, 27, 30-34, 37-41, 43, 45]

P CoA + acetylhistone H3 <4, 8, 11-14, 18> [27, 30-34, 37-39, 41, 43, 45]

S acetyl-CoA + histone H3 peptide <4, 13, 17> (<4,13,17> peptide H3p20 [40,42]; <4,13,17> peptides H3p19, H3p27, H3p11 are substrates for the catalytic domain of Gcn5 and PCAF [40]; <4,13> Lys14 of histone H3 and a peptide containing Lys14 thereof are the preferred substrates for Gnc5 and PCAF protein, as well as Gnc5 and PCAF catalytic domain [37-39]) (Reversibility: ? <4, 13, 17> [37-40, 42]) [37-40, 42]

P CoA + acetylhistone H3 peptide <4, 13> [37-40, 42]

S acetyl-CoA + histone H4 <2, 11> (<11> enzyme form B [27,28]) (Reversibility: ? <2, 11> [27, 28]) [27, 28]

P CoA + diacetylhistone H4 <2, 11> (<11> mono- and diacetylated products [27]; <2> mono-, di- and triacetylated products [28]) [27, 28]

S acetyl-CoA + histone H4 <2> (<2> enzyme form B [28]) (Reversibility: ? <2> [28]) [28]

P CoA + triacetylhistone H4 <2> (<2> mono-, di- and triacetylated products [28]) [28]

S acetyl-CoA + histone H4 <1-5, 8, 11, 13, 14, 17, 18> (<4> SAS complex, native and recombinant, acetylates Lys16 [45]; <4,8,13,18> Gcn5 protein acetylates H4 when purified and presented separately to the enzyme at Lys8 and Lys16 [43]; <13> PCAF protein [40]; <4,17> NuA4-like protein acetylates histone H4 and H2A [35]; <3> enzyme form B nearly exclusively acetylates histones H4 and H2a [23]; <8> histone H4: all of the acetate groups are introduced within the NH_2-terminal amino acids 4 to 17 [20]; <4> enzyme form B has a marked specificity for histone H4 [7]; <4,13> histone H4 is the preferred substrate [10,29,32-34]; <1> histone H4 is a poor substrate [6]; <1,5> enzyme form B is specific for histone H4 [8,21]; <2,8> highly specific for [5,20]; <2,11> enzyme form B [27,28]) (Reversibility: r <4, 8, 13, 17, 18> [43]; ? <1-5, 8, 11, 13, 14, 17> [5-8, 20, 21, 23, 27-35, 40, 45]) [5-8, 20, 21, 23, 27-35, 40, 43, 45]

P CoA + acetylhistone H4 <1-5, 8, 11, 13, 14, 17, 18> (<11> mono- and diacetylated products [27]; <2> mono-, di- and triacetylated products [28]) [5-8, 20, 21, 23, 27-35, 40, 43, 45]

S acetyl-CoA + non-histone chromatin high-mobility group protein <3> (<3> or chymotryptic peptides of [16]) (Reversibility: ? <3> [16]) [16]

P CoA + acetylated non-histone chromatin high-mobility group protein

S acetyl-CoA + poly-L-lysine <1> (<2> not [4]; <1> enzyme form A, not enzyme form B1 and B2 [1]) (Reversibility: ? <1> [1]) [1]

P CoA + N^6-acetyllysine <1> [1]

S acetyl-CoA + protamine sulfate <2, 4> (<2,4> enzyme form A, not enzyme form B [4,7]) (Reversibility: ? <2, 4> [4, 7]) [4, 7]

P ?

S acetyl-CoA + protein p53 <13> (<13> peptide of p53 is a substrate for PCAF catalytic domain [40]; <13> substrate is a DNA-binding transcription activator and a tumor suppressor [31]) (Reversibility: r <13> [43]; ? <13> [31, 40]) [31, 40, 43]

P CoA + acetylprotein p53

S acetyl-CoA + spermidine <1-3> (<2> enzyme form B, not enzyme form A [4]) (Reversibility: ? <1-3> [2, 4, 6, 16]) [2, 4, 6, 16]

P ?

S acetyl-CoA + spermine <1-3> (<2> enzyme form A inactive, enzyme form B active [4]) (Reversibility: ? <1-3> [2, 4, 6, 16]) [2, 4, 6, 16]

P ?

S acetyl-CoA + transcription factor TFIIE <13> (<13> substrate is a basal transcription factor [31]) (Reversibility: ? <13> [31]) [31]

P CoA + acetylated transcription factor TFIIE

S acetyl-CoA + transcription factor TFIIF <13> (<13> substrate is a basal transcription factor [31]) (Reversibility: ? <13> [31]) [31]

P CoA + acetylated transcription factor TFIIF

S Additional information <1, 3, 4, 8, 12, 13, 18> (<4> HeLa nucleosome or core histones are no substrate for recombinant SAS complex [45]; <4> for acetylation activity of Sas2, Sas4 is absolutely required, while Sas5 stimulate [45]; <4, 8, 12, 13, 18> Gcn5 is a coactivator of transcription [43]; <4, 8, 12, 13, 18> Gcn5 and PCAF protein are transcription cofactors [31,37-39,41,43]; <1> diaminodipropylamine and 1,3-propanediamine are no substrates [6]; <3> protamine, bovine serum albumin, and ubiquitin are no substrates [16]) [6, 16, 31, 37, 39, 41, 43, 45]

P ?

Inhibitors

$(NH_4)_2SO_4$ <3> (<3> competitive against both acetyl-CoA and histones [17]) [17]
1-(S-coenzyme A)hex-6-ene <3> [13]
1-(S-coenzyme A)propan-2-one <3> [13]
2-(S-coenzyme A)acetaldehyde <3> [13]
2-(S-coenzyme A)acetic acid <3> [13]
2-(S-coenzyme A)acetic acid thiophenyl ester <3> [13]
Brij-58 <8> (<11> no inhibition [27]) [20]
Ca^{2+} <1, 4, 10> (<4> weak [10]; <1> Mg^{2+} or Ca^{2+} required at low concentration of 5 mM, inhibition at 10-20 mM [19]) [10, 19, 26]
Co^{2+} <1> [19]
CoA <1, 4, 8, 13> (<4> competitive versus acetyl-CoA, Gnc5 protein [39]; <13> product inhibition of PCAF protein, competitive against acetyl-CoA [37]; <1> competitive [14]; <4> reversible [10]) [10, 14, 20, 37, 39]
Cu^{2+} <11> (<11> 5 mM, enzyme form B [27]) [27]
DNA <1, 2, 4, 8, 13> (<13> acetylation of histone H1 by the enzymes PCAF and GNC5 is inhibited in vivo by complexing of H1 with DNA [29]; <8> dsDNA [20]; <1> added DNA forming complexes with the histones inhibits activity [3]; <1> in vitro, enzyme form A inhibited, enzyme form B relatively insensitive [6]; <4> enzyme form A activated by low concentrations, enzyme form B inhibited [7]) [3, 6, 7, 20, 28, 29]
EDTA <1, 3> (<1> high concentrations [6]; <3> weak [2]) [2, 6]
Fe^{2+} <1, 11> (<11> 5 mM, enzyme form B [27]) [19, 27]
HC toxin <11> (<11> cyclic tetrapeptide, decreases enzyme form B expression [36]) [36]
Hg^{2+} <1> [19]
K^+ <6, 10> (<10> strong [26]; <6> 175 mM: 50% inhibition [12]) [12, 26]
Mg^{2+} <1, 4, 6, 10> (<6> 37 mM: 50% inhibition [12]; <4> weak [10]; <3> no inhibition [2]; <1> Mg^{2+} or Ca^{2+} required at low concentration, 5 mM, inhibition at 10-20 mM [19]) [10, 12, 19, 26]
Mn^{2+} <1, 10> (<3> no inhibition [2]) [19, 26]
N-[2-(S-coenzyme A)acetyl]spermidine amide <3> (<3> strong [13]) [13]
N-ethylmaleimide <1, 3, 6> (<3> enzyme form A is more sensitive than enzyme form B [23]) [1, 12, 19, 23, 25]
N^6-acetyllysine <1> (<1> competitive [14]) [14]
Na^+ <3, 6, 10> (<10> strong [26]; <3> competitive against both acetyl-CoA and histones [17]; <6> 160 mM: 50% inhibition [12]) [12, 17, 26]
Ni^{2+} <1> [19]
Nonidet P40 <8> [20]

RNA <8> [20]
Triton X-100 <8> (<11> no inhibition [27]) [20]
Zn^{2+} <1, 11> (<11> 5 mM, enzyme form B [27]) [19, 27]
acetylated histone H3 peptide <4, 13> (<4> noncompetitive versus acetyl-CoA and histone H3 [39]; <13> acetylated at Lys14, product inhibition of PCAF protein, noncompetitive against both substrates [37]) [37, 39]
bisubstrate analogue histone H3-peptide-coenzyme A <13, 17> (<13> peptide consisting of 20 and 7 amino acid residues [42]; <13> potent inhibitor, competitive versus acetyl-CoA, non-competitive versus histone H3-peptide [38]) [38, 42]
bisubstrate analogue methyl-histone H3-peptide-coenzyme A <13> (<13> peptide consisting of 20 and 7 amino acid residues [42]) [42]
desulfo-coenzyme A <4, 13> (<4> dead-end inhibitor, competitive versus acetyl-CoA, Gcn5 protein [39]; <13> dead-end inhibitor, noncompetitive versus histone H3-peptide and competitive versus actyl-CoA [38]) [38, 39]
dimethyl sulfoxide <11> (<11> irreversible at 2% v/v [27]) [27]
dithiothreitol <1> (<1> 10 mM: stimulation, 100 mM: inhibition [19]) [19]
ethanol <11> (<11> irreversible at 2% v/v [27]) [27]
etheno-CoA <4> [39]
histone <1, 8, 10> (<8> histones H2A, H2B, and H3 [20]; <10> histone H1 acetylation is inhibited by all other histone fractions [26]; <1> inhibits spermidine acetylation, enzyme forms A and B [6]) [6, 20, 26]
histone H3-peptide mutant K14A <13> (<13> dead-end inhibitor analogue, mutant histone H3 -peptide consisting of amino acid residues 3-20 K14A [38]) [38]
histone H4 <8> [20]
iodoacetamide <3, 6> (<3> enzyme form A is more sensitive than enzyme form B [23]) [12, 23]
isopropanol <11> (<11> irreversible at 2% v/v [27]) [27]
p-chloromercuribenzoate <1, 3, 4> (<1> strongly inhibits activity of enzyme form B, formation of acyl-enzyme complex [1]; <4> enzyme form B less sensitive than enzyme form A [7]) [1, 2, 7, 23]
polyarginine <8> [20]
polyglutamic acid <8> [20]
polylysine <8> [20]
potassium phosphate <8> (<8> 90 mM [20]) [20]
protamine <8> [20]
sea urchin sperm chromatin <3> [15]
spermidine <1> (<1> inhibits histone acetylation at high concentrations, enzyme forms A and B [6]) [6]
Additional information <3> (<3> monovalent cations cause a 50% decrease in activity at an average concentration of 51 mM, divalent cations at 15 mM [17]) [17]

Activating compounds

DNA <4> (<4> enzyme form A activated by low concentration, enzyme form B inhibited [7]) [7]

Sas4 <4> (<4> absolutely required for acetylation activity of Sas2 in SAS complex [45]) [45]
Sas5 <4> (<4> stimulates acetylation activity of Sas2 in SAS complex [45]) [45]
dithiothreitol <1> (<1> 10 mM: stimulation, 100 mM: inhibition [19]) [19]
spermine <9> (<9> activates acetylation of H3 [24]) [24]
Additional information <4, 10> (<4> phosphorylation of histone H3 enhances Gcn5 enzyme activity [41]; <10> requires free SH-groups [26]) [26, 41]

Metals, ions

Ca^{2+} <1> (<1> Ca^{2+} required at low concentration (5 mM), inhibition at 10-20 mM [19]) [19]
Mg^{2+} <1> (<1> Mg^{2+} required at low concentration (5 mM), inhibition at 10-20 mM [19]) [19]
NaCl <3, 4> (<4> 7fold decrease in activity between 0.15 and 1 M [39]; <3> 10 mM: 100% activity, 100 mM: 50% of activity maximum, 250 mM: less than 10% of activity maximum [25]) [10, 25, 39]
Additional information <3> (<3> addition of Mg^{2+} and Mn^{2+} has no effect on activity [2]) [2]

Turnover number (min^{-1})

0.19 <13> (histone H3-peptide p20, <13> PCAF catalytic domain, recombinant mutant Y638A [42]) [42]
0.7 <13> (acetyl-CoA, <13> PCAF protein [38]) [38]
0.7 <13> (histone H3-peptide, <13> PCAF protein [38]) [38]
1.6 <4, 17> (acetyl-CoA, <4> mutant A190S and A190T of Gcn5 [44]; <17> Gcn5 [44]) [44]
1.6 <4, 17> (histone H3, <4> mutant A190S and A190T of Gcn5 [44]; <17> Gcn5 [44]) [44]
1.7 <4, 13> (acetyl-CoA, <4,13> Gcn5 protein [39,41,44]) [39, 41, 44]
1.7 <4, 13> (histone H3-peptide, <4,13> Gcn5 protein [39,41,44]) [39, 41, 44]
2.3 <13> (acetyl-CoA, <13> PCAF [44]; <13> PCAF catalytic domain [37]) [37, 44]
2.3 <13> (histone H3-peptide, <13> PCAF [44]; <13> PCAF catalytic domain [37]) [37, 44]
2.8 <13> (acetyl-CoA, <13> PCAF catalytic domain [38]) [38]
2.8 <13> (histone H3-peptide, <13> PCAF catalytic domain [38]) [38]
3.5 <13> (histone H3-peptide p20, <13> PCAF catalytic domain, recombinant wild-type [42]) [42]
4 <13> (histone H3-peptide, <13> PCAF catalytic domain [38]) [38]
7.87 <13> (histone H3-peptide p19, <13> PCAF catalytic domain [40]) [40]
10.2 <13> (histone H3-peptide p27, <13> PCAF catalytic domain [40]) [40]
48 <6> (acetyl-CoA, <6> enzyme form type I [9]) [9]
48 <6> (histone, <6> enzyme form type I [9]) [9]
720 <6> (acetyl-CoA, <6> enzyme form type II [9]) [9]
720 <6> (histone, <6> enzyme form type II [9]) [9]
Additional information <4> (<4> Gcn5 protein mutants [41]) [41]

Specific activity (U/mg)

0.0000077 <9> (<9> enzyme form I [24]) [24]
0.0000118 <9> (<9> enzyme forms II and III [24]) [24]
0.00115 <3> (<3> enzyme form A [2]) [2]
0.0014 <3, 10> (<10> purified enzyme [26]; <3> enzyme form B [2]) [2, 26]
0.00186 <1> (<1> enzyme form B [6]) [6]
0.00196 <1> (<1> enzyme form A [6]) [6]
0.0057 <11> (<11> purified enzyme form B [27]) [27]
0.068 <4> (<4> purified enzyme [10]) [10]
Additional information <3, 11> [2, 27]

K_m-Value (mM)

0.0002 <3> (acetyl-CoA, <3> enzyme form B [25]) [25]
0.00023 <3> (acetyl-CoA, <3> enzyme form DB [25]) [25]
0.00058 <17> (acetyl-CoA, <17> Gcn5 [44]) [44]
0.00062 <13> (acetyl-CoA, <13> Gcn5 [44]) [44]
0.0007 <4> (acetyl-CoA, <4> Gcn5 mutant A190T [44]) [44]
0.0009 <4> (acetyl-CoA) [10]
0.00098 <13> (acetyl-CoA, <13> PCAF protein [44]) [44]
0.0016 <13> (acetyl-CoA, <13> recombinant PCAF catalytic domain mutant Y638A [42]; <13> recombinant wild-type PCAF catalytic domain [42]) [42]
0.0021 <4> (acetyl-CoA, <4> Gcn5 mutant A190S [44]) [44]
0.0025 <4> (acetyl-CoA, <4> Gcn5 protein [39,44]) [39, 44]
0.009 <3> (acetyl-CoA) [2]
0.01 <1> (acetyl-CoA, <1> + spermidine, enzyme form B [6]) [6]
0.015 <1> (acetyl-CoA, <1> + spermidine, enzyme form A [6]) [6]
0.046 <13> (acetyl-CoA, <13> PCAF catalytic domain [38]) [38]
0.05 <13> (histone H3-peptide, <13> PCAF catalytic domain [38]) [38]
0.18 <1> (spermidine, <1> enzyme form B [6]) [6]
0.2 <3> (spermidine) [2]
0.27 <1> (spermidine, <1> enzyme form A [6]) [6]
0.273 <4> (histone H3, <4> Gcn5 mutant A190S [44]) [44]
0.352 <4> (histone H3, <4> Gcn5 mutant A190T [44]) [44]
0.357 <17> (histone H3, <17> Gcn5 [44]) [44]
0.471 <13> (histone H3, <13> Gcn5 [44]) [44]
0.49 <4> (histone H3, <4> Gcn5 [44]) [44]
0.49 <4> (histone H3-peptide, <4> Gcn5 protein [39,41]) [39, 41]
0.532 <13> (histone H3, <13> PCAF [44]) [44]
0.75 <13> (histone H3-peptide p27, <13> PCAF catalytic domain [40]) [40]
1.12 <13> (histone H3-peptide p19, <13> PCAF catalytic domain [40]) [40]
Additional information <1, 3, 4, 6, 8, 13> (<4> Gcn5 protein mutants [41]) [12, 14-16, 20, 25, 38, 41]

K_i-Value (mM)

0.00001 <3> (N-[2-(S-coenzyme A)acetyl]spermidine amide, <3> below [13]) [13]
0.000014 <3> (1-(S-coenzyme A)hex-6-ene) [13]
0.00044 <13> (coenzyme A, <13> PCAF catalytic domain [37]) [37]

0.00066 <13> (acetylated histone H3 peptide, <13> PCAF catalytic domain [37]) [37]
0.0013 <3> (2-(S-coenzyme A)acetic acid thiophenyl ester) [13]
0.032 <3> (2-(S-coenzyme A)acetic acid) [13]
1.96 <1> (N^6-acetyllysine, <1> complexed with acetyl-CoA, versus histone [14]) [14]
2.5 <1> (N^6-acetyllysine, <1> complexed with histone, versus acetyl-CoA [14]) [14]
2.77 <1> (coenzyme A, <1> complexed with acetyl-CoA, versus histone [14]) [14]
5.43 <1> (coenzyme A) [14]
8 <3> ($(NH_4)_2SO_4$, <3> versus acetyl-CoA [17]) [17]
12 <3> ($(NH_4)_2SO_4$, <3> versus histone [17]) [17]
30 <3> (Na^+, <3> versus acetyl-CoA [17]) [17]
34 <3> (Na^+, <3> versus histone [17]) [17]
175 <6> (K^+) [12]
Additional information <4, 13> [38, 39]

pH-Optimum

6.9 <1> (<1> assay at [3]) [3]
7.4 <6> (<6> assay at [9,12]) [9, 12]
7.5 <1, 6> [12, 19]
7.8 <1, 3> (<3> assay at [13,22]; <1> enzyme form B, substrate histone [6]) [2, 6, 13, 22]
7.8-8.8 <1> (<1> enzyme form B, substrate spermidine [6]) [6]
7.9-8.6 <2> (<2> enzyme form B [5]) [5]
8 <13, 17> (<17> tGnc5 domain [42]; <13> PCAF protein [38,42]; <13,17> assay at [35,38,42]) [35, 38, 42]
8.1 <1> (<1> enzyme form A [1]) [1]
8.2 <4, 7> (<7> assay at [18]) [10, 18]
8.2-8.5 <11> (<11> enzyme form B [27]) [27]
8.3 <3> (<3> assay at [25]) [25]
8.5 <10> [26]
8.5-8.8 <1> (<1> enzyme form A, substrate spermidine [6]) [6]
8.7 <1, 3> (<1> enzyme form B1 and B2 [1]; <3> spermidine [2]) [1, 2]
8.8 <1> (<1> enzyme form A, substrate histine [6]) [6]
9 <8> [20]
Additional information <1, 3, 13> (<13> reaction kinetic is pH-dependent [37]; <1> pI: 7.8-8.2 [19]) [16, 19, 37]

pH-Range

6-9.5 <13> (<13> PCAF protein [37]) [37]
6.5-9.5 <4> (<4> less than 10% of activity maximum below pH 6.5 and above pH 9.5 [10]) [10]
6.6-8.4 <6> (<6> about 50% of activity maximum at pH 6.6 and pH 8.4 [12]) [12]

Temperature optimum (°C)

4-37 <6> [9]
25 <4> [10]
30 <13, 17> (<17> tGcn5 domain [42]; <13> PCAF protein, assay at [38,42]; <17> NuA4-like protein, assay at [35]; <13> MORF protein, assay at [32]) [32, 35, 38, 42]
37 <1-3, 6, 7, 11> (<2> about [5]; <1, 3, 6, 7> assay at [1, 3, 9, 12, 13, 18, 22, 25]; <11> enzyme form B [27]) [1, 3, 5, 9, 12, 13, 18, 22, 25, 27]

Temperature range (°C)

4-47 <6> (<6> 4-37°C: activity maximum, 47°C: 50% of activity maximum [9]) [9]

4 Enzyme Structure

Molecular weight

70000 <3> (<3> enzyme form A, gel filtration [22]) [22]
80000 <17> (<17> NuA4-like enzyme, sucrose density gradient centrifugation [35]) [35]
90000 <11> (<11> enzyme form B, gel filtration [27]) [27]
92000 <1> (<1> enzyme form B2, gel filtration [1]) [1]
94000 <6> (<6> gel filtration, glycerol gradient ultracentrifugation [12]) [12]
96000 <1> (<1> gel filtration [19]) [19]
98000 <3> (<3> enzyme form B, gel filtration [23]) [23]
99000 <1> (<1> enzyme form A, gel filtration [1]) [1]
110000 <1, 4, 6> (<4> gel filtration [10]; <1> enzyme form B1, gel filtration [1]; <6> enzyme forms I and II, gel filtration [9]) [1, 9, 10]
120000 <2, 3> (<2,3> enzyme form A, gel filtration [4,23]) [4, 23]
130000 <4> (<4> enzyme form B, gel filtration [7]) [7]
150000 <3> (<3> enzyme form A, gel filtration [2]) [2]
160000 <2, 4, 8> (<2> enzyme form B, gel filtration [5]; <4> enzyme form A, gel filtration [7]; <8> nondenaturing PAGE [20]) [5, 7, 20]
170000 <10> (<10> gel filtration, 0.2 M KCl [26]) [26]
175000 <3> (<3> enzyme form B, gel filtration [2]) [2]
200000 <3> (<3> above, enzyme form C, gel filtration [23]) [23]
280000 <10> (<10> 10 mM Tris-HCl, gel filtration [26]) [26]
Additional information <4, 6, 13, 17> (<4,13,17> amino acid sequence comparison, Gcn5 [44]; <4,13> three-dimensional structure of the ternary complex of tGcn5 domain [40,41]; <13> MORF protein contains histone acetyltransferase, and transcription activating and repressing domains [32]; <4,13> model of protein complex composition [31]; <4> enzyme is part of a bridging complex, containing Ada2 and Ada3, for gene activation [31]; <6> amino acid composition and of enzyme forms I and II are indistinguishable, specific antibodies are not cross-reactive [9]) [9, 31, 32, 40, 44]

Subunits

dimer <6, 11> (<11> 1 * 45000 + 1 * 50000, enzyme form B, SDS-PAGE [27,36]; <6> 1 * 50000 + 1 * 40000, enzyme form II, SDS-PAGE [9]; <6> 1 * 45000 + 1 * 62000, enzyme form I, SDS-PAGE [9]) [9, 27, 36]

5 Isolation/Preparation/Mutation/Application

Source/tissue

axis <2> (<2> of embryos [4]) [4]
cell culture <7> (<7> african green monkey kidney cell line CV-1 infected with SV40 [18]) [18]
embryo <8> [20]
gastrula <9> (<9> dormant [24]) [24]
hepatoma cell <8> (<8> tissue culture cell, HTC cells [20]) [20]
kidney <7> [18]
larva <10> [26]
liver <1, 3, 6> [2, 3, 6, 9, 11, 12, 14, 19]
lymphocyte <3> [25]
monocyte <13, 16> (<13,16> leukemic [32]) [32]
plasmodium <5> [8]
root <11> (<11> meristematic part of root of embryos [36]) [36]
seedling <2, 11> (<2> axe [28]; <2,11> i.e. embryo [27,28,36]) [4, 5, 27, 28, 36]
thymus <1, 3> [1, 13, 15-17, 22, 23]
Additional information <11, 13, 16> (<11> preformed enzyme form B is stored during embryogenesis to be available at the very early germination peroid [36]; <13,16> ubiquitous expression of MORF protein in human tissue [32]) [32, 36]

Localization

chromatin <1, 3, 4, 7, 9, 11> (<11> a minor part of enzyme form B is chromatin associated [36]; <3> enzyme form A [25]; <9> enzyme form I [24]; <7> associated [18]) [1, 3, 10, 15, 18, 21, 24, 25, 36]
cytoplasm <1, 2, 5, 11> (<1,2,5,11> enzyme form B [8,21,27,28,36]; <1> enzyme form CI and CII [11]) [8, 11, 21, 27, 28, 36]
nucleosome <3> (<3> closely associated with, enzyme forms B and DB [25]) [25]
nucleus <1, 3, 5-7, 9, 10, 17> (<17> macronucleus, enzyme form A [35]; <3> enzyme forms B and DB [25]; <9> enzyme forms II and III [24]; <6> enzyme forms I and II [9]; <1> enzyme form A and B [6]; <5> enzyme form A1 and A2 [8]; <1> enzyme form A [21]; <1> enzyme forms NI and NII [11]) [1-3, 6, 8, 9, 11, 12, 14, 16, 18, 19, 21, 24-26, 35]
viral capsid <7> [18]

Purification

<1> (partial, enzyme forms A and B [6]) [1, 3, 6, 14, 19]
<2> (partial, enzyme form B [28]; partial enzyme forms A and B [4]; partial, acyltransferase B [5]) [4, 5, 28]
<3> (from calf [2,22,23]; enzyme form A [22]; partial, enzyme forms DB and B [25]) [2, 22, 23, 25]
<4> (native SAS complex and recombinant SAS complex, the latter with Sas2 protein being His-tagged [45]; recombinant catalytic domain of Gcn5 protein [39-41]; recombinant wild-type Sas3 protein and mutants as GST-fusion proteins from E. coli [33]; partial, enzyme form A and B [7]) [7, 10, 33, 39-41, 45]
<5> (partial, enzyme forms A1, A2, and B [8]) [8]
<6> (enzyme forms I and II [9]; partial [12]) [9, 12]
<7> (partial from viral nucleoprotein from infected african green monkey cell line CV-1 [18]) [18]
<8> [20]
<9> [24]
<10> [26]
<11> (enzyme form B [27,36]; recombinant enzyme form B 55000 kDa subunit as His-tagged protein from E. coli [36]) [27, 36]
<13> (recombinant enzyme domains of PCAF and Gcn5 [40]; recombinant His-tagged catalytic domain of PCAF protein from E. coli [38,42]; recombinant catalytic domain of PCAF protein from E. coli [37]; recombinant from Sf9 insect cells via baculovirus infection, FLAG-tagged MORF protein [32]) [32, 37, 38, 40, 42]
<14> (recombinant His-tagged enzyme from E. coli [30]) [30]
<17> (tGcn5 domain, recombinant from E. coli [40,42]) [40, 42]

Crystallization

<4> [41]
<17> (tGcn5 domain cocrystallized with inhibitor histone H3-methyl-CoA-peptide of 20 amino acid residues, hanging drop vapour diffusion method at 20°C, 2.0 M $(NH_4)_2SO_4$, 0.1 M Na cacodylate, pH 6.6, 0.2 M NaCl, CoA is bound via isopropionyl linker to Lys14, structure analysis, modeling of conformational changes [42]) [42]

Cloning

<4> (expression in Escherichia coli of SAS complex, consisting of Sas2, Sas4, and Sas5, Sas2 is His-tagged, coexpression [45]; expression of Gcn5 and mutants thereof in Escherichia coli BL21 (DE3) [44]; overexpression of enzyme domain, yGcn5 consists of amino acid residues 99-262 [40]; expression of Gcn5 protein catalytic domain in Escherichia coli BL21 (DE3) [39,41]; expression of Esa1 gene in Escherichia coli [34]; expression of wild-type Sas3 protein and mutants as GST-fusion proteins in Escherichia coli [33]) [33, 34, 39-41, 44, 45]
<11> (HAT-B-p50 gene, DNA sequence analysis, overexpression of enzyme form B 55000 kDa subunit as His-tagged protein in Escherichia coli [36]) [36]
<13> (overexpression of enzyme domain PCAF amino acid residues 493-658 in Escherichia coli [40]; expression of the catalytic domain of PCAF protein,

amino acid residues 493-658, in Escherichia coli BL21 (DE3) [37,38,42]; expression of MORF protein as FLAG-tagged protein in Sf9 insect cells via baculovirus infection, and as Gal4-fusion protein in 293T cells, amino acid sequence [32]) [32, 37, 38, 40, 42, 44]
<14> (expression of Tip60 protein, which contains an evolutionary conserved domain with yeast silencing factor, as His-tagged protein in Escherichia coli and as GFP-fusion protein in COS cells [30]) [30]
<17> (overexpression of enzyme domain, tGcn5 consists of amino acid residues 48-210 [40]; overexpression of enzyme domain tGcn5 consisting of amino acid residues 48-210 in Escherichia coli 834 (DE3) [42]) [40, 42]

Engineering

A190S <4> (<4> site-directed mutagenesis, mutant of Gcn5, similar kinetics to wild-type [44]) [44]
A190T <4> (<4> site-directed mutagenesis, mutant of Gcn5, decreased K_m for acetyl-CoA [44]) [44]
C303A <4> (<4> site-directed mutagenesis of Sas3, exchange in zinc finger motif, remaining activity: 4.5% of wild-type activity [33]) [33]
C306A <4> (<4> site-directed mutagenesis of Sas3, exchange in zinc finger motif, remaining activity: 11.1% of wild-type activity [33]) [33]
C323A <4> (<4> site-directed mutagenesis of Sas3, exchange in zinc finger motif, no remaining activity [33]) [33]
D214A <4> (<4> site-directed mutagenesis, increased turnover [41]) [41]
E173Q <4> (<4> site-directed mutagenesis, 500-600 decrease in turnover, no effect on substrate binding, Glu173 is the general base catalyst [41]) [41]
G429E <4> (<4> site-directed mutagenesis of Sas3, exchange in acetyl-CoA binding motif, no remaining activity [33]) [33]
G431A <4> (<4> site-directed mutagenesis of Sas3, exchange in acetyl-CoA binding motif, nearly no remaining activity: 0.8% of wild-type activity [33]) [33]
H145A <4> (<4> site-directed mutagenesis, altered peptide binding [41]) [41]
H319A <4> (<4> site-directed mutagenesis of Sas3, exchange in zinc finger motif, nearly no remaining activity: 1.2% of wild-type activity [33]) [33]
K428A <4> (<4> site-directed mutagenesis of Sas3, exchange in acetyl-CoA binding motif, remaining activity: 55.9% of wild-type activity [33]) [33]
Q426A <4> (<4> site-directed mutagenesis of Sas3, exchange in acetyl-CoA binding motif, remaining activity: 38.6% of wild-type activity [33]) [33]
Y430A <4> (<4> site-directed mutagenesis of Sas3, exchange in acetyl-CoA binding motif, remaining activity: 16.7% of wild-type activity [33]) [33]
Y430L <4> (<4> site-directed mutagenesis of Sas3, exchange in acetyl-CoA binding motif, remaining activity: 87.0% of wild-type activity [33]) [33]
Y638A <13> (<13> mutant of PCAF catalytic domain, increased K_m and decreased k_{cat} compared to wild-type [42]) [42]
Additional information <4, 13> (<4> temperature sensitive Esa1 mutant strain shows reduced activity in vitro and needs histone H4 acetylated at Lys5 as substrate in vivo [34]; <4> diverse Esa1 gene mutant strains, over-

view [34]; <13> mutations that reduce histone acetylation activity also lower the transcriptional activation function of a Gal4-CBP fusion protein in vivo [31]) [31, 34]

6 Stability

pH-Stability

6-10 <4> (<4> irreversible loss of activity below pH 6.0 and above pH 10.0 [10]) [10]
6.5 <8> (<8> rapid loss of activity below [20]) [20]

Temperature stability

0-20 <10> (<10> half-life: 20 h [26]) [26]
37 <1, 6> (<1> half-life: 22.2 min [19]; <1,6> glycerol protects against thermal denaturation at 37°C [12,19]) [12, 19]
40 <10> (<10> half-life: 2 min [26]) [26]
45 <1> (<1> half-life: 9.54 min [19]) [19]
60 <4, 8> (<4,8> 1 min, irreversible denaturation [10,20]) [10, 20]
Additional information <11> (<11> enzyme is highly unstable during purification, solubilization at -10°C [27]) [27]

Organic solvent stability

dimethyl sulfoxide <11> (<11> irreversible inactivation at 2% v/v [27]) [27]
ethanol <11> (<11> irreversible inactivation at 2% v/v [27]) [27]
isopropanol <11> (<11> irreversible inactivation at 2% v/v [27]) [27]

General stability information

<1>, dialysis destabilizes [3]
<4>, NaCl, 1.0 M, stable for several days [10]
<4>, sensitive to freezing [10]
<6>, glycerol protects against thermal denaturation at 37°C [12]
<13>, acetyl-CoA stabilizes the enzyme form PCAF [29]

Storage stability

<1>, 0°C, 3 M KCl, 1 month [3]
<3>, -80°C, stable for at least 6 months [22]
<4>, -20°C, 50% glycerol, stable for several months [10]
<4>, 4°C, stable for several weeks [7]
<17>, -70°C, crude extract, 25 mM Tris-HCl, after treatment with DNase I, 10% glycerol, stable for months [35]

References

[1] Gallwitz, D.; Sures, I.: Histone acetylation. Purification and properties of three histone-specific acetyltransferases from rat thymus nuclei. Biochim. Biophys. Acta, **263**, 315-328 (1972)

[2] Libby, P.R.: Calf liver nuclear N-acetyltransferases. Purification and properties of two enzymes with both spermidine acetyltransferase and histone acetyltransferase activities. J. Biol. Chem., **253**, 233-237 (1978)

[3] Fukushima, M.; Ota, K.; Fujimoto, D.; Horiuchi, K.: Nucleosome-histone acetyltransferase from rat liver chromatin. Biochem. Biophys. Res. Commun., **92**, 1409-1414 (1980)

[4] Sendra, R.; Salvador, M.L.; Lopez-Rodas, G.; Tordera, V.; Franco, L.: A plant histone acetyltransferase specific for H3 in nucleosomes. Plant Sci., **46**, 189-194 (1986)

[5] Salvador, M.L.; Sendra, R.; Lopez-Rodas, G.; Tordera, V.; Franco, L.: On the ubiquitous presence of histone acetyltransferase B in eukaryotes. FEBS Lett., **191**, 55-58 (1985)

[6] Libby, P.R.: Rat liver nuclear N-acetyltransferases: separation of two enzymes with both histone and spermidine acetyltransferase activity. Arch. Biochem. Biophys., **203**, 384-389 (1980)

[7] Lopez-Rodas, G.; Perez-Ortin, J.E.; Tordera, V.; Salvador, M.L.; Franco, L.: Partial purification and properties of two histone acetyltransferases from the yeast, Saccharomyces cerevisiae. Arch. Biochem. Biophys., **239**, 184-190 (1985)

[8] Lopez-Rodas, G.; Brosch, G.; Golderer, G.; Lindner, H.; Grobner, P.; Loidl, P.: Enzymes involved in the dynamic equilibrium of core histone acetylation of Physarum polycephalum. FEBS Lett., **296**, 82-86 (1992)

[9] Attisano, L.; Lewis, P.N.: Purification and characterization of two porcine liver nuclear histone acetyltransferases. J. Biol. Chem., **265**, 3949-3955 (1990)

[10] Travis, G.H.; Colavito-Shepanski, M.; Grunstein, M.: Extensive purification and characterization of chromatin-bound histone acetyltransferase from Saccharomyces cerevisiae. J. Biol. Chem., **259**, 14406-14412 (1984)

[11] Yukioka, M.; Sasaki, S.; Qi, S.L.; Inoue, A.: Two species of histone acetyltransferase in rat liver nuclei. J. Biol. Chem., **259**, 8372-8377 (1984)

[12] Kelner, D.N.; McCarty, K.S.: Porcine liver nuclear histone acetyltransferase. Partial purification and basic properties. J. Biol. Chem., **259**, 3413-3419 (1984)

[13] Cullis, P.M.; Wolfenden, R.; Cousens, L.S.; Alberts, B.M.: Inhibition of histone acetylation by N-[2-(S-coenzyme A)acetyl] spermidine amide, a multisubstrate analog. J. Biol. Chem., **257**, 12165-12169 (1982)

[14] Wiktorowicz, J.E.; Campos, K.L.; Bonner, J.: Substrate and product inhibition initial rate kinetics of histone acetyltransferase. Biochemistry, **20**, 1464-1467 (1981)

[15] Wong, L.J.C.: Effect of sea urchin sperm chromatin on histone acetylation. Biochem. Biophys. Res. Commun., **97**, 1362-1369 (1980)

[16] Wong, L.J.C.; Sharpe, D.J.; Wong, S.S.: High-mobility group and other nonhistone substrates for nuclear histone N-acetyltransferase. Biochem. Genet., **29**, 461-475 (1991)

[17] Wong, L.J.; Patton, W.F.: Salt inhibition of nuclear histone acetyltransferase from calf thymus. Int. J. Biochem., **17**, 123-126 (1985)

[18] Otto, B.; Bohm, J.; Knippers, R.: A histone-specific acetyltransferase is associated with simian-virus-40 chromatin. Eur. J. Biochem., **112**, 363-366 (1980)
[19] Wiktorowicz, J.E.; Bonner, J.: Studies on histone acetyltransferase. Partial purification and basic properties. J. Biol. Chem., **257**, 12893-12900 (1982)
[20] Wiegand, R.C.; Brutlag, D.L.: Histone acetylase from Drosophila melanogaster specific for H4. J. Biol. Chem., **256**, 4578-4583 (1981)
[21] Garcea, R.L.; Alberts, B.M.: Comparative studies of histone acetylation in nucleosomes, nuclei, and intact cells. Evidence for special factors which modify acetylase action. J. Biol. Chem., **255**, 11454-11463 (1980)
[22] Belikoff, E.; Wong, L.J.; Alberts, B.M.: Extensive purification of histone acetylase A, the major histone N-acetyl transferase activity detected in mammalian cell nuclei. J. Biol. Chem., **255**, 11448-11453 (1980)
[23] Sures, I.; Gallwitz, D.: Histone-specific acetyltransferases from calf thymus. Isolation, properties, and substrate specificity of three different enzymes. Biochemistry, **19**, 943-951 (1980)
[24] Estepa, I.; Pestana, A.: Isolation and partial characterization of three histone-specific acetyltransferases from Artemia. Eur. J. Biochem., **132**, 249-254 (1983)
[25] Bohm, J.; Schlaeger, E.J.; Knippers, R.: Acetylation of nucleosomal histones in vitro. Eur. J. Biochem., **112**, 353-362 (1980)
[26] Cano, A.; Pestana, A.: Purification and properties of a histone acetyltransferase from Artemia salina, highly efficient with H1 histone. Eur. J. Biochem., **97**, 65-72 (1979)
[27] Eberharter, A.; Lechner, T.; Goralik-Schramel, M.; Loidl, P.: Purification and characterization of the cytoplasmic histone acetyltransferase B of maize embryos. FEBS Lett., **386**, 75-81 (1996)
[28] Mingarro, I.; Sendra, R.; Salvador, M.L.; Franco, L.: Site specificity of pea histone acetyltransferase B in vitro. J. Biol. Chem., **268**, 13248-13252 (1993)
[29] Herrera, J.E.; Bergel, M.; Yang, X.J.; Nakatani, Y.; Bustin, M.: The histone acetyltransferase activity of human GCN5 and PCAF is stabilized by coenzymes. J. Biol. Chem., **272**, 27253-27258 (1997)
[30] Yamamoto, T.; Horikoshi, M.: Novel substrate specificity of the histone acetyltransferase activity of HIV-1-Tat interactive protein Tip60. J. Biol. Chem., **272**, 30595-30598 (1997)
[31] Berger, S.L.: Gene activation by histone and factor acetyltransferases. Curr. Opin. Cell Biol., **11**, 336-341 (1999)
[32] Champagne, N.; Bertos, N.R.; Pelletier, N.; Wang, A.H.; Vezmar, M.; Yang, Y.; Heng, H.H.; Yang, X.J.: Identification of a human histone acetyltransferase related to monocytic leukemia zinc finger protein. J. Biol. Chem., **274**, 28528-28536 (1999)
[33] Takechi, S.; Nakayama, T.: Sas3 is a histone acetyltransferase and requires a zinc finger motif. Biochem. Biophys. Res. Commun., **266**, 405-410 (1999)
[34] Clarke, A.S.; Lowell, J.E.; Jacobson, S.J.; Pillus, L.: Esa1p is an essential histone acetyltransferase required for cell cycle progression. Mol. Cell. Biol., **19**, 2515-2526 (1999)

[35] Ohba, R.; Steger, D.J.; Brownell, J.E.; Mizzen, C.A.; Cook, R.G.; Cote, J.; Workman, J.L.; Allis, C.D.: A novel H2A/H4 nucleosomal histone acetyltransferase in Tetrahymena thermophila. Mol. Cell. Biol., **19**, 2061-2068 (1999)

[36] Lusser, A.; Eberharter, A.; Loidl, A.; Goralik-Schramel, M.; Horngacher, M.; Haas, H.; Loidl, P.: Analysis of the histone acetyltransferase B complex of maize embryos. Nucleic Acids Res., **27**, 4427-4435 (1999)

[37] Tanner, K.G.; Langer, M.R.; Denu, J.M.: Kinetic mechanism of human histone acetyltransferase P/CAF. Biochemistry, **39**, 11961-11969 (2000)

[38] Lau, O.D.; Courtney, A.D.; Vassilev, A.; Marzilli, L.A.; Cotter, R.J.; Nakatani, Y.; Cole, P.A.: P300/CBP-associated factor histone acetyltransferase processing of a peptide substrate. Kinetic analysis of the catalytic mechanism. J. Biol. Chem., **275**, 21953-21959 (2000)

[39] Tanner, K.G.; Langer, M.R.; Kim, Y.; Denu, J.M.: Kinetic mechanism of the histone acetyltransferase GCN5 from yeast. J. Biol. Chem., **275**, 22048-22055 (2000)

[40] Trievel, R.C.; Li, F.Y.; Marmorstein, R.: Application of a fluorescent histone acetyltransferase assay to probe the substrate specificity of the human p300/CBP-associated factor. Anal. Biochem., **287**, 319-328 (2000)

[41] Langer, M.R.; Tanner, K.G.; Denu, J.M.: Mutational analysis of conserved residues in the GCN5 family of histone acetyltransferases. J. Biol. Chem., **276**, 31321-31331 (2001)

[42] Poux, A.N.; Cebrat, M.; Kim, C.M.; Cole, P.A.; Marmorstein, R.: Structure of the GCN5 histone acetyltransferase bound to a bisubstrate inhibitor. Proc. Natl. Acad. Sci. USA, **99**, 14065-14070 (2002)

[43] Hasan, S.; Hottiger, M.O.: Histone acetyl transferases: a role in DNA repair and DNA replication. J. Mol. Med., **80**, 463-474 (2002)

[44] Langer, M.R.; Fry, C.J.; Peterson, C.L.; Denu, J.M.: Modulating acetyl-CoA binding in the GCN5 family of histone acetyltransferases. J. Biol. Chem., **277**, 27337-27344 (2002)

[45] Sutton, A.; Shia, W.J.; Band, D.; Kaufman, P.D.; Osada, S.; Workman, J.L.; Sternglanz, R.: Sas4 and Sas5 are required for the histone acetyltransferase activity of Sas2 in the SAS complex. J. Biol. Chem., **278**, 16887-16892 (2003)

Deacetyl-[citrate-(pro-3S)-lyase] S-acetyltransferase 2.3.1.49

1 Nomenclature

EC number
2.3.1.49

Systematic name
S-acetylphosphopantetheine:deacetyl-[citrate-oxaloacetate-lyase((pro-3S)-$CH_2COO^- \rightarrow$acetate)] S-acetyltransferase

Recommended name
deacetyl-[citrate-(pro-3S)-lyase] S-acetyltransferase

Synonyms
S-acetyl phosphopantetheine:deacetyl citrate lyase S-acetyltransferase
acetyltransferase, citrate lyase
deacetyl-[citrate-(pro-3S)-lyase] acetyltransferase

CAS registry number
42616-18-2

2 Source Organism

<1> *Klebsiella aerogenes* [1]

3 Reaction and Specificity

Catalyzed reaction
S-acetylphosphopantotheine + deacetyl-[citrate-oxaloacetate-lyase((pro-3S)-$CH_2COO^- \rightarrow$acetate)] = phosphopantotheine + [citrate-oxaloacetate-lyase((-pro-3S)-$CH_2COO^- \rightarrow$acetate)]

Reaction type
acyl group transfer

Substrates and products
S S-acetylphosphopantetheine + deacetyl-[citrate-oxaloacetate-lyase((pro-3S)-$CH_2COO^- \rightarrow$acetate)] <1> (Reversibility: ? <1> [1]) [1]
P phosphopantetheine + [citrate-oxaloacetate-lyase((pro-3S)-$CH_2COO^- \rightarrow$ acetate)] <1> (<1> i.e. active form of EC 4.1.3.6 [1]) [1]

Inhibitors

N-ethylmaleimide <1> (<1> preincubation of the acetyl transferase with 0.4 mM N-ethylmaleimide results in large decrease of active citrate lyase formed [1]) [1]

Specific activity (U/mg)

17.4 <1> [1]

pH-Optimum

7.7 <1> [1]

pH-Range

6-8 <1> (<1> 25% of maximal activity at pH 6, 95% of maximal activity at pH 8 [1]) [1]

Temperature optimum (°C)

25 <1> (<1> assay at [1]) [1]

5 Isolation/Preparation/Mutation/Application

Purification

<1> (partial [1]) [1]

6 Stability

Oxidation stability

<1>, O_2-sensitive, DTT stabilizes [1]

General stability information

<1>, DTT stabilizes [1]

References

[1] Singh, M.; Böttger, B.; Stewart, C.; Brooks, G.C.; Srere, P.A.: S-Acetyl phosphopantetheine: deacetyl citrate lyase S-acetyl transferase from Klebsiella aerogenes. Biochem. Biophys. Res. Commun., **53**, 1-9 (1973)

Serine C-palmitoyltransferase 2.3.1.50

1 Nomenclature

EC number
2.3.1.50

Systematic name
palmitoyl-CoA:L-serine C-palmitoyltransferase (decarboxylating)

Recommended name
serine C-palmitoyltransferase

Synonyms
3-oxosphinganine synthetase
SPT
acyl-CoA:serine C-2 acyltransferase decarboxylating
palmitoyltransferase, serine
serine palmitoyltransferase
serine-palmitoyltransferase
Additional information <5> (<5> the amount of enzyme activity in microsomes is inversely correlated with the amount of the major coat Gag protein of the L-A dsRNA virus particles [12]) [12]

CAS registry number
62213-50-7

2 Source Organism

<1> *Hansenula ciferrii* [1-3]
<2> *Rattus norvegicus* [2, 5, 8-10]
<3> *Homo sapiens* (gene hLCB1 and hLCB2 [11]) [4, 11, 18, 19]
<4> *Cricetulus griseus* (mutant defective strain LY-B [15,21]; genetically corrected revertant strain LY-B/cLCB1 of deficient mutant LY-B [15]) [10, 15, 21]
<5> *Saccharomyces cerevisiae* (lcb1 and lcb2 [19]; genes LCB1/TSC2, LCB2/TSC1, TSC3 [16]; gene LCB1/SCS1 [12]) [5, 6, 12, 16, 19]
<6> *Bacteroides melaninogenicus* [5, 7]
<7> *Mus musculus* (gene mLCB2 [11]) [11, 14, 19]
<8> *Homo sapiens* (hLVB1) [11]
<9> *Homo sapiens* (hLVB2) [11]
<10> *Curcurbita pepo* (cv. early profilic straightneck [13]) [13]
<11> *Sphingomonas paucimobilis* (strain EY2395 [17]) [17]

<12> *Homo sapiens* (gene hLCB1 [18]) [18]
<13> *Homo sapiens* (gene hLCB2 [18]) [18]
<14> *Sphingomonas paucimobilis* [20]
<15> *Homo sapiens* (human SPT1 [22]) [22]
<16> *Homo sapiens* (human SPT2 [22]) [22]

3 Reaction and Specificity

Catalyzed reaction

palmitoyl-CoA + L-serine = CoA + 3-dehydro-D-sphinganine + CO_2 (a pyridoxal-phosphate protein; <14> formation of the external aldimine intermediate [20]; <11> ordered Bi-Bi mechanism [17]; <1,3,5> mechanism [1,19])

Reaction type

acyl group transfer
condensation <1-3,5,7,11,14> [2,5,11,16,17,19,20]
decarboxylation

Natural substrates and products

S palmitoyl-CoA + L-serine <1-7, 10, 12-14> (<3,5> involved in cellular stress response [19]; <1-5,7> first step of biosynthesis of sphingolipid bases [1,8-11,18,19]; <3,5,14> rate-limiting enzyme in synthesis of sphingolipids [4,18-20]; <2,3,5,6> initial step of de novo ceramide biosynthesis [5,18,19]) (Reversibility: ir <1, 2, 5, 6, 10, 12, 13> [1, 2, 5, 9, 13, 16, 18]; ? <3-5, 7, 14> [4, 10-12, 15, 19, 20]) [1, 2, 4, 5, 8-13, 15, 16, 18-20]

P CoA + 3-dehydro-D-sphinganine + CO_2 <1-3, 5-7, 10, 14> [1, 2, 4, 5, 9, 11, 13, 16, 20]

Substrates and products

S arachidoyl-CoA + L-serine <11> (<11> 37% activity compared to that with palmitoyl-CoA [17]) (Reversibility: ? <11> [17]) [17]

P CoA + 2-amino-1-hydroxydocosan-3-one + CO_2

S elaidoyl-CoA + L-serine <11> (<11> 39% activity compared to that with palmitoyl-CoA [17]) (Reversibility: ? <11> [17]) [17]

P CoA + 2-amino-1-hydroxy-trans-11-eicosen-3-one + CO_2

S lauroyl-CoA + L-serine <11> (<11> 18% activity compared to that with palmitoyl-CoA [17]) (Reversibility: ? <11> [17]) [17]

P CoA + 2-amino-1-hydroxytetradecan-3-one + CO_2

S myristoleoyl-CoA + L-serine <11> (<11> 46% activity compared to that with palmitoyl-CoA [17]) (Reversibility: ? <11> [17]) [17]

P CoA + 2-amino-1-hydroxy-cis-11-hexadecen-3-one + CO_2

S myristoyl-CoA + L-serine <11> (<11> 75% activity compared to that with palmitoyl-CoA [17]) (Reversibility: ? <11> [17]) [17]

P CoA + 2-amino-1-hydroxyhexadecan-3-one + CO_2

S n-heptadecanoyl-CoA + L-serine <11> (<11> 75% activity compared to that with palmitoyl-CoA [17]) (Reversibility: ? <11> [17]) [17]

P CoA + 2-amino-1-hydroxynonadecan-3-one + CO_2
S oleoyl-CoA + L-serine <11> (<11> 57% activity compared to that with palmitoyl-CoA [17]) (Reversibility: ? <11> [17]) [17]
P CoA + 2-amino-1-hydroxy-11-cis-eicosen-3-one + CO_2
S palmitoleoyl-CoA + L-serine <11> (<11> 80% activity compared to that with palmitoyl-CoA [17]) (Reversibility: ? <11> [17]) [17]
P CoA + 2-amino-1-hydroxy-cis-11-octadecen-3-one + CO_2
S palmitoyl-CoA + L-serine <1-7, 10, 11, 12-14> (<10> optimal palmitoyl-CoA concentration is 0.2 mM [13]; <2,10,11> palmitoyl-CoA is the preferred substrate [5,8,9,13,17]; <2,11> no other amino acids can substitute for serine [5,17]; <2> activities are greatest with palmitoyl-CoA and palmitelaidoyl-CoA, followed by fully saturated homologs, activity considerably diminishes as the alkyl-chain length increases or decreases, or with the presence of a cis-double bond [8]; <3,10> palmitoyl-CoA is used in preference to other saturated or unsaturated acyl-CoA substrates [4,13]) (Reversibility: ir <1, 2, 5, 6, 10, 12, 13> [1, 2, 5, 9, 13, 16, 18]; ? <2, 3-5, 7, 11, 14> [3, 4, 6-8, 10-12, 14, 15, 17, 19, 20]) [1-20]
P CoA + 3-dehydro-D-sphinganine + CO_2 <1-3, 5, 7, 10, 11, 14> (<1, 2> i.e. 2-amino-1-hydroxyoctadecane-3-one, i.e. 3-oxo-dihydroxysphingosine [2]) [2, 11, 13, 16, 17, 19, 20]
S stearoyl-CoA + L-serine <11> (<11> 51% activity compared to that with palmitoyl-CoA [17]) (Reversibility: ? <11> [17]) [17]
P CoA + 2-amino-1-hydroxyeicosan-3-one + CO_2
S Additional information <2, 3> (<3> increasing the acyl-CoA chain length above C_{16} by 1 or 2 carbons is less detrimental to activity than similar decrements in chain length [4]; <2> specificity [8,9]) [4, 8, 9]
P ?

Inhibitors

3-hydroxypropionate <11> (<11> competitive to L-serine [17]) [17]
L-cycloserine <2-5, 7, 10> (<4> cytotoxic, irreversible [15]; <10> mechanism-based inhibition [13]) [4, 5, 11, 13, 15, 19]
O-phosphoserine <11> (<11> competitive to L-serine [17]) [17]
α-methyl-DL-serine <11> (<11> competitive to L-serine [17]) [17]
β-chloro-L-alanine <2-5, 7, 10> (<4> cytotoxic, irreversible [15]; <10> mechanism-based inhibition [13]; <2> stereospecific, no inhibition by β-chloro-D-alanine [10]; <4> in vivo, time dependent [10]; <2> rapid, irreversible and time dependent inhibition at the active site [10]; <2> suicide substrate [10]) [4, 10, 11, 13, 15, 19]
β-haloalanines <2> [5]
cis-4-methylsphingosine <7> (<7> time- and concentration-dependent, causes drastic morphological changes of the cells in vivo [14]) [14]
cis-5-methylsphingosine <7> (<7> weak inhibition [14]) [14]
cis-sphingosine <7> (<7> weak inhibition [14]) [14]
cysteine <11> (<11> competitive to L-serine [17]) [17]
lipoxamycin <3, 7> [11]

myriocin <3-5, 7> (<4> strong inhibition, reversible by sphingosine [15]; <4> i.e. sphingofungin B [15]) [11, 15, 19]
palmitoyl-CoA <2> (<11> no inhibition at concentrations up to 10 mM [17]; <2> 0.5-1.0 mM, substrate inhibition [9]) [9]
serine methylester <11> (<11> competitive to L-serine [17]) [17]
thermozymocidin <4> (<4> strong inhibition, reversible by sphingosine [15]; <4> i.e. ISP-1 [15]) [15]
threonine <11> (<11> competitive to L-serine [17]) [17]
trans-4-methylsphingosine <7> [14]
trans-5-methylsphingosine <7> (<7> weak inhibition [14]) [14]
Additional information <11-13> (<12,13> time-dependent degradation of enzyme mRNA by etoposide, activation of enzyme activity on protein level, time-dependent [18]; <11> not inhibited by halide ions [17]) [17, 18]

Cofactors/prosthetic groups

pyridoxal 5'-phosphate <1-5, 7-11, 14> (<11> bound to ε-amino group of a lysine residue in the active site [17]; <2,10> stimulates [9,13]; <3,5> required for activity [5,19]; <1> a pyridoxal phosphate protein [3]; <2> pyridoxal 5'-phosphate can be resolved from the enzyme by dialysis [5]) [3, 5, 9-11, 13, 17, 19, 20]

Activating compounds

EDTA <10> (<10> required for enzyme activity [13]) [13]
dithiothreitol <2, 10> (<2,10> required for enzyme activity [9,13]) [9, 13]
etoposide <3, 12, 13> (<3> enhanced activity due to chemotherapy with epotoside as therapeutic agent in Molt-4 leukemic cells [19]; <12,13> activates enzyme activity, but decreases mRNA level, time-dependent [18]) [18, 19]
fenretinide <3> (<3> up-regulates enzyme activity in neuroblastoma cells [19]) [19]
retinoic acid <7> (<7> up-regulates enzyme activity in embryonic carcinoma cell ine PCC7-Mz1 [19]) [19]
Additional information <3, 7> (<3,7> enzyme activity is enhanced under stress and in apoptosis, e.g. in pancreatic cells in a model for diabetes, in case of angiotensin II type receptor occupancy [19]) [19]

Metals, ions

Additional information <2> (<2> NaCl, LiCl, $MgCl_2$, $MnCl_2$ have no effect on activity [9]) [9]

Turnover number (min^{-1})

2.33 <11> (palmitoyl-CoA) [17]
3.02 <11> (palmitoyl-CoA, <11> recombinant enzyme [17]) [17]

Specific activity (U/mg)

0.000018 <3> (<3> HEK cell extract [11]) [11]
0.000044 <3, 7> (<3,7> HEK cells transfected with mLCB2, crude extract [11]) [11]
0.000063 <2> [9]
0.00027 <3> [4]

0.057-0.084 <10> [13]
0.7 <11> (<11> purified enzyme [17]) [17]
Additional information <2, 5> (<5> activity in wild-type and tsc3 mutants [16]; <2> assay method [5]) [5, 10, 16]

K_m-Value (mM)
0.1-1 <2> (L-serine) [5]
0.4 <3> (L-serine) [4]
0.67 <2> (L-serine) [9]
0.87 <11> (palmitoyl-CoA, <11> native and recombinant enzyme [17]) [17]
1.8 <10> (L-serine) [13]
4.2 <11> (L-serine, <11> native enzyme [17]) [17]
10.6 <11> (L-serine, <11> recombinant enzyme [17]) [17]

pH-Optimum
7.4 <7> (<7> assay at [14]) [14]
7.5-8 <11> [17]
7.6 <10> [13]
8 <2> [9]
8.2 <3> [4]
8.3 <2, 5, 12, 13> (<2,5,12,13> assay at [8,16,18]) [8, 16, 18]

pH-Range
6.7-8.6 <2> (<2> about 50% of activity maximum at pH 6.7 and pH 8.6 [9]) [9]

Temperature optimum (°C)
25 <2> (<2> assay at [8]) [8]
30 <1> (<1> assay at [1]) [1]
36-40 <10> [13]
37 <2, 5, 7, 11-13> (<2,5,7,11-13> assay at [5,14,16-18]) [5, 14, 16-18]
50 <2> [9]

4 Enzyme Structure

Molecular weight
90000 <11, 14> (<11,14> gel filtration [17,20]) [17, 20]
Additional information <11> (<11> amino acid sequence determination [17]) [17]

Subunits
? <5> (<5> x * 90000, SDS-PAGE [12]) [12]
dimer <11, 14> (<14> 2 * 45041, DNA sequence determination [20]; <11> 2 * 44916, electrospray mass spectrometry [17]; <11,14> 2 * 50000, SDS-PAGE [17,20]) [17, 20]
Additional information <5> (<5> lcb1 and lcb2 are 2 subunits, both necessary for enzyme activity, complexed in ratio 1:1 [19]; <5> Lcbp1 and Lcp2 are 2 subunits of the enzyme, tightly associated and not stable independently, tsc3 protein is loosely associated to them [16]) [16, 19]

Posttranslational modification

no glycoprotein <7> [11]
Additional information <2> (<2> carrier lipid during extraction required for activity [9]) [9]

5 Isolation/Preparation/Mutation/Application

Source/tissue

brain <2> [8]
carcinoma cell <7> (<7> embryonic cell line PCC7-Mz1 [19]) [19]
cell culture <3, 12, 13> (<3> neuroblastoma cell line [19]; <3,12,13> leukemia Molt-4 cells [18,19]) [18, 19]
cerebellum <7, 15, 16> (<15,16> Purkinje cells [22]; <7> granule cells, neuronal cells [14]) [14, 22]
cerebral cortex <15, 16> [22]
endothelium <15, 16> (<15,16> vascular [22]) [22]
epithelium <15, 16> (<15,16> except for ovarian epithelium [22]) [22]
fruit <10> (<10> age-dependent amount of enzyme activity [13]) [13]
keratinocyte <3> (<3> neonatal [4]) [4, 19]
kidney <3, 7> (<3> human embryonic kidney cell line HEK 293 [11]) [11]
liver <2> [2, 9, 10]
lung <7> [11]
macrophage <15, 16> (<15,16> mucosal from colon, lung, prostate, stomach, thyroid, uterus, vascular tissue [22]) [22]
neuron <15, 16> (<15,16> pyramidal, cytoplasm [22]) [22]
ovary <4> (<4> cell culture [10,15,21]) [10, 15, 21]
pancreas <3> (<3> β-cells [19]) [19]
smooth muscle <15, 16> [22]
Additional information <15, 16> (<15,16> not in neuronal cells like astrocytes, microglia, and oligodendritic cells, not in skin and heart [22]) [22]

Localization

cytosol <11, 14, 15, 16> (<15,16> neuronal cytoplasm [22]) [17, 20, 22]
endoplasmic reticulum <1, 3-5, 10> (<4> the N-terminal is orientated to the lumenal side, the C-terminal is oriented to the cytosolic side [21]; <3, 5> cytosolic side of the membrane [19]) [1, 2, 13, 19, 21]
microsome <1-3, 5, 10, 15, 16> (<15,16> recombinant from HEK 293 cells [22]; <5> membrane associated [16]; <5> microsome isolation from mutant strains [12]) [2, 4, 5, 8-10, 12, 13, 16, 19, 22]

Purification

<6> [7]
<11> [17]
<14> (native and recombinant from E. coli, LCB1 and LCB2 co-immunoprecipitate [20]) [20]

Cloning

<3> (hLCB1 and hLCB2 [11]) [11]
<4> (expression of FLAG-tagged LCB1 in the deficient mutant LY-B, overproduction of LCB2 requires co-overproduction of LCB1 [21]) [21]
<5> (genes LCB1 and LCB2, overexpression [19]; gene tsc3 [16]) [5, 6, 16, 19]
<7> (in vitro transcription and translation of mLCB2 [11]; overexpression of mLCB1 and mLCB2 in human embryonic kidney cells, independently and in coexpression with mLCB1 [11]; expression of mLCB1 and LCB2 as glutathione-S-transferase fusion protein in Escherichia coli DH5α [11]) [11]
<8> [11]
<9> [11]
<11> (overexpression in Escherichia coli BL21 (DE3) pLysS, DNA sequence determination [17]) [17]
<12> (RT-PCR from mRNA of Molt-4 cells [18]) [18]
<13> (RT-PCR from mRNA of Molt-4 cells [18]) [18]
<14> (overexpression in Escherichia coli BL21 (DE3) pLysS, DNA sequence determination [20]) [20]
<15> (expression in HEK 293 cells [22]) [22]
<16> (expression in HEK 293 cells [22]) [22]

Engineering

C133V <3> (<3> reduced activity [19]) [19]
C133Y <3> (<3> reduced activity [19]) [19]
V144D <3> (<3> reduced activity [19]) [19]
Additional information <4, 5> (<4> deficient mutant strain LY-B lacks LCB1 and has reduced content in LCB2, the latter is restored upon recombinant expression of LCB2 [21]; <5> tsc3 mutants show reduced enzyme activity, tsc3 is not required for expression, stability and membrane association of Lcb1p and Lcbp2 [16]; <5> natural mutant strains: SCS1/LCB2 knock-out mutant, scs1-1 mutant shows reduced activity, scs1-2 mutant is temperature-sensitive, but shows normal enzyme activity [12]) [12, 16, 21]

Application

analysis <11, 14> (<11,14> useful as a model to elucidate the reaction mechanism [17,20]) [17, 20]

6 Stability

Temperature stability

37 <2> (<2> 30 min, stable [9]) [9]
50 <2> (<2> above, precipitation of enzyme [9]) [9]

Storage stability

<11>, 4°C, Tris-HCl 20 mM, in sterile capped vials, 2 months without loss of activity [17]

References

[1] Brady, R.N.; Di Mari, S.J.; Snell, E.E.: Biosynthesis of sphingolipid bases. 3. Isolation and characterization of ketonic intermediates in the synthesis of sphingosine and dihydrosphingosine by cell-free extracts of Hansenula ciferri. J. Biol. Chem., **244**, 491-496 (1969)

[2] Stoffel, W.; LeKim, D.; Sticht, G.: Biosynthesis of dihydrosphingosine in vitro. Hoppe-Seyler's Z. Physiol. Chem., **349**, 664-670 (1968)

[3] Braun, P.E.; Snell, E.E.: The biosynthesis of dihydrosphingosine in cell-free preparations of Hansenula ciferri. Proc. Natl. Acad. Sci. USA, **58**, 298-303 (1967)

[4] Holleran, W.M.; Williams, M.L.; Gao, W.N.; Elias, P.M.: Serine-palmitoyl transferase activity in cultured human keratinocytes. J. Lipid Res., **31**, 1655-1661 (1990)

[5] Merrill, A.H.; Wang, E.: Enzymes of ceramide biosynthesis. Methods Enzymol., **209**, 427-437 (1992)

[6] Buede, R.; Rinker-Schaffer, C.; Pinto, W.J.; Lester, R.L.; Dickinson, R.C.: Cloning and characterization of LCB1, a Saccharomyces gene required for biosynthesis of the long-chain base component of sphingolipids [published erratum appears in J Bacteriol 1993 Feb;175(3):919]. J. Bacteriol., **173**, 4325-4332 (1991)

[7] Lev, M.; Milford, A.F.: The 3-ketodihydrosphingosine synthetase of Bacteroides melaninogenicus: partial purification and properties. Arch. Biochem. Biophys., **212**, 424-431 (1981)

[8] Merrill, A.H.; Williams, R.D.: Utilization of different fatty acyl-CoA thioesters by serine palmitoyltransferase from rat brain. J. Lipid Res., **25**, 185-188 (1984)

[9] Williams, R.D.; Wang, E.; Merrill, A.H.: Enzymology of long-chain base synthesis by liver: characterization of serine palmitoyltransferase in rat liver microsomes. Arch. Biochem. Biophys., **228**, 282-291 (1984)

[10] Medlock, K.A.; Merrill, A.H.: Inhibition of serine palmitoyltransferase in vitro and long-chain base biosynthesis in intact Chinese hamster ovary cells by β-chloroalanine. Biochemistry, **27**, 7079-7084 (1988)

[11] Weiss, B.; Stoffel, W.: Human and murine serine-palmitoyl-CoA transferase. Cloning, expression and characterization of the key enzyme in sphingolipid synthesis. Eur. J. Biochem., **249**, 239-247 (1997)

[12] Garnepudi, V.R.; Zhao, C.; Beeler, T.; Dunn, T.: Serine palmitoyltransferase (scs1/lcb2) mutants have elevated copy number of the L-A dsRNA virus. Yeast, **13**, 299-304 (1997)

[13] Lynch, D.V.; Fairfield, S.R.: Sphingolipid long-chain base synthesis in plants. Characterization of serine palmitoyltransferase activity in squash fruit microsomes. Plant Physiol., **103**, 1421-1429 (1993)

[14] Van Echten-Deckert, G.; Zschoche, A.; Bar, T.; Schmidt, R.R.; Raths, A.; Heinemann, T.; Sandhoff, K.: cis-4-Methylsphingosine decreases sphingolipid biosynthesis by specifically interfering with serine palmitoyltransferase activity in primary cultured neurons. J. Biol. Chem., **272**, 15825-15833 (1997)

[15] Hanada, K.; Nishijima, M.; Fujita, T.; Kobayashi, S.: Specificity of inhibitors of serine palmitoyltransferase (SPT), a key enzyme in sphingolipid biosynthesis, in intact cells. A novel evaluation system using an SPT-defective mammalian cell mutant. Biochem. Pharmacol., **59**, 1211-1216 (2000)

[16] Gable, K.; Slife, H.; Bacikova, D.; Monaghan, E.; Dunn, T.M.: Tsc3p is an 80-amino acid protein associated with serine palmitoyltransferase and required for optimal enzyme activity. J. Biol. Chem., **275**, 7597-7603 (2000)

[17] Ikushiro, H.; Hayashi, H.; Kagamiyama, H.: A water-soluble homodimeric serine palmitoyltransferase from Sphingomonas paucimobilis EY2395T strain: purification, characterization, cloning, and overproduction. J. Biol. Chem., **276**, 18249-18256 (2001)

[18] Perry, D.K.; Carton, J.; Shah, A.K.; Meredith, F.; Uhlinger, D.J.; Hannun, Y.A.: Serine palmitoyltransferase regulates de novo ceramide generation during etoposide-induced apoptosis. J. Biol. Chem., **275**, 9078-9084 (2000)

[19] Perry, D.K.: Serine palmitoyltransferase: role in apoptotic de novo ceramide synthesis and other stress responses. Biochim. Biophys. Acta, **1585**, 146-152 (2002)

[20] Ikushiro, H.; Hayashi, H.; Kagamiyama, H.: Bacterial serine palmitoyltransferase: a water-soluble homodimeric prototype of the eukaryotic enzyme. Biochim. Biophys. Acta, **1647**, 116-120 (2003)

[21] Yasuda, S.; Nishijima, M.; Hanada, K.: Localization, topology, and function of the LCB1 subunit of serine palmitoyltransferase in mammalian cells. J. Biol. Chem., **278**, 4176-4183 (2003)

[22] Batheja, A.D.; Uhlinger, D.J.; Carton, J.M.; Ho, G.; D'Andrea, M.R.: Characterization of serine palmitoyltransferase in normal human tissues. J. Histochem. Cytochem., **51**, 687-696 (2003)

1-Acylglycerol-3-phosphate O-acyltransferase 2.3.1.51

1 Nomenclature

EC number
2.3.1.51

Systematic name
acyl-CoA:1-acyl-sn-glycerol-3-phosphate 2-O-acyltransferase

Recommended name
1-acylglycerol-3-phosphate O-acyltransferase

Synonyms
1-acyl-sn-glycero-3-phosphate acyltransferase
1-acyl-sn-glycerol 3-phosphate acyltransferase
1-acylglycero-3-phosphate acyltransferase
1-acylglycerolphosphate acyltransferase
1-acylglycerophosphate acyltransferase
acyltransferase, 1-acylglycerol phosphate
lysophosphatidic acid-acyltransferase

CAS registry number
51901-16-7

2 Source Organism

<1> *Escherichia coli* (B [5,6]; K12, strains: LCD45 and CE1061 [2,3]; XL1-Blue cells [28]) [2, 3, 5, 6, 26, 28]
<2> *Saccharomyces cerevisiae* (OC-2 [4]) [4]
<3> *Pisum sativum* (pea, cv. Kleine Rheinlaenderin [7,9,10]; cv. Douce Provence [9]) [7, 9, 10]
<4> *Solanum tuberosum* (potato [9]) [9]
<5> *Spinacia oleracea* (spinach, cv. Mona Lisa [10]) [10, 11]
<6> *Euglena sp.* [20]
<7> *Rattus norvegicus* [1, 8, 12, 14, 17-19, 34]
<8> *Oryctolagus cuniculus* [13, 14]
<9> *Sus scrofa* [15, 17]
<10> *Homo sapiens* [15]
<11> *Bos taurus* (Jersey cow [16]) [16]
<12> *Cocos nucifera* (coconut [21,24,29]) [21, 24, 29]
<13> *Limnanthes douglasii* (clone pLAT2 [22]) [22]
<14> *Limnanthes douglasii* (clone pLAT1 [22]) [22]

<15> *Neisseria meningitidis* [23]
<16> *Neisseria gonorrhoeae* [23]
<17> *Mus musculus* [25, 33]
<18> *Brassica napus* [26, 30]
<19> *Homo sapiens* [27]
<20> *Limnanthes douglasii* [31]
<21> *Brassica napus* [32]

3 Reaction and Specificity

Catalyzed reaction

acyl-CoA + 1-acyl-sn-glycerol 3-phosphate = CoA + 1,2-diacyl-sn-glycerol 3-phosphate

Reaction type

acyl group transfer

Natural substrates and products

S acyl-CoA + 1-acyl-sn-glycerol 3-phosphate <1, 3, 4, 7, 12> (<1,3,4,7> involved in phospholipid biosynthesis [1-3,7-9]) (Reversibility: ? <1, 3, 4, 7, 12> [1-3, 7-9, 21]) [1-3, 7-9, 21]

P CoA + 1,2-diacyl-sn-glycerol 3-phosphate <1, 3, 4, 7, 12> [1-3, 7-9, 21]

S acyl-[acyl-carrier-protein] + 1-acyl-sn-glycerol 3-phosphate <3, 5> (<3,5> involved in phospholipid biosynthesis in chloroplasts [10]) (Reversibility: ? <3, 5> [10]) [10]

P ACP + 1,2-diacyl-sn-glycerol 3-phosphate <3, 5> [10]

S Additional information <3, 11, 12-14, 18> (<3> acyl-CoA thioesters can be considered as the physiological substrates of the microsomal enzyme [7]; <3,5> acyl-ACP rather than acyl-CoA is the physiological acyl donor for glycerolipid biosynthesis in chloroplasts [10]; <11> 1-palmitoyl- and 1-oleoyl-sn-glycerol 3-phosphate are probably the substrates for the enzyme in vivo, since palmitic and oleic acids are the major fatty acid components at the sn-1 position of cow milk triacylglycerols. The specificity of the enzyme plays a major role in determining the fatty acid composition at the sn-2 position of cow milk triacylglycerols, though the influence of other factors such as the in vivo concentration of the acyl-CoAs and the specificities of the monoacylglycerol acyltransferases cannot be ruled out [16]; <12> the enzyme has medium chain length substrate specificity appropriate to the biosynthesis of coconut oil [21]; <13> pLAT2 may be necessary only for specific storage triacylglycerol biosynthesis in developing seed [22]; <14> LPAT1 may be an enzyme required for membrane phospholipid biosynthesis in all tissues [22]; <18> the enzyme controls phosphatidic acid biosynthesis in the plastid [30]) [7, 10, 16, 21, 22, 30]

P ?

Substrates and products

S acyl-ACP + 1-acyl-sn-glycerol 3-phosphate <3-5> (<3,5> catalyzes acylation of position 2 hydroxyl-group, acyl-donor: highly specific for palmitoyl-ACP preferred to oleoyl-ACP, acyl-acceptors are: 1-oleoyl-sn-glycerol 3-phosphate, which is the best and 1-palmitoyl-sn-glycerol 3-phosphate [10]; <3,4> negligible [9]) (Reversibility: ? <3-5> [9, 10]) [9, 10]

P ACP + 1,2-diacyl-sn-glycerol 3-phosphate <3-5> [9, 10]

S acyl-CoA + 1-O-alkyl-sn-glycerol 3-phosphate <17> (<17> the activity is approximately half of that toward 1-acyl-sn-glycerol 3-phosphate [33]) (Reversibility: ? <17> [33]) [33]

P CoA + ? <17> [33]

S acyl-CoA + 1-acyl-sn-glycerol 3-phosphate <1-12, 17, 19> (<1-4> i.e. lysophosphatidic acid, highly specific for 1-acyl-sn-glycerol 3-phosphate, catalyzes acylation of position 2 hydroxyl-group [2,4,7,9]; <2> the enzyme utilizes saturated and unsaturated acyl-CoAs at comparable rates [4]; <2> effect of substrate concentration [4]; <1,7,9> acyl-donors are saturated and unsaturated fatty acyl-CoAs [5,17]; <11> the enzyme is active towards acyl-CoAs of chain length C-8 to C-18 [16]; <9> CoA esters of unsaturated 16- and 18-carbon fatty acids are the best substrates, followed by those saturated 14-, 16-, and 18-carbon acids, and the unsaturated 20-carbon acids [15]; <7,9> C-14 to C-22 [17]; <7> C-16, C-18, C-20 acyl-CoAs and C-22 acyl-CoA [19]; <7> 1 to 6 double bonds [19]; <7-9> with 1 to 5 double bonds [14,15,19]; <7> monoenic and dienic acyl-CoA thioesters [8]; <3,7> unsaturated preferred [7,19]; <3> pronounced specificities and selectivities for unsaturated C_{18}-CoA thioesters [7]; <7,9> ω-3 and ω-6 polyunsaturated fatty acyl-CoAs [17]; <17> recombinant enzyme can catalyse ATP-independent acyl-CoA synthetic activity and CoA-dependent transacylation activity. Acyl-CoA synthetase is not involved in the process [33]) (Reversibility: ? <1-12, 19> [1-21, 27]; r <17> [33]) [1-21, 27, 33]

P CoA + 1,2-diacyl-sn-glycerol 3-phosphate <1-12, 17, 19> (<1-5,7> i.e. phosphatidic acid [1-4,6-9,11,12,19]) [1-21, 27, 33]

S acyl-CoA + 1-arachidonoyl-sn-glycerol 3-phosphate <7> (<7> less efficient than 1-acyl-sn-glycerol 3-phosphate [8,18]) (Reversibility: ? <7> [8, 18]) [8, 18]

P CoA + 1-arachidonoyl-2-acyl-sn-glycerol 3-phosphate <7> [8, 18]

S acyl-CoA + 1-lauroyl-sn-glycerol 3-phosphate <12> (<12> assayed using coconut cDNA encoding a 299-amino acid protein with enzyme activity expressed in canola seeds [29]) (Reversibility: ? <12> [29]) [29]

P CoA + 1-lauroyl-2-acyl-sn-glycerol 3-phosphate <12> [29]

S acyl-CoA + 1-oleoyl-sn-glycerol 3-phosphate <1, 3-5, 7, 9, 11, 15, 16, 18, 19> (<3> 1-oleoyl-sn-glycerol 3-phosphate is the best acceptor [7]) (Reversibility: ? <1, 3-5, 7, 9, 11, 15, 16, 18, 19> [7, 9, 11, 15, 16, 18, 23, 27, 28, 30]) [7, 9, 11, 15, 16, 18, 23, 27, 28, 30]

P CoA + 1-acyl-2-oleoyl-sn-glycerol 3-phosphate <1, 3-5, 7, 9, 11, 15, 16, 18, 19> [7, 9, 11, 15, 16, 18, 23, 27, 28, 30]

S acyl-CoA + 1-palmitoyl-sn-glycerol 3-phosphate <1, 3-9, 11, 17> (Reversibility: ? <1, 3-9, 11> [3, 9-11, 13, 15, 16, 18, 20]; r <17> [33]) [3, 7, 9-11, 13, 15, 16, 18, 20, 33]

P CoA + 1-palmitoyl-2-acyl-sn-glycerol 3-phosphate <1, 3-9, 11, 17> [3, 7, 9-11, 13, 15, 16, 18, 20, 33]

S acyl-CoA + 1-stearoyl-sn-glycerol 3-phosphate <7> (<7> less efficient than 1-acyl-sn-glycerol 3-phosphate [8,18]) (Reversibility: ? <7> [8, 18]) [8, 18]

P CoA + 1-stearoyl-2-acyl-sn-glycerol 3-phosphate <7> [8, 18]

S acyl-CoA + 2-acyl-sn-glycerol 3-phosphate <2> (<2> the rate of CoA-liberation is much lower than with 1-acyl-sn-glycerol 3-phosphate [4]; <1,3,9> the enzyme does not acylate 2-acyl-sn-glycerol 3-phosphate [6,7,15]) (Reversibility: ? <2> [4]) [4]

P CoA + ? <2> [4]

S acyl-CoA + linoleoyl-sn-glycerol 3-phosphate <1, 3-9> (Reversibility: ? <1, 3-9> [3, 9-11, 13, 15, 18, 20]) [3, 7, 9-11, 13, 15, 18, 20]

P CoA + ? <1, 3-9> [3, 7, 9-11, 13, 15, 18, 20]

S acyl-CoA + myristoyl-sn-glycerol 3-phosphate <1, 3-9> (Reversibility: ? <1, 3-9> [3, 9-11, 13, 15, 18, 20]) [3, 7, 9-11, 13, 15, 18, 20]

P CoA + ? <1, 3-9> [3, 7, 9-11, 13, 15, 18, 20]

S arachidonoyl-CoA + 1-acyl-sn-glycerol 3-phosphate <7-9, 17, 19> (<7,8> arachidonoyl-CoA is a poor substrate [8,12,13]) (Reversibility: ? <7-9, 17, 19> [8, 12, 13, 17, 18, 25, 27, 33]) [8, 12, 13, 17, 18, 25, 27, 33]

P CoA + 1-acyl-2-arachidonoyl-sn-glycerol 3-phosphate <7-9, 17, 19> [8, 12, 13, 17, 18, 25, 27, 33]

S arachidoyl-CoA + 1-acyl-sn-glycerol 3-phosphate <19> (Reversibility: ? <19> [27]) [27]

P CoA + 1-acyl-2-arachidoyl-sn-glycerol 3-phosphate <19> [27]

S behenoyl-CoA + 1-acyl-sn-glycerol 3-phosphate <1> (Reversibility: ? <1> [28]) [28]

P CoA + 1-acyl-2-behenoyl-sn-glycerol 3-phosphate <1> [28]

S capryl-CoA + 1-acyl-sn-glycerol 3-phosphate <1, 18> (Reversibility: ? <1, 18> [26]) [26]

P CoA + 1-acyl-2-capryl-sn-glycerol 3-phosphate <1, 18> [26]

S cis-11-eicosenoyl-CoA + 1-acyl-sn-glycerol 3-phosphate <1, 18> (Reversibility: ? <1, 18> [26]) [26]

P CoA + 1-acyl-2-cis-11-eicosenoyl-sn-glycerol 3-phosphate <1, 18> [26]

S eicosatetraenoyl-CoA + 1-acyl-sn-glycerol 3-phosphate <8> (Reversibility: ? <8> [14]) [14]

P CoA + 1-acyl-2-eicosatetraenoyl-sn-glycerol 3-phosphate <8> [14]

S erucoyl-CoA + 1-acyl-sn-glycerol 3-phosphate <1, 18> (Reversibility: ? <1, 18> [26]) [26]

P CoA + 1-acyl-2-erucoyl-sn-glycerol 3-phosphate <1, 18> [26]

S erucoyl-CoA + 1-acyl-sn-glycerol 3-phosphate <8> (<8> erucoyl-CoA is a poor substrate [13]) (Reversibility: ? <8> [13]) [13]

P CoA + 1-acyl-2-erucoyl-sn-glycerol 3-phosphate <8> [13]

S erucoyl-CoA + 1-erucoyl-sn-glycerol-3-phosphate <13> (Reversibility: ? <13> [22]) [22]

P CoA + 1,2-dierucoyl-sn-glycerol 3-phosphate <13> [22]

S gondoyl-CoA + 1-acyl-sn-glycerol 3-phosphate <3> (<3> gondoyl-CoA is utilized less efficiently for the reaction than oleoyl-CoA [7]) (Reversibility: ? <3> [7]) [7]

P CoA + 1-acyl-2-gondoyl-sn-glycerol 3-phosphate <3> [7]

S lauroyl-CoA + 1-acyl-sn-glycerol 3-phosphate <1, 7, 12, 18, 19> (<1,7> lauroyl-CoA is a poor substrate [5,8,12]; <12> transgenic seeds: preferred substrate [29]) (Reversibility: ? <1, 7, 12, 18, 19> [5, 8, 12, 18, 26, 27, 29]) [5, 8, 12, 18, 26, 27, 29]

P CoA + 1-acyl-2-lauroyl-sn-glycerol 3-phosphate <1, 7, 12, 18, 19> [5, 8, 12, 18, 26, 27, 29]

S lignoceroyl-CoA + 1-acyl-sn-glycerol 3-phosphate <1, 19> (Reversibility: ? <1, 19> [27, 28]) [27, 28]

P CoA + 1-acyl-2-lignoceroyl-sn-glycerol 3-phosphate <1, 19> [27, 28]

S linolenoyl-CoA + 1-acyl-sn-glycerol 3-phosphate <1, 8, 19> (Reversibility: ? <1, 8, 19> [14, 27, 28]) [14, 27, 28]

P CoA + 1-acyl-2-linolenoyl-sn-glycerol 3-phosphate <1, 8, 19> [14, 27, 28]

S linoleoyl-CoA + 1-acyl-sn-glycerol 3-phosphate <1, 3, 4, 7, 8, 17, 19> (<7> linoleoyl-CoA is a good substrate but is utilized less efficiently than oleoyl-CoA [12]) (Reversibility: ? <1, 3, 4, 7, 8, 17, 19> [8, 9, 12-14, 18, 27, 28, 33]) [8, 9, 12-14, 18, 27, 28, 33]

P CoA + 1-acyl-2-linoleoyl-sn-glycerol 3-phosphate <3, 4, 7, 8, 17, 19> [8, 9, 12-14, 18, 27, 28, 33]

S linoleoyl-CoA + 1-oleoyl-sn-glycerol 3-phosphate <17> (Reversibility: r <17> [33]) [33]

P CoA + 1-oleoyl-2-linoleoyl-sn-glycerol 3-phosphate <17> [33]

S myristoyl-CoA + 1-acyl-sn-glycerol 3-phosphate <1, 7, 17, 19> (<1,7> myristoyl-CoA is a poor substrate [5,8,12]) (Reversibility: ? <1, 7, 17, 19> [5, 8, 12, 27, 33]) [5, 8, 12, 27, 33]

P CoA + 1-acyl-2-myristoyl-sn-glycerol 3-phosphate <1, 7, 17, 19> [5, 8, 12, 27, 33]

S oleoyl-ACP + sn-1-oleoylglycerol 3-phosphate <1> (<1> microsomal fractions from developing, transgenic seeds but not of untransformed plants [26]) (Reversibility: ? <1> [26]) [26]

P ACP + sn-1,2-dioleoylglycerol 3-phosphate <1> [26]

S oleoyl-CoA + 1-acyl-sn-glycerol 3-phosphate <1-9, 13, 17-19> (<1> high selectivity for oleate [5]; <1> twice as good as palmitoyl-CoA [6]; <3> oleoyl-CoA is the best substrate, both membrane-bound and partially purified enzyme [7]; <7> is the most effective acyl donor [12]) (Reversibility: ? <1-9, 13, 17-19> [3-9, 11, 12, 14, 15, 17, 18, 20, 22, 25-28, 33, 34]) [3-9, 11, 12, 14, 15, 17, 18, 20, 22, 25-28, 33, 34]

P CoA + 1-acyl-2-oleoyl-sn-glycerol 3-phosphate <1-9, 13, 17-19> [3-9, 11, 12, 14, 15, 17, 18, 20, 22, 25-28, 33, 34]

S oleoyl-CoA + 1-oleoyl-sn-glycerol 3-phosphate <18> (<18> assayed with BAT2 protein [30]) (Reversibility: ? <18> [30]) [30]

P CoA + 1,2-dioleoyl-sn-glycerol 3-phosphate <18> [30]

S palmitoleoyl-CoA + 1-acyl-sn-glycerol 3-phosphate <7, 15, 16, 19> (<7> palmitoleoyl-CoA is a good substrate but is utilized less efficiently than oleoyl-CoA [12]; <19> palmitoleoyl-CoA is a good substrate [27]) (Reversibility: ? <7, 15, 16, 19> [8, 12, 23, 27]) [8, 12, 23, 27, 28]

P CoA + 1-acyl-2-palmitoleoyl-sn-glycerol 3-phosphate <7, 15, 16, 19> [8, 12, 23, 27, 28]

S palmitoyl-ACP + 1-acyl-sn-glycerol 3-phosphate <3, 18> (Reversibility: ? <3, 18> [7, 30]) [7, 30]

P ACP + 1-acyl-2-palmitoyl-sn-glycerol 3-phosphate <3, 18> [7, 30]

S palmitoyl-CoA + 1-acyl-sn-glycerol 3-phosphate <1-8, 11, 17-19> (<1> palmitoyl-CoA is the best substrate [5]; <3,4,7> palmitoyl-CoA is a poor substrate [9,12]; <11> maximum activity is obtained with palmitoyl-CoA at a concentration of 0.002-0.003 mM [16]) (Reversibility: ? <1-8, 11, 17-19> [1-9, 11-14, 16, 18-20, 25-28, 33, 34]) [1-9, 11-14, 16, 18-20, 25-28, 33, 34]

P CoA + 1-acyl-2-palmitoyl-sn-glycerol 3-phosphate <1-8, 11, 17-19> [1-9, 11-14, 16, 18-20, 25-28, 33, 34]

S palmitoyl-CoA + 1-oleoyl-sn-glycerol 3-phosphate <18> (<18> assayed with BAT2 protein, palmitoyl-CoA is the preferred substrate [30]) (Reversibility: ? <18> [30]) [30]

P CoA + 1-oleoyl-2-palmitoyl-sn-glycerol 3-phosphate <18> [30]

S stearoyl-CoA + 1-acyl-sn-glycerol 3-phosphate <1, 7, 8, 17, 19> (<7,8> stearoyl-CoA is a poor substrate [8,12,13]) (Reversibility: ? <1, 7, 8, 17, 19> [8, 12, 13, 27, 28, 33]) [8, 12, 13, 27, 28, 33]

P CoA + 1-acyl-2-stearoyl-sn-glycerol 3-phosphate <1, 7, 8, 19> [8, 12, 13, 27, 28, 33]

S Additional information <1-5, 7, 9, 11, 12, 17, 19> (<1> description of a direct nonchromatographic assay for the enzyme [2]; <1> lysophosphatidic acid, synthesized by EC 2.3.1.15, is utilized by the enzyme without prior mixing with the total membrane-associated pool of lysophosphatidic acid, and suggest a close proximity of the two enzymes in native E. coli membranes. This property of the acyltransferases is lost upon separation and reconstitution of enzyme activities [3]; <11> no acyl-donors are butyryl-CoA and hexanoyl-CoA [16]; <11> the chain length specificity of the enzyme in decreasing order is as follows: C_{16}, C_{14}, C_{12}, C_{10}, C_8 [16]; <7> no acceptors are 1-acyl-sn-glycero-3-phosphoethanolamine, 1-acyl-sn-glycero-3-phosphomonomethylethanolamine, 1-acyl-sn-glycero-3-phosphodimethylethanolamine or 1-acyl-sn-glycero-3-phosphocholine [18]; <3-5> no acyl-donors are coenzyme A derivatives [9,10]; <3-5> no acyl acceptor is glycerol 3-phosphate [9,10]; <2> very broad specificity for acyl-CoAs, the enzyme does not seem to discriminate strictly the cis and trans-configurations or the position of double bond in the fatty acyl moiety [4]; <1> the enzyme can utilize various saturated and unsaturated acyl-CoAs at comparable rates, unsaturated fatty acyl-CoAs with cis- or trans-configuration are slightly better substrates than saturated acyl-CoAs [5]; <1> effect of variable ratios of oleoyl-CoA to palmitoyl-CoA on the reaction [5]; <1> 1-acyl-sn-glycerol 3-phosphate is readily acylated by unsaturated and saturated acyl-CoAs [6]; <1> effect of acyl-CoA concentrations on enzyme reaction [6]; <3> the microsomal enzyme displays a specificity for unsaturated C_{18}-CoA thioesters and this fatty-acid specificity is not altered in the course of the purification procedure [7]; <3,4> the enzyme shows higher activities with acyl-CoA than with the corresponding acyl-(acyl carrier protein) thioesters, the enzyme is more active on unsaturated than on saturated acyl-CoA thioesters [9]; <3,5> the enzyme uses preferably ACP-thioesters for the acylation of 1-acylglycerol 3-phosphate, the enzyme does not accept glycerol 3-phosphate as acyl-acceptor [10]; <5> the enzyme uses preferentially those acyl donors and acceptors, which contain oleate, specificity and selectivity experiments [11]; <7> the enzyme exhibits a considerable selectivity for monoenoic and dienoic fatty acyl-CoA thioesters [12]; <7,9> the enzyme system shows relatively broad specificity for saturated and unsaturated fatty acids, with 14- to 20-carbon chains. Time courses fo the acylations of 1-acyl-glycerophosphate in the presence of equimolar mixtures of oleoyl-CoA and arachidonoyl-CoA [17]; <12> acyl-CoAs containing 10:0, 12:0 and 14:0 acyl groups are the preferred acyl-donor substrates, acyl-ACPs are not utilized. There is a slight preference for 12:0-lysophosphatidic acid over 18:1-lysophosphatidic acid as acceptor substrate [21]; <17> the enzyme utilizes both saturated and unsaturated acyl-CoA as an acyl-donor but does not catalyse the transfer of acyl moiety to lysophosphatidylcholine or lyso-platelet-activating factor [25]; <1> an introduced chimeric enzyme gene of E. coli alters the stereochemical composition of transgenic HEAR oil by preferentially directing cis-11 eicosenoic acid to its sn-2 position [26]; <19> the enzyme has affinity for fatty acids of acyl chain lengths from 12 to 18 carbons with a slight dependence on the degree of saturation of the fatty acid. However the enzyme does not incorporate long chain fatty acids like C20:0 and C24:0 unless they are saturated [27]; <12> expression of a 299-amino acid protein with enzyme activity alters the enzyme substrate specificity profile of transgenic canola seeds [29]; <17> lysophospholipids such as 1-acyl-sn-glycero-3-phosphocholine, 1-acyl-sn-glycero-3-phosphoethanolamine, 1-acyl-sn-glycero-3-phosphoinositol and lysoplatelet-activating factor do not serve as acyl acceptor substrates. Free fatty acids do not serve as acyl donors. The enzyme shows a broad specificity for acyl-CoAs [33]; <17> reverse reaction of the enzyme: desulfo-CoA and dephospho-COA do not serve as substrates for acyl-CoA synthesis [33]) [2-7, 9-12, 16-18, 21, 25-27, 29, 33]

P ?

Inhibitors

1-acyl-sn-glycero 3-phosphate <7> (<7> high concentration [19]) [19]
1-palmitoyl-sn-glycerol 3-phosphate <8> (<8> above 0.05 mM [13]) [13]
2-[3,5-di-(t-butyl-4-hydroxyphenyl)thio]hexanate <5> (<5> i.e. DH990, weak [11]) [11]
2-mercaptoethanol <12> (<12> above 5 mM [21]) [21]

CHAPS <3, 12> (<3> above 0.4% w/v [7]; <12> above 0.1% [21]) [7, 21]
Ca^{2+} <12> (<12> 50% inhibition above 10 mM [21]) [21]
$MgCl_2$ <1, 12> (<1> at high concentrations [6]; <12> 50% inhibition above 10 mM [21]) [6, 21]
Mn^{2+} <12> (<12> 50% inhibition above 10 mM [21]) [21]
N-ethylmaleimide <6> [20]
NaCl <12> (<12> above 200 mM [21]) [21]
Triton X-100 <7, 9> (<7> 0.5 mM and above [12]; <9> concentration-dependent inactivation [15]) [12, 15]
sodium deoxycholate <9> (<9> concentration-dependent inactivation [15]) [15]
stearoyl-CoA <11> (<11> strongly inhibitory at 0.002 mM [16]) [16]
Additional information <7, 9, 11> (<7> no inhibition by sulfhydryl reagents, 1-acylglycerol 3-phosphoethanol, 1-acylglycerol 3-phosphoethanolamine, 1-acylglycerol 3-phosphodimethylethanolamine, 1-acylglycerol 3-choline [1]; <9> at low concentrations of enzyme protein, the enzyme is possibly inactivated by the detergent action of substrates [15]; <11> the enzyme can utilize the monomer form of long-chain acyl-CoA but it is inhibited by the micellar form of these substrates [16]) [1, 15, 16]

Activating compounds

bovine serum albumin <7> (<7> induces T-lymphocyte enzyme activity in the presence of palmitoyl-CoA but not oleoyl-CoA [34]) [34]
isopropyl-β-D-thiogalactopyranoside <17> (<17> induction of the enzyme gene [33]) [33]
Additional information <12> (<12> stimulation of the enzyme by incorporation of the enzyme into the artificial membrane vesicles or by stripping of CHAPS from the enzyme into mixed detergent-phospholipid micelles. There is a critical order in which assay components had to be combined to obtain maximum stimulation [21]) [21]

Metals, ions

Additional information <1> (<1> no Mg^{2+}-requirement [6]) [6]

Specific activity (U/mg)

0.0000108 <4> (<4> inner mitochondrial membrane [9]) [9]
0.0000192 <4> (<4> mitochondria [9]) [9]
0.0002742 <4> (<4> outer mitochondrial membrane [9]) [9]
0.0008 <1> [3]
0.0051 <5> [11]
0.0066 <3> [7]
0.44 <7> [8]

K_m-Value (mM)

0.0014 <5> (palmitoyl-CoA) [11]
0.0028 <5> (oleoyl-CoA) [11]
0.003 <5> (1-palmitoyl-sn-glycerol 3-phosphate, <5> with oleoyl-CoA [11]) [11]
0.004 <6> (oleoyl-CoA) [20]

0.0053 <5> (1-oleoyl-sn-glycerol 3-phosphate, <5> with oleoyl-CoA [11]) [11]
0.007 <6> (palmitoyl-CoA) [20]
0.125 <7, 8> (1-acyl-sn-glycerol 3-phosphate, <7,8> in the incorporation of palmitate and linoleate [14]) [14]

pH-Optimum

7-9 <2> (<2> broad [4]) [4]
7.4 <3, 5, 8, 9, 11, 17, 19> (<3,5,8,9,11,17,19> assay at [10,14-16,27,33]) [10, 14-16, 27, 33]
7.5 <1, 12> (<1,12> assay at [5,6,21]) [5, 6, 21]
7.5-8 <1> [6]
8 <18> (<18> assay at [30]) [30]
8-9.5 <9> (<9> broad [17]) [17]
8-10 <4> (<4> broad [9]) [9]
8.5 <7, 9> (<7,9> assay at [17]) [17]
9 <1, 15, 16, 18> (<1,15,16,18> assay at [2,3,23,26,28]) [2, 3, 23, 26, 28]
9-11 <5> (<5> broad, assay at pH 10 [11]) [11]
Additional information <7, 9> (<7,9> pH profiles of the platelet and liver enzyme systems [17]) [17]

pH-Range

6.6-8.6 <1> (<1> about half-maximal activity at pH 6.6 and 8.6 [6]) [6]

Temperature optimum (°C)

22 <9> (<9> assay at [15]) [15]
23-26 <1> (<1> assay at [5]) [5]
25 <3, 5, 7, 9> (<3,5,7,9> assay at [7,11,17]) [7, 11, 17]
30 <1, 8, 11, 12, 15, 16-18> (<1, 8, 11, 12, 15, 16-18> assay at [2, 3, 13, 16, 21, 23, 26, 30, 33]) [2, 3, 13, 16, 21, 23, 26, 30, 33]
35 <1> (<1> assay at [28]) [28]
37 <7> (<7> assay at [8]) [8]

4 Enzyme Structure

Molecular weight

29000 <12> (<12> apparent, gel filtration [24]) [24]
32000 <17, 18> (<18> mature protein, processing of a BAT2 precursor protein, chloroplast-import experiments [30]; <17> Western blotting [33]) [30, 33]
34800 <12> (<12> calculated from amino acid sequence [24]) [24]
38000 <18> (<18> predicted, protein encoded by the BAT2 cDNA [30]) [30]
44000 <12> (<12> estimated, high-speed supernatant, gel filtration [21]) [21]

5 Isolation/Preparation/Mutation/Application

Source/tissue

3T3-L1 cell <7> [1]
T-lymphocyte <7> [34]
blood platelet <9> [17]
brain <17> (<17> analysis of enzyme mRNA expression, transcript of approximately 2 kb length dominant [25]) [25]
embryo <17> (<17> whole, analysis of enzyme mRNA expression, only the transcript of approximately 2 kb length is present. In 7-day embryos both bands of 1.3 and 2 kb are present [25]) [25]
endosperm <12> (<12> immature [21,24]) [21, 24, 29]
erythrocyte <9, 10> (<9,10> mature [15]) [15]
heart <8, 17> (<17> analysis of enzyme mRNA expression, transcript of approximately 2 kb length dominant [25]) [13, 25]
kidney <17> (<17> analysis of enzyme mRNA expression, transcript of approximately 2 kb length dominant [25]) [25]
leaf <3, 5, 13, 14> [9-11, 22]
liver <7, 9, 17> (<17> analysis of enzyme mRNA expression [25]) [1, 8, 12, 14, 17-19, 25, 34]
lung <7, 17> (<17> analysis of enzyme mRNA expression, transcript of approximately 2 kb length dominant [25]) [1, 25]
mammary gland <11> (<11> lactating [16]) [16]
muscle <8> [14]
seed <12, 13, 14> [21, 22, 24]
shoot <3> (<3> etiolated [7]) [7]
skeletal muscle <17> (<17> analysis of enzyme mRNA expression, transcript of approximately 2 kb length dominant [25]) [25]
spleen <17> (<17> analysis of enzyme mRNA expression, transcript of approximately 1.3 kb length dominant [25]) [25]
stem <13, 14> [22]
testis <17> (<17> analysis of enzyme mRNA expression, transcript of approximately 2 kb length dominant [25]) [25]
tuber <4> [9]
Additional information <18> (<18> the BAT2 gene is expressed in a ubiquitous manner in all tissues, including nonphotosynthetic tissues where plastids are present [30]) [30]

Localization

chloroplast <18> (<18> mature protein of 32000 Da processed from the precursor protein of 38000 Da [30]) [30]
chloroplast envelope <3, 5> [10]
endoplasmic reticulum <3, 5, 19> [7, 11, 27]
membrane <1-3, 5, 7, 9, 10, 12, 13, 18> [2-6, 10, 15, 21, 22, 30, 34]
microsome <2, 3, 5-9, 11> (<7> intrinsic membrane protein, cytoplasmic surface [1]; <3> predominantly located in endoplasmic reticulum [7]; <7> not rat liver peroxisomes [1]) [1, 4, 7, 8, 11-14, 16-18, 20]

mitochondrial outer membrane <3, 4> (<3,4> firmly bound [9]) [9]
mitochondrion <2, 3, 4> (<2> the presence of the enzyme in the mitochondrial fraction cannot be explained by the contamination of microsomes in this fraction [4]) [4, 9]
plastid <18, 21> [30, 32]
sarcoplasmic reticulum <8> [14]

Purification

<1> (partial, separated from EC 2.3.1.15 using Triton X-100, the enzyme is not solubilized by Triton X-100 or cholate [3]) [2, 3]
<3> (partial [7,10]; using solubilization with CHAPS at high ionic strength, and anion-exchange chromatography on Mono Q column [7]) [7, 10]
<5> (partial [10]) [10]
<7> (partial [8,12]; using solubilization with Triton X-100 and gel filtration, separation from EC 2.3.1.23 [12]) [8, 12]
<12> (identification of enzyme using solubilization with CHAPS and column chromatography on Red 120 agarose, Bio-Rad HA HT, 12:0-CoA Sepharose 4B, and Superose 12 [24]) [24]

Renaturation

<1> (by mixing 1 vol. of the fraction containing the enzyme with 4 vol. of sonicated phosphatidylcholine/phosphatidylglycerol or total E. coli lipid vesicles [3]) [2, 3]
<12> (reassociation of the solubilized enzyme with soybean phospholipids [21]) [21]

Cloning

<1> (modification of the gene of Escherichia coli which encodes the enzyme by inserting an endoplasmic reticulum retrieval signal to its 3'end and introduction into rapeseed plants under the control of a napin promotor [26]; isolation of the JC201-plsC gene fragment containing the entire sequence of the enzyme. Complementation of the JC201 phenotype: transformation of JC201 cells with pplSC6 or pplsC6.3. Transformation of XL1-Blue cells with the control plasmid pplsC6.3, and the experimental plasmids pplsC6.3A or pplsC6.3L [28];) [26, 28]
<12> (amino acid sequence data used to isolate a full-length clone from a coconut endosperm cDNA library, expression of the cDNA in Escherichia coli [24]; expression of a cDNA encoding protein with enzyme activity in transgenic plants. Expression in canola results in incorporation of laurate into the sn-2 position of triacylglycerol [29]) [24, 29]
<13> (isolation of a potential cDNA for the enzyme by in vivo complementation of the Escherichia coli mutant JC201, clone pLAT2 is an enzyme clone [22]) [22]
<14> (isolation of a potential cDNA for the enzyme by heterologous screening with the maize pMAT1 clone, clone pLAT1 [22]) [22]
<15> (isolation and characterization of a meningococcal mutant, cloning of the meningococcal mutant gene [23]) [23]
<16> (isolation and characterization of a gonococcal mutant [23]) [23]

<17> (expression in Escherichia coli JC201, a mutant strain deficient in the enzyme activity [25]; expression in JC201 Escherichia coli cells [33]) [25, 33]
<18> (isolation of a cDNA encoding a novel enzyme by functional complementation of the Escherichia coli mutant plsC with an immature embryo cDNA library of oilseed rape. Transformation of the acyltransferase-deficient Escherichia coli strain JC201 with the cDNA sequence BAT2 alleviates the temperature-sensitive phenotype of the plsC mutant and confers a palmitoyl-CoA-preferring enzyme activity to membrane fractions. Mapping of the BAT2 genes and sequence analysis of the BAT2 cDNA clone [30]) [30]
<19> (expression in insect and mammalian cell lines. The enzyme gene is located in the MHC class III region [27]) [27]
<20> (transformation of plants of spring high-erocic acid rape, Hero, with a construct carrying the Limnanthes douglasii enzyme, LAT2, under control of the napin promoter. Study of the correlation between 22:1 accumulation at sn-2, production of trierucin and the expression of the transgene as detected by Western blot [31]) [31]
<21> (transformation of a modified plasmid into Escherichia coli JC201 enzyme-deficient strain and transformation of the plasmid DNA containing the cDNA BAT2 encoding the plastidial enzyme into the Escherichia coli XL-1 red mutator strain [32]) [32]

Engineering

C247Y <21> (<21> loss of the ability of the plastidial enzyme to restore growth to JC201, suggests that C247 is essential for activity [32]) [32]
D199A <21> (<21> the mutant fails to grow beyond 0.04 D600 after transfer to the non-permissive temperature. D199 is essential for the ability of the plastidial enzyme to complement the defective acyltransferase activity of JC201 [32]) [32]
E266A <21> (<21> the plastidial enzyme containing this mutation is unable to complement the temperature-sensitive phenotype of JC201, indicating an essential role for this acidic residue [32]) [32]
G140S <21> (<21> loss of the ability of the plastidial enzyme to restore growth to JC201, suggests that G140 is essential for activity [32]) [32]
H194A <21> (<21> the mutant fails to grow beyond 0.04 D600 after transfer to the non-permissive temperature. H194 is essential for the ability of the plastidial enzyme to complement the defective acyltransferase activity of JC201 [32]) [32]
K84Q <21> (<21> greatly impaired growth rate, possibly indicating an importance of the N-terminus of the mature protein in determining enzyme activity [32]) [32]
L203P <21> (<21> loss of the ability of the plastidial enzyme to restore growth to JC201, suggests that L203 is essential for activity [32]) [32]
P187L <21> (<21> loss of the ability of the plastidial enzyme to restore growth to JC201, suggests that P187 is essential for activity [32]) [32]
R92G <21> (<21> greatly impaired growth rate, possibly indicating an importance of the N-terminus of the mature protein in determining enzyme activity [32]) [32]

T122A <1> (<1> the change increases the substrate specificity in vitro for oleoyl-CoA and linoleoyl-CoA [28]) [28]
T122L <1> (<1> the change increases the substrate specificity in vitro for lignoceroyl-CoA [28]) [28]
Additional information <1> (<1> adenine-440 is responsible for the temperature sensitive lesion in the enzyme activity of JC201. Conversion of the mutant adenine-440 back to the native guanine-440 nucleotide restores the enzyme activity at the non-permissive temperature of 42°C [28]) [28]

6 Stability

pH-Stability

6.5-7.5 <12> (<12> lowering the assay pH from 7.5 to 6.5 results in a loss of approximately 20% of the enzyme activity [21]) [21]

Temperature stability

43 <7> (<7> $t_{1/2}$: 62 min [12]) [12]

General stability information

<1>, slow freezing inactivates [6]
<7>, ethylene glycol, 20% v/v, stabilizes [8, 12]
<9>, the long time required for fractionation after hypotonic hemolysis probably contributes to the loss of activity [15]

Storage stability

<1>, -15°C to 4°C, particulate preparation, fast-frozen in liquid N_2, unstable [6]
<1>, -15°C, less than a day [6]
<1>, -80°C, particulate preparation, fast-frozen in liquid N_2, 3 months [6]
<3>, -20°C, particulate preparation, several days [7]
<3>, 4°C, soluble preparation, at least 1 day [7]
<5>, -20°C, microsomal preparation, at least 3 weeks [11]
<12>, 4°C, high-speed supernatant preparation, 50% loss of activity, overnight [21]
<12>, 4°C, high-speed supernatant, reactivated with phospholipids, no loss of activity [21]

References

[1] Bell, R.M.; Coleman, R.A.: Enzymes of triacylglycerol formation in mammals. The Enzymes, 3rd Ed. (Boyer, P.D., ed.), **16**, 87-111 (1983)
[2] Rajasekharan, R.; Ray, T.K.; Cronan, J.E.: A direct nonchromatographic assay for 1-acyl-sn-glycerol-3-phosphate acyltransferase. Anal. Biochem., **173**, 376-382 (1988)
[3] Kessels, J.M.M.; Ousen, H.; Van den Bosch, H.: Facilitated utilization of endogenously synthesized lysophosphatidic acid by 1-acylglycerophosphate

acyltransferase from Escherichia coli. Biochim. Biophys. Acta, **753**, 227-235 (1983)
[4] Yamada, K.; Okuyama, H.; Endo, Y.; Ikezawa, H.: Acyltransferase systems involved in phospholipid metabolism in Saccharomyces cerevisiae. Arch. Biochem. Biophys., **183**, 281-289 (1977)
[5] Okuyama, H.; Yamada, K.; Ikezawa, H.; Wakil, S.J.: Factors affecting the acyl selectivities of acyltransferases in Escherichia coli. J. Biol. Chem., **251**, 2487-2492 (1976)
[6] Okuyama, H.; Wakil, S.J.: Positional specificities of acyl coenzyme A: glycerophosphate and acyl coenzyme A: monoacylglycerophosphate acyltransferases in Escherichia coli. J. Biol. Chem., **248**, 5197-5205 (1973)
[7] Hares, W.; Frentzen, M.: Substrate specificities of the membrane-bound and partially purified microsomal acyl-CoA:1-acylglycerol-3-phosphate acyltransferase from etiolated shoots of Pisum sativum (L.). Planta, **185**, 124-131 (1991)
[8] Yamashita, S.; Hosaka, K.; Miki, Y.; Numa, S.: Glycerolipid acyltransferases from rat liver: 1-acylglycerophosphate acyltransferase, 1-acylglycerophosphorylcholine acyltransferase, and diacylglycerol acyltransferase. Methods Enzymol., **71**, 528-536 (1981)
[9] Frentzen, M.; Neuburger, M.; Joyard, J.; Douce, R.: Intraorganelle localization and substrate specificities of the mitochondrial acyl-CoA: sn-glycerol-3-phosphate O-acyltransferase and acyl-CoA: 1-acyl-sn-glycerol-3-phosphate O-acyltransferase from potato tubers and pea leaves. Eur. J. Biochem., **187**, 395-402 (1990)
[10] Frentzen, M.; Heinz, E.; McKeon, T.A.; Stumpf, P.K.: Specificities and selectivities of glycerol-3-phosphate acyltransferase and monoacylglycerol-3-phosphate acyltransferase from pea and spinach chloroplasts. Eur. J. Biochem., **129**, 629-636 (1983)
[11] Hares, W.; Frentzen, M.: Properties of the microsomal acyl-CoA:sn-1-acylglycerol-3-phosphate acyltransferase from spinach (Spinacia oleracea L.) leaves. J. Plant Physiol., **131**, 49-59 (1987)
[12] Yamashita, S.; Nakaya, N.; Miki, Y.; Numa, S.: Separation of 1-acylglycerolphosphate acyltransferase and 1-acylglycerolphosphorylcholine acyltransferase of rat liver microsomes. Proc. Natl. Acad. Sci. USA, **72**, 600-603 (1975)
[13] Zaror-Behrens, G.; Kako, K.J.: Positional and fatty acid specificity of monoacyl- and diacylglycerol 3-phosphate formation by rabbit heart microsomes. Biochim. Biophys. Acta, **441**, 1-13 (1976)
[14] Waku, K.; Hayakawa, F.; Nakazawa, Y.: Regulation of the fatty acid pattern of phospholipids in rabbit sarcoplasmic reticulum. Specificity of glycerophosphate, 1-acylglycerophosphate and 2-acylglycerophosphorylcholine acyltransferase systems. J. Biochem., **82**, 671-677 (1977)
[15] Mizuno, M.; Sugiura, Y.; Okuyama, H.: Properties of acyl-coenzyme A:1-acylglycerophosphate acyltransferase and lipases in porcine erythrocyte membranes. J. Lipid Res., **25**, 843-850 (1984)
[16] Marshall, M.O.; Knudsen, J.: The specificity of 1-acyl-sn-glycerol 3-phosphate acyltransferase in microsomal fractions from lactating cow mam-

mary gland towards short, medium and long chain acyl-CoA esters. Biochim. Biophys. Acta, **489**, 236-241 (1977)
[17] Inoue, M.; Murase, S.; Okuyama, H.: Acyl coenzyme A:phospholipid acyltransferases in porcine platelets discriminate between ω-3 and ω-6 unsaturated fatty acids. Arch. Biochem. Biophys., **231**, 29-37 (1984)
[18] Miki, Y.; Hosaka, K.; Yamashita, S.; Handa, H.; Numa, S.: Acyl-acceptor specificities of 1-acylglycerolphosphate acyltransferase and 1-acylglycerophosphorylcholine acyltransferase resolved from rat liver microsomes. Eur. J. Biochem., **81**, 433-441 (1977)
[19] Hill, E.E.; Lands, W.E.M.: Incorporation of long-chain and polyunsaturated acids into phosphatidate and phosphatidylcholine. Biochim. Biophys. Acta, **152**, 645-648 (1968)
[20] Hershenson, S.; Ernst-Fonberg, M.L.: Comparison of 1-acylglycerophosphate and glycerophosphate acyltransferases from Euglena microsomes. Int. J. Biochem., **16**, 219-223 (1984)
[21] Davies, H.M.; Hawkins, D.J.; Nelsen, J.S.: Lysophosphatidic acid acyltransferase from immature coconut endosperm having medium chain length substrate specificity. Phytochemistry, **39**, 989-996 (1995)
[22] Brown, A.P.; Brough, C.L.; Kroon, J.T.M.; Slabas, A.R.: Identification of a cDNA that encodes a 1-acyl-sn-glycerol-3-phosphate acyltransferase from Limnanthes douglasii. Plant Mol. Biol., **29**, 267-278 (1995)
[23] Swartley, J.S.; Balthazar, J.T.; Coleman, J.; Shafer, W.M.; Stephens, D.S.: Membrane glycerophospholipid biosynthesis in Neisseria meningitidis and Neisseria gonorrhoeae: identification, characterization, and mutagenesis of a lysophosphatidic acid acyltransferase. Mol. Microbiol., **18**, 401-412 (1995)
[24] Knutzon, D.S.; Lardizabal, K.D.; Nelsen, J.S.; Bleibaum, J.L.; Davies, H.M.; Metz, J.G.: Cloning of a coconut endosperm cDNA encoding a 1-acyl-sn-glycerol-3-phosphate acyltransferase that accepts medium-chain-length substrates. Plant Physiol., **109**, 999-1006 (1995)
[25] Kume, K.; Shimizu, T.: cDNA cloning and expression of murine 1-acyl-sn-glycerol-3-phosphate acyltransferase. Biochem. Biophys. Res. Commun., **237**, 663-666 (1997)
[26] Weier, D.; Luhs, W.; Dettendorfer, J.; Frentzen, M.: sn-1-Acylglycerol-3-phosphate acyltransferase of Escherichia coli causes insertion of cis-11 eicosenoic acid into the sn-2 position of transgenic rapeseed oil. Mol. Breed., **4**, 39-46 (1998)
[27] Aguado, B.; Campbell, R.D.: Characterization of a human lysophosphatidic acid acyltransferase that is encoded by a gene located in the class III region of the human major histocompatibility complex. J. Biol. Chem., **273**, 4096-4105 (1998)
[28] Morand, L.Z.; Patil, S.; Quasney, M.; German, J.B.: Alteration of the fatty acid substrate specificity of lysophosphatidate acyltransferase by site-directed mutagenesis. Biochem. Biophys. Res. Commun., **244**, 79-84 (1998)
[29] Knutzon, D.S.; Hayes, T.R.; Wyrick, A.; Xiong, H.; Maelor Davies, H.; Voelker, T.A.: Lysophosphatidic acid acyltransferase from coconut endosperm mediates the insertion of laurate at the sn-2 position of triacylglycerols in

lauric rapeseed oil and can increase total laurate levels. Plant Physiol., **120**, 739-746 (1999)
[30] Bourgis, F.; Kader, J.C.; Barret, P.; Renard, M.; Robinson, D.; Robinson, C.; Delseny, M.; Roscoe, T.J.: A plastidial lysophosphatidic acid acyltransferase from oilseed rape. Plant Physiol., **120**, 913-922 (1999)
[31] Wilmer, J.A.; Brown, A.P.; Forsyth, K.; Carnaby, S.; Barsby, T.; Slabas, A.R.: Limnanthes douglasii erucic acid-specific lysophospatidic acid acyltransferase activity in oilseed rape: an analysis of biochemical effects. Biochem. Soc. Trans., **28**, 964-966 (2000)
[32] Maisonneuve, S.; Bessoule, J.J.; Lessire, R.; Delseny, M.; Roscoe, T.J.: Mutagenesis of a plastidial lysophosphatidic acid acyltransferase. Biochem. Soc. Trans., **28**, 961-964 (2000)
[33] Yamashita, A.; Kawagishi, N.; Miyashita, T.; Nagatsuka, T.; Sugiura, T.; Kume, K.; Shimizu, T.; Waku, K.: ATP-independent fatty acyl-coenzyme A synthesis from phospholipid: coenzyme A-dependent transacylation activity toward lysophosphatidic acid catalyzed by acyl-coenzyme A:lysophosphatidic acid acyltransferase. J. Biol. Chem., **276**, 26745-26752 (2001)
[34] Jolly, C.A.; Kannan, L.: Albumin stimulates lysophosphatidic acid acyltransferase activity in T-lymphocyte membranes. Lipids, **37**, 475-480 (2002)

2-Acylglycerol-3-phosphate O-acyltransferase 2.3.1.52

1 Nomenclature

EC number
2.3.1.52

Systematic name
acyl-CoA:2-acyl-sn-glycerol 3-phosphate O-acyltransferase

Recommended name
2-acylglycerol-3-phosphate O-acyltransferase

Synonyms
2-acylglycerophosphate acyltransferase
acyl-CoA:2-monoacyl-sn-glycerol 3-phosphate acyltransferase
acyltransferase, 2-acylglycerol phosphate

CAS registry number
51901-17-8

2 Source Organism

<1> *Rattus norvegicus* (Wistar strain, pregnant female [1]) [1, 4]
<2> *Mus musculus* (C57BL strain [1]) [1]
<3> *Saccharomyces cerevisiae* (OC-2 [2]) [2]
<4> *Tetrahymena pyriformis* (WH-14 [3]) [3]
<5> *Spinacia oleracea* (spinach, cv. Subito [5]) [5]
<6> *Pisum sativum* (pea, cv. Kleine Rheinlaenderin [5]) [5]

3 Reaction and Specificity

Catalyzed reaction
acyl-CoA + 2-acyl-sn-glycerol 3-phosphate = CoA + 1,2-diacyl-sn-glycerol 3-phosphate

Reaction type
acyl group transfer

Natural substrates and products
S acyl-CoA + 2-acyl-sn-glycerol 3-phosphate <4> (<4> initial step in glycerophospholipid synthesis [3]) (Reversibility: ? <4> [3]) [3]
P CoA + 1,2-diacyl-sn-glycerol 3-phosphate

Substrates and products

S arachidonoyl-CoA + 2-acyl-sn-glycerol 3-phosphate <1> (<1> poor acyl donor [4]) (Reversibility: ? <1> [4]) [4]
P CoA + 1-arachidonoyl-2-acyl-sn-glycerol 3-phosphate
S lauroyl-CoA + 2-acyl-sn-glycerol 3-phosphate <1> (<1> poor acyl donor [4]) (Reversibility: ? <1> [4]) [4]
P CoA + 1-lauroyl-2-acyl-sn-glycerol 3-phosphate
S linoleoyl-CoA + 2-acyl-sn-glycerol 3-phosphate <1> (<1> poor acyl donor [4]) (Reversibility: ? <1> [4]) [4]
P CoA + 1-linoleoyl-2-acyl-sn-glycerol 3-phosphate
S myristoyl-CoA + 2-acyl-sn-glycerol 3-phosphate <1> (<1> poor acyl donor [4]) (Reversibility: ? <1> [4]) [4]
P CoA + 1-myristoyl-2-acyl-sn-glycerol 3-phosphate
S oleoyl-CoA + 2-acyl-sn-glycerol 3-phosphate <1-4, 5> (<2> preferred substrate [1]; <1,4> poor [3,4]) (Reversibility: ? <1-4, 5> [1-4, 5]) [1-4, 5]
P CoA + 1-oleoyl-2-acyl-sn-glycerol 3-phosphate
S palmitoleoyl-CoA + 2-acyl-sn-glycerol 3-phosphate <1> (<1> poor acyl donor [4]) (Reversibility: ? <1> [4]) [4]
P CoA + 1-palmitoleoyl-2-acyl-sn-glycerol 3-phosphate
S palmitoyl-CoA + 2-acyl-sn-glycerol 3-phosphate <1, 3, 4, 2, 5> (<1> preferred substrate [1,4]; <1> predominant selectivity for saturated acyl-CoA thioesters [4]) (Reversibility: ? <1, 3, 4, 2, 5> [1-5]) [1-5]
P CoA + 1-palmitoyl-2-acyl-sn-glycerol 3-phosphate <3> [2]
S stearoyl-CoA + 2-acyl-sn-glycerol 3-phosphate <1> (<1> most effective acyl donor [4]) (Reversibility: ? <1> [4]) [4]
P CoA + 1-stearoyl-2-acyl-sn-glycerol 3-phosphate

Specific activity (U/mg)

0.0065 <4> [3]
0.073 <1> (<1> partially purified [4]) [4]

Temperature optimum (°C)

25 <3, 4> (<3,4> assay at [2,3]) [2, 3]
37 <1, 2> (<1,2> assay at [1,4]) [1, 4]

5 Isolation/Preparation/Mutation/Application

Source/tissue

leaf <5> [5]
liver <1> [4]
mammary gland <1> (<1> lactating [1]) [1]
seedling <6> (<6> etiolated [5]) [5]

Localization

membrane <3> [2]
microsome <1, 2, 4-6> [1, 3-5]
pellicle <4> (<4> to a lesser extent [3]) [3]

Purification

<1> (partial, solubilized with Triton X-100 [4]) [4]

6 Stability

Storage stability

<3>, -70°C, membrane preparation, at least 2 months [2]

References

[1] Cooper, S.M.; Grigor, M.R.: Fatty acid specificities of microsomal acyltransferases esterifying positions-1 and -2 of acylglycerols in mammary glands from lactating rats. Biochem. J., **187**, 289-295 (1980)

[2] Yamada, K.; Okuyama, H.; Endo, Y.; Ikezawa, H.: Acyltransferase systems involved in phospholipid metabolism in Saccharomyces cerevisiae. Arch. Biochem. Biophys., **183**, 281-289 (1977)

[3] Okuyama, H.; Yamada, K.; Kameyama, Y.; Ikezawa, H.; Fukushima, H.; Nozawa, Y.: Studies on lipid metabolism in Tetrahymena pyriformis: properties of acyltransferase systems. Arch. Biochem. Biophys., **178**, 319-326 (1977)

[4] Yamashita, S.; Hosaka, K.; Numa, S.: Acyl-donor specificities of partially purified 1-acylglycerophosphate acyltransferase, 2-acylglycerophosphate acyltransferase and 1-acylglycerophosphorylcholine acyltransferase from rat-liver microsomes. Eur. J. Biochem., **38**, 25-31 (1973)

[5] Frentzen, M.: Comparison of certain properties of membrane bound and solubilized acetyltransferase activities of plant microsomes. Plant Sci., **69**, 39-48 (1990)

Phenylalanine N-acetyltransferase 2.3.1.53

1 Nomenclature

EC number
2.3.1.53

Systematic name
acetyl-CoA:L-phenylalanine N-acetyltransferase

Recommended name
phenylalanine N-acetyltransferase

Synonyms
acetyl-CoA-L-phenylalanine α-N-acetyltransferase
acetyltransferase, phenylalanine

CAS registry number
9075-16-5

2 Source Organism

<1> *Escherichia coli* (K12 [1]) [1]

3 Reaction and Specificity

Catalyzed reaction
acetyl-CoA + L-phenylalanine = CoA + N-acetyl-L-phenylalanine

Reaction type
acyl group transfer

Natural substrates and products
S acetyl-CoA + L-phenylalanine <1> (<1> enzyme is specific for L-isomer [1]) (Reversibility: ? <1> [1]) [1]
P CoA + N-acetyl-L-phenylalanine

Substrates and products
S acetyl-CoA + L-alanine <1> (<1> 29% of activity with L-phenylalanine [1]) (Reversibility: ? <1> [1]) [1]
P CoA + N-acetyl-L-alanine <1> [1]
S acetyl-CoA + L-histidine <1> (<1> 60% of activity with L-phenylalanine [1]) (Reversibility: ? <1> [1]) [1]
P CoA + N-acetyl-L-histidine <1> [1]

S acetyl-CoA + L-phenylalanine <1> (<1> enzyme is specific for L-isomer [1]) (Reversibility: ? <1> [1]) [1]
P CoA + N-acetyl-L-phenylalanine <1> [1]

Inhibitors

N-ethylmaleimide <1> (<1> 0.1 mM, total inhibition [1]) [1]
iodoacetamide <1> (<1> 0.5 mM, total inhibition [1]) [1]
p-hydroxymercuribenzoate <1> (<1> 0.1 mM, total inhibition [1]) [1]

Specific activity (U/mg)

75 <1> (<1> L-phenylalanine [1]) [1]

K_m-Value (mM)

0.13 <1> (acetyl-CoA) [1]
3.3 <1> (L-phenylalanine) [1]

pH-Optimum

8 <1> [1]

Temperature optimum (°C)

37 <1> (<1> assay at [1]) [1]

4 Enzyme Structure

Molecular weight

18000 <1> (<1> low speed sedimentation without reaching equilibrium [1]) [1]
24000 <1> (<1> gel filtration [1]) [1]

5 Isolation/Preparation/Mutation/Application

Purification

<1> (160fold [1]) [1]

References

[1] Krishna, R.V.; Krishnaswamy, P.R.; Rajagopal Rao, D.: Enzymic synthesis of N-acetyl-L-phenylalanine in Escherichia coli K12. Biochem. J., **124**, 905-913 (1971)

Formate C-acetyltransferase 2.3.1.54

1 Nomenclature

EC number
2.3.1.54

Systematic name
acetyl-CoA:formate C-acetyltransferase

Recommended name
formate C-acetyltransferase

Synonyms
PFL <10, 14, 19> [4, 8, 10, 11, 13, 15-23]
enzyme I <10> [3]
formate acetyltransferase
pyruvate formate-lyase
pyruvic formate-lyase

CAS registry number
9068-08-0

2 Source Organism

<-1> no activity in *Clostridium pasteurianum* (ATCC 6013 [1]) [1, 12]
<1> *bacteriophage T4* [18]
<2> *Clostridium butylicum* (ATCC 14823 [1]) [1]
<3> *Clostridium butyricum* (ATCC 19398 [1,2]; DSM552 [12]) [1, 2, 5, 12, 17]
<4> *Clostridium kluyveri* [1, 12]
<5> *Clostridium pasteurianum* (DSM525 [12]) [12]
<6> *Clostridium pasteurianum* (DSM525 [12]) [12, 20]
<7> *Clostridium sp.* [1, 4, 5, 12]
<8> *Escherichia coli* (PFL wild-type, structure factor amplitude and model coordinate [16]) [16]
<9> *Escherichia coli* (PFL [18]) [18]
<10> *Escherichia coli* (K12 [3,4,6,7,17]; ATCC 10798 [6,17]) [1-23]
<11> *Enterococcus faecalis* (previously Streptococcus faecalis [1,2,4,5]) [1, 2, 4, 5, 12]
<12> *Haemophilus influenzae* [18]
<13> *Lactobacillus bulgaricus* [5]

<14> *Lactococcus lactis* (subsp. cremoris, MG1363 [23]; formerly designated Streptococcus lactis [5,17]; formerly designated Streptococcus cremoris [17]) [5, 17, 20, 23]
<15> *Micrococcus lactilyticus* [1]
<16> *Pasteurella multocida* [18]
<17> *Salmonella typhimurium* [18]
<18> *Streptococcus bovis* (wild-type JB1 pfl [20]) [20]
<19> *Streptococcus bovis* (wild-type JB1 [17,20]) [17, 20]
<20> *Streptococcus mutans* (JC2 [5]) [5, 6, 12, 17]
<21> *Streptococcus salivarius* (ATCC 13419 [12]) [12]
<22> *Streptococcus sanguis* (DSM20066 [12]) [12]
<23> *Vibrio cholerae* [18]
<24> *Yersinia pestis* [18]

3 Reaction and Specificity

Catalyzed reaction
acetyl-CoA + formate = CoA + pyruvate

Reaction type
acyl group transfer

Natural substrates and products
S acetyl-CoA + formate <10> (Reversibility: r <10> [3]) [3]
P pyruvate + CoA
S pyruvate + CoA <1-24> (<10> key role in the metabolism [3]; <10> key enzyme of the glucose fermentation route in anaerobically growing cells [6,7,9,14,15,21]; <10> central reaction in the anaerobic metabolism [4]; <10> acetyl-CoA for ATP synthesis in catabolism [1]; <14> main enzyme competing for pyruvate under anaerobic conditions [23]; <2-4,6> required for synthesis of C_1 units in anabolism [1,2,12]; <10,11,15,20> catabolic function, leading to production of ATP [1,2,12]; <3> physiological function is both anabolic and catabolic [1]) (Reversibility: r <1-5, 7-12, 14, 16-19, 22-24> [1, 3, 4, 7, 18, 23]; ir <10, 11, 15> [1]) [1-23]
P acetyl-CoA + formate

Substrates and products
S 2-oxobutyrate + CoA <2-4> (Reversibility: r <2-4> [1]) [1]
P acetyl-CoA + acetate
S acetyl-CoA + formate <2-4, 10> (Reversibility: r <2-4, 10> [1, 3, 4, 7, 18, 21, 23]) [1, 3, 4, 7, 18, 21, 23]
P pyruvate + CoA
S pyruvate + CoA <1-24> (<10> reverse reaction with only 0.1% velocity of the forward reaction [1]; <11> no report on the reverse reaction [1]) (Reversibility: r <1-5, 7-12, 14, 16-19, 22-24> [1, 3, 4, 7, 12, 14, 16-18, 20, 21, 23]; ir <10, 11, 15> [1]; ? <13, 21> [5, 12]) [1-23]
P acetyl-CoA + formate

S pyruvate + dephospho-CoA <2-4> (Reversibility: r <2-4> [1]) [1]
P dephospho-acetyl-CoA + formate
S pyruvate + dithiothreitol <10> (Reversibility: ? <10> [3]) [3]
P S-acetyl-dithiothreitol + formate <10> [3]
S pyruvate + formate <2-4, 10, 11, 15> (<2-4,10,11,15> carboxyl group exchange reaction [1,4]) (Reversibility: r <2-4, 10, 11, 15> [1, 4]) [1, 4]
P pyruvate + formate <2-4, 10, 11, 15> [1, 4]
S pyruvate + phosphate <15> (Reversibility: ir <15> [1]) [1]
P acetylphosphate + formate <15> [1]
S Additional information <2-4, 10> (<10> not essential for nitrate respiration [11]; <2-4> phosphate cannot substitute for CoA [1]) [1, 11]
P ?

Inhibitors

ADP <2-4> [1]
ATP <2-4, 14, 20> (<2-4> regulates pyruvate cleavage to acetyl-CoA and formate, 10 mM reduces the reactin velocity by 95% [1]) [1, 17]
Cl^- <2-4> [1]
CoA <2-4> (<2-4> inhibits exchange reaction between pyruvate and formate, inhibits formate formation from pyruvate above 0.075 mM [1]) [1]
D-glyceraldehyde-3-phosphate <14, 19, 20> [5, 17, 20, 23]
DEAE-cellulose <2-4> (<2-4> treated extracts do not catalyze formate formation from pyruvate [1]) [1]
HPO_4^{2-}<2-4> [1]
S-adenosyl-L-homocysteine <3, 10> [2, 4]
SO_4^{2-} <2-4> [1]
acetylphosphinate <10> (<10> mechanism-based inactivator [8]) [8]
dihydroxyacetone phosphate <14, 19, 20> [5, 17, 20, 23]
dioxygen <10> [19]
formate <10> (<10> product inhibition [3]) [3]
hypophosphite <2-4, 10> (<2-4> strongly inhibits formate formation from pyruvate [1]) [1, 8]
iodoacetate <10> [6]

Activating compounds

S-adenosylmethionine <2-4, 10, 20> (<2-4,10> obligatory component of the activation reaction [1,4,22]) [1, 2, 4-6, 13, 22]
Y06I <1> (<1> autonomous glycyl radical cofactor, reconstituting the catalytic center of oxygen-fragmented enzyme [18]) [18]
YfiD <9, 10, 12, 16, 17, 23, 24> (<10> autonomous glycyl radical cofactor, reconstituting the catalytic center of oxygen-fragmented enzyme [18]) [18]
ferredoxin <2-4, 10, 20> (<2-4> reactivation of inactive enzyme requires reduced ferredoxin, Fe^{2+}-dithiol or Co^{2+}-mercaptoethanol complexes can substitute for ferredoxin [1]; <20> methylviologen can replace ferredoxin [5]) [1, 2, 5]

oxamate <10, 20> (<10> obligatory component of the activation reaction [4, 22]) [4, 5, 22]
pyruvate <10> (<10> obligatory component of the activation reaction [4, 22]) [4, 22]

Metals, ions
Mg^{2+} <2-4> (<2-4> stimulates pyruvate cleavage about 2fold [1]) [1]

Turnover number (min^{-1})
228 <10> (formate) [3]
260 <10> (formate, <10> backward direction [14]) [14]
660 <10> (pyruvate) [3]
770 <10> (pyruvate, <10> forward direction [14]) [14]

Specific activity (U/mg)
12 <20> [5]
25 <10> [4]
60 <10> [3]
160 <10> [6]

K_m-Value (mM)
0.0068 <10> (CoA) [3, 4]
0.012 <3> (CoA, <3> pyruvate [1]) [1]
0.051 <10> (acetyl-CoA) [3, 4]
0.26 <3> (acetyl-CoA) [1]
0.6 <3> (pyruvate, <3> pyruvate-formate exchange [1]) [1]
1.6 <3> (pyruvate) [1, 17]
2 <19> (pyruvate) [17]
2 <10> (pyruvate) [4]
2.05 <10> (pyruvate) [3]
2.6 <20> (pyruvate) [17]
10 <3> (formate) [1]
20 <3> (formate, <3> pyruvate-formate exchange [1]) [1]
24 <10> (formate) [3, 4]
24.5 <10> (formate) [3]

pH-Optimum
7-7.5 <2-4> (Tris-acetate buffer [1]) [1]
7.5 <2-4, 19, 20> [1, 5, 17]
7.8-8.4 <10> [4]
7.8-8.5 <10> (<10> forward direction [3]) [3]

pH-Range
6.5-8.3 <20> (<20> 40% of activity maximum at pH 6.5, between pH 7.0-8.3 80% of maximum activity [5]) [5]

Temperature range (°C)
22-35 <20> (<20> activity is about 78 times of that at 22°C [5]) [5]

4 Enzyme Structure

Molecular weight

80000 <10> (<10> identification of pfl gene product [6]) [6]
80000-85000 <21, 22> (<21,22> SDS-PAGE [12]) [12]
85120 <10> (<10> amino acid sequence calculation [7]) [7]
140000 <10> (<10> gel filtration, sedimentation velocity [4]) [4]
150000 <10> (<10> gel filtration [3]) [3]
151000 <10> (<10> sedimentation equilibrium [6]) [6]
170000 <10> [8, 10]
170000-180000 <20> (<20> gel filtration [5]) [5]
180000 <19> (<19> TOF-MS flight mass spectrometry [17]) [17]

Subunits

dimer <6, 10, 19, 20> (<6,10> 2 * 85000, homodimer, SDS-PAGE [6,7,9,12,16,19,22]; <10> 2 * 70000 [4]; <10> 2 * 85500, homodimer, SDS-PAGE [6]; <20> 2 * 90000, SDS-PAGE [5]; <19> 2 * 80000-90000, SDS-PAGE [17]; <19> 1 * 87000 + 1 * 83000, TOF-MS flight mass spectrometry, SDS-PAGE [17,20]) [4-10, 16, 17, 20-22]

5 Isolation/Preparation/Mutation/Application

Purification

<10> (partially [4]; recombinant enzyme [16]) [3-7, 9, 16, 19, 22]
<11> [5, 12]
<19> (recombinant enzyme [20]) [20]
<20> [5, 12, 17]

Renaturation

<10> [23]

Crystallization

<10> (space group P4(3)2(1)2, cell parameters a = b = 158.36 A, c = 159.30 A [22]; tetragonal crystals in complex with pyruvate, monoclinic cocrystals with CoA and either pyruvate or oxamate obtained by hanging drop method, space group C222, a = 54.90 A, b = 153.05 A, c = 205.95 [21]; space group P4(3)2(1)2 a = b = 159 A, c = 160 A [16]; space group P4(1)2(1)2 [10]) [10, 16, 18, 19, 21, 22]

Cloning

<6> (pfl gene cloned [12]) [12, 20]
<10> (structural gene pfl cloned and sequenced, homologous expression and overproduction in Escherichia coli K12 [7]; recombinant PFL expressed in Escherichia coli BL21 (DE3) [19,22]; overproducing of the enzyme in Escherichia coli 234M1 transformed with expression vector p153E1 [21]; Escherichia coli SP 264 contains plasmid p29 with pfl gene, identification of pfl gene product as a 80kDa polypeptide [6]; recombinant pfl in Escherichia coli

RM221 [16]; recombinant pfl in Escherichia coli RM220 [20]) [6, 7, 11, 12, 16, 19-22]
<14> (pfl gene expressed under control of different constitutive promoters, PFL-deficient strain CRM40 complemented [23]) [20, 23]
<19> (pfl gene cloned and expressed in Escherichia coli HB101 [17]) [17, 20]

Application
nutrition <14> (<14> plays a significant role in industrial milk fermentation [23]) [23]

6 Stability

pH-Stability
6 <19> (<19> activity decreases to less than 10% of the maximal activity [17]) [17]

Temperature stability
0-30 <10> (<10> stable in anaerobic buffers for days at 0°C, at 30°C for several hours [10]) [10]
30 <10> (<10> activity declines with a half-time of 50 min [4]) [4]

Oxidation stability
<10>, extreme sensitivity towards inactivation by oxygen, admittance of oxygen inactivates completely within a few seconds [3, 4]
<10>, obligate anaerobic enzyme, enzymatic conversion by a unique homolytic mechanism that involves a free radical harbored in the protein structure, protein based organic free radical is essential for catalysis, oxygen destruction of the protein radical [8, 9, 19]
<20>, immediately inactivated by exposure to the air [5]

General stability information
<10>, virtually stable at 0°C and pH 8.0 if kept in media which display a redoxpotential of 0.2 V or below [4]
<20>, unstable even when stored anaerobically [5]

Storage stability
<10>, -20°C, 10 mM MOPS-KOH, pH 7.0, 50 mM KCl, thiols, EDTA in 50% ethyleneglycol, proved stable for months [6]
<10>, -20°C, stable for at least several months [4]
<20>, 4°C, 50 mM potassium phosphate buffer, containing 1 mM dithiothreitol, storage of the purified enzyme in an anaerobic glove box for 2 weeks reduces activity by 50% [5]

References

[1] Thauer, R.K.; Kirchniawy, F.H.; Jungermann, K.A.: Properties and function of the pyruvate-formate-lyase reaction in clostridiae. Eur. J. Biochem., **27**, 282-290 (1972)

[2] Wood, N.P.; Jungermann, K.: Inactivation of the pyruvate formate lyase of Clostridium butyricum. FEBS Lett., **27**, 49-52 (1972)

[3] Knappe, J.; Balschkowski, H.P.; Gröbner, P.; Schmitt, T.: Pyruvate formate-lyase of Escherichia coli: the acetyl-enzyme intermediate. Eur. J. Biochem., **50**, 253-263 (1974)

[4] Knappe, J.; Blaschkowski, H.P.: Pyruvate formate-lyase from Escherichia coli and its activation system. Methods Enzymol., **41B**, 508-518 (1975)

[5] Takahashi, S.; Abbe, K.; Yamada, T.: Purification of pyruvate formate-lyase from Streptococcus mutans and its regulatory properties. J. Bacteriol., **149**, 1034-1040 (1982)

[6] Conradt, H.; Hohmann-Berger, M.; Hohmann, H.P.; Blaschkowski, H.P.; Knappe, J.: Pyruvate formate-lyase (inactive form) and pyruvate formate-lyase activating enzyme of Escherichia coli: Isolation and structural properties. Arch. Biochem. Biophys., **228**, 133-142 (1984)

[7] Rödel, W.; Plaga, W.; Frank, W.; Knappe, J.: Primary structures of Escherichia coli pyruvate formate-lyase and pyruvate-formate-lyase-activating enzyme deduced from the DNA nucleotide sequences. Eur. J. Biochem., **177**, 153-158 (1988)

[8] Ulissi-DeMario, L.; Brush, E.J.; Kozarich, J.W.: Mechanism-based inactivation of pyruvate formate-lyase reaction in Clostridiae. J. Am. Chem. Soc., **113**, 4341-4342 (1991)

[9] Wagner, A.F.V.; Frey, M.; Neugebauer, F.A.; Schäfer, W.; Knappe, J.: The free radical in pyruvate formate-lyase is located on glycine-734. Proc. Natl. Acad. Sci. USA, **89**, 996-1000 (1992)

[10] Knappe, J.; Elbert, S.; Frey, M.; Wagner, A.F.V.: Pyruvate formate-lyase mechanism involving the protein-based glycyl radical. Biochem. Soc. Trans., **21**, 731-734 (1993)

[11] Kaiser, M.; Sawers, G.: Pyruvate formate-lyase is not essential for nitrate respiration by Escherichia coli. FEMS Microbiol. Lett., **117**, 163-168 (1994)

[12] Weidner, G.; Sawers, G.: Molecular characterization of the genes encoding pyruvate formate-lyase and its activating enzyme of Clostridium pasteurianum. J. Bacteriol., **178**, 2440-2444 (1996)

[13] Broderick, J.B.; Duderstadt, R.E.; Fernandez, D.C.; Wojtuszewski, K.; Henshaw, T.F.; Johnson, M.K.: Pyruvate formate-lyase activating enzyme is an iron-sulfur protein. J. Am. Chem. Soc., **119**, 7396-7397 (1997)

[14] Himo, F.; Eriksson, L.A.: Catalytic mechanism of pyruvate formate-lyase (PFL). A theoretical study. J. Am. Chem. Soc., **120**, 11449-11455 (1998)

[15] Kulzer, R.; Pils, T.; Kappl, R.; Huttermann, J.; Knappe, J.: Reconstitution and characterization of the polynuclear iron-sulfur cluster in pyruvate formate-lyase-activating enzyme. Molecular properties of the holoenzyme form. J. Biol. Chem., **273**, 4897-4903 (1998)

[16] Becker, A.; Fritz-Wolf, K.; Kabsch, W.; Knappe, J.; Schultz, S.; Wagner, A.F.V.: Structure and mechanism of the glycyl radical enzyme pyruvate formate-lyase. Nat. Struct. Biol., **6**, 969-975 (1999)
[17] Asanuma, N.; Hino, T.: Effects of pH and energy supply on activity and amount of pyruvate formate-lyase in Streptococcus bovis. Appl. Environ. Microbiol., **66**, 3773-3777 (2000)
[18] Wagner, A.F.V.; Schultz, S.; Bomke, J.; Pils, T.; Lehmann, W.D.; Knappe, J.: YfiD of Escherichia coli and Y06I of bacteriophage T4 as autonomous glycyl radical cofactors reconstituting the catalytic center of oxygen-fragmented pyruvate formate-lyase. Biochem. Biophys. Res. Commun., **285**, 456-462 (2001)
[19] Zhang, W.; Wong, K.K.; Magliozzo, R.S.; Kozarich, J.W.: Inactivation of pyruvate formate-lyase by dioxygen: Defining the mechanistic interplay of glycine 734 and cysteine 419 by rapid freeze-quench EPR. Biochemistry, **40**, 4123-4130 (2001)
[20] Asanuma, N.; Hino, T.: Molecular characterization and expression of pyruvate formate-lyase-activating enzyme in a ruminal bacterium, Streptococcus bovis. Appl. Environ. Microbiol., **68**, 3352-3357 (2002)
[21] Becker, A.; Kabsch, W.: X-ray structure of pyruvate formate-lyase in complex with pyruvate and CoA. How the enzyme uses the Cys-418 thiyl radical for pyruvate cleavage. J. Biol. Chem., **277**, 40036-40042 (2002)
[22] Lehtioe, L.; Leppaenen, V.M.; Kozarich, J.W.; Goldman, A.: Structure of Escherichia coli pyruvate formate-lyase with pyruvate. Acta Crystallogr. Sect. D, **58**, 2209-2212 (2002)
[23] Melchiorsen, C.R.; Jokumsen, K.V.; Villadsen, J.; Israelsen, H.; Arnau, J.: The level of pyruvate-formate lyase controls the shift from homolactic to mixed-acid product formation in Lactococcus lactis. Appl. Microbiol. Biotechnol., **58**, 338-344 (2002)

Kanamycin 6'-N-acetyltransferase 2.3.1.55

1 Nomenclature

EC number
2.3.1.55 (transferred to EC 2.3.1.82)

Recommended name
kanamycin 6'-N-acetyltransferase

Aromatic-hydroxylamine O-acetyltransferase 2.3.1.56

1 Nomenclature

EC number
2.3.1.56

Systematic name
N-hydroxy-4-acetylaminobiphenyl:N-hydroxy-4-aminobiphenyl O-acetyltransferase

Recommended name
aromatic-hydroxylamine O-acetyltransferase

Synonyms
N,O-acetyltransferase
N-hydroxy-2-acetylaminofluorene N-O acyltransferase
NAT <4> [11]
acetyltransferase, aromatic hydroxylamine
aromatic hydroxylamine acetyltransferase
arylhydroxamate N,O-acetyltransferase
arylhydroxamate acyltransferase
arylhydroxamic acid N,O-acetyltransferase
arylhydroxamic acid acyltransferase
arylhydroxamic acyltransferase
Additional information <1, 3> (<1,3> monomorphic enzyme [10]) [10]

CAS registry number
52660-15-8

2 Source Organism

<-1> no activity in *Cavia porcellus* [7]
<1> *Rattus norvegicus* (Sprague-Dawley CD strain [3]) [1-3, 5-7, 9, 10, 13]
<2> *Oryctolagus cuniculus* [7, 10]
<3> *Mesocricetus auratus* [2, 4, 5, 7, 10]
<4> *Homo sapiens* (isoforms NAT1 and NAT2 [11]) [8, 10, 11, 13, 14]
<5> *Mus musculus* [10]
<6> *Rattus norvegicus* (isoform NAT1 [12]) [12]
<7> *Rattus norvegicus* (isoform NAT2 [12]) [12]
<8> *Mesocricetus auratus* (isoform NAT1 [12]) [12]
<9> *Mesocricetus auratus* (isoform NAT2 [12]) [12]

<10> *Canis familiaris* (mongrel dogs [13]) [13]
<11> *Salmonella typhimurium* (Nho gene [15]) [15]

3 Reaction and Specificity

Catalyzed reaction

N-hydroxy-4-acetylaminobiphenyl + N-hydroxy-4-aminobiphenyl = N-hydroxy-4-aminobiphenyl + N-acetoxy-4-aminobiphenyl (<1-3,5> a single cytosolic protein catalyzes N-acetylation, O-acetylation and N,O-acetylation [10]; <1-5> mechanism [10]; <1> ratio of N-acetylation of arylamines, O-aectylation of arylhydroxylamines and N,O-acetylation of arylhydroxamic acids, which are all performed by a single protein, may be highly dependent on the conformation of the enzyme [9])

Reaction type

acyl group transfer

Natural substrates and products

S N-hydroxy-4-acetylaminobiphenyl + N-hydroxy-4-aminobiphenyl <1> (<1> transfers the N-acetyl group of some aromatic acethydroxamic acids to the O-position of some aromatic hydroxylamines [1]; <1> noninducible, no known natural substrate in rat [3]) (Reversibility: ? <1> [1, 3]) [1, 3]
P N-hydroxy-4-aminobiphenyl + N-acetoxy-4-aminobiphenyl
S Additional information <4, 10> (<10> important for the activation of urinary metabolites of carcinogenic arylamines for the initiation of bladder carcinogenesis in the dog [13]; <4> metabolic activation by isoforms 1 and 2 of 3-nitrobenzanthrone and N-acetyl-N-hydroxy-3-aminobenzantrone, 2 compounds with high mutagenic and genotoxic potential, formation of DNA adducts [11]) [11, 13]
P ?

Substrates and products

S 7-acetyl-N-hydroxy-2-acetylaminofluorene + 4-aminoazobenzene <4> (Reversibility: ? <4> [14]) [14]
P 7-acetyl-N-hydroxy-2-aminofluorene + N-acetoxy-4-aminoazobenzene
S 7-bromo-N-hydroxy-2-acetylaminofluorene + 4-aminoazobenzene <4> (Reversibility: ? <4> [14]) [14]
P 7-bromo-N-hydroxy-2-aminofluorene + N-acetoxy-4-aminoazobenzene
S 7-ethoxy-N-hydroxy-2-acetylaminofluorene + 4-aminoazobenzene <4> (Reversibility: ? <4> [14]) [14]
P 7-ethoxy-N-hydroxy-2-aminofluorene + N-acetoxy-4-aminoazobenzene
S 7-ethyl-N-hydroxy-2-acetylaminofluorene + 4-aminoazobenzene <4> (Reversibility: ? <4> [14]) [14]
P 7-ethyl-N-hydroxy-2-aminofluorene + N-acetoxy-4-aminoazobenzene
S N-hydroxy-2-acetylaminofluorene + 4-aminoazobenzene <4> (Reversibility: ? <4> [14]) [14]

P N-hydroxy-2-aminofluorene + N-acetoxy-4-aminoazobenzene

S N-hydroxy-2-acetylaminofluorene + N-hydroxy-4-aminobiphenyl <1, 3, 4, 10> (<1> i.e. N-hydroxy-N^2-fluorenylacetamide, not mammary gland enzyme [1]) (Reversibility: r <4> [8]; ? <1, 3, 10> [1-3, 6, 10, 13]) [1-3, 6, 8, 10, 13]

P N-hydroxy-2-aminofluorene + N-acetoxy-4-aminobiphenyl <1, 4, 10> [8, 10, 13]

S N-hydroxy-2-propylaminofluorene + ? <1> (<1> i.e. N-hydroxy-N^2-fluorenylpropionamide [6]) (Reversibility: ? <1> [6]) [6]

P ?

S N-hydroxy-4-acetylaminobiphenyl + 4-aminoazobenzene <1, 3, 6-9> (Reversibility: ? <1, 3, 6-9> [1, 2, 4, 6, 12]) [1, 2, 4, 6, 12]

P N-hydroxy-4-aminobiphenyl + N-acetoxy-4-aminoazobenzene <1, 3, 6-9> [1, 2, 12]

S N-hydroxy-4-acetylaminobiphenyl + N-hydroxy-4-aminobiphenyl <1-4> (<1-3> S-acetyl-CoA cannot replace N-hydroxy-4-acetylaminobiphenyl [1,7,8]) (Reversibility: r <4> [8]; ? <1-3> [1, 7]) [1, 7, 8]

P N-hydroxy-4-aminobiphenyl + N-acetoxy-4-aminobiphenyl <1-4> [1, 7, 8]

S acetyl-CoA + 4-aminoveratrole <11> (Reversibility: ? <11> [15]) [15]

P CoA + N-acetyl-4-aminoveratrole <11> [15]

S acetyl-CoA + 4-anisidine <11> (Reversibility: ? <11> [15]) [15]

P CoA + N-acetyl-4-anisidine <11> [15]

S acetyl-CoA + 4-iodoaniline <11> (Reversibility: ? <11> [15]) [15]

P CoA + N-acetyl-4-iodoaniline <11> [15]

S Additional information <1, 3, 4> (<4> 7-substituted-N-hydroxy-2-acetylamino fluorenes form adducts with DNA, which is inhibited by addition of ascorbic acid, 4-(2-biphenyl)-2-hydroxytetronic acid, and partly by N-tert-butyl-α-phenylnitrone, triethylphosphite, potassium O-ethylxanthate, thiourea, and N-acetylmethionine, not in case of 7-acetyl-N-hydroxy-2-acetylamino fluorene [14]; <1,3> acetohydroxamic acids with the N-atom of the hydroxamic acid group attached directly to aliphatic or cycloalkyl groups are no substrates or inhibitors [5]) [5, 14]

P ?

Inhibitors

1-(fluoren-2-yl)-2-propen-1-one <1> (<1> irreversible and selective inactivation [10]) [10]

7-substituted N-hydroxy-2-acetamidofluorenes <3> (<3> residues: H, halo, CN, acetoxy, methoxy, ethoxy, butoxy, inhibition kinetics [4]) [4]

N-acetamido-N-hydroxy-7-methoxyfluorene <3> (<3> inhibition kinetics [4]) [4]

N-ethylmaleimide <1> [1]

N-hydroxy-2-acetylaminofluorene <1, 3> (<1,3> irreversible, suicide substrate [2,10]) [2, 10]

N-hydroxy-4-acetylaminobiphenyl <1, 3> [2]

NaF <1> (<1> weak [1]) [1]

PCMB <1> [1]
iodoacetamide <1> [1]
puromycin <1> (weak [1]) [1]
trans-N-hydroxy-4-acetylaminostilbene <1, 3> (<1,3> no protection by GSH, cysteine or N-acetylmethionine [2]) [2]
Additional information <1> (<1> no inhibition by paraoxon [3]) [3]

Activating compounds

reduced pyridine nucleotides <1> (<1> mammary gland [1]) [1]

Specific activity (U/mg)

0.0075 <4> (<4> 7-ethyl-N-hydroxy-2-acetylaminofluorene + 4-aminoazobenzene [14]) [14]
0.025 <7> (<7> recombinant enzyme, N-acetylation of 4-aminoazobenzene [12]) [12]
0.036 <9> (<9> recombinant enzyme, N-acetylation of 4-aminoazobenzene [12]) [12]
0.084 <8> (<8> recombinant enzyme, N-acetylation of 4-aminoazobenzene [12]) [12]
0.097 <6> (<6> recombinant enzyme, N-acetylation of 4-aminoazobenzene [12]) [12]
0.785 <11> (<11> purified enzyme [15]) [15]
Additional information <1, 4> (<4> N-, O-, and N,O-acetyltransferase activity, wild-type and mutants [8]; <1> 14 nmol aminofluorene bound to tRNA per min per mg protein [3]) [3, 8, 14]

K_m-Value (mM)

0.00064 <6> (4-aminoazobenzene) [12]
0.00089 <8> (4-aminoazobenzene) [12]
0.005 <6> (N-hydroxy-4-acetylaminobiphenyl, <6> recombinant enzyme [12]) [12]
0.0063 <1> (N-hydroxy-2-acetylaminofluorene) [3]
0.0068 <8> (N-hydroxy-4-acetylaminobiphenyl, <8> recombinant enzyme [12]) [12]
0.062 <9> (4-aminoazobenzene) [12]
0.134 <9> (N-hydroxy-4-acetylaminobiphenyl, <9> recombinant enzyme [12]) [12]
0.138 <7> (4-aminoazobenzene) [12]
0.356 <7> (N-hydroxy-4-acetylaminobiphenyl, <7> recombinant enzyme [12]) [12]
0.6 <11> (4-aminoveratrole) [15]
1.1 <11> (4-anisidine) [15]
1.3 <11> (4-iodoanisidine) [15]

pH-Optimum

6.8 <1> (<1> liver [1]) [1]
7 <1> (<1> nucleic acid adduct formation assay [3]) [3]
7.5 <1, 11> (<11> assay at [15]; <1> mammary gland [1]) [1, 15]
Additional information <1, 10> (<10> pI: 5.4-5.6 [13]; <1> pI: 4.5 [3]) [3, 13]

pH-Range

5.6-9.6 <1> (<1> about half-maximal activity at pH 5.6 and pH 9.6, nucleic acid adduct formation assay [3]) [3]
6.8-8.2 <1> (<1> about 80% of maximal activity at pH 6.8 and pH 8.2 [1]) [1]

Temperature optimum (°C)

37 <1, 3, 4, 11> (<1,3,4,11> assay at [1-3,8,15]) [1-3, 8, 15]
50 <1> (<1> nucleic acid adduct formation assay [3]) [3]

Temperature range (°C)

25-72 <1> (<1> about half-maximal activity at 25°C and 72°C [3]) [3]

4 Enzyme Structure

Molecular weight

28000 <1> [6]
32000 <1> (<1> gel filtration [9]) [9]
180000 <10> (<10> gel filtration [13]) [13]

Subunits

? <1, 6-9> (<9> x * 29000, isoform NAT2, SDS-PAGE [12]; <6-8> x * 31000, SDS-PAGE [12]; <1> x * 38500, SDS-PAGE [3]) [3, 12]
monomer <1> (<1> 1 * 32000, SDS-PAGE [9]) [9]
trimer <10> (<10> 3 * 60000, SDS-PAGE [13]) [13]

Posttranslational modification

glycoprotein <10> (<10> mannose-rich glycoprotein [13]) [13]

5 Isolation/Preparation/Mutation/Application

Source/tissue

blood <1> [6]
brain <1> [6]
colon <1, 3, 4> [6, 8, 10]
epithelium <10> (<10> of liver and urinary bladder [13]) [13]
esophagus <1> [13]
forestomach <1> [13]
intestine <1, 3> (<1> small and large [13]) [10, 13]
kidney <1> (<1> renal tubules [13]) [1, 6, 7, 13]
large intestine <1> [13]
liver <1-5, 10> (<10> epithelium [13]; <1> noninducible [3]) [1-7, 9-11, 13]
lung <1> (<1> alveolar cells [13]) [13]
mammary gland <1> [1]
muscle <1> (<1> low activity [1]; <1> not [6]) [1]
prostate gland <1> [13]
renal tubule <1> [13]
small intestine <1> [13]

small intestine mucosa <1> [1, 6, 7]
spleen <1> (<1> low activity [1]) [1]
stomach <1> (<1> glandular [13]) [6, 13]
trachea <1> [13]
urinary bladder <10> (<10> epithelium [13]) [13]
Additional information <1-3> (<1> no activity in rat ear duct sebaceous gland, lung, dog urinary bladder mucosa [1,7]; <2> rabbit bladder [7]) [1, 6, 7]

Localization
cytosol <1-5> [1-3, 7-10]
microsome <10> [13]

Purification
<1> (partially [2]) [3, 6, 9]
<3> (partially [2]) [2]
<10> [13]
<11> (recombinant His-tagged protein from E. coli, removal of His-tag [15]) [15]

Crystallization
<11> (sitting and hanging drop methods, 20°C, 10 mg/ml protein solution: 10 mM Tris-HCl, pH 7.9, 0.1 M NaCl, 3 mM dithiothreitol, reservoir solution: 0.8 M Na,K tartrate, 0.1 M MES, pH 7.1, X-ray diffraction [15]) [15]

Cloning
<4> (expression of isoforms 1 and 2 in chinese hamster lung fibroblast derived cell line V79MZ and in TA1538-derived Salmonella typhimurium strains [11]; expression of wild-type and mutants in Escherichia coli strains DH5α and JM105 [8]) [8, 11]
<6> (expression of isoform NAT1 in Escherichia coli [12]) [12]
<7> (expression of isoform NAT2 in Escherichia coli [12]) [12]
<8> (expression of isoforms NAT1 in Escherichia coli [12]) [12]
<9> (expression of isoforms NAT2 in Escherichia coli [12]) [12]
<11> (expression as His-tagged protein in Escherichia coli [15]) [15]

Engineering
G286E <4> (<4> small reduction in activity [8]) [8]
I114T <4> (<4> strong reduction of the acetylation activity [8]) [8]
K268R <4> (<4> slightly increased activity [8]) [8]
R197Q <4> (<4> moderate reduction of acetylation activity [8]) [8]
R64Q <4> (<4> moderate reduction of acetylation activity [8]) [8]
Additional information <4> (<4> nucleic acid substitutions C282T and C481T have no effect on activity [8]) [8]

Application
medicine <4, 10> (<4> bioactivation of N-hydroxy-2-acetylaminofluorenes by N,O-acytransfer resulting in mutagenic and carcinogenic compounds, DNA adduct formation [14]; <10> important for the activation of urinary metabolites of carcinogenic arylamines for the initiation of bladder carcino-

genesis in the dog [13]; <4> metabolic activation by isoforms 1 and 2 of 3-nitrobenzanthrone and N-acetyl-N-hydroxy-3-aminobenzantrone, 2 compounds with high mutagenic and genotoxic potential, formation of DNA adducts [11]) [11, 13]

6 Stability

Oxidation stability

<1>, unstable in the presence of oxygen due to labile SH-groups, dithiothreitol cannot fully stabilize [3]

General stability information

<1>, argon-saturated diphosphate buffer stabilizes [3]
<1>, freezing and thawing inactivates [3]
<1>, lyophilization inactivates [3]
<1>, purification inactivates [3]
<1>, unstable in low salt solutions [9]

Storage stability

<1>, -20°C, up to 3 months in saturating ammonium sulfate sealed under argon [3]
<1>, 0-4°C, about 30% loss of activity after 6 h [1]
<1>, 0-4°C, inactivation after 18 h [1]

References

[1] Bartsch, H.; Dworkin, C.; Miller, E.C.; Miller, J.A.: Formation of electrophilic N-acetoxyarylamines in cytosols from rat mammary gland and other tissues by transacetylation from the carcinogen N-hydroxy-4-acetylaminobiphenyl. Biochim. Biophys. Acta, **304**, 42-55 (1973)
[2] Banks, R.B.; Hanna, P.E.: Arylhydroxamic acid N,O-acyltransferase. Apparent suicide inactivation by carcinogenic N-arylhydroxamic acids. Biochem. Biophys. Res. Commun., **91**, 1423-1429 (1979)
[3] Allaben, W.T.; King, C.M.: The purification of rat liver arylhydroxamic acid N,O-acyltransferase. J. Biol. Chem., **259**, 12128-12134 (1984)
[4] Marhevka, V.C.; Ebner, N.A.; Sehon, R.D.; Hanna, P.E.: Mechanism-based inactivation of N-arylhydroxamic acid N,O-acyltransferase by 7-substituted-N-hydroxy-2-acetamidofluorenes. J. Med. Chem., **28**, 18-24 (1985)
[5] Elfarra, A.A.; Yeh, H.M.; Hanna, P.E.: Synthesis and evaluation of N-(phenylalkyl)acetohydroxamic acids as potential substrates for N-arylhydroxamic acid N,O-acyltransferase. J. Med. Chem., **25**, 1189-1192 (1982)
[6] King, C.M.: Mechanism of reaction, tissue distribution, and inhibition of arylhydroxamic acid acyltransferase. Cancer Res., **34**, 1503-1515 (1974)
[7] Hein, D.W.: Acetylator genotype and arylamine-induced carcinogenesis. Biochim. Biophys. Acta, **948**, 37-66 (1988)

[8] Hein, D.W.; Doll, M.A.; Rustan, T.D.; Fergusion, R.J.: Metabolic activation of N-hydroxyarylamines and N-hydroxyarylamides by 16 recombinant human NAT2 allozymes: effects of 7 specific NAT2 nucleic acid substitutions. Cancer Res., **55**, 3531-3536 (1995)
[9] Land, S.J.; King, C.M.: Characterization of rat hepatic acetyltransferase. Environ. Health Perspect., **102**, 91-93 (1994)
[10] Hanna, P.E.: N-acetyltransferases, O-acetyltransferases, and N,O-acetyltransferases: enzymology and bioactivation. Adv. Pharmacol., **27**, 401-430 (1994)
[11] Arlt, V.M.; Glatt, H.; Muckel, E.; Pabel, U.; Sorg, B.L.; Schmeiser, H.H.; Phillips, D.H.: Metabolic activation of the environmental contaminant 3-nitrobenzanthrone by human acetyltransferases and sulfotransferase. Carcinogenesis, **23**, 1937-1945 (2002)
[12] Jones, R.F.; Land, S.J.; King, C.M.: Recombinant rat and hamster N-acetyltransferases-1 and -2: relative rates of N-acetylation of arylamines and N,O-acyl transfer with arylhydroxamic acids. Carcinogenesis, **17**, 1729-1733 (1996)
[13] Sone, T.; Zukowski, K.; Land, S.J.; Kin, C.M.; Martin, B.M.; Pohl, L.R.; Wang, C.Y.: Characteristics of a purified dog hepatic microsomal N,O-acyltransferase. Carcinogenesis, **15**, 595-599 (1994)
[14] Boteju, L.W.; Hanna, P.E.: Bioactivation of N-hydroxy-2-acetylaminofluorenes by N,O-acyltransferase: substituent effects on covalent binding to DNA. Carcinogenesis, **14**, 1651-1657 (1993)
[15] Sinclair, J.C.; Delgoda, R.; Noble, M.E.; Jarmin, S.; Goh, N.K.; Sim, E.: Purification, characterization, and crystallization of an N-hydroxyarylamine O-acetyltransferase from Salmonella typhimurium. Protein Expr. Purif., **12**, 371-380 (1998)

Diamine N-acetyltransferase 2.3.1.57

1 Nomenclature

EC number
2.3.1.57

Systematic name
acetyl-CoA:alkane-α,ω-diamine N-acetyltransferase

Recommended name
diamine N-acetyltransferase

Synonyms
N1-SAT
N1SSAT
SSAT
acetyl-coenzyme A-1,4-diaminobutane N-acetyltransferase
acetyltransferase, putrescine
diamine acetyltransferase
putrescine (diamine)-acetylating enzyme
putrescine N-acetyltransferase
putrescine acetylase
putrescine acetyltransferase
spermidine N^1-acetyltransferase
spermidine acetyltransferase
spermidine/spermine N^1-acetyltransferase

CAS registry number
54596-36-0

2 Source Organism

<1> *Candida boidinii* (CBS 5777 [1]) [1]
<2> *Candida methanosorbosa* (CBS 6853 [1]) [1]
<3> *Trichosporon melibiosaceum* (CBS 6087 [1]) [1]
<4> *Rattus norvegicus* (female Sprague-Dawley rats [3]; Wistar rats, 2 months old [6]) [2, 3, 6]
<5> *Ascaris suum* (female adults [4,5]) [4, 5]
<6> *Onchocerca volvulus* (female adults [5]) [5]
<7> *Homo sapiens* (recombinant enzyme, expressed in Escherichia coli DH5α [8]; recombinant enzyme, expressed in Escherichia coli CAG2242 [9]; hu-

man cDNA expressed in MCF-7 human breast carcinoma cells [11]) [7-11, 13, 15]
<8> *Cricetulus griseus* [10]
<9> *Mus musculus* (recombinant enzyme, expressed in MCF-7 human breast carcinoma cells [11]; transgenic mouse line overexpressing SSAT [12]; transgenic male CD2F1 mice overexpressing SSAT [14]) [11, 12, 14]

3 Reaction and Specificity

Catalyzed reaction

acetyl-CoA + an alkane-α,ω-diamine = CoA + an N-acetyldiamine (<1-3> sequential mechanism [1])

Reaction type

acyl group transfer

Natural substrates and products

S acetyl-CoA + amantadine <9> (<9> reaction occurs in vivo and in vitro, but only in presence of increased enzyme levels, amantadine may be a specific drug substrate for SSAT [14]) (Reversibility: ? <9> [14]) [14]
P CoA + N^1-acetylamantadine <9> [14]
S acetyl-CoA + putrescine <1> (<1> breakdown of spermidine and putrescine [1]) (Reversibility: ? <1> [1]) [1]
P CoA + acetylputrescine <1> [1]
S acetyl-CoA + spermidine <1, 4, 7> (<1> breakdown of spermidine and putrescine, key agent in the supply of nitrogen to the cell [1]; <4> spermidine and spermine may be physiological substrates, enzyme may play an important role in interconversion of polyamines [2]; <7> spermidine acetylation might be a strategy for inhibiting growth in response to environmental stresses [9]; <7> catabolism of spermidine and spermine [13,15]) (Reversibility: ? <1, 4, 7> [1, 2, 9, 13, 15]) [1, 2, 9, 13, 15]
P CoA + N^1-acetylspermidine <7> [9, 15]
S acetyl-CoA + spermine <4, 7> (<4> spermidine and spermine may be physiological substrates, enzyme may play an important role in interconversion of polyamines [2]; <7> catabolism of spermidine and spermine [13,15]) (Reversibility: ? <4, 7> [2, 13, 15]) [2, 13, 15]
P CoA + N^1-acetylspermine
S Additional information <5-9> (<5> important function in the degradation of diamines of lower eukaryotes [4]; <5,6> enzyme is involved in polyamine degradation and excretion of excessive polyamines, release of N-acetylputrescine, role in the control of polyamine concentrations [5]; <7> rate-limiting enzyme in the degradation and interconversion of polyamines [7]; <7> acetylation is a physiological response to convert excess polyamines to a physiologically inert form which is readily excreted [8]; <7,8> first enzyme in polyamine catabolism [10]; <7,9> enzyme plays an efficient role in maintaining polyamine pool homeostasis during challenges with exogenous polyamines [11]; <7,9> polyamine catabolic en-

zyme [11-15]; <9> SSAT catalyzes together with polyamine oxidase the back-conversion of spermine to spermidine and the latter to putrescine, a function lowering polyamine pools by facilitating their catabolism and excretion [12]; <7> SSAT prevents overaccumulation of higher polyamines from becoming toxic to cell and maintains a balanced ratio of polyamines according to cellular needs [13]; <7> stress-induced enzyme [15]) [4, 5, 7, 8, 10-15]

P ?

Substrates and products

S acetyl-CoA + (S)-2-[(3-aminopropyl)amino]ethylphosphoric acid <7> (<7> WR-2721, about 10% of the rate with spermidine [8]) (Reversibility: ? <7> [8]) [8]

P CoA + N-acetyl-(S)-2-[(3-aminopropyl)amino]ethylphosphoric acid

S acetyl-CoA + (S)-2-[(3-aminopropyl)amino]propylphosphoric acid <7> (<7> WR-44923, about 10% of the rate with spermidine [8]) (Reversibility: ? <7> [8]) [8]

P CoA + N-acetyl-(S)-2-[(3-aminopropyl)amino]propylphosphoric acid

S acetyl-CoA + 1,3-diaminopropane <1, 4, 5> (<4> at a low rate [3]) (Reversibility: ir <5> [4]; ? <1, 4> [1, 3]) [1, 3, 4]

P CoA + N^1-acetyl-1,3-diaminopropane <5> [4]

S acetyl-CoA + 1,5-diaminopentane <1, 5, 6> (<5> cadaverine [4]; <5> at 39% of the rate with putrescine [5]; <6> at 45% of the rate with putrescine [5]) (Reversibility: ir <5> [4]; ? <1, 5, 6> [1, 5]) [1, 4, 5]

P CoA + N^1-acetyl-1,5-diaminopentane <5> [4]

S acetyl-CoA + 1,6-diaminohexane <5, 6> (<5> at 50% of the rate with putrescine [5]; <6> at 74% of the rate with putrescine [5]) (Reversibility: ir <5> [4]; ? <5, 6> [5]) [4, 5]

P CoA + N^1-acetyl-1,6-diaminohexane <5> [4]

S acetyl-CoA + 1,7-diaminoheptane <1> (Reversibility: ? <1> [1]) [1]

P CoA + N^1-acetyl-1,7-diaminoheptane

S acetyl-CoA + 15-deoxyspergualin <7> (<7> antitumor and immunosuppressive agent 15-deoxyspergualin, about 18% of the rate with spermidine [8]) (Reversibility: ? <7> [8]) [8]

P ?

S acetyl-CoA + 2-[(aminopropyl)amino]ethanethiol <7> (<7> radioprotective drug WR-1065, lower affinity than for spermidine, about 10% of the rate with spermidine [8]) (Reversibility: ? <7> [8]) [8]

P CoA + N-acetyl-2-[(aminopropyl)amino]ethanethiol

S acetyl-CoA + 6,6-difluorospermidine <7> (<7> 16.6% of the rate with spermidine [10]) (Reversibility: ? <7> [10]) [10]

P CoA + N^1-acetyl-6,6-difluorospermidine

S acetyl-CoA + 7,7-difluorospermidine <7> (<7> 19.7% of the rate with spermidine [10]) (Reversibility: ? <7> [10]) [10]

P CoA + N^1-acetyl-7,7-difluorospermidine

S acetyl-CoA + N-(n-butyl)-1,3-diaminopropane <7> (<7> weak substrate, lower affinity than for spermidine, 1.3% of the rate with spermidine [8]) (Reversibility: ? <7> [8]) [8]
P CoA + N^1-acetyl-N_3-(n-butyl)-1,3-diaminopropane
S acetyl-CoA + N-acetylputrescine <1> (Reversibility: ? <1> [1]) [1]
P CoA + N^1,N^4-diacetylputrescine
S acetyl-CoA + N^1-acetylspermidine <1> (<1> low affinity [1]) (Reversibility: ? <1> [1]) [1]
P CoA + N^1,N^8-diacetylspermidine
S acetyl-CoA + N^1-acetylspermine <1, 4, 7, 9> (<4> reaction in vitro, but not in vivo [2]) (Reversibility: ir <4> [2]; ? <1, 4, 7, 9> [1, 3, 11]) [1-3, 11]
P CoA + N^1,N^{12}-diacetylspermine <4, 7, 9> [2, 11]
S acetyl-CoA + N^8-acetylspermidine <1> (Reversibility: ? <1> [1]) [1]
P CoA + N^1,N^8-diacetylspermidine
S acetyl-CoA + amantadine <9> (<9> reaction occurs in vivo and in vitro, but only in presence of increased enzyme levels, amantadine may be a specific drug substrate for SSAT [14]) (Reversibility: ? <9> [14]) [14]
P CoA + N^1-acetylamantadine <9> [14]
S acetyl-CoA + β-phenylethylamine <1> (Reversibility: ? <1> [1]) [1]
P CoA + N-acetylphenylethylamine
S acetyl-CoA + histamine <5, 6> (<5,6> at about 30% of the rate with putrescine [5]) (Reversibility: ? <5, 6> [5]) [5]
P CoA + N-acetylhistamine
S acetyl-CoA + histone <1> (Reversibility: ? <1> [1]) [1]
P CoA + N-acetylhistone
S acetyl-CoA + putrescine <1-6> (<5,6> preference for putrescine [5]) (Reversibility: ir <5> [4]; ? <1-6> [1, 4-6]) [1, 4-6]
P CoA + N-acetylputrescine <1-6> [1, 4-6]
S acetyl-CoA + spermidine <1, 3-9> (<4> physiological substrate, higher rate than with spermine [2]; <5,6> very poor substrate, at 2-4% of the rate with putrescine [4,5]) (Reversibility: ir <4> [2]; ? <1, 3-9> [1, 3-5, 7-15]) [1-5, 7-15]
P CoA + N^1-acetylspermidine <1, 3, 4, 7-9> (<1> + N^8-acetylspermidine, ratio N^1-acetylspermidine:N^8-acetylspermidine is 50:45 [1]; <4,7,9> product is exclusively N^1-acetylspermidine [2,3,8,9,11]; <7> N^1-acetylspermidine is the major product [15]) [1-3, 8-13, 15]
S acetyl-CoA + spermine <1, 4-9> (<4> physiological substrate, lower rate than with spermidine [2]; <5,6> very poor substrate, at 2-4% of the rate with putrescine [4,5]; <7> about 40% of the rate with spermidine [8]) (Reversibility: ir <4> [2]; ? <1, 4-9> [1, 3-5, 7-13, 15]) [1-5, 7-13, 15]
P CoA + N^1-acetylspermine <4, 7, 9> [2, 8, 9, 11, 13, 15]
S acetyl-CoA + sym-norspermidine <1, 4, 5> (<1> identical with caldine [1]; <4> high affinity for sym-norspermidine [2]) (Reversibility: ir <5> [4]; ? <1, 4> [1-3]) [1-4]
P CoA + N^1-acetyl-sym-norspermidine
S acetyl-CoA + sym-norspermine <4, 5> (<5> at a much lower rate than with sym-norspermidine [4]) (Reversibility: ir <5> [4]; ? <4> [3]) [3, 4]

P CoA + N^1-acetyl-sym-norspermine

S Additional information <1, 4-7, 9> (<1> not: methylamine, dimethylamine, di-n-butylamine, L-lysine, benzylamine, semicarbazide, L-serine, glyoxylate, oxaloacetate, 4-aminobenzoate, 2-aminobenzoate [1]; <5,6> not: dopamine, serotonin [5]; <4,7,9> not: putrescine [2,3,8,11]; <4-6> not: histone [2-5]; <4> not: sym-homospermidine [3]; <7> not: 1-methylspermidine, WR-2822 [8]; <4,7> enzyme requires a substrate with the structure $H_2N(CH_2)_3NHR$ and acetylates the primary amino group [3,8]; <7> Lys-141 is the first residue in a KRR motif that makes up part of the active site [13]) [1-3, 5, 8, 11, 13]

P ?

Inhibitors

3'-dephospho-CoA <4> (<4> less inhibition than by CoA [2]) [2]
6,6-difluorospermidine <7> (<7> weak inhibition [10]) [10]
CoA <4> (<4> strong inhibitor [2]) [2]
CoA methyl disulfide <4> (<4> very strong inhibitor [2]) [2]
N-acetyl-CoA <4> (<4> at high concentrations slight inhibition [6]) [6]
N-acetylputrescine <5> (<5> slight, competitive inhibition [4]) [4]
N-ethylmaleimide <1> [1]
SKF 525 <4> [6]
amantadine <9> (<9> competitively inhibits spermidine acetylation, at 10 mM: complete inhibition [14]) [14]
berenil <5, 6> (<5,6> competitive with respect to putrescine, i.e. diminazene aceturate [5]) [5]
butan-2,3-dione <1> (<1> at 10 mM, in presence of borate buffer pH 9.0 [1]) [1]
hexamethylenediamine <4> (<4> 30% inhibition [6]) [6]
mescaline <4> [6]
p-chloromercuribenzoate <1, 4> [1, 2]
pentamethylenediamine <4> (<4> 30% inhibition [6]) [6]
pentamidine <5, 6> (<5,6> weak [5]) [5]
phenylglyoxal <1> (<1> at 10 mM, in bicarbonate buffer, pH 9.0 [1]) [1]
spermine <4> [6]
sulfamethazine <9> (<9> inhibits spermidine acetylation [14]) [14]
Additional information <5, 7, 9> (<5> not inhibited by CoA [4]; <7,9> doxycycline suppresses SSAT gene expression [11]; <9> not inhibited by *p*-aminobenzoic acid [14]) [4, 11, 14]

Cofactors/prosthetic groups

Additional information <4> (<4> no known cofactors [3]) [3]

Activating compounds

Additional information <4, 7-9> (<4> enzyme is induced by treatment with hepatotoxins, e.g. carbon tetrachloride [2,3]; <4> enzyme is induced by growth hormone, thioacetamide, dialkylnitrosamines, folic acid and spermidine [3]; <7> enzyme is induced by methylglyoxal bis(guanylhydrazone), potentiated by tetronasin or felodipine, tetronasin is also an active inducer, induction is related to the concentration of intracellular free calcium [7]; <7> enzyme is induced by isopropyl β-D-thiogalactopyranoside [8,9]; <7> N^1,N^{12}-bis(ethyl)spermine is a potent inducer of enzyme in human tumor cells [8]; <7,8> certain stresses induce enzyme by a spermidine-dependent post-transcriptional mechanism, e.g. heat shock, diethyldithiocarbamate and high levels of endogenous spermidine, α-difluoromethylornithine inhibits stress induction [10]; <7,9> N^1,N^{11}-diethylnorspermine induces SSAT mRNA and activity [11]; <9> N^1,N^{11}-diethylnorspermine induces SSAT activity by post-mRNA regulation [12]; <7> enzyme is highly inducible by polyamine analogues such as N^1,N^{12}-bis(ethyl)spermine and N^1,N^{11}-bis(ethyl)norspermine, they prevent enzyme degradation via the Ub/proteasome pathway, mechanism [13]; <7> indomethacin and excess polyamines induce SSAT activity [15]) [2, 3, 7-13, 15]

Specific activity (U/mg)

0.0006 <6> [5]
0.4 <5> [5]
0.4314 <5> [4]
2.45 <1> [1]
8.102 <4> [2, 3]
45 <7> (<7> spermidine [8]) [8]
Additional information <1, 7, 9> [1, 7, 8, 11, 12]

K_m-Value (mM)

0.0015 <4> (acetyl-CoA) [2]
0.0077 <5> (acetyl-CoA) [4]
0.008 <4> (sym-norspermidine) [2]
0.012 <7> (15-deoxyspergualin) [8]
0.014 <1> (acetyl-CoA, <1> putrescine as cosubstrate [1]) [1]
0.025 <7> (spermine) [8]
0.03 <4> (N^1-acetylspermine) [2]
0.035 <4> (spermine) [2]
0.05 <7> (spermidine, <7> wild-type enzyme [13]) [13]
0.06 <7> (spermidine) [8]
0.1 <4> (1,5-diaminopentane) [2]
0.13 <4> (spermidine) [2]
0.25 <5> (putrescine) [4, 5]
0.26-0.44 <1> (putrescine) [1]
0.267 <9> (spermidine) [14]
0.27 <1> (1,3-diaminopropane) [1]
0.33 <6> (putrescine) [5]
0.39 <7> (spermidine, <7> K141R mutant [13]) [13]
0.4 <5> (1,6-diaminohexane) [4]
0.44 <1> (1,5-diaminopentane) [1]
0.5 <1> (sym-norspermidine) [1]
1.25 <5> (1,3-diaminopropane) [4]
1.4 <1> (N^8-acetylspermidine) [1]
1.59 <1> (1,7-diaminoheptane) [1]

1.7 <1> (spermine) [1]
2.4 <1> (acetylputrescine) [1]
2.4 <1> (spermidine) [1]
3 <4> (putrescine) [6]
4.4 <1> (N^1-acetylspermidine) [1]

K_i-Value (mM)
0.0012 <5> (berenil) [5]
0.0042 <6> (berenil) [5]
0.04 <4> (CoA) [2]
0.738 <9> (amantadine) [14]
1.7 <5> (N-acetylputrescine) [4]
3.5 <9> (sulfamethazine) [14]

pH-Optimum
7.7-7.9 <5> [4]
8 <4> [6]
8.5-9 <1> (<1> putrescine [1]) [1]
9-9.5 <1> (<1> spermidine [1]) [1]

pH-Range
7.6-8 <5> (<5> high conversion rates in the range of [4]) [4]

Temperature optimum (°C)
25 <1> (<1> assay at [1]) [1]
30 <4, 7> (<4,7> assay at [2,3,8,13]) [2, 3, 8, 13]
37 <4, 5, 9> (<4,5,9> assay at [4,6,14]) [4, 6, 14]

4 Enzyme Structure

Molecular weight
38000-40000 <5> (<5> gel filtration [4]) [4]
75000 <7> (<7> gel filtration [8]) [8]
114000 <4> [3]
115000 <4> (<4> gel filtration [2]) [2]

Subunits
? <5> (<5> x * ? + x * 13000, catalytic subunit, at least 2 subunits, gel filtration of enzyme precipitated by $(NH_4)_2SO_4$ [4]) [4]
dimer <4> (<4> 2 * 60000, SDS-PAGE [2,3]) [2, 3]
homodimer <7> (<7> 2 * 20000, subunit size of 171 amino acids [13]) [13]
tetramer <7> (<7> x * 21000 + x * 23000, two bands, enzyme is probably a tetramer, recombinant enzyme, SDS-PAGE [8]) [8]

Posttranslational modification
no modification <7> (<7> no posttranslational modification specific to eukaryotes is needed for enzyme activity [8]) [8]
phosphoprotein <7> (<7> purified SSAT is phosphorylated by casein kinase II [13]) [13]

5 Isolation/Preparation/Mutation/Application

Source/tissue

A-549 cell <7> (<7> lung-cancer derived cells [10]) [10]
CACO-2 cell <7> (<7> colon cancer cells Caco-2 and HTC-116 [15]) [15]
CHO cell <8> [10]
HTC-116 cell <7> (<7> colon cancer cells Caco-2 and HTC-116 [15]) [15]
T47D cell <7> (<7> human breast carcinoma cell line [7]) [7]
brain <4> (<4> highest activity [6]) [6]
fetus <9> (<9> primary fetal fibroblasts [12]) [12]
fibroblast <9> (<9> primary fetal fibroblasts [12]) [12]
gut <4> [6]
kidney <4> [6]
liver <4, 9> (<4> carbon tetrachloride induced liver [2,3]) [2, 3, 6, 14]
melanoma cell line <7> (<7> melanoma-derived cells C8146C [10]) [10]
ovary <5> [4, 5]
uterus <5> [4, 5]

Localization

cytosol <4, 5, 7, 9> [2-4, 7, 14]
microsome <4> [6]
mitochondrion <1-3> [1]
nucleus <4> [6]
soluble <7> (<7> recombinant enzyme, expressed in Escherichia coli DH5α, is present at a level of about 2% of the soluble protein [8]) [8]

Purification

<1> (26fold, partial [1]) [1]
<4> (112000fold [2,3]) [2, 3]
<5> (1027fold [4]; partial [4,5]) [4, 5]
<6> (partial [5]) [5]
<7> (62fold purification of recombinant enzyme, expressed in Escherichia coli DH5α [8]) [8]

Cloning

<7> (cloning and expression of cDNA encoding enzyme in Escherichia coli DH5α leads to a significant reduction in the cell growth rate [8]; cloning and conditional expression of cDNA encoding enzyme in Escherichia coli CAG2242 results in a decrease of endogenous spermidine contents and growth rates [9]; complete SSAT cDNA is cloned, tetracyclin-regulated cDNA is expressed in MCF-7 human breast carcinoma cells, conditional overexpression lowers polyamine pools, inhibits cell growth and enhances growth sensitivity to certain analogs [11]; plasmid pSAT9.3, containing SSAT cDNA cloned into Bluescript vector, is used to express the protein from the T7 promoter using rabbit reticulocyte TNT coupled expression system [13]) [8, 9, 11, 13]
<9> (SSAT gene is cloned, tetracyclin-regulated gene lacking the 5′-flanking region is expressed in MCF-7 human breast carcinoma cells, conditional

overexpression lowers polyamine pools, inhibits cell growth and enhances growth sensitivity to certain analogs [11]; full-length cDNA is cloned and expressed in Escherichia coli M15, generation of a transgenic mouse line systemically overexpressing SSAT, overexpression alters polyamine pools and sensitize cells to the antiproliferative activity of N^1,N^{11}-diethylnorspermine [12]; transgenic CD2F1 mice overexpressing SSAT [14]) [11, 12, 14]

Engineering

E152Q <7> (<7> mutation disrupts binding of stabilizing N1,N12-bis(ethyl)-spermine to enzyme [13]) [13]
E170Stop <7> (<7> mutation stabilizes enzyme, due to the lack of ubiquinated complexes [13]) [13]
E171Q <7> (<7> mutation results in a marked stabilization of enzyme, due to the lack of formation of high-molecular-mass complexes with ubiquitin [13]) [13]
E172A <7> (<7> mutation stabilizes enzyme, due to the lack of ubiquinated complexes [13]) [13]
E173A <7> (<7> mutation stabilizes enzyme, due to the lack of ubiquinated complexes [13]) [13]
K111R <7> (<7> mutant with the same activity as wild-type enzyme, but reduced half-life [13]) [13]
K141R <7> (<7> mutant with 35% of activity of wild-type enzyme, increased K_m for spermidine and reduced half-life [13]) [13]
K158R <7> (<7> mutant with 139% of activity of wild-type enzyme [13]) [13]
K161R <7> (<7> mutant with the same activity as wild-type enzyme [13]) [13]
K166R <7> (<7> mutant with 78% of activity of wild-type enzyme [13]) [13]
K22R <7> (<7> mutant with 149% of activity of wild-type enzyme, but reduced half-life [13]) [13]
K26R <7> (<7> mutant with 120% of activity of wild-type enzyme [13]) [13]
K39R <7> (<7> mutant with the same activity as wild-type enzyme [13]) [13]
K3R <7> (<7> mutant with 134% of activity of wild-type enzyme [13]) [13]
K61R <7> (<7> mutant with the same activity as wild-type enzyme [13]) [13]
K87R <7> (<7> mutant with 70% of activity of wild-type enzyme, but significantly longer half-life, suggesting that Lys-87 may be the preferred site for ubiquination [13]) [13]

Application

medicine <7> (<7> induction of enzyme may be critical for the inhibition of tumor cell growth [8]; <7> the ability of polyamine analogues to induce N1SSAT activity correlates with the ability to inhibit growth and viability of human tumor-derived cells [9]; <7> development of SSAT induction strategies to circumvent the ability of tumor cells to escape the antitumor activity of polyamine inhibitors such as α-difluoromethylornithine by salvaging exo-

genous polyamines [11]; <7> logical target for tumor prevention [15]) [8, 9, 11, 15]
pharmacology <7> (<7> compounds capable of potently inducing SSAT and having favorable pharmacological properties in animals are potential anticancer agents [11]) [11]

6 Stability

pH-Stability
7-7.5 <1> (<1> stability maximum, rapid fall on either side [1]) [1]

Temperature stability
35.5 <1> (<1> crude extract, half-life: 0.75 min [1]) [1]
37 <5> (<5> half-life: above 1 h [4]) [4]
45 <4> (<4> half-life: 2.2 min, in 50 mM Tris-HCl, pH 7.5, 10 mM spermidine [2]) [2]
52 <4> (<4> half-life: 0.5 min, in 50 mM Tris-HCl, pH 7.5, 10 mM spermidine [2]) [2]
56 <5> (<5> half-life: less than 0.5 min [4]) [4]
60 <4> (<4> 0.5 min, 90% loss of activity, in 50 mM Tris-HCl, pH 7.5, 10 mM spermidine [2]) [2]
Additional information <4> (<4> very sensitive to heat denaturation [3]) [3]

General stability information
<1>, unstable enzyme, glycerol 10% v/v or bovine serum albumin, 5 mg/ml, partially stabilizes [1]
<4>, sensitive to incubation in dilute solutions, bovine serum albumin, 0.5 mg/ml, stabilizes [3]
<4>, unstable enzyme, spermidine stabilizes [2, 3]
<5>, PMSF and Brij 35 are essential to prevent rapid loss of activity of the enzyme preparation [4]
<7>, native protein conformation is unstable, polyamine analogues such as N^1,N^{12}-bis(ethyl)spermine greatly stabilize, conformational changes caused by their binding prevent the efficient polyubiquination of enzyme and therefore its degradation [13]
<9>, N^1,N^{11}-diethylnorspermine stabilizes [12]

Storage stability
<1>, 25°C, glycerol 10% v/v or bovine serum albumin, 5 mg/ml, more than 22 h, stable [1]
<4>, -20°C, 50 mM Tris-HCl buffer, pH 7.5, 10 mM spermidine, per week, 30% loss of activity [2, 3]
<5>, -22°C, per month, 10-16% loss of activity [4]
<5>, 4°C, per day, 5-9% loss of activity [4]
<7>, -70°C, several weeks, stable [8]

References

[1] Haywood, G.W.; Large, P.J.: The occurrence, subcellular localization and partial purification of diamine acetyltransferase in the yeast Candida boidinii grown on spermidine or putrescine as sole nitrogen source. Eur. J. Biochem., **148**, 277-283 (1985)

[2] Della Ragione, F.; Pegg, A.E.: Purification and characterization of spermidine/spermine N^1-acetyltransferase from rat liver. Biochemistry, **21**, 6152-6158 (1982)

[3] Della Ragione, F.; Pegg, A.E.: Spermidine N^1-acetyltransferase. Methods Enzymol., **94**, 321-325 (1983)

[4] Wittich, R.M.; Walter, R.D.: A novel type of putrescine (diamine)-acetylating enzyme from the nematode Ascaris suum. Biochem. J., **260**, 265-269 (1989)

[5] Wittich, R.M.; Walter, R.D.: Putrescine N-acetyltransferase in Onchocerca volvulus and Ascaris suum, an enzyme which is involved in polyamine degradation and release of N-acetylputrescine. Mol. Biochem. Parasitol., **38**, 13-18 (1990)

[6] Seiler, N.; Al-Therib, J.: Acetyl-CoA:1,4-diaminobutane N-acetyltransferase. Occurence in vertebrate organs and subcellular localization. Biochim. Biophys. Acta, **354**, 206-212 (1974)

[7] Quick, D.M.; Wallace, H.M.: Induction of spermidine/spermine N^1-acetyltransferase in human breast carcinoma cells. A possible role for calcium. Biochem. Pharmacol., **46**, 969-974 (1993)

[8] Parry, L.; Lopez-Ballester, J.; Wiest, L.; Pegg, A.E.: Effect of expression of human spermidine/spermine N^1-acetyltransferase in Escherichia coli. Biochemistry, **34**, 2701-2709 (1995)

[9] Ignatenko, N.A.; Fish, J.L.; Shassetz, L.R.; Woolridge, D.P.; Gerner, E.W.: Expression of the human spermidine/spermine N^1-acetyltransferase in spermidine acetylation-deficient Escherichia coli. Biochem. J., **319**, 435-440 (1996)

[10] Gerner, E.W.; Kurtts, T.A.; Fuller, D.J.M.; Casero, R.A., Jr.: Stress induction of the spermidine/spermine N^1-acetyltransferase by a post-transcriptional mechanism in mammalian cells. Biochem. J., **294**, 491-495 (1993)

[11] Vujcic, S.; Halmekyto, M.; Diegelman, P.; Gan, G.; Kramer, D.L.; Janne, J.; Porter, C.W.: Effects of conditional overexpression of spermidine/spermine N^1-acetyltransferase on polyamine pool dynamics, cell growth, and sensitivity to polyamine analogs. J. Biol. Chem., **275**, 38319-38328 (2000)

[12] Alhonen, L.; Karppinen, A.; Uusi-Oukari, M.; Vujcic, S.; Korhonen, V.P.; Halmekytö, M.; Kramer, D.L.; Hines, R.; Jänne, J.; Porter, C.W.: Correlation of polyamine and growth responses to N^1,N^{11}-diethylnorspermine in primary fetal fibroblasts derived from transgenic mice overexpressing spermidine/spermine N^1-acetyltransferase. J. Biol. Chem., **273**, 1964-1969 (1998)

[13] Coleman, C.S.; Pegg, A.E.: Polyamine analogues inhibit the ubiquitination of spermidine/spermine N^1-acetyltransferase and prevent its targeting to the proteasome for degradation. Biochem. J., **358**, 137-145 (2001)

[14] Bras, A.P.M.; Jänne, J.; Porter, C.W.; Sitar, D.S.: Spermidine/spermine N^1-acetyltransferase catalyzes amantadine acetylation. Drug Metab. Dispos., **29**, 676-680 (2001)
[15] Turchanowa, L.; Dauletbaev, N.; Milovic, V.; Stein, J.: Nonsteroidal anti-inflammatory drugs stimulate spermidine/spermine acetyltransferase and deplete polyamine content in colon cancer cells. Eur. J. Clin. Invest., **31**, 887-893 (2001)

2,3-Diaminopropionate N-oxalyltransferase 2.3.1.58

1 Nomenclature

EC number
2.3.1.58

Systematic name
oxalyl-CoA:L-2,3-diaminopropanoate N^3-oxalyltransferase

Recommended name
2,3-diaminopropionate N-oxalyltransferase

Synonyms
ODAP synthase
oxalyl-CoA:L-α,β-diaminopropionic acid oxalyltransferase
oxalyldiaminopropionate synthase
oxalyldiaminopropionic synthase
oxalyltransferase, 2,3-diaminopropionate

CAS registry number
62213-48-3

2 Source Organism

<1> *Lathyrus sativus* [1]

3 Reaction and Specificity

Catalyzed reaction
oxalyl-CoA + L-2,3-diaminopropanoate = CoA + N^3-oxalyl-L-2,3-diaminopropanoate

Reaction type
acyl group transfer

Substrates and products
- **S** oxalyl-CoA + 2,4-diaminobutyric acid <1> (<1> about 20% of activity with L-2,3-diaminopropionic acid [1]) (Reversibility: ? <1> [1]) [1]
- **P** CoA + N^4-oxalyl-2,4-diaminobutyric acid
- **S** oxalyl-CoA + L-2,3-diaminopropionic acid <1> (Reversibility: ? <1> [1]) [1]
- **P** CoA + N^3-oxalyl-L-2,3-diaminopropionic acid <1> [1]

S oxalyl-CoA + L-homoserine <1> (<1> about 12% of activity with L-2,3-diaminopropionic acid [1]) (Reversibility: ? <1> [1]) [1]
P ?
S Additional information <1> (<1> succinyl-CoA and acetyl-CoA are not accepted [1]) [1]
P ?

Inhibitors
acetone <1> (<1> precipitation with acetone completely inhibits enzyme activity [1]) [1]
ethanol <1> (<1> precipitation with ethanol completely inhibits enzyme activity [1]) [1]

Activating compounds
EDTA <1> (<1> slight activation [1]) [1]

Specific activity (U/mg)
Additional information <1> [1]

K_m-Value (mM)
0.3 <1> (L-2,3-diaminopropionic acid) [1]
0.45 <1> (oxalyl-CoA) [1]

pH-Optimum
7.4-8 <1> [1]

Temperature optimum (°C)
37 <1> (<1> assay at [1]) [1]

5 Isolation/Preparation/Mutation/Application

Source/tissue
seedling <1> [1]

Purification
<1> (60fold [1]) [1]

References

[1] Malathi, K.; Padmanaban, G.; Sarma, P.S.: Biosynthesis of β-N-oxalyl-L-α,β-diaminopropionic acid, the Lathyrus sativus neurotoxin. Phytochemistry, **9**, 1603-1610 (1970)

Gentamicin 2'-N-acetyltransferase 2.3.1.59

1 Nomenclature

EC number
2.3.1.59

Systematic name
acetyl-CoA:gentamicin-C_{1a} $N^{2'}$-acetyltransferase

Recommended name
gentamicin 2'-N-acetyltransferase

Synonyms
AAC(2')-Ia
ACC(2')
acetyltransferase, gentamicin 2'-
gentamicin acetyltransferase II

CAS registry number
50864-40-9

2 Source Organism

<-5> no activity in *Providenicia rustigianii* [4]
<-4> no activity in *Providencia heinbachae* [4]
<-3> no activity in *Morganella morgani* [4]
<-2> no activity in *Proteus vulgaris* [4]
<-1> no activity in *Proteus myxofaciens* [4]
<1> *Streptomyces spectabilis* [1]
<2> *Providencia sp.* (strain 164, clinical isolate [2]) [2]
<3> *Streptomyces kanamyceticus* [1]
<4> *Providencia stuartii* (PR50 [4]) [3, 4, 5]
<5> *Proteus penneri* [4]
<6> *Providencia rettgeri* [4]

3 Reaction and Specificity

Catalyzed reaction
acetyl-CoA + gentamicin C_{1a} = CoA + $N^{2'}$-acetylgentamicin C_{1a}

Reaction type
acyl group transfer

Natural substrates and products

S Additional information <1, 4> (<1>, modification of antibiotics [1]; <4>, enzyme may have a physiological role in secondary metabolism, its activity on aminoglycosides is simply fortouitous [3]; <4>, the enzyme may contribute to the O-acetylation of peptidoglycan [4]; <4>, the enzyme contributes to the O-acetylation of peptidoglycan. Mutants that either underexpress or overexpress the aac(2')-Ia gene are characterized phenotypically to possess either lower or higher levels of peptidoglycan O-acetylation, respectively, compared to the wild-type. Providencia stuartii PR100, with 42-44% peptidoglycan O-acetylation compared to 54% of the wild-type, appears as irregular rods. Strains PR50.LM3 and PR51, with increased levels of peptidoglycan O-acetylation, 63% and 65% respectively, appear as coccobacilli or chain formers [5]) (Reversibility: ? <1, 4> [1, 3, 4, 5]) [1, 3, 4, 5]

P ?

Substrates and products

S N-acetyl-D-glucosamine + tobramycin <2, 4> (<2>, acetylated at 58% the rate of gentamicin C_1 [2]) (Reversibility: ? <2, 4> [2, 3]) [2, 3]

P D-glucosamine + $N^{2'}$-acetyltobramycin

S acetyl-CoA + butirosin <2> (<2>, acetylated at 30% the rate of gentamicin C_1 [2]) (Reversibility: ? <2> [2]) [2]

P CoA + ?

S acetyl-CoA + dideoxykanamycin <2> (<2>, acetylated at 92% the rate of gentamicin C_1 [2]) (Reversibility: ? <2> [2]) [2]

P CoA + $N^{2'}$-diacetyldideoxykanamycin

S acetyl-CoA + gentamicin B <2> (<2>, 7% of the activity with gentamicin C_1 [2]) (Reversibility: ? <2> [2]) [2]

P CoA + $N^{2'}$-acetylgentamycin

S acetyl-CoA + gentamicin C_1 <1, 2> (<2>, best substrate [2]) (Reversibility: ? <1, 2> [1, 2]) [1, 2]

P CoA + $N^{2'}$-acetylgentamicin C1

S acetyl-CoA + gentamicin C_{1a} <1, 2, 3> (<2>, acetylated at 86% the rate of gentamicin C_1 [2]; <1>, acetylates 2'-amino group of hexose moiety [1]) (Reversibility: ? <1, 2, 3> [1, 2]) [1, 2]

P CoA + $N^{2'}$-acetylgentamicin C_{1a} <1> [1]

S acetyl-CoA + kanamycin B <1, 2> (<2>, acetylated at 14% the rate of gentamicin C_1 [2]) (Reversibility: ? <1, 2> [1, 2]) [1, 2]

P CoA + $N^{2'}$-kanamycin B

S acetyl-CoA + kanamycin C <1, 2> (<2>, acetylated at 40% the rate of gentamicin C_1 [2]) (Reversibility: ? <1, 2> [1, 2]) [1, 2]

P CoA + $N^{2'}$-acetylkanamycin C

S acetyl-CoA + neomycin B <2> (<2>, acetylated at 10% the rate of gentamicin C_1 [2]) (Reversibility: ? <2> [2]) [2]

P CoA + $N^{2'}$-acetylneomycin

S acetyl-CoA + paromomycin <1> (Reversibility: ? <1> [1]) [1]

P CoA + $N^{2'}$-acetylparomycin

S acetyl-CoA + ribostamycin <2> (<2>, acetylated at 15% the rate of gentamicin C_1 [2]) (Reversibility: ? <2> [2]) [2]
P CoA + $N^{2'}$-acetylribostamycin
S acetyl-CoA + sisomicin <2> (<2>, acetylated at 71% the rate of gentamicin C_1 [2]) [2]
P CoA + $N^{2'}$-acetylsisomicin
S acetyl-CoA + tobramycin <2, 4> (<2>, acetylated at 58% the rate of gentamicin C_1 [2]) (Reversibility: ? <2, 4> [2, 3]) [2, 3]
P CoA + $N^{2'}$-acetyltobramycin
S acetyl-CoA of soluble fragments of peptidoglycan + tobramycin <2, 4> (<2>, acetylated at 58% the rate of gentamicin C_1 [2]) (Reversibility: ? <2, 4> [2, 3]) [2, 3]
P ?
S Additional information <1, 2> (<2>, no substrates are kanamycin A and 1-hydroxyaminobutyric acid kanamycin A-derivative BB-K 8 [2]) [1, 2]
P ?

pH-Optimum
5.3 <1> [1]
6 <2> (<2>, kanamycin C [2]) [2]
6.6 <2> (<2>, gentamicin C [2]) [2]

pH-Range
5.7-6.5 <2> (<2>, about half-maximal activity at pH 5.7 and 6.5, kanamycin C [2]) [2]
6.4-7.6 <2> (<2>, about half-maximal activity at pH 6.4 and 7.6, gentamicin C [2]) [2]

5 Isolation/Preparation/Mutation/Application

Localization
cytoplasm <1> [1]

Purification
<4> [3]

Application
analysis <1, 3> (<1,3>, rapid analysis of antibiotic concentration in serum [1]) [1]

6 Stability

General stability information
<2>, activity in crude enzyme preparation is stable to repeated freezing and thawing [2]

References

[1] Benveniste, R.; Davies, J.: Aminoglycoside antibiotic-inactivating enzymes in actinomycetes similar to those present in clinical isolates of antibiotic-resistant bacteria. Proc. Natl. Acad. Sci. USA, **70**, 2276-2280 (1973)
[2] Chevereau, M.; Daniels, P.J.L.; Davies, J.; LeGoffic, F.: Aminoglycoside resistance in bacteria medicated by gentamicin acetyltransferase II, an enzyme modifying the 2-amino group of aminoglycoside antibiotics. Biochemistry, **13**, 598-603 (1974)
[3] Payie, K.G.; Clarke, A.J.: Characterization of gentamicin 2'-N-acetyltransferase from Providencia stuartii: its use of peptidoglycan metabolites for acetylation of both aminoglycosides and peptidoglycan. J. Bacteriol., **179**, 4106-4114 (1997)
[4] Clarke, A.J.; Francis, D.; Keenleyside, W.J.: The prevalence of gentamicin 2'-N-acetyltransferase in the Proteeae and its role in the O-acetylation of peptidoglycan. FEMS Microbiol. Lett., **145**, 201-207 (1996)
[5] Payie, K.G.; Strating, H.; Clarke, A.J.: The role of O-acetylation in the metabolism of peptidoglycan in Providencia stuartii. Microbiol. Drug Resist., **2**, 135-140 (1996)

The manufacturer's authorised representative in the EU is Springer Nature Customer Service Centre GmbH, Europaplatz 3, 69115 Heidelberg, Germany. If you have any concerns regarding our products, please contact ProductSafety@springernature.com

Printed and bound by CPI Group (UK) Ltd, Croydon, CR0 4YY
25/09/2026
02222831-0011